Carl Th. Ernst von Siebold

Anatomy of the invertebrata

Carl Th. Ernst von Siebold

Anatomy of the invertebrata

ISBN/EAN: 9783743331372

Hergestellt in Europa, USA, Kanada, Australien, Japan

Cover: Foto ©ninafisch / pixelio.de

Manufactured and distributed by brebook publishing software (www.brebook.com)

Carl Th. Ernst von Siebold

Anatomy of the invertebrata

ANATOMY

OF THE

INVERTEBRATA

BY

C. TH. V. SIEBOLD,

Translated from the German with Additions and Notes

BY

WALDO I. BURNETT, M.D.

BOSTON:
JAMES CAMPBELL, 18 TREMONT STREET.
1874.

PRINTED BY
J. E. FARWELL & CO.,
34 MERCHANTS ROW,
BOSTON.

To

MY ESTEEMED FRIEND,

LOUIS AGASSIZ,

PROFESSOR OF ZOOLOGY, &c.,

IN

Harbard Unibersity,

WHOSE WELL-KNOWN RELATIONS TO COMPARATIVE ANATOMY REQUIRE

NO MENTION HERE, AND WHOSE SPLENDID GENIUS HAS DONE SO MUCH TO AWAKEN,

IN THIS COUNTRY ESPECIALLY,

A LIVELY INTEREST IN OBJECTS OF NATURAL HISTORY,

I Inscribe this Volume,

WITH ADMIRATION AND SINCERE GRATITUDE.

WALDO I. BURNETT.

1*

PUBLISHER'S NOTICE.

For some years there has been a constant and increasing demand for a thorough and reliable text-book on the Anatomy of the Invertebrata; and no other book having appeared upon the subject to meet the requirements and to supersede Dr. Burnett's translation of Von Siebold's *Lerhbuck der vergleichenden Anatomie*, which, together with the translator's valuable and extensive notes, is believed to be incomparably the best and most complete treatise on the science, and the same having been highly commended by Professors Agassiz, Silliman, Hitchcock and others, the publisher feels himself justified in offering this edition to the student, confident that it will supply a want not at present met by any other work in our language.

Boston, May, 1874.

NOTICE OF THE TRANSLATOR AND EDITOR.

In issuing an English translation of the *Lehrbuch der ver-gleichenden Anatomie* of VON SIEBOLD and STANNIUS, any formal account of the work is quite unnecessary. To all Anatomists it is a treatise already well and favorably known, and it has justly been regarded as the most complete and comprehensive work of its kind now extant in any language. The high position and distinguished reputation of its authors have been fully sustained by this portion of their labors.

But there are several features in this work which should be mentioned, since by them it is favorably distinguished from all other treatises of the kind that have preceded it.

In the text will be found a lucid yet succinct exposition of the anatomical structure of organs, arranged as far as practicable under distinct types. The details on which this typical summary is based, are comprised in notes which are as remarkable for their erudition as for their copiousness; indeed, the utmost care has been taken in the literature of the various subjects treated, and the student will here find the most reliable and at the same time the fullest reference to the bibliography of nearly every subject in Comparative Anatomy. In this way, the work as a whole furnishes a complete dictionary of the science, and will prove invaluable even as a work of suggestion and reference, to those who would pursue any special line of inquiry and research in this department.

It may be truly said that the Microscope lies at the foundation of all our best knowledge of anatomy, and especially that of the Invertebrata. This is the case, not only on account of the small size of most of the animals, but because, as *Von Siebold* has said in his preface, the anatomy of these lower forms is scarcely reliable unless based upon histological investigations.

Hence, that part of the work treating of the anatomy of the Invertebrata, by *Von Siebold*, is rich in the results of microscopical researches; and their value in the elucidation of the subject will be readily appreciated. This plan of procedure has not the same urgency with the higher animals, where the character of an organ or part can generally be ascertained from its position, &c.; and, in the second part of the work, on the Anatomy of the Vertebrata, by *Stannius*, details of microscopical structure are comparatively little insisted upon. But, within a few years, the histological composition of organs, even though their character and function is well known, has become of great and increasing interest; and details of this kind, as far as they would be understood without the aid of figures, I have sought to add in their regular order and place.

As to the notes and additions generally, they stand by themselves with ED. affixed, and almost invariably refer to some point treated of in the text or notes of the original, and for the most part relate to the correction, confirmation, or extension of some statements there made. These notes were drawn from all the sources accessible to me; but from the many difficulties in the way of the early receipt of foreign works in this country, they are not as complete a record of the recent progress of the science as would be desired.

As to the translation, I may say, that not being a German scholar, but having read the German language chiefly for scientific purposes, I trust that any inelegances of diction or idiom will be excused. But, throughout, I have endeavored to give a faithful rendering of the author's meaning, and to express this in as simple and terse a form as possible.

In conclusion, I wish to express my gratitude to my friends who have kindly aided me in this work; — prominent among these is *Mr. Edward Capen* of this city, who has been of invaluable assistance to me in the labor of passing the sheets of this volume through the press; — of others, such as *Professors Agassiz, Dana, Leidy*, and *Wyman*, their names will be found honorably recorded by their own important labors in science, to which I have so frequently referred in these volumes.   W. I. B.

BOSTON, *Nov.* 1853.

PREFACE.

As latterly, Zootomists have given much greater attention to the invertebrate animals than formerly; and as, with these investigations they have united, as much as possible, others upon the generation and development of these animals, such a mass of material, composed, in part, of entirely new and very remarkable facts, has accumulated, that the manuals of Zootomy hitherto published are of a scale quite inadequate to receive them. It is unnecessary, therefore, for me to offer further reason for the task I have undertaken of arranging these materials and reducing them to a systematic form. But the order in which I have disposed them may not meet with general approval, for, hitherto, in works of comparative anatomy, the organs, and not the zoological classes, have served as the basis of the order pursued.

But, in the present state of Science, and at least provisionally, it appears to me that the anatomical order should not be followed, for, the types, which, until now, have been recognized in the developmental series of the several organs, appear no longer valid and permanent. Indeed, extended researches made upon a great number of animals, have shown that these types, hitherto regarded as expressive of fundamental laws, may almost be taken as the exceptions. Such genera as *Hydra, Lumbricus, Hirudo, Unio, Astacus*, &c, can now no longer be regarded as the representatives of certain animal classes or orders, for their organization is far from affording the requisite type of that of allied animals. It appears now clearly determined that the types of the development and disposition of the various organs of the Invertebrata are more numerous and varied than hitherto supposed, and that, in this respect, a rule wholly different from that of those of the Vertebrata must here be applied. But as the numberless details which we now possess upon the organization of the Invertebrata, have not been thoroughly worked out and systematized in all the orders, it is really a task too difficult to here distinguish the rule from the exception, and the type from that which is only a secondary modification.

I have especially devoted myself to the collecting and collating as completely as practicable, the numerous new and important facts in the organization of the invertebrate animals, which have as yet been developed. And as occasion presented, I have verified with my own eyes the particular results; and when I have been obliged to refer to the discoveries and observations of others, I have cited exactly their works.

I could not exclude Embryology and Histology from this work, for, in these branches, often lies our only means not only to ascertain the true nature of many larval forms among the lower animals, but also to arrive at the correct interpretation of many organs which, in form, position, and arrangements, have no analogues among the higher animal forms. It is only by the aid of Histology that we are able to show that this or that organ is a branchia, a liver, a kidney, an ovary, or a testicle; while, in the Vertebrata, which are organized after a few principal types, the signification of most of the organs can usually be easily determined by their position and connection.

In order to avoid long descriptions, I have, when practicable, referred to plates and figures; but in so doing I have always endeavored to cite the good and original representations, for I am convinced that many figures which are transferred from one book to another, become, at last, so changed as to be quite dissimilar to the original.

The elaboration of this work having been commenced in 1845, but its completion having been delayed by my change of residence from Erlangen to Freiburg, and partly by a pretty long sojourn of mine on the Adriatic Sea, I have been unable to use the important works which have been published during the last few years, except in the form of a Supplement [additional notes] which will serve to complete, to confirm, or to rectify what has been advanced in the body of the work.

I take this opportunity to publicly express my gratitude to *A. Kölliker*, *H. Koch*, *A. Krohn*, *C. Vogt*, and *H. Stannius*, for the friendly and important aid they have rendered me in the completion of this difficult task — not only by the transmission to me of interesting and rare marine animals, but also in the communication of important manuscripts and letters, the contents of which they have allowed me to freely use for my work.

FREIBURG (IN BREISGAU), *Feb.* 27, 1848.

C. TH. V. SIEBOLD.

TABLE OF CONTENTS.

Introductory Note to the Crustacea.

XII. THE CRUSTACEA.

XIII. THE ARACHNOIDAE.

XIV. THE INSECTA.

CLASSIFICATION

OF THE

INVERTEBRATE ANIMALS.

§ 1.

THE invertebrate animals are organized after various types, the limits of which are not always clearly defined. There is, therefore, a greater number of classes among them than among the vertebrates. But, as the details of their organization are yet but imperfectly known, they have not been satisfactorily classified in a natural manner.

There are among them many intermediate forms, which make it difficult to decide upon the exact limits of various groups.

The following division, however, from the lowest to the highest forms of organization, appears at present the best:

ANIMALIA EVERTEBRATA.

INVERTEBRATE ANIMALS.

Brain, spinal cord, and vertebral column, absent.

FIRST GROUP.

PROTOZOA.

Animals in which the different systems of organs are not distinctly separated, and whose irregular form and simple organization is reducible to the type of a cell.

CLASS I. INFUSORIA.
CLASS II. RHIZOPODA.

SECOND GROUP.

ZOOPHYTA.

Animals of regular form, and whose organs are arranged in a ray-like manner around a centre, or a longitudinal axis; the central masses of the nervous system forming a ring, which encircles the œsophagus.

CLASS III. POLYPI.
CLASS IV. ACALEPHÆ.
CLASS V. ECHINODERMATA.

THIRD GROUP.

VERMES.

Animals with an elongated, symmetrical body, and whose organs are arranged along a longitudinal axis; so that right and left, dorsal and ventral aspects may be indicated.

The central nervous mass consists of a cervical ganglion, with or without a chain of abdominal ganglia.

CLASS VI. HELMINTHES.
CLASS VII. TURBELLARII.
CLASS VIII. ROTATORII.
CLASS IX. ANNULATI.

FOURTH GROUP.

MOLLUSCA.

Animals of a varied form, and whose bodies are surrounded by a fleshy mantle. The central nervous masses consist of ganglia, some of which surround the œsophagus, and others, connected by nervous filaments, are scattered through the body.

CLASS X. ACEPHALA.
CLASS XI. CEPHALOPHORA.
CLASS XII. CEPHALOPODA.

FIFTH GROUP.

ARTHROPODA.

Animals having a perfectly symmetrical form, and articulated organs of locomotion. The central masses of the nervous system consist of a ring of ganglia surrounding the œsophagus, from which proceeds a chain of abdominal ganglia.

CLASS XIII. CRUSTACEA.
CLASS XIV. ARACHNIDA.
CLASS XV. INSECTA.

BIBLIOGRAPHY.

§ 2.

Besides the various ancient and modern works upon general comparative anatomy, — such as those of *Blumenbach*,[1] *G. Curier*,[2] *F. Meckel*,[3] *E. Home*,[4] *Blainville*,[5] *Delle Chiaje*,[6] *Carus*,[7] *Grant*,[8] *Rymer Jones*,[9] *Strauss*

[1] Handbuch der vergleichenden Anatomie. Göttingen, 1824.
[2] Leçons d'Anatomie comparee. Paris, 1799–1805. Translated into German and published with notes and additions by *Meckel* and *Frorirp*. 4 vols. Leipzig, 1809–10. 2nd edit. Paris, 1835–45.
[3] System der vergleichenden Anatomie. 6 vols. Halle, 1821–33.
[4] Lectures on Comparative Anatomy. 6 vols. London. 1814–29.

[5] De l'Organisation des Animaux, ou Principes d'Anatomie comparée. Tom. I. Paris, 1832.
[6] Istituzioni di Anotomia e Fisiologia Comparata. Napoli, 1832.
[7] Lehrbuch der vergleichenden Anatomie. 2nd ed. Leipzig, 1834.
[8] Outlines of Comparative Anatomy. London, 1841.
[9] A General Outline of the Animal Kingdom, and Manual of Comparative Anatomy London, 1841.

Dürckheim,[10] *R. Wagner,*[11]— there exist various contributions upon the relations of these animals in the physiological works of *Treviranus,*[12] *Rudolphi,*[13] *Dugès,*[14] *Burdach,*[15] *J. Müller,*[16] *R. Wagner,*[17] and in the Medical Zoology of *Brandt* and *Ratzeburg.*[18]

The iconographic illustrations by *Carus* and *Otto,*[19] and by *R. Wagner,*[20] contain many plates representing these animals; and in *Guerin's Iconographie,*[21] and *Cuvier's*[22] Règne Animal, edited by several French naturalists, are many illustrations of their internal structure.

The following are some of the anatomical works which treat specially upon these animals :

Schweigger. — Handbuch der Naturgeschichte der skelettlosen ungegliederten Thiere. Leipzig, 1820.

Delle Chiaje. — Memorie su la Storia e Notomia degli Animali senza Vertebre del regno di Napoli. 4 vol. Napoli, 1823–29. 109 tavole.

A second and enlarged edition of this memoir has been published under the following title : Descrizione e notomia degli animali invertebrati della Sicilia citeriore. 1–5, vol. Napoli, 1841. Con tavol. I.–CLXXII.

Sars. — Beskrivelser og Jagttagelser over nogle moerkelige eller nye i Havet ved den Bergenske Kyst levende Dyr af Polypernes, Acalephernes, Radiaternes, Annelidernes og Molluskernes Classer. Bergen, 1835.

Lamarck. — Histoire Naturelle des Animaux sans Vertèbres. Deux. édit., par *Deshayes* et *Milne Edwards.* 11 vols. Paris, 1835–45.

Milne Edwards. — Elémens de Zoologie, ou Leçons sur l'Anatomie, la Physiologie, la Classification, et les Mœurs des Animaux. Deux. édit. Animaux sans Vertèbres. Paris, 1843.

Richard Owen. — Lectures on the Comparative Anatomy and Physiology of the Invertebrate Animals. London, 1843.

H. Frey and R. Leuckart. — Beiträge zur Kenntniss wirbelloser Thiere mit besonderer Berücksichtigung der Fauna des norddeutschen Meeres. Braunschweig, 1847.

These same naturalists have prepared the second part of *Wagner's* Lehrbuch der Zootomie, under the special title of: Lehrbuch der Anatomie der wirbellosen Thiere. Leipzig, 1847.

Stef. Andr. Renier. — Osservazioni postume di Zoologia adriatica pubblicate per cura dell' istituto veneto di scienze, lettere ed arti a studio del *Prof. G. Meneghini.* Venezia, 1847. Con tavol. I.–XVI.

[10] Traité pratique et théorétique d'Anatomie comparée. 2 vol. Paris, 1842.

[11] Lehrbuch der Zootomie. 2nd edit., entirely revised ; or "Lehrbuch der vergleichenden Anatomie." Leipzig, 1842.

[12] Biologie. 6 vol. Göttingen, 1802–22. Also : Erscheinungen und Gesetze des organischen Lebens. 2 vol. Bremen, 1831–33.

[13] Grundriss der Physiologie. 2 vol. Berlin, 1821–28.

[14] Traité de Physiologie comparée de l'Homme et des Animaux. 3 vol. Montpellier, 1838–39.

[15] Die Physiologie als Erfahrungswissenschaft, erste Auflage, mit Beiträgen von *C. v. Baer, Dieffenbach, J. Müller, R. Wagner.* 6 vol. Leipzig, 1826–40. 2 te Auflage, mit Beiträgen von *E. Meyer, H. Rathké, C. v. Siebold* und *G. Valentin.* 2 vol. Leipzig, 1835–37.

[16] Handbuch der Physiologie des Menschen. 2 vol. 4th edit. Coblentz, 1844.

[17] Lehrbuch der Physiologie. 2nd edit. Leipzig, 1843.

[18] Medicinische Zoologie. 2 vol. Berlin, 1829–33.

[19] Erläuterungstafeln zur vergleichenden Anatomie. 6 heft. Leipzig, 1826–43.

[20] Icones physiologicæ. Erläuterungstafeln zur Physiologie und Entwickelungsgeschichte. Leipzig, 1839. Also, Icones Zootomicæ. Handatlas zur vergleichenden Anatomie. Leipzig, 1841.

[21] Iconographie du Règne Animal de G. Cuvier, ou Représentation d'après nature de l'une des espèces les plus remarquables et souvent non encore figurées de chaque genre d'Animaux ; pour servir d'atlas à tous les Traités de Zoologie. 7 vol. avec 450 planches. Paris, 1830–38.

[22] Règne Animal de *Cuvier,* nouvelle édition, accompagnée de planches gravées, &c. &c. Paris 1836–47. Still unfinished.

2*

CONSTANT labors in the whole department of microscopy, and that, too, with greatly improved instruments, during the past few years, have materially changed the face of the class Infusoria since the issue of this work. There have been numerous and signal researches among all the lower forms of animal life; and the imperfect and undeveloped forms of others, which are higher, have been wrought out with an accuracy and detail before unknown.

These movements have all tended to diminish the numbers of the so-called Infusoria, and it remains to be seen how large the proper class will be when these researches shall have been further extended. By some even it is believed that it will be entirely resolved into other classes; this view, however, would appear far from being warranted by our present knowledge; for, while, on the one hand, whole genera have been shown to be only larval worms (*Bursaria*, *Paramœcium*, &c., from Planaria),* yet, on the other, some forms have manifested phenomena and changes leading us to place them almost unhesitatingly among individual animals. In its best aspects, however, the subject has many perplexing points; and, in its present unsettled state, it is almost hazardous for a scientific man to entertain anything like positive views thereon.

I need scarcely allude to the vegetable, algous character which whole sections of the Polygastrica have recently assumed; and the limits of this work will not allow me to discuss in detail this and other interesting points. But there are two or three topics of the highest physiological import, which are prominently introduced by these studies. These are, What is a plant? What is an animal? and, Are the animal and vegetable kingdoms on their lowest confines separate and distinct from each other?

As is well known, all the older criteria by which animals were separated from plants have long since been regarded invalid; and some of those which in late years have been regarded among the most constant, have, quite recently, been declared as equally unsound. Cellulose has been shown to be a component of animal as well as of vegetable structures, and *Kölliker* † has insisted that some forms which have neither mouth nor stom-

---

* *Agassiz*, Ann. Nat. Hist. VI. 1850, p. 156. † *Kölliker. Siebold* and *Kölliker's* Zeitsch. I. 1849, p. 198.

ach, but consist of a homogeneous mass, are true animals. If these premises are correct, nothing will remain, as I conceive, for a distinctive characteristic, but *voluntary motion*. This, when positive, is indubitable evidence of any given form being of an animal character; and it must remain for each individual observer to determine what is, and what is not, voluntary action, in each particular case. Moreover, even should *Kölliker's* view of a stomachless animal prove correct, the inverse condition of a true stomachal cavity being present, must, I think, be regarded as positive evidence of the animal nature of the form in question; for this must always be a distinctive characteristic of the two kingdoms, when present.

In regard to the other point, What constitutes an animal? observers are very far from being agreed. *Siebold, Kölliker*, and others, have taken the ground that individual animal forms may be unicellular; or, in other words, that an animal may be composed of only a single cell.* This view is principally due to *Kölliker's* observations and statements upon Gregarinae.† The facts are indeed striking, but the evidence does not appear to me sufficient, as yet, to settle such a vexed and important question; and more especially so since *Bruch* ‡ has raised the point of their belonging to the Worms. But, aside from such grounds, I was led, some time since, after considerable study of infusoria-forms, to venture an opinion quite at variance with that just mentioned of *Siebold* and *Kölliker*. I then made the following statement: In regard to the question, What characteristic in organic animal matter shall constitute an individual? I feel satisfied of this much, — that cell processes, however closely interwoven they may be with the expressions of individual life, cannot be considered as constituting the ground-work of its definition.§ This statement was made more than two years since; and subsequent observations, some of them of a special character, have not led me to a change of opinion. True individual animal life seems to involve a cycle of relations not implied in simple cells; in other words, these last must always lose their character as such, in a definite form which belongs to the individual.

On this account I regard the Infusoria proper, or those which have been shown to be of an undoubted animal character, as in a completely transition state; and, although it may be well to arrange these forms systematically, for the sake of convenience, yet they cannot be considered as holding fixed zoölogical positions. Further research in this direction, and upon "Alternation of Generation," will, I think, widely clear up this obscure, yet most interesting field of study.   EDITOR.

---

* *Siebold. Siebold* and *Kölliker's* Zeitsch. I. p. 270.

† *Kölliker. Siebold* and *Kölliker's* Zeitsch. I. p. 1.

‡ *Bruch. Siebold* and *Kölliker's* Zeitsch. II. p. 110.

§ *Burnett.* Proceed. Boston Soc. Nat. Hist. V. p. 124.

BOOK FIRST.

INFUSORIA AND RHIZOPODA.

CLASSIFICATION.

§ 3.

THE Infusoria, using this word in a restricted sense, are far from being the highly-organized animals *Ehrenberg* has supposed. In the first place, on account of their more complicated structure, the Rotifera must be quite separated from them, as has already been done by *Wiegmann, Burmeister, R. Wagner, Milne Edwards, Rymer Jones,* and others. The same may be said of the so-called Polygastrica. In fact, a great number of the forms included under Closterina, Bacillaria, Volvocina, and others placed by *Ehrenberg* among the anenteric Polygastrica, belong, properly, to the vegetable kingdom. Indeed, this author has very arbitrarily taken for digestive, sexual, and nervous organs, the rigid vesicles, and the colored or colorless granular masses, which are met with in simple vegetable forms, but which are always absent in those low organisms of undoubtedly an animal nature. Cell-structure and free motion are the only two characteristics in common of the lowest animal and vegetable forms; and since *Schwann* [1] has shown the uniformity of development and structure of animals and plants, it will not appear strange that the lowest conditions of each should resemble each other in their simple-cell nature. As to motion, the voluntary movements of Infusoria should be distinguished from those which are involuntary, of simple vegetable forms; a distinction not insisted upon until lately. Thus, in watching carefully the motions of Vorticellina, Trachelina, Kolpodea, Oxytrichina, &c., one quickly perceives their voluntary character. The same is true of the power of contracting and expanding their bodies.

But in the motions of vegetable forms other conditions are perceived · and there is no appearance of volition in either change of place or form, their locomotion being accomplished either by means of cilia, or other physical causes not yet well understood. Cilia, therefore, belong to vegetable as well as to animal forms, and in this connection it is not a little remarkable that in animals they should be under the control of volition. With vegetable forms these organs are met with either in the shape of ciliated epithelium, as upon the spores of *Vaucheria*, [2] or as long, waving filaments, as upon the earlier forms of many confervæ, [3] in which last can

[1] Mikroskopische Untersuchungen, &c. Berlin, 1839
[2] *Thuret.* Recherches sur les organes locomo-
teurs des spores des Algues. Ann. des Sc. Nat. Botan. 1843, XIX. p. 266. Pl. XI. fig. 29-30.
[3] The same. Pl. X.

often be seen the so-called organization of *Ehrenberg's* Monadina and Volvocina. Until the fact that ciliated organs belong to both animals and vegetables was decided, the real place of many low organisms had to remain undetermined.[4] However, notwithstanding their free motion from place to place by means of cilia, the vegetable nature of many organisms seemed clearly indicated by the rigid, non-contractile character of their forms. It is from a misapprehension of the true nature of these facts, that some modern naturalists have denied the existence of limits between the two kingdoms.[5]

With Bacillareæ and Diatomaceæ, this question has another aspect. Many of these organisms have been taken for animals from their so-called voluntary movements, which truly entirely want the character of volition. In the movements of the rigid Diatomaceæ, for instance, the whole plant has oscillatory motions like a magnetic needle, at the same time slightly changing its place forward and backward. When small floating particles come in contact with such an organism, they immediately assume the same motion. This may be well observed with the Oscillatoria. There are here, undoubtedly, no ciliary organs ; in fact, they could not, if present, produce this kind of motion. According to *Ehrenberg*,[6] the Naviculæ can protrude ciliary locomotive organs through openings of their carapace; but this has not been observed by other naturalists.

§ 4.

The Rhizopoda, whose internal structure is as yet imperfectly known, are closely allied to the Infusoria. Like these last, their bodies are cellular, containing nuclear corpuscles, but no system of distinct organs. These two classes of Protozoa differ, however, in their external form, and the structure of their locomotive organs. The body of the Infusoria, notwithstanding its contractility, has a definite form, and moves chiefly by means of vibratile organs. That of the Rhizopoda, on the other hand, although equally contractile, has no definite form ; their movements also are not due to ciliated organs, but to a change of the form of the body by various prolongations and digitations.

§ 5.

Owing to the present incomplete details upon the organization of these animals, little can here be said about them ; and therefore, instead of devoting to them a separate chapter, it will be proper to treat of them with the Infusoria in general.

As the division of the Polygastric Infusoria, by *Ehrenberg*, into two

---

4 As an example, may be mentioned the various and dissimilar opinions of naturalists upon the question of the animal or vegetable nature of the "red snow ;" a question upon which *Flotow*, after the most careful studies, is still undecided. See *Flotow*, "Ueber Haematococcus pluvialis," in Nov. Act. Acad. Leop. Carol, vol. XX. part ii. p. 18.

5 See *Unger*, Die Pflanze im Momente der Thierwerdung. Wien. 1843.

Also, *Kützing*, Ueber die Verwandlung der Infusorien in niedere Algenformen. Nordhausen, 1844.

In an academic paper (Dissertatio de finibus inter regnum animale et vegetabile constituendis, Erlangae, 1844), I have attempted to show that this confusion between the two kingdoms does not exist.

6 Abhandlungen der Akademie der Wissenschaften zu Berlin, 1836, p. 134, Taf. I. fig. 19, and 1839, p. 102, Taf. IV. fig. 5.

orders, Anentera and Enterodela, appears unfounded, the following classification seems more natural:

PROTOZOA.

CLASS INFUSORIA.
Organs of locomotion chiefly vibratile.

ORDER I. ASTOMA.
Without an oral aperture.

FAMILY: ASTASIAEA.
Genera : *Amblyophis, Euglena, Chlorogonium.*

FAMILY: PERIDINAEA.
Genera: *Peridinium, Glenodinium.*

FAMILY: OPALINAEA.
Genus: *Opalina.*

ORDER II. STOMATODA.
With a distinct oral aperture and œsophagus.

FAMILY: VORTICELLINA.
Genera: *Stentor, Trichodina, Vorticella, Epistylis, Carchesium.*

FAMILY: OPHRYDINA.
Genera : *Vaginicola, Cothurnia.*

FAMILY: ENCHELIA.
Genera: *Actinophrys, Leucophrys, Prorodon.*

FAMILY: TRACHELINA.
Genera: *Glaucoma, Spirostomum, Trachelius, Loxodes, Chilodon, Phialina,
Bursaria, Nassula.*

FAMILY: KOLPODEA.
Genera: *Kolpoda, Paramæcium, Amphileptus.*

FAMILY: OXYTRICHINA.
Genera: *Oxytricha, Stylonychia, Urostyla.*

FAMILY: EUPLOTA.
Genera : *Euplotes, Himantophorus, Chlamidodon.*

CLASS RHIZOPODA.

Organs of locomotion consisting of completely retractile, ramifying prolongations of the body.

ORDER I. MONOSOMATIA.

FAMILY: AMOEBAEA.

Genus: *Amoeba.*

FAMILY: ARCELLINA.

Genera: *Arcella, Difflugia, Gromia, Miliola, Euglypha, Trinema.*

ORDER II. POLYSOMATIA.

Genera: *Vorticialis, Geoponus, Nonionina.*[1]

BIBLIOGRAPHY

O. F. Müller. Animalcula Infusoria. Hafniæ, 1786.

Ehrenberg. Die Infusionsthierchen als vollkommene Organismen. Leipzig, 1838. Also his numerous and important memoirs upon the Infusoria and Rhizopoda in the Memoirs of the Berlin Academy, and its Monthly Bulletin.

Andrew Pritchard. A History of Infusoria, living and fossil, arranged according to the "Infusionsthierchen," of Ehrenberg. Illustrated by nearly 800 colored engravings of these curious creatures, highly magnified. London, 1841.

Kutorga. Naturgeschichte der Infusionsthierchen, vorzüglich nach *Ehrenberg's* Beobachtungen bearbeitet. Calsruhe, 1841.

Dujardin. Histoire Naturelle des Zoophytes. Infusoires. Paris, 1841. This work treats also of the Rhizopoda.

ADDITIONAL BIBLIOGRAPHY.

Besides the various articles quoted in the additional notes I have made the following are among the more important recent writings on this subject:

Cohn. Beiträge zur Entwickelungsgeschichte der Infusorien, in *Siebold & Kölliker's* Zeitsch. III. Hft. 3, and IV. Hft. 3.

Ecker. Zur Entwickelungsgeschichte der Infusorien, in *Siebold & Kölliker's* Zeitsch. III. Hft. 4.

Stein. Neue Beitr. zur Kenntn. d. Entwickelungsg. u. d. feineren Baues d. Infusionsthiere, in *Siebold & Kölliker's* Zeitsch. III. p. 475.

Pritchard. A History of Infusorial Animalcules, living and fossil, &c., with illustrations, new edition. London, 1852.

See also numerous notes in the Annales des Sciences Naturelles, since 1847. — ED.

---

[1] In this table are mentioned the families and genera of those only which have been the objects of anatomical study.

CHAPTER I.

EXTERNAL COVERING.

§ 6.

The PROTOZOA are surrounded by a very delicate cutaneous envelope, which is sometimes smooth,[1] and sometimes covered with thickly-set cilia.[2] Generally these cilia are arranged in longitudinal rows; [3] but in *Actinophrys* they consist of long contractile filaments of a special nature.

CHAPTER II.

MUSCULAR SYSTEM AND LOCOMOTIVE ORGANS.

§ 7.

With the PROTOZOA a distinct muscular tissue cannot be made out, but the gelatinous substance of their body is throughout contractile.

It is only in the contractile peduncle of certain Vorticellina, that there can be perceived a distinct longitudinal muscle, which, assuming a spiral form, can contract suddenly like a spring.[1]

§ 8.

THE VIBRATILE ORGANS on the surface of Infusoria serve as organs of locomotion.

With many species they are found much developed at certain points, and are arranged in a remarkable order and manner.

With *Peridinium*, a crown of them encircles the body; with *Stylonychia*, they are quite long, and surround the flattened body like a fringe; while the Vorticellina have the anterior portion of their body surrounded by retractile cilia, arranged in a circular or spiral manner. In *Trichodina* there is, upon the ventral surface, besides a crown of these cilia upon the back, a very delicate ciliated membranous border, which is attached to a ring which is dentated, and composed of a compact homogeneous tissue. With *Trichodina pediculus* this border is whole and entire; but it is broken or ragged with *Trichodina mitra*.[1]

By means of this organ these animals swim with facility, or invade with skill the arm-polyps and Planaria.[2] With many Infusoria, the vibratile organs are situated at the anterior extremity of the body, as simple or double non-retractile filaments, which move in a manner to produce a vor-

---

[1] *Euglena, Amœba, &c.*
[2] *Trachelius, Paramœcium, Nassula, &c.*
[3] *Amphileptus, Chilodon, Opalina, &c.*
[1] The peduncle is simple with *Vorticella,* but ramified with *Carchesium.* With *Epistylis* it is not muscular.

[1] This Infusorium was discovered by me as a parasite in many Planariæ.
[2] *Ehrenberg* has entirely overlooked the ciliated border of *Trichodina pediculus,* and has regarded the stiff serrations of the ring as movable hooks. See " Die Infusionsthierchen," p. 206.

tical action of the water.[3] But with others the locomotive organ is a long retractile proboscis.[4] With the Oxytrichina and Euplota, there are fleshy movable points (UNCINI) upon the ventral surface, by which these animals move about as upon feet. During these movements with the Oxytrichina, the posterior portion of the body is supported by many setose and styloid processes, which point backward.

The singularly varied and branching locomotive organs of the Rhizopoda are short, and digitated with *Amœba, Difflugia* and *Arcella*.[5] But in the other genera they are elongated and filamentous.[6]

CHAPTERS III. AND IV.

NERVOUS SYSTEM AND ORGANS OF SENSE.

§ 9.

Although the Infusoria clearly evince in their actions the existence of sensation and volition, and appear susceptible of sensitive impressions, yet no nervous tissue whatever has as yet been found in them. If *Ehrenberg* supposed the Polygastric Infusoria to possess a nervous system, he did so because, having decided that the red pigment points of these animals were eyes, he inferred that they necessarily had a nervous ganglion at their base.

§ 10.

With the naked Infusoria the sense of touch exists, undoubtedly, over the whole body. But beside this, it appears specially developed, in many species, in the long cilia forming vibratile circles, or in those movable foot-like and snout-like prolongations of the body. In the same manner, it is probable they have the sense of taste also; for they seem to exercise a choice in their food, although no gustatory organ has yet been found.

All species, whether they have red pigment points or not, seem affected by light. Without doubt, therefore, their vision consists simply in discriminating light from darkness, which is accomplished by the general surface of the body, and without the aid of a special optical organ.

The simple pigment point of many Infusoria,[1] and which *Ehrenberg* has generally regarded as an eye,[2] has no cornea, and contains no body capable of refracting light; there is, moreover, connected with it no nervous substance.

Ehrenberg attaches here too great an importance to the red color of the

---

[3] *Amblyophis, Euglena* and *Peridinium*, have a simple flagelliform cilium, but with *Chlorogonium* it is double.
[4] *Trachelius trichophorus* feels about with a long snout of this kind, without, however, producing a vortical action on the water.
[5] See *Ehrenberg*, "Die Infusionsthierchen," Taf. VIII. and IX.
[6] *Gromia fluviatilis, Miliola vulgaris, Vorticialis strigilata, Euglypha tuberculosa, Trinema acinus,* according to *Dujardin* (Ann. des Sc. Nat. Zool. IV. 1835, p. 343, pl. IX.; also, V. 1836, p. 196, pl. IX. fig. A. See, also, his Histoire des Infusoires, 1841, p. 249, pl. I. fig. 14–17; pl. II. fig. 1, 2, 7—10; pl. IV. fig. 1); *Geoponus stella borealis, Nonionina germanica,* according to *Ehrenberg.* Abhand. d. Berliner Akad. 1839, p. 106, Taf. I. II.
[1] *Amblyophis, Euglena, Chlorogonium, &c.*
[2] Abhandl. d. Berliner Akad. 1831, p. 12; also, "Die Infusionsthierchen," p. 491.

3

pigment,[3] for the blue, violet and green pigments, seen in the eyes of insects and crustacea, show clearly that the red pigment is not essential to the eye.*

CHAPTER V.

DIGESTIVE APPARATUS.

§ 11.

The Infusoria are nourished, either by taking solid food into the interior of their body, or by absorbing by its entire surface nutritive fluids which occur in the media in which they live.

This last mode is illustrated in the Astoma, which have no distinct oral aperture or digestive apparatus. By the ingenious experiment first performed by *Gleichen*,[1] of feeding these animals with colored liquids, no trace of these organs could be found.

Ehrenberg, who also had observed that they did not eat, regarded their internal vesicles as stomachal organs, which were in connection with the mouth by tubes. The correctness of this opinion, however, has not been verified. Indeed, the genus *Opalina*[2] refutes it; here the species are quite large and visible to the naked eye, yet an oral aperture can be detected upon no part of their body, and never do they admit into its interior colored particles. Solid substances found in them cannot be regarded as food. That fluids are here introduced by surface-imbibition is shown by *Opalina ranarum*; this animal is found in bile in the rectum of frogs, and assumes a green color. When *Opalina* requiring only a certain quantity of liquid are placed in water, they quickly absorb it, become greatly swollen, and shortly after die. In such cases, the absorbed liquid is seen as clear, vesicular globules under the surface, and these globules have been taken by *Ehrenberg* as stomachal vesicles (VENTRICULI), and by *Dujardin* as VACUOLAE.

§ 12.

Those Infusoria which are nourished by solid food have a mouth at a certain place, and an œsophagus traversing the parenchyma of the body. Through this last the food is received, and is finally dissolved in the semi-liquid parenchyma of the body, without passing through stomachal or intestinal cavities. In many cases there is at the end of the body opposite the mouth an ANUS, through which the refuse material is expelled. But, when this is

---

[3] "Die Infusionsthierchen," p. 492.

[1] Auserlesene mikroskopische Entdeckungen, 1777, p. 51; also, Abhandlung über die Saamen- und Infusionsthierchen, 1778, p. 140.

[2] The genus *Opalina* was first established by *Purkinje & Valentin*. Many species are found in the rectum of frogs, and it is not rare to meet with them in the alimentary canal of Planaricae.†

* Some recent researches of *Thuret* (Ann. d. Sc. Nat. 3rd ser. XIV. 1850) on the reproductive germs of Algae prove that these bodies have red eye-like specks, resembling those seen in the Polygastrica, but which disappear when the Zoospores attach themselves and germination proceeds. The

fact is a very interesting one in this connection. — Ed.

† [§ 11, note 2.] According to *Agassiz* (Amer. Jour. Sc. XIII. 1852, p. 425), *Opalina* is only a larval form of *Distoma*. — Ed.

wanting, its function is often performed by the mouth. According to *Ehrenberg*, the *Infusoria polygastrica*, such as we have just been describing, differ from the *Infusoria rotatoria*, in having a great number of stomachs, which connect by hollow peduncles with the mouth in the division Anentera, and with the intestine in that of Enterodela. This organization, which, from its high authority, has generally been admitted by naturalists, is not, however, met with in any infusorium.[1]

The vesicular cavities in the bodies of these animals, and which have been regarded by *Ehrenberg* as stomachal-pouches, never have a hollow peduncle, either connecting with the mouth (*Anentera*) or with the intestine (*Enterodela*). Indeed, it is doubtful if a digestive canal can be made out in these Infusoria.

The vesicular, irregular contracting cavities of their body contain a clear liquid, evidently the same as that in which they live, which, with the Astoma, has been absorbed through the surface of the body. But, with those having a mouth and œsophagus, it is received through them, and taken up by the yielding parenchyma of the body.

If the methods of feeding of *Gleichen* and *Ehrenberg* are employed, the colored particles are taken in by a vortical action of the water, caused by the cilia surrounding the mouth. This water, with its molecules, accumulates at the lower portion of the œsophagus, and so distends there the parenchyma as to cause the appearance of a vesicle. Thus situated, the whole has much the aspect of a pedunculated vesicle. But when, from contractions of the œsophagus, this water escapes into the parenchyma, it appears there as an unpedunculated globule, in which the colored particles still float. When the Stomatoda are full-fed in this manner, there appear many of these globules in various parts of the body; and thus substances previously ingested are taken up and disseminated throughout the body.

If the globules thus containing solid particles are closely aggregated, it sometimes happens that they fuse together; a fact which proves that they are not surrounded by a special membrane.

The solid particles of food of the Stomatoda, which are often the lower Algae, such as the Diatomaceæ and Oscillatoria, and often other Infusoria, are sometimes deposited in the parenchyma without being surrounded by a vesicular liquid.*

From observations made upon *Amœba, Arcella* and *Difflugia*, it appears that the Rhizopoda ingest their food like the Stomatode Infusoria.

---

1 *Focke* (Isis, 1836, p. 785) has already raised doubts as to the existence in Infusoria of the stomachs described by *Ehrenberg*. *Ehrenberg* has also opponents in *Dujardin* (Ann. des Sc. Nat. Zool. IV. 1835, p. 364 ; V. 1836, p. 193 ; X. 1838, p. 230; also Hist. Nat. des Infus. 1841, p. 57), in *Meyen* (*Muller's* Arch. 1839, p. 74) and in *Rymer Jones* (Ann. of Nat. Hist. III. 1839, p. 105 ; also, " A General Outline of the Animal Kingdom," 1841, p. 56).

He has attempted to reply to the objections here urged by very detailed illustrations of the organization of the Polygastrica, made by him and *Wer-* neck. (*Muller's* Arch. 1839, p. 80 ; also Monatsbericht der Berliner Akad. 1841, p. 103.) But, detailed as they may be (see *Ehrenberg* Abhandl. d. Ber. Akad. 1830, Taf. III.; 1831, Taf. III.; also "Die Infusionsthierchen," Taf. XXXII. XXXVI. and XXXIX.), they are not representations of nature.

The organ which in *Trachelius ovum* has been taken by *Ehrenberg* ("Die Infusionsthierchen," p. 323, Taf. XXXIII. fig. xiii. 1) for a branching digestive tube, has always appeared to me only as a solid fibrous cord, traversing the soft parenchyma of the body, and by its ramifications presenting a coarse meshed aspect.

* *Bailey* (Amer. Jour. Sc. May, 1853, p. 341) has recently published an account, accompanied with numerous figures, of a new animalcule, which is so remarkable in this connection that I give here his description. He says : " If the reader will imagine a bag made of some soft extensible material, so thin as to be transparent like glass, so soft as to yield readily to extension when subjected to internal pressure, and so small as to be microscopic ; this bag, filled with particles of sand, shells of

§ 13.

If the vesicular cavities containing the liquid and colorless food of the Stomatoda be examined under the microscope by a horizontal central incision, their contents appear colorless; but by changing the focus, viewing alternately the convex and concave surfaces of the vesicle, the points of junction between the colorless globules and the parenchyma appear colored pale-red. This appearance, due to an optical illusion, might easily deceive one into the opinion that the vesicles which are really colorless are colored.

From this it is probable that *Ehrenberg* has described *Bursaria vernalis* and *Trachelius meleagris* as having a red gastric juice.[1]

The violet points which are found upon the back and neck of *Nassula elegans* and *Chilodon ornatus* are only collections of pigment granules, which, in the first case, are often absent, and in the second are often partially dissolved.

This last violet liquid has been regarded by *Ehrenberg*[2] as a gastric juice resembling bile

§ 14.

The solid particles of food, whether surrounded by the parenchyma or enclosed in a liquid vesicle, are moved hither and thither in the gelatinous tissue of the body, during the contracting and expanding movements of the animal. In some, the parenchyma with its contained food moves in a regularly circular manner, like the liquid contained in the articulated tubes of Chara.[1] In *Loxodes bursaria*[2] this circulation is remarkable, and of much physiological interest. Its cause is yet quite unknown, for in no case is it due to cilia, and it may be observed in individuals entirely at rest. *Ehrenberg*,[3] therefore, is incorrect in regarding it as due solely to a contractile power of the parenchyma, displacing the molecules. Much less is his explanation [4] satisfactory, since the digestive tube of an infusorium can be extended at the expense of its stomachal pouches, so as to fill the whole body, giving it the appearance of having a circulation of molecules throughout its entire extent.

---

[1] "Die Infusionsthierchen," pp. 321, 326, 329. *Ehrenberg* has, moreover, in *Trachelius meleagris*, confounded the contractile cavities with those non-contractile, and which receive the food.

[2] Abhandl. d. Berliner Akad. 1833, p. 179 ; also "Die Infusionsthierchen," pp. 319, 338, 339.*

[1] *Vaginicola* and *Vorticella*. See *Focke*, Isis,

1836, p. 786 ; also *Meyen*, *Müller's* Arch. 1839, p. 75.

[2] *Focke* loc. cit.; also *Erdl*, *Müller's* Arch. 1841, p. 278.

[3] Loc. cit. p. 262.

[4] *Müller's* Archiv. 1839, p. 81.

Diatomaceæ, portions of Algæ or Desmidieae, and with fragments of variously colored cotton, woolen, and linen fibres, will give a picture of the animal ; to complete which, it is only necessary to add a few loose strings to the bag to represent the variable radiant processes which it possesses around the mouth." This animal, which is often found with bits of cotton protruding from its mouth, assumes the most bizarre shapes. They appear to multiply by fissuration and gemmation even when filled with these heterogeneous particles, and, on the whole, present characteristics as remarkable as

those of any animalcule with which we are acquainted. — ED.

* [§ 13, note 2.] In this connection should be noticed the experiments of *Will* (*Müller's* Arch. 1848, p. 509). He found evidences of a biliary apparatus, with *Vorticella*, *Epistylis*, and *Bursaria*. These evidences are based on chemical reaction, and he describes no anatomical apparatus. I mention this fact here, although *Vorticella* belongs truly to the Bryozoa, and *Bursaria* to the Planaria. — ED.

§ 15.

The round or elongated oval mouth of Infusoria varies as to its position. Sometimes it is in front, sometimes behind ; and in some cases, near the middle third of the body. Rarely naked,[1] its borders are generally ciliated,[2] and often its circumference is provided with a very remarkable ciliary apparatus. By the aid of this, these animals not only move about, but when quiet produce vortical actions of the water, which are felt at quite a distance; and all minute particles within its reach are quickly drawn towards its mouth, and then swallowed or rejected according to the option of the individual.[3]

It is rare that this oral aperture is provided with a dental apparatus.[4] The oral cavity, generally infundibuliform, extends into a longer or shorter, straight or curved œsophagus, which is lined throughout by a very delicate ciliated epithelium.[5]

The anus, situated usually upon the dorsal surface of the posterior portion of the body, is sometimes, though rarely, indicated by a slight external projection.[6]

CHAPTERS VI. AND VII.

CIRCULATORY AND RESPIRATORY SYSTEMS.

§ 16.

A vascular system entirely distinct by closed walls from the other organs is not found in the Protozoa. But with very many (with all the Stomatoda, without exception) there are contractile pulsatory cavities, the form, number and arrangement of which is quite varied.

They are situated in the denser and outer layers of the parenchyma of the body, and during the diastole they become swollen by a clear, transparent, colorless liquid, which, during the systole, entirely disappears.

[1] *Actinophrys.* The mouth is naked also in the genera *Difflugia* and *Arcella* of the Rhizopoda.*

[2] *Bursaria, Paramecium, Urostyla* and *Stylonychia.* In *Glaucoma scintillans* the ciliated crown of the mouth is replaced by a special semilunar ciliated lobe.

[3] In *Stentor, Vorticella, Epistylis* and *Trichodina,* this apparatus is retractile, and produces in a particular way the vortical actions. In *Spirostomum ambiguum,* there is a long, narrow, ciliated furrow, through which the food is conducted to the mouth, situated at the posterior portion of the body.

[4] *Prorodon, Nassula, Chilodon* and *Chlamidodon.* Here the hair-like teeth are arranged in a cylinder so as to resemble a weir.

[5] The œsophagus is short in *Oxytricha, Stylonychia,* and *Euplotes ;* but is elongated or spiral in *Vorticella, Carchesium* and *Epistylis ;* while it is long and arcuate in *Bursaria truncatella* and *cordiformis.*

[6] The undigested matters accumulate about the anus, and when this opens are expelled from the parenchyma with a certain force. With *Nassula elegans,* the greater or less portions of the *Oscillatoria gracillima* (*Kützing*) upon which it feeds, and which are of a blue-green color, dissolve into granules of this color. But these, during the process of digestion, gradually assume a brown color, and form irregular masses in the posterior portion of the body, and are from time to time expelled as brown fœces. These green granules are not therefore eggs, as *Ehrenberg* (loc. cit. p. 339) has supposed. This *Nassula* when young is perfectly colorless, with the exception of a beautiful blue spot.

* [§ 15, note 1.] *Kölliker* (*Siebold* and *Kölliker's* Zeitsch. I. 1849, p. 198) has given a long and detailed description of *Actinophrys sol.* According to him, it is without mouth or stomach proper, and internally is composed of a homogeneous substance. Yet this remarkable animal lives on other Infusoria, Algae, &c., and avails itself of them by seizing and afterwards invaginating them in its parenchyma, until they finally are included within its interior. — ED.

3*

These movements succeed each other at more or less regular intervals. When these cavities are numerous, a certain order in the succession and alternation of their contractions cannot always be observed. It is very probable that their liquid contained during the diastole is only the nutritive fluid of the parenchyma, and to which it returns during the systole. In this way it has a constant renewal, and all stagnation is prevented. This arrangement constitutes the *first appearance of a circulatory system*, and the *first attempt at a circulation of nutritive fluids*.

From an optical illusion similar to the one mentioned as belonging to the vacuolæ (§ 13) the liquid of these pulsating cavities has a reddish hue. [1]

§ 17.

A round, pulsating cavity is found in the genera *Vorticella*, *Epistylis*, *Loxodes*, and in the following species: — *Amœba diffluens*, *Paramœcium kolpoda*, *Stylonychia mytilus*, *Euplotes patella*, &c. With *Actinophrys*, *Bursaria*, *Trichodina*, there are from one to two; with *Arcella vulgaris*, three to four; with *Nassula elegans*, there are four placed in a longitudinal line on the dorsal surface. With *Trachelius meleagris*, there is a series of eight to twelve upon the sides of the body, and with the various species of *Amphileptus* there are fifteen to sixteen arranged more or less regularly. With *Stentor*, there is a large cavity in the anterior portion of the body, and many similar cavities appear upon the sides, united sometimes into one long canal. A similar canal traverses the entire body of *Spirostomum ambiguum*, and *Opalina planariarum*. With *Paramœcium aurelia*, the two round cavities present a remarkable aspect, being surrounded by five or seven others, small and pyriform, the top of which being directed outward, the whole has a star-like appearance. [1] During the pulsation, often the entire star disappears, sometimes only the two central cavities, and in some cases the rays only.

These cavities, entirely disappearing in the systole, reäppear in the diastole, and usually in the same place and with the same form and number. This would lead us to conclude that they are not simple excavations in parenchyma, but real vesicles or vessels, the walls of which are so excessively thin as to elude the highest microscopic power.

In some individuals, as, for instance, with *Trachelius lamella*, there appear, during the diastole, two or three small vesicles at the extremity of the body, which, after having increased in size, blend into one which is very large. These are probably only globules of nutritive fluid, separated from the parenchyma. Similar phenomena are observed in *Phialina vermicularis* and *Bursaria cordiformis*.

It sometimes happens with these animals that a forcible contraction of the whole body divides an elongated cavity into two spherical portions, as

[1] *Ehrenberg* (loc. cit. p. 321, Taf. XXXIII. fig. viii.), deceived by this illusion, has taken the eight to twelve contractile cavities of *Trachelius meleagris* for stomachal cells, filled with red gastric juice. He has also regarded these cavities, when simple or double, as seminal vesicles. (Abhandl. d. Berliner Akad. 1833, p. 172, — 1835 p. 158.) In species having but few, he has very arbitrarily decided that some are seminal vesicles, others stomachal pouches, as, for example, in *Amphileptus* (loc. cit. p. 355). According to him, the seminal vesicles, upon contraction, pour the sperm upon the eggs contained in the body. It really seems very strange that these animals should practise uninterruptedly these pollutions throughout their entire life. These animals have neither testicles nor ovaries, and the function of these cavities is not, therefore, that assigned to them by *Ehrenberg*, — but is, as I think, with *Wiegmann* (Arch. f. Naturg. 1835, I. p. 12), analogous to that of a heart.

[1] *Dujardin*, Ann. d. Sc. Nat. Zool. tome X. Pl. XV. fig. 3; also, "Infusoires," Pl. VIII. fig. 6. *Ehrenberg's* plates of these star-like vesicles are incorrect.

though it were a drop of oil. The observation of these phenomena would make it doubtful whether or not these cavities are true vesicles or vessels.

These cavities have been met with in only a few of the Astoma, and these are, *Cryptomonas ovata* [2] and *Opalina planariarum*.

§ 18.

The Infusoria appear to respire solely by the skin. In those species whose bodies are covered with vibratile cilia this function is promoted by the vortical action of the water caused by these organs. In others, the contractile cavities just described are situated immediately under the skin, and the opinion may be entertained that the water so communicates with their liquid contents as to perform a respiratory function. In this respect *Actinophrys sol* is quite remarkable, for its contractile cavities are so superficial that when filled they raise the skin in the form of aqueous vesicles,[1] which, however, are so elastic as entirely to disappear in the parenchyma. Here it is plain that a mutual relation between the external water and the contents of these cavities might easily take place.

CHAPTER VIII.

ORGANS OF SECRETION.

§ 19.

No special organ of secretion has been found in the Protozoa; their skin, however, has a power of secreting various materials, which in some species harden and form a carapace, or a head of a particular shape; while in others it serves to glue together foreign particles, forming a case, in which the animal retreats.

Among those having a carapace, may be mentioned *Vaginicola, Cothurnia,* and *Arcella.* This more or less hard envelope does not resist fire, and is probably of a corneous nature. In the Rhizopoda, however, it is usually calcareous, like the shells of Mollusca, and is not affected by heat. The *Difflugiae* carry about with them an envelope of this kind, composed of grains of sand.

---

2 *Ehrenberg,* loc. cit. p. 41, Taf. II. fig. xvii.
1 *Ehrenberg* (Ibid. p. 303, Taf. XXXI. fig. vi. 1) appears to have taken the protrusion of these contractile vesicles for that of a snout.

CHAPTER IX.

ORGANS OF REPRODUCTION.

§ 20.

The Infusoria propagate by *fissuration* and *gemmation*, and never by eggs.[1] They have therefore no proper sexual organs.

This fissuration occurs longitudinally with some,[2] transversely with others,[3] and in many of them by both at once.[4] Gemmation, on the contrary, is very rare.[5]

§ 21.

Nearly all the Infusoria and Rhizopoda have in their interior a nicely-defined body, a kind of a nucleus, which is quite different, in its compact texture, from the parenchyma by which it is surrounded. This nucleus, which, in different species, varies much in number and form, performs an essential part in the fissuration. For, every time the individual divides either longitudinally or transversely, this nucleus, which is usually situated in the middle, divides also. So that, in the end, each of the two new individuals has a nucleus. When an animal is about to undergo fissuration, there is generally first perceived a change in the nucleus. Thus, in *Paramœcium, Bursaria* and *Chilodon,* the nucleus is sulcated longitudinally or transversely, or even entirely divided,[1] before the surface of the body presents any constriction.

This nucleus, which is of a finely granular aspect and dense structure, retains perfectly its form when the animal is pressed between two plates of glass, and the other parts are spread out in various ways. By direct light its color appears pale yellow. It appears to lie very loosely in the parenchyma, and sometimes individuals may be observed turning their bodies around it as it rests motionless in the centre. From all this, it cannot be supposed that this nucleus attaches itself to other parts of the animal, and especially to the pulsatory cavities (*Vesiculæ seminales* of *Ehrenberg*).[2]

§ 22.

A simple, round, or oval nucleus is found in *Euglena, Actinophrys, Arcella, Amœba, Bursaria, Paramœcium, Glaucoma, Nassula* and *Chilodon.* But there are two which are round, and placed one after the other in *Amphileptus anser* and *fasciola,* in *Trachelius meleagris,* and *Oxytricha pellionella.* With *Stylonychia mytilus,* there are four.

---

[1] That which *Ehrenberg* has arbitrarily taken for eggs is sometimes granules of the parenchyma or pigment corpuscles, sometimes bits of food. He did not perceive that these bodies want all that which is necessary to make up an egg, — such as chorion, vitellus, and germinative vesicle and dot. It is on this account that he declares that he never has observed the hatching of young Infusoria. (Abhandl. d. Berliner Akad. 1835, p. 156.)

[2] *Vorticella, Carchesium.*

[3] This may be easily observed with *Stentor, Leucophrys, Loxodes,* and *Bursaria.*

[4] *Bursaria, Opalina, Glaucoma, Chilodon, Paramœcium, Stylonychia* and *Euplotes.*

[5] *Vorticella, Carchesium* and *Epistylis.*

[1] *Ehrenberg,* loc. cit. Taf. XXXVI. fig. vii. 13 to 19, Taf. XXXIX. fig. ix. 4, 5, 11-13.

[2] *Ehrenberg,* from a strange fancy, has taken this nucleus for a seminal gland. (Abhandl. d. Berliner Akad. 1835, p. 163. Also, loc. cit.)

It is not rare that a variable number of these round nuclei, arranged in a row, traverse the body in a tortuous manner. This is so in *Stentor coeruleus* and *polymorphus*, in *Spirostomum ambiguum*, and in *Trachelius moniliger*. In many instances the nucleus has the form of an elongated band, which is slightly curved in *Vorticella convallaria*, *Epistylis leucoa*, *Prorodon niveus* and *Bursaria truncatella*. In *Stentor Raeselii*, it is spiral, and in *Euplotes patella* and *Trichodina mitra*, it is shaped like a horse-shoe. In *Loxodes bursaria*, it is kidney-form, and encloses in one of its extremities a small corpuscle (nucleolus).

The round nucleus of *Euglena viridis* has in its centre a transparent dot. In *Chilodon cucullulus*, the nucleolus has a similar dot, and thus the nucleus as a whole resembles a cell.

§ 23.

These nuclei, which make Infusoria resemble cells, deserve a special attention, since they do not die with the animal. Thus the nucleus of *Euglena viridis*, which, according to *Ehrenberg*,[1] is globular when dying, and surrounded by a kind of cyst, remains unchanged a long time, or even increases in size, having no appearance of a dead body. It may be that the life of this animal, under these circumstances, is not finished, but only assumes another form.[2]

---

1 Loc. cit. p. 110.

2 Perhaps this nucleus, of which the animal is only a temporary envelope, is ultimately developed into a particular animal. Indeed, perhaps this species, as well as many others, are only the larval states of other animals, whose metamorphoses are yet unknown. It may properly be asked, if this nucleus has not, relative to the body containing it, the same signification as have the tubulous larvæ of *Monostomum mutabile* (see below) to the embryos they surround.

That the nucleus contained in Infusoria plays an important part in the propagation of those animalcules, is supported also by a recent observation of *Focke*, who witnessed the development of several young individuals in the nucleus of *Loxodes bursaria*. See Amtl. Bericht über die 22 tr. Versamml. deutsch. Naturforscher. in Bremen, Abth. ii. p. 110.

INTRODUCTORY NOTE TO THE ZOOPHYTA.

WITHIN the past six or seven years the Zoophytes have received more attention from naturalists than any other division of the animal kingdom. The labors of many, if not most of our ablest naturalists, have been directed towards an investigation of the humblest forms of animal life. This fact, combined with the recent improved methods and means for research, would alone be prophetic of the most signal advances in this group; indeed, our knowledge of all these forms has been so modified, as well as increased, that previous writings need rather to be re-written than revised. *Dana, Agassiz, Milne Edwards, Forbes, Dalyell, Müller, Busch,* and others, not to mention the continued labors of older observers, have effected these changes in this group.

The work of *Dana* is most excellent, and will remain a standard of authority in this department for a long time to come. Aside from the many details of structure, in it may be found the first and best philosophical exposition of the relations of organic development with these lower plant-like forms. Had this work been better known in Europe, there would have been saved the constant repetition of the most grave errors. On the labors of *Agassiz* no comment need be made; those who are in this department, whether as minute Anatomists or philosophical Zoologists, will not fail to understand and appreciate him. In the same field is *Busch,* who was extended his brief though excellent labors over the three classes of this whole group; as for the remaining authors mentioned, excepting *Müller,* their position in this department has long been established. *Müller's* researches have been mostly on the Echinoderms, and the careful tracing of the phases of their development and metamorphoses; but where so much has been done, I fear the limits of this book will preclude full details with this class.

This note would be unnecessary, were it not to show that I do not ignore the changes and advance which have been made in this group within the past few years; and more especially so, as I have allowed, in this edition, the classification to stand as in the original. Any great changes of this kind I could not think of making without the consent of the authors, who, although they would undoubtedly fully sanction them, are not sufficiently accessible to me just now, as these pages are going to press. So, however much the present classification may offend the eye of the Zoologist, yet the Anatomist will find under each head the proper details. Thus, he will find as full a description of the anatomical structures of the Bryozoa and Hydroid Polypi, as though they were referred to the Mollusca and Acalephae, where truly they respectively belong.   EDITOR.

BOOK SECOND.

POLYPI.

CLASSIFICATION.

§ 24.

THE POLYPI are either immovably fixed, or seated on a locomotive foot. Their soft body is in part enveloped by a solid support, the polypary. This last is often, for the most part, horny or calcareous; and by it numbers of these animals are united into greater or less groups. The central mouth is always surrounded by a coronet of contractile tentacles. The digestive apparatus is organized after two different types, upon which is based a division of these animals into two orders. The sexual apparatus is always without copulatory organs.

ORDER I. ANTHOZOA.

The digestive canal is without an anus, and opens into the general cavity of the body.

FAMILY : MADREPORINA.

Genera : *Oculina, Millepora, Madrepora, Caryophyllia, Astraea, Desmophyllum, Maeandrina, Monticularia, Agaricia, Favia.*

FAMILY : GORGONINA.

Genus: *Gorgonia.*

FAMILY : ISIDEA.

Genera : *Corallium, Isis.*

FAMILY : TUBIPORINA.

Genus: *Tubipora.*

FAMILY : ALCYONINA.

Genera: *Alcyonium, Lobularia, Alcyonidium.*

FAMILY : PENNATULINA.

Genera : *Veretillum, Pennatula, Virgularia.*

FAMILY: SERTULARINA.

Genera: *Sertularia, Campanularia.*

FAMILY: ZOANTHINA.

Genus: *Zoanthus.*

FAMILY: HYDRINA.

Genera: *Hydra, Eleutheria, Synhydra, Coryne, Syncoryne, Corymorpha.*

FAMILY: ACTININA.

Genera: *Actinia, Eumenides, Edwardsia.*

ORDER II. BRYOZOA.

The digestive canal is closed from the general cavity of the body, and opens behind through an anus.

FAMILY: RETEPORINA.

Genera: *Eschara, Cellepora, Flustra, Bicellaria, Retepora, Telegraphina, Tendra.*

FAMILY: ALCYONELLINA.

Genera: *Cristatella, Alcyonella, Bowerbankia, Vesicularia, Lagenella, Plumatella, Lophopus.*[1]

BIBLIOGRAPHY.

Ellis. Essai sur l'Histoire naturelle des Corallines et d'autres productions marines du même genre. La Haye, 1756.

Pallas. Elenchus zoophytorum. Hagae 1766.

Cavolini. Memorie per servire alla storia dei polipi marini. Napoli, 1785.

Rapp. Ueber die Polypen im Allgemeinen und die Aktinien insbesondere. Weimar, 1829.

Ehrenberg. Die Corallenthiere des rothen Meeres, in the Abhandl. d. Berliner Akad. 1832.

Johnston. A History of the British Zoophytes. Edinburgh, 1838.

Besides the important work of Dana, which will be often quoted in my notes, the additions to the literature of the true polyps have been few since the issue of this work, and have generally been published in the form of articles in the various periodicals, to which reference will be made in my notes. But the Bryozoa have been specially studied, and particularly in the following papers:

1 There are here enumerated only those families whose organization has been specially studied. This remark applies equally to the following classes.

4

Van Beneden. Recherches sur l'Anatomie, la Physiologie et le developpement des Bryozoaires. Mém. Acad. Brux. Tomes XVIII. XIX.
Recherches sur les Bryozoaires fluviatiles de Belgique. Ibid. Tom. XXI.
For further literature on the Bryozoa, see the writings quoted in my
notes, and especially those of *Allman.*   Ed.

CHAPTER I.

CUTANEOUS ENVELOPE AND SKELETON.

§ 25.

The Polypi are composed of either entirely soft parts,[1] or have for their
support a solid frame, which may be calcareous, corneous, or coriaceous.
This frame is always the product of the general skin, and ought therefore
to be compared to a cutaneous skeleton.* This skeleton, known by the
name of *polypary,* is formed partly internally, and partly externally, by
these animals. In the first case it is called an *axial,* and in the second a
tubular polypary.

The axial polypary consists, with some polyps,[2] of a dense substance,
apparently unorganized and composed of carbonate of lime; with others,[3]
of a corneous substance, equally unorganized. When the polypary is
coriaceous, it is often covered by a variable number of calcareous, fusiform
corpuscles, usually bossed or dentated.[4] With some calcareous polyparies[5]
this is also true, and then the corpuscles are arranged in compact reticulated masses. The tubular polyparies serve as a refuge for the animals
living in them, and in many cases, being common to many individuals,
these last are in direct relation to each other by the canals which traverse
the branching tubes. In the axial polyparies there are often cavities or
depressions of a variable size,[6] in which the animals can conceal themselves. When, however, these are wanting,[7] they retire, as is the case
with many soft polyps,[8] beneath their mantle. Sometimes,[9] these cavities
are closed by a movable operculum.

§ 26.

The skin of polyps is very transparent, and should be carefully distinguished from the parenchyma which it envelops. It is smooth, or it is
covered with ciliated epithelium. And, since it has been shown that many

---

1 The Actinina and Hydrina.
2 *Corallium.*
3 The Gorgonina.
4 These corpuscles are easily seen in *Alcyonium*
and *Lobularia.* (*Milne Edwards,* Ann. d. Sc.
Nat., Zool. IV. 1835, pl. XIII. fig. 9 ; Pl. XV. fig.
10–11.) Spicula of this kind are found in the
interior of their tissues, as well as on the surface.
Ehrenberg (Abhandl. d. Berl. Akad. 1841, Th. I.

p. 403, Taf. I.–III.) has described and figured these
spicula under the names of *Spongolithis* and
Lithostylidium.
5 The Madreporina.
6 *Millepora, Madrepora, Oculina* and *Astraea.*
7 *Gorgonia, Isis* and *Corallium.*
8 The *Actiniae.*
9 *Eschara* and *Cellepora.*

* It should here be remarked that the old, and
as now regarded, mistaken view of the formation of
the frame of Polyps is here repeated; for the frame
is generally an internal skeleton, as, for instance,

with *Madrepora, Astraea, &c.* For the formation
of Coral, see *Dana,* loc. cit.; and for the relations of
the corallium carried out in detail, see *Edwards* and
Haime, Ann. d. Sc. Nat. 1849, '50, '51. — Ed.

Anthozoa have the skin, and especially the tentacles, covered with cilia of this nature,[1] these last cannot be regarded as forming a differential characteristic between them and the Bryozoa, as has been done by *Ehrenberg.*[2]

§ 27.

The skin of many polyps is quite remarkable in having nettling or poisonous organs, to which it is only of late that the attention has been directed. They consist of transparent vesicles, having a dense membrane, of a round, oval, or cylindrical form, containing a clear liquid, and a very delicate filament of variable length, which is usually spirally coiled. By the least irritation of the skin, the filament is thrown out of the vesicle, of which it appears to be only a prolongation. These filaments adhere to objects coming in contact with the skin, and in this way the vesicles in question are separated from it.[1] These organs are probably the cause of the nettling sensation felt when certain polyps are handled.

§ 28.

Still more interesting are organs analogous to those just mentioned, and which belong to various species of *Hydra.*[1] They are found not only on the arms, but also upon the skin of the body and foot. They consist of oval vesicles, having a very long and delicate filament, which is slightly swollen and viscous at its free extremity, while the opposite one is directly continuous with the conical neck of the vesicle. The neck of each vesicle is surrounded by three hooks curved backwards. These are always elevated when the skin of the animal is irritated, and especially that of the arms when they seize their prey. This last is then wound about by the free, viscous end of the filament, and the attached vesicle being torn from the body, the whole is often entangled in the arms of adjacent polyps. When this occurs, the vesicles hang by their hooks to the arms of the polyps; and it is this that has given *Ehrenberg* the opinion that the vesicles are detached by their round extremity, that these animals watch their prey with the hooks erected, and that the vesicles and filaments can return into the interior of the arms.[2] But it is probable that they (the hooks) act more as poisonous than as prehensile organs; for if those from the arm of a *Hydra* seize upon a *Nais*, a *Daphnia*, or a larva of *Chironomus*, these last quickly die, even if they escape immediately after being taken.

1 *Erdl* has seen very distinct ciliated epithelium in *Actinia* and *Veretillum.* (See Müller's Arch. 1841, p. 423.)
2 Abhandl. d. Berl. Akad. 1834, p. 255, 377.
1 These nettling organs, which are much more common in the lower orders of the animal kingdom than was at first supposed, are yet quite imperfectly known. *Wagner* first discovered them in the *Actinia*, although he regarded them at first as the spermatic particles of these animals. (*Wiegmann's* Arch. 1835, II. p. 215, Taf III. fig. 7, also 541, I. p. 41 ; Icones Zoot. Tab. XXXIV. fig. 24.) These researches have been extended by *Erdl,* who has shown that they also exist with *Veretillum* and *Alcyonium.* (*Müller's* Arch. 1841, p. 423, Taf. XV. fig. 3--6 and 8, 9.) In *Alcyonium, Erdl* has observed the filament take, on its departure from the vesicle, first a riband-like, and then a spiral aspect. In *Desmophyllum stellaria*

(Ehrenberg), I have seen these cylindrical organs having a long spiral filament. With *Edwardsia, Quatrefages* has found these organs upon the whole surface of the body, as well as upon the arms. (Ann. d. Sc. Nat., Zool. 1842, XVIII. p. 81, Pl. II. fig. 4-6.) For the nettling organs of the *Tubulariae* and the *Actiniae,* see also *Wagner* in *Müller's* Arch. 1847, p. 195, Taf. VIII.
1 These were first described by *Ehrenberg.* (Mittheil. a. d. Verhandl. d. Gesellschaft naturf. Freunde zu Berlin 2 tes. Quartal, 1836, p. 28 ; also, Abhandl. d. Berl. Akad. 1835, p. 147 ; 1836, p. 133, Taf. 11.) They have been carefully studied by *Erdl* (*Müller's* Arch. 1841, p. 429. Taf. XV. fig. 10--13).
2 *Ehrenberg* has figured, ideally (Abhandl. d. Berl. Akad. 1836, p. 133, Taf. 11. fig. 1) an *Hydra* in the act of seizing its prey with extended hooks. In reality this animal is never thus seen.

These poisonous and prehensile organs are destroyed by use, which is also true of the nettling organs. But this loss is probably repaired by their speedy reproduction. This last circumstance may explain the various descriptions given them by different authors, for, probably they have been observed at dissimilar stages of development.[3]

3 *Erdl*, who has discovered a great number of these nettling organs, saw, in some cases, the thread directly continuous with the neck of the vesicle ; in others, these necks appeared furnished with spines directed backwards ; exactly as *Wagner* had before described, and as *Kölliker* had often observed (Beiträge z. Kenntniss d. Geschlechtsverhältnisse u. d. Samenflüssigkeit wirbelloser Thiere, 1841, p. 44, fig. 14). *Erdl* asks if these variations of form are not coincident with an increasing or decreasing activity of the sexual organs (see *Muller's* Arch. 1842, p. 305). *

* [§ 28, note 3.] These nettling organs of the Polypi have recently been very successfully studied by *Agassiz*, who has enjoyed the most enviable advantages with the Polypi and Acalephae of the North American coast. He has changed the entire aspect of the subject, besides almost exhausting it for future research. His special studies were made on the coral polyp of our southern coast, the *Astrangia Danae*, Agass. The complexity of structure of these *lasso-cells*, as he has very appropriately termed them, is truly wonderful for such minute forms. As I have also studied these forms, I will use my own language, in the description of what *Prof. Agassiz* has seen. There are several varieties of these cells or capsules, depending upon the arrangement and structure of the lasso ; sometimes this last is a simple coil, sometimes it is coiled about a staff which is erected from the base, but which is also a part of the projectile apparatus. In the first case, the lasso is much the longer and may be fifty or seventy-five times the length of the vesicle; while, in the second case, it rarely exceeds the length of this last by more than sixteen or twenty times. In all cases, the essential feature of these organs is the lasso or internal coil, which is of a most curious structure. In the first place, it is, in general terms, only an inverted portion of the vesicle or cell itself, an internal instead of an external cilium, coiled up in a regular manner. When thrown out, therefore, it is wholly inverted, and its projection consists of an instantaneous turning of the whole inside out. But the lasso, delicate as it is, has still more delicate structures on its surface. These consist of barbels arranged in regular spiral rows, which extend to the very extremity of the lasso. At this last point, they almost elude the highest and best microscopic powers. These barbels all point backwards when the lasso is extended, and serve, no doubt, as teeth, to prevent it from slipping on the objects over which it is thrown. But these most delicate structures, which in beauty transcend that of all other tissues, can be better appreciated by figures than by the most minute description ; see *Agassiz's* Memoir on Astrangia Danae (forthcoming in the " Smithsonian Contributions to Knowledge"), Pl. VI. These observations, however, were made in 1848 ; see Proceed. Amer. Assoc. Advancem. Sc. 1848, p. 68.

From my own observations there would, indeed, be nothing to add on the special points studied by *Agassiz* ; but a remark or two may be made as to the development of these forms.

The lasso-vesicle is, originally, only an epithelial cell, of a spheroidal shape. It soon elongates, its contents become cloudy, after which, the coil is seen, very faintly marked, lying on the inner wall. It would seem probable, therefore, that its formation was somewhat similar to that of the spiral vessels in plants, although it is true that the lasso-coils and these spiral vessels are analogous only in form and position, and not in structure. The details of the formation are unknown.

These lasso-cells are more widely distributed among the Radiata than hitherto supposed. *Agassiz* (as he has informed me by letter) has observed them on most of the Polypi and Acalephae, and even with some of the Mollusca, and although their general structure is the same, there are points of difference of even a zoological value.

EDITOR.

CHAPTER II.

MUSCULAR SYSTEM AND ORGANS OF LOCOMOTION.

§ 29.

The movements of Polyps are performed, partly by contractions of the sides of their body, in which are found no muscular fibres, and partly by a true muscular tissue. The fibres of this tissue have not regular transverse striæ, although during their contractions there are sometimes, though rarely, seen irregular transverse bands.[1]

§ 30.

In those Polyps having a true muscular system, this tissue is composed of interlaced fibres, forming a layer beneath the skin. A coarse net-work of this kind is seen in the arms of *Hydra*, although in the foot and rest of the body there is scarce anything comparable to muscular fibres.[1] Under the skin of *Synhydra*[2] and in the arms of *Eleutheria*[3] this muscular system is much more apparent. A similar layer, very distinct, is observed in *Actinia*, which, in their mantle, is composed of both longitudinal and circular fibres, the contraction of which draws the tentacles together, and this, combined with that of the radiating fibres of the foot, gives rise to the various forms of these animals.[4]

The Bryozoa have the muscular system more apparent; in the cavity of their body completely isolated fasciculi are seen, composed of parallel fibres, serving especially for the withdrawal of these animals into their cells. These fasciculi arise from the internal surface of the body, and are inserted partly into the base of the tentacles, and partly into the neck and digestive canal, — thus serving almost exclusively as retractors of these last.[5]

---

1 *Milne Edwards*, who declares he has seen striated muscular fibres in *Eschara* (Ann. d. Sc. Nat. VI. 1836, p. 3), must have been deceived. I have been unable to perceive them in *Eschara*, *Alcyonella*, *Cristatella*, and other species. *Nordmann* also has not found them in *Cellaria*. (Observ. sur la Faune Pontique, 1840, p. 679 ; also *Müller's* Arch. 1842, p. ccviii.) The irregular bands appearing during contraction, but afterwards disappearing, have been observed by *Quatrefages* with *Edwardsia* (Ann. d. Sc. Nat. XVIII. 1842, p. 84, pl. II. fig. 7, a–b).*

1 *Corda*, Nov. Act. Acad. C. L. C. Nat. Cur. XVIII. 1839, p. 299. Also Ann. d. Sc. Nat. VIII. 1837, p. 363.

2 *Quatrefages*, Ann. d. Sc. Nat. XX. 1843, p. 288, pl. IX. fig. 3–5.

3 *Quatrefages*, Ibid. XVIII. 1842, p. 281, pl. VIII. fig. 3.

4 *Berthold*, Beitr. zur Anat. u. Physiol. 1831, p. 16 ; also in the body of *Edwardsia*, *Quatrefages* has found longitudinal and circular fibres (Ann. d. Sc. Nat. XVIII. p. 84).

5 Similar muscles have been observed by *Farre* (Phil. Trans. 1837, p. 387) in *Bowerbankia*, *Vesicularia*, *Lagenella* and other Bryozoa. *Milne Edwards* has seen them in *Tubulipora* and *Eschara*. (Ann. d. Sc. Nat. VIII. 1837, p. 324 ; VI. 1836, p. 23, pl. I. fig. 1, c, 1, d ; pl. II. fig. 1, a.) *Coste* has given a very detailed description of the

---

* [§ 29, note 1.] *Busk* has described and figured the striated form of this tissue with *Anguinaria spatulata* and *Notamia bursaria*. (Trans. Microscop. Soc. of London, II.) I have been unable, however, after considerable search upon many Bryozoa, among which were several *Alcyonella*, to detect any appearances of this kind ; and I would venture a pretty confident opinion that in the spe-

cies examined no such form of muscle is present. Quite lately, however, the subject has been carefully examined by *Allman* (Rep. Brit. Assoc. 1850, p. 318), and his descriptions are such as to leave no doubt upon the existence of the striated fibre with the species he has examined, among which are he *Paludicellae*. — ED.

4*

With *Eschara* there are, moreover, two fasciculi in each cell, which move its operculum, and thus close the entrance of this cavity.[5]

§ 31.

Locomotion is performed by the Polyps in various ways.

With the *Hydrae*, by their long-stretching arms; with *Actiniae*, by the contractions of the disc of their foot; [1] while the *Edwardsiae*, having elongated bodies which are not attached by a foot, progress by vermiform movements.[2] With *Cristatella mirabilis*, the whole colony moves itself along by the foot-like basis, like the *Actiniae*.[3]

Some Polyps, at a certain period of their development, move freely in the water by discoid contractions of their body, like the pulmograde Acalephae.[4]

§ 32.

A very remarkable peculiarity is the presence, in certain Bryozoa, of organs shaped like a bird's head, and which swing to and fro at the base of their cells. In some species, these organs have the form of lobster's claws, being composed of both a fixed and a movable piece. This last is corneous, and moved by a muscle which arises from a cavity in the first. It is not yet known by what means either this beak is opened, or the whole organ moves to and fro.[1]

Equally unknown is the function of these singular organs, the movements of which persist after the death of the animal, and of which, therefore, they are independent.[2] They are perhaps organs of defence or prehension, and analogous to the *Pedicellariæ* of the Echinoderms.

muscles of *Plumatella* (Comp. rend. XII. 1841, p. 724 ; *Müller's* Arch. 1842, p. ccx).*
6 *Milne Edwards*, Ann. d. Sc. Nat. loc. cit. p. 24, pl. I. fig. 1, c.
1 *Berthold*, loc. cit. p. 14.
2 *Quatrefages*, Ann. d. Sc. Nat. XVIII. p. 74; also *Forbes*, Ann. of Nat. Hist. VIII. 1842, p. 243.
3 I have been able to confirm the observation of *Dalyell* (*Froriep's* Notizen 1834, No. 920, p. 276) upon this motion in *Cristatella*. *Trembley*, also, has observed that the corallum of *Plumatella cristata* moved half an inch in eight days (see his Mémoire pour servir a l'Hist. des Polypes d'eau douce, 1775, p. 298).
4 See the observations of *Steenstrup* (Ueber d. Generationswechsel, 1842, p. 20) upon *Coryne fritillaria* ; also those of *Van Beneden* (Mém. sur les Campanulaires, 1843, p. 29, or *Froriep's* neue Notizen, 1844, No. 663, p. 38) upon *Campanularia gelatinosa*.

1 These organs were first described by *Ellis* (Essai sur l'Hist. Nat. des Corall. 1756, p. 51, pl. XX. fig. A). *Nordmann* (Observ. sur la Faune Pontique, 1840, p. 679, pl. III. fig. 4) has described and figured them with much accuracy. In *Cellaria avicularis, Bicellaria ciliata* and *Flustra avicularis*, they are formed like lobster's claws. In *Retepora cellulosa* they are pincer-like, and in *Telegraphina* they are articulated stings. See also *Krohn* in *Froriep's* Notizen, 1844, No. 533, p. 70.
For the organs having the form of a bird's head and a lash, and which are present in certain Bryozoa, see also *Van Beneden*, Recherch. sur l'anat. &c., des Bryozoaires, in the Nouv. Mém. de Bruxelles, XVIII. 1845, p. 14, pl II. III., and *Reid* in the Ann. of Nat. Hist. XVI. 1845, p. 385, pl. XII.
2 *Darwin's* Voyage of the Beagle, 1844, pt. I. p. 252.†

*[§ 30, note 5.] *Allman* (Report Brit. Assoc. 1850, p. 314) has described a very complete muscular system in the fresh-water Bryozoa. In the species with bilateral lophophores, there are seven distinct sets : 1. Retractor muscles of the polypide; 2. The rotatory muscles of the crown ; 3. The tentacular muscles ; 4. The elevator muscle of the valve ; 5. Superior parieto-vaginal muscles ; 6. Inferior parieto-vaginal muscles ; 7. Vaginal sphincter. The walls of the stomach also contain circular muscular fibres.

With *Paludicella*, the muscular system is somewhat different ; there are here five sets, — the 1st, 5th, 6th, and 7th of the preceding, and the parietal muscles. But with the 1st there is here only a single instead of a double fasciculus. — ED.

†[§ 32, note 2.] See *Hincks* (Ann. Nat. Hist. VIII. 1851, p. 353), who regards these avicularia as organs of defence, and has observed them seizing and retaining foreign bodies. — ED.

CHAPTERS III. AND IV.

NERVOUS SYSTEM AND ORGANS OF SENSE.

§ 33.

As yet only a very rudimentary and imperfectly distinguished nervous system has been made out in the Polyps; this consists of round masses, which are regarded as composed of nervous matter (ganglia), situated in the parenchyma. A ganglion of this kind has been supposed to have been observed about the mouth.[1]

§ 34.

Investigations upon their organs of sense have not been more successful. However, the sense of touch appears developed over the whole surface of the body, but specially so in the extremely irritable arms and tentacles. But, as yet, no tactile nerves have been found in these parts. In the same manner, light, to which these animals show a greater or less sensibility, is perceived rather by the general surface of the body than by special organs.

There are, however, in some species, at particular stages of development, during which they swim freely about, certain nicely-defined bodies situated upon the sides of the body, and which may be regarded as special organs of light and sound. This is the case with *Syncoryne;*[1] and *Coryne*[2] has in their place four red organs which correspond exactly to those found on the border of the disc of the pulmograde Acalephæ, and which have been regarded as organs of sense.

The organ seen at the base of the six arms of *Eleutheria dichotoma* has quite the appearance of an eye; that is, there can be distinguished in

1 A double œsophageal ganglion has been observed by *Dumortier* (Mém. sur l' Anat. et la Physiol. d. Polypiers composés d'eau douce 1836, p. 41, pl. II. fig. 2) in *Lophopus cristallinus (Plumatella cristata* of Lamarck); and by *Coste* (Comp. rend. XII. 1841, p. 724)in the *Plumatellae* in general. *Nordmann* also has seen a similar ganglion under the mouth of *Plumatella campanulata* (Lamarck) (loc cit. p. 703), and of *Tendra zostericola* (Ann. d. Sc. Nat. XI. 1838, p. 190). According to *Van Beneden*, a nervous ring surrounds the œsophagus of *Alcyonella* (Anu. d. Sc.

Nat. XIV. 1840, p. 222). *Coste* asserts the presence of a nervous system in *Pennatula (Froriep's* neue Notizen, 1842, No. 450, p. 154). That which *Spix* pretended to have discovered in the foot of *Actinia* (Ann. d. Mus. d'Hist. Nat. 1809, p. 443, pl. XXXIII. fig. 4) has been properly rejected by most modern zootomists, as an illusion. See *Berthold,* loc. cit. p. 6.*

1 *Loven, Wiegmann's* Arch. 1837, I. p. 323.
2 *Steenstrup,* Ueber den Generationswechsel, p. 23.

* [§ 33, note 1.] *Allman* has observed with *Cristatella mucedo* a small roundish body situated at the upper end of the pharynx, and which he regards as a nervous ganglion (Rep. Brit. Assoc. Advancem. of Sc. 1846, p. 88). This observation he subsequently confirmed, and has observed with *Plumatella repens* this ganglion (which he terms the great œsophageal ganglion) send off a large filament to each of the tentaculiferous lobes; also a smaller one passing off at each side to embrace the œsophagus, while a very short one was distrib-

nted in the substance of this last organ. And, finally, another set of filaments were distributed to the organs about the mouth. See Report of the same, for 1849, p. 72. According to a late Report, this observer appears to have been able to make out a distinct nervous system in all the fresh-water Bryozoa, except *Paludicella.* He has, however, been able to detect no certain organ of special sense. See report of the same for 1850, p. 319. — ED.

it a cornea, a crystalline lens and a red pigment layer surrounding the whole.[1]

Furthermore, there are upon the border of the disc of the campanulate *Campanularia*, colorless corpuscles, containing a calcareous nucleus, which is transparent as a crystal and soluble in acid.

These organs should probably be regarded as the most simple form of the auditory organs, for they have only a simple vestibule with its single *otolite*.[4]

CHAPTER V.

DIGESTIVE APPARATUS.

§ 35.

The digestive apparatus of Polyps is formed after two different types. With the Anthozoa it consists of a mouth and a simple stomachal sac without an anus. But with the Bryozoa, there is a mouth and anus, and a digestive canal which may be divided into the sections of œsophagus, stomach, small intestine and rectum.

§ 36.

The mouth of Polyps is usually surrounded by a circle of long, very contractile tentacles or arms. These tentacles are tubular, and connect with the cavity of the body.[1] They are simple,[2] or pennate,[3] and may be disposed around the mouth in a single[4] or a multiple[5] circle ; they are also frequently covered with cilia.[6]

Thus, the cylindrical tentacles of *Actinia* are entirely covered by ciliated epithelium. With the Bryozoa, on the contrary, the slightly-flattened ten-

3 *Quatrefages*, Ann. d. Sc. Nat. XVIII. 1842, p 280, pl. VIII. fig. 1, d, d, and fig. 6.

4 See *Krohn* (*Müller's* Arch. 1843, p. 176) and *Kölliker* (*Froriep's* neue Notizen, 1843, No. 534, p. 81). *Van Beneden* has perceived in the campanulate and free individuals of *Campanularia gelatinosa* and *geniculata*, not only eight marginal bodies, each containing a calcareous nucleus, but also four nervous ganglia about the base of the stomach (Mém. sur les Campanulaires de la côte d'Ostende, 1843, p. 24–27, pl. II. III.). I am yet undetermined upon the question whether, as *Van Beneden* thinks, these bodies have sometimes the function of organs of hearing. I am also in doubt as to the opinion of *Huschke* (Lehre von den Eingeweiden und Sinnesorganen, 1844, p. 880), who regards as otolites the calcareous bodies which have been observed in the peduncle of *Veretillum cynomorium*. *Nordmann* (Versuch. einer Monogr. des Tergipes, p. 88) has described as auditory organs the marginal bodies of the free-swimming *Campanularia*.

1 This cavity which is in the arms of most Polyps

does not open outwards at the extremity of these organs. I doubt, in fact, if the Actinia are an exception to this. It therefore appears singular that *Rymer Jones* (A General Outline of the Animal King. p. 41, fig. 13), and *Lesson* (*Duperrey*, Voyage autour du Monde. Zoophytes, p. 82, No. 1, fig. 1), expressly mention and distinctly figure these openings ; the first with an *Actinia*, the second with an *Eumenides*. According to *Van Beneden* (loc. cit. p. 15) the tentacles of *Campanularia* are without these cavities. But this is contradicted by *Lorén* (*Wiegmann's* Arch. 1837, Bd. 1, p. 252). In *Hydra* the cavities open distinctly into the stomach, as is probably the case with many other *Hydrina*. *Frey* and *Leuckart* likewise doubt the constant presence of an orifice at the apex of the tentacles of the *Actiniae*.[*]

2 *Actinia, Hydra, Flustra* and *Campanularia*.

3 *Veretillum, Lobularia, Isis, Gorgonia*, and *Zoanthus*.

4 *Hydra, Flustra, Zoanthus* and *Veretillum*.

5 *Actinia* and *Caryophyllia*.

6 *Veretillum, Flustra, Eschara, Cristatella* and *Tubulipora*.

*[§ 36, note 1.] Subsequent researches have shown that the cavity of the tentacles does open externally through a small papilla. See *Dana*,

Structure and Classification of Zoophytes. Phil. 1846, p. 32. — ED.

tacles have only a single row of cilia, which move regularly and voluntarily, like the rotatory organs of the Rotatoria.

By means of the currents produced by the cilia of their tentacles, many Polyps draw towards their mouth light particles of food; [7] others make use of their ciliated arms to seize larger portions. [8] This act is aided by the nettling and various prehensile organs, which are more usually found upon those Polyp-arms having no cilia. [9] These organs are found upon the tentacles of *Actinia, Edwardsia, Veretillum* and *Alcyonium*, and without doubt serve for the seizing of the prey as well as its retention until death. But these should not be confounded with special prehensile organs found on the tentacles of certain species. These consist of a small coriaceous capsule, from which the animal can project a kind of sting. [10] By means of these organs, the animal can attach itself like a bur to external objects, and not by suction, as is generally supposed.

The circular or oval mouth is always situated in the centre of the anterior extremity of the body; it is often surrounded by a lip formed of circular fibres. [11] In a few species, the mouth projects like a cone at the base of the tentacles. [12] With the *Plumatellae* [13] the mouth is topped by a tonguelet covered with rapidly moving cilia. Some of the Anthozoa, which capture animals of considerable size, can, in swallowing them, dilate their mouth to an astonishing width. [14]

DIGESTIVE CAVITY OF ANTHOZOA.

§ 37.

The simple stomach of Anthozoa, which is of a variable length, opens in general directly external by means of the mouth, [1] and with a few species, only, is there a muscular œsophagus. [2]

With some, the stomach blends with the walls of the body, [3] but usually it is more or less isolated. There remains, therefore, a cavity of the body of variable size, and which is directly continuous with the cavities of the arms. In those Polyps living in colonies, it is prolonged into canals traversing the corallum, and in this way the cavities of the bodies of all the

---

7 Flustra, Eschara, Tubulipora and Cristatella.

8 Actinina.

9 *Hydra, Coryne, Eleutheria, Sertularia, Campanularia* and *Alcyonium.*

10 Such prebensile organs have been observed by *Quatrefages* upon the clavate tentacles of *Eleutheria*. He thinks also he has observed two muscles in their capsules, by which the retractile sting is projected (Ann. d. Sc. Nat. XVIII. 1842, p. 276 and 283, pl. VIII.; or *Froriep's* neue Notizen, 1843, No. 543, p. 230). The oval vesicles which roughen the tentacles of *Campanularia,* and which *Lovén* (*Wiegmann's* Arch. 1837, I. p. 252) has described as small spinous warts, are probably of the same nature. In *Hydra* each hook-organ upon the arm is surrounded by a group of similar vesicles, in the interior of which is a rigid bristle. These organs are here found only upon the arms. They are distinguished from the organs having hooks by their less size, and from their having no projecting filament. *Corda* has not properly distinguished them from the hook-organs, whose fila-

ment is still unprojected (see his Memoir in the Nov. Act. physico-medica XVIII. p. 300, Tab. XV. fig. 5, 9, 10). Perhaps the organs which *Erdl* (*Muller's* Arch. 1841, p. 424, Taf. XV. fig. 3) has seen upon the tactile lobules of *Veretillum cynomorium* are of this kind.

11 *Actinia* and *Edwardsia.*

12 *Hydra, Coryne* and *Campanularia.*

13 *Alcyonella* and *Cristatella.*

14 *Actinia* and *Hydra.*

1 *Veretillum, Alcyonium, Actinia* and *Hydra.*
2 *Edwardsia.* 8 *te Quatrefages* (Ann. d. Sc. Nat. XVIII. pl. I. fig. 2; pl. II. fig. 1, 2).

3 *Hydra.* The stomach of the arm-polyps is not, as has been formerly supposed, a simple excavation in the body. It has proper walls distinct from those of the body, by which, however, they are closely embraced. There is, therefore, in *Hydra* no cavity of the body, and the cavities of the tentacles open directly into the stomach. This is also true of *Eleutheria* (*Quatrefages,* Ann. d. Sc. Nat. XVIII. p. 283).

Polyps are placed in direct intercommunication. It is not rare to find this general cavity divided into chambers by mesenteric membranes stretching longitudinally from it to the external surface of the stomach.[4]

The base of the stomach of many, and perhaps all of the Anthozoa, is pierced by one or more valvular openings, which communicate with the cavity of the body.[5] These animals, by controlling at will these orifices, can allow to pass into the cavity of the body the proper materials, which are probably water and liquid chyle.[6] This digestive apparatus thus communicating with the cavity of the body, reminds one of the organization of the Infusoria.[7]

The cavity of the stomach is lined by very delicate ciliated epithelium, which is continuous through the orifices upon every surface of the cavity of the body and arms, and even into the intercommunicating canals of the corallum.

The color of the walls of the stomach is quite varied, and is due to certain pigment cells which very probably perform the function of a liver; for these animals are entirely wanting in any other glandular appendix of the alimentary canal, analogous to a liver.[8]

4 There are often eight of these longitudinal chambers, as in *Veretillum, Alcyonium* and *Alcyonidium* (see *Icones* zool. Tab. XXXIV. fig. 2 ; also Ann. d. Sc. Nat. IV. 1835, pl. XVI. fig. 3, and pl. XII. fig. 3, 4). In *Actinia* there are seven more. With *Edwardsia* the eighth mesenteric divisions do not reach the sides of the body (*Quatrefages* loc. cit. pl. I. fig. 2).*

5 These orifices were long ago observed by the elder anatomists under various Polyps. Afterwards their existence was incorrectly doubted by other naturalists ; for lately they have been distinctly made out. Thus, in *Veretillum cynomorium* (*Rapp*, Nov. Act. physico-medica XIV. 1829, p. 650), in *Alcyonidium* and *Alcyonium* (*Milne Edwards*, Ann. d. Sc. Nat. IV. p. 325, pl. XV. fig. 6), and in *Edwardsia* (*Quatrefages* Ann. d. Sc. Nat. XVIII. p. 91).

In *Sertularia* and *Campanularia* there are openings between the stomach and the tubulous cavities of the corallum (*Lister*, Phil. Trans. 1834, p. 371, and *Van Beneden*, Mém. sur les Campanulaires, loc. cit. p. 17). There must be direct communication of this kind with the *Actiniae*, since they regularly reject by their mouth nettling filaments, from the chambers of their body. With *Hydra*, the stomach communicates, by an orifice situated at its base, with the narrow tubulous cavity of its cylindrical foot. But at the extremity of this tube there is no oval opening, and the tube itself cannot be regarded as a rectum, for it receives neither faeces, nor fragments of food, and is not affected by the frequent enormous dilatations of these animals from surfeit. *Corda* therefore is incorrect in assigning an anus to these animals. (Nov. Act. physico-medica XVIII. p. 302, Tab. XIV. fig. 2, E.) He appears to have entirely neglected the foot of this animal, which, however, has been well figured by *Ehrenberg* (Abhandl. d. Berl. Akad. 1836, p. 134, Taf. II. fig. 1); and since *Roesel* (Insektenbel. III. Taf. LXXVIII. and LXXIX. fig. 2, and LXXXVI. LXXXVIII. fig. 6) has perceived it in all nonmutilated arm-polyps. *Sars* (Faun. littoral. Norveg, p.

21) has found with a *Lucernaria* a stomach opening inferiorly, and communicating directly with the cavity of the body. This communication has been observed also by *Fry* and *Leuckart* (Beitr. p. 3) with the *Actiniae* and several other Anthozoa.†

6 *Quatrefages* (Ann. d. Sc. Nat. XVIII. p. 87, 91) has seen the stomach of *Edwardsia* entirely filled with *Spirorbis*, and other solid food, without any of it passing into the cavity of the body.

7 With Infusoria, the lower end of the œsophagus is free, so that the food passes directly from it into the parenchyma of the body, where it forms a cavity ; but with the Anthozoa, there is a stomach, from which chyle alone can pass into the cavity of the body.

8 These cells are white in *Edwardsia*, yellow in *Alcyonidium* and *Alcyonium*, and brown in *Veretillum* and *Hydra*. In the last, the brown is distinctly due to irregular pigment granules of that color, floating in the clear liquid of the cell. Probably these cells, by bursting, empty their contents into the stomach ; at least, I have been able to find no excretory duct, such as *Corda* has figured with the *Hydra fusca* (Nov. Act. Acad. physico-medica XVIII. p. 302, Tab. XV. fig. 15—17 ; or Ann. d. Sc. Nat. VIII. p. 366, pl. XIX. fig. 15—17).

In *Hydra viridis*, these brown cells of the stomach can easily be distinguished from the layer of green pigment belonging to the parenchyma of the body. Moreover, if a transverse section of this animal is made, there appears a wide difference of organization between the internal and external surface of the stomach ; the first has ciliated epithelium and hepatic cells, the second a bare skin with prehensile organs. This being so, how can these animals be everted like the finger of a glove, as some naturalists have affirmed, and yet live ? for the two surfaces of the stomach, so different, could not replace each other, and then again the cavities of the arms would open directly outward. Indeed, it is not possible to return unmutilated an everted Polyp, since the inextensible cavity of its foot cannot leave the body with impunity. The gastric

* [§ 37, note 4.] With all the Actinaria the lamellæ of the visceral cavity are the multiples of six ; all the Alcyonaria have eight of these lamellæ. See *Dana* loc. cit. p. 49. — ED.

† [§ 37, note 5.] With the Actinoidea, recent researches have shown that the stomach communi-

cates with the cavity of the body by a *single* orifice only, which may be closed by muscles. See *Dana*, loc. cit. p. 40, 44, pl. XXX. fig. 3, a, b, c, d. It has been since verified by *Cobbold*, Ann. Nat. Hist. XI. 1853, p. 121, with figures. — ED.

DIGESTIVE CAVITY OF BRYOZOA.

§ 38.

The very complicated digestive canal of the Bryozoa floats freely in the spacious cavity of their body. It is composed of an œsophagus which, at its lower extremity, dilates into a round or oval muscular crop; [1] upon this immediately succeeds a cœcal stomach, from the upper portion of which a small intestine arises and passes upwards in front. This, after a course of variable length, ends by a constriction in a short but large rectum, which opens in the vicinity of the mouth, at the external side of the base of the tentacles. [2] The digestive canal here, therefore, is not in communication with the cavity of the body. Its whole inner surface is lined with very active, ciliated epithelium, which keeps its contents in motion, and especially the fæces of the rectum. The sides of the stomach are often colored brown, yellow or green, from the presence of hepatic cells.†

CHAPTERS VI. AND VII.

CIRCULATORY AND RESPIRATORY SYSTEMS.

§ 39.

A vascular system has yet been found only with a few Polyps; but there it is so apparent that its presence in others may be inferred. The blood-vessels exist upon both the sides of the body and of the stomach, and are in part longitudinal, in part circular, ending in a capillary net-work. They are not simple canals excavated in the parenchyma, but have proper walls, and circulate a liquid containing a great number of white (blood) globules. [1]

1 Juice of the Anthozoa must have a very great digestive power, since the *Actinia* eat hard-shelled crustacea, and even the soft *Hydrae* quickly dissolve the larvæ of *Nais* and *Chironomus*. But the indigestible parts of these animals, such as epidermis, bristles, hooks and jaws, are afterwards ejected by the mouth.

1 In *Bowerbankia* (*Farre*, Phil. Trans. 1837, p. 392, Pl. XX. fig. 5; Pl. XXI. fig. 7) this crop is composed of pyramidal corpuscles, with the apices pointing inward, so as to act like teeth. I have observed a very similar structure in *Alcyonella stagnorum*.

2 In *Bowerbankia* and *Vesicularia* the small intestine is very long (*Farre*. loc. cit. Pl. XX. and XXII). I have observed it very short with *Cristatella mirabilis*.*

* [§ 38, note 2.] According to *Allman* (Report Brit. Assoc. 1850, p. 310), the œsophagus succeeds the stomach without the intervention of any distinct crop with all the fresh-water Bryozoa. The stomach is large and thick-walled, and may be divided into a cardiac and a pyloric portion. The pylorus is distinctly valvular, and the intestine,

1 *Milne Edwards* has perceived a vascular net-work of this kind in the sides of the body, with *Alcyonidium elegans*, and *Alcyonium palmatum* and *stellatum* (Ann. d. Sc. Nat. IV. p. 338). Quite recently, *Will* has described the vascular system of *Alcyonium palmatum* (*Froriep's* neue Notizen, 1843, No. 599, p. 68). According to him, white vessels may be perceived, even with the naked eye, upon the longitudinal furrows of this animal. These enter the lobules on the border of the body, and there form a dense net-work, from which a branch is sent to each arm, and this last gives off laterally a twig to each tactile lobule. The principal trunk of the longitudinal vessels continues upon the sides of the stomach to the base of the tentacles. At the point where the bodies of the Polyps continue with the corallum, there are wide at first, passes along the side of the cardiac cavity and œsophagus, and rapidly decreases in diameter, until it terminates in a distinct anus just below the mouth. — ED.

† [Note at end of § 38.] See in this connection my note under § 13, note 2. — ED.

§ 40.

All Anthozoa and Bryozoa have a proper circulation; for there rises
and falls in the cavity of their body a liquid, which is usually clear, and
often contains round and colorless corpuscles. This rises even to the end
of the cavity of the tentacles, and then returns into that of the body
generally. In the colonial Polyps, these currents, by traversing the canals
of the corallum, thereby pass from one animal to another. This movement
is caused by ciliated epithelium, which, as we have just seen, lines all the
cavities of these animals.

With the Bryozoa, the cavity of whose stomach does not communicate
with that of the body, these currents are continuous, regular, and have a
definite direction. But with the Anthozoa they are changed by the
reciprocal action through the stomachic orifices of the liquids of the
stomach and cavity of the body. These currents are perceived in the
arms, even when the cavities of these organs open directly into the
stomach.[1]

§ 41.

Nothing can yet be positively said as to the nature of this circulating liquid,
for it is still doubtful whether this whole phenomenon should be regarded
as an aqueous or a sanguineous circulation. If we refer to the fact that the
Anthozoa can introduce water into the system through the apertures of the
stomach, it should be admitted that this system has an aqueous character,
performing, perhaps, the function of an internal respiratory apparatus,

given off from the eight principal longitudinal vessels numerous lateral branches, which anastomose frequently in the canals of the corallum, and finally form a capillary net-work. The white, semi-transparent corpuscles contained in thin blood have, according to Will, a diameter of about 1-1200 of an inch, and out of the vessels have a globular aspect. According to this same observer, there is a similar vascular system in Actinia.*

1 The circulation in question has been observed by many investigators. Trembley (Mém. pour servir à l'Histoire des Polyps, p. 219) has perceived it in Plumatella cristata. Dumortier (Mém. sur l'Anat. et la Physiol. des Polypes, p. 47) has confirmed this observation. Carolini (see his Memoir on the Anthozoa, p. 56, 87) has seen it in the tubes of several Sertularina. There are various opinions as to the cause of these currents. Gruithuisen (Isis, 1828, p. 506) studied them in the arms of Hydra, and regarded them due to a communication with a circular vessel surrounding the mouth. But, according to the observations of Meyen (Brown's Miscellaneous Botanical writings, IV. p. 490), of Ehrenberg (Mittheil. aus. d. Verhandl. d. Gesellsch. naturf. Freunde z. Berlin, 1836, p. 27) and myself, the cavities of the arms open directly into the stomach.

The movements of the liquid in the arms of Hydra are due not only to the general contractions of the body, as Gruithuisen and Meyen have supposed but also to the cilia covering these parts. This

* [§ 39, note 1.] Subsequent researches have failed to detect any true circulatory system with the real Polyps, and there now can be but little doubt that no such system exists. As with the Acalephs,

was first pointed out by Grant (The new Edinb. Phil. Jour. 1827, p. 107; or Outl. of Comp. Anat. 1841, p. 430), who observed these currents in Flustra, Lobularia, Virgularia and Pennatula. Nordmann, who has examined this circulation in the body and tentacles of Alcyonella diaphana, and Plumatella campanulata, and other Bryozoa, did not find any cilia. He compared the currents to those seen in the joints of Chara (Microg. Beitrag II. p. 75, or Obser. sur la Faune Pontique, p. 709). I feel positive about the presence of cilia in the body of Cristatella mirabilis and Alcyonella stagnorum. Lister has carefully described this circulation with Tubularia, Sertularia and Campanularia; and finding no adequate cause, has likened it to that of Chara (Phil. Trans. 1834, p. 366, et seq.). Ehrenberg (Abhandl. d. Berl. Akad. 1832, p. 299) and Lovén (Wiegmann's Arch. 1837, 1. p. 254) attribute these currents in Sertularia and Campanularia to a peristaltic movement of the canals of the body; which, however, Van Beneden (Mém. sur les Campan. loc. cit. p. 18) has been unable to see in these Polyps. Erdl (Muller's Arch. 1841, p. 426) attributes it, in Veretillum cynomorium, to cilia; and Will (Froriep's neue Notizen, 1843, No. 599, p. 69) has found all the cavities of the body and corallum of Alcyonium palmatum lined with cilia. It is, moreover, certain that the currents observed by Erdl (Muller's Arch. 1841, p. 428) and Dumortier (Mém. loc. cit. p. 52) in the tentacles of Actinia are due to ciliary action.

their nutritive and digestive systems are combined; and, as with them also, the circulating, nutritive liquid is chyme. See also Dana loc. cit. p. 35. — Ed.

while the tentacles, in the cavities of which are regular currents, serve as external organs of respiration, similar to branchiae.

But, if we regard the whole as a true circulation, the contained liquid with its corpuscles will be analogous to blood. But this view is opposed by the fact that, with *Alcyonium*, with *Actinia*, and perhaps many other Polyps, there is a true vascular sanguineous system.[1]

We ought, therefore, to compare the liquid in question to chyle, which passes from the stomach to the general cavity of the body, in the Bryozoa by exosmose, but in the Anthozoa by the orifices of the stomach.[2]

The opinion that these currents form a vascular system, moreover, is not reconcilable with the fact that the Anthozoa can at will empty the contents of their stomach into it, or in the same way shut off from it the water.

We are obliged, then, to regard all these cavities as constituting a vascular aqueous system, performing a respiratory function, by which, in the Anthozoa, all the internal parts are constantly bathed with fresh water. This renewal of water is effected by its alternate ingress and egress through the stomach,[3] during which chyle-corpuscles could easily, by being mixed with water, be carried into this aqueous system.

With the Bryozoa, where this system is, without doubt, equally one of respiration, we shall have to seek for the openings by which this renewal of water takes place. These are situated near the anus, and place the cavity of the body in direct communication with the external water.*[4]

1 See § 39, note 1.

2 *Ehrenberg* and *Loren* regard the canals of the corallum of *Campanularia* and *Sertularia* as direct prolongations of the stomach, and designate them as intestinal tubes, and their contents as chyme.

3 This alternate ingestion and egestion of water has been positively observed by *Lister*, *Loren* and *Van Beneden*, in *Sertularia* and *Tubularia*.

4 By an opening of this kind, *Meyen* (Isis 1828, p. 1228) saw escape the eggs of *Alcyonella stagnalis*, which were free in the cavity of the body. *Van Beneden* (Ann. d. Sc. Nat. XIV. 1840. p. 222) declares that he has observed at the base of the tentacles of *Alcyonella* a series of orifices,

which may be called *aquiferous mouths*, for by them the water enters the cavity of the body. This is perhaps the case with *Actinia*, also; for *Rapp* (Ueb. die Polypen u. die Aktinien, loc. cit. p. 47) has here found numerous small orifices scattered over the whole surface of the body, and through which are emitted jets of water when the animal is squeezed, thus showing that they belong to an aquiferous system. It is quite improbable that the hollow tentacles of *Actinia* are open by an orifice at their apex for the circulation of water, as many naturalists have supposed. *Quatrefages* (Ann. d. Sc. Nat. XVIII. p. 96) is quite opposed to this opinion. See also above § 36, note 1.†

* [End of § 41.] In this connection should be mentioned branchia-like organs, described by *Dana* (loc. cit. p. 42) with the Zoanthina. A pair of them is attached to each of the larger lamellæ. He remarks, "The structure of these organs is such that we can hardly doubt their branchial nature; yet no circulating fluid was detected within them." I find no other mention of these parts, except by *Lesueur* (Jour. Acad. Nat. Sc. Philad. I. 183–185, Pl. VIII. fig. 1, 5, 9), who regarded them as of an hepatic nature. — ED.

† [§ 41, note 4.] The true nature and relations of the respiratory and circulatory systems of the Bryozoa are yet imperfectly understood. There can be but little doubt that water is by some means introduced into the general cavity of the body, and there mingles with the nutritive fluid, which transudes through the walls of the alimentary canal. But the apertures for the introduction of this water have not yet been clearly seen. It is true that *Van Beneden* thinks he has found "Bouches aquifères," as above mentioned, but their existence there has not been fully verified, and is even denied by *Allman*. At present, therefore, it cannot be said that the Bryozoa have a true aquiferous system, like the Anthozoa. The perigastric fluid is, separated from the water, most probably the elaborated product of digestion, and the corpuscles therein contained chyle-corpuscles. *Allman's* view, therefore (Report Brit. Assoc. 1850, p. 319), appears the most correct : "The perigastric circulation, therefore, unites in itself the triple function of a chyliferous, sanguiniferous and respiratory system." — ED.

CHAPTER VIII.

ORGANS OF SECRETION.

§ 42.

Nothing like urinary organs have yet been found in Polyps. Perhaps the borders of the mantles of the cellular Polyps should be regarded as organs of special secretion, since by them the increase and production of these cells take place. [1]

CHAPTER IX.

ORGANS OF GENERATION.

§ 43.

Polyps reproduce by gemmation, fissuration, and by eggs.

1. *Fissuration* is comparatively rare; it takes place nearly always longitudinally, and the division may or may not be complete. [1]

2. *Gemmation* is their most common mode of reproduction. The new individuals may be completely detached, or may remain connected with the parent corallum.

a: In gemmation, complete separation of the young individual is, on the whole, rare. It is best known in *Hydra*, with which the buds always appear upon a certain part of the body, — that is, at its union with the foot. [2] A bud of this kind consists always of a simple fold of the wall of the stomach and the skin, so that the stomach of the young individual is in direct communication with that of the parent, and the chyme can pass freely from one to the other. When the foot of this new being has acquired a proper development, it is completely detached at its inferior extremity.

b: Gemmation without separation of the new beings is quite common with Polyps, and occurs with very various modifications. The buds are formed sometimes upon the sides, sometimes upon the base of the body. In the first case, the coralla have a dendroid aspect; in the second, they are more lamelliform, spherical or lapidescent. These variations are not limited to certain genera or species, being often due to external influences,

---

1 The calcareous tubes of *Tubipora*, and the corneous ones of the Sertularina and other Bryozoa, are, without doubt, secreted by the border of the mantle, as is true of the shells of mollusks.

1 According to *Roesel* (Insektenbelust. III. p. 504, 525. Taf. LXXXIII. fig. 3), fissuration takes place transversely with *Hydra*. Longitudinal fissuration is principally observed with the Madreporina. When it is complete the cells of the corallum are definitely limited, as in *Astraea*, *Favia*,

and *Caryophyllia*; but, when incomplete, the cells are branched, lobulated, and of irregular contour, as in *Agaricia*, *Maeandrina*, and *Monticularia*, &c.

2. *Roesel* (loc. cit. III. Taf. LXXXV. fig. 2, 3, 5, Taf. LXXXVI. and LXXXVIII. fig. g. h. and Taf. LXXXIX. fig. 4). The exceptions to this rule, which are sometimes observed, are probably due to lesions of an accidental nature.

and especially the nature of the soil upon which the colony may have been fixed.* [3]

§ 44.

3. It is probable that all Polyps reproduce by eggs. This requires two kinds of organs, one to produce the egg, the other the semen. Both kinds, *ovary* and *testicle*, have already been described in many species.

Their distribution is quite varied. In some, the sexes are united in the same individual,[1] in others they are distinct;[2] with the colonial polyps the sexes are separate, and each colony[3] may be composed of individuals which are androgynous, or those of one sex alone.[4]

Some species are sexless, and remain so ; but they produce by gemmation individuals of a particular character, which have sexual organs.[5] These last, which have usually either a campanulate or discoid form, are separated from the corallum often before the sexual organs have been formed, and which they do not acquire until an advanced period of their lives. During this time they swim freely about, like the pulmograde Acalephae,[6] for which, as well as for young Polyps, they are often taken.[7]

§ 45.

That the relations just described really exist, may be learned from the following facts : In *Coryne echinata* and *vulgaris*, there are formed at their base, quadrangular and campanulate individuals, which lay numerous eggs.[1] In like manner also, ovigerous capsules are formed about the base of *Syncoryne ramosa*.[2] In *Coryne fritillaria*,[3] the new individuals are completely detached and swim freely about, closely resembling Medusae. In this condition they are developed, and their eggs come to maturity.[4]

3 *Eschara* and *Flustra* have a lamellated form when fixed to stones, shells, or the broad leaves of Algae ; but are tubular when attached to the stems of plants. *Alcyonella stagnorum* undergoes similar changes in the form of its corallum. It divides in a regular dichotomous manner (Eichhorn, Beitr. zur Naturgesch. d. kleinsten Thiere. Taf. IV.; also Roesel, loc. cit. Taf. LXXIII. and LXXIV.), and in this form has been described under the name of *Plumatella campanulata* by *Lamarck*. But when a colony of these Polyps is fixed upon a stone or a sunken root, they commence to be developed in a dichotomous manner. But afterwards they become lapidescent by the branches of both nodes interlacing each other. As the mass becomes more voluminous and dense, the tubes of the dead generation support those of the living. (See *Lamouroux*, Exposit. méthod. des Genres de l'ordre des Polypiers, Pl. LXXVI. fig. 5.) Under this form this Polyp has received the name of *Alcyonella stagnorum* (see *Raspail*, Hist. Nat. de l'Alcyonelle fluviatile).†
 1 *Hydra.*
 2 *Actinia.* ‡
 3 *Alcyonella.*

* [End of § 43.] For a full account of the reproductive process with Polyps, and the most philosophical exposition of the relations of gemmation and its analogies and affinities with other developmental processes, see *Dana*, loc. cit. p. 55. No abstract can be given of such a work. — ED.

† [§ 43, note 3. For full details of the gemmiparous mode of reproduction with the Bryozoa, see *Van Beneden* (Recherch. sur l'organis. des

4 According to *Erdl* (Froriep's neue Notizen, 1839, No. 249, p. 101) the coralla of *Veretillum cynomorium* and *Alcyonium* have always either male or female individuals alone. *Krohn* has perceived the same of *Sertularia* (Muller's Arch. 1843, p. 181).
 5 *Coryne, Syncoryne* and *Campanularia.*
 6 *Coryne* and *Campanularia.*
 7 Very striking, at least, is the resemblance of *Van Beneden's* (Mém. loc. cit. pl. 11.) figure of a free female of *Campanularia gelatinosa* and those of *Sars* (Beskrivelser. loc. cit. p. 28, Taf. VI. fig. 14) of small Acalephae, named by him *Cytaeis octopunctata*, and by *Will* (Horae tergestinae, 1844, p. 68, Taf. II. fig. 5) as *Cytaeis polystyla.*
 1 *R. Wagner.* Isis, 1833, p. 256, Taf. XI.; also Icones zoot. Tab. XXXIV. fig. 16.
 2 *Lowen. Wiegmann's* Archiv. 1837, I. p. 321, Taf. VI. fig. 19–25.
 3 *Steenstrup.* Ueber d. Generationswechsel, p. 20, Taf. I. fig. 41–47.
 4 According to *Sars* (Beskrivelser. loc. cit. p. 6, Taf. I. fig. 3), these remarks are also true of *Corymorpha nutans.*

Laguncula, &c., Mém. Acad. Royale de Bruxelles, XVIII. ; also, Recherch. sur l'Anat. la Physiol. et le développement des Bryozoaires, &c. Ibid. XIX.) See also *Allman*, Report Brit. Assoc. 1850, p. 320. — ED.

‡ [§ 44, note 2.] According to my own observations, the *Actiniae* have both individuals which are hermaphrodites and those of one sex alone. — ED.

The *Campanulariae* and *Sertulariae* produce at the end of their pedicle and branches elongated sexless individuals. But in the angles of these branches cells of another form, and containing many spherical individuals, are developed. In these last sexual organs are formed, which, in *Campanularia geniculata*, occurs without a separation of the new individuals from the corallum, while in *Campanularia gelatinosa* it is after detachment has taken place.[5]

§ 46.

In the eggs of polyps both a germinative vesicle and dot may often be seen. Frequently, however, both disappear at a very early period. The envelopes of the egg are usually of a simple,[1] though sometimes of a complicated structure. The spermatic particles are very active, and in some species are filamentoid, in others composed of a solid body or head, to which is appended a very delicate tail. Water does not appear to affect either their form or motion.[2]

---

5 According to *Krohn* (*Müller's* Arch. 1843, p. 174), it is probable that in *Campanularia* and *Sertularia* both sexes are developed in this way. From *Ellis'* description of *Campanularia dichotoma* (Essai sur l'Hist. Nat. des Corallines, p. 116, pl. XXXVIII. fig. 3), it may be concluded that the females, mistaken by these naturalists for eggs, separate in this way from the corallum. *Meyen* (Nov. Act. physic-medica. XVI. Suppl. I. 1834, p. 195, Tab. XXX. fig. 3, 4) has also taken the medusoid females of this species for spawn.

[Additional note to § 45.] The series of these polyps, the sexless (nurse-like) individuals of which produce self-dependent, medusa-like young, has been increased by several more recent researches. See *Van Beneden*, Rech. sur l'embryol. d. Tubulaires, 1844, pl. 1. IV. (*Tubularia* and *Eudendrium*); *Sars*, Faun. littoral. Norveg. p. 7, Tab. I. (*Podocoryna* and *Perigonimus*); *Dujardin*, Ann. d. Sc. Nat. IV 1845, p. 257, pl. XIV. XV. (various Hydrina). It is true that the development of the genital organs has not been observed in these medusa-like individuals ; but they have indeed in the medusiform individuals of *Syncoryne ramosa* and *Coryne fritillaria*, and therefore it may be proper to infer that the same is true of other Hydrina and Sertularina. If it is correct to regard as the perfect state that in which the individuals resemble Medusae, and as the imperfect state that in which they are polypoid, then should we, as has been done already by many, remove these animals from the class of the Polypi, and place them with the Acalephae.*

1 In most Anthozoa. Eggs of this kind, belonging to *Actinia*, *Coryne* and *Veretillum*, have been figured by *Wagner* (Wiegmann's Arch. 1835, I. Taf. III. fig. 2 ; Prod. Hist. Gener. hom. atque anim. Tab. 1. fig. 1, and Icones zool. Tab. XXXIV. fig. 5, 17, 23).

2 With most Bryozoa the spermatic particles are filamentous. Both from their size and their motions, they have been taken for parasites. *Kölliker* (Beitr. zur Kennt. d. Geschlechtsverhalt. u. d. Stamm. Flüssigkeit wirbellos. Thiere, p. 41, Taf. II. fig. 17) has seen the spermatic particles of a thread-like form, of *Flustra carnosa*, developing in cells, and has seen them moving in the cavity of the body. I have seen similar ones in *Cristatella mirabilis* and *Plumatella campanulata*. Those which were seen by *Farre* (Phil. Trans. 1837, p. 403, pl. XXIII. fig. 5, g) in the cavity of the body of *Valckeria cuscuta*, and were regarded by him as intestinal worms, have an oval body, to which is attached a delicate tail. *Nordmann* (Faune Pontique loc. cit.) has found those of *Cellaria avicularia* having the same form. Those of *Actinia* have also a similar form (see *Erd' Müller's* Arch. 1842, p. 301, and *Kölliker*, loc. cit. p. 44, fig. 13). One should be careful and not confound the spermatic particles with the nettling organs having a similar form ; and especially as the development of these last has apparently some connection with that of the sexual organs (see *Erd'* loc. cit. p. 305). According to *Kölliker*, the spermatic particles of *Alcyonidium gelatinosum* have a lanceolate body, with a hair-like tail (loc. cit. fig. 11).

Spermatic particles of a cercaria-form have been observed by *Wagner* (Icon. zool. Tab. XXXIV. fig. 7, 12) with *Veretillum* and *Hydra*; by *Van Beneden* (Rech. sur l'organisat. d. Laguncula, and Rech. sur l'Annat. d. Bryozoaires, pl. V. in the Nouv. Mém. de Bruxelles, &c. XVIII.), with *Laguncula* and *Halodactylus*; by *Rathke* (Wiegmann's Arch. 1844, I. p. 164, Taf. V. fig. 6) and *Steenstrup* (Untersuch. üb. das Vorkommen d. Hermaphrodit. p. 66, Taf. I. fig. 18, c) with *Coryne*; finally by *Kölliker* (Neue Denkschr. VIII. p. 48, fig. 20, 21, 22, 24) with *Pennaria*, *Eudendrium* and *Sertularia*. In *Crisia*, on the other hand, *Kölliker* found the spermatic particles perfectly filiform. †

---

* [End of additional note to § 45.] The remarkable relations here spoken of, and the conjectures as to the real zoological nature of the animals in question, have been pretty satisfactorily cleared up by the recent researches of *Agassiz*. He has shown that the Hydroid Polyps are not simply a lower form of stemmed animals, producing at a given period more highly-organized Medusae, but that they are themselves, by their structure, real Medusae. See Lectures on Comparative Embryology, 1848; also Proceed. Amer. Assoc. for the Advancement Sc. 1849 ("On the Plan of Structure and Homologies of Radiated Animals"), and Mem. Amer. Acad. loc. cit. p. 225. — ED.

† [§ 46, note 2.] I have been able to trace the development and character of the spermatic particles of many of the true Polyps and the Bryozoa. The development occurs in special daughter-cells,

§ 47.

I. With those polyps which are not sexless, and whose alimentary canal hangs free in the cavity of the body, the sexual organs are situated in this last. They often escape attention, since they are scarcely at all developed except at the sexual epoch. Both ovaries and testicles frequently appear as riband-like bodies, which, being attached by one extremity alone to the stomach, move freely in the general cavity of the body. Sometimes, however, they are attached longitudinally by one of their borders, like a mesentery, the opposite border being free. In other cases, again, they are attached directly to the sides of the body.

The eggs and spermatic particles pass directly from the sexual organs into the cavity of the body. In Coralla having individuals of both sexes, fecundation takes place in the cavities of their bodies, which connect with each other.[1] With the others, however, the individuals of which are of one sex alone, the surrounding water is the medium of fecundation, by transporting the spermatic particles unaffected to the eggs; and this being performed by the aqueous circulation before mentioned, impregnation takes place in the cavity of the body.

§ 48.*

The variations of the internal genital organs in the different families are as follows :

1. With the Bryozoa, a riband-like ovary and testicle are suspended from the extremity of the stomach. In these organs are developed only two to four eggs or fasciculi of spermatic particles, from cells arranged like a string of pearls.[1]

The eggs, of which the germinative vesicle and dot disappear at a very early period, are detached from the ovary before their shell is well formed, and are set in motion by the cilia of the cavity of the body. Usually they are flattened, and at first enveloped by a thin and colorless membrane, which soon becomes thicker and darker, and has upon its borders a clear,

---

1 With *Tendra zostericola*, which is allied to *Flustra*, the Polyps are contained in cells closely bound to each other. But the cells of the males communicate with those of the females by an opening, through which the spermatic particles pass into the cavity of the body of the female (see *Nordmann*, Ann. d. Sc. Nat. XI. 1839, p. 191).

and the particles themselves are the metamorphosed nuclei of these cells, exactly as in other and higher animals. They have invariably, as far as I am acquainted, a cercaria-form consisting of a solid head, to which is attached a most delicate tail. The shape of this head, when studied carefully with the best powers, presents differences of zoological import. Sometimes it is pyriform (*Tubularia, Actinia*), sometimes conical (*Astrangia*), while among the Bryozoa it is long-oblong with *Alcyonella*. I cannot therefore agree with *Kölliker* (Cyclop. Anat. *Art.* Semen. 1849, p. 497) as to the mode of development of these particles with these animals. — Ed.

5*

1 See. for *Alcyonella stagnorum*, *Meyen* (Isis, 1828, Taf. XIV. fig. 1), for *Plumatella cristata*, *Dumortier* (loc. cit. pl. I. fig. 3, u, u) and for *Cellaria avicularia*, *Nordmann* (Obs. sur la Faune Pontique, p. 679, fig. 4, A. u).†

* [§ 48.] In an emendatory note at the end of the volume, the author remarks: "Sections 2d and 3d of this paragraph should be omitted, since the genital organs, with all the Anthozoa, are attached on the internal surface of the visceral cavity. See *Frey* and *Leuckart*. Beitr. &c. p. 13." I have, however, allowed them to remain, for the sake of their notes. — Ed.

† [§ 48, note 1.] My own researches in 1851 have shown me that with *Alcyonella* the sexes are separate. The testicles and ovaries consist of pedunculated sacs, closed at first, but which are ruptured on the mature development of their contents. — Ed.

transparent ring. In *Alcyonella* and *Plumatella*, the eggs are of an oval shape, and of a dark-brown color. In *Cristatella mirabilis, Dal. (Cristatella mucedo, Cuv.*), they are lenticular and clear brown, and have this remarkable peculiarity: [2] Upon both sides of the encompassing ring are a number of double-pointed hooks, which, at first, are imbedded in a gelatinous substance; but as this last is dissolved by water, they become free, and adhere to plants and other bodies. [3]

2. With many Anthozoa, having a cavity of the body, the sexual organs are attached in the form of bands along the external face of the stomach. These are numerous, and during the epoch of reproduction their free borders are often plicated, and have a botryoidal aspect. This form is quite apparent in the *Actiniae*, where these organs are contained in separate chambers of the cavity of the body. [4] The same is true of the *Edwardsiae*. [5] With *Veretillum* [6] and *Alcyonium* [7] these organs form mesenteric divisions which descend deep into the cavity of the body.

3. In *Alcyonidium elegans* [8] and *Tubipora musica* [9] these organs are attached to the internal surface of the cavity of the body, and have a plicated mesenteric form. [10]

§ 49.

The laying of the eggs takes place in different ways with those Polyps having internal sexual organs. With the Bryozoa it probably occurs through the openings near the anus. [1] With the Anthozoa, however, they pass into the stomach through its abdominal orifices, and thence are ejected through the mouth. In the viviparous *Actinia*, the young, developed at the base of the stomach, are expelled in the same manner. [2]

§ 50.

II. Many Anthozoa, which have no general cavity of the body, have *external sexual organs*. This is especially true of *Hydra*, where in the

---

2 *Raspail*, loc. cit. pl. XII. fig. 10–12, pl. XIV. fig. 4–8, and pl. XV. fig. 5.
3 *Turpin* and *Gervais*, Ann. des Sc. Nat. VII. 1837, pl. III. A. fig. 2–4, and pl. IV. A. fig. 1–6.
4 *Wagner. Wiegmann's* Arch. 1835, I. Taf. III. fig. I ; also Icones zool. Taf. XXXIV. fig. 22.
5 *Quatrefages.* Ann. d. Sc. Nat. loc. cit. pl. I. fig. 7, and pl. II. fig. 10.
6 *Carus* and *Otto.* Erläuterungstafeln, Heft. IV. Taf. I. fig. 19 ; also *Wagner*, Icones zool. Taf. XXXIV. fig. 2.
7 *Milne Edwards.* Ann. d. Sc. Nat. loc. cit. pl. XIV. fig. 4; pl. XV. fig. 6, 8, and pl. XVI. fig. 3–5.
8 Ibid. p. 329, pl. XII. fig. 3, pl. XIII. fig. 2, 7.
9 *Rymer Jones.* Outlines, loc. cit. p. 36, fig. 9, after *Lamouroux*.
10 *Külliker's* observation upon the sexual organs

of *Alcyonidium gelatinosum*, Johnst. (*Halodactylus diaphanus* of Farre), is quite remarkable; for he found them wanting in the isolated individuals, but scattered here and there, in the form of small round sacs, in the fleshy substance of the corallum — some being ovaries, others testicles. But he is in doubt whether or not their contents are emptied into the cavity of the body or upon the outer surfaces (Beitr. loc. cit. p. 46). [*]
1 See, for *Alcyonella stagnorum, Meyen* (Isis 1828, p. 1228).
2 *Rathké* has often found spawn in the stomach of *Actinia* (Reise Bemerk. aus Taurien, zur Morph. 1837, p. 10, and Beitr. zur vergleich. Anat. u. Physiol. in der neuesten Schrift. d. naturf. Gesellsch. zu Danzig, III. Hft. IV. 1842, p. 112).

---

*[§ 48, note 10.] With the Actinina, some of the lamellae which partition off the visceral cavity are margined each by a white, capillary, convoluted cord. It is attached to the lamellae by a thin, mesentery-like membrane. These cords are the testicles. Between the spermatic lamellae are others similarly arranged, which are the *ovarian*, on

which are situated the ovaries. With the Zoanthidae the relations are of the same general nature; but with the Tubipora, *Dana* found six spermatic to two ovarian lamellae. See *Dana*, loc. cit. p. 43, pl. XXX. fig. 3, b, c, d, e, f, and pl. LIX. fig. 1, b — Ed.

same individual during the time of heat both ovaries and testicles are developed upon the external surface of the body.

In the place where the eggs are to appear,[1] the transparent and colorless skin rises in the form of swellings, under which the vitelline mass gradually forms. These end each in the form of an excrescence, which, being constricted at its base and rounded, has the shape of an egg. At the point of constriction there is formed from the body of the Polyp a kind of cupel, in the cavity of which the vitellus rests by a small portion of its surface; at this point the skin becomes thin, and ultimately appears like an arachnoid membrane enveloping the egg. In this last neither a germinative vesicle nor dot has been discovered. Its separation is preceded by a thinning of its surrounding membrane, after which the vitellus is immediately clothed by a gelatinous substance. In *Hydra vulgaris* its whole circumference is covered by obtuse prolongations of this kind, which, after an increase in length, divide, each once or more, at their extremity, and so present a dentated appearance.

The arachnoid membrane finally bursting, the detached egg becomes fixed to some body, whilst the gelatinous coat entirely disappears. This is equally true of *Hydra viridis*, with the exception that here the vitelline prolongations are very short and compact.[2]

In these same individuals testicles are developed also. Between the base of the tentacles and the place of the appearance of the egg, there are developed small conical prominences, on the apex of which is a papilla. This has an orifice which leads into an internal cellular cavity. This is the real testicle, wherein are found spermatic particles composed of a body, or head, to which is attached a very movable tail. These particles easily escape through the orifice, and circulate in the water surrounding the Polyps filled with eggs.[3] The number of these testicles in a single individual is not definite.[4] *

1 In the arm-polyps, gemmation always precedes propagation by eggs.
2 The eggs of *Hydra* were long ago observed by *Bernhard Jussieu* (Abhandl. d. schwed. Akad. 1746, VIII. p. 211). But afterwards they were regarded as exanthemata of this animal (see *Roesel*, Insektenbelust. Th. III. p. 500, Taf. LXXXIII. fig. 1, 2). Their true nature was lately first pointed out by *Ehrenberg* (Abhandl. d. Berliner Akad. 1836, p. 115, Taf. II.).
3 The testicles of *Hydra* were known to the elder naturalists, but were taken for an eruptive disease (*Trembley* Abhandl. zur Geschicht. einer Polypenart, p. 264, Taf. X. fig. 4, and *Roesel*, loc. cit. p. 502, Taf. LXXXIII. fig. 4). Latterly this same error has been continued (*Laurent* in *Froriep's neuen Notizen*, 1842, No. 513, p. 104). To *Ehrenberg* is due the first description of their true nature (Mittheil. aus den Verhandl. d. Gesellsch. naturf. Freunde in Berlin, 1838, p. 14).

4 *Wagner*, Icones zoot. Tab. XXXIV. fig. 10, b, b. In *Hydra vulgaris* I have counted fifteen testicles; another individual had seven eggs and eleven testicles; and a third, four eggs and twelve testicles.
[Additional note to § 50.] Other examples of Anthozoa having external genital organs in the form of egg or sperm capsules have been observed by *Van Beneden* (Rech. sur l'embryog. d Tubul. pl. V. VI.), *Rathké* (*Wiegmann's* Arch. 1844, I. Taf. V.), and *Sars* (Faun. littoral. Norveg. p. 7, Tab. II.), with *Hydractinia*, *Coryne* and *Podocoryne*. See also the facts collected by *Frey* and *Leuckart* (Beitr. &c. p. 28). These egg or sperm capsules may, moreover, be regarded as imperfect male or female individuals, and then the porters of these capsules may be considered, being sexless individuals like those mentioned in § 45, in the category of nurse-like generations which, after a more or less complete development, produce generations with sex.

* At end of § 50.] The so-called ova, mentioned above in the text, may be justly questioned as being true ova, for we know of no real ova which do not contain a germinative vesicle. Then, again, simple oval masses of cells as they are, they would exactly resemble the bud-like eggs of Aphides, and the "hibernating eggs" of *Daphnia* and some of the Rotatoria, all of which are properly gemmae, and do not require the agency of the spermatic

particles for their development. It is also worthy of remark, in this connection, that these ova sprout from the same part of the body in which eggs are developed. *Thomson*, however (Edinb. New Philos. Jour. 1847, p. 287), speaks of having observed the granular mass contained within these so-called eggs divide and subdivide like a proper vitellus, and this while still within the capsule, and attached to the parent animal. This does not

§ 51.

III. There are Polyp-colonies which contain two kinds of individuals, those which are sexless, and those having sexual organs only at certain epochs. These last are campanulate or medusoid, and their sexual organs are developed in various parts of their body.

In *Coryne* [1] and *Syncoryne*, [2] the eggs appear upon the external surface of the stomach, then fall into the cavity of the mantle, through the openings on the border of which they escape into the water. In the medusoid individuals of *Coryne fritillaria* and *Corymorpha nutans*, the sexual organs appear to be formed in the angles of the borders of the disc, [3] and in *Campanularia* in the disc itself. *

§ 52.

As to the embryonic developments of Polyps, it is probable that in a great number (perhaps all) there is a *metamorphosis*.

The development commences by the usual segmentation of the vitellus, [1] by which it is ultimately converted into an ovoid, contractile body; this turns upon its longitudinal axis by means of cilia, with which it is entirely covered, swimming about like many Infusoria. These embryos, often developed in the mother, have sometimes been taken for swimming eggs. [2] Afterwards they attach themselves to some body, and usually lose their cilia; the free extremity of their body opens, allowing the escape of the Polyp, which, in the mean while, has been developed in the interior, with its arms in front. Many of the Polyps thus produced multiply by gemmation, and thus become the foundation of new Polyp-colonies. [3]

---

1 *Wagner* (Isis 1833, Taf. XI. fig. 8).
2 *Loven* (*Wiegmann's* Archiv. 1837, I. Taf. VI. fig. 19, 20).
3 *Steenstrup*, Ueber d. Generationswechsel, p. 23, 24.
1 It is indeed singular that with *Hydra* the division of the vitellus takes place before the eggs are either detached from the body, or are surrounded by a dentated envelope. I do not yet know at what epoch the development of the embryo commences, for I have never seen the young come forth. It is impossible for me to say whether or not these Polyps experience a metamorphosis. *Pallas* (Karakteristik d. Thierpflanzen p. 55) has seen the young Polyps come forth from the egg, but he gives no description. *Laurent*, also, only says that the young animal escapes formed from the egg, without describing the embryo (*Froriep's* neue Notizen, No. 513, pl. 101). The segmentation of the vitellus has been observed by *Van Beneden* in the eggs of *Pedicellina*. See his Rech. sur l'anat. d. Bryozoaires (suite) loc. cit. XIX. p. 18, pl. II.
2 As would be inferred from his description, *Cavolini* (loc. cit. p. 47, 50, Taf. IV. fig. 7–10 and 13–15) has observed similar embryos to those of *Gorgonia* and *Madrepora*. His descriptions of various eggs of *Sertularia* leave no doubt that they also

make the matter any more clear; for, even admitting that they are proper ova, it is difficult to conceive how the impregnation (of which the segmentation for a definite result is the sequela) could take place while the ova are thus buried in the capsules.
The subject requires further research. See also *Steenstrup*, Untersuch. üb. Hermaphroditismus, p.

were embryos (Ibid. p. 56, 80 et seq.). *Grant* also has taken for eggs the contractile, ovoid embryos of *Lobularia digitata*, which he has seen issue from the mouth of this animal (*Froriep's* Notizen 1825, No. 440, p. 340). *Meyen* has well described and figured the ciliated epithelium of those of *Alcyonella stagnorum* (Isis 1828, p. 1228, Taf. XIV. fig. 4, 5). *Loven* has observed the elongated embryos of *Campanularia geniculata*, and has taken the division of the vitellus for a spontaneous fissuration of the embryos (*Wiegmann's* Archiv. 1837, I. p. 260, Taf VI. fig. 13, 14). According to *Rathke*, who has seen movable lenticular embryos in the stomachs of *Actinia*, these polyps experience a metamorphosis (Reise Bemerk. aus Taurien zur Morph. p. 10, Taf. 1, fig. 12).
3 This metamorphosis has already been observed by *Cavolini* (loc. cit. p. 261, Taf. VI. fig. 7) with *Sertularia racemosa*, and more lately by *Loven* (loc. cit. p. 261, Taf. VI. fig. 15–17) with *Campanularia geniculata*. There are always developed in the interior of the embryos of *Alcyonella stagnorum* two Polyps, even before the first have escaped from the egg; when the escaped embryo has become fixed, its skin bursts, and the Polyps escape but are able to return again as into a mouth

116, and *Hancock*, Ann. Nat. Hist. 1850, V. p. 282.—ED.

* [End of § 51.] See *Schultze* (*Müller's* Arch. 1850, p. 57), who has found with *Campanularia* seminal capsules corresponding to those for egg-capsules pointed out by *Loven* (loc. cit.). — ED.

This metamorphosis is completed when the skin is covered by a brown and solid layer, and new individuals are developed by gemmation from the two Polyps (See *Meyen*, Isis, loc. cit.). I have seen the development of the coralla of *Cristatella mirabilis* and *Plumatella campanulata* occur in the same way. With the *Cristatellae*, gemmae of new Polyps are often seen to arise from the skin, even after the escape of the two Polyps, and before the young colony has become at all fixed. At this epoch of development these Polyps have been taken by *Cuvier* for a distinct species, and called *Cristatella mucedo*. See *Roesel*, loc. cit. p. 559, Taf. XCI.; and *Turpin*, Ann. d. Sc. Nat. VII. 1837, p. 65, pl. II. and III.

Infusoria-like embryos have been observed also by *Steenstrup* (Untersuch. loc. cit. p. 66, Taf. I. fig. 21) with *Coryne squamata*, and by *Sars* (Faun. littoral. Norveg. p. 7, Tab. II. fig. 7–11) with *Podocoryna carnea*. The round eggs moving about by means of cilia, which *Reid* (Ann. of Nat. Hist. XVI. p. 392, 397, pl. XII. fig. 9, 13) has observed in the visceral cavity of *Pedicellina echinata*, and in special capsules with *Flustra avicularis*, were probably embryos also.*

* [§ 52, note 3.] The embryonic development of the Bryozoa has been carefully wrought out by *Van Beneden*. See Recherch. sur les Bryozoaires, &c., Mém. Acad. Bruxelles. XIX. See also *Allman*, Report, loc. cit. 1850, p. 322 — Ed.

BOOK THIRD.

ACALEPHAE.

CLASSIFICATION.

§ 53.

THE BODY of Acalephae is composed of a transparent, gelatinous substance, quite resembling the *Corpus vitreum* of the eyes of vertebrata. By desiccation it almost entirely disappears, there remaining only a dry cellular tissue, by which the form of the animal is imperfectly preserved. These animals swim freely in the sea after having attained their development.

In the arrangement of their organs in ray-like processes radiating from a common centre or a longitudinal axis, and where also is situated the digestive apparatus, the quaternary system prevails. Copulatory organs are always wanting. The classification is based, according to the system of *Eschscholtz*, upon difference of external form, and upon the structure of their digestive and locomotive organs.

ORDER I. SIPHONOPHORA.

They take in their food by means of numerous tubes, which exist in place of a stomach. Locomotion is aided, generally, by certain cartilaginous capsules.

FAMILY: DIPHYIDAE.

Genera: *Diphyes, Ersaea.*

FAMILY: PHYSOPHORIDAE.

Genera: *Physophora, Stephanomia.*

FAMILY: PHYSALIDAE.

Genus: *Physalia.*

FAMILY: VELELLIDAE.

Genera: *Rataria, Velella, Porpita.*

ORDER II. DISCOPHORA

They have a simple central stomach, and move by means of discoid or campanulate contractions of their body.

FAMILY: AEQUORINA.

Genera: *Aequorea, Polyxenia.*

FAMILY: OCEANIDAE.

Genera: *Oceania, Cytaeis, Thaumantias.*

FAMILY: GERYONIDAE.

Genus: *Geryonia.*

FAMILY: RHIZOSTOMIDAE.

Genera: *Cephea, Cassiopea, Rhizostomum.*

FAMILY: MEDUSIDAE.

Genera: *Pelagia, Cyanea, Chrysaora, Medusa, Aurelia, Ephyra, Sthenonia.*

ORDER III. CTENOPHORA.

Their mouth and stomach is simple and central, and they move by means of cilia arranged in longitudinal rows.

FAMILY: BEROIDAE.

Genera: *Beroë, Lesueuria, Medea.*

FAMILY: MNEMIADAE.

Genus: *Eucharis.*

FAMILY: CALLIANIRIDAE.

Genera: *Cydippe, Cestum.*

BIBLIOGRAPHY.

Eschscholtz. System der Acalephen. Berlin, 1829.

Lesson. Histoire naturelle des Zoophytes. Acalèphes. Paris, 1843.

Will. Horae tergestinae oder Beschreibung und Anatomie der im Herbste, 1843, bei Triest, beobachteten Acalephen. Leipzig, 1844.

Ehrenberg. Ueber die Acalephen des rothen Meeres und den Organismus der Medusen der Ostsee, in the Abhandlungen der Berl. Akad. 1835.

Mertens. Beobachtungen und Untersuchungen über die beroëartigen Acalephen, in the Mémoires de l'Académie des Sciences de St. Petersburg, 6me series, Tom. II. 1833, p. 479. Also, in Isis, 1836, p. 311.

Brandt. Ausfürliche Beschreibung der von C. H. Mertens auf seiner Weltumsegelung beobachteten Schirmquallen, nebst allgemeinen Bemerkung-

en über die Schirmquallen überhaupt, in the Mém. de l'Acad. des Sc. de
St. Petersburg, 6 ser. Tom. IV. 1838, p. 239.

Milne Edwards. Observations sur divers Acalèphes, in the Ann. des
Sc. Nat. 2de Sér. Zoologie. Tom. XVI. 1841, p. 194.

ADDITIONAL BIBLIOGRAPHY.

Forbes. A monograph of the British naked-eyed Medusae, with figures
of all the species. London, Ray Society, 1848. Contains many anatomical details.

Agassiz. Contributions to the Natural History of the Acalephae of
North America.

Part I. — On the Naked-eyed Medusae of the shores of Massachusetts,
in their perfect state of development.

Part II. — On the Beroid Medusae of the shores of Massachusetts, in
their perfect state of development. See the Mem. Amer. Acad. Arts and
Sc. vol. IV. 1850.

Also, Twelve Lectures on Comparative Embryology, delivered before
the Lowell Institute, Boston, 1848–49.

Busch. Beobachtungen über Anatomie und Entwickelung einiger wirbellosen Seetniere. Berlin, 1851.

[The above are among the most important larger works; but see, also,
many papers of great value, to which I have referred in my notes. — EDI-
TOR.]

CHAPTER I.

SKIN AND CUTANEOUS SKELETON.

§ 54.

Generally, the body of the Acalephae is of a gelatinous substance, composed of polyhedral cells. In some species certain parts of the body have
a cartilaginous hardness, but it is only in a few that there is found a cartilaginous or calcareous nucleus, comparable to a rudimentary skeleton.

With the Diphyïdae a large portion of the body has a cartilaginous
density, and with the Physophoridae it is often surrounded by plates of a
similar nature. The Velellidae have a nuclear skeleton, which in *Rataria* is a simple, elongated disc; but in *Velella* this disc, which is horizontal and of an elongated oval form, is surmounted by a vertical crest. The
disc is composed of four pieces joined together by two sutures which cross
each other obliquely. The crest, united to the disc along the whole length
of the two sutures, and resembling the segment of a circle, is composed of
two main pieces, joined in the middle by a third, which is shaped like a
wedge.[1]

The disc situated under the skin of the upper surface of *Porpita*, and

1 *Eschscholtz*, loc. cit. Taf. XV. ; and *Lesson*, Acalèphes, loc. cit. Pl. XII. fig. 1; also,*Duperrey*, Voyage loc. cit. Zoophytes, No. 6. fig. 1, A. A.

which encloses between its two lamellae numerous aërial canals, is said to be of a calcareous nature.[2]

All these discs have upon their surface markings of concentric rings and diverging rays.

§ 55.

The Acalephae are surrounded by a very delicate epidermis. Upon various portions of the body, and especially upon the arms, the tentacles, the prehensile filaments and the cirri, there exist cilia and peculiar nettling and prehensile organs. In those species having active irritating properties the nettling organs are situated in a mass under the epidermis.[1]

§ 56.

These nettling organs are generally composed of an oval capsule, containing a spiral filament which is thrown out from the slightest disturbance, and, together with its capsule, is detached from the skin.[1]

In some species, there exist in place of these nettling organs others of a prehensile nature, consisting of an oval capsule in which is a stiff bristle. These last cause no burning sensation, but are the means by which these animals attach themselves to contiguous objects in a bur-like manner. They are situated, grouped in small masses, under the skin of most of the non-nettling Discophora, and their bristles project upon the cirri situated upon the border of the disc, upon the tentacles, the arms and the sexual organs.[2]

2 *Eschscholtz*, loc. cit. p. 176, and *Lesson*, loc. cit. Pl. XII. fig. 3, also, *Duperrey*, loc. cit. No. 7, fig. 3.

1 *Wagener* (*Muller's* Arch. 1847, p. 183, Taf. VIII. fig. 4, 5) has described the peculiar hair-like productions on the sides of *Beroë* and *Cydippe*. They have, near their free extremity, a multitude of pedunculate small buttons, inserted on a clavate swelling.

1 *Wagner* (Icon. zoot. Tab. XXXIII. fig. 8, 10, 11, A. B. C. and Ueber den Bau der Pelagia noctiluca, 1841. ; also, in *Wiegmann's* Archiv 1841. Th. 1. p. 39) has found in *Pelagia noctiluca* that the nettling capsules are situated among the pigment cells beneath the epithelium of the disc. According to this author, *Oceania*, which has feeble nettling powers, has these capsules only upon the marginal filaments.

Ehrenberg (*Wiegmann's* Archiv 1841, Th. I. p. 71, Taf. III.) has failed to find these organs upon the non-nettling disc of *Cyanea capillata*, although they are found among their prehensile cirri, which have irritating power.

With these, as with the hooked organs of *Hydra*, he thought the capsule was detached before the filament. *Will* (Horæ tergest. pp. 62, 65) did not find these organs in *Cephea*, except on the tentacles of the genital organs ; and in *Polyxenia* only on the marginal filaments. *Külliker* (Beiträge, loc. cit. p. 41) has seen them also about the genitals of *Chrysaora* and *Aequorea*.

The Siphonophora have only the prehensile filaments covered with them. Thus in *Stephanomia*, according to *Milne Edwards* (Ann. d. Sc. Nat. XVI. p. 223, Pl. VIII. fig. 9), they cover the whole surface of these last ; while in *Physophora*, *Diphyes* and *Ersaea*, they exist only upon their enlarged portions, according to *Philippi* (*Muller's* Arch. 1843, p. 62, Taf. V. fig. 9), and *Will* (loc. cit. p. 79, 81, Taf. II. fig. 23-25). *

2 *Siebold* (Beiträge zur Naturgesch. der wirbellosen Thiere, 1839, p. 10, 91, Taf. II. fig. 39) ; also, *Ehrenberg* (Ueber die Acalephen d. rothen Meeres, &c. &c., in th : Abhandl. d. Berl. Akad. 1835, p. 205, Taf. IV-VIII.). He has compared these prehensile organs to suckers.

According to *Milne Edwards* (Ann. d. Sc. Nat. XVI. p. 215), and *Will* (loc. cit. p. 80, Taf. II. fig. 24), they are found also upon the body of *Beroë*, and at the extremity of the prehensile filaments of *Diphyes* and *Ersaea*.

According to *Will*, also (loc. cit. p. 51, Taf. I. fig. 19, A. B.), the prehensile filaments of the Ctenophora have two kinds of capsules ; one, which upon the least touch bursts and discharges a liquid; the other, of a somewhat different appearance, and which contains a delicate, viscous filament. Similar filaments, he says, are found upon the warts on the body of *Eucharis*.

* For these nettling organs and their intimate structure, see my note under § 27, note 1. — Ed.

CHAPTER II.

MUSCULAR SYSTEM AND ORGANS OF LOCOMOTION.

§ 57.

The Acalephae have a distinct muscular system. Their contractile substance is composed of a net-work of elongated, slender filaments and bands; these, in the utriculoid species, are arranged in a longitudinal and annular manner, but in those of a discoid and campanulate form they are disposed in a circular and radiate manner.

In the extremely irritable tentacles and tactile filaments, the longitudinal fibres abound.[1]

Each fibre is smooth when relaxed, but during contraction appears transversely wavy and plicated.[2]

§ 58.

The contractile and aërial natatory vesicles, which are found in the Physophoridae,[1] and the movable lamellae of the Ctenophora, may well be regarded as accessory organs of locomotion. These last, which are arranged in rows upon the sides of the animal, and which by some anatomists have been regarded as respiratory organs, are not simple cutaneous lobes, but are composed of very long cilia closely united together, and the motion of which is voluntary with the animal.[2]

---

1 *Will* (loc. cit. p. 48, Taf. I. fig. 11) has observed in the contractile excrescences of the *Eucharis*, not only circular fibres and numerous longitudinal muscles, but large transversely-flattened ones, which were bound together by oblique bands.

2 *Will*, loc. cit. p. 47, 63, Taf. I. fig. 13. According to *Wagner* (Ueber den Bau, &c.; and Icon. zool. Tab. XXXIII. fig. 30), the muscles of the Discophora have always the transverse striae.

The cartilaginous natatory pieces of the Siphonophora play a completely passive part in the act of locomotion. The swimming is exclusively performed by the energetic contractions of the muscular membrane which lines their cavity, constituting, therefore, a true natatory sac. See *Sars* Faun. littoral. Norveg. p. 42.*

1 Lately, it has been doubted if the Physophoridae can sink and rise in the sea by means of their natatory bladders, because they cannot exhaust the contained air. According to *Olfers* (Abhandl. d. Berl. Akad. 1831, p. 157, 165, Taf. I.), there are two of these bladders in *Physalia*, one of which only has an opening. *Philippi* (Müller's Arch. 1843, p. 63) has found neither internal nor external opening to the bladder of *Physophora tetrasticha*. In *Stephanomia* it would not appear, according to the description of *Milne Edwards* (Ann. d. Sc. Nat. XVI. p. 218, Pl. VIII. fig. 1. b, 2), that this organ had an external opening. *Couch* (Froriep's neue Notizen, No. 275, p. 129) denies that *Physalia* has the power to control the air of its bladder. See also below, § 65.

2 *Grant*, Trans. Zool. Soc. London, I. 1835, p. 9.; *Sars*, Beskrivelser loc. cit. Pl. VIII. fig. 18, c.; *Milne Edwards*, Ann. d. Sc. Nat. XVI. p. 201, 216, Pl. IV. fig. 2, 3, Pl. VI. fig. 1. c.; and *Will*, loc. cit. p. 9, 56, Taf. I. fig. 5.

* [§ 57, note 2.] For the muscular system of the Acalephae, see also *Forbes* (loc. cit. p. 3), and *Agassiz* (loc. cit. p. 256). This last-named author has described this system with full details in many genera. It is much more complex than has hitherto been supposed, and I must refer for the details to the memoir in question.

In regard to the structure of these muscles, *Agassiz* remarks: " With all the power of the best Oberhäuser Microscope, I have been unable to discover the slightest indication of striae on the muscular cells; nevertheless, it cannot be doubted that they are voluntary muscles." To this view I may add my own of the same nature. — ED.

CHAPTER III.

NERVOUS SYSTEM.

§ 59.

A nervous system has been found in many Acalephae. With the Cteno-phora the œsophagus is surrounded by a ring formed of eight ganglia,[1] and at the opposite extremity of the body there is a simple ganglion. Five nervous filaments pass out from these ganglia, and along the sides of the body are nervous fibres, which ultimately divide into delicate threads.[2]

The tentacles of Medusae are supplied with nervous filaments which issue from a ganglion situated at their base.[3]

CHAPTER IV.

ORGANS OF SENSE.

§ 60.

With many Acalephae, there are, upon the borders and extremities of

---

1 These eight ganglia, which are connected together by delicate cords, were first observed by *Grant* (Trans. Zool. Soc. Lond. 1. p. 10) in *Cydippe pileus*. Compare, also, *Wagner*, Icon. zool. Tab. XXXIII. fig. 37, A. B. From each of these ganglia two nerves pass off to the side, while a third, traversing the interior of the body, and having two or three swellings, is finally distributed to the intestine. *Patterson* (The Edin. new Philos. Jour. XX. p. 26), and *Forbes* (Ann. of Nat. Hist. 1839, p. 145), have also observed the œsophageal ring in *Cydippe*, but did not perceive the ganglia.
2 *Milne Edwards* (Ann. des Sc. Nat. loc. cit. p. 206, Pl. IV. fig. 1) has observed at the posterior extremity of the body of *Lesueuria vitrea* (a new Beroid) a ganglionic body which sends

out in front four filaments ; and upon the sides of this animal a nervous cord, from which pass off delicate branches at regular intervals. At the posterior extremity of the body of *Cydippe*, *Eucharis* and *Medea*, *Will* (*Froriep's* neue Notizen, No. 599, 1843, p. 67, and Horæ tergest. p. 44) has likewise observed a round, yellowish ganglion, with four prolongations, from which pass off twenty-five or thirty nerves.
3 *Ehrenberg* has found along the entire border of the disc of *Medusa aurita*, and between each two tactile filaments, a bifid nervous ganglion. He affirms to have seen also two others similar, at the base of each tentacle surrounding the genital organs. See Abhandl. d. Berl. Akad. 1835, p. 203, Taf. IV. fig. 1, x.; and *Muller's* Arch. 1834, p. 571.*

* [§ 59, note 3.] The nervous system of the Acalephae has been successfully studied by *Agassiz* upon several genera (*Hippocrene, Tiaropsis, Staurophora*). His results are new, and different from those of previous observers. I cannot do better than to quote his words : "There is, unquestionably, a nervous system in Medusae, but this nervous system does not form large central masses, to which all the activity of the body is referred, or from which it emanates. There is no regular communication by nervous threads between the centre and periphery and all intervening parts ; and the nervous substance does not consist of heterogeneous elements, of nervous globules and nervous threads, presenting the various states of complication and combination, and the internal structural differences, which we notice in the vertebrated animals, or even in the Mollusca and Articulata."

"In Medusae the nervous system consists of a simple cord, of a string of ovate cells, forming a ring around the lower margin of the animal (Pl. V. fig. 11, 2, 4, 5), extending from one eye-speck to the other, following the circular chymiferous tube, and also its vertical branches, round the upper portion of which they form another circle. The substance of this nervous system, however, is throughout cellular, and strictly so, and the cells are ovate. There is no appearance in any of its parts of true fibres" (loc. cit. p. 232). That this is the nervous system seems placed beyond all controversy ; for, in a private letter, *Agassiz* has informed me that in a new genus (*Rhacostoma*), living on the shores of Massachusetts, he has seen this system at night as an illuminated diagram. — Ed.

their body, button and tongue-like organs, which, as they are connected with neighboring ganglia, may well be regarded as organs of sense.

Their essential structure is a membranous capsule, containing a clear liquid, in which are suspended crystalline corpuscles.

These organs, having sometimes a red pigment, have been taken for eyes; but, as most of them are without pigment, and as the crystalline corpuscles behave in acid like the Otolites of the higher animals, they have more recently been better designated as organs of hearing.

The eight marginal, tongue-like bodies, found upon the disc of *Medusa aurita*, have been regarded as eyes.[1] The sole fact for the support of this opinion is the presence of pigment; for the small hexagonal crystals, irregularly scattered in the interior of these bodies, would scarcely allow them to refract the light like a crystalline lens.

The Ctenophora have only a single organ of this nature, and which is situated near the ganglion at the posterior end of the body. It has been regarded both as an eye and as an organ of hearing.[2]

With many Discophora, these organs appear as pale-yellow, or even colorless marginal corpuscles, having more or less calcareous bodies.[3]

It is yet doubtful whether the otolites of the Acalephae perform the same movements as those of the acephalous and gasteropod mollusca.[4]

1 These marginal corpuscles, already observed in the Medusae by *Gaede* (Beiträge zur Anat. u. Phys. der Medusen, 1810, p. 18, 28), and by *Rosenthal* (Zeitsch. f. Physiol. Bd. 1. Hft. 2, 1825, p. 326), were first described as eyes by *Ehrenberg*. See *Muller's* Arch. 1834, p. 571, and Abhandl. d. Berl. Akad. 1835, p. 190, Taf. IV. V.

2 *Milne Edwards* has called this body, in *Lesueuria vitrea* and *Beroë Forskalii*, "*Organe oculiforme*" (Ann. d. Sc. Nat. loc. cit. p. 206, 211, Pl. IV. fig. 1, k. and Pl. V. fig. 4, i.). According to *Will* (Froriep's neue Not. No. 566, p. 67, and Horae tergest. p. 45, Taf. I. fig. 2, 4, 20, b.), the red pigment of these organs is entirely wanting in *Beroë*, *Eucharis* and *Cydippe*, while the hexagonal calcareous corpuscles are very numerous — a fact leading him to conclude that these organs are auditory vesicles.

3 According to *Wagner* (Ueber den Bau, &c., and Icon. zool. Tab. XXXIII. fig. 31, g. 23, c. and 25), these corpuscles are pale-yellow in *Pelagia noctiluca*, and colorless in *Oceania*, *Cassiopea* and *Aurelia*. In *Cephea*, *Will* has observed only pale-yellow corpuscles, filled with crystals. And, according to him (loc. cit. p. 64, 68), the colorless pedunculated marginal vesicles of *Polyxenia leucostyla* contain, each only a single round otolite, while those of *Cytaeis polystyla* contain numbers, colorless or yellow, and of irregular forms. He has also observed (loc. cit. p. 72, Taf. II. fig 9, 10) that in *Geryonia* the number of these otolites varies from one to nine. *Milne Edwards* (Ann.

d. Sc. Nat. XVI. p. 196, Pl. I⁶ e.) has observed upon the margin of the disc of *Aequorea violacea* vesicles containing two or three spherical corpuscles, and which, probably, are auditory organs. According to *Sars* (*Wiegmann's* Arch. 1841, Th. 1. p. 14, fig. 69), and *Will* (loc. cit. p. 75, Taf. II. fig. 21, A. B.), these marginal corpuscles are found upon young Medusae belonging to *Ephyra*.

4 *Will* has never observed with the Otolites of Acalephae similar movements to those of mollusca. *Kölliker* (Froriep's neue Not. No. 534, p. 82) has observed vibratile cilia upon the inner surface of the marginal corpuscles of *Pelagia*, *Cassiopea*, *Rhizostomum* and *Oceania*, which are pyriform, and contain many calcareous crystals. In the pedunculated vesicles of *Geryonia*, which contain only a single round otolite, these cilia are absent. In none of the Medusae has he found collections of pigment, and in *Oceania* (nov. spec.) only he has observed a mass of brown pigment cells upon the external and superior surface of the base of these corpuscles; in the centre he perceived a round transparent body, and upon the upper surface a circular opening, so that the whole closely resembles an eye, there being, moreover, a kind of pupillary opening, and the traces of an optic nerve from a ganglion.

According to the observations of *Frey* and *Leuckart* (Beitr. &c. p. 39), the group of otolites contained in the auditory organ of a *Cydippe* perform oscillatory movements, due evidently to vibratile cilia situated on the auditive capsule.*

* [§ 560, note 4.] The organs of sense of the Acalephae have been the objects of much study of late, and to *Agassiz* we are indebted for the most minute researches on these obscure points. He has shown the eye-specks to be undoubted organs of sense, from their connection with the nervous system. With the naked-eyed Medusae, he regards them light-perceiving instead of auditory organs. In regard to the single organ found with the Ctenophora, and which *Frey* and *Leuckart* have re-

cently declared to be of an auditory nature, he remarks : " I am inclined to consider this organ, or this speck, as something similar to the central colored speck which occurs in the middle of the disc in Discoid Medusae, and which is particularly distinct in young animals soon after they have been detached from the polyp-like stem on which they grew, as a remnant of the connection which exists between the mother-stem and its progeny in those Medusae which multiply by alternate generations."

CHAPTER V.

DIGESTIVE APPARATUS.

§ 61.

The digestive apparatus of the Acalephae is formed after several very different types. The mouth is sometimes single and central, or there may be many of them. It is often surrounded with arms and retractile filaments, which are endowed with the prehensile and nettling organs just described.

The digestive cavity, which is always lined with ciliated epithelium, has distinct walls, which are united immediately to the parenchyma of the body, leaving, therefore, no surrounding cavity.

With those having a single mouth the stomach is of a variable size, and has often caecal appendages. With *Beroë*,[1] the mouth is very large and free from tentacles, and opens into a very spacious stomach which occupies nearly the whole body. But with *Cestum, Cydippe* and *Lesueuria*, the stomach is small, and appears like a cavity in the body;[2] and with *Cytaeis, Thaumantias* and *Geryonia*, it is likewise small, and has the shape of a tubular projection.[3]

That of *Medusa* has four saccular folds,[4] that of *Pelagia*[5] six, and that of *Cyanea* thirty-two.[6]

When the mouths are numerous, either, as in the Rhizostomidæ,[7] there are many canals which conduct the food through the arms upon which the mouths are situated into the central stomach; or, as in the Siphonophora, each mouth opens into a particular tubular stomach. With these last, however, a certain number of their tentacles are hollow, and have a mouth at the extremity. As it has been observed that these suck in food and digest it, their orifices have been regarded as mouths, and their cavities as stomachs.[8]

---

1 *Milne Edwards*, Ann. d. Sc. Nat. XVI. pp. 5, 6.
2 *Eschscholtz*, loc. cit. Taf. I. II.; and *Milne Edwards*, loc. cit. Pl. III.
3 *Will*, loc. cit. Taf. II.
4 *Baer*, in *Meckel's* deutschs. Arch. VIII. 1823, Taf IV. fig. 2; also, *Ehrenberg* in Abhandl. d. Berl. Akad. 1835, Taf. III. fig. 1.
5 *Wagner*, Icon. zool. Tab. XXXIII. fig. 5.
6 *Gaede*, loc. cit. Taf. II.
7 *Eysenhardt*, Nov. Act. physico-med. X. part II. p. 391, Tab. XXXIV. fig. 1 (*Rhizostomum Cuvieri*).
8 This is so, for examples, in *Diphyes* (*Will*, loc. cit. Taf. II. fig. 22); in *Physalia* (*Olfers* Abhandl. d. Berl. Akad. 1831, p. 162, Taf. I.); in *Stephanomia* (*Milne Edwards*, Ann. d. Sc. Nat. XVI. Pl. VII. IX. X.); and in *Physophora* (*Philippi*, *Muller's* Arch. 1843, Taf. V. fig. I, 4).

Philippi, however, affirms that in this last genus these canals are organs of absorption, and that the true stomach, which has a simple mouth, is concealed at the base of the tentacles (loc. cit. p. 63, Taf. V. fig. 10).

I think, however, that this opening belongs to the respiratory system, as also does a similar opening in *Velella* and *Porpita*, which *Lesson* (Voyage de Duperrey, loc. cit. p. 49, 56, No. 6, fig. B.; and No. 7, fig. C. C.) has regarded as a mouth.

The tubular tentacles of these animals are nothing but stomachs; and *Lesson* himself has called them "*poches stomacales*," since they digest food. It would, moreover, be strange that these organs, which, in *Physalia*, have been admitted to be stomachs, should perform another function in *Physophora, Velella*, and *Porpita*, where their structure is the same. But further researches are

(Loc. cit. p. 316.) On a preceding page he says: "That this may be the case seems probable when we consider the relation of the two sorts of apparatus in the two types. The upper nervous ring in *Sarsia* bears the same relation to the central alimentary cavity, and to the pigmented disc, that the ganglion and eye-speck of *Beroë* bear to the chy-

miferous system, which opens above its gelatinous disc, notwithstanding these openings." (p. 248.) This point, fully as interesting from its zoological importance as from its morphological relations, can be settled only by a knowledge of the embryology of these animals. — ED.

The Acalephæ have no true digestive tube. But, as such, has been regarded a system of vascular canals filled with water, and which, departing from the stomach, traverse the whole body. But these, although sometimes seen to contain fæces, seem to belong more properly to the respiratory system.[9]

In none of the Acalephae has there been found anything like an hepatic organ.[10]

CHAPTER VI.

CIRCULATORY SYSTEM.

§ 62.

Until lately, the longitudinal and circular canals which, in some Acalephae, are spread out through the entire body, have been regarded as belonging to a vascular, sanguineous system. But more recently these have properly been considered as aquatic-respiratory organs, there having been found, moreover, other vessels of exceedingly thin walls, and of a sanguineous nature.

These last constantly accompany and surround in a tubular manner the aquiferous canals; and it is quite rare that small branches are distributed to the general parenchyma.

The delicate walls of these vessels have neither longitudinal nor circular fibres, neither are they lined with ciliated epithelium. They circulate a

required to thoroughly settle this point. See below, the respiratory organs. See also *Hollard*, who unhesitatingly regards the canals, which, with *Velella*, communicate externally by a central opening, as a digestive cavity, and thinks he has observed in their walls brownish spots representing the hepatic cells; see Ann. d. Sc. Nat. III. 1845, p. 240, Pl. IV. bis.

9 The aquiferous canals of the respiratory system having been regarded as intestinal tubes, their orifices, which in the Ctenophora are situated at the extremity of the body, and in the Discophora upon the borders, have been considered as anal openings; and especially so, since in these two orders, accidental fæces in these canals are expelled through these orifices. See *Will*, loc. cit. p. 25,

and *Ehrenberg*, Abhandl. d. Berl. Akad. 1835, p. 189, Taf. I. IV. fig. 2, z.*

10 Acalephæ possess an extraordinary digestive power, which is the more singular as no secretory organ has been found on the sides of their stomach. *Mertens* (Mem. d. l'Acad. de St. Petersburg, loc. cit. p. 490, Taf. I. fig. 5, 6, a.; and p. 518, Taf. VIII. fig. 4, Taf. IX. fig. 1, f.), however, affirms to have seen in *Cestum* and *Cydippe* four vessels in this situation, which are perhaps hepatic organs. The orange-colored cords found upon the sides of the stomach of *Stephanomia*, and which *Milne Edwards* (Ann. d. Sc. Nat. XVI. p. 222, Pl. VII. IX. X.) has taken for genital organs — may they not also be hepatic organs ? |

* [§ 61, note 9.] Upon the nutritive system of the Acalephae, see *Forbes* (loc. cit. p. 4), but especially *Agassiz* (loc. cit.), who has studied the subject with conscientious care. There is no distinction between the alimentary canal proper and the vascular system, for the one opens by large tubes into the other. The Acalephs, therefore, circulate *chyme*, and here we have the rudest form of circulation. If this idea is once well considered, the relations of their nutritive apparatus in general will be quickly appreciated.

The variations in the shape and form of the digestive apparatus are wide and numerous, but

their importance is rather in Zoology. See *Agassiz* for the details of *Sarsia, Hippocrene, Tiaropsis, Staurophora, Pleurobranchia, Bolina.* — Ed.

† [§ 61, note 10.] *Kölliker (Siebold and Kölliker's* Zeitsch. IV. Hft. 3, 4, p. 313) has observed with *Velella* and *Porpita* a glandular mass, corresponding most probably to a liver. It had before been regarded as such by *Delle Chiaje*, but *Kölliker* has given it a special description. It consists of a brown mass which communicates with the bottom of the stomachal cavity by branched, anastomosing ducts. — D.

colored fluid and colored corpuscles; and these corpuscles are not found except in those vessels surrounding the aquiferous canals.

There is no regular circulation, but the shifting motion of the blood hither and thither is due to irregular contractions of various parts of the body.[1]

CHAPTER VII.

RESPIRATORY SYSTEM.

§ 63.

The entire body of the Acalephae is traversed by canals which receive water from the stomach, or directly from without, and which is ejected through openings upon the extremity of the body and on the margin of the disc.

These aquiferous canals are lined with a delicate, ciliated epithelium, by means of which accidental particles of food or faeces are quickly removed. They have been regarded both as digestive and as sanguineous organs. But that they are respiratory organs is highly probable, not only from their structure,—the cilia producing a constant renewal of water,—but also from the fact that they are surrounded by real sanguineous vessels.

This aqueous circulation is oscillatory from one side of the body to the other, being interrupted only by those contractions of the body which occur when fresh water passes from the stomach into the canals.[1]

---

1 These new details upon the sanguineous system of the Acalephae are due to *Will* (Hor. tergest. p. 34, and *Froriep's* neue Not. No. 599, 1843, p. 66). In *Beroe*, he has been able to clearly distinguish the sides of these vessels from those of the aquiferous canals contained in their interior, for the first are covered with numerous red pigment cells.

The blood of this animal has a greenish hue, and contains spherical or slightly elongated red corpuscles, with large nuclei. But, beside these, *Will* has found in *Cydippe* other nucleated cells of a greenish color. In *Polyxenia*, there is no sanguineous system separate from the aquiferous canals, which, in *Cytaeis* and *Geryonia* are quite surrounded by them. The vessels of *Cephea* contain brown corpuscles; and *Will* has concluded that the reddish threads found along the aquiferous canals of this animal, and which *Ehrenberg* (Abhandl. d. Berl. Akad. 1835, p. 195, Taf. VI. fig. 3, 9, and *Müller's* Arch. 1834, p. 568) has taken for striated muscles, are really blood-vessels. Profound researches must decide the real relations of the aquiferous canals to the sanguineous system filled with a violet liquid of *Velella*, as described by *Costa* (Ann. d. Sc. Nat. XVI. p. 188, Pl. XIII. fig. 3). It should be mentioned that the blood-system of the Acalephae,

which *Will* has described with so much positiveness, is not verified either by *Bergmann* or *Frey* and *Leuckart* (Beitr. p. 38), after numerous special researches.[*]

1 If, and especially with the Discophora, these canals have been taken for digestive tubes, it is because faeces and particles of food have been here found, and which have been ejected through the openings on the borders of the body. But the real function of these openings is to discharge the water unfit for respiration; and it is only during the ingestion of this liquid that these foreign particles are thus introduced. This communication between the respiratory and digestive systems reminds one of the Polyps, where (as in the Anthozoa) the openings in the stomach allow its contents to pass into the cavity of the body, which last may be likened to the aquiferous system. On the other hand, the opinion that these canals are blood-vessels would be supported by the Ctenophora, since here they are filled with a red liquid; but, according to *Will* (Hor. tergest. p. 34), this liquid is not in these canals, but in proper blood-vessels surrounding them. He denies, also, that these blood-vessels of the Ctenophora open upon the surface of the body, or that the blood escapes outward mixed with faeces.

* [§ 62, note 1.] A true circulatory system has not been observed also by *Dana* (Struct. and Class. of Zoophytes, 1846, p. 12), by *Forbes* (Brit. Naked-eyed Medusae, 1848, p. 6), by *Agassiz* (Contributions to the Nat. Hist. of the Acalephae of North America, Mem. Amer. Acad. Boston, 1850, p. 260), and by *Busch* (Beobacht. üb. Anat.

u. Entwick. einiger wirbellosen Seethiere, 1851, p. 13). It may, therefore, be concluded that these animals have no system of this kind, and especially so as *Agassiz* failed to notice it after the most intimate research upon the Beröid Medusae (loc. cit. p. 313), which were the objects of *Will's* study. — Ed.

§ 64.

With the Ctenophora, this respiratory system consists of an infundibuliform cavity, communicating with the stomach by two orifices, situated at its base and surrounded by sphincters.

Numerous aquiferous canals pass out of this cavity, traverse the body in a longitudinal direction, and finally anastomose with an annular vessel surrounding the mouth; but, beside these, there are two short canals which pass directly to the posterior extremity of the body, where they open externally.

With *Eucharis* and *Cydippe*, these canals are differently distributed; thus, two go to the tentacles, two to the sides of the stomach, and four to the sides of the body. The same is true with *Beroë*, excepting that those to the tentacles are wanting. The lateral canals divide, at a short distance from the cavity, into as many branches as there are sides. With *Cydippe*, the excretory canals are simple; with *Eucharis* they are provided with vibratile lamellae, and with *Beroë* with branching appendages.[1]

With the Discophora, numerous aquiferous canals pass from the stomach or its appendages, traverse the disc in a radiating manner, sometimes bifurcating, and terminate at the borders of the disc in an annular vessel which opens externally by numerous orifices.

In *Cytaeis*, *Geryonia* and *Thaumantias*, there are four of these canals, arranged in a crucial manner;[2] and in *Aequorea* there are seventy-four disposed in a ray-like way.[3]

In *Medusa aurita*, there pass from the four folds of the stomach sixteen of these canals, eight of which are simple, and eight bifurcating numerously before reaching the marginal vessel of the disc.[4] With *Sthenonia* and *Aurelia*[5] they are very numerous and widely branched.

With *Medusa aurita*, the terminal openings of the annular vessel are eight, and regularly alternate with the organs of hearing there situated.[6] But in *Cephea* these openings are said to be directly beneath these last-named organs.[7]

With the Siphonophora, an aqueous system has not yet been well made out. There is, however, with some, an elongated cavity which is perhaps respiratory, and which, in some species, opens into the stomach, and in others directly upon the outer surface.[8]

---

1 *Will* (Horæ tergest. p. 30, Taf. I.) has made very minute researches upon the aquiferous system of *Eucharis, Cydippe* and *Beroë*. That of *Beroë ovatus, Forskalii*, and of *Lesueuria vitrea*, has been carefully described and figured by *Milne Edwards* as a circulatory system (Ann. d. Sc. Nat. XIII. p. 320; XVI. p. 203, 213, Pl. III.-VI.).

2 *Will*, loc. cit. Taf. II. fig. 5, 7, 8, 14, 16.

3 *Milne Edwards*, Ann. d. Sc. Nat. XVI. p. 197, Pl. I. fig. 1.

4 *Rosenthal*, Zeitsch. f. Physiol. I. Hft. 2, Taf. XI.; also, *Ehrenberg*, Abhandl. d. Berl. Akad. 1835, Taf. I. bis. III.

5 *Eschscholtz*, loc. cit. Taf. IV.; also *Brandt*, Mém. de l'Acad. d. Sc. de St. Petersburg, IV. 1838, Pl. IX. X. XI.

6 *Ehrenberg*, *Müller's* Arch. 1834, p. 566; also, Abhandl. &c. loc. cit. p. 188, Taf. I. fig. 1, w. and Taf. IV. fig. 2, z.

7 *Will*, loc. cit. p. 60.

8 In *Diphyes*, this canal terminates in this way by an oval dilatation, lined with ciliated epithelium, and has perhaps properly been regarded by *Will* (loc. cit. p. 78, Taf. II. fig. 22, a.) as a respiratory organ. A similar cavity, with a cœcal appendage, is found in *Ersaea* (*Will*, loc. cit. p. 81, Taf. II. fig. 27-31, d. e.). If the arms provided with openings, of the *Physophorae*, are really stomachs, then the cavity beneath them, which has a canal passing along the axis of the animal, should be taken as belonging to the aquiferous system, for it receives water by an opening at the base of the anus. This same opening has been taken for a mouth by *Philippi* (*Müller's* Arch. 1843, p. 63, Taf. V. fig. 10). According to *Lesson* (*Duperrey*, Voyage. loc. cit. No. 6, fig. B.), there is between the suckers of *Velella* an orifice which leads from before backward into a large branching canal. This structure, hitherto regarded as a digestive

CHAPTER VIII.

ORGANS OF SECRETION.

§ 65.

The air-cavity of certain Siphonophora, which is surrounded by a double membrane, ought probably to be regarded as an organ of secretion; for, according to many naturalists, the air contained could not have been derived from without, and consequently was secreted by the sides of the internal membrane.[1]

CHAPTER IX.

ORGANS OF GENERATION.

§ 66.

Reproduction by *fissuration* and *gemmation* with the Acalephae has been observed only in the youngest states of certain Medusae.[1] But repro-

cavity, belongs probably to the aquiferous system. That which in *Porpita* has been taken for a mouth, belongs probably, also, to the same system. I would not, however, deny that another signification may be given to the so-called respiratory and digestive organs of the Siphonophora.

If one prefers with *Philippi*, to regard the opening between the tentacles of *Physophora*, *Velella* and *Porpita*, as a mouth, then the cavity of these tentacles should belong to the aquiferous system. Moreover, these tentacles, as to their form and mobility, remind one of the pedicles of the Echinoderms; but it is remarkable that they can absorb food.

Sars (Faun. littoral. Norveg. p. 34, 42, Tab. VI. fig. 3, *gg*. and Tab. VII. fig. 3, *c*.) has observed in the interior of the cartilaginous, natatory pieces of the Physophoridae and Diphylidae, aquiferous canals which are probably of a respiratory nature.

Hollard, likewise, regards the hollow and tubuliform tentacles of *Velella* as aquiferous tubes, and in this way, as the tentacular feet of the Echinoderms, includes them in the aquiferous system. See Ann. d. Sc. Nat. III. 1845, p. 250.

I Many naturalists entirely deny the presence of openings in these aërial cavities, and do not admit that they are filled with gas. Thus *Philippi* (*Muller's* Arch. 1843, p. 63) affirms to have found neither external opening nor air in the pouch at the end of the longitudinal canal of *Physophora tetrasticha*. *Olfers* (Abhandl. d. Berl. Akad. 1831, p. 165) has not been able to find in *Physalia* the opening of the internal sac, said to be near the one of the external sac. In fact, *Bennett* (Proc. Zool. Soc. London, 1837, p. 43; and *Wiegmann's* Arch. 1839, II. p. 332), with the same species,

has not seen an opening of this cavity, and was unable to force air from it. Future researches must determine if these pouches have not a respiratory function.

I See, upon this subject, the Embryology of these animals, below. It is not yet demonstrated that adult Acalephae reproduce by fissuration; and although *Mertens* (Mém. d. l'Acad. de St. Petersburg, II. p. 494, Pl. I. fig. 2–4, and p. 527) has observed detached corpuscles from the body of *Cestum* and *Cydippe* swim freely about, and rapidly enlarge, yet his observations are here limited.

In the same way, *Will* (Horæ tergest. p. 42) has seen analogous bodies detached from *Eucharis*, and has found in the water others supposed to belong to the Ctenophora, but has not traced their further condition.

Propagation by buds has also been found with the Acalephs, through the excellent researches of *Sars* (Fauna littoral. Norveg. p. 11, Tab. IV. fig. 8–12), for this observer has seen on the external surface of the tubuliform stomach of *Cytaeis octopunctata*, and upon the four ovaries of *Thaumantias multicerrata*, small campanuliform Acalephs resembling their parent, in the process of development, and which were finally detached. In the genus *Agalmopsis* which is allied to *Agalma*, *Sars* has observed (Ibid. p. 38, Tab. VI. fig. 14–17) campanuliform bodies sprout out between the prehensile filaments and the tubuliform stomach, and which were finally detached, swimming freely like the Discophora. According to *Sars*, also (Ibid. p. 43, Tab. VII. fig. 11, b. 13, b. and 14), there is, likewise, an analogous mode of propagation with *Diphyes*.*

---

* [§ 66, note 1.] See also *Huxley* (Ann. Nat. Hist. VI. p. 394), who has described the reproductive processes of the Diphyidae, and shown that they multiply by gemmation as well as by ova. See, also, *Muller's* Arch. 1851, p. 380, Taf. XVII. —ED.

duction by eggs, and consequently by the means of proper genital organs, has been observed in all the families.

With the Ctenophora,[2] both sexes are combined in the same individual ; but with the Discophora, the individuals are of one sex alone.[3]

§ 67.

The eggs are spherical, and surrounded by an exceedingly thin envelope. The vitellus is of a whitish violet or yellow color, and contains a germinative vesicle, and germinative dot.[1]

The spermatic particles, which have generally the form of Cercaria (that is, a head and a filiform tail), are very active, and suffer no change in water.[2]

In some Siphonophora, they appear to have a linear form, and attain a very great size.[3]

§ 68.

The genital organs are not developed except at the epoch of procreation, and this period is very brief. On this account, their existence has often entirely escaped the notice of observers.

The male and female organs so closely resemble each other, as to color, form and position, that they are easily confounded. They consist either of elongated pouches, or of riband-like bands, which are situated in different parts of the body. In the first case, the sperm and eggs escape through particular excretory canals ; in the second, they escape directly outwards from the ovaries or testicles, or pass first through large cavities which communicate externally.

As they have no copulatory organs, the water is the medium of fecundation. In this way the unaffected spermatic particles are brought in direct contact with the eggs.

2 *Will, Froriep's* neue Not. No. 599, p. 66.

3 *Siebold, Froriep's* neue Not. No. 1081, 1836, p. 33.*

1 *Wagner* (Prodrom. loc. cit. Taf. I. fig. 2 ; and Icon. zoot. Tab. XXXIII. fig. 15–17) and *Siebold* (Beiträge z. Naturgesch. wirbelloser Thiere. loc. cit. Taf. I. fig. A. B.) have figured the eggs of *Cyanea pelagia,* and of a *Medusa.*

2 The spermatic particles of *Eucharis* and *Beroë* consist of a round body, having a delicate and very movable tail (*Will,* loc. cit. Taf. I. fig. 6, 24). In *Cydippe* they are similar (*Krohn, Froriep's* neue Not. No. 356, 1841, p. 52). This is likewise true of those of the *Discophora*; see *Siebold,* Beiträge loc. cit. Taf. I. fig. c. (*Medusa*); *Kölliker,* Beiträge loc. cit. Taf. I. fig. 8, 9, 10 ; and *Milne Edwards,* Ann. d. Sc. Nat. XVI. Pl. I. fig. 1, d. (*Rhizostomum, Chrysaora* and *Aequorea*) ;

Wagner, Icon. zoot. Tab. XXXIII. fig. 20, and *Will,* Horæ tergest. Tab. II. fig. 12 (*Pelagia* and *Geryonia*).

For the spermatic particles of the Discophora, see also *Kölliker* in the Neue schweiz. Denkschr. VIII. p. ss, Taf. II. fig. 18 (*Cassiopeia*). †

3 It may be that the stout linear and active bodies, seen by *Will* (loc. cit. p. 78, 81, Taf. II. fig. 26) in the respiratory cavity, the stomach and the general cavity of the body of *Diphyes* and *Ersaea,* and which he was inclined to regard as Entozoa, are the spermatic particles of these animals, since they quite resemble those of *Alcyonella* and *Cristatella.*

According to *Sars* (Faun. littor. &c. p. 38), the spermatic particles of *Agalmopsis* have a cercaria-form. ‡

* [§ 66, note 3.] Reproduction by fissuration has been observed with the Discophora by *Kölliker* (*Siebold* and *Kölliker's* Zeitsch. IV. p. 325) ; he witnessed this phenomena with *Stomobrachium mirabile.* It does not appear, however, that he has observed this process with adult forms ; for he remarks that there is reason to believe that this *Stomobrachium* is only a young, imperfect form of his *Mesonema coerulescens.* — Ed.

† [§ 67, note 2.] The spermatic particles of the Acalephae have invariably, I think, a cercaria-

form, like those of the Polyps, and like which, also, they are developed in special daughter-cells. — Ed.

‡ [§ 67, note 3.] These bodies mentioned by *Will* as spermatic particles have since been examined by *Huxley* (loc. cit.), who thinks they are not of this nature, a view which is otherwise probable from the fact that he found no male generative sacs, and also because, as I have shown (see my note after § 46, note 5), these particles with *Alcyonella* have a cercaria-form. — Ed.

§ 69.

The position of the sexual organs varies in the different orders, in the following manner:

1. With the Ctenophora, which are hermaphrodites, they are situated along the sides, under the form of elongated utricles, the testicles being on one side and the ovaries on the other. They have a nodulated appearance, and from the lower part of each passes off an excretory duct, which runs toward the mouth, but the terminal opening of which has not yet been well made out.[1]

2. With many Discophora, these organs are arranged like rays, passing from the centre to the border of the disc. In *Oceania, Cytaeis, Geryonia* and *Thaumantias*, the four saccular ovaries or testicles form at the centre of the disc a cross, which is traversed by four aquiferous canals.[2] Their excretory ducts pass towards the base of the stomach, but their terminal openings are not distinct.[3] In the disc of *Aequorea violacea*, seventy-four ray-like bands are spread out, and the free plicated borders of these hang beneath the inferior surface of the disc, thus permitting the free escape of the eggs and sperm into the water.[4]

3. Another group of the Discophora have at the base of their tentacles four large openings, which lead into as many cavities in the disc.[5] At the base of these cavities, which formerly were regarded as respiratory organs, the genital organs are situated in the form of plicated bands. These as four bands (testicles or ovaries) are bent either into an angle or the arc of a circle, forming sometimes a star with four rays,[6] and sometimes a four-lobed rosette.[7] If these cavities increase in number, the genital organs increase in the same proportion.[8] The border of these organs is generally provided with numerous tentacles which project into the cavity.[9] In the riband-like testicles numerous small sacs are observed; each one of these opens separately into the genital cavity, while the eggs, on the contrary, are separated from the similarly-formed ovary only by a gradual constriction of the latter.[10]

4. With the Siphonophora, all the relations of these genital organs still require much investigation. With the Diphyidae, they consist of sacs communicating with the general cavity of the body.[11] During the epoch

1 *Will*, Horae tergest. p. 38, Taf. I. fig. 22, 23.
2 *Wagner*, Icones. zoot. Tab. XXXIII. fig. 26, a. a.; *Will*, loc. cit. Taf. II. fig. 5, 7, 8, 14, 16; *Blainville*, Manuel d'Actinol. 1834. ¹ᵃ XXXVII. fig. 3; and *Sars*, Beskrivelser loc. cit. Pl. v. ug. 12, 13.
3 *Will*, loc. cit. p. 71.
4 *Milne Edwards*, Ann. d. Sc. Nat. XVI. p. 198, Pl. I. fig. 1, a. b.
5 *Gaede*, Beiträge loc. cit. Taf. I. fig. 1, c. (*Medusa*); and *Lesson* in *Duperrey*, Voyage loc. cit. No. 12, 13 (*Chrysaora*).
6 *Rhizostomum*.
7 *Chrysaora, Medusa, Pelagia* and *Aurelia*. See *Ehrenberg*, Abhandl. d. Berl. Akad. 1835, Taf. I. fig. 1; *Wagner*, Icon. zoot. Tab. XXXIII. fig. 1; and *Brandt*, Mém. de l'Acad. de St. Petersburg, IV. Pl. IX. X. With the male and the female *Cephea*, I have found the testicles and the ovaries disposed exactly as with the Medusae.
8 In *Cassiopea*, these organs are eight in number.
9 *Medusa* and *Pelag* a; see *Ehrenberg*, loc.

cit. Taf. VII.; and *Wagner*, Icon. zoot. Tab. XXXIII. fig. 13.
10 *Siebold*, Beiträge loc. cit. Taf. I. fig. 20, 23; and *Kölliker*, Beiträge loc. cit. p. 40.
11 In *Diphyes* and *Ersaea*, a sac filled with cells opens into the general cavity of the body, and communicates beside with the stomachs and respiratory cavities. *Will* (Horae tergest. p. 78, 81, Taf. II. fig. 23, c.) regards this sac as a sexual organ; and *Meyen* (Nov. Act. physico-med. XVI. Suppl. 1, 1834, p. 214, Tab. XXXVI. fig. 2, h. and fig. 6, 7) asserts to have seen eggs in it. According to *Philippi* (*Muller's* Arch. 1843, p. 63, Taf. V. fig. 10, a. b.), the grape-like clustered genital organs, with *Physophora*, are situated between the prehensile organs; the smallest containing in each lobule six to ten eggs, and the largest a granular liquid (Sperm?).
Hollard (Ann. d. Sc. Nat. III. 1845 ⁿ 251 Pl. IV. bis. fig. 33, 34) has found botryoidal masses of ovaries at the base of the tubuliform tentacles (stomachs). *Sars* loc. cit. p. 37, Pl. V.) has also

of procreation, the females of some Discophora are easily distinguished from the males by the numerous pouches of their tentacles, and in which eggs and newly-hatched young are carried for a short time.[12]

§ 70.

As yet, the development of a few only of the Acalephae has been traced. It is attended by a remarkable metamorphosis.

After the usual segmentation of the vitellus, ovoid embryos resembling infusoria are developed ; these turn freely on their axis, and swim about in the water by means of ciliated epithelium.[1]   Shortly after, they become attached by the anterior extremity to some object.   Upon the opposite free extremity tentacles appear, and between them the mouth.   The animal has then the form of a Polyp.[2]   It is during this period that the young animal reproduces by *gemmation*,[3] and sometimes by transverse *fissuration*. This last mode occurs in the following remarkable manner :

The polyp-like animal increases in length, and its body divides transversely into many segments.   Around each of these segments eight bifid processes are developed ; after this, each segment is successively separated from before to behind, and they float about for a time as eight-rayed Acalephae, but soon attain, however, their adult condition.[4]

seen genital organs of the same form between the tentacles of *Agalmopsis* ; but he found at the same time (loc. cit. p. 38, 43), in the campanuliform individuals produced from buds, testicles with *Agalmopsis*, and ovaries with *Diphyes*. It may therefore be justly supposed that these various Siphonophora are compound, sexless individuals, which, like the Hydrina and Sertularina, reproduce by alternation of generation, — that is, by buds, — individuals having sex.

12 *Medusa aurita* and *Cyanea capillata ;* see *Ehrenberg*, Abhandl. &c. loc. cit. Taf. III. fig. 1, 2, Taf. VIII. fig. 1 ; also, *Sars* in *Wiegmann's* Arch. 1841, I. p. 19.

1 The development and metamorphosis of *Medusa aurita* and of *Cyanea capillata* have been observed by *Siebold* (Beiträge loc. cit. p. 21, Taf. I. II. ; and *Froriep's* neue Not. No. 166, 1838, p. 177 ;) and by *Sars* (*Wiegmann's* Arch. 1841, I. p. 19, Taf. I.–IV.). In the first stage of development (see *Ehrenberg*, Abhandl. &c. loc. cit. Taf. VIII. fig. 15–18 ; also, *Siebold*, Beiträge loc. cit. Taf. I. fig. 17–19 ; and *Sars*, *Wiegmann's* Arch. loc. cit. Taf. I. fig. 1–6), these infusoria-like Medusae have been regarded by *Baer* as the larvae (*Meckel's* Deutsches Arch. VIII. 1823, p. 389).

2 *Siebold*, Beiträge loc. cit. p. 29, Taf. I. fig. 25–33, Taf. II. fig. 34 ; and *Sars*, *Wiegmann's* Arch. loc. cit. Taf. I. fig. 7–31.   During my last visit at Trieste (autumn of 1847), I convinced myself that the young of *Cephea Wagneri* are developed wholly like those of Medusae, by passing from infusoria-like forms to polypoid young animals.*

3 The reproduction of the polyp-form Medusae by buds has been observed by *Sars* in *Cyanea*

capillata.   He has also seen them develop pedicles from the end of which new individuals would appear, which resembled Polyps. See *Wiegmann's* Arch. loc. cit. p. 26, Taf. I. fig. 37, 41, 42, 38, 39, 40.

4 These young Medusae, whilst composed of rings, have been taken for a new genus (*Scyphistoma*) of Polyps by *Sars* (Isis. 1833, p. 222, Taf. X. fig. 2). *Steenstrup* (Ueber d. Generationswechsel, p. 17) has regarded them as nurses of the Medusae. At a latter period, when the rings have been separated (Isis. 1833, p. 224, Taf. X. fig. 4 ; and Beskrivelser, &c., p. 16, Pl. III.) has described them as a new species of Medusae (*Strobila octoradiata*). But lately he has perceived that they are the young of *Medusa aurita* (*Wiegmann's* Arch. 1837, 1. p. 406) ; it did not occur to him, however, that these young constitute, very probably, the genus *Ephyra* of *Eschscholtz* (see *Wiegmann's* Arch. 1841, Th. I. p. 10). It will probably be discovered that many small campanulate or discoid Medusae are only the young of other Acalephae; for it is very likely that they all undergo a similar metamorphosis. It may also prove that many naked Polyps are only transitionary forms of known species of Acalephae. In this connection the observation of *Dujardin* (Comp. rend. 1843, p. 1132) deserves the attention of naturalists. In tracing the development of one of the Discophora allied to *Oceania*, he observed that this animal in its early condition separated from a corallum resembling that of *Syncoryne*, and was of a form quite like an *Eleutheria*. However various these developing forms may be, that one must be regarded as the real one which exists during the development of the testicles and ovaries.†

* [§ 70, note 2.] See, also, for recent researches on the development of *Cephea*, *Ecker*, Bericht üb. die Verhandl. d. naturf. Geselsch. in Basel. VIII. 1849, p. 51 ; *Busch*, Beobachtungen üb. die Anat. &c. Berlin, 1851, p. 30 ; and *Frantzius*, in *Sie-*

bold and *Kölliker's* Zeitsch. f. Zool. IV. p. 118, June, 1852. — ED.

† [§ 70, note 4.] In regard to the development of the Acalephae, it may be mentioned that recent researches, few as they are, have verified some

of the hypotheses suggested in the above note. Hitherto there has been much confusion on this subject, from the want of complete series of observations ; even now the whole class can be regarded only in a somewhat transitionary state, in a zoological point of view. Many genera which have hitherto been regarded good and permanent will no doubt, as *Siebold* has remarked, prove to be only undeveloped forms of well-known species. As already stated, *Agassiz* regards the Hydroid Polypi as true Acalephae, and the analogy which exists between the embryos of Medusae and Polypi may be the foundation of many other important changes. At present, however, broad generalizations must be deferred until we have extensive and serial researches in the embryology of these animals. For separate details on the development of some forms, see *Busch*, loc. cit. (*Sarsia, Lizzia, Cephea, Eudoxia, Diphyes*) ; *Huxley*, loc. cit. (Diphyidae, Physophoridae) ; *Agassiz* and *Desor*, loc. cit. (Medusidae). — Ed.

7

BOOK FOURTH.

ECHINODERMATA.

CLASSIFICATION.

§ 71.

The Echinoderms have a more or less coriaceous envelope, filled with calcareous, reticulated corpuscles. These last are sometimes so numerous that they form a real shell, composed of plates, movable, or tightly bound together. In the ray-like, symmetrically-arranged systems of organs, the quinquenary number prevails.

In many species the digestive canal is asymmetrical. All are marine, and most of them move by means of particular, erectile suckers. Others progress by vermiform motions, and some swim freely by moving their rays like oars. Only a few are stationary. All are without copulatory organs.

ORDER I. CRINOIDEA.

The calcareous shell, composed of movable pieces, forms a true cutaneous skeleton. The body is ray-like; the digestive canal, asymmetrical.

FAMILY: ENCRINIDAE.

Genus: *Pentacrinus.*

FAMILY: COMATULINAE.

Genus: *Comatula.*

ORDER II. ASTEROIDEA.

The calcareous shell, composed of movable pieces, forms an internal skeleton. The cutaneous covering is sometimes coriaceous, and sometimes calcareous. The body is ray-like, and the digestive canal symmetrical.

FAMILY: OPHIURIDAE.

Genera *Astrophyton, Ophionyx, Ophiothrix, Ophiomastix, Ophiocoma, Ophiolepis, Ophioderma.*

FAMILY : ASTEROIDAE.

Genera : *Luidia, Astropecten, Ctenodiscus, Archaster, Stellaster, Astrogonium, Oreaster, Pteraster, Asteriscus, Culcita, Ophidiaster, Chaetaster,
Solaster, Echinaster, Asteracanthion.*

ORDER III. ECHINOIDEA.

The calcareous shell forms a spherical or discoid shield, composed of immovable plates. The digestive canal is asymmetrical.

FAMILY : ECHINIDAE.

Genera : *Echinus, Cidaris.*

FAMILY : CLYPEASTRIDAE.

Genera : *Laganum, Scutella, Encope, Rotula, Lobophora, Echinocyamus,
Mellita, Echinanthus.*

FAMILY : SPATANGIDAE.

Genus: *Spatangus.*

ORDER IV. HOLOTHURIOIDEA.

In place of a calcareous shell, the cutaneous envelope contains a greater
or less number of calcareous reticulated corpuscles. The œsophagus is
surrounded by a calcareous ring, constituting the rudiment of an internal
skeleton. The body is cylindrical. The digestive canal, generally asymmetrical.

FAMILY : HOLOTHURINAE.

Genera : *Holothuria, Pentacta, Bohadschia, Cladolabes.*

FAMILY : SYNAPTINAE.

Genera : *Synapta, Chirodota.*

ORDER V. SIPUNCULOIDEA.

The cutaneous envelope is coriaceous, and free from calcareous corpuscles. There is no calcareous ring about the œsophagus. The body is
cylindrical ; the digestive canal, usually asymmetrical.

FAMILY : SIPUNCULIDAE.

Genera : *Sipunculus, Phascolosoma.*

FAMILY : ECHIURIDAE.

Genera : *Thalassema, Echiurus.*

BIBLIOGRAPHY.

Tiedemann. Anatomie der Röhrenholothurie, des pomeranzenfarbigen Seesterns und Steinseeigels. Landshut, 1816.

Sharpey. Cyclop. of Anat. and Physiol. Art. *Echinodermata,* vol. II. p. 30. London, 1839.

Agassiz. Monographies d'Échinodermes vivans et fossiles. Neuchatel, 1838, 1–3ᵉ, Livr.

Valentin. L'Anat. du Genre Echinus. Neuchatel, 1842. 4ᵉ Livr. des Monographies d'Échinodermes.

Forbes. A History of the British Star-fishes, and other animals of the class Echinodermata. London, 1841.

Müller und Troschel. System der Asteriden. Braunschweig, 1842.

ADDITIONAL BIBLIOGRAPHY.

Müller. Ueber die Larven und die Metamorphose der Ophiuren. In the Transact. Berlin Acad. 1846.

Ueber die Larven und die Metamorphose der Echinodermen. Ibid. 1848.

Ueber die Larven und die Metamorphose der Holothurien und Asterien. Ibid. 1849–50.

Anatomische Studien über die Echinodermen. *Müller's* Arch. 1850. Hft. II.

Berichtigung und Nachtrag zu den Anatomischen Studien über die Echinodermen. Ibid. Hft. III.

Fortsetzung der Untersuchungen über die Metamorphose der Echinodermen. Ibid. Hft. V.

Ueber die Ophiurenlarven des Adriatischen Meeres. Ibid. 1851. Hft. I.

Ueber die Larven und die Metamorphose der Echinodermen, vierte Abhandlung. Read to the Berlin Acad. 7, Nov. 1850; 28 April and 10 Nov. 1851, and published in 1852.

Agassiz. Twelve Lectures on Comparative Embryology, delivered before the Lowell Institute, in Boston, Dec. and Jan., 1848–49. Boston, 1849.

Busch. Beobachtungen über Anatomie und Entwickelung einiger wirbellosen Seethiere. Berlin, 1851.

These writings relate chiefly to development; but, for many special points of Anatomy, see the writings of *Müller, Krohn, Peters,* and others referred to in my notes. — Ed.

CHAPTER I.

CUTANEOUS ENVELOPE AND SKELETON.

§ 72.

With the exception of the apodal Sipunculidae, the Echinoderms have a cutaneous skeleton modified in the different orders in the following manner:

I. In the Holothurioïdea, irregular calcareous corpuscles, which often have reticulated openings, are scattered through the skin.[1]

II. In the Echinoïdea, the calcareous substance is separated from the soft skin, and composed of plates of a definite form, pierced by openings. These plates are immovably united together by means of sutures. These last are easily seen in the Echinidae, but are indistinct in the Clypeastridae; they entirely disappear with age in some species.[2] Among these plates which are arranged in a regular series, those called *ambulacral* should be mentioned; these are perforated, having upon their outer surface the pedicles, and upon their inner the *ambulacral vesicles*. They form, usually, five double rows, so placed between the other plates that their openings form, sometimes five longitudinal rows extending from the mouth to the arm,[3] sometimes a rosette of five lobes[4] on the dorsal surface of the skeleton.[5]

III. The coriaceous skin of the Asteroïdea, like that of the Holothurioïdea, contains numerous calcareous corpuscles, of which the smallest are irregular, the largest porous. But beneath this is a cutaneous skeleton, composed of porous calcareous pieces, movably articulated, and extending on the ventral surface from the mouth to the end of the rays.

In many species, the larger corpuscles, pressed together, form a reticulated support, which is either simple[6] or composed of plates.[7]

With the internal skeleton, each articulation is usually composed of many pieces, the intervening lacunae of which are the *ambulacral pores*. The principal middle pieces unite at an obtuse angle, thus forming an abdominal furrow.[8] The Ophiuridae have also an articulated internal skeleton, but the articulations are simple. But the external envelope of their arms consists of calcareous scales, closely knit together, and which so tightly close up the internal skeleton that the cavity of the body does not extend between the skin and the internal skeleton into these appendages, as in the Asteroidae.

IV. In the Crinoïdea, the skin is soft only on the ventral surface; that of the back is wholly calcareous, and converted into an articulated skeleton, which extends upon the arms and lateral branches. The mobility of these articulations is due to an elastic, interarticular tissue. They constitute discs or short cylinders, which, joined together, form arms, lateral branches (*pinnulae*), cirri, and in some species a peduncle.[9]

There is a canal in axis of all these parts of the skeleton, and upon the

---

1 These irregular and usually perforated calcareous corpuscles are mixed with the sand of the sea, after the death and decomposition of the animal, but can then easily be distinguished with the microscope. *Quatrefages* (Ann. d. Sc. Nat. XVII. 1842, Pl. III. IV.) has figured many of them belonging to *Synapta*. Similar microscopic corpuscles, of various forms, are found in the soft parts of most of the Echinoderms. It is very desirable that, as has already been commenced by *Ehrenberg* (Abhandl. d. Berl. Akad. 1841, p. 408), they should be subjected to careful investigation; for by this way alone can correct views be obtained upon many enigmatical bodies of this kind seen by the naturalist.
For the calcereous corpuscles imbedded in the skin of the Holothurinae, see *Koren* in *Froriep's* neue Not. XXXV. p. 18, fig. 6-9; and in the Arch. skandin. Beitr. f. Naturg. I. p. 449.

2 *Scutella* and *Clypeaster*.
3 *Echinus* and *Cidaris*.
4 *Encope, Rotula, Scutella, &c.*
5 A very detailed description of the shell of *Echinus* will be found in *Meckel's* System der vergleich. Anat. II. Abth. I, 1824, p. 31; and in the monograph of *Valentin*, Anat. du genre Echinus, 1841, p. 5. He has also published very exact researches, with figures, upon the intimate structure of the calcareous plates of this animal (Ibid. p. 17, Pl. 11.).
6 *Asteracanthion, Solaster*.
7 *Asteracanthion, Oreaster, Solaster, &c.*
8 See the figure by *Sharpey*, Cyclop. Anat. and Phys. loc. cit. p. 31, fig. 8, 9; and *Meckel's* vergleich. Anat. II. Abth. 1, p. 19.
9 *Pentacrinus*.

ventral surface of the arms and pinnulae, a furrow, over which the soft skin (*perisoma*) passes in a bridge-like manner.[10]

§ 73.

In many Echinoïdea the buccal cavity is provided with processes pointing perpendicularly into the interior of the shell, and which are the points of attachment of the masticatory muscles and ligaments. This *osseous circle* is most developed in the Echinidae,[1] and is composed of five processes. Between each of these is a smaller one, corresponding to as many ambulacral ones, each of which is perforated by a large opening.[2] In the Clypeastridae, there are five simple processes only ;[3] and in the Spatangidae they are wholly absent.

The sub-cutaneous osseous ring about the œsophagus, in the Holothurioïdea, corresponds probably to this circle. Usually composed of ten pieces, it may be regarded as a rudimentary internal skeleton, for it is the point of attachment of both muscles and tentacles.

In *Holothuria tubulosa* its anterior border is denticulated ;[4] and in *Synapta* it is composed of twelve pieces, five of which have oval openings for the free passage of the aquiferous canals.[5]

§ 74.

The general envelope of many Asteroïdea is more or less covered with various calcareous productions. These have the forms of lamellae, knobs, callosities, granules, immovable rays both sharp and blunt, rough and smooth movable points, double hooks, &c.[1]

In the Echinoïdea, there are points of very variable size united to knobs which are scattered over the external surface of the shell. These points project through the thin skin covering this shell, having at their base a kind of capsular articulation.[2]

Remarkable cutaneous organs are found in *Synapta*. These are small anchor-like hooks, by which these animals attach themselves to objects. Each of them is obliquely inserted under a small sub-cutaneous scale, which is perforated by a canal.[3]

10 In the Crinoïdea, as well as in the Echinoderms generally, the parts of the skeleton have a calcareous, reticulated structure ; see *Müller's* Arch. 1837, p. 93, and *Ueber d. Bau. d. Pentacrinus caput Medusae*, in the Abhandl. d. Berl. Akad. 1841, Taf. I. fig. 3.

1 *Echinus, Cidaris.*

2 *Valentin*, Monogr. loc. cit. Pl. II. fig. 15.

3 *Agassiz*, Monogr. d'Echinodermes, 2te Livr. containing the *Scutellae*, Pl. XIII. fig. 3, Pl. XXVII. fig. 7 (*Lobophora* and *Echinocyamus*).

4 *Tiedemann*, Anat. d. Röhrenholothurie, &c., p. 26, Taf. II. fig. 5 ; also *Wagner*, Icon. zool. Tab. XXXII. fig. 15.

Karen has observed that the osseous ring is composed of ten pieces with *Thyone fuscus* and *Cuvieria squamata* of the Holothurinae.

5 *Quatrefages*, Ann. d. Sc. Nat. XVII. 1842, p. 47, Pl. IV. fig. 5 ; Pl. V. fig. 7, c. c.

1 With *Oreaster* and *Culcita*, the whole body is covered with knobs and granulations. With *Astropecten* and *Stellaster*, you find flattened points and marginal lamellae. Innumerable rays, with

bristled points, project from the surface of *Solaster* and *Chaetaster*. With *Ophiocoma* and *Ophiomastix*, the margins of the arms are covered with smooth points, which in *Ophiothrix* are spinous. In *Ophionyx* these spinous points have movable double hooks ; see the beautiful figures of *Müller* and *Troschel* (System d. Asteriden).

2 The spines of the Echinoïdea have, over their whole extent, numerous, denticulated ribs ; see *Valentin*, Monogr. loc. cit. Pl. III. fig. 26. In *Spatangus* the spines are spatulate, and in the Clypeastridae (*Mellita, Encope, Laganum*) they are clavate. The minute researches of *Valentin* (Monogr. loc. cit. p. 24, Pl. III) have shown the structure of the spines of the Echinoïdea to be very complex.

3 The burr-like roughness of the skin of *Synapta* has already been observed by *Eschscholtz* (Zool. Atlas, Hft. 2, 1829, p. 12). *Jaeger* (De Holothuriis dissertatio, 1833, Tab. I. fig. 3) first figured the cutaneous hooks of *Synapta Beselii*. *Quatrefages* (Ann. d. Sc. Nat. XVII. p. 33, Pl. III.) has given a very exact description of those of *Synapta*

§ 75.

A peculiar calcareous plate (the madreporic plate) is observed upon the cutaneous skeleton of the Asteroïdea and Echinoïdea. In the last it is always situated in the centre of the dorsal surface, but in the first its position varies. In the proper Asteroidae there are often several, having an excentric dorsal situation; while in the Ophiuridae[1] it is found upon the ventral surface, and especially in the angle formed by the junction of the two arms with the tortuous mouth. In some Asteroidae a membranous sac (stony canal), filled with organized calcareous particles, is attached to this plate; in others, an articulated calcareous cord stretches obliquely across the body towards the border of the mouth. The use of these parts is not yet positively known.[2]

CHAPTER II.

MUSCULAR SYSTEM AND ORGANS OF LOCOMOTION.

§ 76.

In the Echinoderms the muscular system is well developed. Its primitive fibres are flat, and without transverse striae.[1]

In the ventral surface, and between each joint of the arms and pinnulae of the Crinoïdea, there are one or two small muscles, antagonistic to which, upon the opposite surface, is an interarticular elastic tissue.[2]

In the Asteroïdea, the interarticular lacunae of the internal skeleton are filled with muscles.[3] The skin of these animals does not aid the motions of the arms, except by its elasticity. But in the Echinidae the skin

---

Duvernaea. The similar hooks found in the sea-mud of Vera Cruz have been taken by *Ehrenberg* for stony concretions belonging to a sponge, and figured and named *Spongolithis anchora* (Abhand. d. Berl. Akad. 1841, p. 323, Taf. III. No. VII. fig. 36). He has also taken the perforated supports of these hooks for an Infusorium with a siliceous carapace, described as *Dictyocha splendens* (Ibid. fig. 319). But, more lately, he has perceived their true nature (Ibid. p. 407, 443). The discovery of analogous cutaneous organs in the marl near Streitberg, by *Count Munster* (Beitr. z. Petrefak. Hft. VI. 1843, p. 92, 96, Taf. IV. fig. 9), is very interesting, since it shows the antediluvian existence of *Synapta*.

Beside the cutaneous corpuscles of carbonate of lime, *Quatrefages* (loc. cit. p. 36, Pl. III. fig. 15) has found others which are of a spherical form in the skin of *Synapta Duvernaea*; and, as they have protractile filaments, he compares them to nettling organs.

1 *Astrophyton.*

2 These parts are found in *Astropecten*. According to *Tiedemann* (loc. cit. p. 54), they furnish the necessary calcareous matter for the skeleton of the Asteroidae. But *Ehrenberg* (*Muller's Arch.*

1834, p. 580) has shown that they do not contain ordinary calcareous matter, but rather that which is organized and perforated in a reticulated manner. A calcareous cord of a special structure is found in *Asteracanthion*; see *Siebold, Muller's* Arch. 1836, p. 291, Taf. X. fig. 14–18; and *Sharpey,* Cyclop. Anat. &c. loc. cit. II. p. 35, fig. 12, 13, s.*

1 According to *Wagner* (*Muller's* Arch. 1835, p. 319), the Echinoderms do not have transversely striated muscles. This has been confirmed by *Muller* (Abhand. d. Berl. Akad. loc. cit. p. 214, Taf. IV. fig. 9) in the genera *Pentacrinus* and *Comatula*. For my own part, I have failed to perceive them in *Echinus, Asterias, Ophiurus, Holothuria,* and *Sipunculus. Valentin* (Monogr. loc. cit. p. 101, Pl. VIII. fig. 153–155) asserts to have seen striæ upon the fibres of the masticatory, spinous and anal muscles of *Echinus*; and *Quatrefages* (Ann. de. Sc. Nat. loc. cit. p. 43, Pl. III. fig. 17) has observed transverse wrinkles during the contraction of the longitudinal muscles of *Synapta*.

2 *Muller,* Abhand. d. Berl. Akad. loc. cit. p. 214, 220, Taf. II. fig. 8, 12.

3 The interarticular muscular layer of the Asteroidae has been accurately described by *Meckel* (System d. vergleich. Anat. III. p. 14).

---

* [§ 75, note 2.] See, for further details on this stone-canal with the Ophiuridae, *Muller,* Arch. 1850, p. 122. — Ed.

covering the shell has distinct muscular bands for the motions of the
points.[4]

In the Holothurioïdea and Sipunculoïdea there is a very thick subcutaneous muscular layer. This is itself composed of two layers, — the first
and upper being made up of circular, the second and lower of longitudinal fibres. In the Holothurioïdea,[5] these fibres form five large, thick,
widely-spread bundles, which are inserted into the osseous ring. In the
Sipunculoïdea, these bundles are more numerous, but more compactly
bound together.[6]

The muscles of mastication, of the digestive canal, and of the tentacles,
will be treated hereafter.

§ 77.

With the exception of the Synaptinae and Sipunculoïdea, the Echinoderms
have special, tentacular, locomotive organs (ambulacra). These are hollow
and very contractile prolongations of the skin, and communicate through
the ambulacral pores with small contractile sacs (ambulacral vesicles), found
upon the internal surface of the coriaceous or calcareous envelope of the
body. The ambulacra and their vesicles have transverse, longitudinal fibres,
and contain a clear liquid, which, from contractions, oscillates from one to
the other through the pores. In this way the ambulacra are capable of
erection and elongation, and the animal uses them as feelers to find a
proper object of attachment; and on this account, also, they have in some
species a suctorial extremity.

These organs, which are sometimes locomotive, sometimes prehensile,
have the following variations of structure and form:

I. With the Crinoïdea they are small, delicate and cylindrical, and are
found upon the borders of a furrow, which runs from the mouth along the
soft perisoma covering the arms and pinnulae. Each one of them is covered with small cylindrical, clavate tentacles.[1]

II. The Ophiuridae have upon their arms, and between the plates, pores
which connect with small cylindrical ambulacra; these last, from numerous
small warts, present a studded aspect.[2]

III. With the Asteroidea they are situated in a double or quadruple
row, in the ventral furrows which extend from the mouth to the end of the
rays. They form compact cylinders of considerable size, the acute or
truncated extremity of each of which has a sucker.[3]

IV. With the Echinoïdea they are situated upon an elongated stalk,
and have a sucker. They are found both upon the ambulacral plates and
immediately around the mouth.[4] Being extremely movable, they are

4 *Valentin*, Monogr. loc. cit. p. 35, Pl. III. fig. 39.
5 The cutaneous muscular system of *Holothuria* has been described by *Tiedemann* (loc. cit. p. 27, Taf. II. IV.); and that of *Synapta* by *Quatrefages* (Ann. d. Sc. Nat. loc. cit. p. 41).
6 For the muscular system of *Sipunculus nudus*, see *Grube*, in *Müller's* Arch. 1837, p. 240, Taf. XI. fig. 1.
1 The ambulacra of *Comatula*, which have active vermicular movements, have no opening at their free extremity; see *Müller*, Abhandl. d. Berl. Akad. loc. cit. p. 222, Tab. IV. fig. 13, 14.
2 By these the very active arms of the Ophiuri-

dae are attached to surrounding objects; see *Erdl* in *Wiegmann's* Arch. 1842, I. p. 58, Taf. II. fig. 1, a.
3 Beside the very correct description given of these organs by *Tiedemann* (loc. cit. p. 56), see *Rymer Jones* (A Gen. Outl. of the Anim. King. p. 148, fig. 65). It appears that in *Astropecten* the extremity of the ambulacra can be inverted, thus compensating for the sucker found in *Echinaster*, *Asteriscus*, and *Asteracanthion*.
4 With *Echinus* the suckers, which exactly resemble the ambulacra, are fixed upon the contractile membrane surrounding the mouth. With *Spatangus* and *Echinanthus* there is

chiefly locomotive; for from them numerous points are prolonged, by which they adhere to objects, and to which they become afterwards fixed by their sucker. They are covered with ciliated epithelium, and their suckers are made firm by a coarse calcareous network. Elongated calcareous corpuscles of the same nature are found also in their walls, — some branching and others hook-like.[5]

V. With those Holothurioïdea which have them, they have a more or less complete sucker, and are scattered irregularly over the entire surface of the body, or disposed in regular rows. Usually very short, they can be retracted deeply in the skin; but they are capable of equal prolongation, and thus perform well the function of suckers.[6]

The ambulacral vesicles, which are intimately connected with the circulatory and respiratory systems, will be fully treated hereafter.

§ 78.

With the Echinoïdea, and Asteroidae, there are other movable organs (pedicellariae), which, scattered over the surface of the body, are prehensile, and used in a pincer-like manner. With the Asteroidae, they usually consist of two delicate forceps-like pieces (pedicellariae forcipatae), or of two large valvular flaps (pedicellariae valvulatae). Generally they are not pediculated.[1] Those of the Echinoïdea have been carefully studied in *Echinus*. They are numerous, and occur for the most part about the mouth, presenting three different forms: 1. Those composed of three short, lenticular pieces (pedicellariae gemmiformes). 2. Those formed of three long delicate pieces, laterally denticulated (pedicellariae tridactyli). 3. Those with three laterally denticulated spoon-like pieces (pedicellariae ophiocephali). They are supported by a base of calcareous, reticulated substance; and in the Echinoïdea, always rest upon a stalk, the lower part of which contains a cylindrical, calcareous nucleus, while the remaining portion is soft, and capable of a spiral contraction.[2] Here also they are covered with ciliated epithelium, and can, by means of movable processes, seize hold of objects, which, being passed along, may be conveyed even from the dorsal surface to the mouth.

near the mouth, and opposite the ambulacral rosette, a row of ambulacra having special pores.

5 See *Valentin*, Monogr. loc. cit. p. 37, Pl. IV. V., and *Erdl* in *Wiegmann's* Arch. 1842, I. p. 55, Taf. II. fig. 10. The corpuscles found by *Ehrenberg* (Abhandl. d. Berl. Akad. 1841, p. 324, Taf. III. No. VII. fig. 37, a. b.) in the marine sand of Vera Cruz, and figured under the name of *Spongolithis uncinata*, are only the cruciform parts of the skeleton of *Echinus*. This will be evident from comparing them with the calcareous corpuscles figured by *Valentin* (Monogr. loc. cit. Pl. V. fig. 55).

6 Catalogue of the Physiol. series of Comp. Anat. contained in the Royal Coll. of Surgeons, London, IV. 1838, p. 196, Pl. XLIX. fig. 3–5.

1 With *Luidia*, there are, however, three tongue-like pedicellariae. In *Asteracanthion*, they have a soft pedicle. In *Asteropsis, Stellaster*, and

Astrogonium, they are valvular and without a pedicle; see *Müller* and *Troschel*, loc. cit. p. 10, Taf. VI. fig. 5–6.

2 The pedicellariae of *Echinus* were, at first, taken for parasitic Polypi by O. F. *Müller* (Zool. Dan. I. 1777, p. 16, Tab. XVI.). See *Lamarck*, Hist. Nat. des Anim. sans Vertèbres, II. p. 75. More recently, *Agassiz* (*Valentin's* Monogr. loc. cit. p. 51) has expressed the opinion that they were young individuals. The researches of *Delle Chiaje* (Memor. sulla storia e notom. degli Anim. senza Vertebr. II. 1825, p. 324, Tab. XXIII.) and of *Sars* (Beskrivelser, &c., p. 42, Tab. IX.) upon *Echinus, Cidaris*, and *Spatangus*, have dispelled all doubts as to the real nature of these organs. Very correct descriptions of them have lately been published by *Valentin* (Monogr. loc. cit. p. 46, Pl. IV.), and by *Erdl* (*Wiegmann's* Arch. loc. cit. p. 49. Taf. II. fig. 1–9).*

*[§ 78, note 2.] See *Adams* (Ann. of Nat. Hist. VIII. 1851, p. 237), who has found what he regards as Pedicellariae on the skin of *Voluta*

vespertilio; he thinks, therefore, that they are independent parasitic organisms. — Ed.

CHAPTER III.

NERVOUS SYSTEM.

§ 79.

The central portion of the nervous system consists of a ring which is usually pentagonal, and surrounds the commencement of the œsophagus. The main nervous branches are given off from this, and pass to the other end of the body along the median line of the rays, or their corresponding parts. The form of this ring is mainly due to that of the mouth; and therefore, with the reniform mouth of *Spatangus*, it is unequally pentagonal.[1] Ganglia have not yet been found in it. But in *Echinus* and *Holothuria*, the nerves passing from it have between their fibres, violet, green, or red pigment granules.[2]

§ 80.

The principal nervous trunks have a longitudinal furrow, as if composed of double cords, and give off from each side, during their course, branches which go to the ambulacra.[1]

With the Crinoïdea, a nervous cord passes beneath the furrow formed by the perisoma on the ventral surface of the arms; this has a slight swelling opposite each pinnula, to which it sends off a branch.[2] With the Asteroidae, the nervous trunks which pass off from the œsophageal ring are lodged in the ventral furrows of the rays.[3] But in the Ophiuridae, they pass in a canal, concealed by the ventral plates of the arms. The five nerves, analogous to those of the Echinoïdea, pass along the internal surface of the ambulacral plates, between the vesicles, even to the centre of the dorsal region. In *Echinus*, there are, moreover, special nerves directly from the œsophageal ring, for the organs of mastication and digestive canal.[4] In *Holothuria*, this ring is situated directly on the anterior border of the osseous circle, and sends off five nerves which pass along the median line of the longitudinal muscles, even to the end of the body;[5] it sends off also special nerves to the oral tentacles.[6]

---

1 *Krohn* (*Müller's* Arch 1841, p. 8, Taf. I. fig. 3, 4).

2 *Krohn*, loc. cit.

1 *Krohn*, ibid. p. 4, 10.

2 *Müller* (Abhandl. d. Berl. Akad. loc. cit. p. 233, Taf. IV. fig. 11, i. ; Taf. V. fig. 16).

3 The nervous system of the Asteroidae was first clearly shown by *Tiedemann* (loc. cit. p. 62, Taf. IX. and *Meckel's* Deutsch. Archiv. I. 1815, p. 69, Taf. III. fig. 1). This anatomist, like *Krohn* (loc. cit. p. 4), did not perceive the ganglia of the œsophageal ring, observed by *Wagner* (Vergleich. Anat. 1834, p. 372).

The ganglia and nerves that *Spix* (Ann. du Mus. d'Hist. Nat. XIII. 1809, p. 439, Pl. XXXII. fig. 3, 6) and *Konrad* (De Asteriarum fabrica dissert. 1814, p. 13, fig. 3, 6.) affirm to have seen on the internal (dorsal) surface, opposite the ventral furrows of the articulations of the rays, in *Asteracanthion rubens*, and *glacialis*, are probably only tendinous fibres.

4 *Krohn*, who has studied the nervous system of *Echinus* and *Spatangus*, has traced the filaments given off from the main trunks, across the ambulacral pores, to the suckers of the ambulacra. See also *Valentin's* figures of this system, in *Echinus* (Monogr. loc. cit. p. 98, Pl. VIII. IX.).

5 The œsophageal ring of *Holothuria*, observed by *Krohn* (*Müller's* Arch. 1841, p. 9, Taf. I. fig. 5), sends off its principal nerves across the fissures of the dentations of the five great pieces of the osseous rings. Their lateral filaments, going to the ambulacral vesicles, are so fine that *Krohn* could scarcely find them.

6 *Grant*, loc. cit. p. 184.*

---

* [§ 80, note 6.] *Müller* has furnished some valuable contributions on the nervous system of the Holothurioïdea ; see Arch. 1850, p. 226. He makes this statement, which is worthy of remem-

With the Sipunculidae, as with the other worm-like Echinoderms which approach the Annelids, the arrangement of the nervous system is quite different. Here, the nervous ring is a simple, aganglionic thread extending to the posterior end of the body, and may be regarded as the first trace of a ventral cord.[7]

CHAPTER IV.

ORGANS OF SENSE.

§ 81.

The sense of *touch* is well developed with the Echinoderms, and seems to have its seat in the oral tentacles, the ambulacra, and pedicellariae.

With the Asteroïdea, and Echinoïdea, no organs of *vision* have yet been found. As such, however, have been regarded the red pigment dots situated, with the former, at the extremity of their rays,[1] and with the latter, in the middle of the dorsal region upon five ocellary plates which alternate regularly with those of the genital organs.[2] These ocellary plates are perforated each by a very fine canal, through which passes a delicate filament from the main nerve for the pigment dot.[3] Although these pigment dots have thus a nervous connection, no proper organ to refract the light has yet been found in them.[4]

---

7 According to *Krohn* (*Müller's* Arch. 1839, p. 348), the œsophageal ring of *Sipunculus nudus* has two super-œsophageal ganglia blended together. These had already been observed by *Delle Chiaje* (Memor. loc. cit. I. p. 15, Tav. I. fig. 6. i.); but more lately *Grube* had taken them for cartilaginous rudiments of the osseous circle (*Müller's* Arch. 1837, p. 244). He has also confounded with the muscular system the two lateral nerves of this ring, and its abdominal branch which in its course sends off laterally branches to the muscular layer and to the skin, and terminating at the end of the body in a swelling. Then, on the other hand, the filaments surrounding the digestive tube, and taken by him for nerves, appear to be only cellular fibres (loc. cit. p. 244, Taf. XI. fig. 4).
According to *Forbes* and *Goodsir* (*Froriep's* neue Not. No. 392, 1841, p. 279), the nervous system of *Echiurus* is composed of an œsophageal ring, with an abdominal cord, from which pass off asymmetrical branches.

brance : " It is a noticeable fact that the nervous trunks of these animals throughout are contained in a sheath, which, after the maceration of its contents, has exactly the aspect of a blood-vessel." The nervous system of these animals cannot, therefore, be properly studied from alcoholic specimens. — Ed.

* [§ 80, note 7.] See also *Blanchard* (Ann. d. Sc. Nat. 1849, XII. p. 57), who has well made out the nervous system with *Sipunculus rufo-fimbriatus*. It consists of two cerebral ganglia

According to *Quatrefages*, *Echiurus Gaertneri* has an abdominal cord which possesses ganglia, and by this character the Echinidae approach the Annelida ; see Ann. d. Sc. Nat. VII. 1847, p. 332, Pl. VI. fig. 4.*
1 In the Clypeastridae and Echinidae.
2 These dots, which *Vahl* (*Müller* Zool. Dan. Tab. CXXXI.) had already observed in *Pteraster militaris*, were first regarded as eyes by *Ehrenberg* (*Müller's* Arch. 1834, p. 577, and Abhand. d. Berl. Akad. 1835, p. 209, Taf. VIII. fig. 11, 12). He has seen in *Asteracanthion violaceus*, a small swelling at the extremity of the nerve of the ocellary dot. *Forbes* (Hist. of the Brit. Star-fishes, 1841, p. 152) first noticed these dots in *Echinus*, and their presence has been confirmed by *Agassiz* and *Valentin* (Monogr. loc. cit. p. 10, 100, Pl. II. fig. 12, Pl. IX. fig. 188, 189).
3 *Valentin*, loc. cit. Pl. IX. fig. 190.
4 *Valentin* has failed to discover in these organs a crystalline lens. Although in *Echinus* they are upon the back, and therefore favorable to vision ;

united so as to form a single cordiform mass — the brain, which is situated under the muscles of the proboscis. From this brain passes off a cord on each side, forming a collar about the œsophagus ; these unite below, and then continue as a ventral cord to the posterior extremity of the body. This cord has slight swellings along its course, which may be regarded as ganglia ; they send nerves to the integuments.
This anatomist has also observed here a very distinct splanchnic system of nerves. — Ed.

CHAPTER V.

DIGESTIVE APPARATUS.

§ 82.

The alimentary canal is situated in the cavity of the body, isolated, but is retained in its place by a kind of mesentery which is composed of fibres,[1] or of a thin membrane.[2]

The mouth, which is usually central, is often surrounded by a circle of tentacles.[3] In the Asteroïdea, the digestive canal is a large central pouch, an anus and appendages extending into the rays being present in some[4] and wanting in others.[5] In the other Echinoderms, the digestive canal has usually thin walls, is of a variable length, and tortuous quite to the anus.

The position of the anus is quite varied. In the Echinidae, and Asteroïdae, it is in the centre of the back, exactly opposite the mouth. In the Holothurioïdea, it is at the posterior end of the body; while in the Clypeastridae, and Spatangidae, it opens laterally upon the margins of the shell. In the Crinoïdea, it is near the mouth upon the ventral surface, and in the Sipunculoïdea, it has a similar position.

The internal surface of this canal has generally been found lined with ciliated epithelium.[6]

§ 83.

With the Asteroïdea, and Echinoïdea, the pedicellariae already described, are used to seize the food and convey it to the mouth. Their ambulacra are perhaps sometimes used in the same way. In the Crinoïdea, the furrow of the tentacles, aided by the tentacles themselves, serves well to conduct the food from the arms and pinnulae to the mouth.[1]

In the Holothurioïdea, and Sipunculidae, there are completely retractile tentacles of a special nature. In the first, they are hollow, pinnated or branched, and, arranged in a circle around the mouth, are attached by their base to the osseous circle and to the elongated vesicles which project into the cavity of the body. These tentacular vesicles contain a liquid,

---

and, in the Asteroïdae, where they are upon the ventral surface at the end of the furrows, the rays bend round to the dorsal surface; and again, although *Tiedemann* (*Meckel's* Deutsch. Arch. loc. cit. p. 175) thinks these last can distinguish light from darkness, yet it is doubtful if these animals can really see by these organs. They appear, like many other inferior animals, to perceive the light by its action as an excitant upon their skin, and in this way can, like plants, seek the sunlight. The account which *Forbes* (Hist. of British Star-fishes, p. 139, and *Froriep's* neue Not. No. 420, 1841, p. 26) has given of *Luidia fragilissima*, which, having made its escape by the loss of an arm, looked with scornful eyes upon its persecutor, is pleasant to read, but is far from settling this question.

1 Asteroïdea, Echinoïdea, and Sipunculoïdea.

2 Holothurioïdea.
3 Holothurioïdea and Sipunculoïdea.
4 Asteroïdae.
5 Ophiuridae.
6 According to *Sharpey* (Cyclopædia, &c., loc. cit. 1. p. 616) and *Valentin* (*Wagner's* Handwörterbuch der Physiol. 1. 1842, p. 433), the internal surface of the stomach and its appendages, of the Asteroïdae, has a ciliary movement. *Valentin* (Monogr. &c. p. 79) has also found ciliated epithelium in the entire digestive canal of *Echinus*.

With *Phascolosoma*, where I have found cilia upon the tentacular apparatus, and with *Comatula*, where *Muller* (Abhandl. d. Berl. Akad. 1841, p. 233) has found them in the anus, they extend probably through the intestine.

1 *Muller*, Abhandl. d. Berl. Akad. 1841, p. 222.

which, by their contraction, is pressed into the cavity of the tentacles for lubrication.[2]

The retraction of the tentacles is due in part to their own contractility, and in part to the numerous muscles, which, arising from the internal surface of the cavity of the body, are inserted into the osseous circle. By these means, it, together with the tentacles, can be retracted into the body.[3] With the Sipunculidae the tentacular apparatus consists of a fringed border on the margin of the mouth, which is also provided with vesicles.[4] In *Sipunculus*, and *Phascolosoma*, there are four long muscles, which, arising from the internal surface of the body, pass on to the mouth, and are retractors of the tentacular membrane.[5] It is possible that these oral tentacles serve not only as prehensile organs of food, but also as those of locomotion and respiration.[6]

§ 84.

The *mouth* of the Comatulinae presents nothing remarkable; but with the Asteroidae, it is covered with hard papillae, projecting from its corners and angles. In the Ophiuridae, the inverted angles are covered with hard papillae, while the everted ones have calcareous teeth, between which are concealed soft cylindrical tentacles. Immediately behind all of these, the entrance of the stomach is indicated by a membranous sphincter. In the Asteroidae, however, this is wanting, there being a short œsophagus leading directly into the stomach.

With the Echinoïdea, and Holothurioïdea, the mouth has a soft circular lip, between which, with the Echinidae, and with the Clypeastridae, project the points of enamelled teeth.

The mouth of the Echinidae, and Clypeastridae, has a very remarkable masticatory apparatus. In the first, the calcareous basis which supports the teeth has long been known as *Aristotle's lantern*. This conical basis is divided into a base and summit; the first being the superior part of the animal itself, while the second is formed by points of teeth projecting from the mouth. It is, moreover, composed of fifteen pieces, five of which are three-sided, hollow pyramids, and so adjusted that they touch each other by their plane surfaces, presenting externally the third surface which is convex. This last has internally a longitudinal furrow, in which is fitted a very long, narrow and slightly-curved tooth. Beside these five principal pieces, which form the jaws of *Echinus*, there are two other kinds, much

2 These vesicles are found in *Holothuria* and *Chirodota*; see *Tiedemann*, loc. cit. Tab. II. fig. 4, e. 6, i.; also the Catalogue of the Museum, London, &c., IV. Pl. XLIX. fig. 1, 2 (*Holothuria tubulosa*); and the Atlas Zool. du Voyage de l'Astrolabe. Zoophytes, Pl. VIII. fig. 3 (*Chirodota fusca*). In *Pentacta doliolum*, I have found only a single cylindrical vesicle fixed to the circle of tentacles. In *Synapta Duvernaea (Quatrefages* Ann. d. Sc. Nat. loc. cit.), these vesicles are entirely wanting.

Cuvier (Anat. Comp. V. p. 454) and other anatomists (see *Grant*, Outlines, &c., p. 383) have erroneously taken these parts for salivary organs. They do not communicate with the digestive canal, but connect freely with the circulatory and respiratory systems, — a point, therefore, to which we shall further allude hereafter.

3 In *Pentacta*, there are five large cylindrical muscles arising from the subcutaneous longitudinal ones, and inserted into the osseous circle; — they are special retractors of the tentacles; see *Meckel*, System d. vergleich, Anat. IV. p. 62.

4 I am inclined to regard as tentacular the two vesicles of Poli, in *Sipunculus*; and of which *Delle Chiaje* (Memor. &c. Tav. I. fig. 6, d.) perceived only one, although *Grube* (*Müller's* Arch. 1837, p. 251, Taf. XI. fig. 2, P.) has seen them both fixed in a space circumscribed by the tentacular membrane.

5 *Grube*, Ibid, p. 241, Taf. XI. fig. 1, n. 2, m. m.; and *Delle Chiaje*, Memor. &c. Tav. I. fig. 3.

6 The oral tentacles of *Synapta Duvernaea*, which, according to *Quatrefages* (loc. cit. p. 63, Pl. IV. fig. 1), have suckers on their internal surface, are certainly used as locomotive organs.

8

smaller. Of these, five are elongated quadrilateral plates, placed at the base of the lantern, between each two pyramids. The other five, smaller and longer, are curved upon the first.

All these pieces are united by many tendons and muscles to each other, and to the neighboring osseous circle which projects inwards from the shell.

The muscles of mastication are in ten pairs; five of these arise from the longest processes of the osseous circle, and are inserted on the pyramids below the summit of the lantern. The other five, on the other hand, pass from the shortest processes of this circle to the base of the pyramids.

By this arrangement, when the first five contract and separate the summits of the pyramids together with their teeth, the second five, contracting also, carry the points of the teeth again together, by separating the bases of the pyramids.[1]

In the Clypeastridae, the masticatory apparatus is more simple. It is composed of ten unequal, triangular pieces, joined together, V-form, two and two. Each of these pieces has in its projecting angle, a furrow in which a tooth is fitted. These five jaws are so arranged around the mouth that their angles and the points of their teeth meet together in its centre.[2]

§ 85.

The digestive cavity of the Ophiuridae is only a simple stomachal sac, occupying the centre of the hollow disc of their body.

It is divided by walls projecting inwardly, into many caeca, which never extend into the rays.[1]

There are usually ten of these caeca, which in *Astrophyton* are subdivided into numerous smaller caeca.[2]

With the Asteroidae, the stomach is large and has a similar situation; but it sends off radial caeca into the rays.

In those species which have an anus, the digestive canal may be divided into three parts. The stomach is separated into two chambers by a circular, projecting fold. The first of these is the true stomach, and the second sends off the radial caeca. A narrow, short rectum, passing off from the stomach, forms the third part of this canal, and terminates in an anus, situated upon the back of the animal and concealed among points, callosities, &c. This rectum has folds which, of a variable length and sometimes branched, are called the inter-radial caeca, and are situated between instead of in the rays.[3]

In the Comatulinae, this canal consists of a coecum situated at the end of a short œsophagus, and which, after a spiral course about the axis of the body, terminates in an anus having the form of a short tube projecting from the ventral surface not far from the mouth.[4]

In *Comatula europaea*, the axis, around which the digestive canal passes

1 This apparatus has been minutely described by *Tiedemann* (loc. cit. p. 72, Taf. X. fig. 1, 2), by *Meckel* (syst. d. vergleich, Anat. IV. p. 56), and by *Valentin* (Monogr. &c. p. 64, Pl. V.). See also the beautiful figure by *Rymer Jones* (Outline of the Anim. King. &c. p. 167, fig. 70, 71).

2 *Agassiz*, Monogr. &c. 2e Livr. Scutelles. p. 15, Pl. XII. XIII. XIV. &c.

3 *Konrad*, De Asteriarum fabrica, fig. 5.

2 *Meckel*, Syst. d. vergleich, Anat. IV. p. 50.

3 See also *Tiedemann* (loc. cit. Taf. VII.), whose beautiful figures have been copied everywhere; and the original designs of the digestive cavity of *Asteracanthion, Archaster,* and *Culcita,* by *Muller* and *Troschel* (loc. cit. Taf. XI. XII.).

4 Upon the digestive canal of *Comatula,* see *Heusinger,* Zeitschr. f. d. organische Physik. III. 1829, p. 371, Taf. X. XI.

spirally, consists of a spongy substance, from which projects a lamina like the *lamina spiralis* of the conch of a snail shell.[5]

In *Spatangus*, the toothless mouth opens into a delicate œsophagus which passes insensibly into a long tube of nearly the same size. This last makes two convolutions in its course, and sends off at about its anterior fourth a very long caecum. The digestive canal, situated between the origin of this caecum and the œsophagus, is of a dark color and has transverse plicae, while the remaining portion below is smooth and of an orange hue.[6]

In the Clypeastridae, the numerous spiral turns of this canal are supported by many calcareous laminæ situated upon the interior of the shell.[7]

In many species of *Clypeaster*, this canal has at its commencement, transverse folds, and further on numerous lateral caeca, which are separated from each other by laminæ like those just described.[8]

In the Echinidae the pharyngx has very thick muscular walls, and is surrounded by masticatory organs. Upon it succeeds a proper œsophagus, which, after a few convolutions, passes to the anus situated in the centre of the back. The digestive canal is a caecum given off by this last, and has many spiral turns in the cavity of the body.[9]

In the Holothurioïdea, the very muscular pharynx is surrounded by the osseous circle. In the Holothurinae, the intestinal canal, which is long and equal throughout, has many turns from behind forwards, ending at last in a large cloaca situated at the posterior part of the body. But in the Synaptinae, it is short and nearly straight, and terminates in an anus having no cloaca.[10]

In the Echiuridae [11] this canal closely resembles that of the Synaptinae.

In the Sipunculidae it is long, making its first turn about the middle of the body, and its second near the posterior extremity. The ascending and descending portions of this last pass spirally around each other on their way to the anus which is situated on the ventral surface of the body.[12]

§ 86.

As to the glandular appendages of the alimentary canal, the *salivary organs* are perhaps entirely wanting in these animals.

In the Holothurinae alone, are there particular appendages opening into its anterior portion, which could be regarded as organs of this nature. In the different genera, species, and even individuals of this family, these appendages widely vary as to form and number.

5 *Müller*, Abhandl. d. Berl. Akad. 1841, p. 230, Taf. V.

6 See *Meckel*, Syst. d. vergleich, Anat. IV. p. 55, and *Delle Chiaje*, Memor. &c. Tav. XXV. fig. 12; also *Carus* and *Otto*, Erläuterungstafeln z. vergleich. Anat. IIft. IV. Tab. I. fig. 25, and *Wagner*, Icon. zool. Tab. XXXII. fig. 8. The nature of the canal figured by *Delle Chiaje* is yet unknown. It arises from the first portion of the intestine, and returns to it at its middle portion. It has not been mentioned by *Meckel*.

7 *Agassiz*, Monogr. des Scutelles, p. 14, Pl. III. fig. 19, a.

8 Ibid. p. 17, Pl. XXII. fig. 23 (*Laganum* and *Mellita*).

9 See *Tiedemann* and *Valentin*, loc. cit.

10 The digestive canal of the Holothurinae was first figured by *Delle Chiaje* and by *Tiedemann*,

loc. cit.; afterwards by *Quoy* and *Gaimard* (Atlas zool. du Voyage de l'Astrolabe. Zoophytes, Pl. VI. fig. 2, Pl. VII. fig. 3). The cloaca is always wholly attached to the skin by numerous tendinous fibres. In *Chirodota fusca* the intestine is spiral (Atlas zool. &c. Pl. VIII. fig. 3); but in *Synapta Duvernaea* it is nearly straight (*Quatrefages* Ann. d. Sc. Nat. loc. cit. Pl. II.).

11 See the remarks of *Forbes* and *Goodsir* upon the Anatomy of *Thalassema* and *Echiurus* (*Froriep's* neue Notizen, No. 392, p. 273, fig. 12).

12 The alimentary canal of *Sipunculus nudus*, and of *Echinorhyncus*, has been faithfully described by *Delle Chiaje* (Memor. &c. 1. p. 9, Tav. I. fig. 5, 6; p. 126, Tav. X. fig. 11) and *Grube* (*Müller's* Arch. 1837, p. 245, Taf. XI.). I have found a similar intestine in *Phascolosoma granulatum*.

In *Holothuria tubulosa*, they are cylindrical, pure white, and very numerous, being united in bundles which are attached to the digestive canal near the pharynx by short white pedicles.[1]

In *Pentacta doliolum*, there is usually only one of these organs, — a small, white, curved horn, which sends to the pharynx a very tortuous canal, which is widely removed from the excretory duct of the genital organs.

The whiteness of these organs in Holothurinae is due to a reticulated calcareous skeleton in their walls.[2]

The radial caeca of the Asteroidae ought probably to be regarded as hepatic organs. They are often quite developed, extending as a double canal from the stomach into each ray. Their walls have numerous small botryoidal vesicles, which secrete a yellow liquid. Usually each of these ten liver-like organs arises from the stomach by a proper canal ;[3] but in some, two of them connect with this organ by a single canal.[4]

With those Asteroidae which have an anus, there is another series of glandular appendages, the inter-radial caeca, which pass off from the rectum. Their function is not yet known. They contain a brownish liquid, in which, with *Asteracanthion rubens*, no uric acid has been found. In *Astrogonium, Solaster*, and *Asteracanthion*, these organs are branched, and only two in number.[5] In *Archaster*, and *Culcita*, there are five ; but in *Culcita coriacea*, each of these is divided dichotomously into two other long botryoidal caeca, which, separated by a septum, are spread out between the rays.[6]

In *Astropecten*,[7] which is without an anus, there are sometimes found two short, analogous caeca, which open into the base of the stomach by a common orifice. But in *Luidia*, which is also without an anus, these organs are entirely absent.[8]

In the other Echinoderms, which are entirely without these glandular appendages, the walls of the alimentary canal probably secrete the fluid requisite for digestion, and thus supply also the want of the hepatic organ.[9]

1 It has already been shown that the cylindrical vesicles of *Holothuria* taken by *Cuvier* and other naturalists for salivary organs do not communicate with the digestive canal, but rather with the tentacles. The white appendages of *Holothuria tubulosa* were first described as testicles by *Delle Chiaje* (Memor. &c. I. p. 97. Tav. VIII. fig. 6.), and *Tiedemann* (loc. cit. p. 29, Taf. II. fig. 6, p.) assigned to them the same function. It is certain that they have no testicular character, although I cannot affirm that they are salivary organs. They have been figured, in *Holothuria atra*, by *Jaeger* in his dissertation : De Holothuriis, Tab. III. fig. 2, c. c.

2 This calcareous tissue has been observed by *Jaeger* (loc. cit. p. 38, Tab. III. fig. 7), by *Wagner* (*Froriep's* neue Not. No. 249, 1839, p. 99), and by *Krohn* (Ibid. No. 356, 1841, p. 53). This last observer, who affirms that these organs are in connection with the great circulatory vessel surrounding the digestive canal, compares them to the stony canal of the Asteroidae.

3 In *Astropecten aurantiacus*, according to

Tiedemann (loc. cit. Taf. VII. or, *Wagner*, Icon. zool. Tab. XXXII. fig. 1). It is the same, also, in *Archaster, Culcita*, and *Luidia* ; see *Muller* and *Troschel*, loc. cit. p. 132, Taf. XI. fig. 2 ; Taf. XII. fig. 1.

4 *Asteracanthion* ; see *Konrad*, De Asteriarum fabrica, fig. 1 ; and *Muller* and *Troschel*, loc. cit. Taf. XI. fig. 2.

5 See *Muller* and *Troschel*, loc. cit. p. 132, Taf. XI. fig. 1 (*Asteracanthion rubens*) ; an entire group of these rectal caeca of *Asteracanthion glacialis*, has been figured by *Konrad*, loc. cit. fig. 1, d.

6 *Muller* and *Troschel*, loc. cit. p. 132, Taf. XI. fig. 2, Taf. XII. fig. 1.

7 *Tiedemann*, loc. cit. Taf. VII.

8 *Muller* and *Troschel*, loc. cit. p. 132.

9 According to *Valentin's* figure of the intimate structure of the digestive membranes of *Echinus*, they are lined with hepatic epithelium, like that of the Lumbricinae, and that of the Polyps, already mentioned (Monogr. &c. Pl. VII. fig. 120, 131, 133).

CHAPTER VI.

CIRCULATORY SYSTEM.

§ 87.

The vascular, sanguineous system of these animals is yet imperfectly known. The constant confusion and imperfection of its descriptions are probably due to the fact that it has not been carefully distinguished from the respiratory system; and also, as was true of the Acalephae, because it has been confounded with the aquiferous system, which is usually present.[1]

From all the old and new researches upon this subject, it is evident that all the Echinoderms have an isolated system of this kind, composed usually of both an arterial and venous trunk, between which there is, in some species, an organ like a heart.

§ 88.

In the Crinoïdea, there is, at the base of the calyx, a heart-like sacculus, from which pass off vessels into the central cavity of the arms, the cirri, and the pedicle when it is present. From its centre, another vessel is given off for the spongy axis of the cavity of the body.[1]

The Asteroïdae have three vascular rings, one of which is under the skin of the back, while the other two are beneath, around the mouth. Between these vascular rings there is a long muscular heart, which, united to the calcareous pouch or cord, extends from the madreporal plate to the mouth.

It is probable that the Asteroïdae, which have many of these plates, have also many calcareous cords and hearts.[2] From these vascular rings numerous other vessels are sent off, some to the stomach and its appendages, and the genital organs, and others to the ambulacra and their vesicles.[3]

---

1 The extended, and in some respects contradictory works of *Tiedemann* and *Delle Chiaje* (loc. cit.; see, also, *Meckel*, Syst. d. vergleich. Anat. V. p. 25; and *Sharpey* Cyclopæd. &c. II. p. 41) have not, for reasons which may be stated, cleared up this point. The same may be said of what relates to the blood of these animals, for it has been confounded in part with the ambulacral liquid belonging to the aquiferous system. See *Wagner*, Zur vergleich Physiol. der Blutes, 1833, p. 28.

The observations of *Delle Chiaje* (Memor. &c. II. p. 345) and of *Carus* (Analekt. zur Natur. u. Heilkunde, 1829, p. 132, and Lehrb. d. vergleich. zoot. 1834, p. 673) do not give correct ideas upon the mode and direction of the circulation of these animals; for it is evident that they did not see it, but only the vibratile phenomena of the aquiferous system.

1 The vascular system of *Comatula* and *Pentacrinus* has become known through *Heusinger* (Zeitsch. f. organisch. Physik. III. 1828, p. 373, Taf. X. XI.) and *Muller* (Abhandl. d. Berl. Akad. 1841, p. 198, 236, Taf. V.). The membranous canal, situated beneath the nervous branches of the arm, and directly above the calcareous articulations, and the passage of which through the arm into the calyx *Muller* (loc. cit. p. 233) has not been able to clearly make out, is probably a blood-vessel. It is yet unknown how the blood of these vessels is distributed to the organs.

2 As in *Echinaster solaris*, and *Ophidiaster multiforis*; see *Muller* and *Troschel*, loc. cit. p. 134.

3 According to *Tiedemann* (loc. cit. p. 49, Taf. VIII.), the lower extremity of the heart of *Astropecten aurantiacus* opens into the vascular ring which surrounds the mouth. This last sends arterial branches to the stomach, the cœca, and the genital organs; the superior extremity of the heart communicates in like manner with another vascular ring upon the back, and which receives the veins of the organs just mentioned. From a third and reddish vascular ring, situated directly under the skin of the mouth, *Tiedemann* has seen pass into each ray a vessel placed superficially in the furrow of the ambulacra, but he did not ascertain

In the Echinidae, the heart is long,[4] and attached to the œsophagus.

In *Echinus*, it has several saccular enlargements, and internally has a cavernous aspect, due to numerous irregularly arranged septa. At each of its extremities there are two vascular rings. The two below are situated on the top of the lantern and surround the œsophagus, while the two above surround the anus; all belong probably to the arterial and venous systems. One of these last sends off five branches to the genital organs, while the other receives one of the two trunks which pass along the whole length of the intestinal canal. Two longitudinal vessels, which send off branches right and left, pass between each of the five pairs of ambulacral organs. These are, probably, a branchial artery and vein.[5]

In the Holothurinae, the vascular system, which is without a heart, is very distinct. An aortal trunk arises from the vascular ring, which surrounds the œsophagus, and ramifies upon the intestine and the genital organs. By a reünion of these ramifications, a second trunk like a vena cava, is formed. This divides into two arteries, which ramify upon the branchiae, and from which arise two branchial veins, which return to the aorta.[6] With the Sipunculidae, and Echiuridae, there is a main vascular trunk, which, after sending off laterally small branches, passes along the ventral median line, above the digestive canal.[7]

---

its relations with the rest of the vascular system. Moreover, he has taken for an isolated, special sanguineous system belonging to the ambulacra, the aquiferous system, which communicates directly with the ambulacra, and which forms a third ring, situated between the two sanguineous ones of the mouth.

Volkmann's description (Isis 1837, p. 513) is wholly different. According to him, the vascular trunks of the superficial ring, and which are located in the furrows of the arms of *Asteracanthion violaceus*, send off laterally ambulacral branches; the oral ring, situated more deeply, sends off branches, which, passing through the cavity of the body, go to the rays and ambulacra, and *freely communicate with the cavity of these last*. This same ring has also an anastomotic connection with that of the back. According to this, the circulation occurs, he thinks, in the following manner : The heart sends the blood into the superficial oral ring ; thence it passes by the vessels in the furrows of the arms into the cavity of the ambulacra ; these last, acting as venous hearts, send it, by the vessels in the interior of the rays, to the second oral ring, from which it passes to the third and dorsal ring, and thence to the heart.

It is evident that *Volkmann* has taken a part of the aquiferous system for that of the sanguineous one ; and it is probable that he did not observe the second oral ring. No correct idea can be formed of the distribution of the arteries and veins of the Asteroidae, or of their vascular system in general, except by carefully separating it from the aquiferous system, and considering the fact that the blood-vessels do not open into the ambulacral vesicles, but probably are spread as a capillary net-work upon their walls.

4 The heart of *Echinus*, which is accurately described by *Valentin* (Monogr. &c. p. 92, Pl.

VIII.), is attached to the œsophagus by a kind of mesentery.

5 These details are supported by *Valentin* (loc. cit. p. 93), who has already added much to the labors of *Tiedemann* and *Delle Chiaje* upon the sanguineous system of *Echinus*, although, like his predecessors, he has been deceived as to its connections.

The received opinions upon the circulation of these Echinoderms are, therefore, hypothetical. The nature of the five glandular organs, which *Valentin* has seen communicate with one of the two vascular rings situated upon the lantern, is very problematical (Monogr. &c. p. 95, Pl. VII. fig. 119, i. 120).

6 See *Tiedemann*, loc. cit. p. 15. The sanguineous system of *Synapta Duvernaea*, as described by *Quatrefages* (loc. cit. p. 58), corresponds, properly, to the aquiferous system of *Holothuria*, which *Tiedemann* also has taken for a special sanguineous system of the skin and ambulacra. Hereafter we shall notice further both of these systems.*

7 For the sanguineous vascular system of *Sipunculus*, and *Echiurus*, see *Grube* and *Krohn* (*Müller's* Arch. 1837, p. 248 ; 1839, p. 350), also *Forbes* and *Goodsir* (*Froriep's* neue Not. No. 392, loc. cit.). The vascular trunk embraces there the nerve so closely, that care is necessary not to overlook one, or confound both together.

Quatrefages has found in the anterior part of the body of *Echiurus Gaertneri* three heart-shaped swellings of the blood system, namely, a *ventral heart* upon the ventral vessel, a *dorsal heart* upon the dorsal vessel, and a *mesenteric heart* situated beneath the digestive tube. This last communicates with the ventral heart by a flexuous vascular canal, and with the dorsal vessel by a small vascular ring ; see Ann. d. Sc. Nat. loc. cit. p. 324, Pl. VI. fig. 4.

---

* [§ 88, note 6.] See, for the vascular system of the Holothurioidea, *Müller* (Arch. 1850, p. 229), who has carefully studied it with the larger Synaptinae. He confirms *Tiedemann's* observa-

tion above quoted as to the general distribution of the vessels, and especially as to the presence of a splanchnic system, which, as is well known, *Quatrefages* has supposed to be wanting. — ED.

CHAPTER VII.

RESPIRATORY SYSTEM.

§ 89.

The respiration of the Echinoderms is performed in various ways. These are: 1. By exclusively respiratory branchiae. 2. By organs serving at the same time other functions. 3. By means of water passing through the openings of the skin into the cavity of the body, and aërating the blood through the capillary vessels of the viscera.

With the Asteroïdea, Synaptinae, Sipunculidae and Echiuridae, every individual has always two of these modes of respiration, and sometimes all three, as with the Echinidae and Holothurinae.

§ 90.

I. Organs which are exclusively respiratory are found in the Echinidae, Holothurinae, and Echiuridae. They consist of external branchiae in the first, and internal in the last two.

The external branchiae of the Echinidae are situated upon the soft membrane of the mouth, being formed of five pairs of arborescent, hollow lobules.[1] They are contractile, but cannot be retracted within the body. They are covered both internally and externally with ciliated epithelium.

The cavity of each communicates with that of the body by a large orifice situated on the internal surface of the oral membrane.[2] By this means they are bathed with water upon both of their surfaces. Their walls contain a coarsely reticulated calcareous skeleton,[3] and without doubt, also a capillary net-work belonging to the branchial vessels.

The internal branchiae of the Holothurinae arise as two tubes from the cloaca of the intestinal canal, and send off, through the whole cavity of the body, numerous coecal branches.[4] In *Holothuria tubulosa*, one of these tubes is closely connected with the turns of the intestine, while the other is attached to the inner walls of the body. With the first, especially, may be perceived the ramifications of the branchial vessels. They are also covered with ciliated epithelium, and their contractile and expan-

---

1 The ramified organs of the Echinidae, already known by *Tiedemann* (loc. cit. p 78, Taf. X. fig. 5, d. d.) and *Delle Chiaje* (loc. cit. II. p. 338), have been more exactly described by *Valentin* (Monogr. &c. p. 82, Pl. IV. fig. 57 ; Pl. VIII. fig. 42), and by *Erdl* (*Wiegmann's* Arch. 1842, I. p. 50, Taf. II. fig. 12, 13).
2 *Valentin*, loc. cit. Pl. VII. fig. 135, l.
3 *Valentin*, loc. cit. fig. 143 ; and *Erdl*, loc. cit. fig. 13.*
4 The branchiae of *Holothuria tubulosa* have

been very well described by *Tiedemann* (loc. cit. p. 11, Taf. 11. or *Wagner* Icon. zool. Tab. XXXII. fig. 9), and by *Delle Chiaje* (loc. cit. Tav. VIII. IX.). See also Atlas Zool. du Voyage de l'Astrolabe. Zoophytes, Pl. VII. fig. 2, 9, p. (*Holothuria ananas*) and Pl. VII. fig. 3, e (*Cladolabes spinosus*). *Pentacta doliolum* has similar organs. According to *Cuvier* (Anat. Comp. VII. 1840, p. 536) there is only a single branchia in the other remaining Holothurinae.

---

* [§ 90, note 3.] See, in this connection, *Müller* (Arch. 1850, p. 122), who has confirmed *Valen-* tin's observations as to the structure of the external gills. — ED.

sive power, united with the action of the cloaca, enables them to receive into and expel from their interior the water of the sea.[5]

The internal branchiæ of the Echiuridae consist of branchless tubes. In *Echiurus vulgaris*, the two branchiæ, which are very movable and open into a kind of cloaca, have, on their exterior, infundibuliform, ciliated protuberances; and to each of these there is internally a corresponding ciliated sac, capable of being inverted. The very bright-red, vascular network which is spread over these branchiæ, communicates with the great ventral vessel at the posterior extremity.[6]

§ 91.

II. Among the organs which are not exclusively respiratory, are the ambulacra of the Echinodermata pedata, and the oral tentacles of the Holothurioïdea and Sipunculidae, — organs which are used also for prehension and locomotion.

These ambulacra and tentacles have always a cavity which communicates directly with the proper vascular, aquiferous system. Their whole interior is covered throughout with ciliated epithelium.

This aquiferous system has, until recently, been taken by anatomists as a special vascular one, or confounded with it. Its water serves partly to distend the ambulacra and tentacles, as shown above (§ 77), and partly for respiration, which is performed by the vesicles over which ramify the branchial vessels. These vesicles are therefore like internal branchiæ, their vessels being bathed by the water of the sacs, and that of the cavity of the body. Usually this system consists of a ring situated between the vascular rays of the mouth, which sends canals to the oral tentacles and to the sides of the body. These canals always pass along by the rows of ambulacral vesicles, with which they communicate by lateral branches.

§ 92.

In the Echinodermata pedata, this aquiferous system has the following modifications:

In the Crinoïdea, and Ophiuridae,[1] only traces of it have been found.

In the first, there is an apparently aquiferous canal for the tentacles, situated directly under their furrow. This may be regarded as forming a part of such a system. In *Pentacrinus*, it is simple, but in *Comatula*, it is divided at several points by simple septa.[2]

In the Asteroidae, this system is highly developed, the central ring being provided with pediculated and often elongated vesicles.[3] The main

---

5 There is found, but inconstantly, it would appear, upon the trunk of the branchiæ of some Holothuriæ particular pedunculated cœca, which in *Bohadschia marmorata* have been regarded as urinary organs by *Jaeger* (De Holothuriis, &c., Tab. III. fig. 9, g.). But they require further investigation.*

6 *Forbes* and *Goodsir* (*Froriep's* neue Not. No. 392, p. 277, fig. 12, c. — 19).

1 From the figures of *Delle Chiaje* (loc cit. Tav.

XXI. fig. 17) it would appear that *Ophiurus* has an aquiferous system.

2 *Muller*, Abhandl. d. Berl. Akad. 1841, p. 234.

3 These pyriform vesicular appendages are always situated between the principal vessels of the rays, varying both as to number and volume, and being sometimes entirely wanting. *Astropecten bispinosus* has only five ; *Asteriscus verruculatus*, *Astropecten pentacanthus*, and *Asteracanthion glacialis*, have ten, in pairs. In this

* [§ 90, note 5.] For many new details upon the respiratory system of the Holothurioïdea, see

Muller, Arch. 1850, p. 129-155 (*Synapta, Chirodota*, and *Molpadia*). — ED.

trunks from this oral ring pass along the furrows of the rays close to their external surface. The ambulacral vesicles into which their lateral branches open, are sometimes simple,[4] or, from a kind of sulcation, have a heart-like form.[5]

In the Echinoïdea, the oral ring wants the pyriform appendages,[6] and its main trunks pass along the internal wall of the shell. The ambulacral vesicles of the oral membrane are conical; but the others are flattened, overlap each other in a tile-like manner,[7] and have a distinct branchial, vascular network.[8]

The aqueous oral ring of the Holothurinae has hollow appendages (tentacular vesicles) projecting into the cavity of the body.[9] It has also, in many species, a larger, longer, and sometimes double, cœcal vessel (*Ampulla Poliana*).[10] Opposite the tentacular vesicles, the ring sends off to the oral tentacles, vessels which are often arborescent and comparable to external branchiae;[11] while, between these vesicles, arise five other vessels which descend along the internal surface of the body. As usual, they send off lateral branches to the generally very small ambulacral vesicles.[12]

In a few species only of the Synaptinae, the aquiferous ring has hollow appendages.[13] From it pass off vessels both to the tentacles and to the sides of the body. As the ambulacra are here absent, the five main trunks do not give off lateral branches.[14]

In the Sipunculoïdea, the aquiferous system is least developed. As yet there has been found only a liquid moved by vibratile cilia in the doubly-laminated cavity of the lobulated tentacles of the Sipunculidae. With this cavity, two vesicles of Poli communicate, thus indicating the presence of an aquiferous system.[15]

last species they are only slightly developed; in *Astropecten aurantiacus* there are three to seven vesicles, opening by a common duct into each of the five angles of the aqueous vascular ring; see *Delle Chiaje*, loc. cit. II. p. 296; *Tiedemann*, loc. cit. p. 52, Taf. VIII.; *Konrad*, loc. cit. fig. 3; and *Meckel*, Syst. d. vergleich. Anat. V. p. 32. Here should be mentioned also the glandular corpuscles which are attached to the aqueous vascular ring, and which resemble in some respects the glandular organs of the vascular sanguineous rings of *Echinus*, pointed out by *Valentin*; see *Delle Chiaje*, loc. cit. II. Tav. XXI. fig. 12, 14; *Tiedemann*, loc. cit. Taf. VIII. o. o., or *Wagner*, Icon. zool. Tab. XXXII. fig. 2, m.

4 *Ophidiaster*, *Asteracanthion*, *Luidia*; see *Muller* and *Troschel*, loc. cit. Taf. XI. fig. 4.

5 *Astropecten*; see *Konrad*, loc. cit. fig. 4. I am not yet settled upon the question whether the aquiferous system of the Asteroïdea is filled by the extremity of the ambulacra, or by the oral ring. I have not been able to convince myself of the presence of an opening at the extremity of these first.

6 *Delle Chiaje* (loc. cit. Tav. XXVI.) has given very detailed figures of the aquiferous system of *Echinus* and *Spatangus*; but he has confounded it with the sanguineous vessels of the intestinal canal.

7 *Valentin*, Monogr. &c. Pl. CXXXIV.—CXXXVI.

8 The branchial vessels ramifying upon the flattened ambulacral vesicles appear to have been seen by *Monro* (Vergleichung des Baues und der Physiol. der Fische, 1787, p. 91, Taf. XXXIII. fig. 13–15; or Cyclopædia of Anat. and Physiol. II. p. 35, fig. 14). *Krohn* (*Muller's* Arch. 1841, p. 5) has accurately described them. It is affirmed that the ambulacra of *Echinus* can be filled with water through an opening of the sucker at their

extremity, and that it is discharged from the aquiferous system through ten openings between the teeth; see *Tiedemann*, loc. cit. p. 81; *Valentin*, Monogr. &c. p. 84, or Repertor. f. Anat. 1843, p. 237; and *Monro*, loc. cit. p. 92.

9 *Tiedemann*, loc. cit. Taf. II. fig. 4, e. e. fig. 6, m., and *Delle Chiaje*, loc. cit. Tav. VIII. IX.

10 *Tiedemann*, loc. cit. Taf. II. fig. 4, a. a. fig. 6, g.; *Delle Chiaje*, loc. cit. Tav. IX. fig. 6, f. (*Holothuria tubulosa*).

11 The position of the tentacular vesicle seems exactly adapted to enable them to force, during their contraction, their water into the tentacles, thus causing the prominence and development of these last. I am yet uncertain if they are not aided by the vesicles of Poli. With some Holothurinae, as with *Cladolabes spinosus* (Atlas zool. du Voyage de l'Astrolabe. Pl. VII. fig. 3, f.), and with *Pentacta doliolum* according to my own observations, the aquiferous ring has only one vesicular appendage, and it would be questionable whether this is analogous to a tentacular vesicle, or to one of Poli.

Thyone and *Cuvieria* have, according to *Koren* (loc. cit. p. 20, 36, fig. 2, 11), only a single large, vesiculiform appendage upon their aqueous ring.

12 See *Delle Chiaje*, loc. cit. Tav. IX. fig. 6 (*Holothuria tubulosa*); but here also the aquiferous is confounded with the sanguineous system.

13 In *Chirodota Doreyana*, and *fusca*, these hollow tentacular vesicles are very apparent; see Atlas zool. du Voyage, &c., loc. cit. Pl. VII. fig. 16, Pl. VIII. fig. 3.

14 *Quatrefages*, loc. cit. p. 58, Pl. IV. fig. 1, Pl. V. fig. 5.

15 That the tentacular membrane of the Sipunculidae has the function of a branchia, is indicated

§ 93.

III. In nearly all the Echinoderms, as has been seen, all the viscera are bathed with water which certainly affects their delicate blood-vessels. It is very probable that from ciliated epithelium covering the entire cavity of the body and the viscera this water circulates in a definite manner. It is rejected at last through many respiratory openings, through which also fresh water is introduced.

In the Ophiuridae, there are in each inter-radial space two or four large openings of this kind, leading into the cavity of the body.[1]

In the Asteroidae, water passes freely in and out the cavity of the body, through small contractile trachean tubes, which have been known for a long time, and which are very numerous upon the back. They are covered within and without with ciliated epithelium, and have an opening at their extremity.[2] As yet it is unknown how the cavity of the body of the Echinoïdea and Holothurioïdea receives the water. Only in *Synapta Duvernaea*, have there been found proper respiratory openings; these are four or five papillæ, covered with cilia, concealed at the base of the oral tentacles, and connecting with the cavity of the body through a narrow canal.[3] In the Sipunculidae, the water is received through an opening at the posterior end of the body.[4]*

CHAPTER VIII.

ORGANS OF SECRETION.

§ 94.

The Echinoderms appear to have special organs of secretion. In different parts of the body there are glandular organs, the real nature of which, however, has not yet been determined.[1]

by the presence of delicate and tortuous vessels, observed by *Grube* (*Müller's* Arch. 1837, p. 253) upon that of *Sipunculus nudus*. The same conclusion might be drawn from the liquid moved by cilia observed by myself in the interior of the tentacular lobules of *Phascolosoma granulatum*. *Grube* (*Müller's* Arch. 1837, p. 251, Taf. XI. fig. 2, P.) has seen in *Sipunculus nudus* the two vesicles of Poli, communicating with the cavity of the tentacular membrane.

1 *Müller* and *Troschel*, loc. cit. Taf. IX. X.

2 *Ehrenberg*, Abhandl. d. Berl. Akad. 1835, Taf. VIII. fig. 12, e.; and *Sharpey*, Cyclopædia of Anat. &c. I. p. 615, fig. 238, C.

3 *Quatrefages*, Ann. d. Sc. Nat. loc. cit. p. 64, Pl. V. fig. 7, f.

4 The manner in which the water enters into the interior of the Echinuridae is not quite clear to me from the description of *Forbes* and *Goodsir* (*Froriep's* neue Not. No. 382, p. 277).

1 The attention has already been directed to these glandular organs, when speaking of the parts to which they are attached. The calcareous sac, or stony canal as now understood, of certain *Asteriae*, can scarcely be regarded as organs of secretion.

* [End of § 93.] In *Echinarachnius* and *Clypeaster Agassiz* has observed that trachean tubes, similar to those of the Asteroidae, perform the function of carrying the water in and out of the body. They are situated chiefly along the margin of the disc, emptying first into a circular tube, analogous to the circular tube of the Discophora, from which extend ramifications into the main cavity of the body ; see Compt. rend. 1847. — ED.

CHAPTER IX.

ORGANS OF GENERATION.

§ 95.

Although most Echinoderms have extraordinary powers of reproduction, yet this, apparently, is not for the multiplication of the individuals, for they do not reproduce either by fissuration or by buds.

The Holothurioïdea alone, perhaps, form the exception.[1] All propagate by the sexual organs of separate male and female individuals, and hermaphroditism is very rare.

The eggs which are usually round, are covered by a thin chorion, and contain beside a little albumen, a variously colored vitellus with its germinative vesicle and dot.[2] The sperm is always milky, and the spermatic particles which are unaffected by sea-water, are nearly always composed of a round or oval, rigid body, to which is attached a delicate, very active tail.[3]

§ 96.

Externally, the organs of both sexes exactly resemble each other, and especially during the interval of procreation; but at the sexual epoch they often differ in color. Their situation is very varied, and they are composed of simple or branched tubes, with proper excretory ducts. These last, however, are sometimes wanting, and then the contents of the former escape by rupture, and, falling into the cavity of the body, pass out through the respiratory openings.

Here, as in the Polyps and Acalephs, the copulatory organs being absent, the water is the medium of the fecundation of the eggs, by bringing the spermatic particles in contact with them.

---

1 The *Holothuria*, which, when captured, discharge all their viscera through the mouth, can, according to *Dalyell* (*Froriep's* neue Not. No. 331, p. 1), not only reproduce all these, but also can divide spontaneously into two or more parts, each of which becomes a complete individual. This multiplication by fissuration occurs also, perhaps, with *Synapta Duvernœa*; see *Quatrefages*, loc. cit. p. 26.

2 See the eggs of *Comatula Europaea* (*Muller*, Abhandl. d. Berl. Akad. 1841, Taf. V. fig. 17), of *Asteracanthion violaceus* (*Wagner*, Prodromus, &c., Tab. 1. fig. 3, or *Carus* and *Otto*, Erläuterungstafeln, Hft. V. Taf. I. fig. 1), of *Echinus lividus* and *sphaera* (*Valentin* Monogr. &c. fig. 167, 169), of *Holothuria tubulosa* (*Wagner*, Icon. zool. Tab. XXXII. fig. 12), and of *Synapta Duvernaea* (*Quatrefages*, loc. cit. Pl. V. fig. 1).

3 See, for the spermatic particles of *Asteracanthion*, *Solaster*, and *Echinus* (*Kolliker*, Beiträge, loc. cit. fig. 1-4, and *Valentin*, Monogr, &c. fig. 168), of *Holothuria* and *Synapta* (*Wagner*, Icon. zoot. Tab. XXXII. fig. 13, and *Quatrefages* loc. cit. Pl. V. fig. 2). Those of similar form have been seen in *Comatula* by *Muller* (Monatsbericht d. Berl. Akad. 1841, p. 189, or the Abhandl. of the same, loc. cit. p. 235). According to *Valentin* (Repertorium, 1841, p. 301), those of *Spatangus violaceus* have an elongated body, pointed in front, with a very delicate hair-like tail. Those of *Ophioderma longicauda*, and *Ophiothrix fragilis*, according to my own observation, have a round body, with an equally delicate hair-like tail.*

---

* [§ 95, note 3.] The spermatic particles of the Echinoderms are developed, like those of the other Radiates, in special cells, and like them also have, I think, invariably a cercaria-form. The differences in the shape of the head of these particles are wide, and of zoological import. Thus it is sometimes round (*Asterias*, *Urastes*), sometimes pyriform (*Echinocidaris*), and sometimes long-conical (*Mellita*). — Ed.

§ 97.

In the Crinoïdea, these organs, in the form of tubes, are situated under the soft perisoma of the pinnulae, and probably are without proper excretory ducts.[1]

In the Ophiuridae, they consist of lobular, pedunculated sacs, which are suspended in pairs in the inter-radial spaces of the disc.

These ten organs are usually deeply fissured, and the lobules thus formed appear as so many proper sacs attached to the peduncle.[2] These last are sometimes subdivided also.[3]

Sometimes each organ, divided in its whole length into lobules, is turned in the shape of a ram's horn.[4] The peduncle of these organs is directed towards the mouth, but it is yet uncertain whether their contents escape this way or fall into the cavity of the body. In the first case, the peduncle would be the excretory duct;[5] and in the second, the eggs and sperm would escape through the respiratory openings.[6]

In the Asteroidae these organs consist of varicose lobular sacs, situated in the angles of the inter-radial spaces.[7] In those species which are without an anus, there are no proper genital openings;[8] these openings are also wanting in those Asteroidae which have an anus.[9] In these last, the sperm and very small eggs pass into the cavity of the body, and probably have their escape through the respiratory openings.[10]

But in some species,[11] there are upon the back and near each angle of the inter-radial spaces two small approximated plates, perforated by small openings (*Laminae cribrosae*). These are the simple openings of these organs, which here consist of multi-ramose sacs, situated all along each side of the inter-radial septa, to the common duct which opens through one of the plates.

The number of these genital sacs varies widely in the different genera of the Asteroidae. In many, a single trunk of them hangs on each side of the inter-radial septa;[12] in others, there is a whole row of them;[13] and in others still, there are two rows attached to the dorsal surface of the cavity of the body, and extending into the rays.[14]

In the Echinoïdea, these organs descend along the internal surface of

---

1 The development of the genital organs of *Comatula* was first observed by *Dujardin*, who asserts that the red vesicles situated on both sides of the tentacular furrows secrete, during the epoch of rut, a very beautifully red liquid (L'Instit. No. 119, p. 268, or *Wiegmann's* Arch. 1836, II. p. 207). *Thompson* has seen the eggs of *Comatula* escape in clusters through the openings of the pinnulae (Edinb. New Philos. Jour. No. XX. p. 295, or *Froriep's* neue Not. No. 1057, 1836, p. 4, fig. 8); while, according to *Muller*, they escape by rupture (Abhandl. d. Berl. Akad. 1841, p. 234, Taf. V. fig. 17, 18).

2 *Ophioderma longicauda*, and *Ophiolepis scolopendrica*; see *Rathké*, *Froriep's* neue Not. No. 269, p. 65; and, Neueste Schrift. d. Naturforsch Gesellsch. in Danzig. III. Hft. IV. 1842, p. 116, Taf. II. fig. 3, 4.

3 *Ophiocoma nigra*; see *Rathké*, Danzig. Schrift. &c. loc. cit. Taf. II. fig. 5–7.

4 *Ophiothrix fragilis*.

5 *Rathké*, loc. cit.

6 *Muller* and *Troschel*, loc. cit. p. 133.

7 *Muller* and *Troschel* have very interesting details upon the various arrangements of the genital organs of the Asteroidae (loc. cit. p. 132).

8 As in *Astropecten* and *Luidia*.

9 As in *Ophidiaster*.

10 According to *Sars*, the ventral surface of the disc and arms of the female *Echinaster sanguinolentus* and *Asteracanthion Mulleri* have at certain times a kind of incubating cavity, in which the eggs remain during their development. He thinks they get there from the cavity of the body, through particular openings upon the ventral surface of this last; see *Wiegmann's* Arch. 1844, I. p. 169, Taf. VI. fig. 1, 2.

The genital parts of *Echinaster sanguinolentus* have been described with much detail by *Sars*, Faun. littor. Norveg. p. 48.

11 *Asteracanthion rubens*, and *Solaster papposus*; see *Muller* and *Troschel*, loc. cit. Taf. XII. fig. 2–4.

12 *Echinaster*, *Astrogonium*, *Asteriscus*, and *Ctenodiscus*.

13 *Astropecten*, *Oreaster*, and *Culcita*; see *Tiedemann*, loc. cit. p. 61, Taf. VIII. L. L.

14 *Archaster*, *Chaetaster*, *Luidia* and *Ophidiaster*; see *Muller* and *Troschel*, loc. cit. Taf. XII. fig. 5.

the shell, filling the empty spaces between the double rows of ambulacral vesicles.

They consist of widely ramified, deeply interlocked cœca, having always proper excretory ducts, which open upon the genital plates of the back of the shell.[15] There are here always five of these organs, and the genital plates, alternating with the ocellary ones, surround the anus.[16] In some species of the Clypeastridae, and Spatangidae, there are, perhaps, only four of these organs, judging from that number of the plates.[17] In the Holothurinae, these organs have a very different arrangement. They consist of widely-branched cœca,[18] floating, as loose clusters, freely in the cavity of the body, and opening through a single common excretory duct, situated below the osseous circle, and between the oral tentacles.

The testicle, which is of a whitish color, consists of a cluster of cylindrical sacs, branched and interlocked with each other.[19] But the ovary is pale red, very long, branched, a little flattened, and extends even to the posterior end of the body.[20]

As the only exception among these animals, the Synaptinae are hermaphrodites. But it should be stated that we know of them only through *Synapta Duvernaea*. It is said that here the testicles and ovaries are united in one and the same organ.[21] Three or four long cylindrical sacs float in the cavity of the body, and have an excretory duct which opens back of the osseous circle. At the epoch of procreation, vesicular prolongations appear on their interior surface, in which are formed spermatic particles. The spaces between these prolongations are filled by a pultaceous mass, in which appear eggs.[22]

In the Sipunculidae, and Echiuridae, there are only two or four simple cylindrical contractile pouches attached to the ventral wall. It is yet undetermined whether their contents escape by rupture, or through special openings.[23]

---

15 The separate sexes of *Echinus* were first shown by *Peters*; see *Müller's* Arch. 1840, p. 143.

16 See *Tiedemann*, loc. cit. p. 85, Taf. X. fig. 1, 4, 8; and especially *Valentin*, Monogr. &c. p. 103, Pl. VIII.

17 With *Echinanthus, Mellita, Rotula, Scutella* (see *Agassiz*. Monogr. des Scutelles), and *Spatangus arenarius*, I can count only four genital plates, while in *Encope*, and *Clypeaster*, I find five; yet *Valentin* (Repertorium, 1840, p. 301) expressly speaks of five genital organs in *Spatangus violaceus*.

18 *Wagner* and *Valentin* were the first who noticed the sexual differences of *Holothuria tubulosa*; see *Froriep's* neue Not. No. 243, p. 93.

19 See *Wagner*, Icon. zoot. Tab. XXXII. fig. 11 (*Holothuria tubulosa*). I have already remarked (§ 86), that the white cylindrical pedicellae, taken by some zootomists as testicles (*Delle Chiaje*, loc. cit. I. p. 97, Tav. VIII. fig. I. o.), are distinct from the genital organs, and communicate directly with the intestinal canal.

20 See the Catalogue of the Physiol. Series, &c., loc. cit. IV. Pl. XLIX. fig. 1. c. (*Holothuria tubulosa*).

21 *Quatrefages*, Ann. d. Sc. Nat. loc. cit. p. 66, Pl. IV. fig. 1, q. Pl. V. fig. 1.

22 This deep confusion of the organs of two sexes is something so remarkable, that one cannot but believe that *Quatrefages* has here taken the parent sperm cells for the eggs.

23 In *Sipunculus*, and *Phascolosoma*, there is observed on each side, a little front of the anus, a sac attached to the side of the body (see *Delle Chiaje*, loc. cit. Tav. I. fig. 5, s. s. and *Grube*, *Müller's* Archiv. 1837, Taf. XI. fig. 1. v°).

These have been regarded as genital organs. In *Sipunculus nudus*, Grube has found eggs not only in these sacs, but in the cavity of the body also. It may, therefore, be questioned if the eggs escape from the sacs into the cavity of the body, whence they are expelled through an opening at its posterior extremity, or if they are accidentally introduced from without with the water, during respiration. In this last case, these sacs should have excretory ducts; and there are, indeed, in *Sipunculus nudus*, two external fossae opposite the point of insertion of the sacs (see *Delle Chiaje*, loc. cit. Tav. I. fig. 2, f.), and in which it is said, there are two very small openings. According to *Forbes* and *Goodsir*, the genital sacs of the male *Echiurus vulgaris* contain a seminal liquid, with very active spermatic particles, while those of the female are filled with eggs; see *Froriep's* neue Not. loc. cit. p. 28] fig. 20, 22, 12, f. f.*

* [§ 97, note 23.] For the sexual organs of *Sipunculus*, see *Peters* (*Müller's* Arch. 1850, p.

§ 98.

The few observations hitherto made upon the embryology of the Echinoderms belong solely to the Asteroidae. Here, the vitellus undergoes the usual segmentation, and then is changed into a long, cylindrical, infusorial embryo, covered with cilia.

A few days after, four papillae are formed upon the anterior part of the body, and by these the embryo is attached to the walls of the incubating cavity (Bruthöhle). It then begins to be flattened laterally, and upon one of these lateral surfaces, ray-like tentacles appear, while the margin of the body forms five angles, upon the extremity of each of which is a red pigment dot. Then the cilia upon its surface disappear, and the young individual, deprived of its papillae and set free, moves about by its ambulacra.[1]

1 These interesting observations of *Sars* (*Wiegmann's* Arch. 1837, I. p. 401, 1844, I. p. 164, Taf. VI. fig. 4–22) were made upon *Echinaster sanguinolentus*, and *Asteracanthion Mulleri*. He has also observed that during the development, the point of attachment is gradually changed, until it reaches the back ; thus supporting the view that the madreporal plate is the relic of this last, which, in *Comatula*, has been well compared by *Muller* and *Troschel* (Syst. d. Asteriden, p. 134), to a button, since from it the young individuals are attached by a pedicle, as *Thomson* has shown upon (formerly) *Pentacrinus Europaeus ;* see Zeitsch. f. die Organisch. Physik. 1828, p. 55, and the Edinb. new Philos. Jour. 1836, p. 296, or *Froriep's* neue Not. No. 1057, 1836, p. 1. The assertion of *Sars* (*Wiegmann's* Arch. 1844. I. p. 176) that the animal which he formerly called *Bipinnaria asterigera* (Beskrivelser, &c., p. 37, Tab. XV. fig. 40) is probably only a developing Asteroid endowed with a great swimming apparatus, deserves to be considered. The remark of *Dalyell* (*Froriep's* neue Not. No. 331, p. 2) that the young of *Holothuria* are of the size of barley-corns, and resemble white maggots, is not one that affords us any data upon the development of these animals. There remains, therefore, a vast field open to observers concerning the development of the Echinoderms.

Sars (loc. cit. p. 47, Taf. VIII.) has furnished numerous data on the development of *Echinaster.* It appears, moreover, that all the Asteroidae are not developed after this type; for, *Koren* and *Danielssen* (Ann. d. Sc. Nat. VII. 1847, p. 347, Pl. VII. fig. 7–9) have shewn that *Bipinnaria asterigera* first observed by *Sars*, is a young Asteroid which moves by means of a particular

appendage, which is very complicated, and provided with numerous oars,— an appendage which is subsequently detached, but which continues then to execute natatory movements. There were, perhaps, similar appendages detached from young Asteroids that *Muller* and *Wagner* found at Helgoland, and which they have described and figured under the name of *Actinotrocha branchiata* ; see *Muller's* Arch. 1846, p. 101, Taf. V. fig. 1, 2, and 1847, p. 202, Taf. IX. fig. 1–6.

Various naturalists have noticed interesting facts on the development of the Echinidae in endeavoring to produce artificial fecundation. In the first of these experiments, by *Baer*, in 1845 (Bull. de. la Classe physico-math. de l'Acad. des Sc. de St. Petersburg, V. p. 231, *Froriep's* neue Not. XXXIX. p. 36), the eggs of *Echinus esculentus*, and *lividus*, thus fecundated, were transformed, after a complete segmentation of the vitellus, into a round, infusoria-like body, covered with cilia. *Dufossé* and *Derbès* (Ann. d. Sc. Nat. VII. 1847, p. 44, and VIII. p. 80, Pl. V.) followed still further, with *Echinus esculentus*, the development of these infusoria-like embryos. They gradually became pyriform, and acquired a peduncle at their smaller anal extremity ; while at the larger, oral end, tentacles and several long calcareous spines were developed. At the same time the digestive canal was formed in the interior of the body.

A small marine animal, first described by *Muller* (Arch. 1846, p. 108, Taf. VI. fig. 2, 3, and 1847, p. 160) under the name of *Pluteus paradoxus*, has been recently found by this same naturalist to be the young of an *Ophiura*. This animal swims by means of vibratile cilia, and is supported by a frame composed of ten diverging, calcareous prolongations, resembling a painter's easel.*

382, Taf. IV. fig. A—H), and *Krohn* (Ibid. 1850, p. 368, Taf. XVI.).

Peters has found that the fine whitish line described by *Grube* as lying contiguous with the bloodvessel of the intestine is an oviduct, being filled with ova, which move along by the action of the cilia with which it is lined. Connecting with this oviduct are botryoidal appendages, situated on the intestine, and filled with eggs ; these are the ovaries. The eggs, when matured, escape into the general cavity of the body, and thence are transferred outwardly through two brownish tubes, which open externally, and whose internal extremity is not closed, as has hitherto been supposed, but opens

into the general cavity of the body. These tubes, or oviducts, have been regarded hitherto as respiratory or secreting organs. *Krohn's* observations confirm those of *Peters* on this point. — Ed.

* [§ 98, note 1.] The development of the Echinoderms has been much and successfully studied of late, and chiefly by *Muller*, who, by several successive researches (see loc. cit.), has changed the zoological face of this class, beside making himself the great authority on all that relates to its embryology. The writings of *Agassiz* and others furnish also many details, but in any account I may give I shall depend mainly on the first-mentioned authority.

The first condition of every Echinoderm is the same, — an oval, ciliated body, resembling an infusorial animalcule, and without external organs, or distinction of parts. This is the starting-point, and upon it succeed variations according to the different families. Upon this ciliated body are developed, at one part, peduncles for its attachment to other bodies, while the rest of the germ increases in size, and assumes a star-fish form.

The larvæ thus formed may be divided into two groups :

1. Those of the Ophiuridae and Echinidae.
2. Those of the Asteroidae and Holothuridae.

The first are somewhat hemispherical bodies, with one edge of their truncated side prolonged into a single flat and wide process, which carries the mouth and œsophagus ; while from the opposite extremity project rods, of four, eight or more in number, and which form the internal skeleton. (See Ueb. d. Ophiurenlarven d. Adriat. Meeres. Taf. I. II.) These larvæ have a globular stomach in their hemispherical portion, and from which proceeds a short intestine terminating in a circular anus. They have, moreover, a ciliated fringe, which consists of a ridge covered with large cilia, passing above the mouth and before the arms, completely encircling the body in an oblique manner.

With the second group there is no internal calcareous skeleton, and they form *Muller's Auricularia* (of the Holothuridae), and *Bipinnaria* (of the Asteroidae).

The first of these are concavo-convex bean-shaped bodies, with an irregular transverse fissure answering to the hilum of the bean, in which the mouth is placed. The margins of this fissure are ciliated ; the anus opens on the ventral surface.

The *Bipinnaria* closely resemble these last, but they have a distinct ciliated circle in front of the mouth ; as they increase in size, the anterior part of their body is covered with long processes, which vary according to different forms.

Out of these larvæ, all of which have a strictly bilateral symmetry, the more or less radiate adult Echinoderms are developed by a process which is a sort of internal gemmation.

The changes and variations of this metamorphosis I will give in *Muller's* own words :

" 1. The change of the bilateral larva into the Echinoderm takes place when the larva yet remains an embryo, and is universally covered with cilia, without a ciliated fringe. A part of the body of the larva takes on the form of the Echinoderm ; the rest is absorbed by the latter (a part of the Asteroidae, *Echinaster, Asteracanthion,* Sars).

" 2. The change of the bilateral larva into the Echinoderm takes place when the larva is perfectly organized ; that is, possesses digestive organs and a special ciliated fringe.

" The Echinoderm is constructed within the Pluteus like a picture upon its canvas or a piece of embroidery in its frame, and then takes up into itself the digestive organs of the larva. Hereupon, the rest of the larva vanishes (*Ophiura, Echinus*), or is thrown off (*Bipinnaria*).

" 3. The larva changes twice. The first time it passes out of the bilateral type with lateral ciliated fringe into the radial type, and receives, instead of the previous ciliated fringe, new locomotive larval organs, the ciliated rings. Out of this pupa-condition, the Echinoderm is developed, without any part being cast off (*Holothuria,* some Asteroidae).

" If we call embryonic type the condition in which the animal leaves the egg, and when the internal organs are not yet developed, we have four stages or types, — the embryonic type, the larval type, the pupa type, and the Echinoderm type. The animal may pass from either of the first three forms into the Echinoderm, or may run through them all." See Ueber. d. Larven u. d. Metamorph. d. Holoth. u. Aster. p. 33. See, also, a review of *Muller's* researches, by *Huxley* (Ann. Nat. Hist. VIII. 1851, p. 1), and by *Dareste* (Ann. d. Sc Nat. XVII. 1852, p. 349).

These results are highly interesting in both a zoological and a physiological point of view, and I need only suggest their important relations to the doctrine of " alternation of generations."

In this connection, it may be proper to allude to another point. It is well known that *Vogt* (Naturgesch. d. lebend, u. untergegang. Thiere. I. Liefer. 3, p. 254) has removed the Beroïd Medusae from the Acalephae to the " Molluscoida," regarding them bilateral animals. In a private letter from *Agassiz,* there is a passage bearing directly on this point. He says : " The young Echinoderms are structurally and morphologically homologous with Beroïd Medusae, showing that Beroïds are genuine Radiates, and truly belong to the class of Acalephae, and cannot be referred to the Molluscoids. These relations will be plain by comparing Taf. I. fig. 6, of *Muller's* Larven und d. Metamorph. d. Ophiuren und Seeigel. 1848, with the figures of Pl. VIII. of *Agassiz'* Memoir on the Beroïd Medusae, in the Mem. of the Amer. Acad. of Arts and Sc. Vol. IV."

For further writings on the development of the Echinoderms, see *Muller's* papers, published in his Arch. 1848, p. 113 ; 1849, p. 84, 364 ; 1851, p. 1, 272, 353 ; but these papers are all included in his large memoirs already given. See, also, *Krohn,* Beiträg. zur Entwickelungsgeschichte der Seeigellarven, 1849, and in *Muller's* Arch. 1851, p. 338, 344, 368 ; and *Desor, Muller's* Arch. 1849, p. 79. — Ed.

BOOK FIFTH.

HELMINTHES.

CLASSIFICATION.

§ 99.

It is very difficult to characterize the class Helminthes, for it contains animals having widely dissimilar organization. On this account, the separation of its groups, and their distribution among the other classes of the invertebrata, has been attempted. But such various difficulties have arisen from this, that for the present, it is best that all these animals should remain together. If a common character is not furnished by their structure, it must be sought for in their manner of life; for nearly all are parasites,[1] and during their whole life, or at least during some of its periods, seek their abode and nourishment in or upon other living animals.

ORDER I. CYSTICI.

The body is swollen in the form of a bladder, and filled with a serous liquid. Digestive and genital organs are wanting.[2]

Genera: *Echinococcus, Coenurus, Cysticercus, Anthocephalus.*

ORDER II. CESTODES.

The parenchymatous body is riband-like, having often incomplete transverse fissurations; often it is wholly divided transversely into rings. Digestive organs are wanting. The genital organs of both sexes are combined in the same individual, and generally are often repeated. Copulatory organs are present.

Genera: *Gymnorhynchus, Tetrarhynchus, Bothriocephalus, Taenia, Triaenophorus, Ligula, Caryophyllacus.*

---

1 *Anguillula* is the only exception to this.
2 The head of the sexless Cystici, as to its form, its hook and suckers, strikingly resembles that of some Cestodes; from which it might be inferred that they are only the larval forms of these last.

ORDER III. TREMATODES.

The body is parenchymatous, and usually flattened. The intestinal canal, which is often branching, has a mouth, but nearly always is without an anus. The genital organs of both sexes are combined in the same individual. Copulatory organs are present.

Genera : *Gyrodactylus, Axine, Octobothrium, Diplozoon, Polystomum, Aspidocotylus, Aspidogaster, Tristomum, Monostomum, Holostomum, Gasterostomum, Pentastomum.**

ORDER IV. ACANTHOCEPHALI.

The sack-like body is flattened, transversely striated, and swollen cylindrically by the absorption of water. Digestive organs are wanting. The genital organs are situated in separate individuals. Copulatory organs are present.

Genus : *Echinorhynchus.*

ORDER V. GORDIACEI.

The body is filiform and cylindrical. The digestive organs are without an anus. The genital organs are situated upon separate individuals. Copulatory organs are sometimes present.

Genera : *Gordius, Mermis.*

ORDER VI. NEMATODES.

The body is sack-like and cylindrical. The digestive canal has a mouth and an anus, and passes in a straight line through the cavity of the body. The genital organs are situated upon separate individuals. Copulatory organs are present.

* In this connection, and especially in reference to the remarks made by the author under § 99, it may be well to notice that *Van Beneden* does not regard the *Linguatulae* as true Helminthes, but that they belong rather to the division of articulated animals, — coming nearest to the *Lerneae.* His reasons are the following :

"These animals, on their extrication from the egg, are provided with two pairs of articulated feet terminated by hooks.

"The nervous system differs from that of the *Lerneae* only in having two cords which form the ganglionic chain, separated throughout their whole length, whilst in the *Lerneae* they are separated for only half their length.

"In both cases the males are comparatively very small. The ovisacs of the females are equally bulky ; but in the *Lerneae* which live in water they project externally, whilst in the *Linguatulae,* which always live in a different medium, they remain in the interior.

"Besides the ring of nerves, the sub-œsophageal ganglion, and the cords which represent the ganglionic chain, the *Linguatulae* are provided with different ganglia representing the great sympathetic. I detected four perfectly distinct ganglia spread over the sides of the lower surface of the œsophagus in the new species from the Mandrill. In another species *M. Blanchard* detected these ganglia and stomato-gastric nerves; but he referred them to the system of the nerves of relation or those of animal life, judging, at least, from the name which he has assigned to them.

"Another point, which, however, had not escaped the attention of naturalists, is that the muscles exhibit in their primitive fibres the transverse lines which are not met with in the lower animals."

See Bull. de l'Acad. Royale de Belgique, 1848, XV. No. 3. See also *Blanchard,* Comp. Rend. 1850, XXXI. p. 629. — ED.

9*

Genera : *Sphaerularia, Trichosoma, Trichocephalus, Filaria, Anguillula, Physaloptera, Liorhynchus, Lecanocephalus, Cheiracanthus, Gnathosoma, Ancyracanthus, Spiroptera, Hedruris, Strongylus, Cucullanus, Oxyuris, Ascaris.*

BIBLIOGRAPHY.

Goeze. Versuch einer Naturgeschichte der Eingeweidewürmer. Blankenburg, 1782.

Zeder. Erster Nachtrag zum vorhergehenden Werke. Leipzig, 1800.

Brera. Vorlesungen über die vornehmsten Eingeweidewürmer des menschlichen lebenden Körpers. From the Italian by Weber. Leipzig, 1803.

Rudolphi. Entozoorum Historia Naturalis. Amstelaedami, 1808–10, and Entozoorum Synopsis. Berolini, 1819.

Bremser. Ueber lebende Würmer im lebenden Menschen. Wien, 1819. Translated into the French under the title : " *Traité zoologique et physiologique sur les Vers intestinaux de l'Homme,* par Bremser, traduit par Grundler, revu et augmenté de notes par De Blainville. Paris, 1824." To this Leblond has added a new Atlas. Paris, 1837.

Bremser. Icones Helminthum. Viennae, 1824.

Cloquet. Anatomie des vers intestinaux Ascaride lombricoïde et Échinorhynque géant. Paris, 1824.

Creplin. Observationes de Entozois. Gryphiswaldiae, 1825 ; and Novae Observationes de Entozois. Berolini, 1829.

Mehlis' excellent remarks upon this work in the Isis, 1831, p. 68, 166.

Bojanus. Enthelminthica, in the Isis, 1821, p. 162.

Nitzsch. In *Ersch's* and *Gruber's* Encyclopaedie, articles : Acanthocephala, Acephalocystis, Amphistoma, Anthocephalus, Ascaris, &c.

Creplin. In the same work, articles : Distomum, Echinococcus, Echinorhynchus, Eingeweidewürmer, Enthelminthologie, &c.

Delle Chiaje. Compendio di elmintografia umana. Napoli, 1833.

Leuckart. Versuch einer naturgemässen Eintheilung der Helminthen. Heidelberg, 1827 ; and his Zoologische Bruchstücke, Hft. I. Helmstädt, 1820, and Hft. III. Freiburg, 1842.

Baer. Beiträge zur Kenntniss der niederen Thiere, in the Nov. Act. Acad. Leop. Carol. Vol. XIII. p. 525.

Nordmann. Micrographische Beiträge zur Naturgeschichte der wirbellosen Thiere. Berlin, 1832.

E. Schmalz. XXIX. Tabulae Anatomiam Entozoorum illustrantes. Dresdae, 1831. These contain mostly copies.

C. Th. E. Siebold. Helminthologische Beiträge und Jahresberichte über die Helminthen, in *Wiegmann's* Arch. für Naturgeschichte.

Diesing. His excellent Monographs in the Annalen des Wiener Museums.

F. J. C. Mayer. Beiträge zur Anatomie der Entozoen. Bonn, 1841.

R. Owen. His excellent article, Entozoa, in the Cyclopaedia of Anatomy and Physiology.

Dujardin. Histoire naturelle des Helminthes. Paris, 1845.

ADDITIONAL BIBLIOGRAPHY.

The following are among the more important contributions to the Anatomy of the Helminthes which have been published since the issue of the original work. I should mention, however, that I have not had very much access to recent German contributions in this department, from the tardiness with which such matters reach this country. However, I am happy in not being ignorant of the late publications of *Siebold*, who is truly at the head of Helminthology.

Blanchard. Recherches sur l'organisation des Vers., in the Ann. d. Sc. Nat. VII. 1847, p. 87, VIII. 1847, p. 119, 271, X. 1848, p. 321, XI. 1849, p. 106, XII. 1849, p. 1.

Van Beneden. Recherches sur l'organisation et le developpement des Linguatules (Pentastoma, *Rud.*), &c. in the Mém. de l'Acad. de Bruxelles, 1848 ; also, in Ann. d. Sc. Nat. 1849, XI. p. 313.

Note sur le developpement des Tétrarhynques, in the Bull. de l'Acad. de Belgique, XVI. 1849.

Recherches sur les Vers Cestodes, in the Mém. de l'Acad. de Belgique, 1850, XXV.

Siebold. Ueber den Generationswechsel der Cestoden nebst einer Revision der Gattung Tetrarhynchus, in *Siebold* and *Kölliker's* Zeitsch. II. 1850, p. 198.

Ueber die Verwandlung des Cysticercus pisiformis in Taenia serrata Ibid. IV. p. 400.

Ueber die Verwandlung der Echinococcus-Brut in Taenien. Ibid. IV. 1853, p. 409.

See, also, various valuable though small contributions, in the form of letters to *Siebold*, in *Siebold* and *Kölliker's* Zeitsch. IV. p. 52, 116, 451, 454; as well as the references in my notes. — ED.

CHAPTER I.

CUTANEOUS SYSTEM.

§ 100.

The body of the Helminthes is generally surrounded by a firm skin, which may be separated into a thin epidermis, and a pretty hard dermis. The epidermis of the adults is never ciliated ; but not unfrequently it has horny spines pointing backwards, which sometimes are limited to the anterior part of the body, and sometimes spread over a large surface, in transversely serrated rows.[1]   In the first case, the spines serve to attach them

---

1 In many Nematodes, Acanthocephali, and Trematodes, the epidermis is spinous like a rasp. These spines are simple in *Liorhynchus denticulatus*, *Lecanocephalus spinulosus* (according to *Diesing*, Annalen des Wiener Museums, II. Abth. 2, 1839, Taf. XIV. fig. 14–20), *Echinorhynchus pyriformis*, and *hystrix* (*Bremser*, Icon. Helm nt. Tab. VII.), *Distomum lima*, *maculosum*, *scabrum*, *ferox*, and *perlatum* (Ibid. Tab. X. and *Nordmann*, Micograph. Beiträge. Hft. I. Taf. IX.), and *Pentastomum denticulatum* (*Diesing*, loc. cit. I. Abth. 1, Taf. III. fig. 10–13). But they are polydenticulated in *Cheiracanthus* (*Diesing*, loc. cit. II. Hft. 2, Taf. XIV. XVI. XVII.).

to other animals, and therefore will be specially described with the locomotive organs.

With most of the Nematodes, the epidermis has very fine and closely approximated transverse folds, which are but occasionally so prominent that the body appears annulated.[2] Sometimes, but rarely, the body is also plicated in a longitudinal manner.[3] The dermis has a fibrous structure, consisting of two fibrous layers, — one longitudinal and the other transverse, — which cross each other at right angles; and of two other layers, which intersect each other more acutely.[4] The skin of these animals has a great absorptive power which during life is voluntary, but which continues to a certain extent after death, so that then these worms often swell enormously, and sometimes burst.[5]

§ 101.

Directly beneath the skin of the Cystici, and Cestodes, are found hard corpuscles containing carbonate of lime, and which may be regarded as the vestige of a cutaneous skeleton. But, as they are scattered here and there more deeply in the parenchyma, they certainly may be compared to the spicula and calcareous net-works found in the skin of many Polyps and Echinoderms. Oval or discoid, they are usually of equal size in the same individual. Sometimes, however, they present irregular and unequal forms. Always colorless and transparent, and composed of concentric layers, they refract the light like small vitreous bodies.

In *Taenia, Triaenophorus, Bothriocephalus,* and the young of *Echinococcus,* they are subcutaneous, and more or less scattered; but in the wrinkled and vesicular body of *Coenurus,* and *Cysticercus,* they are so very abundant that they form quite thick layers. They are absent in the caudal vesicle of *Cysticercus,* but in *Coenurus,* and *Echinococcus,* they are found in the vesicular walls beneath the delicate epithelium which lines the interior of the body.[1]

2 This is so, for instance, with the anterior extremity of *Liorhynchus denticulatus,* and *Strongylus annulatus,* mihi (from the trachea of the wolf).

The epidermis of *Ascaris nigrovenosa* has such long and loose folds that its body, seen laterally, has a fringed appearance.

3 Excepting the longitudinal folds of the epidermis, which form lateral wings of variable form and length at the cephalic extremity of the Nematodes, or on both sides of the extremity of the tail of many males of this order (*Bremser,* Icon. Helminth. Tab. IV. fig. 20–24), I have as yet found the epidermis longitudinally plicated over the whole body only with *Strongylus striatus,* and *inflexus.*

4 These different dermic layers are distinct, especially with *Gordius* and *Mermis*; see *Dujardin's* figure in the Ann. d. Sc. Nat. XVIII. 1842, Pl. VI. I have found this structure also in *Ascaris mystax, microcephala, Distomum echinatum, hians, lineа,* and in *Monostomum verrucosum.*

In *Amphistomum giganteum, Diesing* (Annal. d. Wiener Museums, I. Abth. 2, p. 239, Taf. XXII. fig. 1, c, d), has regarded these layers as muscular. The same is true of *Bojanus* (Isis, 1821, p. 166, Taf. II. fig. 12), and *Laurer* (De Amphistomo conico, p. 6, fig. 15).

But the structure of the skin of *Echinococcus* is quite different. Here no epidermis can be separated from the dermis or the sac of the body; and the whole is a thick membrane, resembling congulated albumen and composed of numerous very thin layers, tightly bound together.

5 This absorbent power of the skin is particularly prominent with the Acanthocephali. It is here really a vital act; for *Echinorhynchus,* which naturally absorbs only a little liquid into its constantly flattened and wrinkled body, swells and relaxes alternately when in contact with water. This has been observed with many species by *Creplin* (Nov. Observ. de Entozois. 1829, p. 44, and in *Ersch* and *Grube's* Encyclopaedie XXX. 1838, p. 381), by *Mehlis* (Isis, 1831, p. 167), and by myself. With the Nematodes it is otherwise. These cannot voluntarily govern this absorbing power, and when, therefore, they are put in water, they swell to bursting and die. With the Gordiacei this power is purely physical, so that the dead and dried individuals of *Gordius aquaticus,* when placed in water, quickly become round again, and perform very active hydroscopic motions.

1 These calcareous corpuscles, which are always without an envelope and are scattered through the whole body of these Helminthes, have been taken by *Pallas, Goeze, Zeder,* and by most Helminthologists until lately, for eggs, and as such were often figured.

CHAPTER II.

MUSCULAR SYSTEM AND ORGANS OF LOCOMOTION.

§ 102.

The muscular system is well developed with the Helminthes; its primitive fibres are flattened, and never transversely striated. In the Cystici, and Cestodes, the muscles are least distinct, although in *Cysticercus* there can be no question as to the muscular fibres which traverse in every direction the walls of the caudal vesicle.[1] Equally distinct is a subcutaneous layer of longitudinal fibres in the rings of *Bothriocephalus* and *Taenia.*[2] Moreover from the great contractility of the rings, and especially those of the cephalic portion of the Cystici and Cestodes, there must be muscular fibres concealed in the parenchyma, but which from their tenuity escape our observation. In the Trematodes, having also an extreme contractility, a large portion of the parenchyma of the body is composed of a muscular reticulated tissue, the transverse and longitudinal muscles of which embrace the various organs in a retiform manner.[3] In the Acanthocephali, the Gordiacei, and Nematodes, the general movements of the body are due to a subcutaneous muscular layer, which surrounds the visceral cavity in a sac-like manner. Its longitudinal and transverse muscles are quite distinct from each other; and their fibres, although parallel, communicate with each other by angular anastomes, and in this way form a net-work.[4]

In most of the Nematodes, the longitudinal muscles form four, large bands, two upon the ventral, and two upon the dorsal surface. These

In the Cestodes, this error is unnecessary, for in the posterior portions of their body the eggs can easily be distinguished from the corpuscles; moreover, these last are the most numerous about the neck and anterior rings, — localities where the genital organs are scarcely and sometimes not at all developed. It may be added, also, that these bodies dissolve in a weak acid with the escape of gas, while the eggs of *Taenia* under the same circumstances remain unaffected. In the Cystici, which are sexless, and where therefore eggs are vainly sought for, these corpuscles, as to their structure, chemical composition, and position, so closely resemble those of the Cestodes, that it appears strange that they have always been taken for eggs. *Eschricht* (Nov. Act. Acad. Leopold Carol. Vol. XIX. Suppl. alter. 1841, p. 59, 103), not having perceived that they contain carbonate of lime, has described them as elementary granules, and thinks that they have a nutritive function analogous to that of the blood and lymph corpuscles.

Gulliver (Med.-Chir. Trans. VI. London, 1841, p. 1; see *Wiegmann's* Arch. 1841, II. p. 314) has given an exact description of those of *Cysticercus*, but he also has taken them for eggs. In *Taenia filum, linea, serrata,* and *infundibuliformis,* they are spherical or oval; and in the first two species, *Goeze* (Versuch. einer Naturgesch. d. Eingeweidewürmer, p. 393, Taf. XXXII. A. fig. 6, 7, 12) has taken them for eggs, and the concentric rings of the calcareous layers for the coils of the embryo. With those of *Cysticercus cellulosae,* and *pisiformis,* the discoid form prevails; I have often seen here four to six calcareous layers about

the nucleus; and sometimes there are two nuclei thus enclosed, and then the corpuscles have exactly the aspect of the precious stones of Imatra.

Those of *Taenia cucumerina, Bothriocephalus solidus,* and *Cysticercus fasciolaris,* are usually of an oval form, sometimes irregular, and of a variable size. *Tschudi* (Die Blasenwürmer, 1837, p. 24, Taf. II. fig. 21) has figured those of the last species as eggs.

[1] I have easily seen these muscular fibres in the caudal vesicle of *Cysticercus cellulosae,* and *tenuicollis.* But they are wholly absent in the parent-vesicle of *Echinococcus hominis,* and *veterinorum.* This vesicle, therefore, has probably no spontaneous movements, whilst the embryos it contains at certain times have distinct locomotive organs.

[2] The longitudinal fibres of the subcutaneous muscular layer, have been observed in *Bothriocephalus latus,* by *Eschricht* (loc. cit. p. 55) and in *Taenia angulata, lanceolata, nasuta,* and *villosa,* by myself.

[3] The reticulated muscular parenchyma of the Trematodes (*Amphistomum giganteum*) has been represented by *Diesing* very beautifully (Ann. d. Wiener Museums I. Abth. 2. Taf. XXII. fig. 4–8).

[4] In *Ascaris lumbricoides,* as in most Nematodes, the muscular fibres are so closely approximated that the meshes of their net-work are not seen except by tearing asunder the muscles; see *Bojanus,* Isis 1821, Taf. III. fig. 48. The reticulated form of the longitudinal muscles is very distinct in *Cheiracanthus gracilis;* see *Diesing,* Annal. d. Wiener Museums, II. Hft. 2, p. 225, Taf. XVII. fig. 1, 2.

bands are separated by the same number of longitudinal lines, the two narrowest of which are above and below; while the others, which are large and riband-like, are on the sides.[5]

In the Acanthocephali the transverse muscles are more superficial than the longitudinal,[6] while in the Nematodes and Gordiacei the inverse is true.[7]

§ 103.

There are with the Helminthes a great variety of organs for the movements of the body. With the Cystici, Cestodes, and the Trematodes, there are often sucking-cups and cavities; the first of these are more or less alveolate, being formed of numerous layers of circular and radiated muscular fibres,[1] while the second are only excavations in contractile parenchyma of the body, and are divided into many chambers by septa, or have very variable lobular appendages.[2] Many of these suctorial organs have, — some at their bottom, others on their borders, hooks with a horny support, by which these animals can firmly attach themselves to objects.[3]

5 *Bojanus*, Isis, 1821, p. 186, Taf. III. fig. 49, 55, B. (*Ascaris lumbricoïdes*).

6 In *Acanthocephalus* the transverse muscles intercommunicate with each other by short and narrow anastomoses, and form a complete ring, which surrounds the longitudinal ones like a large girdle. See, in reference to this, *Echinorhynchus gigas*. In *Echinorhynchus gibbosus* these annular muscles have been found only above the swelling of the body.

7 The transverse muscular bundles of the Nematodes, which are not so closely united as the longitudinal ones, do not form closed rings, but produce four segments, which are separated from each other by the crossing over of the longitudinal muscles. At least, this is so in *Ascaris lumbricoïdes*, *Strongylus gigas*, and most of the species of this order. *Bojanus* (Isis, 1821, p. 187, Taf. III. fig. 51, 54) and *Cloquet* (Anat. des Vers intestin. p. 26, Pl. II. fig. 3) have taken these transverse muscles for vessels ; and *Diesing* has made the same mistake with the ramified muscles of *Cheiracanthus* and *Ancyracanthus* (Ann. d. Wiener Museums II. Abth. 2, Taf. XVI. fig. 1, and Taf. XVIII. fig. 2). In *Ascaris inflexa*, and *Filaria attenuata*, I have seen the transverse muscles ramified in the same manner. In *Ascaris spiculigera* they have a peculiarity ; their more or less long fibres pass off from the longitudinal muscles at a right angle, and are inserted into one or the other of the two narrow longitudinal rays. In the Gordiacei, the longitudinal layer is not broken by any ray of this kind, but forms continuous tubes which have thin walls and a satin aspect, and where the flattened and riband-like fibres are bound together by their faces, and at the same time anastomose with each other.

This would at least appear to be so, judging from the net-work with long meshes which is produced by a little traction. I have not found the transverse muscles in *Gordius* ; but in *Mermis nigrescens*, there is, under the longitudinal muscular layer, a net-work like the preceding, but with very large meshes. *Dujardin* appears to have observed it, but he regarded it as connected with the eggs of this worm (Ann. d. Sc. Nat. XVIII. 1842, Pl. VI. fig. 13).

1 The young *Echinococcus*, the *Cœnurus*, *Cysticercus*, and *Tænia*, have usually upon the cephalic extremity four imperfect cup-like cavities, which can serve only as suckers. It must have been an oversight of *Nitzsch* (*Ersch* and *Gruber's* Encyclopædie XII. 1824, p. 95), who regarded these in *Tænia* as so many oral orifices

leading into the alimentary canals. It is only with *Distomum*, *Amphistomum*, *Polystomum*, and other Trematodes, that this sucker, which is situated in front, is perforated at its bottom, and serves also the function of mouth.

The ventral sucker of *Distomum*, and that found at the posterior extremity of *Amphistomum*, and *Polystomum*, as well as the numerous analogous organs upon the back of *Monostomum verrucosum*, and upon the terminal dilatation of *Aspidocotylus mutabilis* (*Diesing*, Ann. d. Wiener Mus. II. Abth. 2, p. 234, Taf. XV.), are all imperforate. That upon the posterior extremity of *Amphistomum subclavatum*, and *unguiculatum*, is remarkable ; it has a small duplicate at its bottom, which *Diesing* (loc. cit. I. Abth. 2, p. 254, Taf. XXIV.) has erroneously regarded as the opening of the genital organs. In *Polystomum*, six large muscles pass from the interior of the body, and are spread upon the convex surfaces of as many suckers, situated at the posterior extremity, and which they properly move during the animal's creeping.

2 Upon the head of *Bothriocephalus*, *Tetrarhynchus*, and *Anthocephalus*, there are two to four simple cup-like fossæ ; with *Tristomum*, *Polystomum*, and some other Trematodes, there are two on each side of the mouth, and with *Axine*, *Octobothrium*, and *Diplozoon*, there are two which are subcervical and behind the mouth.

With *Bothriocephalus tumidulus* (*Bremser* Icon. Helminth. Tab. XIII. fig. 21, or *Leuckart*, Zool. Bruchstücke. Hft. I. Taf. I. fig. 4, 5), there are four, which are divided into chambers by many septa ; and with *Aspidogaster* (*Baer*, Nov. Act. Acad. Leop. Carol. Vol. XIII. pt. 2, Tab. XVIII.), the whole ventral disc is divided by septa into quadrangular suctorial fossæ. The head of *Bothriocephalus auriculatus* has a singular aspect, due to numerous partly crenulated lobes, which flank its four suckers (*Bremser*, loc. cit. Taf. XIII. fig. 17, 19, and *Leuckart*, loc. cit. Taf. I. fig. 6–11). A very simple structure is found upon the head of *Bothriocephalus tetrapterus*, mihi (from the intestine of the seal) ; here the points of junction of the fossæ are prolonged into four triangular lobes, by which the animal can adhere tightly to its object. *Holostomum*, which lives in the intestines of birds and mammals, has analogous appendages around the cavity which is situated at the anterior extremity, and which it fastens to the intestinal villosities ; see *Nitzsch*, in *Ersch* and *Gruber's* Encyclop. III. p. 393, IX. 1822, fig. 1.

3 This condition of things is found especially in

The young of *Echinococcus, Coenurus, Cysticercus,* and many of the *Taenia,* have their head armed with a circle of single or double hooks, which were known to the oldest Helminthologists. Each hook consists of a strongly-curved point, situated upon a round, straight pedicle, of variable length. At the point where the curve ceases, there is, upon the concave side of the organ, a small conical process. When this circle of hooks is unfolded, the points project around the anterior part of the head, whilst the pedicles point towards the inner and their processes towards the posterior portion of the body, and are buried in the parenchyma. Both are surrounded by muscular substance. When, therefore, the muscles of the pedicles contract, the hooks are drawn downwards and outwards, and their points are brought together upon their convex surface in the long axis of the head; but when, on the other hand, the muscles of the processes contract, these last are depressed, the pedicles are again elevated, and the hooks project outward. With many Cestodes, this circle of hooks is situated upon a particular proboscis (rostellum), which can be retracted into a sheath which is concealed between the four suckers of the head.[4]

With *Anthocephalus, Gymnorhynchus,* and *Tetrarhynchus,* there are upon the head four long and completely retractile proboscees, which are armed with an extraordinary number of small, backwardly-curved hooks which are attached by a large base to the external surface of the organ, and are without special muscles; by these, these animals can penetrate the most compact animal tissues. Each proboscis is a hollow muscular tube, which can be voluntarily retracted within a sheath of the same nature, and then the hooks, with their points directed in front, are drawn together in its axis. The length of the sheath, which is usually enlarged at its base, depends upon that of the proboscis. In many species of *Tetrarhynchus,* they reach far into the neck of the animal.[5]

The Acanthocephali have only one of these organs, and the hooks, which are without special muscles, form rows arranged one after another. Both the number of these rows and the form of the hooks vary in different species. Usually their size decreases from before backwards, so that those

---

the suctorial apparatus of the Trematodes. In *Tristomum hamatum* (see *Rathke*, Nov. Act. Acad. Leop. Carol. XX, 1843, p. 241, Taf XII. fig. 11), several sharp points project from the bottom of the sucker at the posterior extremity. With *Polystomum appendiculatum* (*Nordmann,* Micrograph. Beiträge, IIft. I. p. 82, Taf. V. fig. 6, 7), the borders of the six suckers at the posterior extremity are armed with a sharp claw. The disc of *Gyrodactylus* (Ibid. Taf. X.) has its borders provided with six horny points, and its base is supported by two sides of the same nature, curved like an arc.

A very complicated support, formed of horny arches and ridges, sustains the eight suckers at the posterior extremity of *Octobothrium sagittatum, Merlangi,* and of *Diplozoon paradoxum;* an analogous support wholly surrounds the large foot at the end of the body of *Axine* (*Leuckart,* Zool. Bruchstücke IIft. 3, Taf. II. and *Nordmann,* Micogr. Beitr. IIft. 1, Taf. VII.; also *Diesing,* Nov. Act. Acad. Leop. Carol. XVIII. pt. 1, Tab. XVII.).

The four fossae found on each side of the mouth of *Pentastomum* contain simple and double very curved hooks, which the animal can erect at will (*Diesing,* Ann. d. Wiener Mus. I. Abth. 1, Taf. III. IV.). A remarkable exception among the Nematodes is found with *Hedruris androphora,*

of which *Nitzsch* (*Ersch* and *Gruber's* Encyclopaed. VI. p. 49, IX. Taf. II. A.) has made a separate genus; the females have a protrusive sting in the sucker situated at the posterior extremity.

4 With *Echinococcus, Coenurus,* and *Cysticercus,* the number of hooks is twenty to thirty; and I have seen as many with *Taenia scolecina,* and *infundibuliformis;* but I have found only eighteen with *Taenia angulata,* ten with *Taenia setigera,* and eight with *Taenia lanceolata.*

Taenia scolecina, crassicollis, and *Cysticercus,* have an equal number of large and small hooks alternating with each other, and, at a certain point, forming a double circle.

With *Taenia,* and especially those which have these organs on the proboscis, they may be partly or even wholly detached.

Rudolphi has regarded *Taenia gracilis, angulata, infundibuliformis, setigera,* and *stylosa,* as naturally without these organs, but I have often found them having a complete circle.

With *Taenia cucumerina,* the structure is different; its seven rows of hooks are in all respects like those of *Echinorhynchus.*

5 See *Leblond,* Ann. d. Sc. Nat. VI. 1836, Pl. XVI. fig. 5, 6, 7; and *Goodsir, Froriep's* neue Notiz. 1841, No. 429, fig. 18; also *Mayer, Müller's* Arch. 1842, Taf. X.

of the last row are only rudimentary. The sheath of the proboscis is very muscular, and terminates behind in a caecum; it extends across the neck of the animal even into the cavity of the body, and its movements are aided by some special muscles. In all the species whatever, there are three muscles which act as retractors of the sheath and neck. Two of these arise as delicate cords at the anterior extremity of the body from both sides of the internal surface of the subcutaneous muscular sac; they traverse thence the cavity of the body obliquely, and are inserted, in *Echinorhynchus acus, angustatus, fusiformis,* and *proteus,* upon the sides of the sheath; but in *Echinorhynchus gigas, haeruca, polymorphus, hystrix,* and *strumosus,* the insertion is at its inferior extremity. Between these two muscles, and below their points of origin, there is a third, which divides from the subcutaneous muscular sac; this is simple, riband-like, and is inserted at the lower extremity of the sheath. In *Echinorhynchus polymorphus,* and *proteus,* its form is pyramidal. In *Echinorhynchus gigas,* and *gibbosus,* two thin muscles arise from the anterior extremity of the body, and are inserted upon the sides of the sheath; they serve, probably, for the protrusion of this organ and the neck.[6]

There are, moreover, upon the different parts of the body of some Helminthes, horny hooks and spines, which serve for their creeping about and permanent attachment to objects.[7]

CHAPTER III.

NERVOUS SYSTEM.

§ 104.

The apparently quite feebly-developed nervous system of the Helminthes is yet but very incompletely known.

Our whole knowledge is limited to that of a small obscure ganglion found in some species, which, as it sends off several nerves, may be regarded as

---

6 Helminthologists are not yet agreed as to the number and arrangement of the proboscideal muscles of the Acanthocephali; see *Nitzsch,* in *Ersch* and *Gruber's* Encyclop. 1. 1818, p. 242; *Bojanus,* in *Isis,* 1821, Taf. III. fig. 34; *Westrumb,* De Helminthibus Acanthocephalis 1821, p. 50; and *Cloquet,* Anat. des Vers. intestin. p. 76, Pl. VII. *Mehlis* (Isis 1831, p. 82) has taken the proboscideal sheath for an oesophageal organ, and its two muscles for vessels. *Burow* (Echinorhynchi strumosi Anatome, 1836, p. 16, fig. 1, e) has fallen into a similar error, in regarding these same muscles as intestinal tubes.

7 With many Trematodes, as, for example, with *Polystomum, Octobothrium, &c.* (see *Baer,* Nov. Act. Acad. Leop. Carol. XIII. pt. 2, Tab. XXXII. fig. 7, f. and *Mayer,* Beiträge, &c., Taf. III. fig. 3, m. m. fig. 8), there are found between the suckers at the posterior extremity, special hooks, and to which, with *Polystomum,* I have seen proper muscles proceed from the interior of the body. With certain *Cercariae* (larvae of *Distomum)* one can distinctly observe the use which they make of

a sting which projects from the back above the oral sucker (and not from the mouth itself, as *Wagner* has supposed, Isis. 1834, p. 131), and which serves to open a passage through the parenchyma of the animals they infest. An entire group of *Distomum,* as *Distomum echinatum, militare, uncinatum* (*Bremser,* Icon. Helminth, Tab. X. fig. 5), which *Rudolphi* has designated as Echinostomata, have around their oral sucker an annular collar, upon which are numerous straight spines arranged in a circular manner. An armature of this kind is found upon a *Cercaria.* These spines are as easily detached as the hooks of the armed *Taeniae.* With *Spiroptera crassicauda,* I have found on each side of the mouth a doubly-pointed sting pointing backwards, and behind this two others three-pointed. A still more remarkable form is seen in the four penniform stings, which project behind the mouth of *Ancyracanthus pinnatifidus* (see *Diesing,* Ann. d. Wiener Mus. II. Abth. 2, Taf. XIV. XVIII.). These Nematodes undoubtedly use these instruments for piercing the stomachal membranes of the animals they infest.

a central nervous organ. There are, however, various other parts which
have been taken for nerves, but some of these, certainly, do not belong to
this system.

In the Cystici, no nervous system has yet been found, and the researches
made upon the Cestodes have ended equally unsatisfactory. A single
observation upon a *Tetrarhyncus* would lead us to think that in these last
the nervous system is situated at the cephalic extremity.

In *Tetrarhynchus attenuatus*, there is a small flattened swelling between
the sheaths of the four probosces, and from which pass off filaments to both
of these organs.[1]

The observations upon the nervous system of the Trematodes are more
numerous and positive. Immediately behind the oral sucker, and upon
the sides of the œsophagus, are two nervous swellings, connected by a
transverse cord, which passes beneath this canal. Among the branches
given off in all directions from these, there are two, large and long, extending from each side of the body to its extremity, and which give off in their
course many lateral branchlets.[2]

In *Pentastomum*, the central portion of this system consists of a single
large ganglion, sub-œsophageal, and due perhaps to the fusion of two lateral
ganglia. From this, filaments pass off in every direction ; two of these
surround the œsophagus in a ring-like manner, while two others, analogues
of the two main trunks of the Trematodes, pass to the very extremity of
the body, giving off on their way, very fine filaments.[3]

1 *Muller*, not without reason, regards this organ
as the nervous system of *Tetrarhynchus* (Arch.
1836, p. CVI.). New observations are needed to
decide if, as *Lereboullet* (Institut. 1839, No. 812,
p. 115) supposes, there can be included in this system the two longitudinal stripes, which, with *Ligula simplicissima*, extend along both sides of the
ventral surface, and from which I, at least, have
seen pass no filaments.*

2 Our very exact knowledge of the nervous system of *Amphistomum subtriquetrum*, and *conicum*, and of *Distomum hepaticum*, we owe to the
researches of *Bojanus* (Isis 1821, p. 168, Taf. II.
fig. 14, 15, 19), of *Laurer* (De Amphistomo conico, p. 12, fig. 21, 26), and of *Mehlis* (De Distomate hepatico, p. 22, fig. 13).

By continuing the methods of these helminthologists, this system will undoubtedly be found in
other Trematodes. *Diesing* (Ann. d. Wiener Mus.
I. Abth. 2, p. 246, Taf. XXII. fig. 9) has found in
Amphistomum gigantéum, and I have done the
same in *Distomum duplicatum* (which is properly only a larva of a species of this genus) the
same disposition noticed in *Amphistomum coni-

cum*. In *Distomum holostomum*, I have found
also a similar structure, except that the two œsophageal ganglia are widely separated, and united
by a very long cord-like commissure. *Laurer*
alone affirms to have seen enlargements upon the
principal nervous trunks of the Trematodes. But
their existence may be yet doubtful for no other
anatomist has mentioned them, and in no case have
I myself been able to see them.

3 *Miram* (Nov. Act. Acad. XVII. pt. 2, p. 652.
Tab. XLVI. fig. 8) did not, apparently, notice in
Pentastomum tarnioides the nervous ring which
surrounds the œsophagus ; although it had already
been noticed by *Curier* (Règne Anim. III. 1830,
p. 254), and by *Nordmann*, in a work in common
with *Mehlis* (Microgr. Beitr. Hft. 2, p. 141). The
existence of this ring has been placed beyond a
doubt by the figures of it as found in *Pentastomum tarnioides*, and *proboscideum*, given by
Owen (Trans. of the Zool. S c. I. p. 325, Pl. XI.
fig. 13, of Cyclop. Anat. and Phys. II. p. 130, fig.
78), and *Diesing* (Ann. d. Wiener Mus. I. Abth.
I, p. 13, Taf. I. II.).†

* [§ 104, note 1.] *Blanchard* (Ann. d. Sc. Nat.
1848, X. p. 338) appears to have distinctly made
out a nervous system in *Taenia*. With *Taenia
serrata*, there are directly behind the proboscis two
small medullary nuclei united by a commissure ;
from these pass off on each side a nerve which is
distributed to the lateral parts of the head, and
connects with a ganglion situated at the base
of each sucker, which sends filaments to the
muscles of this last. Posteriorly there are given
off filaments which run parallel to the intestinal
tubes. This, however, has not been confirmed by
other observers, and *Agassiz* has made a statement
in a private letter to me which is worthy of notice.

He says : " I believe the nervous system described
by *Blanchard* to be bands of muscular fibres which
cross each other between the fossae of the proboscis : at least, this is so in the new species of *Taenia*
from *Amia calva* which was observed alive for several hours ; and I could discover no nervous
threads, but only muscular fibres, which had exactly the arrangement of *Blanchard's* nervous system." See, however, *Valenciennes'* report to the
Acad. des Sc. in the Comp. Rend. 1847, XXIV. p.
1054, also *Blanchard's* response to *Dujardin*,
Ibid. 1849, XXIX. p. 60. — ED.

† [§ 104, note 3.] *Blanchard* has found with
Linguatula another ganglion above the œsopha-

The central nervous system of the Acanthocephali is very distinct. It is always concealed at the bottom of the sheath of the proboscis, which this last, being never in a state of complete retraction. does not fill. It consists of a dense mass of ganglionic, cellular globules blended together, and here and there may be seen through the cell-membranes their nuclei and corpuscles. This comparatively large mass sends off nerves in every direction, but the tenuity of these prevents their being traced, especially after they have entered the muscular walls of the proboscideal sheath.[4]

With the Gordiacei,[5] and Nematodes, a nervous system has been found with certainty only in *Strongylus gigas.* Here a cord arises from a swelling in the head, traverses the whole length of the body upon the median ventral line, and terminates at the posterior end of the body in another swelling. It sends off in its course lateral filaments, thus resembling the nervous system of the Sipunculidae.[6]

4 I have thus found the nervous system of the Acanthocephali in *Echinorhyncus gigas, angustatus, haeruca,* and *proteus.* It can be easily observed by carefully pressing or tearing the proboscideal sheath. In thus tearing, you sometimes completely expose the ganglionic mass with the roots of the nerves. In no species that I have dissected have I been able to find the ganglionic ring mentioned by *Henle* (*Froriep's* neue Not. No. 285, p. 330, and *Muller's* Arch. 1840, p. 318) as found about the genital orifice of *Echinorhyncus nodulosus.*
Dujardin also (Hist. Nat. d. Helm. p. 495, 491, Pl. VII. fig. D. 4), has not observed it, but he distinctly perceived the central mass at the base of the proboscis, and has figured and named it as *un corps glanduleux ou ganglionaire.*
5 As yet no nervous system has been found in the Gordiacei. *Berthold* (Über den Bau des Gordius aquaticus, 1842, p. 12) has been inclined to regard as nerves two delicate filaments which traverse the cavity of the body of *Gordius*; but, as these give off no lateral branches, this opinion cannot be admitted.*
6 Many Helminthologists have erroneously taken for nerves the delicate projecting lines which, situated directly subcutaneous and often blended with the skin, traverse the whole length of the body of many Nematodes, and have been called the ventral and dorsal lines. Their lateral branches, as already observed, are only transverse muscular bands. Quite different from these is the longitudinal cord, which *Otto* (Magaz. d. Geselich. naturf. Freunde zu Berlin, 7th Jahrg. 1816, p. 225, Taf. V.) has described and figured as belonging to the nervous system; a view which I am disposed to adopt, in spite of *Nitzsch* (*Ersch* and *Gruber's* Encyclop. VI. 1821, p. 45) and other Helminthologists.
In a large female *Strongylus gigas,* now under my eyes, there is a simple longitudinal cord beneath the muscular envelope, and therefore in direct connection with the skin, and which extends along the ventral surface. In its course it sends off numberless lateral branches, which in their intimate structure are quite different from the transverse muscular bands. But neither here nor upon the nerves of other worms have I ever seen the enlargements spoken of by *Otto.* Grant's figure of a double nervous filament traversing the body of *Ascaris* is probably imaginary; see Outlines of Comp. Anat. p. 186, fig. 82, A.

gus, which he regards as a brain; these observations have since been confirmed by *Van Beneden* (Ann. d. Sc. Nat. XI. 1849, p. 319), who, however, regards this mass as belonging to the sympathetic system. But, however viewed, an oesophageal collar has been distinctly made out, thus confirming the views of *Cuvier.*
In regard to the splanchnic system of nerves with these animals, *Van Beneden* (loc. cit.) describes it as consisting of two ganglia lying on the oesophagus back of the oesophageal collar, and from which pass off two filaments, which run along the oesophagus, and enter the collar laterally. He thinks the two ganglia are united by a transverse commissure. Further behind is another and larger ganglion on each side, and from which pass off filaments to the digestive cavity. See also my note under § 99. — ED.

* [§ 104, note 5.] This view of *Berthold* is supported also by *Blanchard* (Ann. d. Sc. Nat. 1849, XII. p. 6), who affirms that he has observed on both sides of the body a double longitudinal cord, which is usually very distinct. This, examined microscopically, appeared to be composed like the nerves of the other Helminthes. *Blanchard,* however, did not succeed in tracing these cords to any cephalic centres. Nothing of special value, therefore, is known on this subject. — ED.

CHAPTER IV.

ORGANS OF SENSE.

§ 105.

The sense of touch is probably the only one well developed with the Helminthes. The granulations, warts, papillae, filaments, and retractile lobes, found upon the head of some species,[1] are, without doubt, the organs of this function. The red and black points upon the back of many, both adults and larvae, and which have been regarded by some naturalists as organs of vision, appear to be only pigmentary spots; for they contain nothing like a light-refracting body.[2]

CHAPTER V.

DIGESTIVE APPARATUS.

§ 106.

The digestive organs with the Helminthes have a variable degree of development in the different orders.

In the Cystici, Cestodes, and Acanthocephali, neither mouth nor alimentary canal is perceived. In the first two orders, there is, however, a system of vessels which may be regarded as a digestive apparatus; but these are designed for circulation, rather than for digestion, since their walls are complete throughout and have no openings, as has erroneously been supposed, which communicate with the suckers of the head; and their contained nutritive material is received by them through the skin in an endosmotic manner.[1]

---

1 These tactile granulations are found with many species of *Ascaris*, as, for instance, in *Ascaris osculata*, between the large oral collars; in *Physaloptera alata*, they surround the oral extremity of the body as a single row; but they form a double one in *Ascaris trunculata*. With *Distomum laureatum*, and *nodulosum*, they are found upon the borders of the oral sucker. With *Holostomum excavatum*, and *podomorphum*, there are two retractile lobules protruding from the sides of the mouth; and in *Holostomum alatum*, these have antenna-like filaments; see *Nitzsch's* figures of *Holostomum*, in Ersch and *Gruber's* Encyclop. III. p. 399. IX.

2 These dark pigment-dots upon the infusoriform embryos of many Trematodes when they escape from the egg, and of which there is only one upon the neck of *Distomum nodulosum*, and *hians*, and two upon *Monostomum mutabile*, have been taken for eyes by *Nordmann* (Microgr. Beitr. Hft. 2, p. 139), and formerly by myself also (*Wiegmann's* Arch. 1835, I. p. 69, Taf. 1. fig. 3, 4, 5). Three of these dots have been observed upon a larva of a *Monostomum* which *Nitzsch* (Beitr.

zur. Infusorienkunde, p. 29, Taf. I.) has described in *Cercaria ephemera*; I have seen only two upon the back of many cercarian larvae. Of this same nature are the two red dots of *Scolex polymorphus* (*Muller*, Zool. Danica. Tab. LVIII. fig. 16, 17), as also the brown ones upon the neck of *Gyrodactylus auriculatus* (*Nordmann* Microgr. Beitr. Hft. 1. p. 108. Taf. X. fig. 4). Finally may be mentioned *Amphistomum subclavatum*, which has two large oval black dots upon its neck. These pigment-cells are physiologically, without doubt, simply colored spots, which in *Polystomum integerrimum* are highly developed, forming a widely-spread subcutaneous net-work. Sometimes, and especially in the various *Cercariae*, and in many individuals of *Amphistomum subclavatum*, these dots have a very effaced aspect; this is probably due to a dissolution of the walls of the cells, — the pigment-granules being then scattered through the skin.

1 It has already been observed that the four suckers of *Taenia*, regarded by *Nitzsch* as oral orifices, are imperforate at their bottom. *Owen* (Cyclop. Anat. &c. II. p. 131) has fallen into a

The food enters the cavity of the body of *Echinorhyncus* probably in the same manner, for their skin has great power of absorption.[2]

The Acanthocephali have this peculiarity, that between the skin and the muscular walls of the cavity of the body there is a thin layer of finely-granulated parenchyma, often of an orange or yellow color, which is traversed by longitudinal and transverse canals.

These canals, having no proper walls, form a continued vascular system, and contain a liquid filled with granules and vesicles. As this system is completely closed, and cannot therefore receive nutritive substances from without, it must be regarded as nutritive or circulatory, and not digestive, as it has been by many naturalists.

§ 107.

In the other groups of the Helminthes the digestive organs are pretty generally well developed.

The Trematodes have a mouth situated usually upon the border of the cephalic extremity, and where there is a sucker occupying its bottom. From this there passes along the middle line of the neck a thin-walled œsophagus, which is often of an S-like form. Directly behind the mouth or oral sucker, but sometimes a short distance removed from it, the œsophagus is surrounded by a round or oval muscular pharynx.[1] From the extremity of this pass off, usually, two blind intestinal tubes, which, passing along both sides of the body, extend generally to its posterior extremity.[2] The other forms of the digestive canal are as follows: in *Monostomum mutabile,*[3] and *flavum,* the two intestinal tubes, instead of ending coecally, form the arc of a circle;[4] in *Aspidogaster,* a simple and uniform intestine succeeds upon the pharynx, and ends in a coecum at the posterior extremity of the body;[5] in *Gasterostomum fimbriatum,* this canal is very short, and terminates in the same way, but there is a mouth in the middle of the ventral surface; in *Bucephalus polymorphus,*[6] the structure is similar; and in *Pentastomum,*

similar error in regarding these organs as mouths, not only in *Taenia* and *Cysticercus,* but also in *Bothriocephalus.* I have been unable to find a mouth upon the cephalic extremity of the Cestodes, as has *Mehlis* (Isis, 1831, p. 131), or upon that of *Taenia solium,* as has *Owen* (Lect. on the Comp. Anat. &c. p. 48, fig. 21, a.). The fossa sometimes found upon this last, is due to the retraction of the circle of hooks, or of the proboscis, within the sheath.

2 Most Helminthologists admit that *Echinorhynchus* receives its food through a small orifice at the extremity of the proboscis, the sheath of the last aiding in suction and deglutition. I have been unable to convince myself of the existence of this orifice, and never have found food in the cavity of the sheath. On the other hand, I have often, like *Creplin* and *Mehlis,* seen *Echinorhynchus* receive and reject liquids through the skin.

1 With *Distomum globiporum,* the pharynx is somewhat removed from the oral sucker; see *Burmeister,* in *Wiegmann's* Arch. 1835, II. fig. Taf. 1, 3. In *Distomum echinatum, militare* and allied species, the œsophagus is usually very long. But in *Distomum oxycephalum,* it is very short; and in *Distomum appendiculatum,* it is entirely wanting, and consequently the intestinal bifurcation is directly behind the pharynx.

2 In *Monostomum, Amphistomum, Holosto-*

mum, Distomum, and *Polystomum,* the intestinal bifurcation extends to the posterior extremity of the body. With *Distomum chilostomum,* and many other species of this genus living in the Neuroptera, the whole intestine is reduced to two short right and left coeca, which are given off from the end of the œsophagus.

3 *Creplin,* Nov. Observ. de Entozoïs, fig. 10, 11.

4 This arrangement has been also, but erroneously, assigned to *Distomum tereticolle;* see *Wagner,* Lehrbuch der vergleichenden Anat. 1834. p. 75, and *Creplin,* in *Ersch* and *Gruber's* Ency clop. XXIX. 1837, p. 314.

This error is probably due to the inaccurate copying of figures; see Ann. d. Sc. Nat. II. 1824, p. 493, Pl. XXIII. fig. 4, 5; and *Schmalz,* Tabulae Anat. Entozoorum, Tab. VIII. fig. 2, 3. By referring to the original figure in the Memoir of *Jurine* (Mém. de la Soc. de Phys. et d'Hist. Nat. de Genève, II. pt. 1, 1825, p. 149, fig. 4, 5), from which these have been copied, there is found no trace of a closed, arcuate intestinal canal behind. Moreover, *Jurine* expressly says that he has seen the intestinal tubes of *Distomum tereticolle,* as coeca.

5 *Baer,* in Nov. Act. Acad. &c. XIII. pt. 1, p. 536, Taf. XXVIII.; also *Diesing,* Med. Jahrbuch. d. k. k. österreichischen Staates. XVI. 1834, p. 423, fig. 8–11.

6 *Bucephalus polymorphus* is probably a larval

this canal is simple, straight, and ends posteriorly in an anus.[7] In many Trematodes, the intestinal tubes have in all their course simple or ramified caeca, and in some, these caeca are so fully developed that the intestinal canal appears to fill the whole body.[8] The intestinal walls here are very thin, but this does not prevent peristaltic and anti-peristaltic movements, by which their contents move backwards and forwards, and are often rejected through the mouth.[9]

§ 108.

In the Nematodes, and Gordiacei, the intestinal canal passes straight from the mouth which is at the anterior extremity, through the cavity of the body to the anus, which, in the first, opens front of the caudal extremity.[1] In very many Nematodes, the mouth has nodosities and swellings, but it is seldom that its cavity has horny, tooth-like processes.[2]

From the mouth extends a long and very muscular œsophagus, which is usually dilated claviform at its lower extremity. When the œsophagus is very long, it has one or more constrictions.[3] It is nearly always composed of three longitudinal muscles which are united by longitudinal seams. The triangular cavity circumscribed by these muscles is lined by a very firm epithelium, which is sometimes horny, and in some species so thickly set in the clavate dilatation that it resembles a masticatory apparatus.[4] The intestine consists of a straight tube, with thin walls and without dilata-

---

Gasterostomum; and the species above mentioned I have discovered in the intestinal canal of *Perca fluviatilis,* and *Lucioperca.*

7 See *Miram, Owen,* and *Diesing,* loc. cit. The opening at the posterior extremity of many Trematodes, and by many Helminthologists taken for an anus, belongs to a special secretory organ, which will be mentioned hereafter.

8 In many species allied to *Monostomum trigonocephalum,* the two intestinal tubes have simple caeca upon both sides of their entire length. In *Octobothrium lanceolatum,* the structure is the same; see *Mayer,* Beitr. p. 21, Taf. III. fig. 3. These lateral caeca are more or less ramified in *Octobothrium palmatum, sagittatum, Merlangi, Polystomum appendiculatum,* and *Tristomum elongatum* (*Leuckart,* Zool. Bruchstücke, Hft. 3, p. 26, 54, Taf. I. fig. 4, c. h. Taf. II. fig. 5, d. ; *Nordmann,* Microgr. Beitr. Hft. 1, p. 79, 81, Taf. VII. fig. 2, Taf. V. fig. 6 ; and *Baer,* Nov. Act. Acad. Leop. XIII. pt. 1, p. 665, Taf. XXXII. fig. 2). With *Distomum hepaticum,* these ramifications are very fully developed ; see *Mehlis,* Observ. de Distomate, fig. 1, 2, 7, 8. In the very remarkable genus *Diplozoon,* the digestive canal consists of a single tube which traverses the whole body upon the median line, and sends off laterally ramified caeca, while at the point of junction of the two bodies of the animal it dilates into a stomachal cavity ; see *Nordmann,* loc. cit. Hft. I, p. 67, Taf. V. fig. 2. The blackish ramifications of *Polystomum integerrimum,* and which have been regarded by *Baer* (Nov. Act. Acad. Leop. loc. cit. p. 682, Taf. XXXII. fig. 7, 8) and other authors as a digestive canal, belong to the subentaneous pigmentary net-work already mentioned.

9 The digestive canal of Trematodes is usually partly filled with blood which they have absorbed, and partly with brown or yellowish chyme ; it is therefore evident how, from the thinness of its walls, it would, when empty, entirely escape the observation.

1 Among the Nematodes, and Gordiacei, there

are, moreover, species which have very rudimentary digestive organs. In *Sphaerularia bombi,* there is neither mouth nor anus, and in the place of the intestinal canal there is a row of long sacs clinging together, and around which the genital organs are coiled (*Wiegmann's* Arch. 1838, I. p. 305). In *Filaria rigida,* living in the intestines of *Aphodius fimetarius,* I have found no digestive canal whatever (*Muller's* Arch. 1836, p. 33). In the various species of *Mermis,* there is a distinct mouth, œsophagus and intestine, but this last ends in a caecum. I have been unable as yet to positively determine a mouth with *Gordius aquaticus ;* the anus is certainly wanting, and it might be questioned if the two tubes which traverse the body should be regarded as an intestine ; see *Wiegmann's* Arch. 1843, II. p. 305.

2 With *Strongylus armatus, hypostomus, dentatus,* and *tetracanthus,* the entrance of the mouth is provided with a circle of horny teeth, which are moved by special muscles ; see *Mehlis,* Isis. 1831, p. 78, Taf. II. fig. 5, 6. With *Spiroptera strongylina,* I have seen the entire internal surface of the mouth provided with a spiral, horny swelling. In *Cucullanus,* there is a very complicated apparatus for opening and closing the mouth, composed of solid, horny pieces.

3 With *Anguillula fluviatilis, Oxyuris vermicularis, Ascaris acuminata, brevicaudata, dactyluris, oxyura,* and *vesicularis,* the œsophagus has this enlargement. But it is divided into two portions by a prominent constriction with *Cucullanus elegans, Physaloptera alata, Spiroptera anthuris, europtera, obvelata,* and *crassicauda.* In *Trichocephalus,* it is very long, and has behind very many constrictions, which are successive at short intervals ; see *Mayer,* Beitr. &c. Taf. I. II. With *Trichosoma falconum,* it is equally long and divided into many sections, which give it an articulated aspect.

4 By many Helminthologists this tube has been called *œsophagus,* and its dilatation *stomachus.*

tions, and which terminates in a short muscular rectum. The proper intestine is of a brown, greenish, or dirty yellow color, which is due to its walls being formed of compact cells filled with colored granules. The loose and cellular walls, having very feeble peristaltic movements, are surrounded externally by a kind of dense peritoneum, and lined internally by a very fine epithelium.[5] In some species of *Ascaris*, the intestine is lengthened into a caecum at its junction with the œsophagus.[6]

§ 109.

There are observed, here and there, only traces of appendant organs of the digestive canal.

In many Trematodes, there are upon each side of the neck, two more or less developed cords or canals, of a cellular aspect, and of a pale yellow color by direct light. They pass towards the mouth, open perhaps into its cavity, and have a function, probably, like that of *salivary organs*.[1] In many Nematodes, two or four caeca extend from the cephalic extremity along the œsophagus, and as they open distinctly into the oral cavity, it is, therefore, the more probable that they should be regarded as salivary organs.[2] The same signification should be given to the coecal appendage found in many species of *Ascaris*, which extends from the constriction of the œsophagus to the beginning of the intestine.[3]

Hepatic organs have been found nowhere but in the Nematodes; but it may be that the granular cells in the thick walls of the intestinal canal, take their place.

5 This epithelium has sometimes special inequalities, which, with *Ascaris osculata*, and *spiculigera*, form a regular zig-zag series, resembling the valves of the intestinal mucous membrane of some vertebrates. With *Ascaris aucta*, they have the form of long, sharp villosities.

6 This coecal appendage, accompanied usually with a constriction of the posterior end of the œsophagus, was first observed by *Mehlis* (Isis. 1831, p. 91, Taf. II. fig. 16, 17, 18). It is found with many *Ascaris*, but its length is very variable. In *Ascaris heterura, semitera*, and *ensicaudata*, it is very short, and protrudes scarcely beyond the œsophageal constriction; while in *Ascaris depressa, aucta, angulata*, and *mucronata*, it reaches to the middle of the œsophagus, and in *Ascaris spiculizera, osculata*, and the species described as *Filaria piscium*, it extends nearly to the cephalic extremity.*

1 These glandular-like organs are often very distinct in the cercarian larvae of the Trematodes, and in many adults of *Monostomum*, and *Distomum* ; see *Wiegmann's* Arch. 1843, II. p. 322.

2 *Mehlis* (Isis, 1831, p. 81, Taf. II. fig. 6) has observed with *Strongylus armatus*, an annular vessel surrounding the mouth, which communi-

cates with it directly, and also with two cords accompanying the œsophagus. According to him, there is also a similar disposition with *Strongylus hypostomus*, and *tetracanthus*.

Similar appendages, analogous to salivary organs, occur, according to *Owen*, in the new genus *Gnathosoma*, as four caeca surrounding the œsophagus, and opening into the mouth (*Wiegmann's* Arch. 1838, I. p. 134). With *Cheiracanthus*, and *Ancyracanthus*, there are four similar organs, and *Diesing* is certainly in error in regarding them as analogous to the ambulacral vesicles of the Echinoderms (Ann. d. Wiener Mus. II. Abth. 2, p. 224, 226, 228, Taf. XVII. fig. 8, 9, Taf. XVIII. fig. 3). I am disposed to regard as salivary organs, also, the two long caeca which pass from the mouth along the œsophagus of *Strongylus striatus*.

3 I have discovered a similar œsophageal appendage in a group of *Ascaris* known as *Filaria piscium* (*Wiegmann's* Arch. 1838, I. p. 309) ; such are, *Ascaris mucronata, angulata, osculata, spiculizera, aucta, acus*, and *labiata*. It is remarkable that with the exception of the last two, all these have also a caecum upon the intestine.

* [§ 108, note 6.] See, for the alimentary canal of *Ascaris infecta, Leidy* (A Flora and Fauna within living animals, Smithsonian Contrib. V. Art. 2, p. 43, Pl. VI. fig. 1–7). He divides it into a strongly muscular gizzard, a cylindroid intestine lined with hexahedral epithelium, and a pyriform rectum.

See also his description of that of *Streptosomum, Thelastomum*, &c. (Ibid. p. 49). In *The-*

lastomum appendiculatum, there is this peculiarity, that the intestine commences by a broad, deeply sinuate, cordiform dilatation, which rapidly narrows to a short, cylindroid portion, and then sends off a long, capacious, gourd-form receptacle, or diverticulum, and afterwards proceeds backwards to the rectum, and in its course, in the vicinity of the generative aperture, performs a single short convolution. — Ed.

CHAPTER VI.

CIRCULATORY SYSTEM.

§ 110.

Most of these animals have a vascular system. The circulating liquid is usually wholly colorless, and often contains vesicular or granular corpuscles, which are difficult to perceive from their delicacy and transparency. The circulation is due to the general contractions of the body or of the walls of the vessels.

In the Acanthocephali, the vessels have no proper walls, but are spread out, as has already been said (§ 106), in the subcutaneous parenchyma. There are two larger, lateral canals, which pass from the neck to the caudal extremity, sending off laterally numerous small canals, which anastomose with each other. A similar net-work is found in the proboscis through its whole length.[1] These two canals connect also with the *lemnisci*, upon each side of the neck. These last, of which there are always two upon the sides of the proboscis, passing from the neck to the cavity of the body, are usually riband-like, and composed of a finely-granulated parenchyma, which, like the cutaneous one, has a system of vascular canals.[2]

In most species of *Echinorhynchus*, this system consists of a main canal upon the border of the lemniscus, from which are sent off inwardly, numerous small branches. These last form the net-work which fills the parenchyma of the proboscis.[3]

In many,[4] the lemnisci are surrounded by muscular fibres, which, converging to the posterior extremity of these organs, form two short muscles, which, in their turn, are blended with those passing obliquely to the proboscideal sheath. The point of junction is at a short distance from the place where they are detached from the subcutaneous muscular layer. Each lemniscus is constricted into a narrow neck at its base, which passes into the skin at the base of the proboscis. The junction of the cutaneous with the lemniscian vascular system occurs at this point, as is indicated by the contained liquid passing backwards and forwards between the two from

---

1 This vascular system, taken by many Helminthologists for a digestive canal, has been figured by *Westrumb* (De Helminth Acanthocephalis Tab. II. fig. 10, 111. fig. 10, 12, 21), and *Burow* (Echinorhynchi strumosi Anat. 1836, fig. 1, 8). The movements of the nutritive liquid may be distinctly seen by placing these animals alive and undilated as natural under the microscope. One will then be quickly convinced that the circulation is due to the general movements of the body. If *Echinorhynchus* is placed in much water, the absorption distends not only the body, but the canals of the vascular system are so filled that the subcutaneous parenchyma is swollen, and the skin is raised here and there into vesicles.

2 With *Echinorhynchus angustatus, acus, fusiformis, proteus,* and *polymorphus*, the two lemnisci have a riband-like form. In *Echinorhynchus gigas*, they are very long; and in *Echi-*

norhynchus claviceps, they are longer than the body, and lie coiled in its cavity. In *Echinorhynchus gibbosus, hystrix*. and *strumosus*, they are discoid and very short.

3 *Echinorhynchus angustatus, hacruca, polymorphus, proteus,* and *gibbosus*. As a wide exception, the principal canal occupies the median line of the lemnisci, and sends off laterally small branches, with *Echinorhynchus gigas*. Here and there its course is broken by oval, voluminous, transparent and apparently vesicular bodies ; see *Westrumb* loc. cit. Tab. II. fig. 7. Similar bodies in the lemnisci and subcutaneous parenchyma, are found with *Echinorhynchus claviceps* ; see *Muller*, Zool. Danica. Tab. LXI. fig. 3. These bodies are, moreover, regular neither as to their number nor position, and I have not learned their nature.

4 *Echinorhynchus acus, angustatus, fusiformis,* and *proteus.*

the peristaltic actions of the body and the alternate retraction and protraction of the proboscis.[5]

In the Gordiacei, and Nematodes, no vascular system has as yet been
found. Only in a group of species described as *Filaria piscium*, has there
been found a riband-like organ concealed in the cavity of the body, and
traversed by a net-work of canals, which resemble those of the lemnisci of
the Acanthocephali.[6]

§ 111.

In the Cystici, Cestodes, and Trematodes, the vascular system is well
developed. Its canals have proper walls, the contraction of which produces the circulation. In the first two orders, it consists of two pairs of
longitudinal canals, which pass along the sides of the body and head, and
intercommunicate occasionally, by transverse canals. These four vessels
open, in the head, into an annular ring which surrounds the proboscideal
sheath; there is here, therefore, a completely isolated system.[1] In the
Trematodes, this system consists of a contractile net-work spread over
the whole body; and in which are two larger trunks, which pass along
the sides of the neck and body.[2]

5 *Mehlis* (Isis, 1831, p. 82) affirms to have seen
on the neck of *Echinorhynchus gigas* two small
orifices by which the lemnisci open outwards. But
I have been unable to see them in this species, or
others of this same genus. If they really exist,
they will shed light upon the doubtful functions of
these organs. From what we know of their structure, it is not improbable that they belong to the
nutritive system, and transude a liquid which
bathes and nourishes the organs in the cavity of
the body.*

6 With the Nematodes, the liquid appears to
transude through the walls of the intestine into the
cavity of the body, and there bathe, without a vascular system, all the organs. The riband-like organ
found in the *Filaria piscium* (see *Wiegmann's*
Arch. 1838, I. p. 310), and which I have also found
in *Ascaris osculata*, has the same vascular ramifications as the lemnisci of *Echinorhynchus gigas*, and the vesicle-like bodies are not wanting
upon the course of the principal canal. Perhaps
they also transude the nutritive liquid, for I have
not found any communication between them and
the intestinal canal.

The two lateral enlargements also, which, as
already mentioned (§ 102), are extended between
the longitudinal muscles of the skin, have often
been regarded as sanguineous vessels; but I have
observed with them neither longitudinal nor lateral
canals.†

1 These lateral vessels, regarded by some Helminthologists as intestinal tubes, give off in their
course no lateral branches, except these transverse
canals. With the articulated Cestodes, these last
are always situated at the posterior extremity of
the articulations, thus giving a ladder-like aspect
to the entire vascular system. They are also
found, however, in *Caryophyllaeus mutabilis*,
which is not articulated.

*[§ 110, note 5.] The observations of *West-
rumb* and *Burow* on the circulatory system of the
Acanthocephali, have recently been thoroughly
verified by *Blanchard*, who has illustrated it with
excellent figures; see Ann. d. Sc. Nat. 1849, XII.
p. 21, and Règne animal, nouv. Édit. Zoophytes,
Pl. XXXV. fig. 2. — Ed.

Platner (*Müller's* Arch. 1838, p. 572, Taf.
XIII. fig. 4, 5) affirms to have seen semilunar valves
at the orifices of the transverse canals of *Taenia
solium*.

The four lateral cervical vessels which I have
observed not only in *Taenia*, but also in *Bothriocephalus*, and *Cysticercus*, may be traced with
perfect distinctness in *Taenia cyathiformis*, and
serrata, to the vascular ring which surrounds the
proboscideal sheath. With *Caryophyllaeus mutabilis*, and *Taenia ocellata*, which are without a
proboscis, this vascular ring does not exist any
more than with *Bothriocephalus*; here also the
four lateral vessels widely ramify in the head, and
form by anastomoses, a distinct net-work. *Bothriocephalus claviceps* has a similar organization.
It should, moreover, be here observed that from
the contraction of its very thin walls the vascular
system will easily elude the observer.

2 The vessels of the Trematodes are remarkable
for their prominent flexures; see *Distomum cirrigerum*, *tereticolle*, *duplicatum*, and the various
species of *Diplostomum* (*Nordmann* Microgr.
Beitr. Hft. 1, Taf. 11. fig. 8, IV. fig. 5, 6). One
should not confound with the sanguineous vessels,
as has often been done, the very finely-ramified
canals of the excretory organ, which will hereafter
be mentioned. Thus I think that the vascular
net-work of *Distomum hepaticum* described by
Bojanus (Isis, 1820, p. 305, Taf. IV.) belongs to
this excretory organ. *Laurer* also (de Amphistomo conico. p. 10, fig. 22), has not carefully distinguished them; and *Nordmann* appears to have
fallen into the same error (loc. cit.).

With *Diplostomum*, the vessels open each side
into a large reservoir situated at the extremity of
the body. Between these two receptacles, the
excretory organ passes to the extremity of the
body, and *Nordmann* has taken its orifice as

†[§ 110, note 6.] *Berthold* (Ueber den Bau
des Wasserkalbes, &c. loc. cit.) has described a
vascular system with the Gordiacei; but *Blanchard*
(Ann. d. Sc. Nat. 1849, XII. p. 7) has failed to
confirm his statements after very careful research.
— Ed.

CHAPTER VII.

RESPIRATORY SYSTEM.

§ 112.

A respiratory system has not yet been found with certainty in the Helminthes.

The pedunculated vesicles of many Nematodes, situated under the skin, and projecting into the cavity of the body, and which have great absorptive power, have been compared to trachean pouches and branchiae; but their structure is so little known, that any opinion as to their function ought to be deferred.[1]

A remarkable fact is the presence in some Trematodes of extremely active vibratile lobules, situated intermittingly on the inner surface of the walls of the vessels.[2] It may be questioned if these vessels have a special function, different from that of the others. They somewhat resemble the aquiferous system of the Polyps, Acalephs, and Echinoderms, and like it, belong, perhaps, to the respiratory system. They differ, however, in not having openings which communicate outwardly; but, probably, they receive by endosmosis, water absorbed by the skin.[3] But another objection to this view, is, that in this order there has been found nothing like blood-vessels,

---

belonging to the nutritive vessels. The nutritive liquid of the vascular system differs from the coarsely-granulated excretion of the excretory organ, by its homogeneous and colorless aspect.

It is remarkable that in *Distomum tereticolle* this liquid has a reddish color, which, in the finest capillaries has a yellowish cast ; see *Wiegmann's* Arch. 1835, I. p. 59.

H. Meckel, likewise, thinks that the above-described vascular system of the Trematodes, is in direct communication with the secreting organ peculiar to these Helminthes ; see *Muller's* Arch. 1846, p. 2, Taf. I. fig. 2.*

1 *Bojanus* (Isis, 1821, p. 187, Taf. III. fig. 51–55) affirms to have observed in *Ascaris lumbricoides* these pedunculated vesicles, which are found also in *Ascaris depressa*, and *Strongylus gigas*, in connection with the lateral swellings ; but this throws no light upon the nature of these vesicles, for we are yet ignorant of that of these swellings. The stigmata which he affirms (loc. cit. p. 187, Taf. III. fig. 56) to have observed upon these lines with *Ascaris acus*, are, according to my own observations, only subcutaneous cell-like bodies.

2 I have quite distinctly seen these vessels with *Diplozoon paradoxum*, *Aspidogaster conchicola*, *Distomum echinatum*, and an allied species of this last from the intestine of *Falco apivorus*.

*[§ 111, note 2.] *Van Beneden* (Ann. d. Sc. Nat. 1852, XVIII. p. 23) has recently expressed doubts upon the presence of a circulatory system

I am yet uncertain if the vibratile organs found in the neck of *Distomum globiporum* and *nodulosum* (*Wiegmann's* Arch. 1836, I. p. 218), and in the parenchyma of *Distomum duplicatum* behind the ventral sucker, are of the same nature.

Ehrenberg (*Wiegmann's* Arch. 1835, II. p. 128) was the first who observed this ciliary movement in the vessels of *Diplozoon*. When the motions of these lobules are free, there is a rapid current of the liquid, as *Nordmann* has remarked (Microgr. Beitr. Hft. I. p. 69). But if an animal is compressed between two plates of glass, and their motions thus impeded, it will be quickly seen that these last are the cause of the circulation ; in fact, when the lobules cease moving, the colorless, homogeneous, and, without doubt circulatory liquid, is no longer perceived.

3 *Burmeister* (Handbuch d. Naturgesch. 1837, p. 528) compares, not without reason, this system to the trachean system of insects, the first being *aqueous*, and the second *aerial* respiratory organs, thus confounding this vascular system of Helminthes with the excretory organ and duct found in most Trematodes. There may be, however, a comparison between these two systems, if we except the insects with stigmata, and take those which are aquatic and have a completely closed trachean apparatus (see below), admitting no air from without.

with the Cestodes and Trematodes, but see the beautiful plates of *Blanchard*, Ann. d. Sc. Nat. 1848, X. Pl. XI. — ED.

CHAPTER VIII.

ORGANS OF SECRETION.

§ 113.

No organs of secretion have been found, except in the Trematodes and Nematodes. In most of the Trematodes, there is, upon the median line of the posterior part of the body, a contractile sac, which usually opens outwards,[1] at the caudal extremity, and seldom at the posterior part of the back.[2] This sac is single,[3] bifurcate,[4] or multiramose. In the last case, its branches are spread usually over the whole body.[5] Its walls are quite thin, and therefore, it is seen with difficulty when wholly contracted or empty. It contains a colorless liquid filled with numerous granules or vesicles, which, during the contractions, pass up and down, or escape through the external opening.[6] This organ is sometimes so crowded with clear, solid corpuscles, composed apparently of earthy matter, that examined by reflected light, it has a cretaceous aspect.[7]

In many Nematodes, there is on the ventral surface and at a variable distance from the head, a small oblique opening surrounded by a sphincter. In some species, two canals pass from it and run backwards on each side of the intestinal canal; and in others, there are also two other canals which extend forwards in the same way. The use of the colorless and homogeneous secretion of these organs is yet unknown.[8]

1 This opening, known as the *Foramen caudale* with *Distomum, Holostomum, Monostomum, Aspidogaster,* and *Diplostomum,* has formerly been compared to an anus by Nardo (*Heusinger's* Zeitsch. für organische Phys. 1827, I. p. 68), and by Baer (Ibid. II. p. 197). Mehlis (Observ. de Distomate, p. 16) having shown that it belonged, in *Distomum hepaticum,* to a particular organ which is ramified like a vessel, has properly rejected this analogy; see Isis,1831, p. 179. With the larvae of Trematodes, known as *Cercaria, Bucephalus,* and *Distomum duplicatum,* the base of the tail is thrust into the excretory opening of this organ, and its contents cannot escape until the animal has lost the tail.
2 *Amphistomum.*
3 *Monostomum faba, Distomum cirrigerum, Gasterostomum fimbriatum,* and *Bucephalus polymorphus.*
4 *Distomum chilostomum, clavigerum, lima, maculosum, terreticolle, variegatum,* and many species of *Monostomum,* — where the two closed ends of the sac often extend to the cephalic extremity. With *Distomum appendiculatum,* the two branches of the excretory organ unite directly behind the oral sucker. With *Aspidogaster conchicola,* it divides into two canals near the *Foramen caudale,* which extend to the anterior extremity. In *Amphistomum,* two similar canals wind from the head along each side of the body, to the middle of the posterior back, where they open outwards, after having formed by reunion a pyriform reservoir. Laurer (De Amphistomo conico, p. 10, fig. 24) has given a figure of this reservoir, in which he has confounded the secretory canals with the nutritive vessels.

5 Beside *Distomum hepaticum, Holostomum urnigerum,* the *Distoma* also with a spinous head, have a widely-ramified excretory organ; see Mehlis, Isis,1831, p. 182.
6 With the spinous-headed *Distomum militare,* and *echinatum,* this organ is often so reduced in substance, that here and there are perceived only isolated groups of the ramified canals.
7 The solidity of these corpuscles may have been the reason why Ehrenberg (Symb. Physic. Anim. Evertebr. Ser. I. Phytozoa entozoa) has taken those of *Cercaria ephemera* for eggs, and the two canals of the excretory organ for ovaries; and why Nordmann (Microgr. Beitr. Hft. 1, p. 54, Taf. I. fig. 7) has regarded their escape from the body with *Distomum annuligerum,* as an act of oviposition.

The corpuscles of this kind found in the excretory organ of certain Trematodes, as for instance in a larva of *Monostomum* known as *Cercaria ephemera,* remind one from their aspect, of the small calcareous subcutaneous bodies of many *Taeniae,* and it may be asked if they are not an effete material, which, not being contained in proper organs, is with these Helminthes thus subcutaneously deposited.
8 This organ, to which I first called the attention in the dissertation of Rugge (De evolutione Strongyli auricularis et Ascaridis acuminatae, 1841, p. 13), is composed of two canals which run backwards in *Strongylus auricularis, Ascaris brevicaudata,* and *acuminata* (Rugge, loc. cit. fig. 30, A. B.); and in *Ascaris dactyluris,* and *paucipara,* mihi (from the intestine of *Testudo graeca*), of two anterior and posterior canals, the common opening of which is near the middle of the body.

CHAPTER IX.

ORGANS OF GENERATION.

§ 114.

Although most of the Helminthes propagate by means of genital organs, yet there are a few species which multiply by *fissuration* and *gemmation*. The *fissuration* is always transverse, and differs from that of the Protozoa and Zoophytes in the fact that complete individuals are not produced, there being only a separation of certain organs from the perfect animal, as, for instance that of the segments of the body in the Cestodes. This fissuration is complete or incomplete. In the first case, occurring in the *Taenia*, the segments are detached from the body, and continue to live independently, without, however, ever forming a new individual.[1]

Gemmation has been observed in the sexless *Coenurus* and *Echinococcus*. In *Coenurus cerebralis*, it is incomplete. The buds are formed on the internal surface of the parent-vesicle, and never separate from it, nor become perfect individuals. They have only a head and neck which project outwardly after the complete development. In *Echinococcus*, however, the gemmation is complete. The buds appear as in *Coenurus*, but the young animals are sooner or later detached and fall into the liquid of the parent vesicle. When completely developed, this vesicle bursts, and they are set at liberty. That their development occurs in this way is shown by their hanging by a cord, which, like the tail of *Cercaria*, is inserted into a fossa at the posterior extremity of the body. Like this last, also, this cord subsequently disappears, and the young animal moves freely about, by the aid of its double circle of hooks and its four suckers.[2]

§ 115.

In those species which reproduce by male and female genital organs, these last are sometimes upon a single animal, and sometimes upon two separate individuals. The eggs and spermatic particles are formed after very different types. In all, the copulatory organs are extraordinarily developed. The Cestodes and Trematodes are hermaphrodites.[1] The structure of

---

1 The imperfect fissuration with *Ligula* and *Triaenophorus* is limited almost to a constriction of the lateral borders. With *Bothriocephalus punctatus*, it is only here and there that a ring is detached, and over most of the body the transverse and opposite sulcations do not extend near to the median line. With *Bothriocephalus tetrapterus*, the fissuration is more complete; but even here, there are only some incompletely limited rings among numerous others which are completely so. Of all Helminthes the *Taeniae* have the most complete fissuration; here not only is the separation of the rings indicated by a complete furrow, but the rings are sometimes detached and live thus independently. The separated rings of *Taenia solium, cucumerina*, and others, move freely, and are so individualized, that they resemble some Trematodes.

2 See *Chemnitz*, De Hydatibus Echinococci hominis commentatio, 1834 ; *Muller*, in his Arch. 1836, p. CVII. ; and *Siebold*, in *Burdach's* Physiol. II. 1837, p. 183.
1 According to *Nordmann* (Microgr. Beitr. Hft. 2, p. 141), *Diesing* (Ann. d. Wiener Mus. I. Abth. 1, p. 9), and *Miram* (Nov. Act. Acad. XVII. pt. 2, p. 636), the male and female genital organs of the genus *Pentastomum*, classed by many modern Helminthologists among the Trematodes, are situated upon different individuals. But *Owen* affirms to have observed the opposite (Trans. of the Zool. Soc. of London, 1835, I. p 325). The only way to settle this point is by analyzing accurately the contents of these organs ; a method pursued by *Valentin* (Repertorium III 1837, p. 135), who found filamentoid spermatic particles in the organs of an apparently female

the genital organs of the first is yet imperfectly known; while that of those of the second is well understood. The female apparatus of the Trematodes consists of a germ-forming organ (ovary), with its excretory duct; then, two others for forming the vitellus, which have also excretory ducts; and then a simple uterus with its vagina. The male apparatus consists of testicles with their excretory canals, an internal seminal vesicle, a cirrhus-sac, an external seminal vesicle, and a penis.[2]

The ovary consists of a round or pyriform [3] reservoir, situated, usually, upon the median line of the body,[4] from which it is distinguished by its pale color and transparency. It is filled with simple round cells — the egg-germs. The nucleus of these cells is the germinative vesicle, and the nucleolus, the germinative dot.[5]

The short and small excretory duct of the ovary opens at the commencement of the uterus. The organs which secrete the vitellus are two in number, of variable length, and situated upon each side of the body near the dorsal surface; they occupy either the cervical, the central, or the posterior portion of the animal, and sometimes extend over them all. They are nearly always composed of ramified caeca filled with white, granular, vitelline corpuscles. By reflected light these caeca appear through the skin as a white, ramified, botryoidal mass,[6] and from each of them, pass off inwardly, numerous excretory ducts, which reünite opposite the ovary into two common canals. These last approach each other transversely, and form a single canal upon the median line, which, after a short course, opens at the bottom of the uterus by an orifice which is common to it and the ovary.[7]

Pentastomum taenioides, organs which are regarded by *Diesing* as caeca for secreting the envelope of the eggs.

Since all the parts of the genital organs of *Pentastomum* have not been examined with this same precision, I can give no opinion as to their use.*

2 See *Siebold*, in *Wiegmann's* Arch. 1836, I. p. 217, Taf. VI., and in *Muller's* Arch. 1836, p. 232, Taf. X. fig. 1.

3 The ovary here is always smaller than the testicle, and sometimes as to form very closely resembles it, as in *Distomum globiporum*, and *longicolle*, mihi (from the urinary bladder of *Cottus gobio*); consequently it may easily be taken for a third testicle.

4 With *Monostomum*, it lies wholly at the posterior extremity.

5 In *Polystomum*, *Octobothrium* and *Diplozoon*, the germs are so large that they may easily be taken for perfect eggs.

There is here, moreover, between the cell-wall and the nucleus (the germinative vesicle), quite a thick layer of albuminous substance, somewhat

representing a vitellus. But in the other Trematodes it is so thin as scarcely to be perceived.

6 With the following Trematodes there is a wide deviation from this usual arrangement. In *Distomum longicolle* the organs producing the vitellus are two simple round caeca located behind the ventral sucker; in *Distomum cygnoides*, they are two very small deeply-fissured bodies; and in *Distomum gibbosum*, there is one only, which is star-shaped and located at the middle of the body.

7 These organs, until now regarded as ovaries, secrete only vitelline cells. With most Trematodes their nuclei are clear, and have been taken for eggs. In eggs recently formed, one can always distinguish these cells from the germs. In passing the excretory canals they are compressed and elongated, but never run into each other. When these canals are crowded, they have the aspect of white cords, which have often been taken for nerves. But when they are empty, they, as well as the vitellus-secreting organs, are almost invisible.†

* [§ 115, note I.] See upon this subject *Van Beneden* (Ann. d. Sc. Nat. XI. 1849, p. 326), who has described in detail the sexual organs of *Linguatula Diesingii*, and has shown the sexes to be separate. See also my note under § 99. — ED.

† [§ 115, note 7.] To say that certain organs secrete vitelline cells, is a little obscure, and no doubt *Siebold* intended to convey the meaning that they secreted the plastic material out of which these cells are formed. I make this perhaps seemingly unnecessary reference to the matter, since it concerns the subject of the development of the ovum. In the *Ascaris*, where the origin and development of the ovum can be satisfactorily

studied, you first notice the germs as nucleolated cells, of which the nucleus is the future germinative vesicle and the nucleolus the germinative dot. These cells increase in size, and as they move along there appear in the liquid which lies between the nucleus and the cell-wall minute granules which ultimately become cells; in this way the vitellus is formed, the formation being endogenous and not exogenous. These special organs or tubes therefore are vitellus-forming organs, in virtue of their secreting the formative material out of which the vitellus is formed *within* the original, nucleolated germ-cell. — ED.

The neck of the internal seminal vesicle (*Vesicula seminalis interior*), discharges its contents at this same place into the uterus, through a special *Vas deferens* from one of the testicles. The *Uterus* commences as a narrow tube, which may be regarded as a *Tuba Fallopii*. Its dilated portion, which has powerful peristaltic motions notwithstanding its thin walls, is throughout of nearly an equal diameter. It winds through a large portion of the body and terminates in a narrow, more or less straight, muscular vagina, which always opens externally by the side of the penis.[8]

The testicles, of which there are usually two,[9] are generally of a round or oval form,[10] and located in the posterior region of the body, nearly always one before the other.[11] They are transparent and colorless, and the filiform spermatic particles are extremely small and active.[12] The two *Vasa deferentia* open into the cirrhus-sac, which is perforated at its bottom to communicate with the *Vesicula seminalis exterior*.[13] From each testicle there passes off, also, a third *Vas deferens* which opens into the neck of the *Vesicula seminalis interior*.[14] The cirrhus-sac is pyriformly elongate, or round,[15] and the *Vesicula seminalis exterior* is always situated at its base. This last is prolonged, opposite the openings of the vasa deferentia, into usually a very long, tortuous *Ductus ejaculatorius*, which opens into a tubular penis.[16] There is one common genital opening for the penis and vagina which are usually side by side, and out of which the penis often considerably projects.[17] In most Trematodes, these two organs are located at the anterior extremity of the body, and only in *Holostomum*, and *Gasterostomum*, are they removed to the other extremity.[18]

---

8 The length of the uterus varies very much in different genera and species, and its coils are always irregular. With *Monostomum mutabile*, and *verrucosum*, the oviduct arising in the posterior extremity, passes in front with numerous transverse coils.

9 I have found one testicle only, in *Amphistomum subclavatum*, and *Aspidogaster conchicola*, although I have seen three or four in *Distomum appendiculatum*, and *cygnoides*.

10 With *Distomum ovatum*, the two testicles are side by side behind the ventral sucker ; with *Distomum chilostomum*, they are on each side of this sucker, and with *Distomum crassum*, mihi (from the intestine of *Hirundo domestica*), they are in front of it, on each side of the neck.

11 With *Distomum longicolle*, *lanceolatum*, *oxyurum*, *echinatum*, *globiporum*, and *Amphistomum conicum*, the testicles have many depressions ; see *Bojanus*, Isis, 1821, Taf. II. fig. 25–27 ; *Burmeister* and *Siebold*, in *Wiegmann's* Arch. 1835, II. Taf. II. 1836, I. Taf. VI. ; also *Laurer*, De Amphistomo conico. fig. 21, 24, 25. With *Amphistomum subtriquetrum*, *giganteum*, and *Distomum hians*, the number and depth of these depressions gives the testicle the aspect of a bundle of cæca ; see *Bojanus*, loc. cit. Taf. II. fig. 14–17, and *Diesing* Ann. d. Wiener Mus. I. Abth. 2, Taf. XXII.

12 In the testicles of the Trematodes, the development of the spermatic particles occurs after the usual mode.

The bundles which they form are separated in their passing the vasa deferentia, and they collect into irregular masses in the seminal vesicles.

Their extremely active movements cannot be perceived unless they are quite isolated. When put in water they become twisted together, and assume a loop-like arrangement, — their motions instantly ceasing.

For the development of the spermatic particles of the Trematodes, see *Kölliker*, Die Bildung du Saamenfäden in Bläschen, loc. cit. p. 44, fig. 31.*

13 These two vasa deferentia are sometimes blended together before reaching their destination ; this is so in *Distomum variegatum*, and *longicolle*.

14 The internal seminal vesicle is so extraordinarily large in *Distomum variegatum* that it exceeds that of the ovary and two testicles.

15 This cirrhus-sac, together with the penis, is very long with *Distomum lima*, *maculosum*, *variegatum*, and *ovatum* ; but it is especially so with *Aspidogaster conchicola*, and *Monostomum verrucosum*.

16 The protruding cirrhus or penis of *Distomum holostomum* is provided with small bunches ; and that of *Monostomum verrucosum* with numberless little warts.

17 When the penis is protruded, it may then be seen how the contents of the vagina are emptied at its base. When the common genital opening is closed, the very flexible penis can be turned into the vagina and there discharge its contents, and in this way the self-impregnation of these animals may occur.

18 The common genital opening is usually situated on the middle of the neck, and with *Distomum*, it is directly in front of the ventral sucker.

With *Distomum clavigerum*, and *ovatum*, it is upon the sides of the neck, and with *Distomum caudate*, and holostomum, exceptionably, it is on

---

ured the spermatic particles of *Polystomum appendiculatum* as Cercaria-form. — Ed.

In the terminal, constricted portion of the uterus, eggs, vitelline cells, and spermatic particles are often found mixed together. It is probably here that the eggs are formed, their fecundation occurring without copulation, and by means of the *Vesicula seminalis interior.* The succeeding folds of the uterus contain already, nicely-defined, oval eggs containing a germ and many vitelline cells. Their recently-formed envelope is still colorless, and so thin and flexible, that the peristaltic contractions of the uterus give it a variety of forms. But in passing from the uterus they lose this flexibility; their envelope becomes more solid, — of a yellow and then a brown color; and the whole, at the same time, undergoes a decrease in size, due probably to a condensation of their substance. The eggs of most of the Trematodes have an opercular opening at one extremity.[19]

In the Cestodes, the walls of the genital organs are so very thin, and so intimately blended with the parenchyma of the body, that their structure and relations have not yet been well made out.

With the exception of in *Caryophyllaeus,*[20] these organs are repeated many times one after another, having in the same individual different degrees of development. They are always most complete in the posterior portion of the body, being only rudimentary near the neck, while in the neck itself they do not exist at all. In the articulated Cestodes, each ring contains both male and female sexual organs; and in their two Groups, the arrangement of these is the same as in the Trematodes. It is probable that the ovaries and the secreting organs of the vitellus are separate.[21] In *Ligula, Triaenophorus,* and *Bothriocephalus,* the uterus consists, exactly as in the Trematodes, of a very tortuous tube filled with oval eggs.[22] But in

---

the posterior extremity of the body. Its position is indicated, even when the penis is not protruded, by a small papilla.

With *Octobothrium,* and *Polystomum,* there is a round muscular sac concealed directly behind this opening, which contains a circle of delicate horny ribs, the lower extremities of which are bifid and form a support like a bownet. *Mayer* (Beitr. loc. cit. p. 21, Taf. III. fig. 3, 6) has seen ten similar ribs with *Octobothrium lanceolatum.* I have found eight with *Polystomum integerrimum,* and forty with *Polystomum ocellatum.* Their use is wholly unknown to me.

19 The eggs of the Trematodes have apparently only a single envelope. Among the normal eggs in the uterus may often be found others which are malformed, also very irregular bodies of a yellowish or brown color, formed almost entirely of the substance of these envelopes. These bodies were most probably secreted by the walls of the uterus (the *Tuba Fallopii*) at a time when the ovaries and the secreting organs of the vitellus were inactive, so that the substance of the envelopes' was hardened before receiving their usual contents. With *Amphistomum subclavatum, Octobothrium lanceolatum, Polystomum integerrimum,* and *ocellatum,* and *Diplozoon paradoxum,* the eggs are very large, and in the last-named species their extremities are narrowed and lengthened into a spiral filament, wherefore one of these eggs has been taken for a testicle and penis; see *Nordmann* Microgr. Beitr. Hft. 1, p. 73, Taf. V. VI. fig. I, b.; also *Vogt,* in *Müller's* Arch. 1841, p. 34, Taf. II. fig. 11.

The eggs of *Monostomum verrucosum,* and some other species of this genus which live in the intestine of *Chelonia esculenta,* have a very dif-

ferent form; they are oval and colorless, and at each extremity have two papillae, which are gradually developed into very long, sharp appendages; see *Dujardin,* Hist. Nat. d. Helminth. Pl. VIII. fig. G, B. 3.*

20 With *Caryophyllaeus mutabilis,* there is only a single cirrhus-sac upon the ventral surface of the posterior body, and from which a delicate long penis often protrudes.

21 I think I have seen an ovary in each of the segments of *Bothriocephalus punctatus,* and *Taenia ocellata.* As such, ought, perhaps, to be regarded those organs which *Eschricht* (Nov. Act. Acad. Leop. XIX. Suppl. 2, Tab. I. fig. 2, c, c) has considered with *Bothriocephalus latus* to be ovaries. The organs secreting the vitellus are a mass of irregularly arranged granulations situated upon both the dorsal and the ventral surfaces, and which have very fine excretory ducts. This mass, called by *Eschricht* (loc. cit. p. 25, Tab. I. fig. 5) the ventral and dorsal granules, cannot, together with its excretory ducts, be made out, except when filled with the vitelline substance. With *Taenia ocellata,* the vitelline organs are limited to the sides of each segment, at the anterior border of which two main excretory ducts are easily seen; these form a single short canal in the middle of the body. In this same place are two transversely-placed oval sacs, and which are probably the two ovaries.

22 The uterine convolutions are generally in the middle of the body, and when filled with mature eggs, appear through the skin as a brown rosette; see *Eschricht* loc. cit. Tab. I. II. (*Bothriocephalus latus*).

---

* [§ 115, note 19.] See also for the structure of the genital organs *Thaer, Müller's* Arch. 1850,

p. 602, Taf. XX. fig. 17 (*Polystomum appendiculatum*). — ED.

Taenia, it is a reservoir, composed of numerous ramified cocca, and intimately blended with the parenchyma of the body.[23] The vagina is a narrow, muscular canal, which usually opens close to the penis by a special orifice (*Vulra*), or by a common genital opening (*Porus genitalis*).

It is difficult to decide whether the testicles, which always form the middle layer of the body, consist of a collection of inter-opening cacca, or of a single spirally-rolled tube. The cirrhus-sac with the *Vas deferens* opening at its bottom, is always very distinct. As in the Trematodes, it has a *Vesicula seminalis*, with a *Ductus ejaculatorius* and a muscular penis.[24] The contents of the different canals, the seminal vesicle and the ejaculatory duct, are always very active, filiform spermatic particles.[25] The genital openings are upon the middle of the ventral surface, or on the lateral borders of the body ; but in those species where the sexual openings are separate, they are lateral for the male, and ventral for the female.[26]

The eggs of the Cestodes, situated like those of the Trematodes in a spiral, pouch-like uterus, have also a similar structure. Their simple, oval, brownish-yellow envelope, has also, sometimes, an operculum. The eggs of *Taenia* have a very different structure ; the envelope is colorless, and of a very variable, and sometimes quite remarkable form.[27]

23 With most *Taeniae* the borders of the cellular uterus are very difficult to distinguish. But its lateral cacca with *Taenia ocellata*, and its arborescent divisions with *Taenia solium*, are very easily seen ; see *Delle Chiaje*, Compendio di Elmintografia umana, Tav. III. fig. 10.

24 The cirrhus-sac is either short and pyriform, or very long. With very many *Taeniae*, as with *Taenia amphitrica*, *lanceolata*, *multistriata*, *scolecina*, and *setigera*, the penis has numerous small spines pointing backwards ; see *Dujardin*, Hist. d. Helm. Pl. IX.-XI. That of *Taenia infundibuliformis* is surrounded with very large bristles ; and according to *Dujardin* (loc. cit. Pl. IX. B. 210) this is also true with *Taenia sinuosa*.

25 By very slight pressure, the spermatic particles contained in the *Vesicula seminalis* of the cirrhus-sac are pressed out through the penis ; this is so with *Bothriocephalus punctatus*, *latus*, *Taenia cucumerina*, *planiceps* (from the intestine of *Hirundo urbica*), *inflata*, *pectinata*, *serpentulus*, and *villosus*. As with the Trematodes, the spermatic particles here cease to move when put in water, and are twisted into loops.*

26 With *Ligula*, *Bothriocephalus nodosus*, *latus*, *claviceps*, *ditremus*, *punctatus*, and *tetrapterus*, the two genital openings are situated on each side of the ventral surface, while the penis protrudes from a special opening directly in front of the vulva ; see *Mehlis* in Isis, 1831, Taf. I. fig. 1, 2, and *Eschricht*, loc. cit. Tab. I. fig. 5. With *Bothriocephalus punctatus*, there are two pairs of these openings upon each segment, one under the other, but in *Bothriocephalus tetrapterus*, these are side by side. With *Triaenophorus*, *nodulosus*, and *Taenia ocellata*, the vulva is upon the ventral surface, and the penis upon the lateral border. With *Bothriocephalus fra-*

gilis, *proboscideus*, *rugosus*, and with most *Taeniae*, the cirrhus-sac and the vagina open by a common genital orifice upon the lateral border, and usually through a papilla. With *Taenia cucumerina*, and *bifaria*, mihi (from the intestine of *Anas leucophthalmus*), I have found an orifice of this kind upon the two lateral borders of each segment, and behind which were the genital organs.†

27 Although I have not seen either the germinative vesicle or dot in the eggs of the Cestodes, probably from their delicacy, yet I do not for a moment doubt their presence there, since *Kölliker* (*Muller's* Arch. 1843, p. 92, Taf. VII. fig. 44) has seen them in the eggs of a *Bothriocephalus*. Many species of this genus produce oval eggs which have a simple brown envelope. Of an oval form, but colorless, are those of *Caryophyllaeus*, *Ligula*, *Triaenophorus*, *Taenia literata*, and *scolecina*. Those of *Taenia amphitricha*, *bifaria*, *macrorhyncha*, *serpentulus*, and *serrata*, are round, and have two colorless envelopes ; this is true also of the oval eggs of *Taenia angulata*, *villosa*, &c. There are three of these envelopes with the round or oval eggs of *Bothriocephalus infundibuliformis*, *proboscideus*, *Taenia porosa*, *lanceolata*, *ocellata*, *setigera*, and *solium*. With *Taenia infundibuliformis*, and *planiceps*, each extremity of the envelope is lengthened into a long and delicate appendage. Two similar but fibrillated appendages exist upon those of *Taenia variabilis*. With *Taenia cyathiformis*, the external pyriform envelope of the eggs has, at its attenuated extremity, two round, bladder-like appendages. *Dujardin* (Hist. d. Helm. Pl. IX.-XII.) and I (*Burdach's* Physiol. 1837, II. p. 201) have seen many other forms with the eggs of *Taenia*. The round and doubly-enveloped eggs

* [§ 115, note 25.] I have observed the development of the spermatic particles with *Taenia*. They are developed in special cells, and before they have escaped, are therein coiled up resembling those of the coleopterous insects. They are simply filiform. — ED.

† [§ 115, note 26.] The Cestodes have been the objects of much careful study during the last few

years, by *Blanchard* (Ann. d. Sc. Nat. X. 1848, p. 321) and *Van Beneden* (Mem. Acad. Belgique, 1850, XXV.) and the sexual parts pretty clearly made out. They both agree that, internally, the male and female organs are wholly distinct, and therefore that impregnation of the ova must be by self-copulation. — ED.

§ 116.

In the Acanthocephali, the genital organs occupy a large portion of the cavity of the body. They arise in the posterior portion, and are supported by a *Ligamentum suspensorium*, which extends from this last to the base of the proboscideal sheath.

In the females, there are neither proper ovaries, nor an uterus; but in their place there are numerous oval, or round, flattened bodies of considerable size, which float freely in the liquid of the cavity of the body; they have nicely-defined borders, and are composed of a vesicular, granular substance, and, as eggs are formed within them, they may be regarded as so many loose ovaries.[1]

When the eggs have reached a certain size, they fall from the ovaries into the cavity of the body. At this time they are ovo-elongate, have only a single envelope, and contain both a vesicular and a finely-granular substance, but no trace of a germinative vesicle. They continue to increase in size, and two new envelopes are formed about them.[2] The muscular canal which passes off from the simple vulva which is situated at the posterior part of the body, may be regarded as a uterus.

At the point where it is attached to the *Ligamentum suspensorium*, it becomes a campanulate or infundibuliform organ, whose borders float freely in the cavity of the body, and thus the whole is comparable to a *Tuba Fallopii*. The bottom of this bell-shaped organ communicates with the superior extremity of the uterus by a narrow, valvular opening, which presents a lateral, semilunar fissure.

This whole organ is endowed with very active peristaltic motions, by which the loose contents of the cavity of the body are absorbed; and while the larger ovaries are thrown out, the little immature eggs are returned into the cavity of the body by the lateral fissure, — the more mature ones only, reaching the uterus.[3] This uterus, which is of variable length, opens outwardly through a very short and narrow vagina.

The males of *Echinorhynchus* have usually two oval or elongated testicles, one before the other, and attached to the *Ligamentum suspensorium*.

---

of *Taenia cucumerina* (Creplin, Observ. de Entozois fig. 12, 13) and *crateriformis*, have the remarkable arrangement of being grouped in tens to twenties, and each group is surrounded by a gelatinous envelope.*

1 The ovaries of *Echinorynchus* were formerly taken both for mature eggs, and for cotyledons; and to this is due the very inaccurate figures of them by *Westrumb* and *Cloquet* (loc. cit.). *Dujardin*, however (Hist. d. Helm. Pl. VII. fig. D. 6), perceived their true nature.

A state of development which I have observed with many females of *Echinorhynchus gibbosus*, would appear to throw some light upon the question as to the part of the body where the ovaries are first formed. Here the *Ligamentum suspensorium* had, over most of its extent, large granular globules, while the cavity of the body contained neither ovaries nor eggs. I think, therefore, that this ligament is the elementary material from which the ovaries are developed under the form of globules, which, being subsequently detached, continue their development in the liquid of the cavity of the body.

* [§ 115. note 27.] See *Van Beneden* (loc. cit. p. 67), who has observed the eggs of the Cestodes

2 The long eggs of many *Echinorhynchi* are formed by the same process. They are all colorless, and may be distinguished by the peculiar aspect of their middle envelope which at both extremities is constricted like a neck. But those of *Echinorhynchus gigas* form an exception; for they are shorter and oval, their middle envelope is yellowish, and, like the two others, has externally numberless small obtuse spines. With *Echinorhynchus strumosus, hystrix, angustatus,* and *proteus,* the external envelope of the eggs presents the peculiar phenomenon that when pressed between two plates of glass, it separates into very fine fibrillae.

3 The nature of this campanulate *Tuba Fallopii* has been wholly mistaken by *Bojanus, Westrumb* and *Cloquet*. *Burow* (Echinorhynchi strumosi Anat. p. 22, fig. 1. g, fig. 6) was the first to describe it, without however conveying the correct idea. See my description (*Burdach's* Physiol. loc. cit. p. 197), which has been confirmed since by *Dujardin* (Hist. d. Helm. p. 495, Pl. VII. fig. D. 5).

composed like those other animals,— with a germinative vesicle, &c. — ED.

They send off two varicose *Vasa deferentia* to the posterior portion of the body, where, after uniting very probably with the neck of an odd elongated vesicle (*Vesicula seminalis?*), they are prolonged into a copulatory organ.[4] There are six pyriform bodies,which secrete a finely-granular substance, and are attached behind the testicles to the *Vasa deferentia*. Their six excretory ducts successively unite, ending finally in two which open into the copulatory organ.[5] The penis is usually folded inward, but when projecting outwardly, it is a muscular, cup-shaped appendage, whose fossa receives the posterior portion of the body of the female during copulation.[6]

The spermatic particles are developed after the usual mode; they are filiform and very active, and quickly die in water, interlooping and twisting together.[7]

The very adhesive, viscous, yellowish-brown wax-like substance, often found about the vulva, is apparently the secretion of the pyriform bodies during copulation.[8]

§ 117.

With the Nematodes, the genital organs consist of a long, simple or partly double caecal tube, which winds around the straight intestine.

In the female it has the following parts: *Ovarium, Tuba Fallopii, Uterus,* and *Vagina;* and in the male, *Testes, Vas deferens, Vesicula seminalis,* and *Ductus ejaculatorius.*

With *Trichosoma, Trichocephalus,* and *Sphaerularia,* the genital tube is simple in the females, and usually so in the males. But in *Filaria, Ascaris, Strongylus, Spiroptera, Oxyuris,* and *Anguillula,* the ovary, Fallopian tube, and uterus, are double.[1] In the females, the ovary is the posterior portion of this genital tube, and in its terminal portion are small round

4 With *Echinorhynchus strumosus,* these two round testicles are side by side. Having always found the odd, long vesicle empty, I cannot decide whether or not it serves the function of a seminal vesicle.

5 These six pyriform bodies were formerly taken for seminal vesicles; see *Westrumb,* de Helminth. Acanthocephalis, p. 55. Tab. III. fig. 24; and *Nitzsch,* in *Ersch* and *Gruber's* Encyclop. VII. 1821, plate for the Acanthocephala, fig. 2, 3, i. With *Echinorhynchus claviceps,* I have found only one of these bodies.

6 The copulatory organ, which protruded has mostly an oblique direction, has been very exactly figured by *Dujardin* (Hist. d. Helm. p. 493, Pl. VII. fig. D, 1, D, 2).

7 For the spermatic particles of the Acanthocephali, see my observations in *Müller's* Arch. 1836, p. 252.

8 This waxy substance incrusts sometimes the whole caudal extremity of females; this is so with *Echinorhynchus gigas,* and *globocaudatus;* see *Cloquet* (Anat. &c. &c. p. 100, Pl. VIII. fig. 4, 5) and *Nitzsch* (*Wiegmann's* Arch. 1837, I. p. 64.*

1 For the simple genital tube with its various parts of the female of *Trichocephalus dispar,* see *Mayer,* Beitr. &c. Taf. II. With *Filaria rigida,*

* [§ 116, note 8.] For some further details on the genitalia of the Acanthocephali, see *Blanchard* (Ann. d. Sc. Nat. 1849, XII. p. 23), and *Règne*

11*

and *Ascaris pauciparo,* I have found the female organs likewise simple. When these organs are double, either one uterus with its ovary and oviduct passes in front from the simple vagina, while the other passes behind, as is the case with *Ascaris brevicaudata, nigrovenosa, Oxyuris vermicularis, Spiroptera anthuris. Strongylus auricularis,* and *striatus;* or both pass side by side behind, as in *Ascaris aucta, mystax, lumbricoides* (*Cloquet,* Anat. &c. Pl. I. fig. 2) and *ausculata.* With *Cucullanus elegans,* and *microcephalus* (from the intestine of *Emys lutaria),* the uterus alone is double; one horn terminating posteriorly in a caecum without an ovary or Fallopian tube, while the other, which has these parts, passes in front. There are, moreover, species of *Ascaris* into whose vagina open three or four genital tubes. Thus with *Ascaris microcephala,* I have seen the uterus divide upon reaching the vagina into three tubes, each having an ovary and oviduct. According to *Nathusius* (*Wiegmann's* Arch. 1837, I. p. 57), the uterus of *Filaria labiata,* which is at first simple, divides at its posterior extremity into five tubes.

The double uterus of *Strongylus inflexus* has, posteriorly, numerous constrictions, giving it a moniliform aspect.

animal nouv. edit. Zoophytes, Pl. XXXV. fig. 3b, 3¹, 3¹, 3², 3². — ED.

cells; in the anterior portion, these cells are more numerous and begin to be surrounded by a granular vitelline substance, in which the primitive nucleated cells are still seen; these cells therefore, ought perhaps to be regarded as germinative vesicles. In front, these eggs, which are of a discoidal form, are arranged in a row, or are grouped closely around a rachis which traverses the axis of the ovary. In the Fallopian tube, which may be known by its less diameter, the eggs become more mature, and, having been surrounded by a double colorless envelope, pass into the base of the uterus.[2] This last is the largest portion of the genital tube, and is distinguished by its well-marked power of peristaltic action. The vagina, which is distinguished from the uterus by its narrowness and its muscular walls, opens at very different points of the body. Generally, as for instance in *Ascaris, Spiroptera, Strongylus, Oxyuris, Cucullanus,* and *Trichocephalus,* the *Vulva,* consisting of a transverse fissure, and often surrounded by a very remarkable fleshy swelling, is situated either a little in front of, or near the middle of the body; but sometimes it opens just in front of the anus.[3] The sperm is usually so accumulated in the bottom of the uterus, that this is probably the locality of fecundation.[4]

In the males, the posterior portion of this tube is the testicle; another portion of it, which is short and constricted, is the *Vas deferens,* which passes into a dilated portion,— the *Vesicula seminalis.* Usually this last is separated by a constriction from the *Ductus ejaculatorius,* which opens into another muscular tube (sheath of the penis).[5] At the anterior portion of this last, is a horny, copulatory apparatus. The simple or double penis is of variable length, and is protruded by the muscular contractions of its sheath through the external opening, which is always situated at the poste-

---

2 The formation of eggs in various Nematodes has been described by *Siebold* (*Burdach's* Physiol. loc. cit. p. 208), by *Bagge* (Dissert. de. Strongylo, &c., fig. 1–5), and *Kölliker* (*Müller's* Arch. 1843, p. 69, Taf. VI. fig. 20). I have found a rachis in the ovaries of *Ascaris aucta, lumbricoides, mystax, osculata, Cucullanus elegans,* and *Strongylus inflexus.* The eggs of these, while yet immature and flattened, have a point on one of their extremities by which they are attached to the rachis.

With those of *Ascaris lumbricoides,* this point is very long during a certain period of development, and the opposite end has many deep sulcations, giving it a remarkable appearance; see *Heute,* in *Müller's* Arch. 1835, p. 602, Taf. XIV. fig. 11.

In the mature eggs, which are nearly always oval, it is rare that the double colorless envelope can be clearly perceived. With *Trichosoma,* and *Trichocephalus,* there is a short diverticulum at each extremity of the egg. But in *Ascaris dentata,* there is at this same place a long fibrillated filament; see *Mayer,* Beitr. Taf. II. fig. 8, and *Kölliker,* in *Müller's* Arch. 1843, Taf. VI. fig. 16–19.*

3 With *Ascaris dactylaris, Cucullanus elegans, Strongylus nodularis,* and *striatus,* the borders of the vulva appear quite swollen. With *Trichosoma,* this swelling is so attached to the vulva as to resemble a prolapsus of the vagina (*Dujardin,* Hist. d. Helm. Pl. I.).

With *Filaria attenuata, inflexo-caudata,* mihi (from the pulmonary cysts of *Delphinus phocaena*), and *papillosa* (see *Leblond,* Quelques matériaux pour servir à l'histoire des Filaires et des Strongles, 1836, Pl. II. fig. 1), the vulva is at the side of the mouth.

With *Strongylus paradoxus,* it is swollen to the form of a bladder, and is situated near the caudal extremity; while that of *Ascaris paucipara* is directly upon the anus.

4 See *Bagge,* loc. cit. p. 12; and *Kölliker,* in *Müller's* Arch. 1843, p. 72.

5 For the male genital tube, see *Mayer,* Beitr. Taf. I., and *Cloquet,* Anat. &c. Pl. I. fig. 5, Pl. II. fig. 8. As yet I have observed only a few exceptions to this typical form with male Nematodes.

With *Filaria attenuata,* the posterior portion of the testicle is bifurcate, and with *Ascaris vesicularis,* there are two moderately large caecal prolongations which arise from the *Vesicula seminalis* at the place where it empties into the *Vas deferens.*

---

* [§ 117, note 2.] Primitively, the ova of *Ascaris* consist of nucleolated cells, which are polyhedral from mutual pressure. These increase in size gradually, in their passage down towards the oviduct, and the granules of the liquid lying between the nucleus or germinative vesicle and the cell-wall become developed into cells, and in this way the mature ova are formed. Probably no better opportunity is afforded to perceive that morphologically the ovum is at first only a nucleolated or nucleated cell; see *Leidy,* loc. cit. p. 43, Pl. VII. fig. 14, c. — ED.

rior portion of the body.[6]   It has a great variety of forms, and from its sheath arise two antagonistic muscles, which are inserted at its base.[7] The spermatic particles, which are always motionless, have usually a cell-form, or, at least, are never filiform corpuscles.[8]   For aiding the union of the sexes during copulation, the males have lobular appendages, papillae. and suckers, situated about the genital opening. Without doubt, the spiral posterior extremity also of the animal, is often used for the same purpose. Moreover, in many instances, there is secreted a wax-like substance intended to fasten the two sexes together.[9]

6 According to *Leblond* (loc. cit. p. 20, Pl. III. fig. 1), both the male and female genital openings with *Filaria papillosa* are quite near the oral orifice. I have been unable to confirm this observation, at least with *Filaria attenuata*, *inflexo-caudata*. and another species found in the thoracic cavity of *Sturnus vulgaris*.

7 With *Trichocephalus*, and *Trichosoma*, the penis is simple and very long, and, beside the muscular sheath, has another which is membranous, and sometimes covered with small spines pointing backwards. This sheath, being folded outwards when the penis is protruded, is comparable to a *Praeputium* ; see *Mayer*, Beitr. loc. cit. Taf. I., and *Dujardin*, Hist. d. Helm. Pl. I.-III. With nearly all the other Nematodes the penis is double. It is very long with *Ascaris acuminata, brevicaudata, depressa, spiculigera,* and *Strongylus paradoxus* ; but is very short with *Ascaris ensicaudata, semiteres, Cucullanus elegans, Filaria attenuata, inflexo-caudata, Spiroptera anthuris,* and *Strongylus inflexus.* With *Spiroptera*, the two penises are of unequal length, and with *Ascaris paucipara, brevicaudata,* and *Strongylus*, there is an additional horny piece like a third penis.

With most Nematodes, the penises are sulcated, and those of *Strongylus* have a singular form due to the presence of numerous appendages. The two delicate, retractor muscles of this organ, arise from the internal surface of the cavity of the body, and when the penis is double there are two pairs.

With *Ascaris osculata, vesicularis,* and *spiculigera*, I have found these four muscles very long. See upon the penis of the Nematodes, *Mayer*, Beitr. Taf. I., and *Dujardin*, Hist. d. Helm. Pl. I.-VI.

8 For the spermatic particles of the Nematodes, see *Bagge*, Dissert. de Strongylo, &c., p. 12, fig. 27, 28. The development of these cell-like spermatic particles may be easily observed with *Ascaris paucipara*, where the parent-cells are very large. In the posterior end of the testicle the

nuclei with their nucleoli are first formed ; afterwards these nuclei are surrounded by a finely-granular substance around which the cell-membrane is formed.

In this state the testicle exactly resembles an ovary filled with germinative vesicles and eggs. Still later, the parent-cell membrane increases more and more, and the granular substance is found only upon the internal surface of the cell. During these changes, the nucleus which resembles a germinative vesicle, is transformed into a long, solid, and neatly-circumscribed corpuscle. With *Strongylus auricularis*, the spermatozoal [daughter ?] cells are pyriform ; and with *Oxyuris ambigua* their form is similar (*Külliker*, loc. cit. p. 73, Taf. VII. fig. 26).

It is very probable that *Mayer's* assertion (Neue Untersuch. aus dem Gebiete der Anat. u. Physiol. 1842, p. 9) that he had seen thread-like spermatic particles with *Oxyuris vermicularis*, has led *Külliker* to regard these pyriform cells as so many bundles of filamentoid spermatic particles. But never have I seen filaments of this kind in the Nematodes.

The pyriform spermatic particles of *Strongylus auricularis*, which have a short peduncle, as well as the round, cell-like, and nucleated ones of *Ascaris acuminata*, have been figured by *Reichert* (Beitr. zur Entwickel. der Saamenkörp. bei den Nematoden). This same naturalist has shown that these spermatic particles arise by endogenous generation, by fours in each cell ; see *Muller's* Arch. 1847, p. 88, Taf. VI.*

9 The large caudal valve of the male *Strongylus*, and the spiral tail of the male *Spiroptera*, may be here instanced. With very many male *Ascaris*, there are two rows of papillae upon the sides of the genital opening, and with *Ascaris vesicularis*, and *inflexa*, I have found a copulatory sucker directly in front of this opening. The male of *Hedruris androphora* winds himself about the female during copulation, and the caudal valve of the male *Strongylus trachealis* glues itself so

* [§ 117, note 8.] The statement here made that *Reichert* has observed the development of the spermatic particles of an *Ascaris* by fours in each cell, deserves attention from its histological relation. According to my own observations, the histological formative conditions of the development of the spermatic particle are exactly analogous to those of the development of the embryo. The nucleus of the sperm-cell divides or segments like the vitellus of the ovum, and this process continues until the sperm-cell which has now attained a large size, is filled with numerous small nucleated cells (daughter-cells) ; and the nucleus of these last is changed into the spermatic particle.

I think, therefore, that, invariably, the spermatic particle is only a metamorphosed nucleus of a

daughter-cell (see my Memoir, The Origin, Development, and Nature of the Spermatic Particles in the four classes of Vertebrata, in the Mem. Amer. Acad. of Arts and Sc. V. 1853). The view of *Reichert*, therefore, that four spermatic particles are here formed in one cell, does not appear to me admissible, although I have no observations upon the instance in question. It appears to me explicable in this way : the nucleus of the parent sperm-cell underwent here only a second segmentation, thereby only four daughter-cells being produced. The nucleus of each of these became a spermatic particle, and these four particles passed into the cavity of the parent-cell. *Reichert* therefore, probably saw four spermatic particles in a parent and not in a daughter cell. — Ed.

'The few observations hitherto made upon the genital organs of the Gordiacei have shown that they are wholly tubular as in the Nematodes. But their intimate structure, and the development of their spermatic particles are so strikingly different, that this point alone would justify their separation from the Nematodes. [10]

<h2 style="text-align:center">§ 118.</h2>

With the exception of the Nematodes, and Gordiacei, the development of all Helminthes, which reproduce by means of genital organs and eggs, is metamorphotic. A complete series, from beginning to end of these metamorphoses has yet never been observed with any species. From the separate parts of it here and there which have been observed, there appears the remarkable fact, that the embryos after escaping the egg, are not always changed at the end of the metamorphosis, into individuals like the parent, but appear as larva-like animals, capable in their turn of producing other larvae. These last larvae alone, change into individuals, which are like the parent.

This particular kind of transformation and development which is quite common among the Trematodes, has received the name of *Alternate Generation*. [1] Whether it occurs among the Cestodes and Acanthocephali, cannot now be stated positively, for as yet we are unacquainted with the first period of their metamorphosis, — the embryo as it escapes from the egg. [2] In many Cestodes and Trematodes, the embryos are developed before the eggs are cast, and in some of the last order, they make their escape while the eggs are in the uterus.

The development of the Cestodes occurs as follows: After the disappearance of the germinative vesicle, large, transparent embryonic cells appear in the midst of the vitellus, which undergoes fissuration. These multiply by division, increasing at the expense of the vitellus, which in the

tightly to the vulva of the female in this act, that they cannot disengage themselves (*Siebold* and *Nathusius*, in *Wiegmann's* Arch. 1836, I. p. 105, Taf. III. 1837, I. p. 60, 66). With many other species of *Strongylus*, and *Ascaris*, it is not rare to find a brownish gum about the vulva, and in which there is, sometimes, the very distinct impress of the male caudal valve (*Mehlis*, Isis,1831, p. 87).*

10 In the genus *Mermis* formed by *Dujardin*, the tubular uterus, the muscular vagina, and the vulva situated far from the caudal extremity, — all remind one much of the Nematodes. The eggs of *Mermis nigrescens*, like those of *Ascaris dentata*, have long fibrillated appendages (*Dujardin* Ann. d. Sc. Nat. 1842, XVIII. p. 134, Pl. VI., and *Siebold*, in *Wiegmann's* Arch. 1843, II. p. 300); and at the caudal extremity of the males of *Mermis albicans*, mihi (Entom. Zeit. 1843, p. 79), there are, as in most Nematodes, two horny penises.

But with *Gordius*, the structure of the genital organs is very different (see *Siebold* and *Dujardin*, loc. cit.). In both sexes the cavity of the body is completely filled with a double genital tube, straight, and simple posteriorly, the sides of which are formed of large cells. The genital

opening is always at the posterior extremity of the body. The testicular tubes of *Gordius aquaticus* contain anteriorly, cell-like bodies; but posteriorly there are others, staff-like, and which, being found among the eggs in the uterine tube, I have regarded as perfect spermatic particles. The genital opening of the male *Gordius* is between the two more or less prominent lobes of the caudal extremity, and is without copulatory organs. The simple, round, colorless eggs, are bound together at the posterior part of the uterus by an albuminous substance, and are deposited in a very long row. It is this row of eggs which *Leon Dufour* has described as *Filaria filariae* (Ann. d. Sc. Nat. XIV. 1828, p. 222, Pl. XII. fig. 4).

1 See *Steenstrup*, Ueber den Generationswechsel, &c., 1842.

2 In various marine fish there is a trematode larva of a *Tetrarhynchus* (*Miescher*, Bericut ueber die Verhandl. d. Naturforsch. Gesellsch. in Basel. 1840, p. 29, and in *Wiegmann's* Arch. 1841, II. p. 302), which would lead one to conclude that alternate generation exists also with the Cestodes.]

* [§ 117, note 9.] For many details of the reproductive organs of *Ascaris infecta*, with beautiful illustrative figures, see *Leidy*, A Flora and Fauna, &c., loc. cit. 4 B. Pl. VII. 14, 16, b. 19. — ED.

†[§ 118, note 2.] The view here suggested of

the alternating generation of the Cestodes, has recently been confirmed most thoroughly by *Siebold*, who has treated the subject in a most comprehensive manner, in a Memoir in *Siebold* and *Kölliker's* Zeitsch. II. 1850, p. 198. — ED.

end they completely replace. When this has taken place, there is a mass of extremely small cells, which, being covered with a delicate epithelium, form a round or oval embryo, upon one extremity of which there are gradually formed six small horny hooks.[3]

The embryos of the Acanthocephali are perhaps developed in the same manner, but they have only four hooks.[4]

The Trematodes are developed exactly like the Cestodes, excepting that their oval embryos have usually ciliated epithelium, and there is an oral sucker in place of the hooks.[5]

Beside this first period of development, or embryonic state, there are other more advanced or larval states, during which many Helminthes have been described and figured as separate species in the science.[6] Among these may be especially noticed two forms of the Trematodes — the cylindrical and the cercarian larvae. The first (the germinative tubes of Baer), form one of the phases of the alternate generation, and have a more or less complete organization. In the cavity of their body, germinative corpuscles are formed; these consist of a vesicular, granular substance, and resemble eggs neither by their structure nor mode of development.

These corpuscles produce larvae of a cylindrical or cercarian form, which, deprived of their tail, are changed into perfect animals which have genital organs; and thus the series of metamorphoses is terminated.[7]

3 For the embryonic development of *Bothriocephalus*, and *Taenia*, see Siebold (Burdach's Phys. loc. cit. p. 200), *Dujardin* (Ann. d. Sc. Nat. X. 1838, p. 29, Pl. I. fig. 10, also XX. 1843, p. 341, Pl. XV., and his Hist. d. Helm. Pl. IX.-XII.), and *Kölliker* (*Muller's* Arch. 1843, p. 91, Taf. VII. fig. 44-56).

The small hooks which the cestoid embryos so actively protrude and retract, somewhat resemble those which are circularly arranged with the adult *Taenia.*

4 As yet, with *Echinorhynchus gigas* alone have I succeeded in liberating the embryos from the egg by compression. The four hooks of these embryos resemble, by their form and position, those of the Cestoid embryos. It does not appear, however, that the embryos of all *Echinorhynchus* have them; at least *Dujardin* has not found them with those of *Echinorhynchus transversus*, and *globocaudatus* (Hist. d. Helm. Pl. VII.).

5 For the embryonic development of *Monostomum*, and *Distomum*, see *Siebold* (Burdach's Phys. loc. cit. p. 206), and *Kölliker* (*Muller's* Arch. loc. cit. p. 99). The embryos which swim about like Infusoria by means of ciliated epithelium, and which escape the egg while yet in the uterus, have been observed of *Distomum hians*, by *Mehlis* (Isis 1831, p. 190); of *Distomum nodulosum* and *globiporum*, by *Nordmann* and *Creplin* (Microgr. Beitr. Hft. 2, p. 139, and in *Ersch* and *Gruber's* Encyclop. XXIX. 1837, p. 324); of *Distomum cygnoides*, *longicolle*, *Amphistomum subclavatum*, and *Monostomum mutabile*, by myself (*Wiegmann's* Arch. 1835, I. p. 66, Taf. I.). See also *Dujardin*, in the Ann. d. Sc. Nat. VIII. 1837, p. 303, Pl. IX. fig. 3. I have seen the embryos of *Distomum tereticolle*, and *Aspidogaster conchicola*, without the ciliated epithelium.

Those of *Distomum longicolle*, *cygnoides*, *Monostomum mutabile*, and *Aspidogaster conchicola*, have an oral sucker. In this last species, there is another sucker also, at the posterior extremity of the body (*Dujardin*, Hist. d. Helm. p. 325).

6 In this category are the genera *Cercaria*, *Histrionella*, *Bucephalus* and others, which as yet have been founded only upon different species of Trematode larvae. The Helminth described by *Leblond* (Ann. d. Sc. Nat. VI. 1836, p. 289, Pl. XVI. fig. 3) as *Amphistomum ropaloides*, is only a larva of a *Tetrarhynchus*. The species forming the genus *Scolex* are certainly only imperfect *Bothriocephalus*; and the *Gryporhynchus pusillus* of *Nordmann* (Micr. Beitr. Hft. 1. p. 101, Taf. VIII. fig. 6, 7), is probably only a young *Taenia*. There may also be a doubt here, if the Cystici can be considered as real species.

It is very probable that they are imperfect Cestodes whose genital organs are to be afterwards developed, as with *Cysticercus fasciolatus*, while the Rodents in which it lives are devoured by carnivorous animals. *Taenia crassicollis* is, perhaps, to *Cysticercus fasciolaris*, what *Bothriocephalus nodosus* is to *Bothriocephalus solidus*; see *Creplin*, Nov. Observ. &c. p. 90.

7 The cylindric larvae of the Trematodes have been termed by *Steenstrup* (loc. cit. p. 50) *nurses* (Ammen). They are yet known only as living parasitically upon Mollusks, as for instance, upon *Paludina*, *Lymnaeus*, *Planorbis*, *Ancylus*, *Succinea*, *Anodonta*, and *Unio*; also upon *Helix pomatia*, and *Tellina baltica*, according to *Bojanus*, *Baer*, *Carus*, *Steenstrup*, and myself. The cylindric larvae of *Bucephalus polymorphus*, are very long tubes, varicose here and there, sometimes ramified, and which do not exhibit any

* [§ 118, note 3.] The history of all our best embryological studies shows that the segmentation of the vitellus is the invariable preface to the beginning of development with all true ova. In the case of the Cestodes, if, as above mentioned, there is no such process, it is highly probable that such

development is not from true eggs but rather from buds, a view which is the more worthy of attention from the recent developments made by *Siebold* with *Gyrodactylus*; see below, my note under § 118, note 7. — Ed.

§ 119.

With the Nematodes, of which very many are viviparous, the embryos are developed within the egg in two different ways: Either the embryo-

movements (*Baer*, in the Nov. Act. Acad. Leop. XIII. pt. 2, p. 570. Tab. XXX.). Those of *Distomum duplicatum* have simple, oval, and rigid germinative utricles (*Baer*, Ibid. p. 558, Tab. XXIX.). Those of *Cercaria ephemera*, are also very simple, but of a cylindrical form (*Siebold*, in *Burdach's* Phys. loc. cit. p. 187, and *Steenstrup*, loc. cit. p. 78, Taf. III. fig. 1–6). Those of *Cercaria furcata* are simple and cylindrical, but very long and endowed with quite active peristaltic motions (*Baer*, loc. cit. p. 626, Tab. XXXI. fig. 6). The curious animal, *Leucochloridium paradoxum*, consisting of only a cylindrical sac with a tail, is only a trematode larva (*Carus*, in the Nov. Act. Acad. Leop. XVII. pt. 1, p. 85. Tab. VII.). With the slow-moving, cylindrical, orange-colored nurses of *Cercaria ephemera*, there may be easily seen a mouth, a pharynx, and a simple cœcal intestine (*Siebold*, in *Burdach's* Phys. loc. cit. p. 187). Those of *Cercaria echinata*, are similar, but they have also two short oblique prolongations in front of the obtuse caudal extremity (*Baer*, loc. cit. p. 629, Tab. 31, fig. 7, and *Steenstrup*, loc. cit., p. 51, Taf. II. fig. 2–4). The germinative bodies from which *Cercaria* is developed, have nothing comparable to a chorion or germinative vesicle. Their larvae have always a tail, which is simple (*Cercaria armata, ephemera, Distomum duplicatum*), or bifurcated (*Cercaria furcata*), or double (*Bucephalus polymorphus*). The movements of this tail are very slow with *Distomum duplicatum*, but extremely lively and vertical with *Cercaria*. With *Bucephalus*, the two filiform tails lengthen and shorten considerably, at the same time jerking all about.

When the larvae are developed, they leave the corpuscles and pass into other animals to complete their final metamorphoses. Many *Cercariae* appear to prefer the larvae of insects whose bodies they enter by means of their cephalic hooks. In this way I have seen the *Cercaria armata* easily enter the larvae of *Ephemera, Nemura*, and *Perla*. By the aid of its sting it can perceive the intersegmental membrane of these larvae. Frequently it loses its tail in passing through a narrow opening it has made.

Immediately upon reaching the cavity of the body of the larva, it is surrounded by a vesicular membrane, in which the sting is rejected, and the animal enters upon its final metamorphosis. But I have a doubt whether it is there completed, for among the numerous similar parasites which I have found in the most different insects whose larvae are

aquatic, as of *Libellula, Agrion, Ephemera*, and *Phryganea*, I have never met with one whose genital organs were in a state of advanced development.

The full development of these organs, the delicate contours of which may be seen while the parasites are in the bodies of these animals, is not perhaps attained, until the insects have been swallowed by birds and other animals, — being thereby furnished with more proper conditions for their complete formation.

Some *Cercariae* lose their tail and are surrounded with a capsule without leaving the Mollusks which are their first habitat. This is probably so, because these Mollusks are liable to be eaten by aquatic birds, in which these parasites may properly reach their final development. It should, however, be remarked that when these larvae become chrysalides, their investing capsule or cyst, is a secretion from their bodies, and not a product of the animals in which they live. It is probable that very many of these larvae never attain a perfect state, for, in their migrations, they fail to reach their destined and final habitat.

These migrations undoubtedly occur with many Cestodes while young; at least *Miescher* (loc. cit.) has observed it with *Tetrarhynchus*. But although we have followed these in their migrations, and the transformation of many of them into *Monostomum* and *Distomum* has been observed, and therefore the completion of their metamorphoses, yet we are but slightly informed as to their beginning by the alternation of generation.

There is yet little known as to the manner in which these embryos are changed into the cylindric nurses. There are now only two isolated facts throwing light upon this point. According to my own observations (*Wiegmann's* Arch. 1835, I. p. 75, Taf. l.), each embryo of *Monostomum mutabile* contains a germinative tube, which, at the death of the embryo, is freed and quite resembles the nurse of *Cercaria echinata*. I have also observed in the embryos of *Amphistomum subclavatum* a tubular body, but I could not satisfy myself of its germinative nature. According to *Steenstrup* (loc. cit. p. 98), there is an animal like a *Paramaecium*, and probably an embryo of a *Distomum*, living in Muscles, and which finally is deprived of its epithelium, and changed into the rigid, germinative tube of *Distomum duplicatum*; see upon this, my Jahresbericht in *Wiegmann's* Arch. 1843, II. p. 300.[*]

---

* [§ 118, note 7.] In this connection should be noticed the remarkable phenomena of reproduction with *Gyrodactylus* as recently observed by *Siebold* (*Siebold* and *Kölliker's* Zeitsch. I. 1849, p. 345). Individuals are here developed viviparously as in the so-called alternating generations, and *Siebold* has observed a mother in which was a daughter and in this last a grand-daughter, the series being therefore three-fold. These viviparous individuals contain no sexual organs proper, but the new individual is developed out of a group of cells situated within the body. The whole reproductive conditions which *Siebold* has detailed with his usual care appear to me to closely resemble those of the viviparous Aphides which I have

recently investigated; and I believe this mode of reproduction to be only a peculiar form of gemmiparity or budding suited to some ulterior, economical purpose of the animal's life. On a future page I shall speak more fully on this point and attempt to show that the whole set of phenomena known under the name of "Alternation of Generations" is, when divested of its paraphernalia, only a kind of Gemmiparity.

See also for further details on that curious animal *Leucochloridium paradoxum*, *Piper*, in *Wiegmann's* Arch. 1851, I. p. 313, but especially *Siebold*, in *Siebold* and *Kölliker's* Zeitsch. IV. 1853, p. 425, Taf. XVI. B. This last-named observer has shown that this animal form is only a

nic cells present the same successive phases as in the Cestodes and Trematodes, without the appropriated vitellus undergoing any segmentation ; or, the whole vitellus after a complete segmentation, is changed into an embryo.[1]

In both cases, the embryo has the parent's form. A muscular œsophagus and straight intestine appear in its body in the midst of the refuse vitelline granules ; and thus the young animal attains its perfect state by simple increase and by the development of its genital organs, but without any metamorphosis.[2]

From the few observations hitherto made upon the development of the Gordiacei, it appears that the embryos exactly resemble the parents.[3]

---

1 *Kölliker* was the first to call the attention to these two types of development with the Nematodes (*Müller's* Arch. 1843, p. 68, Taf. VI. VII.). With *Ascaris dentata, Oxyuris ambigua,* and *Cucullanus elegans,* free embryonic cells are formed in the vitellus without its fissuration. But there is a complete segmentation with *Ascaris nigrovenosa, acuminata, succisa, osculata, labiata,* and *brevicaudata, Strongylus auricularis, dentatus, Filaria inflexo-caudata, rigida,* and *Sphaerularia bombi.* After I had already noticed this vitelline segmentation with the Nematodes (*Burdach's* Phys. loc. cit. p. 211), which *Bagge* (Dissert. loc. cit.) described very fully, *Kölliker* (loc. cit.) attempted to reconcile it with the cell-theory, by regarding the cells which appear in the segmented, vitelline globules, as the embryonic cells, and in the multiplication of which by segmentation, the enveloping vitellus participates.

2 It appears that, as with the Trematodes, so in the Nematodes, a migration of the young precedes their complete development.

In the tissues of the most different insects and vertebrates, there are found small Nematodes without genital organs, and contained in a cyst. They could not get there except by a migration, and they cannot attain the full development of

their genital organs or their bodies in general, except through a transplantation upon other animals ; exactly as occurs with the trematodal larvae. (See the observations of *Creplin* and myself upon the sexless Trematodes, in *Wiegmann's* Arch. 1838, I. p. 302, 373.)

The *Trichina spiralis* of man is undoubtedly an encysted and imperfect form of one of the Nematodes, and in which one may seek in vain for genital organs. Some of these Nematodes appear to increase in their cysts without their genital organs being developed in the same proportion. Thus, the *Filaria piscium* are sometimes found very large, while their genital organs are very little developed ; and these last do not probably attain their perfect state, until, as with *Bothriocephalus solidus,* these worms have passed into other animals. For the same reason, I agree with *Steenstrup* (loc. cit. p. 113), who doubts that the *Filaria piscium* become, as *Miescher* has affirmed (loc. cit. p. 26), a globular capsule out of which there afterward appears an animal at first resembling a Trematode, but which finally becomes a *Tetrarhynchus.**

3 See *Dujardin* (Ann. d. Sc. Nat. XVIII. loc. cit. Pl. VI. fig. 15, 16) upon *Mermis nigrescens,* researches which I have been able thoroughly to confirm.†

---

kind of nurse of a *Distomum,* containing peculiar germ-bodies which are developed into *Distomum.* But the most important result obtained is that all *Distomum* are not developed by means of a cercarian, larval stage,—the economy of some making it seemingly requisite that the developmental process should be more direct. — ED.

* [§ 119, note 2.] In regard to *Trichina spiralis,* the various researches upon its structure, made in England and America, would show that it is a true animal having genital organs. The following are some of the references upon this subject : *Owen,* London Med. Gaz. April and December, 1835, or Transact. Zool. Soc. London, IV., or Cyclop. Anat. and Phys. Art. *Entozoa; Wood,* London Med. Gaz. May, 1835 ; *Farre,* Ibid. December, 1835 ; *Harrison,* Report of the Brit. Assoc. for the Advancem. of Sc. 1835 ; *Knox,* Edinb. Med. and Surg. Jour. 1836, XLVI. p. 86 ; *Hodg-*

kin, Lect. on Morbid Anat. of Serous and Mucous Membranes, I. p. 212 ; *Curling,* London Med. Gaz. February, 1836 ; *Bowditch,* Boston Med. and Surg. Jour. April, 1842 ; *Luschka, Siebold* and *Kölliker's* Zeitsch. III. 1851, p. 69, Taf. III., and *Gairdner,* Edinb. Monthly Jour. of Sc. May, 1853. The subject is one that deserves especial attention from Helminthologists. — ED.

† [§ 119, note 3.] *Grube (Wiegmann's* Arch. für Naturgesch. 1849, p. 358) and *Leidy* (Proc. Acad. Sc. Philad. V 1850, p. 98) have observed the development of *Gordius.* It corresponds pretty closely with that of *Ascaris* as described by *Bagge ;* but the embryo on escaping from the egg is annulose and tentaculated, and differs much from the adult form. Nothing is known of the history of the animal between these two conditions. — ED.

BOOK SIXTH

TURBELLARIA.

CLASSIFICATION.

§ 120.

THE TURBELLARIA receive their name from the ciliated epithelium, which covers their whole body. Their flattened, or cylindrical, non-articulated body, is formed of a loose parenchyma, in which lie hid the viscera. The nervous system appears very little developed, and when visible, consists only of a cervical ganglion, from which there never extends a ventral cord. The multiramose intestinal canal is always without an anus. The genital organs are either very much developed, or entirely absent.[1] In the first case, these animals are always hermaphrodites, and have copulatory organs.

The Turbellaria have been shifted from one zoological system to another, but their organization has sufficient peculiarities to entitle them to a special class by themselves.

Ehrenberg was the first to found the group Turbellaria; but he has included therein many different animals; and we are, therefore, indebted to *Örsted*, for a late revision of this group.

ORDER I. RHABDOCOËLI.

The alimentary canal is simple and cylindrical; the œsophagus, non-protractile; locomotion, mostly natatory.

Genera : *Vortex, Derostomum, Gyratrix, Strongylostomum, Mesostomum, Typhloplana, Macrostomum, Microstomum.*

ORDER II. DENDROCOËLI.

Intestinal canal dendritically ramified ; œsophagus completely protractile ; locomotion reptatory.

---

1 I cannot here omit the question, if these small sexless Turbellaria, as for example, *Derostomum*, and *Microstomum*, really constitute distinct genera, and if they are not rather the larvae of other inferior animals.

Genera: *Polycelis, Monocelis, Planaria, Leptoplana, Eurylepta, Planoce-ra, Thysanozoon.*

BIBLIOGRAPHY.

Baer. Ueber Planarien, in the Nov. Act. Acad. Leop. XIII. 1826, p. 691.
Dugès. Recherches sur l'organisation et les mœurs des Planariées, in the Ann. d. Sc. Nat. XV. 1828, p. 139; and XXI. 1830, p. 72. See Isis 1830, p. 169, or *Froriep's*, Not. 1829, No. 501.
Mertens. Ueber den Bau verschiedener in der See lebender Planarien, in the Mém. de l'Acad. de St. Petersbourg, 6ᵉᵐᵉ Sér. Tom. II. 1833, p. 1.
Ehrenberg. Phytozoa turbellaria, in the Symbolae physicae, Ser. I. 1831.
Focke. Ueber Planaria Ehrenbergii. in the Ann. des Wiener Mus. I. Abth. 2, 1836, p. 193.
F. F. Schulze. De Planariarum vivendi ratione et structura penitiori nonnulla. Dissertatio. Berolini, 1836.
A. S. Örsted. Entwurf einer systematischen Eintheilung und speciellen Beschreibung der Plattwürmer. Copenhagen, 1844.

ADDITIONAL BIBLIOGRAPHY.

Beside the writings referred to in my notes, see the following:
Schmidt. Die rhabdocoëlen Strudelwürmer des süssen Wassers. Jena, 1848.
Neue Beiträge zur Naturgesch. der Würmer. Jena, 1848. Erster Abschnitt, Turbellarien.
Handbuch d. vergleich. Anatomie, 1849, p. 294.
M. S. Schultze. Ueber die Microstomeen, eine Familie der Turbellarian, in *Wiegmann's* Arch. 1849, p. 280, Taf. VI.
Beiträge zur Naturgeschichte der Turbellarien. 1851. — ED.

CHAPTER I.

CUTANEOUS SYSTEM.

§ 121.

The whole body of the Turbellaria is covered with ciliated epithelium, under which lies a loose cellular parenchyma. In this parenchyma, and directly beneath the epithelium, there are found, in many species, particular cell-like bodies, which sometimes remind one of the nettling organs of certain zoophytes, and sometimes exactly resemble the prehensile organs

12

of the arms of Polyps.[1]   These bodies contain six or eight, or even more, staff-like, colorless corpuscles, which are parallelly arranged side by side, or curved a little spirally.   With their further development, the envelope disappears, and they then remain free under the skin, but sometimes projecting through it.[2]

CHAPTER II.

MUSCULAR SYSTEM AND LOCOMOTIVE ORGANS.

§ 122.

Although their parenchyma is extremely contractile, yet the Turbellaria have only a very feebly-developed muscular system.

In many small species of the Rhabdocoëli, the parenchymal muscles may be made out; and in the larger *Planariae*, when the muscles are visible, their fibres appear unstriated.

The small Rhabdocoëli swim by means of their ciliated epithelium, like many Infusoria, their bodies revolving on its longitudinal axis; while the flattened Dendrocoëli crawl along like the Gasteropoda.[1]   Many larger species of the first order,[2] appear to float from place to place by means of their epithelium, thus really neither creeping nor swimming.

---

1 With *Microstomum lineare, Örsted*, these prehensile organs so closely resemble those of *Hydra* that they need not be described. According to *Örsted* they are urn-shaped glands in the centre of which are parabolic bodies which are constantly in motion (loc. cit. p. 73, Taf. II. fig. 18). But had he pressed these organs between two plates of glass, he would have seen the protruding filament, together with its double hooks.

2 I have seen these corpuscles protruding through the lateral border of the body of *Planaria lactea*. In the dorsal papillae of *Thysanozoon Diesingii*, a part of these corpuscles are contained in cells; but the others are free and often protrude through the skin. With *Mesostomum Ehrenbergii*, and *rostratum*, they are arranged in rows in the anterior half of the body, forming striae, which quickly catch the attention. *Örsted* has taken these corpuscles for as many muscular columns (loc. cit. p. 70, Taf. II. fig. 26, 37). The spines which, according to him (loc. cit. p. 72, Taf. II. fig. 29, 34) cover the entire surface of *Macrostomum*

hystrix, are probably of the same nature, as may also be said of the delicate short bristles found everywhere under the skin of *Derostomum leucops, Dugès*.

Quatrefages, in his monograph on marine *Planariae* (Ann. d. Sc. Nat. IV. 1845, p. 146, Pl. VIII. fig. 9, 10), also mentions various formations which, partly as spines, partly as nettling organs, are found in the skin of certain Dendrocoëli.

1 The mode of locomotion by which these animals move over solid bodies, or upon the surface of the water, has not yet been satisfactorily explained. The ciliated epithelium cannot here be the principal agent. According to *Schulze*, loc. cit. p. 32, the staff-like corpuscles projecting from the back of these animals, and which he terms bristles, are used as oars.

According to *Mertens* (Mém. de l'Acad. de St. Petersbourg, 6eme, sér. II. 1833, p. 5), *Planaria lichenoides* moves by means of the protruded lobes of its pharynx.

2 For example, *Mesostomum*.

CHAPTERS III. AND IV.

NERVOUS SYSTEM AND ORGANS OF SENSE.

§ 123.

The nervous system with the Turbellaria, is quite indistinct, for it has not yet been observed in the small species, and in the larger ones its disposition is yet doubtful. A double ganglion in the cervical region appears to form its central part, and from this nerves pass off in different directions.[1]

§ 124.

Among the organs of sense, those of vision are the most developed with very many species.

The red, brown, or black spots on the anterior extremity, two or more in number, are not always simple pigment cells,[1] but may be regarded as eyes, for they have a cornea, — a light-refracting body surrounded with pigment, and a nerve-bulb.[2]

As to the sense of touch, no special tactile organs have yet been found, but the whole surface reäcts sensitively from the lightest contact; and this sensibility appears particularly prominent at the anterior extremity, which, with many Dendrocoëli, is furnished with lobular and other appendages.[3]

---

1 *Ehrenberg* has seen two disconnected ganglia with *Planaria lactea* (Abhand. d. Berl. Akad. 1835, p. 243). With other Dendrocoëli, as with *Planocera sargassicola*, and *pellucida*, these two ganglia are blended into one ; at least, the organ which *Mertens* has here described as a heart, has exactly the appearance of two united ganglia (loc. cit. Taf. I. fig. 6, Taf. II. fig. 3, m. or Isis 1836, Taf. IX. fig. 3, c. m.). The light pulsations which this author affirms to have here observed, are perhaps, as *Ehrenberg* has supposed (loc. cit. p. 244), due to the contractions of neighboring organs. According to *Schulze* (loc. cit. p. 39), with *Planaria torva*, the double central ganglion gives off two nerves, which pass backwards on both sides of the intestine.

This double ganglion, situated in the cervical region, and the nervous filaments which it gives off, have been demonstrated by *Quatrefages* (loc. cit. p. 172, Pl. IV.,-VI.).

1 Most commonly there are two eye-dots. With *Planocera*, and *Leptoplana*, there are many which are grouped together, and with *Polycelis nigra*, the whole anterior part of the body is covered with them. In many small species, they appear to be of a simple pigmentary nature.

2 With *Planaria lactea*, there is, between the cornea-like bulging of the skin, and a semilunar, pigment layer, a small, conical, transparent body, corresponding exactly to a crystalline lens ; see *Ehrenberg*, loc. cit. p. 243, and *Schulze*, loc. cit.

* [§ 124, note 2.] See also *Leidy* (Proc. Acad. Nat. Sc. Philad. III. 1848, p. 248) on the eye-specks of *Phagocata gracilis*, a sub-genus made by him from *Planaria*, and *Schmidt* (Die Rhabdocoëlen Strudelwürmer, &c., p. 7, and Neue Beiträge Zur Naturgesch, der Würmer, &c., p. 11). Both of these observers agree in considering these parts in

p. 37. With *Monocelis*, these organs are very remarkable, being composed of two eyes blended into one, and the simple and spherical ball of the eye is filled, according to *Örsted*, with a transparent vitreous body, in which two conical crystalline lenses are buried with their apices pointing inwards (loc. cit. p. 6, 56, Taf. I. fig. 1, 2, and in the text, fig. 10). *Örsted* has distinctly seen two optic nerves passing laterally to this organ. It is quite remarkable that with one of the three known species, the *Monocelis unipunctata*, the eye is entirely without pigment. *Ehrenberg* affirms that he has observed with *Polycelis*, many star-like ganglia in the middle of the anterior part of the body, which are for the long row of eye-dots (loc. cit. p. 243).

For the eyes of the marine *Planariae*, see also *Quatrefages*, loc. cit. p. 178, Pl. III. The organ which with *Monocorlis* has been taken for an eye by *Örsted*, appears to be, according to the researches of *Frey* and *Leuckart* (Beitr. p. 83, Taf. I. fig. 18), an auditory organ. That which *Örsted* regarded a vitreous body, is an otolite, and his two crystalline lenses, are two semicircular prolongations attached loop-like to the otolite. *Frey* and *Leuckart* are also convinced that *Convoluta paradoxa Örst.*, has a single auditive capsule, situated on the median line of the cervical region, and containing an otolite which floats in a lilac-colored fluid ; see Beitr. loc. cit. p. 82, Taf. I. fig. 17.*

3 There are contractile and antenniform append-

question as visual organs. *Schmidt* has often failed to find anything like an otolite ; but, on the other hand, has often found with various *Derostomum* a complete visual apparatus. This point, therefore, is still unsettled, unless, as *Schmidt* ingeniously suggests, it may be that one organ serves the functions of two separate senses. — ED.

CHAPTER V.

DIGESTIVE APPARATUS.

§ 125.

With the two orders of Turbellaria, this apparatus is formed upon very different types. But in both orders, the location of the mouth varies so much, that it serves as the basis of genera, according as it is at the anterior extremity, or a little behind it, — or, at the middle of the belly, or a little behind that also. Th. ..alls of the intestinal canal are always intimately blended with the pa. .   .'' the body.

With the Rhabo.. ..i, the mouth leads to a muscular œsophagus, which is either an annular sphincter, or a longer or shorter tube, but which, in no case, can be everted from the mouth. The intestinal canal is a simple caecum extending from the œsophagus to the posterior extremity ; but with those species which have the mouth situated more or less posteriorly, it stretches forward as a coecum to the anterior portion of the body.[1] With the Dendrocoëli the mouth opens into a large throat, containing a protractile and very movable deglutitory organ (*Pharynx*).

This organ, which can be protruded entirely out of the throat while the animal is eating, is either a tube composed of longitudinal and transverse muscles, or a collection of lobular and ramified tentacles circularly arranged about the mouth.

Its base is prolonged into the proper intestine, whose dendritic ramifications extend over the whole body.[2]

Scarcely a trace of salivary or hepatic organs have here been found with these animals.[3]

---

ages on the anterior part of the body of *Planaria tentaculata*, and *Eurylepta cornuta*, and upon the neck of *Planocera*. With the last, they support a part of the eye dots.

1 The mouth and cylindrical œsophagus of *Gyratrix hermaphroditus*, and *Vortex truncata*, are at the cephalic extremity (*Ehrenberg*, Abhandl. d. Berl. Akad. 1835, p. 178, Taf. I. fig. 2, 3). But the mouth and annular œsophagus of *Derostomum* is situated just back of this extremity, into which, however, the coecal intestine extends. The œsophagus is also annular with *Mesostomum*, and *Typhloplana*. In the first, the mouth is at the middle of the ventral surface ; and in the last, a little behind this point, while the intestine projects coecally far into the anterior extremity (*Örsted*, loc. cit. Taf. II. fig. 26, 31, and *Focke*, loc. cit. Taf. XVII.).

2 The genus *Planaria* has become famous for its movable organ of deglutition, which, being separated from the body, still continues for a while to swallow all presented to its mouth (*Baer*, loc. cit. p.

* [§ 125, note 2.] With *Phagocata (Planaria) gracilis*, *Leidy* (Proceed. Acad. Nat. Sc. Phil. III. 1848, p. 248) found, instead of a single sucker, twenty-three, in the full-grown animal. These are all protruded when the animal feeds, but when not in use, are closely packed together within the animal. They all connect separately with portions of the dendritic alimentary cavity. — ED.

716, Tab. XXXIII. fig. 8–11, and *Dugès*, loc. cit. XV. p. 152, Pl. IV. fig. 18, 19).

The large and plicated œsophagus of *Planaria tremellaris*, constitutes the transition to the tentacular form of the deglutitory organs (*Dugès*, loc. cit. XV. Pl. IV. fig. 20, 21). Fully ramified tentacles are found with *Planocera sargassicola*, *pellucida*, and *Leptoplana lichenoides*. When collected in the throat, they present exactly the aspect of a ramified intestine (*Mertens*, loc. cit. Taf. I. fig. 2, 3, 6, Taf. II. fig. 3, 4, and the Isis, 1836, Taf. IX. fig. 3, b. 3, c.). The ramified intestine of many Dendrocoëli has been figured by *Baer*, *Dugès*, and *Mertens*, in their works already cited.*

3 *Focke* (loc. cit. p. 196, Taf. XVII. fig. 11, c. f.) is inclined to regard as salivary and hepatic organs, two large lateral vessels, and a glandular organ which he has discovered near the œsophagus and intestine of *Mesostomum Ehrenbergii* ; but he himself admits that this view is not yet well founded.]

† [§ 125, note 3.] *Will* (*Müller's* Arch. 1848, p. 508) has shown that the brownish layer covering the whole extent of the intestine of *Planaria* is composed of hepatic glands (*Dendrocoelum lacteum*, *Planaria torva*, and *nigra*). — ED.

CHAPTERS VI. AND VII.

CIRCULATORY AND RESPIRATORY SYSTEMS.

§ 126.

As yet, only a very imperfect vascular system has been observed in the parenchyma of these animals. With the Dendrocoëli, there are constantly two principal vessels, extending along each side of the body, which give off many lateral branches and anastomose together at their two extremities.

This system has no central heart-like organ, and the walls of the vessels not being contractile, the circulation is probably effected through the general contractions of the body.[1] The contained homogeneous and colorless liquid ought therefore to be considered as a nutritive fluid.

With the Rhabdocoëli, the disposition is different. In many there are one or two vessels which traverse the body and loop at its extremities, without either giving off branches or diminishing in size. The movement of their colorless liquid is due to isolated vibratile lobules situated here and there in the vessels.

This organization reminds one more of an aquiferous than a sanguineous system.[2]

Special respiratory organs are here wholly absent, if we do not regard as such the aquiferous system just mentioned. There remains, therefore, only the conjecture that the ciliary epithelium upon the entire surface of the body is subservient to a general cutaneous respiration, by constantly bringing the water in contact with the skin.

---

1 *Dugès* has described and figured very completely the vascular system of *Planaria* (loc. cit. XV. p 160, Pl. V. fig. 1, 2, XXI. p. 85, Pl. II. fig. 24, 25). The cordiform organ which *Mertens* (loc. cit. p. 12, Taf. I. fig. 6, Taf. II. fig. 3) refers to the vascular system of *Planocera sargassicola* and *pellucida*, is probably, as seen above, the central part of the nervous system. *Dugès* is the only observer who affirms to have seen with the Dendrocoëli proper movements of the vessels; while *Mertens*, *Ehrenberg* (Abhandl. d. Berl. Akad. loc. cit. p. 243), *Schulze* (loc. cit. p. 18), and *Örsted* (loc. cit. p. 10), have observed only the contrary.

2 With *Derostomum leucops*, *Dug.*, I have seen two intertwined vessels of equal size throughout, extending from the caudal extremity to the head where they form a simple loop. At the cau-

dal extremity, they approach so near to the cutaneous surface that it is impossible to decide whether they terminate there by a loop, or open externally. *Ehrenberg* (Abhandl. d. Berl. Akad. loc. cit. p. 278, Taf. I. fig. 2) has figured two pairs of such vessels with *Gyratrix hermaphroditus*, and which loop at the posterior extremity, but in front terminate indistinctly. The trembling in the interior of these vessels observed by *Ehrenberg*, indicates certainly the presence of vibratile lobules, and which *Örsted* (loc. cit. p. 17, Taf. III. fig. 48) has distinctly found in the vessels of *Mesostomum Ehrenbergii*, while *Focke* (loc. cit. p. 200) could see only their effects. These are the very vessels which this author supposes connect with the pharynx; but this is not so according to my own observations.*

* [§ 126, note 2.] See for these two systems, *Schmidt*, Die Rhabdoc, Strudelw. &c., p. 11, and

Neue Beitr. zur Naturgesch. d. Würmer, &c., p. 15. - ED.

CHAPTER VIII.

ORGANS OF SECRETION.

§ 127.

No special organs of secretion have yet been found with the Turbellaria, although these animals, and especially the Dendrocoëli, secrete from their cutaneous surface an extraordinary quantity of mucus.[1]

CHAPTER IX.

ORGANS OF GENERATION.

§ 128.

The Turbellaria propagate by transverse fissuration, and by the means of genital organs.

In the smaller Rhabdocoëli, which have no trace of genital organs, the transverse fissuration is the rule.[1] It is, however, probable that at certain epochs of their lives, genital organs are developed, and therefore, that they multiply also by eggs.[2]

With both the larger Rhabdocoëli, and the Dendrocoëli, the genital and copulatory organs of both sexes are situated upon one and the same individual, so that they are capable of self-impregnation ; but there is generally a reciprocal copulation.[3] This genital apparatus is very complex, and as the contents of its various parts have not yet been subjected to a careful analysis, it is not positively certain that the right interpretation of them is given.

1 It is yet undecided whether the subcutaneous cell-like bodies of the Dendrocoëli have any relation to this secretion.

1 *Dugès* (Ann. d. Sc. Nat. XV. p. 169, Pl. V. fig. 15) has observed a voluntary transverse fissuration with *Derostomum leucops*. I have been able to follow the very regular fissuration of *Microstomum lineare*, where each unseparated half of the body began to halve again, and then these four pieces also each divided, and so finally the body appeared worked by seven transverse furrows, into eight divisions.

I must here remark, to prevent an error, that I, contrary to *Örsted* (loc. cit. p. 73), regard these two mentioned species as distinct ; for *Derostomum leucops*, *Dug.*, is without the reddish brown eye-dots and the prehensile organs, which are found with *Microstomum lineare*, *Örst.* The wonderful reproductive power of the sexless *Planariae*,

and which can be multiplied artificially by divisions in all directions, would lead us to infer that they propagate also from accidental divisions, to which their vulnerable nature is constantly exposed.*

2 *Örsted* (loc. cit. p. 21, Taf. III. fig. 53) and *Ehrenberg* (Abhandl. d. Berl. Akad. loc. cit. p. 178, Taf. I. fig. 2, 3) affirm to have seen ovaries, testicles, copulatory organs, and eggs with *Microstomum lineare*, and many other allied Rhabdocoëli, such as *Gyratrix*, *Vortex*, and *Strongylostomum* ; but the details they have given are too imperfect to allow definite opinions upon this organization. I must here ask if these animals have not been confounded with the sexless larvae which multiply by fissuration like those of Medusae.

3 Coition has often been observed with *Planaria* and *Mesostomum*, and has been figured by *Baer*, *Dugès*, and *Focke*.

* [§ 128, note 1.] See *Leidy* (loc. cit.) ; he found that with *Phagocata* (*Planaria*) *gracilis*, this subdivision could not be carried successfully beyond three or four parts. — ED.

The following are the parts usually found : an ovary or organ of vitelline secretion, which is double, and, extending into the parenchyma of the body, opens by a common excretory duct into a large cavity, — a vagina or oviduct; a double testicle sends its seminal liquid, full of filamentoid and motionless spermatic particles, into the seminal vesicle through two tortuous vasa deferentia; to this seminal vesicle is attached a very erectile penis, situated by the side of the vagina. There is a common genital opening, situated always behind the mouth, for the protrusion of this penis and the escape of the eggs.

With *Planaria*, there are, beside, two special, hollow organs, with narrow excretory ducts, which open into the vagina. Of these, one very probably secretes the envelope of the egg, while the other serves as a *Receptaculum seminis.*[4]

§ 129.

The embryonic development of the Turbellaria is yet unknown except with the *Planariae*.

It differs wholly from anything yet known with other Invertebrates. Many of these embryos are developed, always simultaneously, in one large egg; but it is impossible at first to determine their number, since

4 See, for the genital organs of *Mesostomum Ehrenbergii*, *Focke* (loc. cit.); for those of *Planocera*, and *Leptoplana*, *Mertens* (loc. cit.); and for those of *Derostomum*, and *Planaria*, *Dugès*, *Baer*, and *Örsted* (loc. cit.). But the interpretation here given of the different parts of these organs must be much changed. For, to speak here only of the genus *Planaria*, what *Baer* has regarded as the ovaries and oviducts, are certainly the two testicles with their vasa deferentia, since I have always found them filled with spermatic particles (loc. cit. Tab. XXIII. fig. 18, a. b.). The two seminal canals open into a hollow, flask-shaped body like a *Vesicula seminalis* or a *Ductus ejaculatorius*, the neck of which is continuous with a very contractile and erectile tube (*Penis*). This penis is in a cavity separated by a septum from the large vulva, with which, however, it communicates by a special orifice, and consequently can be protracted through the common genital opening. There is, beside the intestinal canal, another ramified organ in the body of *Planaria*, and which very probably is an ovary, or at least a vitellus-secreting organ. But its caeca contain only simple vesicular bodies, which have no germinative vesicles. The canal which *Dugès* (loc. cit. XV. Pl. V. fig. 4, b.) has taken for an oviduct, belongs probably to the ramifications of this organ. The other two organs which this author (Ibid. Pl. V. fig. 4, 8, c.) has described as *Vésicule copulatrice ou reservoir du sperme et des oeufs*, do not appear to me to exist in all *Planariae*. They consist of two hollow, pyriform organs, not blended together as *Dugès* has figured them, but distinct ; one opens by a long, and the other by a shorter canal, into

the vulva. As I have found many spermatic particles in the first of these, I am led to regard it as a *Receptaculum seminis*. But in the other, which *Baer* (loc. cit. Tab. XXXIII. fig. 18, e.) has taken for a penis, I have never found either eggs or germs, but always only a granular substance ; from this I am inclined to think that this organ secretes the material which envelops the vitelline cells grouped in the vulva. With the *Planariae*, one egg at a time is always formed in the round vagina ; this is very large, and when it is deposited others succeed it in the same way. This is not true, however, with *Mesostomum Ehrenbergii*; here the vagina is short and narrow, and receives various organs whose nature is not yet well determined. One of these contains, according to my own researches, a confused mass of active, filamentoid spermatic particles, and may therefore be regarded as a *Receptaculum seminis*. Two canals which pass off right and left from the vagina, bifurcate into two simple coeca, one of which passes forwards, and the other backwards, and in which very large eggs remain for a long time. This therefore may be regarded as an uterus. See *Focke*, Taf. XVII. fig. 1, 11, g. g.

According to the very minute researches of *Quatrefages* (loc. cit. p. 163, Pl. IV.–VIII.) made upon various marine *Planariae*, both the male and the female organs of these Dendrocoeli have two distinct orifices situated in the ventral region, one behind the other. The posterior is a vulva and opens into a more or less long coecum (vagina or copulatory pouch) upon which are laterally inserted two oviducts. The anterior orifice is for the protrusion of the protractile penis.*

* [§ 128, note 4.] See, for many details on the sexual organs of the Turbellaria, and illustrated with figures, *Schmidt*, loc. cit. (*Protostomum*, *Vortex*, *Hypostomum*, *Derostomum*, *Mesostomum*, *Opistomum*, *Macrostomum*, *Microstomum*, *Stenostomum*, *Schizostomum*, *Typhloplana*; according to this author, *Dinophilus vorticoides* is separate-sexed, — the exceptional instance among the Rhabdocoeli. The subject of the

spermatic particles of the *Planariae* is little understood. They probably have not a hair-like form as mentioned in the preceding note, but are Cercaria-like ; see *Kölliker*, loc. cit., *Quatrefages*, loc. cit. Pl. VIII. fig. 5–9, and *Schmidt*, Die Rhabdoc. Strudelwürmer, &c., p. 16 ; this author, however, describes those of *Opistomum pallidum* as somewhat different, there being a filament beyond the head (Taf. V. fig. 14b). — Ed.

their chorion contains only loosely-arranged vitelline cells, among which there is seen no trace of one or more germinative vesicles. The vitelline cells always contain, beside a finely-granular albuminous substance, a round nucleus which has a nucleolus. Both the nucleus and the granular substance are shifted from one side to the other of the cell by the very remarkable peristaltic movements of the cell-membrane. After a time, these movements cease, the cell-membrane disappears, and the contents mix with those of other cells which have been affected in the same way: by these means, little collections of vitelline substance here and there are formed, which increase by the addition of other cells, — and finally are transformed into roundish, nicely-defined embryos which become covered with ciliated epithelium. From this time the embryos do not increase as before by the external fusion of cells, but there is a muscular, discoid œsophagus formed upon their periphery, and through this the remaining cells are ingested and assimilated within the animal.

Still later, the embryo, hitherto spherical, becomes flat and elongated at two opposite points ; — ultimately, and upon the appearance of the eye-specks, it assumes exactly the form of the adult *Planariae*.

The size of the young *Planariae* depends upon the number of embryos developed in the same egg, for the smaller this number, the larger the embryos at the time of their hatching, and *vice versa*.

The cause regulating the number of embryos in an egg is yet unknown.[*] [(1)]

---

1 See my details upon this subject in the Bericht. ueber die Verhandl. d. Berl. Akad. 1841, p. 83. During the development of *Planaria*, one can, after a while, ascertain the number of vitelline cells assimilated by fusion and deglutition, by counting their nuclei which are easily seen in the parenchyma of the body. According to *Focke* (loc. cit. p. 201), the eye-specks, and the œsophagus are developed very early in the young *Mesostomum Ehrenbergii ;* — a species with which each egg contains a single embryo only, and which is developed while the egg is in the uterus.

The remarkable movements of the vitelline cells in the eggs of the *Planariae*, and which I was the first to observe, have since been confirmed by *Kölliker*, with *Planaria lactea* ; see *Wiegmann's* Arch. 1846, I. p. 291, Taf. X. I am unable to say whether or not the spontaneous movements observed by *Quatrefages* (loc. cit. p. 169, Pl. VII. fig. 6–9) upon the larger portions of the vitellus of *Polycelis pallidus* while in the oviducts, are of the same nature ; this naturalist himself supposes that these portions were the embryos of this *Planaria*.[†]

* [End of § 129.] Recent embryological studies have thrown some light upon this point — the so-called plurality of embryos in a single egg. The undoubtedly an ovarian sac, in which are developed many germs; some of these germs may perish, and the fewness of those remaining would give the appearance of an egg with many germs. — ED.

† [§ 129, note 1.] The development of *Planaria* has been also observed by *Schmidt*. Die Rhab-

doc. Strudelwürmer, &c., p. 17; by *Agassiz* (Proc. Amer. Ass'c. Advancem. of Sc. 2ᵈ meeting, 1849, p. 438), who made the interesting observation that the Infusoria-genera, *Kolpoda* and *Paramaecium*, are only larvæ of *Planaria* ; by *Girard* (Ibid. p. 398), and by *Muller* (*Müller's* Arch. 1850, p. 485). *Muller* has here some interesting remarks on the relations of the study of these forms to the class Infusoria. — ED.

BOOK SEVENTH.

ROTATORIA.

CLASSIFICATION.

§ 130.

THE body of the ROTATORIA is covered with a smooth, hard epidermis, and, from transverse incisions, at least at its posterior portion, usually appears articulated; while its anterior portion has vibratory retractile parts — the so-called rotatory organs. The very indistinct nervous system is almost wholly comprised in a cervical ganglionic mass. The fully-developed digestive canal lies in the large cavity of the body, and its anterior portion is provided with masticatory organs, while posteriorly, it terminates in an anus. Female genital organs alone have as yet been found with certainty.

No one would deny that the Rotatoria, whose organization is so high, ought to be separated from the Infusoria, whose structure is scarcely advanced above that of a simple cell-nature. One can be in doubt only as to their other and proper place in the animal kingdom; — whether, with *Burmeister*, they are to be placed among the Crustacea; or with *Wiegmann, Wagner, Milne Edwards, Berthold*, and others, among the Worms. But the choice here between these two classes will not be difficult, for, as will soon be shown, they differ widely from the Crustacea. Aside from the absence of a ventral cord and of striated muscular fibres, these animals have vibratile organs upon the surface of their body, as well as upon their respiratory and digestive organs — a structure not found with the Crustacea, nor with the Arthropoda in general. Their development is non-metamorphotic, and they do not have articulated feet when they escape from the egg; while the Crustacea, and even those which, from a retrograde metamorphosis, become vermiform, have at least three pairs of articulated legs when hatched. On the other hand, they have, in common with most worms, an articulated body, internal and external vibratile organs, absence of a ventral cord, and, with all, the want of articulated feet.

Although the uniformity of their organization does not admit of these animals being divided into orders, they can at least be considered as a separate class in the great section of Worms.

FAMILY: MONOTROCHA.

Genera: *Ptygura, Ichthydium, Chaetonotus, Oecistes, Conochilus.*

FAMILY: SCHIZOTROCHA.

Genera: *Megalotrocha, Tubicolaria, Stephanoceros, Lacinularia, Melicerta, Floscularia.*

FAMILY: POLYTROCHA.

Genera: *Enteroplea, Pleurotrocha, Hydatina, Notommata, Synchaeta, Polyarthra, Diglena, Triarthra, Eosphora, Cycloglena, Theorus, Mastigocerca, Euchlanis, Salpina, Stephanops, Squamella.*

FAMILY: ZYGOTROCHA.

Genera: *Rotifer, Actinurus, Philodina, Noteus, Anuraea, Brachionus.*

BIBLIOGRAPHY.

See the works already cited under Infusoria.

ADDITIONAL BIBLIOGRAPHY.

Besides the writings of *Brightwell, Huxley, Leydig,* and others, quoted in my notes, see the following:

O. Schmidt. Versuch einer Darstellung der Organisation der Räderthiere, in *Wiegmann's* Arch. 1846.

Frey. Ueber die Bedeckungen der wirbellosen Thiere. Göttingen, 1848.

D'Udekem. Bull. de 'l 'Acad. Roy. des Sc. de Belgique, XVIII. 1, 1851.

See also the new edition of *Pritchard's* Infusoria, given under Book first. — ED.

CHAPTER I.

CUTANEOUS SYSTEM.

§ 131.

Nearly all the Rotatoria are covered with a smooth, hard skin,[1] which is thrown into folds by the contractions of the subcutaneous parenchyma; at the anterior extremity only, it is very delicate, and covered with vibratile organs, which also move to and fro with the parenchyma. With many,

---

[1] With *Chaetonotus*, and *Philodina aculeata*, the structure of the skin is quite different from this; for its surface bristles with stiff points and spines. With *Noteus*, and *Anuraea*, there are species whose faceted skin is roughened by innumerable granulations.

the annular sulcations of the skin, partial, or over its whole extent, give the body an articulated aspect.[2] Many others have a skin so hard and stiff as to be like a carapace.[3]

CHAPTER II.

MUSCULAR SYSTEM AND LOCOMOTIVE ORGANS.

§ 132.

The muscular system of the Rotatoria is quite distinct in many parts of the body. There can at once be observed, distinctly separated from the general parenchyma, unstriated muscles, of which some are transversely annular, and many others narrow and longitudinal.[1] The first, subcutaneous and widely separated from each other, are usually upon the borders of the segments of the body. The second, divisible into dorsal, ventral and lateral portions, arise from the internal surface of the skin, and are inserted at the cephalic or opposite extremity.[2]

The posterior extremity of those species which move freely, has two stiff points of variable length, which are moved as tentacles by two cylindrical, or clavate, caudal muscles. Some have long, movable bristles or pedicles, by which they row along or move by quick leaps.[3]

§ 133.

The prominent characteristic of the Rotatoria is the retractile, vibratile apparatus at their cephalic extremity, known as the rotatory organs. By these, they swim freely about, revolving upon their axis, or, when at rest, produce vortex-like motions of the water. The form, number, and arrangement of these organs varies much according to the genera, and may be used even to characterize families.

The rotatory organ is either single, double, or multiple. Often it consists of a disc, supported by a pedicle of variable length, upon whose borders are successive rows of regularly-arranged cilia, the motion of which gives the appearance of rotation to the disc itself. This apparent motion

---

2 With *Conochilus, Megalotrocha, Lacinularia, Brachionus, Noteus, Squamella, Notommata,* and *Stephanops,* the tail is transversely marked or articulated. With many species of *Hydatina, Rotifer, Philodina, Actinurus,* and *Eosphora,* not only the caudal extremity, but the whole body, is regularly segmented, and capable, especially at the posterior extremity, of being intussuscepted or drawn out, like a telescope.

3 A solid carapace, like the shell of *Daphnia,* is found with *Brachionus, Anuraea, Noteus, Sulpina* and *Euchlanis.*

1 The muscles are smooth when at rest, but when contracted, they appear more or less distinctly plicated transversely. The assertion of *Ehrenberg* is therefore remarkable, that the longitudinal muscles of *Euchlanis triquetra* are transversely striated like those of the higher animals (Die Infusionsthierchen, p. 462, Taf. LVII. fig. 8).

2 For the muscles of the Rotatoria in general, see *Ehrenberg,* loc. cit. and his description of the *Hydatina senta,* in the Abhandl. d. Berl. Akad. 1830, p. 47.

3 Many Rotatoria use their caudal pincers as a fulcrum when creeping along. *Philodina* moves along in a leech-like manner, using its mouth and tail as suckers. *Polyarthra* has many bundles of bristles upon the sides of its body, which it uses as oars. *Triarthra* has under the throat and at the posterior extremity of the body, long stiff bristles, articulated with the body, and by which these animals can leap like a flea.

is quite remarkable with those species whose single or double disc is not crenulate, but entire.[1] With those whose organs are more numerous, but smaller, this appearance is not observed.[2]

With *Floscularia*, and *Stephanoceros*, the rotatory organs have quite a different form. With the first, there are five or six button-like processes about the mouth, covered with very long bristles; these bristles produce usually but very feeble motions, and rarely give rise to vortexes. But *Stephanoceros* reminds one much of the Bryozoa, for its rotatory apparatus consists of five tentacle-like processes covered with vibratile cilia[3] The rotatory organs differ, moreover, from the ordinary vibratile cilia of epithelium, in being under the animal's control, — that is, moved or kept at rest, at will.[4]

CHAPTERS III. AND IV.

NERVOUS SYSTEM AND ORGANS OF SENSE.

§ 134.

Notwithstanding the transparency of the Rotatoria, and the distinctness with which their organs are separated from each other, yet their nervous system has not yet been made out with certainty, for their bodies are so small that their peripheric nerves elude the microscope, and their principal nerves and ganglia cannot be distinguished from the muscular fasciculi, the ligaments, and the contractile parenchyma of the body.

It appears certain, however, that in all, there is, as a nervous centre, a group of cervical ganglia, from which pass off threads in various directions.[1]

---

1 *Conochilus, Philodina,* and *Actinurus.*

2 *Hydatina, Notommata, Synchaeta,* and *Diglena.*

3 See *Ehrenberg,* Die Infusionsthierchen, Taf. XLV.

4 According to *Ehrenberg,* there are, at the base of each cilium of the rotatory organs, many striated muscles, which, acting antagonistically, produce the motion (Abhandl. d. Berl. Akad. 1831, p. 34). But neither *Dujardin* (Infusoires, loc. cit. p. 579), nor *Rymer Jones* (Compar. Anat. &c. p. 120), has been able to perceive this apparatus. The contractile parenchyma on which the vibratile discs are situated, appears to be destined only for the protrusion and retraction of the rotatory organs.*

1 *Ehrenberg,* to whom we are indebted for our chief information upon the nervous system of these animals, first took for a cerebral ganglion the gland-

uliform body found upon most Rotatoria, and in the neck of *Hydatina senta,* and *Notommata collaris* (Abhandl. d. Berl. Akad. 1830, p. 52, Taf. VIII. 1833, p. 189, Taf. IX., and, Die Infusionsthierchen, p. 386, &c.). Besides this ganglion, he has mentioned with *Hydatina, Synchaeta,* and *Diglena,* many others scattered through the anterior part of the body, and connecting with the cerebral one by nervous filaments. Likewise, with *Enteroplea, Hydatina, Notommata,* and *Diglena,* he has regarded as a nervous loop, the two filaments which pass off from the cerebral ganglion, and go to the cervical respiratory orifice. Finally, he refers to the sensitive system, a white sacculus, single or double, and situated behind the cerebral ganglion, with *Notommata, Diglena,* and *Theorus* (Die Infusionsthierchen, p. 425). *Grant's* description of the nervous system of the *Hydatina,* as being composed of many ganglia and a ventral cord,

---

* [§ 133, note 4.] *Dobie* (Ann. of Nat. Hist. 1848) speaks of two kinds of cilia with *Floscularia;* "one of the usual short vibratile kind, covering the interior of the alimentary tube; the other extremely

long and filiform, of uniform thickness, and not vibratile under ordinary circumstances." They are slowly moved, being spread out by the contractile substance of the lobes of the rotatory organ. — ED.

§ 135.

Beside the sense of touch, apparently located chiefly in the rotatory organs and their tentaculiform processes,[1] these animals have also an organ of vision. Usually this consists of a single or double eye-speck upon the neck; and sometimes, though rarely, of three or four red specks upon the forehead.[2] These specks are usually very small, but nicely defined, and covered by a kind of cornea. They are situated immediately upon the cerebral ganglion, or are directly connected with it, by nervous filaments.[3]

CHAPTER V.

DIGESTIVE APPARATUS.

§ 136.

The digestive apparatus is well developed with the Rotatoria, and has the following parts:

The mouth opens into a muscular pharynx which has two horny, masticatory organs, which move laterally upon each other. Succeeding this pharynx is a narrow œsophagus of variable length, which leads to a stomachal

---

(Outlines, &c., p. 88, fig. 82, B.), is founded, undoubtedly, upon supposition, and not upon real observation.*

1 The vibratile disc of *Conochilus* has upon its centre, four cylindrical processes, terminating usually by a bristle, and quite resembling antennæ. The two or four styles projecting from the front of *Synchaeta*, are probably of the same nature.

2 The eye-speck is simple with *Euchlanis*, *Notommata*, *Synchaeta*, *Cycloglena*, and *Brachionus*; double with *Conochilus*, *Megalotrocha*, *Diglena*, *Rotifer*, and *Philodina*; with *Eosphora*, there are three, and with *Squamella*, four; while *Hydatina*, *Enteroplea*, *Ptygura*, *Tubicolaria*, and the adult *Floscularieae*, have none at all.

3 *Ehrenberg*, who was the first to regard these red dots as eyes, has given their intimate structure in none of his writings; this is the more to be

regretted since *Dujardin* has not regarded them as visual organs (Infusoires, p. 591). He supports this view by the fact that they disappear with the adult individual; but this objection will appear valueless when it is remembered that this is also true of certain parasitic Crustacea. At all events, the small ocular dots of *Conochilus*, *Rotifer*, and *Philodina*, are nicely-defined organs surrounded with a solid capsule, and appear to me wholly different from the diffused masses of red pigment which *Ehrenberg* has erroneously taken for eyes with the Infusoria. The disproportionate size of the red dots which *Ehrenberg* (Die Infusionsthierchen, Taf. LI. LIII. LVI.) has figured with *Notommata forcipata*, *Synchaeta baltica*, *Cycloglena*, and *Eosphora*, lead one to suppose that they are only collections of pigmentary granules.

* [§ 134, note 1.] *Gosse* (Ann. Nat. Hist. 1850, p. 21) describes the nervous system of *Asplanchna priodonta* as follows : " Each of the three eyes rests on a mass that appears ganglionic ; the clubbed masses at the lateral apertures are probably of the same character ; and the interior of the body contains a number of very delicate threads, floating freely in the contained fluid, which have thickened knobs here and there, especially where they anastomose."

Leydig (Zur Anat. und Entwickelungsgeschichte der Lacinularia socialis, in *Siebold* and *Kölliker's* Zeitsch. Feb'y, 1852, p. 457) describes a very peculiar nervous system with *Lacinularia*, consisting of : " 1. A ganglion behind the pharynx, composed

of four bipolar cells with their processes. 2. A ganglion at the beginning of the caudal prolongation, similarly composed of four larger ganglionic cells and their processes." But, that these parts belong to the nervous system, appears by no means positive ; for, as, this observer candidly observes, and it is, I think, a capital comment on this whole class of study : " That these cells, with their radiating processes, are ganglion globules and nerves, is a conclusion drawn simply from the histological constitution of the parts, and from the impossibility of making anything else out of them, unless indeed, organs are to be named according to our mere will and pleasure." — ED.

dilatation. This dilatation is continuous into an intestine which opens externally by an anus.

The mouth is always between the rotatory organs, so that it receives what is drawn in by their vortical action, — the animal swallowing or rejecting the particles at will.[1]

The pharyngeal masticatory apparatus is round, and composed of two jaws having one or several teeth, which are brought together laterally by the action of special muscles.[2]

Usually these jaws are formed of two knee-shaped divisions (*Processus anterior* and *posterior*). The posterior division gives insertion to the masticatory muscles, but the anterior terminates with a tooth,[3] or as a multidentate apophysis.[4] With some which have this last arrangement,[5] the two jaws are formed of three horny arches, and noted for their stirrup-like form. Two of these arches (*Arcus superior* and *inferior*), form the arched portion of the stirrup, pointing inwards, while its base is formed by the third arch (*Arcus externus*), pointing outwards. The masticatory muscles are inserted upon the inferior arch, and move against each other — the transversely-arranged teeth passing over the other two.

With the multidentate Monotrocha, and Zygotrocha, the pharynx rests always in the same locality; but with the unidentate Polytrocha, it can move up and down, and even be protruded through the mouth. In this last case, the teeth serve as pincers for the seizure of food. The intestine usually traverses the cavity of the body in a straight line, rarely looping,[6] and is lined throughout with ciliated epithelium.

From the stomachal dilatation to a point near the anus, its walls are very thick. The walls of the stomach and intestine are formed of large cells with a colorless nucleus, and which, as they contain a brownish or greenish granular substance, are of an hepatic nature.

With most species, two caeca, rarely more, with thick walls and lined with ciliated epithelium, open on the right and left of the beginning of the stomach. Their walls are also composed of large cells, which, as they differ widely from the hepatic ones by their colorless contents, may perhaps serve the function of salivary glands or pancreas.[7]

The term Rectum has been given to a short and terminal portion of the intestine, which has thin walls, capable of being widely distended by faeces. Its orifice is excretory not only of the faeces, but also of the contents of the genital organs and of the aquiferous system — and may therefore be regarded as a cloacal as well as an anal opening. It is nearly always at the base of the caudal extremity.

1 The tentaculiform, rotatory organs of *Stephanoceros*, are also used for the seizure of food ; see *Ehrenberg*, Abhandl. d. Berl. Akad. 1832, Taf. XI. fig. 1, e., also, Die Infusionsthierchen, Taf. XLV. fig. 11. 5.

2 For the structure of the teeth, see *Ehrenberg*, Abhand. d. Berl. Akad. 1831, p. 46, Taf. III. IV.

3 *Pleurotrocha*, *Furcularia*, and many species of *Notommata*, and *Diglena*.

4 *Hydatina*, *Euchlanis*, *Salpina*, *Anuraea*, *Brachionus*, and many species of *Notommata*, and *Diglena*.

5 *Philodina*, *Lacinularia*, *Melicerta*, and *Conochilus*.

6 With *Euchlanis*, and *Brachionus*, the stomach is separated from the intestine by a constriction, and with *Philodina*, the intestine is of equal

size throughout, except the rectum which is dilated. But it is coiled, especially with those which are enclosed in a carapace, as with *Tubicolaria*, and *Melicerta*, since here the anus is far in front.

7 These two pancreatic caeca are nearly always present, being wanting only with some species of *Ichthydium*. With *Notommata clavulata*, and *Diglena lacustris*, there are, besides these caeca which are long, attached to the stomach many smaller sacs, which are colorless and perhaps of the same nature.

With *Megalotrocha albo-flavicans*, there are also two like caecal appendages entering the base of the stomach, and which are independent of the short pancreatic ones of the same locality ; see *Ehrenberg*, Abhandl. d. Berl. Akad. 1831, Taf. III. and, Die Infusionsthierchen, Taf. L. LIV.

CHAPTERS VI. AND VII.

CIRCULATORY AND RESPIRATORY SYSTEMS.

§ 137.

As no sanguineous system has yet been found with the Rotatoria, it must be admitted that all the organs are bathed directly by the nutritive liquid which transudes through the intestine.[1]

§ 138.

The vessels observed with the Rotatoria belong probably to the aquiferous system, which, from its structure and limited distribution, must be regarded as of a respiratory nature. In most species, a straight and riband-like organ is seen upon each side of the body, which contains a stiff, tortuous, vasculiform canal. At the anterior extremity of these two lateral bands, their canals connect with many short lateral vessels which open into the cavity of the body, — their orifices being furnished each with a very active, vibratile lobule.[1]

These lateral orifices have the appearance of pyriform, or oval corpuscles, in the interior of which, the vibratile lobule, produces the aspect, when its motions are diminished by pressure between plates of glass, of a small, flickering flame.

The number of these organs varies with the species, and also, it would appear, even with different individuals of the same species. Usually there

---

1 The sanguineous vessels which *Ehrenberg* has frequently described and figured, have not appeared as such to *Dujardin* (Infusoires, p. 589), *Rymer Jones* (Comp. Anat. p. 125), *Doyère* (Ann. d. Sc. Nat. XVII. 1842, p. 201), and myself.

The so-called annular vessels encircling the body of many species at regular and wide distances, and which, as he himself avows (Die Infusionsthierchen, p. 415), are not connected by longitudinal vessels, are undoubtedly only the transverse sulcations, or muscles. From their extreme tenuity, it is difficult to determine the nature of the other filiform organs in the body of the Rotatoria, and which *Ehrenberg* has also referred to the sanguineous system. But, equally well might they be taken for muscular fasciculi, ligaments or nerves.*

1 *Ehrenberg* was the first to point out these vibratile organs, and designated them as the internal gill-like respiratory organs (Abhandl. d. Berl. Akad. 1833, p. 183).

* [§ 137, note 1.] *Dalrymple* (Phil. Trans. 1849, p. 334) has described with *Asplanchna Brightwellii* what he regards as a peculiar circulatory system. It "consists of a double series of transparent filaments (for there is no proof of their being tubes or vessels), arranged, from above downwards, in curved or semicircular form ; symmetrical when viewed in front. These filaments, above and below, are interlaced loop-like ; while another fine filament passes in a straight line, like the chord of an arc, uniting the two looped extremities. To this delicate filament are attached tags, or appendices, whose free extremities are directed towards the interior of the animal, and are affected by a tremulous, apparently spiral motion, like the threads of a screw. This is undoubtedly due to cilia arranged round these minute appendices. The tags are from eight to twelve, or even twenty, in number, varying in different specimens." He thinks these organs fulfil their function by the ciliated tags producing currents in the fluid which fills the body of the animal.

These observations are curious and deserve further attention. — ED.

are two or three on each side, and sometimes there are from five to eight pairs,[2] but rarely more.[3]

The lateral bands approach each other at the posterior extremity, and their canals join in a common, highly-contractile vesicle with thin walls, which empties externally its aqueous contents through the cloacal opening.[4]

An orifice, situated usually upon the neck, and sometimes pedunculated, serves probably to introduce the water into the cavity of the body. This water enters the aquiferous system through the lateral vessels which float free in this cavity, and at last is expelled through the contractile vesicle. In this way, a constant renewal of water can occur, and the opening upon the neck may therefore be properly termed a respiratory orifice or tube.[5]

There can be but little doubt that the rotatory organs also, have a respiratory function, for their surface is covered with thin epithelium, and their cilia produce a constant change of the water.

CHAPTER VIII.

ORGANS OF SECRETION.

§ 139.

Some of the Rotatoria secrete a gelatinous substance, which, hardening, forms the cells and tubes into which they can partly or wholly withdraw themselves. The organ of this secretion is yet unknown; but the secretion appears to be derived from the posterior extremity, and especially from the cloacal opening.[1]

---

2 *Notommata copeus*, and *syrinx.*
3 With *Notommata clavulata*, and *myrmeleo*, the number of these organs is remarkable; each lateral band has thirty-six to forty-eight; see *Ehrenberg*, Die Infusionsthierchen, Taf. XLIX. L.
4 *Ehrenberg* was the first to direct the attention of naturalists to these two lateral bands and their contractile vesicles; but he regarded them as two testicles with their vesiculae seminales (Abhandl. d. Berl. Akad. 1830, p. 51). The incorrectness of this opinion, and which he has maintained in his grand work, cannot be doubted, if it is considered that these two bands with their appendages are already developed and in activity with the young animals, and this even before they have escaped the cavity of the parental body.
In all *Ehrenberg's* published figures, one notices nothing of the flexuous canals of these organs, and which, therefore, he does not appear to have observed.
5 The respiratory orifice is cervical with *Enteroplea, Hydatina, Diglena*, and many species of *Notommata*; but, with *Rotifer, Philodina,*

Brachionus, and some species of *Salpina, Euchlanis*, and *Notommata*, it is replaced by a tube.
With *Actinurus*, exceptionally, a simple respiratory tube is placed under the throat; and with *Tubicolaria*, and *Melicerta*, there are two in the same region.
1 With *Conochilus*, and *Lacinularia*, where several individuals are attached by their tails around a common centre, the nucleus of one of these colonies is formed by a loose, gelatinous substance, in the cells of which these animals can partially withdraw themselves. With *Oecistes, Tubicolaria, Stephanoceros, Flosculuria*, and *Limnias*, each individual occupies an isolated and more or less hard gelatinous tube (*Ehrenberg*, Die Infusionsthierchen). The tubes of *Melicerta*, of which *Schäffer* has given an excellent figure (Die Blumen-polypen der süssen Wasser 1755, Taf. I. II.), are very remarkable, and according to *Ehrenberg*, are composed of brown polygonal cells which are excreted through the cloacal opening and glued together (Die Infusionsthierchen, p. 406).

CHAPTER IX.

ORGANS OF GENERATION.

§ 140.

Although it is certain that the Rotatoria propagate only by genital organs, yet the female organs only are yet well known. These consist of a single or double ovarian tube of variable length, situated upon the sides of the intestinal canal at the posterior part of the cavity of the body, and opening into the cloacal cavity through a short oviduct. These ovaries never develop but a few eggs at a time. The mature eggs are always oval and surrounded by a simple, solid, colorless envelope. They contain a finely granular and usually colorless vitellus, in which there is a distinct germinative vesicle. Many species are ovigerous, but a few only are viviparous. [1]

It would be naturally supposed that these animals, which have such distinct female organs, would have also those of the other sex. But as yet the most minute researches have failed to detect them. It is therefore doubtful whether these animals are hermaphrodites or of separate sexes.† [2]

---

1 For the various forms of the ovaries see the classical works of *Ehrenberg*. With *Philodina roseola, Brachionus rubens*, and *Mastigocerca carinata*, the vitellus of the eggs as well as the parenchyma of the body is of a reddish color. With those species which live in the tubes, the eggs are usually deposited in the cavity of these last. But with *Triarthra, Polyarthra*, and *Brachionus*, they remain glued to the cloacal opening.

With *Philodina*, the young are often hatched in the cavity of the parental body, and are, according to *Ehrenberg* (Die Infusionsthierchen p. 483), always surrounded with an extensible membrane of the ovary (uterus). But it has always appeared to me that the mature eggs of the viviparous *Philodinae*, are detached from the ovaries and fall into the cavity of the body, where afterwards the hatched young move about. Perhaps oviducts are here wanting and the young escape from their parent through an orifice near the cloacal opening.*

2 Admitting that there are here male genital organs, the respiratory tube upon the neck of

---

* [§ 140, note 1.] The view here expressed that the young of the viviparous *Philodinae* may find their escape from the body of the parent through an opening near the anus — the oviducts being perhaps wanting — is probably correct, since, in the viviparous Aphides, where the processes of reproduction occur likewise by a kind of gemmiparity, there are, according to my observations, no oviducts proper, but the young, having fallen into the abdominal cavity, thence escape through a *Porus genitalis* situated near the anus. — ED.

† [End of § 140.] The discovery of distinct males with the Rotatoria is due to *Brightwell* (Ann. Nat. Hist. Sept. 1848) who has positively determined it with *Asplanchna*. Here it is about half the size of the female, being also of a different form; it is exceedingly transparent and easily eludes observation. The testis appeared as a round vessel situated at the bottom of the body on one side, and filled with spermatic particles. This author thinks also that he observed a well-defined intromittent organ connected with the testis, and a passage for its extension from the body of the animal.

In verification of this observation it may be mentioned that *Brightwell* observed the actual coitus between the sexes, and *Gosse* (loc. cit. p. 22) has witnessed the development of the males from the ovum.

Huxley, on the other hand (Quat. Jour. Mic. Sc. No. I. Oct. 1852, p. 1), has found with *Lacinularia* no trace of a male individual, but in some specimens he observed singular bodies which answered precisely to *Kölliker's* description of the spermatic particles of *Megalotrocha*. He says, "They had a pyriform head about 1-1000 in. in diameter by which they were attached to the parietes of the body, and an appendage four times as long which underwent the most extraordinary contortions, resembling however a vibrating membrane

§ 141.

.Their embryonic development occurs, as in most invertebrate animals, through a complete segmentation of the vitellus; and the embryonic cells then appear in the segmented portions.

The newly-hatched embryo has already rotatory and masticatory organs, eye-specks, &c., and the general form of the adult animal.† [1]

many species was formerly taken for a penis. But the incorrectness of this view has since been seen, for no one has here observed the copulatory act. According to *Ehrenberg*, who regards these animals as hermaphrodites, certain parts of the aquiferous system represent the male organs. He regards the two lateral bands as testicles, and their inferior extremities as vasa deferentia, while the contractile vesicle is the vesicula seminalis. But these organs contain only an homogeneous aqueous fluid, in which there is at no time anything like spermatic particles ; moreover they are fully developed in the young individuals which then have no trace of female organs.

It would be wholly anomalous that these animals should constantly secrete sperm during their whole life. One would therefore wholly assent to the doubts of *Dujardin* (Infusoires, p. 587), upon this view of *Ehrenberg*, and some contradictions into which this last has fallen upon this subject, have been noticed by *Doyère* (Ann. d. Sc. Nat. XVII. 1842, p. 199). *Kölliker* has also thought this view unfounded, and has sought to remove the doubts by a search after the spermatic particles. He regards as such, with *Megalotrocha albo-flavicans*, the peculiar trembling bodies which he has seen in the cavity of the body, since they are composed of a pyriform body, to which is attached a movable tail. These bodies he affirms are developed in round cells, often nucleated, and he has often counted ten to twenty in the same individual. As he also asserts to have seen eggs at this time in the same individual, this would certainly be a proof of the hermaphroditism of these

more than the tail of a spermatozoon." He very justly concludes that they cannot at present be definitely regarded as spermatic particles. — ED.

* [§ 140, note 2.] The subject of the form and character of the spermatic particles of the Rotatoria is quite interesting, as it may perhaps throw some light on the position of these animals in the animal kingdom. As yet, however, we have very few observations, and even these are not fully definite.

Schmidt (Vergleich. Anat. &c. p. 268, note) speaks of the spermatic particles of *Euchlanis macrura*, as being cercaria-form.

Leydig (*Siebold* and *Kölliker's* Zeitsch. III. Hft. 4, p. 471) has given those of *Lacinularia* as composed of a nuclear body from which radiate many tails, like these particles with the Decapods. See Taf. XVII. fig. 2. — ED.

† [End of § 141.] We are indebted to *Leydig* (Zur Anat. u. Entwickelungsgesch. d. Lacinularia socialis, in *Siebold* and *Kölliker's* Zeitsch. III. p.

animals (*Froriep's* neue Not. No. 28, 1843, p. 17). But this whole observation is somewhat suspicious, for *Kölliker* has very probably confounded the vibratile lobules of the aquiferous system with the spermatic particles, and of which there are four with *Megalotrocha* in the anterior extremity.

The observation of *R. Wagner* (Isis, 1832, p. 386, Taf. IV. fig. 1, 7) is particularly worthy of attention, for followed out, it might lead to the discovery here of male genital organs. He has described peculiar eggs, found frequently by him with *Hydatina senta*, and whose whole surface is covered with very fine, thickly-set hairs. He has regarded these as in their first stages of development, although *Ehrenberg* (Abhandl. d. Berl. Akad. 1835, p. 154, and, Die Infusionsthierchen p. 415), has taken this villous envelope for an alga of the genus Hygrocrocis. But these villous envelopes have always reminded me of the masses of spermatic particles in the testicles of leeches and which have been figured by *Henle* as whitish felt-like globules (*Müller's* Arch. 1835, p. 584, Taf. XIV. fig. 6. a).

[*Additional Note*.] *Kölliker* (Neue Schweiz. Denkschr. VIII. Taf. II. fig. 31, a.) having since figured the spermatic particles of *Megalotrocha albo-flavicans*, my former view that he had confounded these with vibratile organs, is incorrect.*

1 *Kölliker* was the first to observe the complete segmentation of the eggs, with *Megalotrocha* (*Froriep's* neue Not. loc. cit.). It wholly escaped the observation of *Ehrenberg* amid his numerous researches upon the eggs of these animals ; see Abhandl. d. Berl. Akad. 1835, p. 152.‡

452) and to *Huxley* (loc. cit. p. 11–15), for extending our knowledge in this direction. They have carefully observed the development of *Lacinularia*, and the phases correspond exactly with those of *Megalotrocha* as described by *Kölliker*. But beside this ordinary mode of reproduction, they have observed another which is a sexual and analogous if not identical with what has been observed with some of the lower Crustacea (see infra § 292), — propagation by the so-called hibernating eggs. Their observations throw light on the whole of this interesting subject, and have fully confirmed me in my previous conjectures that these "Ova" are only gemmæ having their exact representative in the bud-like eggs of the viviparous Aphides. — ED.

‡ [§ 141, note 1.] *Kölliker's* observation above-mentioned on *Megalotrocha*, has since been confirmed by *Leydig* (Isis, 1848, p. 170) who has observed it likewise with *Notommata* and *Euchlanis*. — ED.

BOOK EIGHTH.

ANNELIDES.

CLASSIFICATION.

§ 142.

THE ANNELIDES are distinguished from all other worms by their ventral, ganglionic cord, and by their annulated body, at the two extremities of which there is a mouth and anus. They resemble the Arthropoda, but at the same time differ from them in having a completely closed vascular system, and in wanting articulated, locomotive organs. The epithelium of their body is not ciliated except where it covers the external branchiae.

The Nemertini, which have hitherto been classed among the Turbellaria, belong more properly to the Annelides, since their body is more or less distinctly articulated, and its parenchyma closely resembles that of the Hirudinei. Moreover, the power which many of them have to divide spontaneously into many segments, is another affinity with various Annelides. It will therefore appear proper to unite the Nemertini with the other Annelides in the following manner : [1]

ORDER I. APODES.

Body without bristles.

SUB-ORDER I. NEMERTINI.

Posterior extremity of body without a sucker; cephalic extremity often provided with lateral respiratory fossæ.

---

1 Since *Kölliker* (Verhandl. d. Schweiz. naturf. Gesellsch. zu Chur. 1844, p. 89) and *Quatrefages* (Ann. d. Sc. Nat. VI. 1846, p. 173) have published their researches on the anatomy of the Nemertini, I have, also, during my last visit at Trieste in 1847, been convinced that these animals should be classed among the Turbellaria, and that they especially deserve this name since their entire body is covered with very distinct vibratile cilia. — *Additional note.*

Genera: *Tetrastemma, Polystemma, Micrura, Notospermus, Meckelia, Polia, Nemertes, Borlasia.*

SUB-ORDER II. HIRUDINEI.

Posterior extremity of body provided with a sucker.

Genera: *Branchiobdella, Piscicola, Clepsine, Nephelis, Haemopis, Aulacostomum, Sanguisuga, Pontobdella.*

ORDER II.   CHAETOPODES.

Body provided with bristles.

SUB-ORDER III.   LUMBRICINI (ABRANCHIATI.)

Body without feet.

Genera: *Chaetogaster, Enchytraeus, Naïs, Lumbriculus, Euaxes, Saenuris, Lumbricus, Sternaspis.*

SUB-ORDER IV.   CAPITIBRANCHIATI.

Body provided with feet; branchiae situated upon the cephalic extremity.

Genera: *Siphonostomum, Chloraema, Amphicora, Serpula, Sabella, Amphitrite, Terebella.*

SUB-ORDER V.   DORSIBRANCHIATI.

Body provided with feet; branchiae situated upon its segments.

Genera: *Arenicola, Ammotrypane, Chaetopterus, Aricia, Aricinella, Cirratulus, Peripatus, Glycera, Goniada, Nephtys, Alciopa, Syllis, Phyllodoce, Hesione, Lycastis, Nereïs, Oenone, Aglaura, Lumbrinereis, Eunice, Amphinome, Sigalion, Polynoë, Aphrodite.*

BIBLIOGRAPHY.

Pallas. Miscellanea zoologica. Hague, 1766, p. 72.
O. F. Müller. Von den Würmen des süssen und salzigen Wassers.
Copenhagen, 1771.
Savigny. Description de l'Egypte. Histoire Naturelle. Tom. XXI.
1826. Annelides. Also, Isis, 1832, p. 937.
Moquin-Tandon. Monographie de la famille des Hirudinées. Paris,
1827.
Morren. De lumbrici terrestris historia naturali, nec non Anatomia.
Bruxelles, 1829.

Andouin et Milne Edwards. Classification des Annélides, et description des espèces qui habitent les côtes de la France, in the Annales des Sciences. Tom. XXVII.–XXX. 1832–33, but published separately under the title, *Recherches pour servir à l'histoire naturelle du littoral de la France.* Tom. II. Paris, 1834.

Ehrenberg. Symbolae physicae. Phytozoa turbellaria.

Milne Edwards. Cyclopaedia of Anatomy and Physiology, vol. 1, 1836. Art. *Annelida.*

Grube. Zur Anatomie und Physiologie der Kiemenwürmer. Königsberg, 1838; also, Aktinien, Echinodermen und Würmer des Adriatischen und Mittelmeers. Konigsberg, 1840.

Örsted. Grönlands Annulata dorsibranchiata. Kjöbenhaven, 1843. Annulatorum Danicorum conspectus. Fasc. I. Maricolae. Hafniae, 1843. Entwurf einer systematischen Eintheilung und speciellen Beschreibung der Plattwürmer. Copenhagen, 1844.

Hoffmeister. De vermibus quibusdam ad genus lumbricorum pertinentibus. Berolini, 1842.

Rathké. Zur Fauna der Krim. St. Petersbourg, 1836, p. 117. De Bopyro et Nereïde. Rigae, 1837. Beiträge zur vergleichenden Anatomie und Physiologie; in the Neuesten Schriften der Naturforschenden Gesellschaft in Danzig, III. Hft. 4, 1842, p. 56. Beiträge zur Fauna Norwegens; in the Nov. Act. Acad. Nat. Cur. XX. pt. 1. 1843, p. 149.

ADDITIONAL BIBLIOGRAPHY.

Besides the references in my notes, see the following writings:

Moquin-Tandon. Monographie de la Familie des Hirudinées. Nouv. Ed. revue et augmentée, accompagnée d'un Atlas de 14 planches gravées et coloriées. Paris, 1846.

Schmidt. Neue Beiträge zur Naturgeschichte der Würmer. Jena. 1848. (zweiter Abschnitt, Ringelwürmer).

Quatrefages. Etudes sur les types inférieurs de l'embranchement des Annéles, containing: Memoire sur la Familie des Hermellicus, Ann. d. Sc. Nat. 1848, X. p. 1.; Sur la Circulation des Annélides, Ibid. XIV. 1850, p. 281; Sur la Respiration des Annélides, Ibid. XIV. 1850, p. 290; Mémoire sur le Système Nerveux des Aunélides, Ibid. XIV. 1850, p. 329; Mémoire sur le Système Nerveux, et les affinities et les analogies des Lombries et des Sangsues, Ibid. XVIII. 1852, p. 167; Mémoire sur le Branchellion de D'Orbigny, Ibid. XVIII. 1852, p. 279. Note sur le Système Nerveux et sur quelques autres points de l'Anatomie des Albiones, Ibid. XVIII. 1852, p. 328.

———— Mémoire sur l'Embryogénie des Annélides. Ibid. X. 1848, p. 153.

Leydig. Zur Anatomie von Piscicola geometrica mit theilweiser Vergleichung anderer einheimischer Hirudineen, in *Siebold* and *Kölliker's* Zeitsch. I. 1849, p. 103.

———— Anatomisches über Branchellion und Pontobdella. Ibid. III. 1851, p. 315.

Leidy. Descriptions of some American Annelida abranchiata, in the Jour. Acad. Nat. Sc. Philad. 1850, II. p. 43 [contains many anatomical details]. — Ed.

CHAPTER I.

CUTANEOUS SYSTEM.

§ 143.

The skin of the Annelides consists of a very thin, non-ciliated epidermis,[1] and a more or less compact dermis composed of solid, but delicate fibres obliquely intertwisted.

The iridescence and often splendid colors of many of the Chaetopodes, are not due to a pigment, but to an optical effect produced by the reticulated union of the dermic fibres.

But the Apodes, on the other hand, owe their many colors to a pigment net-work and cells. Usually the epidermis is separated with difficulty from the dermis, but with many Capitibranchiati, and Dorsibranchiati, the opposite is true. With the Apodes, the skin is closely united with the subjacent muscular layer.

With many Dorsibranchiati, the skin has filiform or lamellar appendages sometimes so much developed that they overlap each other like scales.[2] With some Chaetopodes, there are, beside the bundles of locomotive bristles and hairs, numerous appendages of this kind covering most of the body.[3]

CHAPTER II.

MUSCULAR SYSTEM AND LOCOMOTIVE ORGANS.

§ 144.

The muscles of the Annelides, although highly developed, are never striated.

The whole body is enveloped by a subcutaneous muscular layer divisible into three sheets : an external, of circular fibres ; an internal, of longitudinal fibres ; these are the most developed. Then a middle one composed of obliquely intertwisted fibres ; this is less distinct, and sometimes entirely wanting.[1]

---

1 The external respiratory organs alone are covered with ciliated epithelium. *Örsted* (Beschreib. d. Plattwürmer, loc. cit. p. 77) however, affirms that the body of the Nemertini is provided with vibratile cilia. But this is to me improbable at least with the large species of *Borlasia, Nemertes,* and *Polia.*

At all events this statement of his requires new proof. See additional note under § 142.

2 Scaly appendages of this kind cover the back of *Aphrodite, Polynoë,* and *Sigalion.* With *Polynoë squamata,* they are very easily detached.

3 The back of *Aphrodite hystrix* has numerous bristles and hairs ; with *Aphrodite aculeata,* these hairs are so thickly set that they conceal the back by a kind of felt.

1 The middle muscular sheet is found with the Hirudinei and Lumbricini ; see *Brandt* and *Ratzeburg,* Med. Zool. II. p. 244, Taf. XXIX. fig. 1, 2, and *Morren,* loc. cit. p. 83.

But with the Nemertini it is wanting ; see *Rathké,* Neueste Schrift. d. naturf. Gesellschaft, in Danzig. loc. cit. p. 95.

With the Apodes, this muscular envelope so closely embraces the viscera, that the cavity of the body is made very small. But with the Chaetopodes, this cavity is larger.

With many of the Branchiati, the muscular fibres form distinct fasciculi — so that instead of a common muscular envelope there are longitudinal and annular muscles distinct from each other.[2]

With many Chaetopodes, the internal surface of this envelope sends off annular muscular septa into the cavity of the body, at the junction of the segments, — thus dividing this last into as many chambers as there are segments; sometimes these septa bind the intestinal canal so closely, as to regularly constrict it.[3]

§ 145.

Besides the common subcutaneous muscles, which produce the vermicular motions of the body, there are other groups: 1st. For the auxiliary locomotive organs, and 2nd, for many other organs.

1. The Hirudinei are distinguished, as is well known, by a sucker situated at their posterior extremity, which contains both circular and radiating muscular fibres. This sucker serves both to move and to attach the body.

All the Chaetopodes have short, horny stings (aciculi), and long bristles (setae), united in fasciculi of various forms, which they use as fulcra when they creep, or as oars when they swim.

With the Branchiati, these organs are most fully developed, and are nearly always situated laterally upon a double row of fleshy knobs; and those of the two inferior rows may be regarded as rudimentary feet.

The Lumbricini have short and usually S-shaped stings which are arranged in many rows upon the belly, and may be wholly withdrawn into the abdominal cavity.

Beside these last, Naïs has also a row of bristles each side of the body.[1]

2 These separate muscles are found in *Aphrodite*, *Polynoë*, and *Nereis*, with which the longitudinal ones especially, are seen separated into dorsal, ventral and lateral layers. See for the subcutaneous muscles of the Branchiati in general, *Rathké*, De Bopyro et Nereïde, p. 29, Tab. II., and in the Danzig. Schrift. loc. cit. p. 62, Taf. IV. fig. 6; also *Grube*, Zur Anat. und Physiol. d. Kiemenwürmer, p. 4. et seq.

3 When these septa are largely developed, and embrace closely the digestive canal, as in *Lumbricus*, *Sabella*, *Serpula*, and *Eunice*, there are always foramina in these diaphragms or septa, through which the contents of the cavity of the body can pass from one chamber into another.*

1 The stings and bristles of the Abranchiati, upon whose various forms see *Örsted* (Conspectus generum specierumque Naïdum, in *Kröyer's* Naturhistor. Tidskrift. IV. 1842, p. 128, Pl. III.), are easily lost from use, but are as easily reproduced.

* [§ 144, note 3.] The development and intimate structure of the muscles of the Annelides has been carefully studied by *Leydig* (Siebold and *Kölliker's* Zeitsch. I. 1849, p. 103) upon *Piscicola*, *Clepsine*, *Nephelis*, and other Hirudinei. The muscular fibre is here developed as in the higher animals out of large nucleated cells arranged in rows, and the adult fibre often shows the relics

The number of these organs may therefore vary very much upon the different segments of the same individual. It is remarkable that with the Lumbricini the stings are often detached interiorly, and falling into the cavity of the body form there tough masses which are glued together by a viscous substance lodged in the posterior chambers of the body; see *Hoffmeister*, De vermibus quibusdam loc. cit. Tab. II. fig. 3, and in *Wiegmann's* Arch. 1843, I. p. 196. These agglutinated masses in which are lodged usually various kinds of vibrioid parasites, have been taken by *Montègre* (Observ. sur les Lombries, in the Mém. du Museum I. p. 246, fig. 5, 6, g) for the eggs and fœtuses of the Lumbricini. *Morren* (loc. cit. p. 195, Tab. XXV.-XXIX.) has gone even further, by taking these stings for the chrysalids, and their enclosed vibrios for the embryos of these animals.

of these elementary parts. The fibre is not transversely striated, and is composed of a structureless envelope or sheath which is filled with a fine granular substance; see loc. cit. Taf. VIII. fig. 13-23.

See also *Holst*, De struct. Muscul. in genere et annulat. musculis in specie, Diss. Dorpate 1846. — ED.

With the Branchiati, these organs are often of a cultrate, lanceolate, or sagittate form. Often too, they are denticulated, or barbed upon one or both of their sides, and sometimes they appear articulated.[2] These stings and bristles are moved by a special, muscular apparatus, consisting of many short muscles which arise from the internal surface of the cavity of the body and pass obliquely front and behind to the bases of these organs. These bases project into the cavity of the body, and as their fasciculi are surrounded by a common membranous sheath, when all the muscles contract at once, these organs are thrust out; but they move in various ways, when the muscles contract separately. The other transverse muscles which pass either from the median line of the belly, or from the anterior and posterior parts of the body, and are inserted at the base of these organs, retract them anew into the cavity of the body.[3]

2. With many Branchiati, there is a group of longitudinal muscles at the cephalic extremity, which, arising from the internal wall of the anterior segments, act as elevators and depressors of the œsophagus,[4] as well as retractors and protractors of the fasciculi of the cephalic bristles and tentacles.[5]

CHAPTER III.

NERVOUS SYSTEM.

§ 146.

The nervous system is highly developed in all the Annelides excepting the Nemertini.

The central is distinctly separated from the peripheric portion. The first is usually composed of a row of ganglia, joined together by nervous cords upon the median line of the body.

The most anterior ganglion, and which in some respects may be considered analogous to the brain of the higher animals, rests upon the œsophagus, although the rest of the ganglionic chain which is situated on the median line under the digestive canal, may be regarded as a ventral cord. This cerebral ganglion differs from the others in its larger volume, and appears to be the product of a fusion of two or more symmetrically-arranged unequal-sized ganglia. These ventral ganglia are of uniform size, although not always of the same number with the segments of the body. Strictly, each of them is composed of two ganglia blended together either very perfectly, or very incompletely. The cerebral ganglion is joined to the first of the ventral chain by two cords which surround the œsophagus

------

2 For the nearly inexhaustible variety of form of these horny locomotive organs, which, when cultrate, or lanceolate, are used as weapons of defence, see *Audouin* and *Milne Edwards*, Classification des Annélides, loc. cit. XXVII. p. 370, and *Oersted*, Grönlands Annulata und Annulatorum Danicorum Conspectus, fasc. I. Pl. 1.

3 See *Rathké*, De Bopyro et Nereide, p. 31, Tab. II. fig. 7, 12 ; *Grube*, Zur Anat. d. Kiemenwürmer p. 5 ; and *Gruithuisen*, Anat. d. gezüngelten Naïde, in the Nov. Act. Acad. XI. p. 240, Tab. XXXV.

4 *Aphrodite, Nereis,* and *Arenicola.*

5 *Amphitrite,* and *Siphonostomum.*

laterally, thus forming a ring (the œsophageal ring) through which the œsophagus passes.

§ 147.

The histological elements of the nervous system of these animals, are arranged in the following manner : [1]

The central mass of the nervous system is enveloped by a fibrous tissue (*Neurilemma*), of longitudinal and transverse fibres which are often covered with special pigment cells.

The nervous cords and filaments are composed of extremely fine, primitive fibres, between which in the ganglia are situated various-sized cell-like ganglionic globules.[2] Some of these primitive fibres pass from the cerebral ganglion through all the ventral ganglia, while others pass off from the central ganglia to the peripheric nerves. Many of the ganglionic globules of the brain and abdominal cord are remarkable for their longer or shorter prolongations which may be traced even into the roots of the nerves.[3]

§ 148.

The nerves are given off usually from the ganglia, and rarely from the interganglionic cord. The cerebral ganglion sends off nerves to the organs of sense in the head, and to the labial, proboscideal, and masticatory organs about the mouth. Its development therefore corresponds exactly to the more or less complicated condition of the cephalic extremity.

The ventral ganglia send off from each side usually two or three symmetrically-arranged main nerves to the muscles and skin.

It is with the Annelides that there have been found the first traces of a vegetative or splanchnic nerve (*Nervus splanchnicus*). This consists of delicate filaments which are distributed upon the intestinal canal, with here and there enlargements, and which anastomose, some directly with the œsophageal

1 Our knowledge of the intimate structure of the nervous system of the Annelides is as yet based upon researches of that of the Hirudinei only : see *Helmholtz*, De fabrica systematis nervosi evertebratorum dissertatio. Berol. 1842, p. 12 ; *Hannover*, Recherches microscopiques sur le système nerveux. Copenhague, 1844, p. 72; *Will*, Vorläufige Mittheilung über die Struktur der Ganglien und den Ursprung der Nerven bei wirbellosen Thieren, in *Müller's* Arch. 1844, p. 82 ; *Ehrenberg*, Beobachtung einer auffallenden, bisher unerkannten Struktur des Seelenorgans bei Menschen und Thieren, in the Abhandl. d. Berl. Akad. 1834, p. 720, Tab. VI. fig. 7 ; and *Valentin*, Ueber den Verlauf und die letzten Enden der Nerven, in the Nov. Act. Acad. XVIII. 1836, p. 202, Tab. VIII.

2 *Valentin* declares that he has seen in the brain

and ventral ganglia of the leech so regular and symmetrical an arrangement of the ganglionic globules, that those of the two lateral halves corresponded exactly as to number, volume and position : see *Valentin*, loc. cit. p. 208, Tab. VIII. fig. 62, &c. This symmetry must appear highly astonishing.

3 These prolongations give the ganglionic globules a clavate aspect, as already seen and figured by *Ehrenberg* (loc. cit. Tab. VI. fig. 7, and fig. 7.11., 7.12).

Further researches must decide if these prolongations are really continuous with the primitive nervous fibres, as *Helmholtz* (loc. cit. p. 15), and *Hannover* (loc. cit. p. 73, Tab. VI. fig. 78), affirm to be the case ; for, *Valentin* in his apparently so careful researches, has never seen any trace of a clavated or pedunculated ganglionic globule.*

* [§ 147, note 3.] This alleged relation of the elements of the nervous tissue is a point of no little histological importance and I shall give it a special consideration in noticing the minute structure of this tissue with the higher animals. I have made no observations on the animals in question, but

Leydig, an excellent observer, confirms the general view here advanced of the direct connection of the ganglionic globule with the nerve-tube ; see loc. cit. p. 130, Taf. X. fig. 67 (*Piscicola*). See also *Bruch*, *Siebold* and *Kölliker's* Zeitsch. 1848, p. 175, Taf. XII. fig. 7, 8, 9. — ED.

14

ring, and others with the cerebral ganglion by means of other small ganglia near the cephalic extremity.[1]

With the different orders and sub-orders of the Annelides, the nervous system has the following modifications:

1. The Nemertini differ remarkably from the other Annelides in this respect; for their ventral cord is without ganglionic enlargements, and composed of two separated cords, one on each side of the body, which send off, right and left, lateral branches along their course. These two cords arise at the anterior extremity in two ganglia blended together above the œsophagus, which represent the cerebral ganglion, and send off many nervous branches in front.[2]

2. With the Hirudinei, the ventral ganglia are much fewer than the segments of the body, and are bound together by two contiguous cords. The first and last of these ganglia are remarkable for their size. The first sends filaments to the lips, the second to the caudal sucker.[3]

The *Splanchnic system* is composed of a small ganglion situated in front of the cerebral one, and with which it is connected by two filaments. By its side are two others, which are also small and connect with the cerebral by delicate threads. All three send branches to the oral parts, while a delicate filament goes to the inferior surface of the intestinal canal, and represents an inner splanchnic nerve.[4]

3. The ventral medulla of the Lumbricini consists of two nervous cords

---

1 *Brandt*, Bemerkungen über die Mundmagenoder Eingeweide-nerven der Evertebraten. Leipzig 1835, p. 37.

2 *Rathké* (Danzig. Schrift. loc. cit. p. 100, Taf. VI. fig. 10, 11) has thus described the nervous system of *Borlasia striata*.

He has seen particularly two pairs of cephalic nerves arise from the cerebral ganglion. One and the larger of these is principally distributed to the respiratory fossae of the head, while the other, the smaller, passes directly in front, probably for the vermiform organ upon the cephalic extremity. *Örsted* (Beschreib. d. Plattwürmer loc. cit. p. 5, 18), appears to be wrong in suspecting that *Rathké* has taken the vascular for the nervous system, for *Quatrefages* (Icon. du Règne anim. de Cuvier. Zoophytes, Pl. XXXIV. fig. 1) has figured by the side of the vascular system, the nervous system of *Nemertes Camillæ*, exactly as it is described by *Rathké*.

* [§ 148, note 3.] *Leydig* (loc. cit. p. 129) has found the structure of the cerebral nervous centre of *Piscicola*, quite different from that as described by *Leo*. The cerebral mass is composed of capsules containing ganglionic globules; these capsules are symmetrically situated on each side of the median line, but are connected by a well-marked transverse commissure composed of nerve-fibres; see Taf. X. 67, 68, 69. See also upon the nervous system of the Leeches, *Bruch* (Ueber das Nervensystem des Blutegels: Ein Beitrag zur topographischen Histologie des Nervensystems, in *Siebold* and *Kölliker's* Zeitsch. 1849, p. 164). This memoir is principally histological, and bears upon that disputed point, — the alleged direct connection between the ganglion-corpuscles and the nerve-fibres. It has, however, some topographical anatomical details, and the accompanying figures would make the whole subject very clear. — ED.

3 See *Brandt* and *Ratzeburg*. Med. Zool. II. p. 250, Tab. XXIX. B. (*Sanguisuga medicinalis*), and *Leo*, in *Müller's* Arch. 1835, p. 422, Taf. XI. fig. 10 (*Piscicola geometra*). *Wagner* has found an arrangement quite different from the above, in *Pontobdella muricata* (Isis, 1834, p. 131, Taf. 1. fig. 3). He saw here the ventral ganglia united by a single cord which sends off from each side only a single nerve. This nerve, after a short course, has a ganglion, and then divides into lateral branches. According to *Stannius*, these lateral branches are not united together by longitudinal cords as is the case with the Amphinome.*

4 See *Brandt*, Med. Zool. II. p. 251, Tab. XXIX. B. fig. 7, and, Bemerk. über die Mundmagennerven loc. cit. p. 39 (*Sanguisuga medicinalis*).†

† [§ 148, note 4.] See, for some further remarks on the Splanchnic system of the Hirudinei and Lumbricini, *Quatrefages*, Ann. d. Sc. Nat. VIII. 1847, p. 36. According to him that of the Hirudinei resembles that of the Insects, and is composed of a chain of ganglia from which pass off filaments, some to the abdominal chain, others to the jaws, and others still to the walls of the œsophagus. There is also a frontal ganglionic chain which forms in front a real arcade, and from which filaments are given off anteriorly.

With the Lumbricini it is considerably different, and he thinks unlike that which has yet been described of all the other Annelides.

See further, Ann. d. Sc. Nat. XIV. 1850, p. 282. and XVIII. 1852, p. 167.

See also *Leydig*, Siebold and *Kölliker's* Zeitsch. III. Hft. 3, p. 315, and *Quatrefages*, Ann. d. Sc. Nat. XVIII. 1852, p. 316 (*Branchellion*). — ED.

which are nearly blended together into one, and whose closely-successive ganglia correspond numerically with the segments of the body.[5]

4. With those Chaetopodes which have external branchiae, the nervous system is most highly developed, but has wide variations as to its whole or its details, according to the more or less complicated structure of the cephalic extremity and segments of the body. With those species which are without antennae and eyes, the ventral medulla is composed of two contiguous cords the enlargements of which are indistinct and not sharply defined.[6] These two cords are separated at the cephalic extremity, and terminate either, by a ganglion on each side without apparently forming by a commissure an œsophageal ring,[7] or by encompassing the œsophagus, and forming a ring through a ganglion lying upon it.[8]

With some, the two parallel cords are without ganglia but are reünited at each segment of the body by two transverse threads.[9] With others, this connection occurs through transverse threads and ganglia.[10] There are many Branchiati with which the two cords are so closely contiguous that they are separated only by a longitudinal furrow. Their round or elongated ganglia are then common, and succeed each other at longer or shorter intervals.[11] With an entire series of the Dorsibranchiati, the ventral ganglia are so closely approximated that the interganglionic cords appear wholly wanting.[12]

The brain is composed of only two ganglia, which are more or less blended into one with the Capitibranchiati, and with those Dorsibranchiati whose head is very slightly developed;[13] while with the other Dorsibranchiati whose head is distinct and the eyes and tentacles very much developed, it is the product of the fusion of many ganglia.[14]

---

5 See *Gruithuisen*, in the Nov. Act. Acad. XIV. 1828, p. 412, Tab. XXV. fig. 3-5 (*Chartogaster diaphanus*); *Hente*, in *Muller's* Arch. 1837, p. 85, Taf. VI. fig. 2, 3, 8, x, y (*Enchitraeus*); *Roth*, De Animalium invertebratorum systemate nervoso. Wirceburg, 1825, fig. 3 ; and *Morren*, loc. cit. p. 117, Tab. XIX.-XXIII. (*Lumbricus terrestris*).

In the common *Lumbricus*, two pairs of nerves (*Nervi annulares*) pass off laterally from the centre of the ganglionic enlargements ; and between every two ganglia, exceptionally, there passes off another pair (*Nervi interannulares*) which are distributed to the transverse muscular septa ; see *Morren* loc. cit. The nervous system of *Sternaspis thalassemoides* is quite different, and appears retrograded to the type of that of the Sipunculidae, for the ventral medulla consists only of a simple cord which is enlarged at the caudal extremity ; see *Will*, in *Muller's* Arch. 1842, p. 427.

6 *Arenicola, Ammotrypane*, and *Terebella*.

7 *Arenicola*; see *Grube*, Zur. Anat. d. Kiemenwürmer, p. 17, Tab. I. fig. 7 ; and *Stannius*, in *Muller's* Arch. 1840, p. 379, Taf. XI. fig. 15.

8 *Ammotrypane* ; see *Rathké*, in the Nov. Act. Acad. XX. p. 197, Tab. X. fig. 14, 19.

9 *Sabella*; see *Wagner*, Isis, 1832, p. 657, Taf. X. fig. 14 ; and *Grube*, Zur. Anat. d. Kiemenwurmer, p. 30, Taf. II. fig. 16.

10 *Phyllodoce*. Here, the transverse threads commence only at the border of the 7th or 9th ganglion. They alternate regularly with these and disappear towards the last segments of the body ; see *Quatrefages*, Ann. d. Sc. Nat. II. 1844, p. 95, Pl. II. fig. 2, 3.

11 *Siphonostomum, Amphitrite, Amphinome, Aricinella, Polynoe*, and *Aphrodite*. With *Si-*

phonostomum, the ventral ganglia are very long ; see *Rathke*, Danzig. Schrift. loc. cit. p. 90, Taf. VI. fig. 3. Here, the peripheric nerves are given off from the interganglionic cords and not from the ganglia themselves.

With *Amphitrite*, the ventral ganglia are long also, but from the fifth segment of the body they alternate with others that are round, so that each segment has two ganglia. Both of these ganglia furnish exclusively the peripheric nerves, but in front where the round ganglia are wanting, they are furnished also by the interganglionic cords ; see *Rathke*, loc. cit. p. 75, Taf. V. fig. 7, 15. With *Aricinella (Quatrefages*, loc. cit. p. 96, Pl. II. fig. 5), and *Amphinome (Treviranus*, Beobacht. aus d. Zoot. u. Physiol. 1839, p. 83, Taf. XI. fig. 72), the ganglia are very closely set together. With *Aphrodite*, and *Polynoe*, the number of ventral ganglia exceeds that of the segments of the body ; see *Grube*, loc. cit. p. 66.

12 *Nereis, Eunice, Glycera* ; see *Wagner*, in Isis, 1834, p. 135, Taf. I. fig. 11 ; *Muller*, in the Ann. d. Sc. Nat. XXII. 1831, p. 22, Pl. IV. fig. 19 ; *Rathke*, De Bopyro et Nereide, p. 41, Tab. II. fig. 13 ; *Grube*, Zur. Anat. d. Kiemenwürmer, p. 43, Taf. II. fig. 9 ; and *Quatrefages*, Ann. d. Sc. Nat. loc. cit. Pl. I. fig. 1, 2, 3.

13 *Amphitrite, Siphonostomum (Rathké*, Danzig. Schrift. loc. cit. Taf. V. fig. 7, 14, Taf. VI. fig. 5), and *Glycera (Quatrefages*, Ann. d. Sc. Nat. loc. cit. p. 96, Pl. I. fig. 3).*

14 *Nereis, Eunice*, and *Phyllodoce* ; see *Muller*, Ann. d. Sc. Nat. loc. cit. Pl. IV. fig. 10 ; *Rathke*, De Bopyro et Nereide, p. 43, Tab. II. fig. 4, 5, 13 ; and *Quatrefages*, loc. cit. p. 81, Pl. I. fig. 1, 2, Pl. II. fig. 1.

---

* [§ 148, note 13.] See *Quatrefages*, Ann. d. Sc. Nat. X. 1848, p. 47 (*Hermella*) and XII. 1849, p. 300 (*Chloraema*). — ED.

With the Dorsibranchiati, the many delicate threads which arise from
the cerebral ganglion by special roots and pass to the different portions of
the digestive canal with a ganglion here and there upon their course, may
be regarded as splanchnic nerves.[15]

With the Amphinomae, Euniceae, Nereïdeae and Ariciae, there arise
from the posterior border of the cerebral ganglion two roots which may be
regarded as *Nervi pharyngei superiores*, and which unite near their origin
into a *Ganglion pharyngeum superius*. From this last pass off posteriorly
delicate threads which form many ganglia upon the œsophagus, and afterwards spread over probably many other parts of the digestive canal.

Beside this *Plexus splanchnicus superior*, there is sometimes a *Plexus
splanchnicus inferior* — formed by other roots which pass off inferiorly
from the brain; part of these form under the œsophagus a *Ganglion pharyngeum inferius*, while others, passing backwards, constitute *Nervi pharyngei* and *œsophagei*.[16]

The Amphinomae have on each side of the abdomen a very remarkable
ganglionic chain. Their ganglia intercommunicate, not only by longitudinal,
but also by transverse anastomoses, with the central mass of the nervous
system. Among these last, those which are given off from the anterior
lateral ganglia, join the connecting filaments of the œsophageal ring; while
the others, arising from the posterior lateral ganglia, go to the various
ganglionic enlargements of the ventral cord.[17]

It has not yet been possible to ascertain the signification of these lateral
ganglionic chains.

CHAPTER IV.

ORGANS OF SENSE.

I. Organs of Touch.

§ 149.

With the Annelides, the sense of touch is particularly developed at the
cephalic extremity.[1] With some Lumbricini, this extremity is prolonged
into a kind of tentacular proboscis.[2] The Branchiati have special and

---

15 *Cuvier* (Leç. d'Anat. Comp. II. 337) has noticed with *Aphrodite* two nerves passing backwards which ought to be regarded as of a splanchnic nature; but *Grube* (Zur. Anat. d. Kiemenwürmer, p. 58,) has been unable to find them even
in the same species.

16 *Stannius* (Isis, 1831, p. 986, Taf. VI. fig. 8, r.
r.), and *Grube* (De Pleïone carunculata, 1837, p 9,
fig. 5, r.), have seen with certain species of *Amphinome* the two roots of the *Plexus splanchnicus
superior*, but were unable to trace them further.
However, with *Eunice Harassii*, *Grube* (Zur.
Anat. d. Kiemenwürmer, p. 43, Taf. II. fig. 9, i.) has
found beside these two roots, the *Ganglion pharyngeum superius* which they form, and the nervous
filaments which pass off from this last. Quite lately,
Quatrefages has given very exact and detailed
descriptions and figures of the expansions of the
Plexus splanchnicus superior and *inferior*, with
Eunice Nereis, *Glycera*, *Phyllodoce*, and *Aricinella*; see Ann. d. Sc. Nat. 1844, 11. p. 81, Pl. I. II.

17 These two ganglionic chains were first described by *Stannius* with *Amphinome rostrata*
(Isis, 1831, p. 986, Taf. VI. fig. 4). He saw three
ganglia connect with the œsophageal ring on each
side. But *Grube* (De Pleïone carunculata, p. 10,
fig. 5) has seen six on each side with *Amphinome
carunculata*. These lateral ganglia, moreover,
remind one of those described by *Wagner*, a already noticed with *Pontobdella muricata*.

1 According to *Rathke* (Danzig. Schrift. loc. cit.
p. 94, 100), the two cephalic and respiratory fossae
with the Nemertini, are the seat of a most delicate
sense of touch; and their white, long and protractile proboscis is also a tactile organ. But other
naturalists attribute wholly different functions to
these organs.

2 The proboscis is non-articulated with *Nais proboscidea*, and *Enaxes filiformis* (Grube, Wiegmann's Arch. 1844, I. p. 204, Taf. VII. fig. 1).
But it is articulated with *Rhynchelmis* (Hoffmeister, Ibid. 1843, I. p. 192, Taf. IX. fig. 8).

often very prominent tactile organs, in the form of processes of variable number and shape, which are situated principally though not entirely upon the cephalic extremity of the body. Those upon the head have been named *Antennae*, and the others *Cirri*. These last are often very numerous upon the first segment of the body. Both are contractile and usually unarticulated, though sometimes having very distinct joints.[3] The antennae receive their nerves directly from the cerebral ganglion,[4] while those of the cirri of the first segment, are given off from the base of the two lateral cords of the œsophageal ring, and from the first ventral ganglion.[5]

II. Organs of Vision.

§ 150.

With nearly all the Capitibranchiati,[1] and with many Nemertini, and Lumbricini, the eyes are wanting. But, as visual organs, have been regarded the brown or black dots, which are two in number with many *Naïs*, four with *Tetrastemma*, but are innumerable and arranged irregularly or in rows upon the neck, with *Polystemma* and *Nemertes*. But these are scarcely more than simple pigment dots.[2]

With the two to ten eye-specks of the Hirudinei,[3] however, the structure is quite different. Here the eye[4] is composed of a transparent cylindrical body, a little attenuated and rounded at its inferior extremity, while the opposite one causes the skin to bulge out like a cornea. Its remaining portion is

3 The antennae of the Annelides have been distinguished from those of insects by being termed *Tentacula*; for they are non-articulated, while those of insects are articulated. But this distinction is not valid, for, with the Branchiati, there are insensible transitions from the non-articulated tentacula to the articulated antennae. But another and more essential difference is, that those of the Annelides are contractile, while those of insects are not. These organs are articulated with *Eunice*, *Peripatus*, and *Syllis*. In this last it is true of the cirri also. The modifications and varieties of the antennae and cirri belong, however, to the province of Zoology.

4 With *Nereis*, four nerves pass off from the anterior portion of the brain to the four antennae; the two external as gustatory nerves and which go to the larger antennae, are largely swollen at their extremity; see Rathké, De Bopyro et Nereïde, p. 43, Tab. II. fig. 4, 5.

5 See Rathké, Ibid. Taf. II. fig. 18, d. d. and in the Danzig. Schrift. loc. cit. p. 76, Taf. V. fig. 14, d. d.

1 A remarkable exception to this occurs with *Amphicora Sabella* as described by Ehrenberg (Mittheil. aus d. Verhandl. d. Gesellsch. naturf. Fr. nde zu Berlin, 1836, p. 2). It has, it would

appear, two eyes not only at the cephalic extremity, but at the opposite one also.*

2 *Gruithuisen* (Nov. Act. Acad. Nat. Cur. XI. p. 242) has described the two eyes of *Naïs proboscidea* as particles of pigment enveloped by a sensitive parenchyma. But this is not based upon observation, and is an hypothesis only, as *Muller* has very judiciously remarked (Ann. d. Sc. Nat. XXII. 1831, p. 20). The assertion of *Quatrefages* is of more weight (Comp. rend. XIX. 1844, p. 195). He affirms that the pigment specks of many Nemertini and of a marine species allied to *Naïs*, contain really light-refracting bodies, and connect with the nervous centre by particular nerves. The last of these Annelides has similar pigment specks also upon each side of the segments of the body, which receive each a distinct nerve from the ventral marrow. Is not this species identical with the *Naïs picta* described by Dujardin (Ann. d. Sc. Nat. 1839, XI. p. 263, Pl. VII. fig. 9)? †

3 *Clepsine* has two, four or six eyes; *Nephelis*, eight; and *Haemopis* and *Sanguisuga*, ten; while with *Branchiobdella*, they are wanting. With this family (the Hirudinei), these organs are always symmetrically arranged upon the neck.

4 At least with *Sanguisuga officinalis*.

* [§ 150, note 1.] *Quatrefages* (Ann. d. Sc. Nat. X. 1848, p. 48, Pl. II. fig. 10, y.) describes two colored points situated on the middle of the brain of *Hermella*, as eyes. They are composed of pigment and rest directly on the nervous substance. — Ed.

† [§ 150, note 2.] *Quatrefages* (Compt. rend.

Dec. 31, 1849) has found very perfect eyes with *Torrea vitrea*, consisting of a crystalline lens, a choroid coat, a vitreous humor, a transparent cornea, &c. He thinks also that he has discovered with *Sabella* eyes situated on the branchiae! — Ed.

enveloped with a layer of black pigment.[5]   Each of these bodies receives a nervous filament from the cerebral ganglion.   Undoubtedly, these filaments are optic nerves, and the cylindrical bodies are light-refracting and light-concentrating organs.[6]

Many of the Dorsibranchiati are entirely without eyes, having only the eye-specks ; but others, belonging to the Amphinomae, Nereïdeae, Euniceae, and Aphroditae, have two to four very distinct eyes.[7]   In these, there is an eye-ball invested with a black or brown pigment layer : and this layer often has, above, a very distinct round pupillary opening, covered by the skin, which bulges out like a cornea.   At the central portion of this layer, there is concealed a transparent body, which is very probably surrounded by a retina-like expansion of the optic nerve.   The optic nerves which are given off usually from the upper surface of the brain have, after a short course, and before entering the pigment layer of the eye, an enlargement.   It is said that with some the light-refracting body and the pupillary opening are wanting.   In such cases, the eyes could only distinguish light from darkness.[8]

III. Organs of Hearing.

§ 151.

Although it has never been doubted that the Annelides can perceive sounds, yet it is of late only that the attention has been directed to the locality of the auditory organs.   The two vesicles, which, with some Chaetopodes, are situated near the œsophageal ring, and contain crystalline bodies, may be regarded as simple *Vestibula*, containing many otolites.[1]

5 *Weber* was the first to show that the black specks of *Sanguisuga officinalis* were really eyes (*Meckel's* Arch. 1827, p. 301, Taf. III. fig. 24). This has been confirmed by *Brandt* (Med. Zool. I. p. 251, Taf. XXIX. A. fig. 10–12), and more recently, *Wagner* has discovered in the interior of the pigment layer, a transparent body, composed, he thinks, of two parts, a crystalline lens and a vitreous portion ; see *Wagner*, Lehrbuch, d. vergleich. Anat. 1835, p. 428 ; also Lehrb. d. speziellen Phys. 1843, p. 383, and, Icon. physiol. 1839, Tab. XXVIII. fig. 16.

6 *Brandt* has been able to trace the ten optic nerves of *Sanguisuga officinalis* from the brain even to the eyes (Med. Zool. loc. cit. p. 250, Taf. XXIX. II. fig. 2).*

7 With *Glycera, Aricia, Arenicola*, and *Cirratulus*, the eyes are wanting. With *Goniada*, and *Nephtys*, there are only simple pigment specks upon the head. With *Eunice, Phyllodoce*, and *Alciopa*, there are two eyes ; and four with *Nereis, Syllis, Hesione*, and *Amphinome*. The genus *Alciopa* is well suit d, from its large size, for the dissection of these organs.

8 For a most detailed description of the eyes of *Nereis*, we are indebted to *Muller* (Ann. d. Sc. Nat. XXII. 1831, p. 22, Pl. IV. fig. 6 10), and *Wag-*

ner (Lehrb. d. Physiol. p. 383, and Icon. physiol. Tab. XXVIII, fig. 15). *Wagner*, who, formerly (Zur vergleich. Physiol. d. Blutes, 1833, p. 55), could not, any more than *Muller*, perceive the light-refracting body, has at last seen it distinctly. For my own part, I can confirm its presence in the two eyes of *Eunice gigantea*, which have a circular pupil. According to *Rathké* (De Bopyro et Nereïde, p. 44, Tab. II. fig. 4, 5) the eyes of *Nereis pulsatoria* and *lobulata* want the pupil, although it is present with those of *Nereis Dumerilii*. According to *Wagner*, the pupil is wanting with the two posterior eyes, but is present with the two anterior ones, with most of the Nereïdeae.

1 I have compared the swellings noticed upon the œsophageal ring of *Arenicola*, by *Grube* and *Stannius* (see *Wiegmann's* Arch. 1841, I. 166), to the auditory vesicles of Mollusks, and their contents to otolites ; since then, *Quatrefages* has recognized the presence of similar auditory organs containing many otolites with two species allied to *Amphicora* (Compt. rend. XIX. 1844, p. 195, and Ann. d. sc. Nat. 1844, II. p. 94). *Frey* and *Leuckart* (Beitr. &c. p. 81), after a very careful examination of the organs of *Arenicola*, which I have regarded as auditive, have confirmed this opinion.

* [§ 150, note 6.] For further details on the ocular organs of the Hirudinei, see *Moquin-Tandon*, loc. cit. Ed. 1846, p. 80, Pl. VIII. fig. 11. According to him, they contain neither a lens nor a vitreous humor, and are only light-perceiving organs. See also *Leydig* (loc. cit. p. 129) who makes the following statement upon the nature of these bodies

with *Piscicola :* "They receive no nerve, neither do they contain a light-refracting body. I regard them as simple ornaments, wholly analogous to the corresponding pigment dots on the pedal shield, with which they also correspond in color and distribution."— ED.

CHAPTER V.

DIGESTIVE APPARATUS.

§ 152.

The digestive canal of the Annelides, which is organized after very different types, opens always at the anterior part of the body by a mouth, and at the posterior part by an anus. It is situated upon the axis of the body, and is usually straight, rarely having convolutions. Often it is divided into many sections, to which the names of pharynx, œsophagus, stomach, and intestine, may be given. The mouth is usually surrounded with thick lips, and, with many Capitibranchiati, it has very erectile tentacles and cirri, which may be not only tactile but prehensile organs.[1] With others of this group, the food is taken in by the action in the water of the ciliated branchial rays which surround the mouth in an infundibuliform or spiral manner.[2] But usually the food, both soft and solid, is seized by the protuberant lips, and swallowed by the very muscular pharynx. Many Annelides can also suck in liquid food through their organs of deglutition.[3] The stomach and intestine is lined with ciliated epithelium. The intestinal canal, whose walls are in general very thin, is either closely embraced by the parenchyma of the body,[4] or, when there is a cavity of the body, is supported and constricted by numerous muscular septa.[5]

I. Organs of Deglutition and Mastication.

§ 153.

The mouth of the Nemertini is situated upon the ventral surface, and usually at some distance from the cephalic extremity. It is a longitudinal orifice opening into a long, muscular and very spacious pharyngeal tube.[1] This tube is intimately united with the parenchyma of the body, and after passing a short distance backward, joins directly with the intestinal canal.[2] With many Hirudinei, the mouth is at the anterior extremity. Its anterior border projects so as to form a kind of lip, which the animal can voluntarily change into a sucker. Other species have a complete oral sucker,

---

1 *Terebella, Amphi .rite,* and *Siphonostomum.*
2 *Sabella,* and *Serpula.*
3 Many Hirudinei.
4 This is true of the Hirudinei, and many Nemertini.
5 With the Chaetopodes.

1 See *Delle Chiaje,* Memorie loc. cit. Tav. LXXVIII. fig. 8. b. (*Polia geniculata*); *Huschke,* Isis, 1830, Taf. VII. fig. 2 (*Notospermus drepanensis*); *Grube,* Aktinien, Echinod. und Würmer, &c., loc. cit. fig. 7, a. (*Meckelia annulata*); *Rathké,* Danzig. Schrift. loc. cit. Taf. VI. fig. 8. b. (*Borlasia striata*); und *Ehrenberg,* Symbol. Physic. Phytozoa Turbellaria, Tab. IV. fig. 4. g. (*Micrura fasciolata*). *Ehrenberg,* moreover, was deceived in regarding this mouth as the opening of the genital organs, and in taking the proboscideal organ of this species, for the true mouth. There is yet in this respect much contradiction among naturalists.

Thus *Dugès,* with *Polystemma (Prostoma) armatum* (Ann. d. Sc. Nat. XXI. 1830, p. 74, Pl. II. fig. 5), and *Quatrefages,* with *Nemertes mandilla* (Icon. du Règne anim. de Cuvier, Zooph. Pl. XXXIV. fig. 2), regard the long canal which opens at the cephalic extremity, as the pharyngeal tube, and the spines at its base as masticatory organs; while *Orsted* (Beschreib. d. Plattwürmer, p. 22, Taf. III. fig. 41, 49, 50) regards this whole apparatus with *Tetrastemma* as a copulatory organ (see below). In my opinion, the animals here cited do not belong even to the Nemertini.

2 *Borlasia* (*Rathké,* loc. cit. p. 96, Taf. VI. fig. 10, 11) and *Polia* (*Delle Chiaje,* loc. cit. II. p. 407, Tav. XXVIII. fig. 3, j., or Isis, 1832, p. 648, Tat. X. fig. II. 3, j.). With *Meckelia annulata,* I have found the pharyngeal tube arranged in the same way.

entirely distinct from the rest of the body.[3] These suckers serve not only as locomotive organs, as the one, for instance, which is situated at the posterior end of the body, but also for the drawing in of liquid food, and particularly blood. For this purpose, many Hirudinei have a short and spacious pharynx, possessed of muscular walls, which are blended with the parenchyma of the body, and which are armed with horny teeth, by which they cause the wounds necessary for sucking the blood. With *Branchiobdella*, the pharynx has horny upper and lower jaws, of a pyramidal form.[4] With *Sanguisuga*, and *Haemopis*, on the contrary, the base of the pharynx has three fleshy swellings, the projecting arciform border of which is edged with bicuspid teeth.[5] In this respect, *Clepsine* is quite different. The pharyngeal tube is very long, and from its base a movable fleshy tube can be protruded out of the mouth, and which the animal can use as a proboscis.[6] With the Abranchiati, and Capitibranchiati, the pharynx is simple, short and muscular, and presents nothing remarkable. With the Dorsibranchiati, it is very muscular, of variable length, and stretches freely into the cavity of the body. By the aid of special muscles, it may be folded upon itself, and project far out of the mouth.[7] With many Annelides, the pharynx has a horny, masticatory apparatus of sometimes a very complicated structure, and which, when the pharynx is protruded, often extends out beyond it, and serves as a prehensile organ.[8] These two, four, seven, eight or nine jaws always move laterally upon each other. They are usually curved like hooks, and denticulated upon their concave side. When numerous, they are of dissimilar forms with the same individual.[9]

II. Intestinal Canal.

§ 154.

The intestinal canal of the Nemertini passes directly from the mouth to the anus, without forming a stomachal dilatation. Its walls are closely united with the parenchyma of the body, and its internal surface throughout is thickly set with annular folds, which, projecting far into the canal, form there pouch-like divisions.[1]

3 *Piscicola*, and *Pontobdella*.
4 See *Henle, Muller's* Arch. 1835, p. 575, Taf. XIV. fig. 1.
5 See *Moquin-Tandon*, Monog. des Hirud. p. 43, Pl. I. fig. 2. 11, Pl. IV. V. ; *Brandt*, Med. Zool. II. p. 245, Taf. XXIX. A. fig. 13-18, 21, Taf. XXIX. B. fig. 13-17. The swellings of these leeches are carried in front during suction, so as to resemble a three-rayed star — the form of the wound which they produce.
6 See *Moquin-Tandon*, loc. cit. Pl. IV. This proboscis quite reminds one of the pharyngeal tube of the *Planariae*, which also can be protruded from the mouth, but without being reversed.
7 This pharyngeal tube is short with *Amphinome, Nereis, Eunice*, and *Peripatus*; but very long with *Aphrodite, Polynoë, Hesione, Phyllodoce, Glycera*, and *Goniada*; see *Audouin* and *Milne Edwards*, Recherches, &c., loc. cit. That of *Aphrodite, Polynoë, Amphinome*, and others, has been regarded as a stomach ; see *Treviranus*. In Tiedemann's Zeitsch. f. Phys. III. p. 161. Taf. XII. fig. 9. 10, k ; *Grube*, Zur Anat. d. Kiemenwürmer p. 54, et seq. and *Stannius*, Isis, p. 982. But the position, structure and muscular apparatus of this organ are against this view, and quite in favor of its being a pharynx.
8 The jaws are wanting with *Amphinome, Phyllodoce, Aricia, Chaetopterus*, and *Arenicola*.
9 There are two strongly-curved jaws with *Nereis, Lycastis*, and *Peripatus*; four with *Polynoë, Aphrodite*, and *Glycera*; and eight with *Lumbricnereis*. Of the seven with *Eunice*, there are four, one on one side, and three on the other. The same asymmetry exists with those of *Aglaura* and *Oenone* ; see *Audouin* and *Milne Edwards*, Recherches, &c., loc. cit.
1 According to *Rathké* (Danzig. Schrift. loc. cit. p. 96), these transverse folds do not exist with *Borlasia striata*, except when the body is shortened by contraction, and they disappear when it is again extended. But it did not appear thus to me with the numerously folded intestine of *Meckelia annulata. Delle Chiaje* had already observed these folds with *Potia sipunculus*, but figured them as isolated pouches (Memorie, loc. cit. II. p. 407, Tav. XXVIII. fig. 3, 6, or Isis, 1832, Taf. X. fig. 11. 3, 4). According to *Quatrefages* (Ann. d. Sc. Nat. VI. 1846, p. 245), the intestinal canal of the Nemertini, which occupies the axis of the body,

With the Hirudinei, the intestinal canal varies very much, especially as to the number and volume of its appended cocca.[2] Its very narrow anal opening is upon the back directly above the pedal sucker.[3] With *Nephelis*, the canal is simple and gradually enlarges from before backwards, but has no cocca.

With *Branchiobdella*, it is deeply constricted in several places.[4] With *Pontobdella*, it is simple with its two anterior thirds, but there is a caecum on each side of its remaining portion.[5] This last is also true of its posterior third with *Haemopis*, *Clepsine*, and *Sanguisuga*.[6] With this last genus, the other portions of the canal are divided by ten or eleven constrictions into as many parts which send off on each side short cacca;[7] while that of *Clepsine* has on each side five or six cocca, all of which may be ramified. There is a kind of valve directly behind the last two cacca, and so the part of the intestinal canal in front of this may be regarded as a stomach and a small intestine, while the remaining portion behind it, represents the rectum.[8]

With the Abranchiati, the intestinal canal is short, and its œsophagus which is usually narrow passes into a muscular pharynx, which leads into a stomachal dilatation. Upon these parts follow the remaining portions of the intestine which are separated from each other by the transverse septa of the body and often resemble the stomach. With a few species only, the stomach is remarkable for its thick, muscular walls.[9]

With some of the Capitibranchiati, the digestive canal arises directly behind the œsophagus and has bulging portions like those of the colon,[10] assuming, posteriorly, sometimes a spiral form.[11] With others, the œsophagus is continuous directly into the intestinal canal, which, free and unattached by diaphragmatic septa, makes many turns in the cavity of the body, and by constrictions is divided into a stomach, small intestine, and rectum.[12]

With many Dorsibranchiati, the intestine follows directly upon the œsophagus, and is either straight and divided by constrictions,[13] or assumes a spiral form[14] or is without constrictions and irregularly tortuous.[15] With others, the portion of intestinal canal between the pharynx and in-

---

forms with the buccal orifice, a cavity distinct from that of the abdomen, and its anus has a kind of sphincter. But this is certainly an erroneous view of the organization of these worms : the contents of the cavity are sufficient alone to confute it.

2 *Moquin-Tandon*, loc. cit. Pl. I.–IV.

3 With *Piscicola*, exceptionally, the anus is upon the ventral surface of the last segment of the body ; see *Leo*, in *Muller's* Arch. 1835, p. 420.

4 *Henle*, *Muller's* Arch. 1835, Taf. XIV. fig. 1.

5 *Wagner*, Isis, 1834, p. 130, Taf. I. fig. 1, 2.

6 *Brandt* and *Ratzeburg*, Med. Zool. II. p. 246, Taf. XXIX. B. fig. 12.

7 Ibid. Taf. XXIX. A. fig. 19, 20, 55.

8 With *Clepsine marginata*, this rectum has coccal appendages also ; see *F. Muller*, in *Wiegmann's* Arch. 1844, I. p. 371, Taf. X. fig. 14.*

9 With *Lumbricus*, the stomach is very muscular ; see *Morren*, loc. cit. Tab. XI.–XIV. This is also true of *Nais proboscidea*, but not with *Lumbriculus*, and *Enchitraeus*.

10 *Terebella*, and *Sabella* ; see *Grube*, Zur Anat. d. Kiemenwürmer, p. 20, 27, Taf. II. fig. 12, and *Milne Edwards*, Ann. d. Sc. Nat. X. 1838, Pl. X. XI.

11 *Sabella* ; see *Carus* and *Otto*, Erläuterungstaf. Hft. IV. Taf. III. fig. 4, 6, and *Wagner*, Icon. zoot. Tab. XXVII. fig. 21.

12 *Amphitrite*, and *Siphonostomum*. With the first, the stomach is long, spiral, and divided into an ascending and a descending portion ; see *Rathké*, Danzig. Schrift. loc. cit. p. 64, 86, Taf. V. VI.

13 *Amphinome*, *Arenicola*, *Eunice*, and *Nephtys* ; see *Stannius*, Isis, 1831, Taf. VI. fig. 10 ; *Milne Edwards*, Ann. d. Sc. Nat. X. 1838, Pl. XII. XIII. ; *Grube*, Zur Anat. d. Kiemenwürmer, Taf. I.

14 According to *Grube* (Ibid. p. 34), the intestine of *Cirratulus* is spiral like that of *Sabella*.

15 *Ammotrypane* (according to *Grube*, see *Rathké*, Nov. Act. Acad. Nat. Cur. XX. p. 197, Tab. X. fig. 13).

* [§ 154, note 8.] For many special details illustrating as well the histology as the anatomy of the intestinal canal of the Hirudinei (*Piscicola*, *Clepsine*, *Nephelis*), see *Leydig*, loc. cit. p. 110, Taf. VIII. IX. fig. 24–37. — Ed.

testine, receives the excretory ducts of glandular appendages and is therefore, more properly a stomach than an œsophagus.[16] With many, the
stomach and its appendages are wanting, but then the entire canal stretching directly across the cavity of the body has on both sides long analogous
appendages which sometimes consist of dilated sacs, so that these appendages have wholly the aspect of caeca.[17]

III. Glandular Appendages.

§ 155.

The glands appended to the digestive canal of the Annelides may be divided into the salivary and hepatic organs. The first of these are sometimes absent, but the last are never wanting.

The organs regarded as salivary glands are attached either to the pharynx or to the beginning of the intestinal canal. With the Nemertini,
they are absent. But with *Sanguisuga*, as abdominal salivary glands,
may be regarded the many groups of round corpuscles which surround the
commencement of the intestine, and whose excretory ducts open into it by
many orifices, after anastomosing together.[1] With *Lumbricus*, there is
a long lobular body on each side of the pharyngeal tube which secretes a
whitish liquid, and which is analogous perhaps to an oral salivary gland.[2]
The four pairs of transparent vesicles, which, with *Enchytraeus*, open at
the inferior extremity of the œsophagus, are possibly of the same nature.[3]
With *Siphonostomum*, there are two riband-like caeca which pass along the
œsophagus and open separately into the oral cavity.[4] With many Dorsibranchiati, the commencement of the intestine has two glands of probably
a pancreatic nature.[5] It is difficult to decide as to the hepatic or salivary nature of the numerous and usually white appendages, which belong
to both sides of the whole alimentary canal of the Aphroditae. With *Polynoë*, these consist of six cylindrical, caecal, and sometimes bifid tubes,
lying between the muscles of the walls of the body.[6]

With *Aphrodite hystrix*, there are twenty of these tubes on each side

---

16 *Nereis*; see *Rathké*, De Bopyro et Nereïde p.
35, Taf. II. fig. 7, 8.
17 With *Aphrodite hystrix*, and *aculeata*, the
intestine has on each side twenty glandular appendages with long peduncles. In this last species,
these appendages are caeca also, for they have at
their extremities saccular dilatations filled with
chyme ; see *Pallas*, Miscell. Zool. p. 85, Tab. VII.
fig. 11 ; *Treviranus*, in his Zeitsch. f. Physiol.
III. p. 162, Taf. XII. fig. 9, 10 ; and *Milne Edwards*, in Cyclop. Anat. and Phys. I. p. 169, fig.
70.
1 *Brandt*, Mediz. Zool. II. p. 247. Taf. XXIX.
A. fig. 22, 23.*
2 *Morren*, loc. cit. p. 129, Tab. X. XI. (*Lumbricus terrestris*).
3 *Henle*, in *Muller's* Arch. 1837, p. 79, Taf. VI.
fig. 6, d. d.

4 *Rathké*, Danzig. Schrift. loc. cit. p. 87, Taf. V.
fig. 5, c. c.
5 With *Nereïs*, these two salivary glands communicate by two narrow ducts with that portion of
the intestinal canal which should be regarded as a
stomach ; see *Rathké*, De Bopyro et Nereïde, p.
58, Tab. II. fig. 7, g. 8. *Grube* has found these
two appendages at the beginning of the intestinal
canal with *Arenicola* (Zur Anat. d. Kiemenwürmer, p. 6, Taf. I. fig. 1, 5, h.), and with *Ammotrypane* (Nov. Act. Acad. XX. p. 197, Tab. X. fig.
13, 19, h.). See also *Milne Edwards*, in the Ann.
d. Sc. Nat. X. 1838, Pl. XII. fig. I, j. (*Nereis*), and
Pl. XIII. fig. I, c. c. (*Arenicola*); and *Wagner*,
Icon. zoot. Tab. XXVII. fig. 18, g. g. (*Nereïs*).†
6 *Grube*, Zur. Anat. d. Kiemenwürmer, p. 62,
Taf. II. fig. 13 (*Polynoë squamata*).

* [§ 155, note 1.] For the salivary glands of
Hirudinei, see *Moquin-Tandon*, loc. cit. Edit.
1846, p. 108, Pl. X. fig. 4 (*Hirudo medicinalis*),
Pl. VI. fig. 11 (*Haemopis*), Pl. I. fig. 5 (*Branchellion*).— Ed.

† [§ 155, note 5.] For the salivary glands of
Branchellion, see *Leydig*, *Siebold* and *Kölliker's*
Zeitsch. III. Hft. 3, p. 315, and *Quatrefages*, Ann.
d. Sc. Nat. XVIII. 1852, p. 296, Pl. VI. fig. 3, a.
c. — Ed.

of the intestinal canal ; these are narrow and their botryoidal extremities lie in the interstices of the dorsal wall of the body.

With *Aphrodite aculeata*, the structure is analogous but differs in that these appendages have more the aspect of cœca with thin walls, and have not the ramified diverticula except in their central part and between the already-mentioned saccular dilatations.[7]

As an hepatic organ may be regarded with more certainty a particular tissue colored in part brownish yellow, and partly greenish yellow, which closely surrounds the whole intestinal canal of most Annelides. Carefully examined, this tissue is found composed of closely-aggregated glandular sacs which empty their contents into the intestine either directly, or by many common excretory ducts.[8] This contained liquid is, with most species a transparent fluid in which are suspended brown granules, and it resembles the bile of the higher animals.

CHAPTER VI.

CIRCULATORY SYSTEM.

§ 156.

This system is highly developed with the Annelides. The blood is usually colored, and the vascular system, remarkable for many peculiarities, is complete and closed.

This system may be divided into a central and a peripheric part.

The first consists of large contractile vessels taking the place of a Heart. There are also various heart-like organs in the shape of varicose dilatations upon the course of the contractile vessels. The principal vessels have a longitudinal course, occupying the whole length of the median line of the body, — one as a dorsal, and the other as a ventral vessel.

With many Hirudinei, there are also lateral vessels. The dorsal and ventral vessels unite at both extremities, beside anastomosing by transverse branches in the separate segments.

When there are lateral vessels, these also connect with the median ves-

---

7 See *Pallas*, *Treviranus*, *Milne Edwards*, loc. cit., and *Grube*, loc. cit. p. 54.

8 According to *Henle* (*Muller's* Arch. 1837, p. 81, Taf. VI. fig. 2), this glandular envelope forms a villous envelope about the intestine. This is also true of *Lumbricus*, *Lumbriculus*, *Nais*, and *Chaetogaster*. The glandular sacs are greenish with *Branchiobdella* (*Henle*, loc. cit. 1835, p. 575), yellowish with *Amphitrite* (*Rathke*, Danzig. Schrift. loc. cit. p. 65). With *Sanguisuga*, the excretory ducts of the hepatic sacs inter-anastomose and form a kind of net-work around the stomach and its cœca ; see *Brandt*, Med. Zool. p. 247, Taf.

XXIX. A. fig. 28, 29. With many, this hepatic layer envelops also the blood-vessel upon the dorsal surface of the intestine. It is possible that the yellow canal described by *Morren*, with *Lumbricus terrestris* as Chloragogena, is only this hepatic mass (loc. cit. p. 142, Tab. XV. XVI.). Another canal which is traversed by blood-vessels and closed at both extremities, and which is contained in a longitudinal enlargement upon the internal surface of the intestinal canal, and is called by *Morren*, Typhlosolis (loc. cit. p. 138, Tab. XI. XII. XVI. XVII.) may perhaps be regarded as a receptacle of chyle.*

* [§ 155, note 8.] The hepatic organs with the Annelides have been successfully studied by *Will* (*Muller's* Arch. 1848, p. 508), who has used chemical tests. He has found the glandular layer (*Lumbricus*, *Nais*), and the long, thread-like and cœcal glands (*Hirudo*, *Haemopis*, *Aulacostoma*, *Helluo*, *Piscicola*, *Clepsine*) which surround the intestinal canal, to be organs of this nature. — ED.

sel by transverse anastomoses. The peripheric vessels arise by means of a capillary net-work, from the most various points of the longitudinal and transverse vessels. The circulation has on the whole a determinate direction, — the dorsal vessels force by a kind of peristaltic movement the blood from behind forwards into the ventral vessel, which returns it into the dorsal vessel. The blood can, however, pass from the dorsal to the ventral vessel by a much shorter way, — by traversing the capillaries, or directly through the transverse anastomoses. It is, moreover, very probable that the course of the blood in the transverse vessels is not always in the same direction, and that it may under certain circumstances pass from the ventral into the dorsal vessel. This makes it difficult to decide which of these vessels are arteries, and which veins. The respiratory organs, which usually form the limit be ... the venous and arterial systems, are, with most Annelides connec.. .. '.. transverse vessels, and therefore throw no light upon this doubtful po .. From the multitude of these transverse anastomoses, it must appear impossible to distinguish the arterial from the venous blood, and the distinction of veins from arteries with most Annelides must be wholly arbitrary.

The blood of the Annelides, although red like that of the vertebrates, is, however, quite different. It is composed of a li uid containing globules. These last, which are always colorless, of unequal size, and of a spherical form, are granulated on their surface.[1] The blood liquid is either colorless, or contains a coloring matter, which is usually red, but sometimes yellow or green.

§ 157.

With the Nemertini, whose blood is red [1] but as yet imperfectly known, the circulation appears to be due to two cardiac dilatations concealed in the cephalic extremity.[2]

The Hirudinei have, beside the two median vessels, two lateral ones also, which intercommunicate by very numerous transverse vessels.[3]   From the contractions of these vessels, the blood is driven sometimes forwards, and

1 For the blood-globules of the Annelides, see *Wagner*, Zur vergleich. Physiol. d. Blutes, Hft. I. p. 23, Hft. II. p. 39. According to him, those of *Terebella* (Ibid. Hft. I. fig. 8) are pale red, circular discs. Here the exception is remarkable, supposing there was not an error of observation. It appears that beside the blood which circulates in the vessels the fluid contained in the visceral cavity of the Chaetopodes plays also an important part in the act of nutrition, for the eggs and the spermatic particles which with these animals are often detached from the ovaries and testicles at a time when still quite imperfect, attain their complete development while remaining in the visceral cavity, probably by means of this nutritive fluid. See *Quatrefages*, Ann. d. Sc. Nat. V. 1846, p. 379.

1 According to *Milne Edwards* (Ann. d. Sc. Nat. X. 1838, p. 197), the blood of the Nemertini is colorless.

2 The vascular system of *Polystemma* has been distinctly seen by *Dugès* (Ann. d. Sc. Nat. XXI. 1830, p. 75, Pl. II. fig. 6), and by *Orsted* (Beschreib. d. Plattwürmer, p. 17). It is composed of many longitudinal vessels, which intercommunicate not by transverse ones, but by arcuate anastomoses at the cephalic extremity, and by two hearts in the cervical region. According to *Orsted*, these

hearts are divided into two chambers, the anterior having deep-colored blood, while that of the posterior one is more clear. This arrangement has led this naturalist to regard as hearts the bodies described by *Rathke*, with *Borlasia striata*, as cerebral ganglia, and as blood-vessels, the nerves which are given off from them laterally (see above, § 148, note 2). But if *Quatrefages'* figures of the nervous and vascular systems of *Nemertes mandilla* are examined (Règne anim. de Cuvier éd. illustr. Zoophytes, Pl. XXXIV. fig. 1), it will be seen that there are here three main trunks, a median and two lateral. These last accompany the lateral nerves, while a bifurcating vessel which passes from the median to the two lateral trunks, embraces closely, in a loop-like manner, the two cerebral ganglia, so that they easily escape observation. This is perhaps true also of *Borlasia*.

3 The sanguineous system of *Sanguisuga* has been very carefully described by *Brandt* (Med. Zool. II. p. 247, Taf. XXIX. B.); see also *Bojanus*, in the Isis, 1818, p. 2089, Taf. XXVI. fig. 3. 4. With *Nephelis*, there are only two lateral vessels and an abdominal one, lying along the ventral medulla; see *Muller*, in *Meckel's* Arch. 1828, p. 24, Taf. I. fig. 1.

sometimes backwards and oscillates from one side to the other, through the transverse vessels.[4] With most genera, the blood is red, being colorless with a few only, and it is always poor in corpuscles.[5] The Chaetopodes have no lateral vessels. Their circulation is often due to pulsatory organs, and there is a great variety in the disposition of their vascular trunks and sinuses.

With the Abranchiati, the dorsal vessel lies close upon the intestinal canal, and is almost wholly enveloped in the hepatic tissue. At the anterior extremity, it divides in many bifurcating branches, which, after encompassing the pharynx, unite below it, and form the ventral vessel.[6] This vessel accompanies the ventral cord to the posterior extremity, and connects with the dorsal vessel by bifurcating branches, as before.[7] The transverse anastomoses connecting the dorsal and ventral vessel, form at each segment simple, or torose canals.[8] With the small Lumbricini, these are usually

---

4 The irregularity of the blood-currents has, undoubtedly, given rise to the numerous different opinions upon the circulation of these animals ; see *Dugès*, Ann. d. Sc. Nat. XV. 1828, p. 308 ; *Weber*, in *Meckel's* Arch. 1828, p. 399 ; *Müller*, Ibid. p. 24 ; and in *Burdach's* Physiol. IV. 1832, p. 143 ; and *Wagner*, Isis, 1832, p. 635. If the valves which *Leo (Müller's* Arch. 1825, p. 421, Taf. XI. fig. 9) has found in the dorsal and ventral vessels of *Piscicola*, should be found also with other Hirudinei, it would throw some light upon the real course of the circulation.*

5 With *Sanguisuga, Haemopis, Pontobdella, Nephelis, Piscicola*, and others, the blood is red ; it is colorless with some *Clepsine*, according to *Filippi* (Lettera sopra l'Anat. e lo sviluppo delle Clepsine, 1839, Pavia, p. 11) ; it is also brown, violet or red, according to the species. He also declares (loc. cit. p. 8), that with *Clepsine* and *Piscicola*, which live wholly upon the blood of the lower animals, the

* [§ 157, note 4.] The memoir of *Gratiolet* (Mém. sur. l'Organisation du système vasculaire de la Sangsue médicinale et de l'Aulostome vorace, pour servir à l'histoire des mouvements du sang dans les Hirudinées bdelliennes, in extract in the Comp. Rend. 1850, XXXI. p. 699), is worthy of a special reference in this connection. He says : " The lateral vessels, whose walls are very muscular, are the principal organs for the movement of the blood ; they contract alternately, as has been well observed by *Dugès, Weber* and *Müller*, and their contained blood moves in a circular manner, sometimes one way, sometimes the opposite.

" The branches given off by these lateral vessels are of two kinds :

" A. Those destined for the skin, and which are ramified in the respiratory net-works ; they never anastomose with those of the opposite side. Before their final and minute ramifications, they form a large varicose net-work under the skin, which hitherto has been regarded as a plexus of hepatic vessels, but which is positively an interlacement of blood-vessels.

" B. The other branches are destined for the small intestine, and its spiral valve for the testicles, the copulatory apparatus, to the loops and to the muciparous vesicles.

"All these branches arise from the branches or the large arches which form a free anastomosis between

vascular system communicates directly through small canals with the coeca of the digestive canal, so that the contents of this last may pass into the blood without being changed.†

6 See *Henle*, in *Müller's* Arch. 1837, p. 83, Taf. VI. fig. 5 (*Enchytraeus*), and *Hoffmeister*, De vermibus quibusdam, &c., loc. cit. p. 14, Taf. II. fig. 4 (*Saenuris variegata*).

7 With *Lumbricus*, there are, beside the principal ventral vessel, three others smaller, and in direct connection with the ventral cord. Two of these pass off laterally, and the third underneath ; see *Leo*, De Structura Lumbrici terrestris, p. 27; *Dugès* Ann. d. Sc. Nat. XV. 1828, p. 298; and *Morren*, loc. cit. p. 152, Tab. XXI.–XXIV. fig. 5, who especially has carefully described the vascular system of *Lumbricus terrestris*.

8 The transverse anastomoses are simple with *Lumbricus*, but torose with *Saenuris* ; see *Hoffmeister*, loc. cit.

the two lateral vessels. The consequences of this form of structure may be easily summed up. The blood oscillates from the alternate contractions from one pulmonary net-work to another. It *circulates* in the principal organ of the intestinal absorption, in the testicles, and in the muciparous glands.

" This circulation, very different from that which *Dugès* admits in the alleged pulmonary vesicles, shows how various are the means employed by nature. Here she determines the course of the blood by means of valves and stoppers ; while elsewhere she accomplishes the same end by causing certain blood-currents to prevail over others."

The valvular structure of the vessels with *Piscicola*, as mentioned by *Leo*, has since been confirmed by *Leydig* (loc. cit.), who has found it also with *Clepsine*. *Leydig* calls the attention to another kind of circulatory system in *Piscicola* ; see loc. cit. p. 116. But this point has not yet been well made out ; see also *Moquin-Tandon*, loc. cit. p. 133, Pl. X. fig. 10, 15, 16, and Pl. XII, fig. 13. — ED.

† [§ 157, note 5.] The recent observations of *Leydig* (loc. cit. p. 119), have shown the blood of *Piscicola* to be always colorless. This view is probably the correct one, since it better accords with the histological relations of the blood of these animals. — ED.

only in the anterior segments of the body.[9]   With the genus *Lumbricus*,
the cardiac organs consist of five to nine pairs of moniliform, transverse
canals, situated above the stomach, and whose pulsations are very distinct.[10]   With all the Abranchiati yet examined, the blood is red.

With the Capitibranchiati, there are often two dorsal vessels, one immediately subcutaneous, the other lying, as usual, on the intestine.[11]   This
duplicity of the dorsal vessel is observed particularly with those species
which have a coiled intestinal canal.[12]   In this case, there is also a second
ventral vessel accompanying the coils of the intestine.   All these longitudinal vessels interanastomose very frequently, and send many transverse
branches to the intestine and the walls of the body, where they blend with the
capillary system.   Not unfrequently, the dorso-intestinal vessel is dilated
at its anterior extremity, above the pharynx, into a large, pulsatory, heart-
like canal, which sometimes has two lateral arcuate, sinuses situated at the
commencement of the intestine.[13]   The extremity of this vessel sends off,
right and left, many branches to the branchiae, which are situated in this
region.   Leaving these organs, these vessels are distributed, some in front
to the tentacles, and to the other organs surrounding the mouth ; while
others pass below to unite with the ventral vessel.   As the blood is thrown
from behind forwards in the dorsal vessel, and thence passes into the branchiae, this vessel may be called a dorsal vein, and its dilatation a branchial
heart ; while the ventral vessel, which receives the returning blood from the
branchiae, would be an abdominal aorta.   But there are other reasons for
this view.   The dorso-intestinal vessel, from its intimate connection with the
liver, might well serve the function of a *Vena portarum*, while the close
union of the ventral vessel to the ventral cord, is undoubtedly for the purpose
that the latter, as a central nervous mass, may receive arterialized blood directly from the branchiae.   With these animals (the Capitibranchiati), the
blood is red in some, and green in others.[14]

The Dorsibranchiati often have double dorsal and ventral vessels, two of
which belong to the intestinal canal, and two to the walls of the body.[15]
With some, these longitudinal vessels are divided into two or three branches.[16]   The principal dorsal vessel is sometimes dilated at its anterior extremity, above the pharyngeal tube, into a cardiac sinus, to which, at the
beginning of the intestine, there are added two lateral, arcuate dilatations.[17]

---

9 *Enchytraeus, Chaetogaster,* and *Naïs.* The
vascular system of *Euaxes* and *Lumbriculus* is
very remarkable in this respect. Instead of transverse anastomoses, there are, in each segment of the
body, two vessels which pass off from the dorsal
trunk, and divide into many coecal branches ; see
Treviranus, Beobacht. aus d. Zool. loc. cit. p. 60 ;
and *Grube,* in *Wiegmann's* Arch. 1844, I. p. 205,
Taf. VII, fig. 1, 2, d.

10 See *Dugès,* loc. cit. Pl. VIII. fig. 1, and *Morren,* loc. cit. p. 162, Tab. XX.–XXIII., XXI.–
XXIV. fig. 1.

11 *Milne Edwards* has made very beautiful researches upon the vascular system of the Capitibranchiati ; see Ann. d. Sc. Nat. X. 1838, p. 193,
Pl. X. XI.

12 *Amphitrite* and *Siphonostomum ;* see *Rathké.* Danzig. Schrift. loc. cit. p. 76, 88, Taf. V. fig. 4,5.

13 With *Terebella,* there is a vascular heart and
two lateral sinuses; see *Milne Edwards,* loc. cit. Pl.
X. XI. fig. 1.   With *Siphonostomum,* there is a
similar cardiac dilatation upon the pharynx, and it
is divided into two chambers by a well-marked

constriction at its posterior part ; see *Rathké,* loc.
cit. p. 89, Taf. VI. fig. 5, f. g.

14 With *Terebella, Amphitrite,* and *Serpula,*
the blood is red ; with *Siphonostomum, Chloraema,* and some species of *Sabella* and *Serpula,* it is
green.

15 We are indebted to *Milne Edwards* for very
detailed accounts of the vascular system of the Dorsibranchiati ; see Ann. d. Sc. Nat. loc. cit. Pl. XII.
XIII.; see also, for that of *Arenicola, Stannius,*
in *Müller's* Arch. 1840, p. 357.

16 With *Eunice sanguinea,* there is a double
dorso-intestinal vessel (*Milne Edwards,* loc. cit.
Pl. XII. fig. 2, 3 ·) ; and a double ventral one
with *Nephtys Hombergi.* With *Arenicola,* there
are three ventral vessels accompanying the ventral
cord (*Müller,* in *Burdach's* Phys. loc. cit. p. 147).
and with *Amphinome, Grube* has found three
dorsal ones beside, all widely separated from each
other.

17 *Eunice ;* see *Milne Edwards,* loc. cit. Pl.
XII. fig. 2. The vascular system here resembles
that of *Terebella.*

These last are sometimes found alone.[18] With many of these Annelides, the transverse vessels are dilated, before branching, into real branchial hearts.[19] As their branchiae are variously situated among the transverse anastomoses, the distinction between the arterial and venous blood is not as marked as with the Capitibranchiati; it must be arbitrary, as with the Hirudinei and Abranchiati. The blood is usually red, but sometimes is yellow or nearly colorless.[20]

CHAPTER VII.

RESPIRATORY SYSTEM.

§ 158.

With the various families of the Annelides, the respiratory organs are formed after wholly dissimilar types.

With the Nemertini, they are least developed, for, excepting two longitudinal fossae upon the sides of the cephalic extremity,[1] there are no organs which can be regarded as of this nature.

These two respiratory cavities are of variable depth, and their lateral borders are so approximated as to have the aspect of a longitudinal opening, and with some they are situated so far out on the cephalic extremity as to be blended together.[2] They are lined with a delicate ciliated epithelium, quite different from that covering the rest of the body,[3] and by the vortex actions of which, fresh water is brought constantly in connection with the blood.[4] Considering the smallness of these organs, it is very probable that the whole skin has also a respiratory function.

---

18 *Arenicola;* see *Milne Edwards*, loc. cit. Pl. XIII.

19 *Eunice;* Ibid. Pl. XII. fig. 2.

20 With *Eunice, Nephtys,Glycera,* and *Arenicola,* the blood is red ; with *Phyllodoce,* it is yellow ; and it is nearly colorless with *Aphrodite, Polynoe,* and *Sigalion ;* see *Milne Edwards,* loc. cit. p. 196.*

1 See *Müller,* Zool. Danica. Tab. LXVIII. fig. 1–4 (*Tetrastemma (Planaria) viride*) ; *Delle Chiaje,* Memor. loc. cit. Tav. LXXVIII. fig. 8, a (*Polia geniculata*); *Quoy* and *Gaimard,* Atlas Zool. de l'Astrolabe Zooph. Pl. XXIV, fig. 10 (*Borlasia viridis*) ; and the Dict. d. Sc. Nat. LVII. Art. *Vers,* p. 574, Pl. Parentomozoaires, Nemertés, fig. 1, 2 (*Borlasia Angliae,* and *Cerebratulus bilineatus*); also *Huschke,* Isis, 1830, Taf. VII. fig. 1–3. *Notospermus drepanensis.*

2 *Tetrastemma viride, Polia geniculata,* and *Micrura fasciolata* (*Ehrenberg,* Symb. phys. Phytozoa Tab. IV. fig. 4. e. i. g.).

3 See *Quatrefages,* Règne anim. illustr. Zooph. Pl. XXXIV. fig. 1, h. b. (*Nemertes Camillae*).

4 *Rathké* (see above, § 149, note 1) is of the opinion that these two cephalic fossae are the seat of touch ; but the view of *Orsted* (Beschreib. d. Plattwürmer, p. 18, 77), who thinks them of a respiratory nature, is, perhaps, the more correct. In support of this last, is the fact of the presence of ciliated epithelium, and of a very large blood-vessel directly beneath them (see *Quatrefages,* loc. cit. Pl. XXXIV. fig. 1, g. g. (*Nemertes Camillae*)) and which, in many Nemertini, is clearly seen through the thin epithelium ; see *Müller,* Zool. Dan. Tab. LXVIII. (*Tetrastemma virire*); *Delle Chiaje,* Memor. Tav. LXXVIII. fig. 8 (*Polia geniculata*), and Isis, 1830, Taf. VII. (*Notospermus drepanensis*).

* [§ 157, note 20.] See also for the blood of the Annelides, *Quatrefages,* Ann. d. Sc. Nat. XIV 1850, p. 287. — ED.

§ 159.

With the Hirudinei, and Lumbricini, the peculiar canals found in the abdominal cavity may be regarded as internal branchiae, or as aquiferous vessels.

The intimate structure of this aquiferous system is difficult to unravel with the Hirudinei. It is most easily observed with the Branchiobdella; here there are only two pairs of curved canals whose inner surface is ciliated. One of these pairs opens upon the ventral surface at the beginning of the second third of the body, while the other opens at the extremity near the median line. Each of these four canals is dilated just before its external opening into a round, yellow cavity, from which pass off many loop-like vessels.[1] With the other Hirudinei, these organs in pairs are more numerous, and situated one after another from the second third to the extremity of the body.

It is remarkable that the ciliated epithelium lining these canals with *Branchiobdella*, is absent in all the other species.[2]

The structure of the respiratory system of the Lumbricini is not less difficult to be understood. With all the genera there are, at the commencement and on each side of the intestine, very tortuous canals which open upon the ventral surface, by a narrow orifice near the median line. These canals are lined with long cilia which have an undulatory movement;[3] they also are colorless and sometimes have dilatations before opening externally, but they never contain air, so that the terms tracheae or pulmonary cells, have been erroneously applied.[4] Often they float loosely in the cavity of the body, and their free extremity has an orifice surrounded by long vibratory cilia.[5] With some, however, they terminate by thickly-

---

1 See *Henle*, in *Müller's* Arch. 1835, p. 576, Taf. XIV. fig. 1. This epithelium would undoubtedly favor the constant renewal of water in these canals.

2 With *Sanguisuga*, there are seventeen pairs of these organs. They have been taken by *Brandt* (Med. Zool. II. p. 251, Taf. XXIX. A. fig. 55-58) for organs of special secretion, since he has seen a whitish liquid escape from their ventral orifices. The riband-like organ of these Annelides is not, moreover, as is usually supposed, a simple canal, but is composed of numerous interwoven and frequently interanastomosing canals, having no trace of ciliated epithelium.

From *Dugès'* remark (Ann. d. Sc. Nat. XV. 1828, p. 308, Pl. VIII. fig. 2), I think it probable that this net-work is formed of blood-vessels which are occasionally empty, for I have always found them colorless.

In this case, the real aquiferous canals are probably concealed in the net-work, and from their want of ciliated epithelium not easily seen.

With *Nephelis vulgaris*, I have seen the same number of internal branchiae as with *Sanguisuga*. Here, the aquiferous system appears as a knot of colorless, non-ciliated canals connecting with a vesicular pouch which is filled with red blood ; so that in the posterior two-thirds of the body there is a double row of seventeen sanguineous sinuses, inside the lateral vessels. These sinuses, already carefully described by *Muller* (*Meckel's* Arch. 1828, Taf. I. fig. 1), take no part in the pulsations of the main vessels, and are not alternately emptied and filled during the transverse circulation of these animals.

A very interesting fact to me, is the existence with this *Nephelis*, of a multi-lobulate, rosetted, ciliated, colorless organ in the interior of these sanguineous sinuses.*

3 See *Henle*, in *Müller's* Arch. 1837, p. 84, Taf. VI. fig. 7, 8, v. w (*Enchytraeus*) ; and *Gruithuisen*, Nov. Act. Acad. Nat. Cur. XI. 1823, p. 238, Tab. XXXV. fig. 1, x, XIV. 1828, Tab. XXV. fig. 5 (*Nais* and *Chaetogaster*).

4 An aqueous respiratory fluid circulates in these canals undoubtedly by the aid of cilia. The terrestrial Lumbricini which live only in the damp earth obtain this fluid therefrom.

5 I have observed this with *Saenuris variegata, Lumbriculus variegatus, Nais elinguis*,

* [§ 159, note 2.] See, upon the circulatory system of *Branchellion, Quatrefages* (Ann. d. Sc. Nat. XVIII. 1852, p. 314). According to him, the general cavity of the body here, is represented by a collection of canals which are lacunae. This forms a vascular lymphatic system which circulates chyle to the branchiae ; these last are, therefore, "branchies lymphatiques."

He also states as conclusions :

1. The non-communication of the abdominal vessels with the branchiae.

2. The existence of a subcutaneous lymphatic vessel.

3. The origin from this vessel of trunks which go to the branchiae. — ED.

arranged loops.[6]   With *Lumbricus*, these aquiferous canals are surrounded by a very distinct vascular net-work, which has a botryoidal aspect from its numerous pedunculated, vesicular dilatations which are filled with blood.[7]

§ 160.

With most of the Capitibranchiati, and Dorsibranchiati, the respiratory organs consist of external branchiae, which are very apparent, although having variations in their development.   They always consist of lobules or filaments covered with ciliated epithelium, and in which are very considerable vessels as branchial arteries and veins.[1]   The branchiae are here always situated between the venous and arterial systems, so that a portion only of the whole blood is made to pass through the respiratory organs.

It is possible that the two bundles of tentacles which are found with many Capitibranchiati, as infundibuliform,[2] or spiral[3] tufts, are also respiratory; for their vibratory organs not only draw in food, but also produce a constant change of the water.

Other Capitibranchiati have distinct and exclusively respiratory organs in the cervical region, which are either dendritic,[4] or semi-pinnate.[5]

With the Dorsibranchiati, nearly every segment of the body has branchiae upon both sides of its dorsal surface.   These are so simple and rudimentary with the Ariceae, and Nereïdeae, that they consist only of simple lobules, exactly resembling the cirri of the feet.[6]

Enchytraeus albidus, and others. Those canals thus situated remind one of the trembling organs of Rotatoria, connecting the two lateral canals with the cavity of the body ; see above, § 138.

6 With *Lumbricus terrestris*, and its allied species, I have as yet been unable to find any orifices of the aquiferous canals. *Henle* also (*Muller's* Arch. 1835, p. 580) has always found them looped upon themselves. They may however exist, although they have eluded the notice of *Henle* and myself, for the respiratory organs of *Lumbricus* are so difficult to study that there is yet no description or figure giving any idea of their complexity ; see the poor figures of *Lumbricus terrestris* by *Leo* (loc. cit. p. 25, Tab. I. fig. 4), and *Morren* (loc. cit. p. 53, 148, Tab. XIV. XV.). Those of *Hoffmeister*, although more detailed, are scarcely less unsatisfactory (loc. cit. p. 15, Tab. I. fig. 35, 36). It now remains to inquire as to the relations existing between these aquiferous canals and the glands at their base which have been taken by many for mucous pouches. I cannot, for my own part, perceive that these glands with the *Lumbricus terrestris*, excrete any liquid whatever upon the ventral surface. On the other hand, I have often seen escape from the back of this animal, a watery liquid which was only the contents of the cavity of the body, issuing through small orifices upon each side of the median line between the segments of the body. Although I do not know, yet I suppose, that similar orifices exist with the other Lumbricini, and thus, by these orifices and by those of the internal branchiae, the necessary renewal of water for these last, can take place. This hypothesis appears admissible since the cilia of the aquiferous canals always move in the same way.

*[§ 159, note 7.] See, for many details on these parts, *Gegenbaur* (Ueber d. sogenannten

7 These vesicular dilatations do not pulsate, and are undoubtedly analogues of the simple sinuses which communicate with the aquiferous canals of *Nephelis vulgaris.*

1 These respiratory organs may be taken as analogous to the aquiferous vessels of the Lumbricini, which can be everted so that the internal ciliated surface becomes external, and the external blood-vessels internal.

2 *Serpula*, and *Protula*.

3 *Sabella*.

4 *Terebella*; see *Delle Chiaje*, Mem. loc. cit. Tav. XLIII. fig. 1-5, Tav. XLV. fig. 2, 10 ; and *Milne Edwards*, Ann. d. Sc. Nat. X. 1838, p. 200, Pl. X. XI. fig. 1. There are here on each side of the neck three multiramose, contractile branchiae which are placed close together. Into these a large portion of the blood of the median dorsal vessel enters by six lateral branches, while the remaining portion passes on through the dorsal vessel to the tentacles and the borders of the lips.

In each branchial tuft there are a simple artery and a vein placed side by side, which anastomose at its extremity in an arcuate manner.

The returning blood from the six branchiae passes by as many veins into the median dorsal vessel, and the frequent strong contractions and dilatations of the branchiae, certainly very much aid the current.

5 *Amphitrite* ; see *Pallas*, Miscell. zool. p. 120, Tab. IX. fig. 1, 5, 6, 8, e, c ; *Rathke*, Danzig. Schrift. loc. cit. p. 5J, Taf. V. fig. 1, 3. Here the four semi-pinnate branchiae are upon both sides of the second and third rings of the body, and each lamella contains a tightly-closed vascular network.

6 The filaments of these branchiae are very short

Respirationsorgane d. Regenwurms, in *Siebold* and *Kölliker's* Zeitsch. IV. p. 221). — ED.

It is remarkable that the branchiae are perhaps entirely wanting, with the Aphroditae,[7] while they are often highly developed, partly in a pectinate and partly in a fasciculate manner, with the Euniceae, Amphinomae, and Arenicolae.[8]

with *Glycera*, *Nereis*, *Lycastis*, *Nephtys*, and others ; but with *Cirratulus*, they are very long. With *Phyllodoce*, a ad *Alciopa*, there are flattened lobules. But with *Lumbrinereis*, *Aglaura*, and some other allied genera, these are wholly wanting ; see *Milne Edwards*, Classif. loc. cit. The question here arises if the Dorsibranchiati which have atrophied branchiae, have not therefore internal respiratory organs. It is at least probable that the two pairs of remarkable networks surrounding the pharynx of *Nereis* and which have given rise to various interpretations (see *Rathke*, De Bopyro et Nereide, p. 48, Tab. II. fig. 5, bb, fig. 8, f, g, h, and Tab. III. fig. 14 ; also *Milne Edwards*, Ann. d. Sc. Nat. X. 1838, p. 210, Pl. XII. fig. 1, o, p) are properly internal branchiae. They receive the blood from the dorsal vessel through two lateral vessels, and it is returned to the median ventral vessel by two others which are also lateral. Moreover, according to *Rathke* (loc. cit. p. 40), there is, between every two feet upon both sides of the segments of the body, a small orifice opening into the cavity of the body and through which water for respiration can pass.

7 Different observers have equally different opinions upon the branchiae of these animals. For my own part, I have found no trace of these organs, either internal or external with *Aphrodite aculeata*, and *hystrix*. I suppose, therefore, that water enters the cavity of the body by orifices which are very small and difficult to be seen, and comes in contact with the entire vascular system. *Milne Edwards* (Règne anim. illustr. Annelides, Pl. XVIII. fig. 2[a], c) has figured rudimentary branchial lobules with *Aphrodite aculatea*, which are crenulated and concealed between the scales, and are, perhaps, invisible, when the animal is in a fresh state. Moreover, *Sharpey* (Cyclop. Anat. and Phys. I. p. 618), having observed with the same species a very active ciliary motion especially upon the external surface of the intestine and its coeca, it is very probable that here, as with

the Asteroïdae, the respiration occurs by water entering the cavity of the body and bathing the intestine.

8 With *Onuphis*, and *Eunice*, the branchiae are pectinate or semi-pinnate ; see *Milne Edwards*, Classific. loc. cit. With *Diopatra*, and *Chloeia*, each branchia consists of a single ramified fasciculus ; but with the Amphinomae, and Arenicolae, there are several fasciculi ; see *Milne Edwards*, loc. cit. and his plates annexed to Règne anim. de Cuvier, Annelides ; also *Stannius*, Isis, 1831, Taf. VI. With *Eunice*, the blood of the median dorsal vessel passes first into the inferior lateral vessels which have the form of cardiac sinuses, and by the pulsations of which it passes into the branchial vessels, whence it returns into the other two dorsal vessels by the superior lateral ones ; see *Milne Edwards*, Ann. d. Sc. Nat. X. 1838, p. 207, Pl. XII. fig. 2.

With *Amphinome*, there is at the base of each branchial fasciculus, in the cavity of the body, a *Plexus branchialis*, closely resembling the wonderful net-works of *Nereis*, and from which the blood passes into two lateral vessels which here exist ; see Catal. of the Physiol. Ser. &c. II. Pl. XIV. fig. 10, or *Rymer Jones*, Outlines, &c., p. 218, fig. 93.

With *Arenicola piscatorum*, only the thirteen middle segments of the body have branchial fasciculi. These communicate with the ventral and dorsal vessels by simple lateral vessels.

As there exist here at the extremity of the body between the two dorsal and ventral vessels, two cardiac sinuses, it is probable that these force the blood from before backward into the ventral vessel, and thence by the lateral vessels into the branchiae ; so that these inferior lateral vessels would be called arteries, and the superior lateral vessels which return the blood to the dorsal vessel, veins ; see *Milne Edwards*, Ann. d. Sc. Nat. loc. cit. p. 215, Pl. XIII.*

---

* [§ 160, note 8.] The respiratory organs of the Annelides have been much studied by *Quatrefages* (Ann. d. Sc. Nat. XIV. 1850, p. 290), and the following is his résumé :

" 1. The respiration is at first general and entirely cutaneous (*Lumbrinereis*, *Lysidice*, *Hesione*, &c).

" 2. It is still cutaneous, but is confined or concentrated upon particular rings of the body (*Chaetopterus*).

" 3. It is localized upon certain points of each ring, without the structure of these points being sensibly modified (*Nereis*).

" 4. The first degree of the specialization of the respiratory organ appears under the form of a simple cul-de-sac, or an ampulla into which the blood flows (*Glycera*).

" 5. The branchiae become gradually character-

ized by the formation of a canal which communicates with the more or less spacious lacunae.

" 6. These true branchiae may be distributed all along the body (*Eunice sanguinea*).

" 7. They may be concentrated upon a certain number of rings situated near the middle of the body (*Eunice Bellii*, *Arenicola*, *Hermella*, *Polydora*).

" 8. They may be concentrated towards the anterior extremity of the animal, and occupy only a few rings (*Terebella*, *Pectenaria*).

" 9. Finally, they may be located wholly at the extremity of the body, and form only a double tuft (*Sabella*, *Serpula*).

" 10. In considering sometimes the entire body, sometimes each ring separately, a real distinction between the venous and the arterial system may nearly always be made out."— ED.

CHAPTER VIII.

ORGANS OF SECRETION.

§ 161.

Many Annelides are covered with a mucus which is secreted by small, simple follicles situated in the skin.[1]

The calcareous tubes of the Serpulini, appear to be secreted by a collar surrounding the first segment of the body.[2]　It is not yet decided that the leathery tube in which many other Branchiati are concealed,[3] is secreted by an analogous organ.

Those Capitibranchiati which form tubes with grains of sand, bits of shells, &c., have, perhaps all, an opening close behind the mouth upon the ventral surface.　This opening is in connection with many glands situated at the anterior extremity of the body, which probably secrete a substance for the gluing together of the materials of these tubes.[4]

CHAPTER IX.

ORGANS OF GENERATION.

§ 162.

The Annelides reproduce partly by a transverse fissuration, and partly by a sexual apparatus.

1 Similar muciparous follicles are arranged in curved rows with the Hirudinei, upon both the ventral and dorsal surfaces, giving the skin a granulated aspect ; see *Brandt*, Med. Zool. II. p. 244.　I have seen similar groups of follicles with the larger Lumbricini.*

2 The secretion of the calcareous matter occurs here probably as upon the border of the mantle of Mollusks.

3 *Sabella, Onuphis*, and *Chaetopterus.*

4 As secreting organs of this glue, *Rathké* has correctly described four yellowish glands situated, with *Amphitrite*, upon the ventral surface of the first and second segment of the body, and opening by a common canal at the first segment ; see Dauzig. Schrift. loc. cit. p. 71, Taf. V. fig. 6, aa. fig. 2, d.　With *Terebella*, and *Sabella*, the two glands near the cephalic extremity, are perhaps of the same nature.　*Grube*, however, thinks them male genital organs ; see, Zur Anat. d. Klemenwurmer, p. 31, Taf. II. fig. 12, y.; and *Milne Edwards*, Ann. d. Sc. Nat. X. 1838, Pl. X. n. Pl. XI. fig. 1, h. fig. 2, f.

* [§ 161, note 1.] *Leydig (Siebold* and *Kölliker's* Zeitsch. 1849, p. 103) has described with *Piscicola, Clepsine, Nephelis*, and other Hirudinei, cutaneous glands. These consist of an infundibuliform sac, which exactly resembles a nucleated cell, from which passes off a long, tortuous duct.

With *Piscicola*, these are situated in the cephalic and pedal shield, but in *Clepsine*, and *Nephelis*, they are also present in the skin throughout ; see loc. cit. Taf. VIII. fig. 23. This structure is remarkable from its resemblance to some of the cutaneous glands of the higher animals. — ED.

Spontaneous transverse fissuration occurs particularly with the Abranchiati,[1] but has also been observed with the Nemertini,[2] and Branchiati.[3]

It occurs usually at the middle portion or at the border between the second and third segments of the body. Very often there may be perceived at this point, when this process has somewhat advanced, the place where, with the two future individuals, there will be a new fissuration. If the animal has a proboscis, tentacles, or eyes, these organs are developed with the posterior individual before its final separation.[4]

These animals have no trace of genital organs, while this process of division lasts. The individuals thus produced, re-divide, and this division continues until a certain time of the year. It then ceases, and genital organs being developed, reproduction takes place by eggs.

The extreme vulnerability and reproductive power of many Chaetopodes, give rise to their frequent multiplication by artificial and accidental division. The fragments thus produced are finally developed, and the mutilated animal ultimately regains its lost parts.[5] Some have the power of voluntary division from the least handling of their body,[6] and these separated parts are probably developed to new individuals.

§ 163.

Most of the Annelides reproduce by sexual organs, and the few Lumbricini which, as just observed, multiply by fissuration, have probably, like their allied species, genital organs at certain seasons of the year.[1]

The eggs of the Annelides present nothing remarkable; they are always spherical, and have a chorion and thin vitelline membrane containing a finely-granular vitellus with a germinative vesicle and dot.[2] This

1 *Lumbriculus, Nais, Chaetogaster* and *Aeolosoma.*
2 See *Johnston,* in the Mag. of Zool. and Bot. I. 1837, p. 534.
3 With the Nereideae.
4 Fissuration with many species of *Nais,* has already been noticed by O. F. *Muller* (Naturgesch. einiger Wurm-Arten des süssen und salzigen Wassers. Taf. II. &c.). For that of *Nais proboscidea* and *Chaetogaster diaphanus,* see *Gruithuisen,* Nov. Act. Acad. Nat. Cur. XI. p. 243, Tab. XXXV. fig. 1, 3; XIV. p. 412, Tab. XXV. fig. 2. For that of *Aeolosoma,* see *Orsted* in *Kröyer's* Naturhist. Tidskrift. IV. Pl. III. fig. 7; and for that of *Nereis prolifera,* see *Muller,* Zool. Dan. II. p. 16, Tab. LII. fig. 6. This last species is a very young *Nereis.* It is probable that many other Branchiati multiply in the same way. *Quatrefages* (*Froriep's* neue Not. No. 726, 1845, p. 344) has recently recognised a *Syllis* in *Nereis prolifera.* *Sars* (Faun. litt. &c. p. 87, Taf. X. fig. 18, 19) has observed multiplication by transverse division with *Filograna implexa,* a young animal detaching itself from the caudal end of this *Serpula.* He observed a like division with a *Protula,* a genus allied to *Serpula.* According to *Milne Edwards* (Ann. d. Sc. Nat. III. 1845, p. 180, Pl. XI.) a single individual of *Myrianida fasciata,* which is allied to *Phyllodoce,* produces six young by as many successively disposed divisions. According to *Frey* and *Leuckart* (Beitr. &c. p. 94, Taf. II. fig. 1), there are with *Syllis prolifera* also, several young developed simultaneously, one after the other, at the caudal extremity.*
5 See the experiments upon this subject with the Lumbricini by *Réaumur, Bonnet, Trembley,* and *Roesel.* *Dalyell* (*Froriep's* neue Not. No. 331, 1840, p. 1) has observed a similar mode of reproduction with *Sabella.*
6 This has been observed by *Grube,* with *Polia delineata* (Zur Anat. d. Kiemenwürmer, p. 58). *Meckelia annulata* has also the same property.
1 *Aeolosoma.*
2 See *Wagner,* Prod. Hist. gener. loc. cit. Tab. I. fig. 9, 10 (*Sanguisuga* and *Nephelis*); *Stannius,* in *Muller's* Arch. 1840, Taf. II. fig. 1, 2 (*Arenicola piscatorum*); *Milne Edwards,* Ann. d. Sc. Nat. III, 1845, Pl. V. fig. 2, 3, Pl. IX. fig. 43, 44 (*Terebella* and *Protula*); and *Sars,* in *Wiegmann's* Arch. 1845, 1. Taf. I. fig. 13 (*Polynoë cirrata*). If the bodies which *H. Meckel* has figured (*Muller's* Arch. 1844, p. 481, Taf. XIII, fig. 13-23) as the eggs of *Lumbricus terrestris,* are really such, which I think is doubtful, they differ much

* [§ 162, note 4.] See in this connection, *Schultze* (Ueber die Fortpflanzung durch Theilung bei Nais proboscidea, in *Wiegmann's* Arch. 1850, p. 293). He has carefully described this form of multiplication with this animal, and according to him it is a true *fissuration,* and not a *gemmation,* as that of *Syllis,* described by *Frey* and *Leuckart.* See further *Leuckart,* Ueber die ungeschlechtliche Vermehrung bei Nais proboscidea, in *Wiegmann's* Arch. 1851, p. 134, Taf. II. fig. 1.-III.; and *Krohn,* Ueber die Erscheinungen bei der Fortpflanzung von Syllis prolifera und Antolytus prolifer. Ibid. 1852, p. 66. — ED.

vitellus is usually whitish or yellowish, but rarely of a more marked color.[3]

With the Hirudinei and Lumbricini, the spermatic particles are filamentoid and very active, while with the other Annelides they have the form of *Cercariae.*[4]

§ 164.

With the Hirudinei, and Lumbricini, the two sexes are always united in the same individual. The sexual organs consist of testicles, vasa deferentia, and vesiculae seminales; then, ovaries, oviducts, and the male and female copulatory organs. The female copulatory organs are upon the ventral surface of the anterior part of the body and behind the male organs — so that two individuals by placing together their anterior ventral surfaces in an inverse position, can be mutually impregnated.[1]

The excretory ducts of both sexes are often lined with a very delicate ciliated epithelium.

from the eggs of other Annelides, in containing between the vitellus and vitelline membrane a layer of caudate cells. These cells, of variable number and size, but always of uniform size in the same egg, have often been compared, from their form, to Naviculacae; see *Henle*, in *Muller's* Arch. 1835, p. 591, note, and *Hoffmeister*, De vermibus quibusdam, &c., Tab. II. fig. 14–17.
3 The vitellus is rose-colored, or greenish, with *Clepsine*, and violet with *Polynoë.*
4 The development of the spermatic particles of the Hirudinei and Lumbricini is very remarkable. The cell-membrane of the parent cells, in which the spermatic particles are usually developed, disappears before these last are developed. There are then small cells grouped around a discoid nucleus. These cells lengthen out, and finally become spermatic particles, and they remain attached to the disc until fully developed. If a bundle of these is placed in water, they separate and become intertwisted in the usual manner; see *Henle*, in *Muller's* Arch. 1835, p. 584, Taf. XIV. fig. 4, 6, 7, 9; *Külliker*, Beitr. zur Kenntn. d. Geschlechtsverhältnisse, p. 17, Taf. II. fig. 16, 18, 19; *H. Meckel, Muller's* Arch. 1844, p. 477, Taf. XIII. fig. 2–10 (*Sanguisuga, Pontobdella,* and *Branchiobdella*); and *Hoffmeister,* De vermibus quibusdam, &c., Tab. II. fig. 6–10. From *Stannius'* description and figures of the sperm of *Arenicola* (*Muller's* Arch. 1840, p. 375, Taf. XI. fig. 3–6); and *Rathké* of that of *Amphitrite auricoma* (Danzig. Schrift. loc. cit. p. 67, Taf. V. fig. 13); and *Quatrefages,* of that of *Nemertes mandilla* (Règne anim. illustr. Zooph. Pl. XXXIV. fig. 3–5), we can conclude that the spermatic particles of the other Annelides are de-

veloped like those of the Hirudinei and Lumbricini. While in the excretory ducts of the sperm, the spermatic particles are found in bundles; and when, as at the procreative period, many of these bundles are collected together, their very active, undulatory movements give a most wonderful appearance beneath the microscope; see *Morren,* loc. cit. p. 178, Tab. XXIV.–XXVIII., and myself in *Muller's* Arch. 1836, p. 42. Among the filamentoid spermatic particles of Hirudinei, those of *Branchiobdella* are worthy of special mention. One of their extremities is delicate and spirally turned (see my observations, *Muller's* Arch. 1836, p. 42, Taf. II. fig. 8), and terminates, according to *Külliker,* by a small vesicle (loc. cit. p. 18, Taf. II. fig. 16, f.). With the Branchiati, the cercarian-form predominates, according to *Quatrefages* (Comp. Rend. XVII. 1843, p. 424). With the Nemertini, they are either simply filamentoid (*Notospermus,* according to *Örsted,* Entwurf. einer Einth. d. Plattwürm. loc. cit. Taf. III. fig. 54) or more cercarian-form (*Nemertes,* according to *Quatrefages,* Règne anim. illustr. Zooph. Pl. XXXIV. fig. 6; and *Külliker,* Verhandl. d. schweiz. naturf. Gesellsch. bei ihrer Versammlung zu Chur. 1844, p. 91). For the spermatic particles of the Annelides, see especially *Külliker* in the Neue schweiz. Denkschr. VIII. p. 33.*
1 See *Bojanus,* Isis, 1818, Taf. XXVI. fig. 1; *Brandt,* Med. Zool. II. Taf. XXX. fig. 25 (*Sanguisuga medicinalis*); *Leo, Muller's* Arch. 1835, Taf. XI. fig. 3 (*Piscicola geometra*); *Morren,* loc. cit. Tab. XXVII.–XXXI.; and *Hoffmeister,* De vermibus quibusdam, &c., Tab. I. fig. 29, 30 (*Lumbricus* and *Enchytraeus*).†

* [§ 163, note 4.] For the spermatic particles of the *Hermella,* see *Quatrefages* (Ann. d. Sc. Nat. X. 1848, p. 167); he describes them as being of a cercaria-form. My own results on the spermatic particles of the Annelides and their development, do not agree with the view above expressed. Here, as elsewhere, I have found them to be the metamorphosed nucleus of the daughter-cells. It is true that with the Lumbricini they present some peculiarities, but these are apparent only. The mulberry-like mass to which they are here found adherent, is composed of the remains of the development, and the spermatic particles which seem to radiate from it in all directions present this appear-

ance because they are then just escaping from the daughter-cells, and the more or less adherence of the membrane of these last to the particles, gives the appearances above mentioned in the note. I have observed the same appearances with some of the Coleopterous insects, where the development occurs unmistakably in special cells. These particles are, according to my own observation, hair-like with the Hirudinei, and Lumbricini, but are pin-shaped with some of the Capitibranchiati.—Ed.

† [§ 164, note 1.] See also *Leydig, Siebol d* and *Külliker's* Zeitsch. III. IIft. 3, p. 318, and *Quatrefages,* Ann. d. Sc. Nat. XVIII. 1852, p. 299 (*Branchellion*).—Ed.

With the Nemertini, and Branchiati, the sexes are upon separate individuals, and the genital organs are composed simply of testicles and ovaries.

§ 165.

The structure of the genital organs of the Nemertini is yet quite obscure. The few researches hitherto made only furnish the general result that the sexes are separate.

There are numerous glandular follicles situated laterally in the parenchyma of the body between the skin and the intestinal canal, which are closely aggregated and serially arranged.

With some individuals, these follicles contain eggs, and with others, sperm. They ought, therefore, to be regarded as ovaries and testicles. Each follicle opens separately upon the surface of the body.[1]   There are very contradictory statements as to whether these animals have, or have not, copulatory organs.

According to some Naturalists, the worm-like organ, concealed in a canal extending along the back, and which, with both sexes, is often protruded and moved actively about, ought to be regarded as an excitatory organ, — although no connection between it and the testicles or ovaries, has as yet been found.   According to others, it is a proboscis unconnected with the genital organs.[2]

§ 166.

The disposition of the genital organs of the Hirudinei and Lumbricini, is essentially different.

The first have only two simple genital openings, — one male, the other female, both situated, one after the other, upon the median line of the ven-

1 See *Dugès*, Ann. d. Sc. Nat. XXI. 1830, p. 76, Pl. II. fig. 5 (*Polystemma (Prostomum) armatum*); *Johnston*, Mag. of Zool. I. p. 532, Pl. XVII. fig. 2°*, 6°*, Pl. XVIII. fig. 3° (*Nemertes* and *Borlasia*); *Örsted*, Entwurf. einer Beschreib. d. Plattwürm. p. 22, Taf. III. fig. 41 (*Tetrastemma varicolor*); *Kölliker*, Verhandl. d. schweiz. naturf. Versamml. zu Chur. p. 91 (*Nemertes*); and *Rathké*, Danzig. Schrift. loc. cit. p. 98 (*Borlasia striata*). This last author has not seen the orifices of the genital organs. *Quatrefages* (Règne anim. illustr. loc. cit. Pl. XXXIV. fig. 1, n. n.) did not see them with *Nemertes Camilla*, and *Johnston* is also silent on this subject. According to *Örsted* (Entwurf. &c. loc. cit. p. 25, Taf. III. fig. 47, of *Notospermus flaccidus*) the Nemertini secrete from the whole surface of their body, a gelatinous mucus, which surrounds the eggs, and thus forms an envelope into which they can draw their bodies. Something similar to this occurs with the Lumbricini and Hirudinei. See below.

2 The Nemertini being of distinct sexes, this organ can be regarded neither as a penis, nor as an everted spermatic vessel, as *Huschke* has done (Isis, 1830, p. 682, Taf. VII. fig. 5). More properly could it be considered, with *Örsted* (Entwurf. &c. p. 25), as an excitatory organ; although *Rathké* (Danzig. Schrift. loc. cit. p. 100, and Nov. Act. Acad. Nat. Cur. XX. p. 233) regards it as of a tactile, and *Kölliker* of a prehensile nature (Verhandl. d. schweiz. p. 90). Other observers agree with *Ehrenberg* (Symb. phys. loc. cit.) that it is an intestine and an everted œsophagus, its orifice being a mouth; but this is undoubtedly erroneous. With

Polystemma armatum (*Dugès*, Ann. d. Sc. Nat. loc. cit. p. 75, Pl. II. fig. 5) *Tetrastemma varicolor* (*Örsted*, Entwurf. &c. p. 23, Taf. III. fig. 41), and *Nemertes* (*Johnston*, Mag. of Zool. I. p. 530, fig. 2; *Quatrefages*, Règne anim. illustr. loc. cit. Pl. XXXIV. fig. 2, and *Kölliker*, Verhandl. d. schweiz. &c.) there is at the centre of this organ a dart pointing forward, which is horny, according to *Dugès*, and calcareous, according to *Örsted*. On each side of this dart, there is a reservoir of many others, smaller and yet imperfect, destined, probably, to replace the former when lost. *Dugès*, *Johnston*, and *Quatrefages*, who regard this organ as an intestinal canal, and *Kölliker*, who considers it prehensile, all regard these darts as a kind of teeth; but *Örsted* thinks they serve to excite the genital organs. For my part, they involuntarily remind me of the darts of the Helicina.

[Additional Note.] — I have now satisfied myself upon living individuals of *Tetrastemma*, that the eggs can escape from the visceral cavity through numerous lateral openings in the wall of the body. I am also satisfied that with the Nemertini, the walls of the digestive canal (the middle body-cavity according to *Quatrefages*) are not the points of departure of the genital organs, as *Quatrefages* thinks, and who also would regard as a digestive tube the snout of these animals, an organ which is yet enigmatical. The very detailed figures which this naturalist has given (loc. cit.) of the walls of the digestive canal of these animals, present nothing like an ovary, and show no trace of the presence of germs.

tral surface. The posterior opening connects with a short muscular canal which may be regarded as a reservoir of eggs.

From the base of this reservoir, a narrow spiral canal passes off, and bifurcating into two oviducts, terminates with two round ovaries.[1]

From the anterior opening, a long filiform penis may be protruded, which, when not erected, lies spirally concealed in a bulbous muscular sheath. A *Ductus ejaculatorius* extending from the seminal vesicles, opens into each side of this sheath. These seminal vesicles are formed each by a kind of continuation of the *vas deferens* into a varicose tortuous canal, which lies in the midst of a dense cellular tissue. The *Vasa deferentia* are narrow, and passing backwards along the sides of the body, receive upon their internal surface the short excretory ducts of the five, nine, or twelve pairs of round isolated testicles, which form a double row near the ventral cord.[2]

With many Hirudinei, a portion of the skin is connected with the sexual function. Such is the case with *Nephelis*, with which numerous cutaneous glands are developed upon the back and belly near the female genital opening. The skin soon has a bloated, transparent appearance, so that the animal appears to have a girdle about its anterior extremity. Before the deposition of the eggs, these glands secrete a substance which hardens in water, and surrounds the body of the animal like a horny belt. This belt is filled with a greater or less quantity of eggs; the animal then withdraws, or slips out from it, while its two extremities are closed up by its own elasticity; but the embryos developed in this egg-capsule are not thereby prevented from making their escape.[3]

The *Sanguisugae* form cocoons in a similar manner; but they are surrounded with a very thick, spongy substance.[4] The various species of *Clepsine* form sac-like capsules for their eggs, and which they usually carry about with them, attached under their belly, — shielding them with their body at the approach of danger.[5]

1 See *Brandt*, Mediz. Zool. II. p. 252, Taf. XXIX. A. fig. 45, 46; *Moquin-Tandon*, Monogr. loc. cit. p. 80, Pl. I.-III.; *Leo, Muller's* Arch. 1835, p. 424, Taf. XI. fig. 10 (*Sanguisuga, Aulacostomum, Nephelis, Pontobdella,* and *Piscicola*).

According to the careful researches of *Filippi* (Lettera sopra l'anat. e lo sviluppo delle Clepsine, p. 16. Tav. I. fig. 5), *Grube* (Untersuch. üb. d. Entwickl. d. Clepsinen. p. 6, Taf. III. fig. 3), and *Fr. Muller* (Muller's Arch. 1846, p. 138, Taf. VIII.), the two ovaries of *Clepsine* and *Nephelis*, consist of long flexuous cords surrounded by two more or less long muscular sheaths, which are uninterruptedly continuous into the oviducts; they receive the eggs as they are detached from the ovaries, and pass them along by peristaltic movements.

2 *Sanguisuga* has nine pairs of testicles (*Brandt* Med. Zool. II. p. 252, Taf. XXIX. A. fig. 32-44). The *Vasa deferentia* of the seven pairs with *Piscicola* are dilated before reaching the two seminal vesicles into two long and very flexuous tubes (*Epididymis*, according to *Leo*, loc. cit. 1835, p. 423, Taf. XI. fig. 10). With *Pontobdella*, there are

five pairs of these organs; with *Haemopis*, eight; and with *Aulacostomum*, twelve (*Moquin-Tandon*, Monogr. loc. cit. Pl. III. fig. 8; Pl. I. fig. 3, Pl. II. fig. 10). With *Nephelis*, the arrangement is different, there being on each side of the posterior part of the body, numerous testicular vesicles united in a botryoidal manner; see *Moquin-Tandon*, Monogr. loc. cit. Pl. III. fig. 4.*

3 See *Rayer*, Ann. d. Sc. Nat. IV. 1824, Pl. X. fig. 1-6, and *Moquin-Tandon*, loc. cit. Pl. VI. fig. 4, c-h. These cocoons are often found as brown scales, glued to aquatic plants. *Piscicola* forms similar cocoons, but they never have more than one egg each; see *Leo*, loc. cit. p. 425, Taf. XI. fig. 6; and *Brightwell*, Ann. of Nat. Hist. IX. 1842, p. 11.†

4 See *Rayer*, loc. cit. Pl. X. fig. 10, and *Moquin-Tandon*, loc. cit. Pl. V. According to *Brdeke* (Froriep's neue Not. No. 452, 1842, p. 183), the medicinal leech ejects from the mouth as a scum, the spongy envelope of these cocoons.

5 See *Grube*, Untersuch. über die Entwick. d. Clepsinen, 1844, p. 1.

* [§ 166, note 2.] For many valuable details on the genitalia of the Hirudinei, see the often-quoted and valuable works of *Leydig*, loc. cit. p. 120. It contains histological, as well as anatomical results. According to him, *Piscicola* has six, and not seven (*Leo*) pairs of testicles. — Ed.

† [§ 166, note 3.] See, for an histological examination of these genital glands in *Piscicola, Leydig*, loc. cit. p. 122, Taf. IX. fig. 43, e. 49, a. b. c. — Ed.

§ 167.

The genital organs of the Lumbricini are very difficult of dissection: for often those of both sexes are intimately united together into a common mass. It is certain, however, that the male and female orifices are always in pairs and situated at the anterior extremity of the body, near the ventral median line.[1]

These orifices communicate with more or less numerous glands, sacs, and pyriform or cylindrical vesicles.

Their nature as testicles, ovaries or sperm-receptacles, is known only by their contents. As yet it has been possible to trace only very imperfectly their excretory ducts. With some, two of these caecal organs have been observed intersuscepted in each other. The internal one contained spermatic particles, and should therefore be regarded as a testicle; while the outer one contained at its base, eggs and egg-germs, and ought therefore to be taken for an ovary.[2]

The larger Lumbricini appear to be without copulatory organs, the collar situated back of the genital orifices, taking their place. With many, it is situated chiefly on the back, but terminates on the belly with two long lateral swellings, which, during coition, seize those of the other individual.[3] This collar, moreover, is composed of a mass of glandular follicles, which copiously secrete, during the sexual period, a white, viscous liquid. It is then very fully developed, but at other periods it is scarcely visible. The belt which is developed near the genital openings of the smaller Lum-

---

1 With *Lumbricus terrestris*, the two anterior genital openings are male, and the two posterior, female. These have been figured by *Montègre* (loc. cit. fig. 2, a. c.), *Leo* (De Struct. Lumbr. terrestr. Tab. I. fig. 2), and by *Morren* (loc. cit. Tab. III. fig. 2). With *Saenuris*, and *Nais*, I have also found these two pairs of genital openings.

2 This invagination of the testicle in the ovary has been distinctly observed by me with *Saenuris variegata*, and *Nais proboscidea*. From *H. Meckel's* late researches upon the very complicated genital apparatus of *Lumbricus terrestris*, it appears that there are three pairs of seminal vesicles and testicles; these last being intimately joined with as many ovaries (*Muller's* Arch. 1844, p. 480, Taf. XIII. fig. 12). It is probable that here the testicles and ovaries are also invaginated, and that the vesicles which have usually been taken for testicles are only vesiculae seminales; see *Morren*, loc. cit. p. 175, Tab. VII.-X. and *Treviranus*, in his Zeitsch. f. Physiol. V. p. 154, Taf. VII. However, as yet I have been unable to trace to their termination the excretory ducts of the testicles and ovaries which are invaginated together. This point is all the more difficult, for, as *Dugès* appears to represent (Ann. d. Sc. Nat. XV. 1828, p. 328, Pl. IX. fig. 2, or 1838, 1830, Taf. III. Tab. 9, fig. 2), the vas deferens is probably invaginated in the oviduct. Many observers have gone so far as to think that the eggs having escaped from the ovary, and fallen into the cavity of the body, pass gradually to its posterior portion, and are there evacuated through invisible openings. On this account, several of them have taken for eggs and embryos, the collections of horny spines, and vibrios, which are often found in these animals. See above, § 145, note 1; and *E. Home*, Lect. on Comp. Anat. IV. 1823, Pl. CXLIX.

I have always been astonished that, at the epoch of procreation with *Saenuris*, *Euaxes*, and *Nais*, the two anterior genital openings should communicate with two caeca which contain sperm and long

bundles of spermatic particles, but never their cells of development. *Dugès* has made a similar observation with his *Nais filiformis* (Ann. d. Sc. Nat. loc. cit. p. 320, Pl. VII. fig. 2), only he does not specify the contents of the organs. *Menge*, also, has observed these two caeca with *Euaxes*, but he unhesitatingly regards them as testicles (*Wiegmann's* Arch. 1845, I. p. 32, Taf. III. fig. 2, aa. fig. 3). Never having seen any connection between these caeca and the testicles behind them, I am disposed to think that the two posterior genital openings of some Lumbricini, are the common orifices of the invaginated testicles and ovaries, while the anterior caeca, which are filled at certain times with sperm, are two isolated *Receptacula seminis*. During the mutual copulation, the sperm will pass from the testicles into these reservoirs, in order to be used during the subsequent deposition of the eggs.

From *Hoffmeister's* description (Die bis jetzt bekannt. Arten aus der Familie der Regenwürmer, 1845, p. 15) of the copulatory act with *Lumbricus agricola*, it would appear that the sperm remains equally distant from the female organs, being received into special fossae, which correspond perhaps to the *Receptacula seminis*. *Nais proboscidea*, although having a pair of genital openings, has only one testiculo-ovarian canal, both of which although invaginated, have a very active and independent peristaltic action. They are bifurcated anteriorly. See *Gruithuisen* (Nov. Act. Acad. Nat. Cur. XI. p. 246, Tab. XXXV. fig. 4, 5), who has very correctly perceived the eggs in the bottom of the ovarian sac, but not the nature of the invaginated testicular canal.

3 With *Lumbricus olidus*, the two copulating individuals seize each other so slightly by their collars, that each of these animals completely envelops the other by this organ; see *Hoffmeister*, in *Wiegmann's* Arch. 1843, I. p. 190, and, De vermibus quibusdam, Tab. I. fig. 30.

bricini at this period, is of an analogous nature. It is also composed of numerous cutaneous glands, closely aggregated, and extending completely over many segments of the body.[4] The secretion of this collar is like that observed with the Hirudinei, probably for the formation of cocoons. But these cocoons differ from those of the Hirudinei in having the place of their opening prolonged into a long, narrow neck.[5]

§ 168.

The Branchiati resemble the Arthropoda in their annulated body, their distinct head endowed with organs of sense, the structure of their nervous system, and the development of their locomotive organs; but, from the simplicity of their locomotive apparatus, and the complete absence of copulatory organs, they would be carried towards the Zoophytes.

Here the sexes are separate, and the genital organs of both the Capitibranchiati, and Dorsibranchiati, appear as simple glandular bodies, ovaries or testicles, which project from the ventral surface into the cavity of the body between the fasciculi of the cutaneous muscle.[1] At the sexual period, they are filled with eggs, or spermatic particles, although at other times they can scarcely be seen.[2]

Neither the testicles nor the ovaries have excretory ducts which open upon the surface of the body. The sperm and eggs escape into the cavity of the body, which, during this period is thereby filled throughout.[3]

It is possible that the scarcely visible orifices said to be concealed be-

4 *Saenuris*, *Nais*, &c.; see *Gruithuisen*, loc. cit. Tab. XXXV. fig. 5, b.b.

5 With the large species of *Lumbricus*, each cocoon has from one to six eggs (*L. Dufour*, Ann. d. Sc. Nat. XIV. 1828, p. 216, Pl. XII. B. or, *Froriep's* Notiz. No. 472, 1828, p. 149, fig. 13–16; and *Hoffmeister*, De vermibus quibusdam, Tab. I. and Die Arten aus der Familie, &c., p. 16, 25, 42). With the smaller Lumbricini, as with *Saenuris*, *Euaxes*, *Nais*, &c., the cocoons contain nearly always five to eight eggs (*Dugès*, loc. cit. XV. Pl. VII. fig. 5, *Nais*). Most of these cocoons have appendages by which they are attached to vegetables and other bodies. *Hoffmeister* (Die Arten aus der Fam. &c. p. 42, fig. 9, c.) has figured a very remarkable husk-shaped cocoon of a new species, *Criodrilus lacuum*.

1 See *Treviranus*, Zeitsch. f. Physiol. III. 1827, p. 165, Taf. XIII. fig. 17, 18 (*Aphrodite*); *Rathké*, De Bopyro et Nereide, p. 39, Tab. II. fig. 12, 1. (*Nereis*), and Danzig. Schrift. loc. cit. p. 66, Taf. V. fig. 6, hh. fig. 11, aa. (*Amphitrite*); *Grube*, Zur Anat. d. Kiemenwürmer, p. 16, Taf. I. fig. 1, 2, m. (*Arenicola*), p. 44, Taf. II. fig. 6, y. z. (*Eunice*); also, Nov. Act. Acad. XX. p. 201, Tab. X. fig. 13, 15, m. (*Ammotrypane*). *Rathké's* and *Grube's* opinion upon the presence of both male and female organs with the same individual is only an uncertain supposition, founded upon no histological examination of the parts.

2 This condition of the genital glands after the procreative season, is the reason why, as yet, we possess so few facts as to their structure. Most observers, and among them *Rathké* and *Grube*, are

of the opinion that the Branchiati, like the Lumbricini, are hermaphrodites. But *Quatrefages*, from his knowledge of the development of the spermatic particles, has recognized separate sexes with the most different species, thus : *Terebella*, *Sabella*, *Aricinella*, *Nephtys*, *Syllis*, *Glycera*, *Eunice*, *Sigalion*, *Phyllodoce*, *Nereis*, and *Aphrodite* ; see Comp. Rend. XVII. 1843, p. 423. But before this, *Stannius* had concluded that the sexes were separate with *Arenicola*, from a difference in the contents of the cavity of the bodies of different individuals (*Muller's* Arch. 1840, p. 375). The glands at the cephalic extremity of the Branchiati which live in cases, and which *Grube* has regarded as male genital organs, are certainly not such, for they occur with both sexes, and do not change in size during the procreative season. (See § 161, note 4.) *

3 According to *Quatrefages* (Compt. rend. XVII. 1843, loc. cit.), the parent sperm-cells leave the testicle before the formation of the spermatic particles, which occurs in the cavity of the body. This is confirmed with *Arenicola*, by *Stannius* (*Muller's* Arch. 1840, loc. cit.). According to *Krohn* (*Wiegmann's* Arch. 1845, I. p. 182), the eggs and the spermatic particles, with *Alciopa*, are developed free in the visceral cavity, without the intervention of special organs, ovaries and testicles. *Frey* and *Leuckart* (Beitr. &c. p. 88) think they have observed the same fact with *Nereis*, *Syllis*, *Phyllodoce*, *Aonis*, *Ammotrypane*, *Ephesia*, *Hermella*, *Vermilia*, *Fabricia*, and *Spirorbis* ; they speak of the presence of ovaries and testicles in certain Annelides (*Aphrodite*, *Arenicola*) as the exception.

* [§ 168, note 2.] According to *Quatrefages* (Ann. d. Sc. Nat. X. 1848, p. 46) the sexes are separate with the *Hermella*. Both the testicle and the ovary consist of a delicate areolar tissue adherent to the inferior internal surface of the general cavity of the body. These genital organs are evidently tem-

porary, for they are not found in many individuals, having, probably, quite disappeared from atrophy after the procreative period. This fact should be remembered in the study of the genitalia of other Annelides. — ED.

tween the feet of many Branchiati, serve for the escape of the sperm and eggs.[4] With others, the cavity of the body opens outwardly, probably by a loss of the last segment, especially with those which are viviparous.[5]

The water is undoubtedly the medium of fecundation, and receives the sperm from the males, probably through orifices like those which serve for the escape of eggs with the female.

With the viviparous Branchiati, water filled with sperm can enter the body and fecundate the eggs through these same openings.

§ 169.

The development of the Annelides as far as yet known, occurs after two different types; but it always commences with a complete segmentation of the vitellus.

I. With the Hirudinei, after the vitellus has divided into many large cells, a central one becomes distinguished from the others by its still further division; this becomes the digestive tube. The others, still dividing, form a primitive embryonic part in which appears the future ventral and nervous portion.

The embryo is at first spherical, and ultimately is covered with a delicate ciliary epithelium. A kind of sucker is then developed upon a certain point of its surface; this connects with the stomach, and through it is received, for food, the albumen surrounding the embryo. It then gradually lengthens, and, losing its ciliary epithelium before the escape from the egg, a sucker appears upon the posterior extremity, and it finally becomes fully developed without a Metamorphosis.[1]

II. With the Branchiati, there is a complete metamorphosis. The segmentation of the vitellus is uniform throughout, and this last is finally changed into a round embryo — which, escaping from the egg, swims freely about like an Infusorium, by means of the ciliated epithelium which covers its whole body. The embryo then lengthens, and the epithelium disappears

---

4 According to *Milne Edwards'* observations upon several Capitibranchiati, as *Terebella, Serpula, Protula,* &c., the eggs are glued together in masses by an albuminous substance, and attached to the stones of the anterior border of their cases; see Ann. d. Sc. Nat. III. 1845, p. 148, 161, Pl. V. fig. 1, Pl. VII. fig. 28, Pl. IX. fig. 42. With *Polynoë cirrata,* on the other hand, masses of eggs are attached and borne about on the scales of their body; see *Sars,* in *Wiegmann's* Arch. 1845, I. p. 13, Taf. 1. fig. 12. With the females of *Exogone* and *Cystonereis,* the eggs are situated in longitudinal rows upon the ventral surface; see *Örsted,* in *Wiegmann's* Arch. 1845, I. p. 21, Taf. II. fig. 4, and *Külliker,* in an as yet unpublished memoir for the Helvetic Society, titled: *Einige Worte zur Entwickelungsgeschichte von Eunice, von H. Koch* in Trieste, mit einem Nachwort von *Külliker.*

[Additional note.] The often-quoted memoir of *Koch* and *Külliker* on the development of the An-

nelides has recently appeared in the Neue Schweiz. Denkschr. VIII.*

5 According to my friend *H. Koch* of Trieste (in the MS. just indicated), the eggs of a species allied to *Eunice sanguinea,* are developed in the cavity of the female body, whence the young escape through a rupture of its posterior extremity.

1 See *F. de Filippi,* Lettera sopra l'Anatomia, e lo sviluppo delle Clepsine, Pavia, 1839, Tav. II.; *Grube,* Untersuch. über die Entwick. d. Clepsine, p. 15, Taf. I., and *Frey,* Zur Entwickel. von. Nephelis vulgaris, in *Froriep's* neue Not. No. 807, 1846, p. 228. The old observations of *E. H. Weber* (*Meckel's* Arch. 1828, p. 366, Taf. X. XI.) and *R. Wagner* (Isis, 1832, p. 398, Taf. IV.) agree very well with those of *Filippi*

As yet, we possess nothing upon the development of Lumbricini, whose young, as is known, like those of the Hirudinei, leave their cocoons without undergoing any metamorphosis.†

---

* [§ 168, note 4.] According to *Felix Dujardin* (Ann. d. Sc. Nat. XV. 1851, p. 298) *Exogone pusilla* is androgynous. Beside the well-known pediculated ovarian sacs on the ventral surface, each segment of the body, except the first two, has, with this species, a dorsal, fusiform cirrus, in which are developed spermatic particles. This ob-

servation, from its singularity, requires confirmation. — ED.

† [§ 169, note 1.] For the embryology of *Nemertes,* see *Desor,* Boston Jour. Nat. Hist. VI. p. 1. The general facts accord with those mentioned in the text. — ED.

except upon the belt-like parts of the two extremities. The future anterior extremity is directed in front during the motions of the animal, and eyes appear upon it; while the other extremity is gradually divided into segments upon which bristles and feet appear.[2]

While the embryos are thus acquiring the adult form, there appear upon the cephalic extremity and upon the sides of the body, tentacles, cirri, and branchiae, of forms which vary according to families, genera and species. The development of the digestive and circulatory organs occurs also with equal pace.[3]

2 See *Lorén* in *Wiegmann's* Arch. 1842, I. p. 302, Taf. VII. (*Nereis*); *Sars*, Ibid. 1845, I. p. 12, Taf 1. fig. 1–21 (*Polynoë*); *Örsted*, Ibid. p. 20. Taf. II. (*Exogone*); and *Milne Edwards*, Ann. d. Sc. Nat. III. 1845, p. 145, Pl. V.–IX., or, *Froriep's* neue Not. No. 721, p. 257 (*Terebella, Protula*, and *Nereis*). *Kölliker* (in MS. already cited) has also observed the development of an *Exogone*, and of a *Cystonereis*, an allied genus. Here the embryo is not formed through a complete and uniform segmentation of the vitellus, but, as with the Hirudinei, the formation is preceded by an irregular division of that portion to be the ventral and nervous parts. He, at the same time, calls the attention to a figure of *Milne Edwards*, representing the development of *Protula*, from which it would appear that other Branchiati also are developed like the Hirudinei; see Ann. d. Sc. Nat. loc. cit. Pl. IX. fig. 47.*

3 One ought therefore to be careful about forming distinct genera from these larval Branchiati. Thus, *Sabellina brachycera*, described by *Dujardin* (Ann. d. Sc. Nat. XI. 1839, p. 291, Pl. VII.

fig. 6), is only a larval *Terebella*, as will be seen by referring to *Milne Edwards'* figures of the development of *Terebella nebulosa* (Ann. d. Sc. Nat. loc. cit. Pl. VII. fig. 24, 25). *Anisomelus luteus*, of *Templeton* (Transact. Zool. Soc. II. 1841, p. 27, Pl. XV. fig. 9–14), is perhaps only a young *Serpula*.

The absence of branchiae and blood-vessels which *Quatrefages* has noticed with many small Branchiati of which he has made new genera (as *Aphlebine*, and *Doyeria*, &c.), would lead one to suspect that they are only larvae; see Ann. d. Sc. Nat. I. 1844, p. 18, or *Froriep's* neue Not. No. 726, p. 341. *H. Koch* (see above, § 168, note 5) has lately observed that the young individuals found in the body of *Eunice* are identical with the *Lumbrinereis* of *De Blainville*.

The new animal described by *Muller* and *Busch* (*Muller's* Arch. 1846, p. 104, Taf. V. fig. 3–5, and 1847, p. 187, Taf. VIII. fig. 1–3) under the name of *Mesotrocha sexoculata*, appears likewise to be only a young larva of an Annelid.†

* [§ 169, note 2.] For the embryology of *Polynoë*, see *Desor*, loc. cit. p. 12. It agrees closely with that of *Nemertes*; see also *Max Muller*, in *Muller's* Arch. 1851, p. 323. — ED.

† [§ 169, note 3.] See *Quatrefages* (Sur l'Embryogenie des Annélides, in Ann. d. Sc. Nat. X. 1848, p. 153). – ED.

BOOK NINTH.

ACEPHALA.

CLASSIFICATION.

§ 170.

The Acephala are principally characterized in having a headless body, and a very large mantle, which so envelops the body, that there is a spacious and more or less closed cavity in which the oral and anal orifices are often entirely concealed.

Their body is either wholly asymmetrical, or divided into a right and a left side. In this last case, the organs, excepting the digestive canal, are in pairs; and the two sides are perfectly symmetrical, or one is developed at the expense of the other. All Acephala are aquatic; many are permanently attached during life; others creep about, and a few only can swim freely. Copulatory organs are wanting throughout.

ORDER I. TUNICATA.

Body wholly asymmetrical and so enclosed in the mantle, that there are only two narrow openings.

Family: Ascidiae.

Genera: A. Compositae.

Didemnum, Diazona, Aplidium, Botryllus, Botrylloïdes, Leptoclinum, Eucoelium, Synoecium, Polyclinum, Sigillina, Perophora, Pyrosoma.
B. Simplices.

Clavelina, Phallusia, Rhopalaea, Boltenia, Cynthia, Chelyosoma.

Family: Salpinae.

Genus: Salpa.

ORDER II. BRACHIOPODA.

Animals which are symmetrical and bivalved, and whose widely-open mantle encloses two fringed, arm-like, protractile tentacles.

Genera: *Orbicula, Terebratula, Lingula.*

ORDER III. LAMELLIBRANCHIA.

Animals which are symmetrical and bivalved, and whose more or less
closed mantle encloses two pairs of lamelliform tentacles and branchiae.

SUB-ORDER I. MONOMYA.

FAMILY: OSTRACEA.

Genera: *Ostrea, Anomia.*

FAMILY: PECTINEA.

Genera: *Pecten, Spondylus, Lima.*

FAMILY: MALLEACEA.

Genera: *Malleus, Perna, Crenatula.*

SUB-ORDER II. DIMYA.

FAMILY: AVICULACEA.

Genera: *Avicula, Meleagrina, Pinna.*

FAMILY: ARCACEA.

Genera: *Arca, Pectunculus, Trigonia, Nucula.*

FAMILY: NAIADES.

Genera: *Anodonta, Unio.*

FAMILY: MYTILACEA.

Genera: *Mytilus, Modiola, Lithodomus, Tichogonia.*

FAMILY: CHAMACEA.

Genera: *Chama, Isocardia.*

FAMILY: CARDIACEA.

Genera: *Cardium, Lucina, Hiatella, Cyclas, Piscidium, Tellina, Psammobia, Venus, Cytherea, Venerupis, Mactra, Lutraria, Ungulina.*

FAMILY: PYLORIDAE.

Genera: *Mya, Solen, Solenomya, Panopaea.*

SUB-ORDER III. INCLUSA.

FAMILY: TEREDINA.

Genera: *Pholas, Teredo.*

16*

FAMILY: ASPERGILLINA.

Genera: *Aspergillum, Claragella.*

BIBLIOGRAPHY.

Poli. Testacea utriusque Siciliae eorumque historia et anatome. 1791–95.

J. Rathke. Om Dammuslingen, in the Skrivter af Naturhistorie-Selskabet, IV. Kjöbenhavn, 1797, p. 139.

Cuvier. Mémoire sur l'animal de la Lingule, in the Ann. du Mus. d'Hist. Nat. I. 1802, p. 69. Mémoire sur les Thalides et sur les Biphores, Ibid. IV. 1804, p. 360. Both of these are included in his Mém. pour servir à l'histoire et à l'anatomie des mollusques. Paris, 1817.

Schalk. De Ascidiarum structura, Dissert. Hal. 1814.

Savigny. Mémoires sur les animaux sans vertèbres, Pt. II. 1816. Recherches anatomiques sur les Ascidies composées et sur les Ascidies simples. Also, Isis, 1820, lit. Anz. p. 659, Taf. XI.–XXI.

Carus. Beiträge zur Anatomie und Physiologie du Seescheiden (Ascidiae), in *Meckel's* deutsch. Arch. 1816, p. 569, and Nov. Act. Acad. Leop. Carol. X. 1821, p. 423, Tab. XXXVI. XXXVII.

Cuvier. Mémoires sur les Ascidies et sur leur Anatomie; in the Mém. du Mus. d'hist. Nat. II. 1815, p. 10; also Isis, 1820, p. 387, Taf. 8, 9.

Chamisso. De animalibus quibusdam e classe vermium Linnaeana. Fasc. I. De Salpis. 1819.

Bojanus. Ueber die Athem-und Kreislaufwerkzeuge der zweischaligen Muscheln. Isis, 1819, p. 42, Taf. I. II., 1820, p. 404, and 1827, p. 752, Taf. IX.

Eysenhardt. Ueber einige merkwürdige Lebenserscheinungen an Ascidien, in the Nov. Act. Acad. Leop. XI. 1823, p. 250, Tab. XXXVI. XXXVII.

Pfeiffer. Naturgeschichte deutscher Land-und Süsswasser-Mollusken. Abth. II. 1825.

Unger. De Anodonta anatina. Dissert. Vindobon 1827.

Carus. Neue Untersuchungen über die Entwickelungsgeschichte unserer Flussmuschel, in the Nov. Act. Acad. Leop. XVI. 1832, pt. I. Tab. I.–IV.

Meyen. Beiträge zur Zoologie. Abhandl. I.; Ueber die Salpen. Ibid. p. 363, Tab. XXVII.–XXIX.

Owen. On the Anatomy of the Brachiopoda, in the Trans. Zool. Soc. I. 1835, p. 145, Pl. XXII. XXIII.; also, in Isis, 1835, p. 143, and Ann. d. Sc. Nat. III. 1835, p. 52.

Deshayes. Conchifera. Cyclop. Anat. Phys. I. p. 694. London, 1836.

Eschricht. Anatomisk-physiologiske Undersögelser over Salperne. Kjöbenhavn, 1840; also, in Isis, 1842, p. 467, Taf. II. III., and: Anatomisk Beskrivelse af Chelyosoma Mac-Leayanum. Kjöbenhavn, 1841.

Milne Edwards. Observations sur les Ascidies composées. Paris, 1841.

Garner. On the Anatomy of the Lamellibranchiate Conchifera, in the Trans. Zool. Soc. of London, II. 1841, p. 87, Pl. XVIII.–XX.

Neuwyler. Die Generationsorgane von Unio und Anodonta, in the Neuen Denkschrift. der allg. schweizerischen Gesellsch. f. die gesammten Naturwissensch. VI. 1842, p. 1, Taf. I.–III.

Vogt. Anatomie der Lingula anatina. Ibid. VII. 1843, p. 1, Taf. I. II.

Van Beneden. Mémoire sur l'Embryogénie, l'Anatomie et la Physiologie des Ascidies simples, &c., in the Bullet. de l'Acad. royale de Belgique, XIII. No. 2.

ADDITIONAL BIBLIOGRAPHY.

Kölliker. Ueber das Vorkom. d. Holzfas. im Thierreich., in the Ann. d. Sc. Nat. 1846, p. 193, Pl. V.–VII.

Van Beneden. Recherches sur l'Embryogénie, l'Anatomie, et la Physiologie des Ascidies simples, in the Mém. de l'Acad. Roy. de Belgique, XX. 1847.

Frey and Leuckart. Beiträgen zur Kenntniss der wirbellosen Thiere mit besonderer Berucksichtigung der Fauna des Norddentschen Meeres. Braunschweig, 1847, p. 46, Anatomie des Pfahlwurmes (*Teredo navalis*).

Deshayes. Exploration scientifique de l'Algérie, pendant les années 1840, 1841, 1842. Histoire naturelle des Mollusques, avec un Atlas de 117 Planches. Paris, 1847.

Ed. Forbes and Hanley. A History of British Mollusca and their Shells. 4 vol. London, 1853. [Contains many anatomical details.]

Dalyell, T. G. Rare and remarkable animals of Scotland, represented from living subjects, with practical observations on their nature. Vol. II. London, 1848, p. 138–173, Pl. XXXIV.–XLIII. (Ascidiae).

Lovén. Om utvecklingen af Mollusca acephala, Oversigt af k. Vet. Akad. Förhandl. 5te Argängen, Dec. 1848. Stockholm, 1849, p. 233–257 ; or, its translation in *Müller's* Arch. 1848, p. 531 ; or, in *Wiegmann's* Arch. 1849, p. 312.

Quatrefages. Mémoire sur le Genre Taret (*Teredo* Lin.), in the Ann. d. Sc. Nat. XI. 1849, p. 19.

——— Mémoire sur l'embryogénie des Tarets. Ibid. p. 102.

T. Rupert Jones. Cyclop. Anat. and Physiol. IV. p. 1185, Art. *Tunicata.*

G. A. F. Keber. Beiträge zur Anatomie und Physiologie der Weichtheire, Königsberg, 1851. [Devoted to the nervous, circulatory, and respiratory systems of the fresh-water Bivalvia.] — ED.

CHAPTER I.

CUTANEOUS SYSTEM.

§ 171.

The body of the Acephala is enveloped in a special mantle, which, with the Tunicata, is composed of a leathery, cartilaginous, or gelatinous substance, scarcely at all irritable.[1] But with the Lamellibranchia, and Brachiopoda, it is composed of a contractile, fleshy membrane. With the Tunicata, it com-

---

1 The mantle is leathery with *Cynthia*, cartilaginous and hard with *Phallusia*, cartilaginous and soft with *Salpa*, and gelatinous with *Clavelina*, *Diazona*, *Aptidium*, *Botryllus*, and *Pyrosoma*.

pletely surrounds the body and has only an oral and anal opening; [2] and with the compound species, it is continuous with the common substance which contains the individuals and binds them into more or less regular groups, and is, therefore, analogous to a corallum. With the Lamellibranchia, and Brachiopoda, it is more or less open, or even may be wholly divided into halves; [3] it has here the property, especially upon its borders, of secreting calcareous matter for the formation of the shell.

§ 172.

With the Tunicata, the mantle is remarkable both for its histological structure, and its chemical composition. Recent investigations have shown that, with the Ascidiae and Salpinae, it is composed of *Cellulose* and therefore of a non-azotized substance. [1]

Its anatomical structure is quite complicated. Usually it can easily be separated into two or three layers, the internal one of which is composed, in some species, of a lamellated epithelium formed of a single layer of polygonal nucleated cells. [2]

Its principal mass in both the compound and simple forms of this order, is formed of a single, or a double confluent layer of a homogeneous transparent substance, through which are scattered granules, nuclei, groups of pigment molecules, cells, fibres, and crystals of carbonate of lime, — all varying according to genera and species, and often differently arranged in one and the same species. [3] But in each species, they are variously arranged in the inner portion of this mantle-substance. [4] In some species

2 These openings are properly only simple orifices of the cavity of the body, and correspond to the respiratory tubes of certain Lamellibranchia ; see below, § 190.

3 With *Mya, Panopaea, Pholas, Teredo, Aspergillum*, the mantle is almost entirely closed, but it has two long fissures at each extremity with *Solen, Cyclas, Tellina, Mytilus, Lithodomus* and others ; with the Ostracea, Pectinea, Arcacea, Naïades, and Brachiopoda, it is entirely open.

1 This important fact was first stated by *Carl Schmidt* (Zur. vergleich. Physiol. d. wirbellosen Thiere. 1845, p. 61), with *Cynthia mamillaris*, and has subsequently been confirmed by *Lowig* and *Kölliker*, after the most careful investigations upon the entire order of Tunicata (Compt. rend. 1846, p. 38). These two authors found this non-azotized substance, particularly in the different species of *Phallusia, Cynthia, Clavelina, Diazona, Botryllus, Didemnum, Aplidium, Salpa*, and *Pyrosoma*; but not with the other Mollusca, nor with the Annelides, the Helminthes, the Echinodermata, the Acalephae, and the Polypi. It is certainly wanting in the true Infusoria, for *Frustulia salina*, which *Carl Schmidt* cites as belonging to this order and as containing cellulose, is evidently a vegetable. *Lowig* and *Kölliker* justly fear, moreover, that this discovery will be quickly seized by those who deny that there is any limit between the animal and vegetable kingdom (see loc. cit. p. 8). They seek, therefore, to oppose this view by insisting upon the circumstance that this cellulose is never found in a pure state in the mantle of the Tunicata, but always combined with other substances, and that, moreover, no animal has as yet been found entirely composed of this substance.

2 *Phallusia mamillaris, sulcata, Cynthia papillata, pomaria*, and *Salpa bicaudata*.

3 This basement substance is homogeneous, and has the same chemical properties as cellulose.

4 *Kölliker* has made very detailed investigations upon the structure of this mantle. He has kindly allowed me to communicate his results, and authorized me to make use of them without waiting for the publication of his work in common with *Lowig* (Ueber das Vorkommen von Holzfaser im Thierreich). According to them, the middle layer of the mantle of *Phallusia monachus*, and *sulcata, Clavelina lepadiformis*, and *Aplidium gibbulosum*, contains numerous nuclei and starlike crystals lodged in a transparent structureless substance. But the external layer of this organ is filled with very large round cells with very thin walls, containing no nucleus, but filled with a transparent liquid. With *Clavelina lepadiformis*, the peduncle and branches of the whole mantle are so crowded with non-nucleated cells, some round and others elongated, that the basement-substance is apparently absent. It has therefore quite the aspect of a vegetable tissue. With *Aplidium gibbulosum*, and *Botryllus violaceus*, the cells of the external layer contain carbonate of lime which ultimately so increases that it gives them a petrified aspect. With *Didemnum candidum*, these petrified cells have calcigerous rays and are so numerous that the whole mass of this compound Ascidian appears filled with white starlike corpuscles.

According to *Milne Edwards*, this is true also of *Leptoclinum maculosum* (Observ. sur les Ascidies composées, p. 81, Pl. VIII. fig. 2).

With *Diazona violaceum, Pyrosoma giganteum, Botryllus polycyclus, Salpa maxima*, and *bicaudata*, the mantle is without these elegant cells, and in the basement-substance are found only granules and nuclei, and with *Diazona*, in addition, are pigment-granules, and crystalline points, or calcareous concretions.

of this order, the mantle receives, moreover, numerous blood-vessels, or ramified prolongations of the body of the animal.[5]

These last are spherical or star-like, with *Salpa maxima*, and dendritic with *Salpa bicaudata*; with these species they are not soluble in hydrochloric acid, and are therefore probably composed of silex. In the mantle of *Botryllus*, there are, in certain places, peculiar flexuous fibres running in all directions. If these are treated with potassa they will appear evidently composed of celluloso.

According to *Kölliker*, the structure of the mantle of *Cynthia papillata* is still more complicated. Its middle layer is composed of longitudinal and circular flexuous non-azotized fibres. Between these lie granules, nuclei, crystals and cells; these cells are nucleated, and contain, sometimes pigment granules, and sometimes daughter-cells which gives them the appearance of those of cartilage.

Kölliker was unable to determine the structure of the third and external layer which is horny, for he had at his disposal only alcoholic specimens of this *Cynthia*. He saw however that it united with the middle layer to form the spines which project from the surface of the skin.

In the mantle of *Cynthia pomaria*, the longitudinal fibres predominate, and between them lie crystals, round pigment-cells, and other cells which are peculiar and filled with yellow corpuscles; and finally, a third variety arising from the transformation of the pigment-cells, whose walls are gradually thickened and ultimately split up into filaments, forming concentric layers around the cell-cavity. When subjected to potassa, these cell-membranes are decomposed, like the principal fibres, into an insoluble, non-azotized substance, while all the other elements of the mantle entirely disappear under the action of this agent.

The researches of *Kölliker* and *Löwig* upon the mantle of the Tunicata, have been recently published in the Ann. d. Sc. Nat. V. 1846, p. 193, Pl. V.–VII.*

5 Blood-vessels are found in the mantle of various *Phallusia*; they are spread out in a reticulated manner, especially in the external layer. See *Cuvier*, Mém. sur les Ascidiés, &c., p. 16, Pl. III. fig. 1 (*Phallusia mamillaris*); *Savigny*, Mém. &c. p. 102, Pl. IX. fig. 1. B. (*Phallusia sulcata*); and *Delle Chiaje*, Descrizione e notomia degli animali invertebrati della Sicilia citeriore Tom. III. 1841, p. 33, Tav. LXXXIV. fig. 2 (*Phallusia monachus*).

---

* [§ 172, note 4.] The presence of cellulose in animal tissues is a fact of no inconsiderable importance in animal and vegetable physiology. The subject has recently received much attention from *Schacht* (*Müller's* Arch. 1851, p. 176), and his conclusions are sufficiently interesting to be presented in full.

"1. In the mantle of the Ascidiae there is a substance insoluble in caustic potass, but soluble in sulphuric acid, which is turned to a beautiful blue by iodine and sulphuric acid, and which therefore consists entirely of cellulose. This substance constitutes the interstitial substance of the cells; in the mantle of *Phallusia* it is homogeneous, but in Cynthia it occurs for the most part in a fibrous form.

"2. The mantle of the Ascidiae contains beside this cellulose, another material which is soluble in caustic potass, but insoluble in sulphuric acid, and not colored blue by iodine and sulphuric acid, and which consequently is not cellulose; in the mantle of *Phallusia* it is only sparingly present, but in Cynthia and the new Chilian Ascidian, it is much more abundant and alone constitutes the corneous epidermis of their mantle.

"3. The membrane of the cells in the mantle of *Phallusia* does not consist of cellulose, it is colored brown by iodine and sulphuric acid; it is soluble in caustic potass, and behaves exactly like an animal membrane as do the nuclei and vessels.

"4. In the mantle of *Phallusia*, cells abound in a homogeneous, interstitial substance composed of cellulose; it is only at the inner margin of the mantle that fibres composed of cellulose, with nuclei among them, make their appearance. In Cynthia, &c., there are scarcely any traces of cells, while the nuclei and cellulose fibres abound.

"5. A tessellated epithelium, containing no cellulose, covers the inner surface of the three Ascidiae which I examined; the outer surface of the mantle of *Phallusia* appears to have a similar epithelium.

"6. There are two essential points of difference between the modes in which cellulose occurs in the Ascidiae and in the vegetable kingdom:

"(1.) In *Phallusia*, the cellulose constitutes the inter-cellular substance, but does not, as in plants form an integral part of the cell-wall itself.

"(2.) In *Cynthia* and other species, the cellulose forms free fibres, a form in which it is never observed in the vegetable kingdom.

"7. The substance of the mantle of the Ascidiae is not disintegrated by boiling with caustic potass and nitric acid, like the vegetable cellular tissue, into its elementary parts; there is in it none of the inter-cellular substance universally present in vegetable tissues, and by which the cells are connected but which inter-cellular material is never composed of cellulose, as it resists sulphuric acid, but is soluble in caustic potass, as well as by maceration;" see loc. cit. p. 197, 198. This valuable paper is accompanied with three colored plates representing sections, &c., of the mantle-tissues, drawn by the camera lucida.

From this it is clear that this discovery of cellulose in animals is very far from confounding the animal and vegetable kingdoms, for whatever else may be said, the previously established law that the animal cell-membrane always contains nitrogen, retains its force.

See, also, the report of *Payen* on *Kölliker* and *Löwig's* paper, before the Institute, in the Compt. Rend. 1846, XXII. p. 581.

But see for some dissenting views on this subject, *Huxley* (Quarterly Jour. of Microscop. Sc. No. 1, Oct. 1852, p. 22). — ED.

§ 173.

With the Bivalvia, the mantle exhibits (especially near its free borders), contractile motions upon the slightest touch. These are due to numerous muscular fibres which traverse in every direction its granular parenchyma, but are most abundant in the borders. It contains here, moreover, nerves, blood and aquiferous vessels, and in some species, even genital organs. The borders of the mantle of the Lamellibranchia are often provided with very sensitive contractile tentacles; [1] these are rarely wanting around the anal opening, — an orifice which serves also for the respiration. [2] In many, this anal opening is divided by a septum into a round, superior and inferior orifice. [3] The borders of these two orifices are often prolonged each into a longer or shorter fleshy tube (*Sipho*). These two tubes, which are often blended together, project considerably out beyond the mantle and shell, but usually can be wholly withdrawn. [4]

With the Brachiopoda, the border of the mantle has, instead of retractile tentacles, — hyaline, radiating filaments, which are hollow and deeply inserted in the substance of the mantle. [5]

With the Lamellibranchia, and Brachiopoda, the internal surface of the mantle is covered with ciliated epithelium, which extends also upon the abdomen, foot, oral tentacles, and branchial lamellae.

This epithelium is of great importance, since it constantly directs currents of water into the mantle, and thereby food is brought to the mouth, fresh water to the branchiae, the eggs and sperm are carried away from the genital openings, and the faeces are rejected outwardly. The existence of this epithelium makes it clear how these animals can continue to live when buried in wood or stone.

§ 174.

The mantle of the Bivalvia is covered by two shells, whose infinite variety of form serves for their zoological classification into genera and species. These shells are composed for the most part of carbonate of lime so closely

With many of the compound Ascidiae, the body sends fleshy ramified prolongations into the mantle. These have been regarded as blood-vessels by *Savigny* (Mém. &c. p. 47, (*Diazona* and *Botryllus*)), and *Delle Chiaje* (Descriz. &c. III. p. 34, Tav. LXXXIII. fig. 13, 15 (*Polyclinum viride*)); but *Milne Edwards* (loc. cit. p. 41, Pl. VII. fig. 1, 1b. 1c. 5d.) has regarded them with *Botryllus rotifera*, and *Didemnum gelatinosum*, as hollow prolongations,—a view entirely assented to by *Kölliker*.

1 With *Avicula*, *Anomia*, *Pecten*, and *Spondylus*, there are two or three rows of cylindrical tentacles along the border of the mantle; with *Lima*, these tentacles are highly developed, and are situated upon the convex edge of the fold of the mantle. With *Mytilus edulis*, they are peculiar, being flattened and digitiform.

2 With the Naiades (*Unio* and *Anodonta*), there are no tentacles around the anal fissure, while the principal mantle-orifice which is separated from this last by only a narrow isthmus, has them quite numerously upon its borders; see *Pfeiffer*, Naturg. deutsch. Land-und Susswasser Mollusken, Abth. II. Taf. I. fig. 2, 5, 9, p. b. These Naiades have also a third fissure, which is dorsal and situated quite distant from the anal one; it was first pointed out by *Bojanus*. I am yet unsettled as to its nature. See *Pfeiffer*, loc. cit. Taf. I. fig. 5, t.

3 *Isocardia*, *Tridacna*, and *Chama*.

4 With *Psammobia*, *Tellina* and *Venus* the siphon is double and very protractile. With *Cyclas*, and *Teredo*, the two respiratory tubes are more or less blended together at their base; and they are united so as to appear as a single organ with *Mactra*, *Mya*, *Panopaea*, *Solen*, *Pholas*, *Lutraria*, *Clavagella*, and *Aspergillum*.

In these two last genera, the mantle is prolonged directly into a siphon without any appreciable line of separation. It is almost entirely closed, and beside the siphon and the narrow anterior opening, there is in the middle of its ventral border, a very small aperture, whose nature is yet with me doubtful; see *Ruppell* and *Leuckart*, Neue wirbellose Thiere des rothen Meeres. p. 41, Taf. XII. fig. 4, a; and *Owen*, On the Anatomy of Clavagella, in the Transact. of the Zool. Soc. London, 1. p. 270, Pl. XXX. fig. 13, 14, or the Isis, 1836, p. 440, 1837, Tab. 11. fig. 13, 14.

5 These filaments appear to be composed of a horny substance. They are smooth and very small with *Terebratula*, and very long and jointed with *Orbicula* and *Lingula*.

With *Orbicula*, each article of the filament is surrounded with short bristles; see *Owen*, Trans. Zool. Soc. p. 147, 154, Pl. XXII. XXIII.; or the Isis 1835, p. 144, 151, Taf. V. VI.; or in the Ann. d. Sc. Nat. III 1835, p. 55, 66, Pl. I. II.; and *Vogt*, Neue Denkschriften der allg. schweizerischen Gesellschaft für du gesammten Naturwissenschaften, loc. cit. p. 3, Taf. 1.

blended in a homogeneous organic base, that this last is not apparent except by the aid of acids. In a few only, does this organic base predominate over the calcareous matter.[1]

The intimate structure of shells is quite varied,[2] but nearly always an external fibrous, and an internal lamellated layer may be distinguished by aid of a simple lens. The external layer appears to have a crystalline texture, being composed of thickly-set, calcareous prisms, attached perpendicularly or obliquely upon the internal layer. These prisms, however, are not the result of a crystallization, but, as is shown from their development,[3] are only cells filled with lime, and if dissolved in acid, delicate prismatic cells remain as the organic base. The internal layer is made up of numerous superposed, non-cellular lamellae composed of the organic base, and arranged intricately in various ways. To the plicae thus formed, and between which the carbonate of lime is deposited, is due the pearly aspect of this internal layer. The relative thickness of these layers varies, sometimes one, and sometimes the other, being the greater.[4] The external layer is undoubtedly secreted by the borders of the mantle, while the internal is formed by a secretion of its external surface.

The growth of the shell is not continuous, but occurs only at certain periods of the year; hence the formation of concentric lines and furrows upon its surface, analogous to the yearly rings of trees.

The external layer is often colored, either uniformly throughout, or only in spots; while the internal one rarely contains any pigment. By examining the cicatrized wounds which these animals accidentally present, it will appear plain that this pigment is secreted by the borders of the mantle. For, if these wounds are situated at a distance from these borders, the shell is never filled except by a layer of colorless matter.[5]

In the shells of some Bivalvia there are, moreover, special, narrow canals, which are either simple and traverse the shell obliquely from within outwards, or branched in a reticulated manner throughout its whole extent.[6]

The shells are not attached to the animal except by muscular insertions along their borders, and by an epidermis belonging to the borders of the mantle. This epidermis, composed of a horny, yellowish-brown substance, stretches from the borders of the shell over its whole external surface,[7] and

1 The shells of *Lingula* contain very little lime, and there is even still less in the flexible valves of *Orbicula*.

2 The microscopic structure of shells has of late been studied by several naturalists; see *Deshayes*, Cyclop. of Anat. &c. 1. p. 797; *Shuttleworth*, ueber den Bau d. Schalen, &c., in the Mittheil. d. naturforsch. Gesellsch. in Bern 1843, p. 43; and *Carpenter*, Annals of Nat. Hist. XII. 1843, p. 373, Pl. XIII. XIV. and especially the Rep. of the Brit. Assoc. 1844, p. 1, with many figures.*

3 *Mya arenaria* forms an exception to this; the tooth of its shell contains true prismatic crystals bound together in a star-like manner; see *Carpenter* Annals of Nat. Hist. loc. cit. Pl. XIV. fig. 8.

4 These two layers, of which the outer one quite resembles the enamel of the teeth, are very distinctly seen with *Malleus*, *Perna*, *Crenatula*, *Avicula*, *Meleagrina*, *Pinna*, *Anodonta*, *Unio*, &c. With *Ostrea*, and *Chama*, they alternate with each other several times. In many Pectines,

and *Cardiacea*, and with *Anomia*, the fibrous layer appears to be wholly absent.

5 The formation of pearl occurs only upon the inner surface of the mantle. It has, therefore, the same lamellated structure and iridescent property, as the natural layer of shells.

6 With *Terebratula*, these canals are quite distinct — occupying the whole thickness of the shell. I have observed the same arrangement with *Cyclas*, while with *Lingula*, they are confined to the internal layer. By direct light they appear black. I am yet uncertain whether this color is due to their extreme tenuity, or to calcareous matter in their interior. If the first, they would be comparable to the canaliculi of the dentine of teeth; but if the second, to the corpuscles of bone. *Carpenter* (Annals of Nat. Hist. loc. cit. p. 384, Pl. XIII. fig. 5), has observed that in the shells of *Lima rudis*, those canals are divided and form a kind of network.

7 See *Mytilus*, *Anodonta*, *Unio*, *Solen*, *Lutraria* and *Mya*.

* [§ 174, note 2.] For the complete labors of *Carpenter* in this direction, see Cyclop. Anat. and Physiol. Art. *Shell*, IV. p. 556. It is replete with figures. — ED.

in some species covers even the whole of the siphon.[9]   Very often, however, this epidermis is worn away upon old portions of the shell, which
is quite striking with those which have lamelliform or pilous prolongations
around the borders of the shell.[9]

The two shells are joined together partly by a hinge (*Cardo*), and partly
by an elastic tissue (*Ligamentum*).[10]   This last, either external or internal, is antagonistic to the adductor muscles of the shell.   It is composed
of elastic fibres, the internal of which, when the shell is closed, are compressed between the borders of the hinge, while those which are external
are lengthened out.   In both cases, their natural action is to open the two
shells.[11]

§ 175.

The *Terebratulae* have a very remarkable internal calcareous support situated upon the inner surface of the two shells.   It consists, first of two delicate
outwardly curved peduncles, which arise from the sides of the two cardinal
teeth situated upon the non-perforated valve ; then there are two other peduncles which are shorter, and arise from a longitudinal ridge upon the
centre of the same valve ; these pass in front and unite in an arcuate manner.   The two branches thus formed are abruptly recurved after a short
course, and unite, forming a common arc behind the centre of the shell.[1]
With many, this structure is much more simple, consisting only of a median
apophysis, from which pass off two alar prolongations which are curved at
their extremity.[2]   This structure serves principally for the insertion of
the tentacles.[3]

CHAPTER II.

MUSCULAR SYSTEM AND ORGANS OF LOCOMOTION.

§ 176.

The muscles of the Acephala are composed of simple, smooth fibres.

8 With *Mya*, and *Lutraria*, the epidermis forms
a complete sheath around the siphon.

9 *Mytilus hirsutus*, *Arca barbata*, *lacerata*,
and *ovata*.

10 I must omit a description of the various forms
of the hinge and ligament, for they belong properly
to the department of Zoology.   The hinge is wholly
wanting with the *Inclusa*, and the ligament is absent with the Brachiopoda ; and with *Orbicula*,
and *Lingula*, both are absent.

The Aspergillina are distinguished from all the
other Acephala, by a singular disposition of their
valves.   Their mantle ceases early to secrete the
matter for the shell-formation.   The two valves are
then joined at a point, where, most probably, they
would have been articulated, while the mantle
which has only two small openings, and its long
siphon with a double canal, is covered with a calca-

reous formation which constitutes, with the two
united valves, the singular tube of these animals.

11 The ligament is internal with *Pecten*, *Spondylus*, *Mya*, *Lutraria*, and *Pholas* ; but external
with the Chamacea, Cardiacea, Arcacea, and Naïades.   It is half external and half internal with
Malleus, and many other species.

1 *Terebratula chilensis, dorsata, dentata,* and
Sowerbyi ; see *Owen*, loc. cit. Pl. I. fig. 4.

2 *Terebratula rubicunda,* and *psittacea*.

3 According to *Owen*, in those species which
have this apparatus highly developed and bent
backwards, these arches, notwithstanding their calcareous nature, are somewhat elastic ; and when
the valves are closed, they are slightly depressed,
and thus may serve in the absence of the elastic
ligament, for the opening of the shell.

But *Salpa* presents a remarkable exception to this, for here the fibres are striated.[1]

With the Tunicata, the muscular system is most simple, being limited to a subcutaneous layer, which, with the Ascidiae, envelops like a sac the body of each individual, and is attached to the skin only at the two openings of the cavity of the body. It is formed of numerous circular and longitudinal interlaced muscles, among which there are, here and there, oblique fasciculi.[2]

With *Salpa*, this cutaneous muscle consists only of a few isolated bands bound together by a thin, homogeneous membrane. These bands, which vary much in number, distance apart, and direction, surround the cavity of the body, usually in a belt-like manner. They are sometimes straight, sometimes curved, and their extremities never meet upon the ventral surface so as to form a complete belt, but terminate loosely, or are blended by anastomoses with adjoining bands. Around the two openings of the body, they form real sphincters.[3]

By means of this muscle, the Tunicata can enlarge or diminish the cavity of the body, and thus cause the necessary renewal of water for nutrition and respiration, beside ejecting the faeces and products of generation. The *Salpa*, by rhythmical contractions of their body (its anterior superior opening, being closed by a membranous valve), eject water through its posterior opening, and thus are propelled along.

§ 177.

With the Bivalvia, the muscular system is much more complicated. Not only are muscular fibres scattered through nearly the whole body, but in certain points, they are so aggregated as to form distinct isolated muscles.

The largest of these muscles are the Adductores of the valves. With the Lamellibranchia, these consist of a single or a double mass of thickly-set, parallel fibres, the ends of which are inserted at opposite points of the two valves. Those species which have two of these muscles are called Dimya ; here one of these muscles is anterior, and the other, larger, posterior. With the Monomya, there is one muscle alone ; this is large and situated near the centre of the valves.

With Brachiopoda, these muscles are more complicated, there being four pairs. Part of these, only, are doubly inserted to the valves,[1] while the rest, which arise from one of the valves, are inserted upon the peduncle.

1 See *Eschricht*, Over Salperne, &c., p. 64, Tab. III. fig. 16. These striae are due to a zig-zag plication, as I have satisfied myself from a specimen of *Salpa zonaria* preserved in alcohol. *Will* has observed the same in the muscles of other invertebrates (*Muller's* Arch. 1843, p. 359). The muscular fibrillae of *Salpa* are bound together in primitive riband-like fasciculi which are plicated during contraction like the frill of a shirt ; this is easily seen when one of these fasciculi is observed in an edgewise position.

2 See *Savigny*, Mém. &c. Pl. V. fig. 1, 2 (*Boltenia* and *Cynthia*) ; *Delle Chiaje*, Descriz. &c., III. p. 23, Tav. LXXXIV. fig. 3, 5 (*Phallusia*) ; and the Catal. of the physiol. Series, &c., I. Pl. V. (*Phallusia*).

3 *Salpa cordiformis* and *zonaria* have from five to seven isolated and equi-distant muscular girdles ; see *Eschricht*, loc. cit. Tab. I. III.

Salpa cylindrica has ten or eleven of these girdles, the anterior of which converge upon the back and are curved from before backwards ; see *Cuvier*, Mém. sur les Thalides. loc. cit. fig. 9 ; and *Savigny*, Mém. loc. cit. Pl. XXIV. fig. 1. With *Salpa mucronata*, and *maxima*, these girdles are blended together upon the back ; see *Meyen*, Ueber die Salpen, loc. cit. Tab. XXVIII. fig. 5, Tab. XXIX. fig. 2. *Salpa pinnata (cristata)* is remarkable for having numerous anastomoses uniting the girdles upon the sides of the body and presenting a trellis-like aspect ; see *Chamisso*, De Salpa, fig. 1, G. II., and *Cuvier*, loc. cit. fig. 1, 2.

1 Several of these muscles do not always arise directly from the shell, but from the visceral sac ; so that here their action is not solely for displacing the viscera, but also for the movement of the valves, to which this sac is attached.

As the points of the insertion of these muscles do not always correspond with the two valves,[2] their direction is often oblique, or, they sometimes even cross each other.[3]   *Orbicula* and *Lingula*, which want both the ligament and hinge, have this disposition of the adductors which terminate usually by a delicate tendon, and the contraction of either of these muscles alone, produces the lateral movements of their valves.

The spirally-pointed tentacles of the Brachiopoda are moved by a particular apparatus.   The fringes of these organs are inserted upon a cartilaginous tubular prolongation which tapers to a point.   This is closed at both extremities and contains a liquid, which, by the contractions of the circular muscular fibres, is propelled from the base to the extremity, thereby unrolling the spiral turns.[4]   These tentacles here certainly take the place of the elastic ligament of the bivalves, for their extension probably tends to slightly open the valves.

With the Lamellibranchia, the tentacles which are not rolled, are arranged quite differently.   Like their branchiae, they are scarcely at all irritable, — this being due to the fewness of their muscular fibres.   But in the mantle these fibres are very abundant, and especially near their free borders.   This is true also of the Siphon, in which both longitudinal and circular fibres can be easily seen ; here, two very distinct flattened muscles arise from the base of the siphon and are inserted upon the two valves, external to the posterior adductor muscle : these serve as a *Retractor siphonis*.

§ 178.

Very many of the Lamellibranchia have a highly-developed organ of locomotion, — the *foot*.[1]   This is a muscular prolongation from the ventral surface, which passes obliquely forward to be inserted upon the internal surface of the back of the shell, by four, rarely more, tendinous cords.[2]

These cords surround the abdominal viscera, and becoming gradually thicker and more muscular, finally blend with numerous, interlaced muscular fasciculi which compose the foot.   This last varies considerably as to its size and form, and can be protruded a long way out through the open shells, but may also be wholly withdrawn.[3]

2 *Lingula* forms an exception.   Here the principal adductor is a short, solid muscle, which stretches straight across from one shell to the other, at their posterior extremity.

3 This muscular apparatus has been particularly described and figured by Owen, and *Vogt* (loc. cit.) with *Terebratula*, *Orbicula*, and *Lingula*.

With the *Terebratula*, two pairs of muscles arise from each valve.   The two anterior ones arising from the imperforate valve, are the longer, and have their origin back of its centre.   After the crossing of their delicate tendons, they pass through the opening of the valve and terminate in the pedancle, together with the two posterior ones which are short and fleshy, and which arise at the base of the hinge.   Of those of the perforated valve, the two posterior ones only pass to the peduncle — the two anterior being attached to the base of the other valve.   With *Orbicula*, there are two posterior, and two anterior fleshy muscles, all of which pass obliquely from one valve to the other, while the anterior ones sometimes send off fibres to the short peduncle.   In the space circumscribed by these four muscles, lie four others, which are small and interlaced, and extend from the

visceral sac to the valves.   With *Lingula*, there are, beside the principal adductor, four pairs of interlaced muscles, which pass obliquely through the centre of the cavity of the valves, and are attached by their two ends to the visceral sac.

4 Owen, loc. cit. ; and *Vogt*, Anat. d. Lingula, p. 8, Tab. 11. fig. 16-18.

1 The foot is absent, particularly with those mollusks which are fixed to rocks and other solid bodies, by a calcareous cement.

2 Usually one pair of these delicate cords passes above, and another below, and are inserted upon the valves, quite near the four points of insertion of the two adductor muscles.   This is so with *Anodonta*, *Unio*, *Cardium*, &c.   With *Isocardia*, I have found a third pair of cords inserted upon the posterior extremity of the summits of the shell.   These serve not only as *Retractores* of the foot, but when this last is fixed to some point, draw the animal towards it.

3 The laterally-compressed foot of *Anodonta*, and *Unio*, arises, by a large base, from the abdomen, and has carinated borders.   That of *Pectunculus* and *Venus*, is quite similar, but its free border is hollowed by a furrow, and is therefore bi-carinated.

Most of these animals use this organ to dig in the sand, or to creep along on soft surfaces. For this purpose, they reach it out in front, and then by alternate contractions and elongations, drag their body after it. Some species can in this way glide freely along like the Gasteropoda, or even seize hold of aquatic plants.[4] Sometimes this foot is truncate and hollow at its extremity, and probably, therefore, acts like a sucker.[5]

§ 179.

With many of the Lamellibranchia,[1] the foot appears imperfectly developed, and has a secretory organ of the *Byssus*, a part by which these animals are attached to wood, stone, and other bodies. In this case the foot is a delicate, protractile, tongue-like body,[2] capable of a stiffness sufficient for creeping, but used chiefly as a feeler to find the points of attachment by the byssus.[3] It always points towards the oral extremity, and upon its inferior surface there is a longitudinal furrow which has a cavity at its base. The walls of this furrow and cavity secrete the byssus. From their glandular aspect, they differ much from the rest of the organ, which is formed of numerous interlaced muscular fibres.[4] The bottom of this cavity from which the furrow arises, is regularly divided by numerous delicate, parallel lamellae, from which arises the compact root of the byssus.[5] This byssus is, therefore, inserted into the base of the cavity as are our finger-nails into their matrix. Its base has a fibrous, or lamellated structure, and passes into a longer or shorter trunk composed of numerous cylindrical, or flattened filaments,[6] whose extremities are sometimes discoid.[7]

§ 180.

Many Bivalvia, which are likewise wanting in locomotive organs, and have, moreover, no organs of the byssus, attach themselves to bodies in another and peculiar way. Thus, with *Anomia*, one of the valves is marked by a deep fissure, across which, like a short peduncle, a portion of the ad-

With *Tellina*, *Donax*, and *Cyclas*, it is very long and more or less ridged, and often quite small at its base.

With *Cardium*, *Nucula*, *Trigonia*, *Mactra*, and *Isocardia*, it is curved like a hook or knee from behind in front. With *Solen*, it is very long, straight and nearly cylindrical.

4 *Cyclas* and *Pisidium*. It is probable that those species whose foot is furrowed upon its inferior border (*Pectunculus*), or bent in front (*Nucula* and *Trigonia*), can also creep like the Gasteropoda.

5 *Pholas*.

1 The Malleacea, Aviculacea, Mytilacea, with *Pecten*, *Lima*, *Arca*, *Tridacna*, &c. Quite singularly there exists with *Anodonta*, *Unio*, and *Cyclas*, when hatched, a secretory organ of the byssus ; see below, § 197, note 13.

2 For the byssus-forming organ, see *Deshayes*, Cyclop. of Anat. &c. I. p. 702 ; and especially *A. Muller*, De Bysso Acephalorum, Dissert. Berolini. 1836 ; or, his Memoir in *Wiegmann's* Arch. 1837, I. p. 1, Taf. I. II.

3 The manner in which *Mytilus* and *Tichogonia* act in spinning their byssus has been described by *Marion de Procé* in the Ann. d. Sc. Nat. XVIII. 1842, p. 59 ; and by *A. Muller*, loc. cit.

4 I do not yet clearly understand the true nature of the walls which secrete the byssus. *A. Muller* has designated them as *Glandula byssipara*, composed of round cells. He affirms to have seen at the base of the furrow of *Mytilus edulis*, orifices of the excretory ducts of this gland ; see *Wiegmann's* Arch. loc. cit. Taf. I. fig. 6. On the other hand, neither *J. Muller* (De glandul. structura, p. 39), with *Tridacna* ; nor *R. Wagner* (Lehrb. d. vergleich. Anat. 1835, p. 271), with *Arca*, and *Pinna*, has been able to find these glands.

5 See *A. Muller*, in *Wiegmann's* Arch. loc. cit. Taf. I. fig. 5, c. (*Tichogonia*), and *Poli*, loc. cit. II. p. 152, Tab. XXIV. fig. 5–7 (*Arca*).

6 For the intimate structure of the byssus see the Memoir of *A. Muller*, loc. cit. With *Arca*, its form is very remarkable, consisting of a solid, laterally-compressed trunk, carinated above and below, and having filaments upon no portion. With that of *Pinna*, on the contrary, its filaments remain ununited even to the very root.

7 *Avicularia* and *Mytilus* ; see *Poli*, loc. cit. Tab. XXXI. (*Mytilus edulis*), and Tab. XXXIV. fig. 2 (*Pinna muricata*).

ductor muscle of the other valve[1] passes, in order to be attached to foreign bodies by its smooth, calcareous extremity.

With the Brachiopoda, there is a real peduncle which constantly projects through an opening near the hinge. It is a soft tendinous or muscular tube, which is, perhaps, only a prolongation of the mantle.[2]

CHAPTER III.

NERVOUS SYSTEM.

§ 181.

The nervous system, which has been observed in all the orders of the Acephala, consists of a central and a peripheric portion.

The first is composed of one, or several (usually three) ganglia; the second consists of nervous trunks of variable size, which pass off in the most different directions. When the number of these ganglia is considerable, they are arranged in pairs which are situated more or less near the median line, according to the different regions of the body. The ganglia of each pair intercommunicate by a transverse commissure of variable length. They connect, moreover, with others, which are even far removed. by anastomosing filaments. It is difficult to decide which of these ganglia is the brain. Many species want a complete ganglionic ring surrounding the buccal cavity.

§ 182.

From its extreme softness, the internal structure of the nervous system of the Acephala is very difficult of study. Its primitive fibres are very delicate, and are surrounded, in the nervous trunks, by a distinct and very thin neurilemma. In the ganglia, through which orange-colored granules are usually scattered,[1] these fibres pass into a very loose tissue composed of small transparent vesicles, which probably take the place of the ganglionic globules which are so distinct with other invertebrates.[2]

§ 183.

1. The nervous system is most simple with the Tunicata. It here consists of a single ganglionic mass, which is subcutaneous, and. situated between the two respiratory tubes.

---

1 Carefully examined, *Anomia* will be found to have three unequal adductor muscles arising from the imperforate valve. The largest of these, together with one of the others, passes into the fissure of the other valve; while the third is inserted upon the same valve.

2 It has already been remarked (§ 177), that with the Brachiopoda the peduncle receives muscles both from the body and from the valves. It has, moreover, muscles of its own, and ought, therefore, to be contractile. This contractility is quite prominent with the very large and long peduncle of *Lingula*, especially in comparison with

the very short, sucker-like one of *Orbicula*. Externally, it is composed of a thick cartilaginous tissue, while its interior is occupied by a hollow, muscular cord, composed of longitudinal filaments; see *Owen*, loc. cit. (*Terebratula*), and *Vogt*, loc. cit. Tab. I. fig. 1–6 (*Lingula*).

1 These orange-colored ganglia are quite distinct with *Unio*, and *Anodonta*.

2 Although the Naïades have very large ganglia, yet their microscopic examination has furnished no further results, for neither by the compressorium, nor by chemical means, can these globules be separated from the intervening tissue.

With the Salpinae, the central nervous mass is upon the dorsal surface in front of the middle of the body. It consists of many closely-aggregated, yellowish ganglia, from which nerves pass off in all directions.[1]

With the Ascidiae, it consists of a single large ganglion, which is easily found within the muscular envelope, in an angle formed by the oral and anal tubes. The nerves which pass off in different directions from this ganglion, belong chiefly to the muscular envelope. Some of them, however, pass to the organs of sense situated near the two respiratory tubes, and form around the orifice of that one of them which is buccal also, a complete circle which corresponds perhaps to an œsophageal ring.[2]

2. The nervous system of the Brachiopoda is as yet little known. However, from the presence, in some species, of two or three ganglia about the œsophagus, it may be concluded that it is analogous to that of the Lamellibranchia.[3]

3. With the Lamellibranchia, the nervous system is the most distinct.[4] Its very symmetrical arrangement is prominent, except in the unequivalved species.

1 *Meyen* was the first to describe with care this nervous mass with *Salpa*; for, before him, many other parts of the animal had been erroneously taken by *Savigny* (Mém. &c. II. p. 127), and *Chamisso* (De Salpa, &c., p. 5), for the nerves and ganglia; see *Meyen*, Nov. Act. Acad. Leop. loc. cit. p. 394, Tab. XXVII. fig. 5, d. 18 (*Salpa pinnata*), Tab. XXVIII. fig. 5, h k. 12 (*Salpa mucronata*). *Quoy* and *Gaimard* have also observed a central nervous system in the dorsal region of several *Salpa*; see Voyage de l'Astrolabe, Zool. III. p. 559, and the Atlas zoologique of the same, Mollusques, Pl. LXXXVI.; or Isis, 1836, p. 113, Tab. VI.

Eschricht's description is still more minute; but he has taken for the ventral surface that portion of the body in which is lobulated ganglionic mass is situated; see his Memoir, Over Salperne, &c., p. 12, Tab. II. fig. 8. 10, u, v. (*Salpa cordiformis*), and Tab. III. fig. 22 (*Salpa zonaria*). See, also, *Delle Chiaje*, Descriz. &c. III. p. 45, Tav. LXXVIII. fig. 3, n. 12 (*Salpa maxima*). I am yet undecided whether the nervous ring, which, according to *Eschricht*, is formed by the junction of the two nerves surrounding the anterior respiratory orifice, really corresponds to the œsophageal ring.

2 A very detailed description accompanied with figures of the nervous system of the simple Ascidiae, may be found in the works of *Cuvier* (Sur les Ascidiés, &c., loc. cit. p. 24, Pl. II. fig. 2 c. 5, g. III. fig. 2, 3 c. (*Cynthia* and *Phallusia*)); *Eschricht* (Beskrivelse af Chelyosoma, loc. cit. p. 8, fig. 4, c.); *Delle Chiaje* (Descriz. &c. III. p. 25, Tav. LXXXII. fig. 2, and LXXXIV. fig. 3, 5, (*Phallusia*)); and *Savigny*, who has included also the compound Ascidiae (Mém. &c. p. 32, Pl. IX. fig. 2², XI. fig. 1², D'. (*Phallusia*); also Pl. XXI. fig. 1², XXII. fig. 1⁴, XXIII. fig. 1⁶, D'. d'. (*Botryllus* and *Pyrosoma*)).

The principal ganglionic mass is always nearer the anal than the oral tube, and does not give off branches except at its extremities. The nervous ring of the Ascidiae, has been observed by both *Cuvier*, and *Delle Chiaje*.

The last of these authors has also mentioned a particular ganglion which he has observed with *Phallusia mamillaris*, in this ring, and which he

regards as the brain; while he has given the name of *Ganglion sympathicum* to a principal ganglionic mass, lying near the anal tube.

3 *Cuvier* (Sur la Lingula, loc. cit. p. 8) thinks he has observed two ganglia at the base of the arms, but which give off no nerves. *Owen* (loc. cit.) has found two ganglia in front of the œsophagus of *Orbicula*, and one behind it; and also two nerves arising from the two anterior ganglia attended with two arteries which go to the two hearts.

4 The nervous system of the Lamellibranchia was discovered by *J. Rathké*. In 1797, he had well represented the anterior pair of ganglia of *Anodonta* (loc. cit. p. 162, Tab. IX. fig. 10, 11). *Poli*, it is true, had already figured the nervous system of several species of this order (loc. cit. Tab. XXXVI. fig. 1, n. (*Pinna*); Tab. VIII. fig. 1, l. (*Pholas*); Tab. IX. fig. 10, a. (*Unio*); Tab. X. fig. 15, Tab. XI. fig. 1, Tab. XIII. fig. 6, (*Solen*); Tab. XXV. fig. 1, (*Arca*); Tab. XXXII. fig. 18, r. (*Mytilus*)). But he erroneously took it for a lymphatic system.

The following works may be consulted upon this system: *Mangili*, Nuove ricerche zool. sopra alcune specie di conchiglie bivalvi, Milano, 1804 (translated in *Reil's* Arch. IX. 1809, p. 213, Taf. x². (*Anodonta*)); *Brandt*, Medizin. Zool. II. p. 310, Taf. XXXVI. fig. 10,–12 (of the Oyster); *Garner*. On the Nervous System of Molluscous Animals, in the Trans. of the Linn. Soc. XVII. 1837, p. 485, Pl. XXIV. (*Ostrea, Pecten, Modiola, Mactra, Mya*, and *Pholas*); and, On the Anatomy of the Lamellibranch. loc. cit. p. 89, Pl. XIX. fig. 5 (*Venerupis*); *Keber*, De Nervis Concharum, Diss. Berolini, 1837; *Duvernoy*, Sur l'animal de l'Unguline, in the Ann. des. Sc. Nat. XVIII. 1842, p. 118, Pl. V. B. fig. 8; and, Sur le Système nerveux des Mollusques Acéphales bivalves, in the Comp. rend. 1844, Nos. 22, 25, 1845, No. 731; *Blanchard*, Observ. sur le Système nerveux des Mollusques Acéphales testacés ou Lamellibranches, in the Ann. des Sc. Nat. III. 1844, p. 321, Pl. XII. and in *Froriep's* neue Not. No. 741 (*Solen, Mactra*, and *Pecten*); and *John Anderson*, Art. *Nervous System*, in the Cyclop. of Anat. III. p. 604.*

* [§ 183, note 4.] See, also, *Duvernoy* (suite) Compt. Rend. XXXIV. 1852, p. 665, and XXXV. 1852, p. 119; also, *Frey* and *Leuckart*, loc. cit.

p. 46; *Deshayes*, loc. cit. p. 69, Pl. VIII. IX.; and *Quatrefages*, loc. cit. p. 63, Pl. I. fig. 3, 6 (*Teredo*). — ED.

A. The central nervous mass is composed of three pairs of principal ganglia, as follows: *Par anterius* or *labiale, Par posterius,* and *Par inferius* or *abdominale.* This last pair is extraordinarily developed with those species which have a foot, and has, therefore, received also the name of *Par pedale.*

The ganglia of the *Par anterius* are situated one on each side of the digestive canal, and are connected by a filament which extends arcuately over the circumference of the oral cavity.[5] They send off two long nerves, which pass along the back to the *Par posterius,* which is the largest of all and usually situated upon the anterior surface of the posterior adductor muscle. Its two ganglia are either blended together, or connected by a transverse commissure.[6] In this way, the anterior and posterior pairs with their commissures form a kind of œsophageal ring which surrounds the base of the abdomen.

The inferior or pedal pair is situated at the point where the base of the foot is joined to the abdomen. Its ganglia are contiguous upon the median line, or are blended together into one.[7] They connect also with the labial ganglia by two nerves, thus forming a second œsophageal ring.[8]

Besides these principal ganglia, there are others, smaller, and situated in various parts of the body. But these are not constant, for their presence is always due to an unusual development of the muscular system.

B. The peripheric nerves arise almost exclusively from the three pair of principal ganglia, for the nerves of connection do not usually give off branches. The few and very delicate filaments which sometimes pass off from these last, belong, probably, to the splanchnic system, for the principal ganglia appear to furnish only sensitive and motory nerves; these are usually distributed in the following manner:

The *Par anterius* sends nerves to the anterior part of the mantle,[9] to the anterior adductor muscle, and to the tentacles of the mouth and its circumference.

The *Par posterius* sends two very large trunks to the branchiae; also other nerves to the lateral and posterior part of the mantle,[10] to the posterior adductor muscle, and delicate filaments to the heart and rectum.

---

5 The length of the arc of this filament of connection depends upon the position of the two labial ganglia. Thus with *Pecten,* where these ganglia are situated unusually in the rear, it is very long and very arched ; while with *Pholas,* and *Solen,* where they are close upon the oral opening, it is short. With *Venus,* and *Mactra,* these ganglia lie so close to each other, that this anastomotic filament is replaced by a very short, transverse commissure.

6 The *Par posterius,* which, from its relations to the branchiae, is also called the *Par branchiale,* is blended into a single ganglion in those species whose branchiae are united at their lower part ; as *Unio, Anodonta, Mactra, Mya, Solen,* and *Pholas.* On the other hand these ganglia are separate, and connected simply by a transverse commissure, with those whose branchiae are isolated ; as *Ostrea, Pecten, Avicula, Mytilus, Lithodomus, Modiola,* and *Arca.*

7 According to the earlier Zootomists, the *Par pedale* was wanting in those species which have no foot, although careful investigation has shown that there is a pair corresponding to the *Par inferius.* I refer to that found with *Ostrea* (Brandt, loc. cit. Tab. XXXVI. fig. 11, a. o.), directly behind the labial ganglia, and with *Pecten* (Grube, Mül-

ler's Arch. 1840, p. 33, Taf. III. fig. 3, g. ; and *Blanchard,* loc. cit. p. 336, Pl. XII. fig. 3, a. b.), between these ganglia with which it is in connection by commissures.

8 Of these three ganglia, the *Par anterius* has often been regarded as the brain. But others have rather taken the *Par posterius* for the principal nervous mass. For my own part I think that all three, together with their commissures, correspond to the pharyngeal system of Gasteropoda.

9 The anterior nerves of the mantle of *Solen,* which is prolonged far beyond the oral opening, and is strengthened by a muscular mass — have ten to twelve ganglia lying along the border of the mantle. With *Pecten,* the mantle-nerve has also a small ganglion upon the muscular mass which is found upon each side of the anterior border of this organ ; see *Blanchard,* loc. cit. p. 333, Pl. XII. fig. 1, f. (*Solen*), fig. 3, c. (*Pecten*).

10 In the mantle of *Ostrea, Spondylus, Pecten, Lima,* and in general those species in which its borders have numerous sensitive organs, the branches of the anterior and posterior mantle-nerves unite and form a common marginal nerve whose size depends upon the number of the sensitive organs to which it sends filaments.

The siphon and its muscular apparatus receive their nerves also from this same pair. [1]

The nerves of the *Par inferius* being destined chiefly for the foot, correspond in number and size with the degree of development of this organ. This number, however, varies between two and six for each side.

§ 184.

The Acephala have, certainly, a Splanchnic nervous system, but as yet it has been found only with the Lamellibranchia; [1] and even here it is seen with difficulty and imperfectly on account of the extreme tenuity of its filaments.

With some species, delicate, lateral filaments pass off from the nerves of communication, which connect the *Par gangliorum inferius* and *posterius* with the *Par anterius;* these may be properly termed sympathetic nerves, for they are distributed partly to the walls of the digestive canal, and the heart, and partly to the liver, the gland of *Bojanus*, and the genital organs. [2]

CHAPTER IV.

ORGANS OF SENSE.

§ 185.

Of the organs of sense with the Acephala, those of Touch are the most highly developed. They usually consist of conical, or flattened, protractile prolongations of the skin, which are extremely irritable, covered with ciliated epithelium, and often of a deep color.

11 When the two retractor muscles of the siphon are large, as is the case with *Solen, Mactra, Venus*, and *Cytherea*, their two nervous trunks have several ganglionic enlargements along their course, connected by transverse filaments; see *Blanchard*, loc. cit. p. 333, Pl. XII. fig. 1, 2, d. (*Solen* and *Mactra*).*

1 With the simple Ascidiae, as a sympathetic system may perhaps be considered the ganglion, which, according to *Schalk* (loc. cit. p. 9, fig. 4, g. q.) is concealed between the intestinal convolutions, at the posterior extremity of the body of *Phallusia*, and send off filaments in various directions. But, as yet, the existence of this ganglion needs confirmation.

2 *Garner, Duvernoy*, and *Blanchard* have seen the filaments, which issue from the principal ganglia, enter the vegetative organs ; but as they could not further trace them, they hesitate to regard them as organic nerves. *Keber* is more positive in favor of the existence of a sympathetic system with the Lamellibranchia. He has observed (loc. cit. p. 15) that the commissural filaments, which pass into the *Par posterius*, give off branches to the intestinal canal, to the liver, and gland of *Bojanus;* and that those of the *Par pedale* give off similar branches to the genital organs ; and also, that these nerves form several *Plexus* between these organs, and from which are given off filaments to the heart. From this disposition, he ought to conclude that these are real organic nerves.

If this is so, the same signification would be given to the nervous filaments which *Blanchard* (loc. cit. p. 335, Pl. XII. fig. 1, e.) has seen arise with an *Arca*, and a *Solen*, from the two small ganglia which belong to the commissures of the *Par posterius*. More profound researches upon the destination of their nerves, must determine whether the two ganglia situated between the labial ganglia, with the apodal Lamellibranchia (see above § 183 note 7), really correspond to the *Par pedale*, or do not rather belong to the sympathetic system.

* [§ 183, note 11.] See *Quatrefages* (Mém. sur le genre Taret. in Ann. d. Sc. Nat. 1849, XI. p. 63, Pl. I.), who has described in detail this system with the Teredina. — ED.

With both the simple and the compound Ascidiae, there are, at the base of the oral tube and at the entrance of the respiratory cavity, numerous filiform and sometimes fringed tentacles inserted upon a kind of ring.[1] With the Lamellibranchia, there are often conical tentacles around the respiratory and anal openings of the mantle,[2] and the orifice of the siphon.[3] Among those which have an open mantle, there are many the borders of whose mantle, either wholly, or only posteriorly,[4] are provided with thickly-set conical tentacles.[5] These receive all their nerves from those of the mantle.

Instead of these retractile tentacles, the Brachiopoda have long radiating bristles upon the borders of their mantle.[6] These project a considerable way beyond the borders of the valves, and having perhaps sensitive nerves at their base, they are thus tactile organs like the vibrissae of some Mammalia.

The oral opening of all the Lamellibranchia is provided, moreover, with two pairs of contractile, foliated lobes, pointing backwards, which are perhaps oral tentacles.[7] Each pair is composed of two lobules, an internal and an external, which are united at their base, and whose surfaces lie against each other. Behind, the border of these four lobes is somewhat thinned, while in front, the two on the same side usually pass into each other, the external being above, and the internal below, the oral opening.[8] The free surfaces of the lobules are smooth and covered with a very thin epithelium, while the other and opposite surfaces are furrowed transversely throughout, and the borders of these furrows are fringed with very large vibratile cilia.[9]

As tactile organs, may be mentioned the two remarkable arms which, with the Brachiopoda, are spirally rolled up near the oral opening. The long, pectinate fringes upon their borders are united at their base by a soft, hollow membrane which is probably contractile, and is provided with vibratile cilia.[10]

---

1 See the figures in *Savigny*, Mém. &c. loc. cit.

2 *Cardium, Chama, Tridacna*, and *Isocardia*.

3 *Solen, Pholas, Aspergillum, Mactra, Venus, Donax*, &c. With *Donax trunculus*, the respiratory tube is remarkable for its ramified tentacles; see *Poli*, loc. cit. Tab. XIX. fig. 15–20.

4 *Unio, Anodonta.*

5 With *Donax, Mactra*, and *Tellina*, this row of tentacles is single; but it is multiple with *Avicula, Anomia, Ostrea, Pecten, Spondylus*, and *Lima.*

6 See *Owen*, and *Vogt*, loc. cit.

7 As to the oblong organ which, with *Salpa cordiformis*, projects into the cavity of the body as two parallel cutaneous folds between the anterior respiratory opening and the central mass of the nervous system, I am yet undetermined whether or not it corresponds to the tactile lobes of the Lamellibranchia. It appears smooth upon its free border, and receives, at its transversely striated base, two nerves from the principal ganglia. With *Salpa zonaria*, a similar organ lies directly in front of the central nervous mass; see *Eschricht*, Over Salperne, loc. cit. p. 14, fig. 8, 10, 22, t. With *Salpa mucronata*, this singular organ is situated in front of the nervous centre, and has been taken by *Meyen* for a male genital organ; see Ueber die Salpen, &c., p. 397, Tab. XXVIII. fig. 5–10.

b *Avicularia, Isocardia, Pinna, Cardium,*

Pectunculus, Mactra, Anodonta, Aspergillum, &c. But *Spondylus* and *Pecten* form, in this respect, an exception. Here, the lobes upon each side, instead of being continuous, are separated by numerous curiously-branched tentacles which surround the oral orifice and strikingly resemble those surrounding the mouth of certain Holothurioidae when contracted; see *Poli*, loc. cit. Tab. XXII. fig. 8, 13, 14, XXVII. fig. 6, 10.

9 The branchial lamellae of the Lamellibranchia have these furrows upon all their surfaces, and in their outward aspect closely resemble these tactile lobes. It is therefore probable that, like the oral tentacles of the Polyps and Holothurians, they have a varied function. Thus, they could serve not only as gustatory organs for the food entering the mouth, but also as those of ingestion, beside taking a part also in the respiration.

10 The researches of *Cuvier, Owen*, and *Vogt* (loc. cit.) upon the arms of the Brachiopoda, were made upon specimens preserved in alcohol.

The relations, therefore, of these organs and their fringes during life are not known. *Muller* also (Zool. Danica, I. p. 4), and *Poli* (loc. cit. II. p. 190, Tab. XXX. fig. 22, 23), say nothing upon the motions of the fringes of *Orbicula* and *Terebratula*. If they are really contractile and ciliated, the whole apparatus is quite analogous to that of the *Alcyonellae*.

§ 186.

As yet, organs of hearing with the Acephala have been found only among the Lamellibranchia. They are here feebly developed, consisting only of two simple round capsules filled with a transparent liquid. Their very thick and somewhat solid walls are homogeneous and transparent; they enclose a vitreous spherical otolite, of a crystalline structure,[1] and composed of carbonate of lime. These otolites constantly keep up very singular swinging and rotatory motions, which instantly cease, however, when the capsule is ruptured.[2] These auditory capsules when present, are situated in the foot in front of the pedal ganglia with which they always communicate,[3] either contiguously, or by two auditory nerves which they receive.[4]

§ 187.

Organs of vision are very common with the Acephala, and always many in number. With some, they occupy a large portion of the borders of the mantle; with others, they are confined to the external orifices of the longer or shorter mantle-tubes.[1]

---

1 These organs were first noticed by me with the Naïades, Cardiacea, and the Pyloridae; but were regarded as of a doubtful nature. Since then, after comparing them with the auditory organs of the embryos of fish, I am satisfied that they are really very simple organs of hearing; see *Muller's* Arch. 1838, p. 49, and *Wiegmann's* Arch. 1841, I. p. 148, Taf. VI. fig. 1, 2 (*Cyclas cornea*); or, the Ann. d. Sc. Nat. X. 1838, p. 319, XIX. 1843, p. 193, Pl. II. B. It appears, moreover, that similar corpuscles are found in other orders of these animals. Thus, *Delle Chiaje* mentions with *Salpa neapolitana*, an organ situated above the nervous centre which exactly resembles the auditory capsules I have discovered in the foot of *Cyclas*. Unfortunately he has neither figured nor carefully described this organ (Descriz. &c. III. p. 45, Tav. LXXVI. fig. 1, 1.). *Eschricht* (Anat. Beskriv. af Chelyosoma Macleayanum, p. 9, fig. 4, 6, d. γ. and fig. 5) has also regarded as an auditory organ a remarkable apparatus which he found near the nervous centre of a simple Ascidian. This consists of a pyriform vesicle filled with whitish matter, and of a clavate body which has, upon its large end, a fissure and two lateral depressions. *Delle Chiaje's* figure (Descriz. &c. III. Tav. LXXXII. fig. 4.), of the principal ganglionic mass of *Cynthia papillata*, reminds me of the clavate body of *Chelyosoma* and leads me to think that this author has confounded it with the nervous centre. I think that this organ exists generally with both the simple and compound Ascidiae, for *Savigny* has noticed with *Cynthia, Phallusia, Aplidium, Polyclinum, Botryllus, Eucoelium, Synoecium, Pyrosoma*, &c., two tubercles near the nervous ring which surrounds the respiratory tube (*Tubercule antérieur et postérieur*). And, to judge from his figure (Mém. &c. Pl. VI. fig. 12, 2, 4, h. Pl. VII. fig. 21), of one of these tubercles, with *Cynthia*, these organs appear analogous to the clavate body just mentioned. At all events, these tubercles deserve, with Zootomists, more attention than has hitherto been given them.

2 These motions are probably due to the ciliated epithelium lining the cavity of the capsules; see, below, the auditory organs of the Gasteropoda.

3 I have been unable as yet to find these capsules with the apodal Lamellibranchia, — at least, with *Tichogonia*, and *Mytilus*. They appear to exist, however, for recently *Deshayes* has found them both in *Teredo*. Here they were situated at the extremity of the septa lying between the pericardium and the elevator of the anus, and upon which the anterior extremity of the branchiae is inserted; see Comp. rend. 1846, XXII. No. 7; or *Froriep's* neue Not. No. 813, p. 323.

4 With *Cyclas*, and *Tellina*, the auditory capsules are contiguous with the ganglia of the *Var pedale*. With *Anodonta, Unio, Cardium*, and *Mya*, they are a little removed.

It is remarkable that these organs appear very early in the embryos of certain Lamellibranchia (*Cyclas*), while in others (*Anodonta* and *Unio*), no trace of them is seen during the embryonic life.

1 *Pali* (loc. cit. II. p. 153, 107, Tab. XXII. fig. 1, 4; and Tab. XXVII. fig. 5, 14, 15), was the first to compare to human eyes these remarkable bodies, which, brilliant as diamonds, lie upon the borders of the mantle of *Pecten* and *Spondylus*, with this expression: *Ocelli smarazdino colore coruscantes*. Nevertheless, it is only of late that these organs have received much attention. *Garner* (On the Anat. of the Lamellibr. Conchifer. &c. Pl. XIX. fig. 1, c. 3) was the first to notice anew the *Ocelli* of *Pecten*. *Grant* (Outlines, &c., p. 258) has described those of *Pecten* and *Spondylus* as organs long known. *Grube* (*Muller's* Arch. 1840, p. 24, Taf. III. fig. 1, 2). and *Krohn* (Ibid. p. 381, Taf. IX. fig. 16) have described the structure of these organs, and, quite recently. *Will* (*Froriep's* neue Not. 1844, No. 622, 623) has treated this subject most profoundly.

Deshayes is not satisfied of the existence of organs of vision with the Pectinea, while *Duvernoy* regards as such the bodies situated on the border of their mantle (Instit. 1845, p. 52, 88). It is astonishing that *Deshayes* should have denied eyes to the Pectinea, where they are so complete. He could have better denied them to *Phallusia, Arca, Ostrea*, and other Acephala. During my last visit at Venice and at Trieste, I examined living individuals of the genera *Arca, Ostrea, Pinna*, as well as other Lamellibranchia and various Ascidiae; but with all possible care, I was unable to verify *Will's* description (loc. cit.) of the eyes of these animals. In most cases, the bodies which he has described as eyes, have appeared to me only as simple excrescences of the mantle, which are variously colored, but are wholly without the indispensable optic apparatus for a visual organ.

Each eye is composed of a ball formed of a fibrous *Sclerotica*, which is situated upon a small eminence, or is sunken in a contractile prolongation of the mantle from which projects a *cornea*, covered by the general skin. Within the sclerotica there is a reddish-brown pigment which is continuous in front into a brownish or bluish-green *Iris* which has a circular pupil; while behind, at the base of the eye, it has the appearance of a kind of *Tapetum*. This is composed of staff-like corpuscles, which produce that beautiful emerald-green appearance of the eyes of certain species.

The *Retina* surrounds a vitreous body, composed of non-nucleated cells, and which receives in front a very flattened crystalline lens. The optic nerves which enter the eye-ball at its posterior part, are, together with those of the tentacles, received from those of the mantle, and especially from the marginal branches.[2]

The following are the modifications which have already been observed with the eyes of these animals:

With the Ascidiac, there are eight eyes at the entrance of the respiratory tube, and six of a deep-yellow color at the entrance of the anal tube. They are situated in the special fissures around the openings, and in the midst of a mass of orange-colored pigment.[3]

With *Pholas*, *Solen*, *Venus*, and *Mactra*, these organs are very numerous and non-pedunculated, and are situated at the base of the tentacles surrounding the two orifices of the siphon. With *Cardium*, the borders of the orifices of the short siphons have an extraordinary number of protractile tentacles which can be protruded through the open valves, each of which bears an eye of diamond brilliancy.[4]

With *Tellina*, the two borders of the mantle have small, reddish-yellow, pedunculated eyes, which are quite numerous at the posterior portions.

With *Pinna*, the anterior part of the mantle near the adductor muscle has, on each side, about forty brownish-yellow eyes situated upon short peduncles. But with *Arca*, and *Pectunculus*, the numerous reddish-brown eyes, usually sessile, are scattered irregularly over the borders of the mantle.[5]

Anomia has about twenty brownish-yellow sessile eyes concealed among the tentacles, upon each border of the mantle. With *Ostrea*, the number is still larger; for, for more than a third of the length of the mantle, there is a very small short-pedunculated yellowish-brown eye between every second tentacle.

But the beautiful emerald-green eyes of the Pectinea are the most remarkable. They are pedunculate and situated between the tentacles of the marginal fold of the mantle, being very much more numerous upon the side of the plane, than upon that of the convex valve.[6]

---

2 See *Garner*, loc. cit. fig. 3; *Krohn*, loc. cit. fig. 16, and *Grube*, loc. cit. fig. 2.

3 *Phallusia*, *Cynthia*, and *Clavellina*, according to *Will*, loc. cit. No. 623, p. 102. *Grant* (Outlines, &c., p. 361) has seen, at least with *Phallusia*, these fourteen eyes.

4 See *Will*, loc. cit. p. 100. The color of the eyes appears reddish blue with *Mactra*, and of a yellowish brown with others.

5 See *Will*, loc. cit. The pupil is an elongated oval with *Pinna*. With *Pectunculus pilosus*, the very numerous eyes are partly isolated, and partly grouped in twenties and thirties.

6 Beside the figures already cited, all of which belong to *Pecten* and *Spondylus*, see also those which *Delle Chiaje* (Descriz. &c. Tav. LXXV. LXXVI.) has given of the eyes of *Pecten*. In this same genus, *Will* has seen sixteen to twenty-four of these organs upon the convex portion of the mantle, and thirty-five to forty-five upon the plano portion; and with *Spondylus gaederopus*, sixty upon the convex, and ninety upon the plano side.

CHAPTER V.

§ 188.

The digestive canal of the Acephala is formed, throughout the class, upon a single plan. It always consists of irregular convolutions which are separated with difficulty, for their walls are generally not covered by a peritoneal envelope, but are intimately blended with contiguous organs and especially the liver and genital gland. The oral and anal openings, which are always present, are not upon the surface of the body, but are situated in a cavity circumscribed by the mantle.[1]

The mouth has always tumid lips and often tentacular appendages. Its cavity has neither distinct muscular walls, nor any trace of a masticatory apparatus. It passes either directly, or by a short œsophagus, into a kind of stomach which gradually contracts into a longer or shorter intestine, scarcely different from it in its intimate structure. The extremity of the intestine often projects into the cavity of the body, as a kind of papilla, upon the end of which the anus is situated. Internally, this canal is lined throughout with a very distinct, ciliated epithelium.

The food of these animals, which consists of slime and small organized bodies, is taken into the cavity of the body with the water, and is conducted to the mouth by the ciliated epithelium which lines this last. In a similar manner the fæces are rejected with the refuse water.

§ 189.

The very feebly-developed digestive canal of *Salpa* consists only of a small knob (*Nucleus*) situated in the posterior part of the cavity of the body. It connects with a furrow formed by two narrow folds situated along the ventral median line.

This furrow may become a canal by the joining of its borders, and its posterior extremity, which is a little lateral, opens directly at the entrance of the intestinal canal which is surrounded with a lip, and ought therefore to be regarded as a mouth. The folds of this furrow arise directly behind the anterior respiratory orifice, and are very probably covered with cilia, by which, solid particles of food taken into the body during respiration, are borne towards the mouth.[1]

---

1 With many Acephala, as with the Ascidiae and Salpinae whose mantle is entirely closed with the exception of the two respiratory orifices, it is only in an improper manner that the terms oral and anal can be given to these orifices.

1 With *Salpa cordiformis*, and *maxima*, I have seen this furrow quite distinctly. It appears to be present in all species. *Cuvier* has already mentioned and figured it (Mém. sur les Thalides, &c., p. 12, fig. 1, 2, 3, &c., φ.), and it has also been noticed by *Savigny* (Mém. &c. p. 124, Pl. XIV. fig.

1, 2, l.), and *Eschricht* (Over Salperne, p. 26, fig. 4, 8, 18, m.); but they describe it as a dorsal furrow and a dorsal fold, for they have taken the abdominal cavity of these animals for the back. That of *Salpa gibbosa* is quite distinctly figured in the Catalogue of the Physiological Series, &c., I. Pl. VII. fig. 1, k. This furrow corresponds, probably, less to an open œsophagus, than to the tentacle-furrow which, with all the Lamellibranchia, is situated upon the two sides of the mouth.

The intestinal canal is short, without a distinct stomach, and somewhat spirally convoluted. [2]   Its extremity opens by a large anal orifice near the mouth. [3]

With the Ascidiae, the intestinal canal is quite distinct.   The mouth is situated in the respiratory cavity, far removed from the so-called oral tube, or more properly speaking, the respiratory orifice.   It is surrounded with thick lips, and has at its posterior extremity with many species, a semi-canal closely resembling, and undoubtedly of the same signification as the ventral furrow of *Salpa*.   This canal is formed by two narrow folds arising below the circle of tentacles which surround the interior of the oral cavity ; it passes along the large curvature of the respiratory cavity, and rising upon its opposite side, ends, after a longer or shorter course, below the oral cavity. [4]   The mouth opens into a short œsophagus, and this last ends in a long or round stomach, which is often quite circumscribed and plicated longitudinally on its inner surface. [5]   The intestine passes first towards the base of the body by a short arch, then by a longer one it rises towards the mouth, and thence passes to the anal tube, opening, by a fringed anus, sometimes close behind the mouth, and sometimes further below it. [6]

With the Brachiopoda, the mouth is simple and concealed between the base of the two tentacular arms.   With *Terebratula*, the œsophagus is very long and curving, opening into a large stomach ; [7] but with the other Brachiopoda, the stomachal dilatation is wanting, and the intestine is simply convoluted. With *Orbicula*, and *Terebratula*, the intestine is short, and has only a single convolution which passes to the right and terminates in a lateral anus hidden between the lobes of the mantle.   But with *Lingula*, it is much longer, and its turns are quite numerous ; the anus here is lateral also, and opens through a small papilla which projects from the cavity of the body into that of the mantle. [8]

With the Lamellibranchia the intestinal canal is highly developed, but always buried in the midst of other abdominal viscera.   The mouth, situated at the bottom of the cavity of the mantle, and beneath the anterior adductor muscles, is surrounded by two pairs of tentacles in the form of tactile lobes ; these often form a furrow leading to the mouth, and along which pass the particles of food drawn in by the cilia. [9]   The mouth opens, either

2 See *Home*, Lect. on Comp. Anat. II. Pl. LXXII. (*Salpa Tilesii*).

3 For the oral and anal orifices of *Salpa*, see the figures of Cuvier, and Savigny, loc. cit. Some species however differ from the descriptions here given. Thus, according to a preparation in *Hunter's* Museum, the intestine of *Salpa gibbosa* has two caecal appendages (*Home*, Lect. &c. Pl. LXXI. fig. 2, 3, and Catalogue of the Phys. Series, 1. p. 132, Pl. VII. fig. 1, 2, i. i.). The intestinal canal of *Salpa pinnata* presents a still more remarkable exception. No nucleus is formed, but the mouth opens directly into the stomach which is curved and sends off an intestine in front, and the anal orifice is situated near the anterior extremity of the ventral groove ; see *Cuvier*, loc. cit. p. 11, fig. 2 ; *Home*, loc. cit. Pl. LXXIII. fig, 2, and the Catalogue of the Phys. Series, 1. Pl. VI. fig. 4.

4 *Savigny* has described this canal with the most different Ascidians as a *Sillon dorsal* ; see the figures (loc. cit. Pl. VI. &c.) of *Cynthia*, *Phallusia*, *Diazona*, *Synoicum*, *Aplidium*, *Eucoelium*, *Polyclinum*, *Botryllus*, *Pyrosoma*, &c. *Carus* also has called the attention to this canal with *Cynthia microcosmus* (Nov. Act. Acad. Physico-Med. loc. cit. p. 452, Tab XXVII. fig. 1, 2, &c.).

With *Phallusia intestinalis*, there is, opposite this canal and upon the side of the respiratory cavity corresponding to the anal tube, a longitudinal row of very long thickly-set filaments, extending even to the oral aperture. *Eschricht* has seen a row of similar tentacles with *Chelyosoma* ; see loc. cit. p. 10, fig. 4, 6, z.

5 The stomach is elongated with *Boltenia*, *Phallusia*, *Cynthia*, *Sigillina*, and spherical with *Aplidium*, *Eucoelium*, &c. Its longitudinal folds are often very distinctly marked externally by deep grooves, as is the case with *Sigillina*, *Aplidium*, and *Botryllus* ; see for this, *Savigny*, loc. cit. According to him also there is a small caecum at the base of the stomach with *Botryllus Schlosseri*, and *polycyclus* ; see *Mém.* &c. p. 201, Pl. XX. fig. 5², Pl. XXI. fig. 14, c.

6 Upon the course of the intestine with the Ascidiae, see Cuvier, Savigny, and *Home*, loc. cit. Pl. LXXIV. and the Catal. of the Phys. Ser. 1. Pl. V. (*Phallusia*).

7 See the figure given by *Owen*, loc. cit.

8 For the intestinal canal of several Brachiopoda, see *Cuvier*, Owen, and *Vogt*, loc. cit.

9 With *Cardium*, *Isocardia*, *Avicula*, &c., these two pairs of gustatory lobules are very distinctly seen passing towards the mouth by as many lateral

directly, or by a short œsophagus,[10] into a large stomach lined with numerous papillae and apparently perforated by many biliary canals. The intestine, when short, forms a single arch only; but when long, it has many convolutions; it terminates in a rectum which lies along the dorsal surface of the abdomen,[11] and passes between the lobes of the mantle, under the hinge and above the posterior adductor muscle, finally terminating above in a ciliated anus, situated upon a small prominence.[12] With the majority of this order, the rectum traverses the heart.[13] There is often, near the pylorus, a long caecum[14] extending between the convolutions of the intestine to the lower extremity of the abdomen, and which contains, through its whole extent, a cylindrical transparent cartilaginoid body — the so-called crystalline-stalk.[15] A longitudinal fold extends along the inner surface of the entire intestine and a large part of that of the rectum, and thereby the intestinal surface is increased.

§ 190.

The anterior portion of the digestive canal of the Acephala is entirely without a *Salivary gland*.[1] The *Liver*, however, is always present; it is

grooves, whose borders as already mentioned (§ 185) are blended above and below with the oral orifice. With *Pectunculus*, and *Arca*, there is a still more remarkable arrangement.
Their lobules of this kind consist only of two narrow folds upon each side of the mouth, and between which is a transverse furrow, resembling the ventral-groove of *Salpa*, or the semi-canal of the Ascidiae. The important part which this apparatus serves in the prehension of food, can be seen by covering those of *Anodonta* and *Unio* with a powdered colored substance.
This powder is carried by cilia from the surface to the borders of the tentacles, thence upon their transversely grooved internal surfaces even into the angle formed by these last, thence into currents of the grooves, and so direct into the mouth.
10 A distinct but short œsophagus is found with *Arca*, *Chama*, *Pinna*, *Cardium*, and *Mactra*.
11 The intestine is short and has a single arch with *Spondylus*, *Pecten*, *Arca*, and *Chama*. It is long and has many turns with *Pholas*, *Tellina*, *Cardium*, *Mactra*, *Pinna*, *Ostrea*, &c.
12 The anus is short and situated directly behind the anal fissure of the mantle with *Unio*, *Anodonta*, *Cardium*, *Isocardia*, &c.; while with *Aspergillum*, *Lutraria* and *Solen*, it is situated far removed from the siphon. With *Arca*, *Pectunculus*, *Pinna*, and *Avicula*, the rectum passes around a large portion of the adductor muscle and ends in front in a papilla, which, in the last two genera is quite long. With *Lima*, it ascends a little way along the anterior surface of the adductor muscle, and with *Pecten* and *Ostrea*, it leaves the median line upon the back of this muscle and passes obliquely towards the smaller valve.
13 To this, *Arca*, *Ostrea*, and *Teredo*, form an exception, and especially the last, where the intestinal canal is distinguished for several other peculiarities. Thus, the stomach is double and anteriorly divided to its base by a longitudinal septum; see *Home*, Lect. &c. Pl. LXXX., and *Deshayes*, Comp. Rend. 1846, XXII. No. 7; or *Froriep's* neue Not. No. 813.
14 For the caecum of *Solen*, *Mactra*, and *Cardium*, see the figures of *Garner*, On the Anat. of the Lamellibr. &c. Pl. XVIII. fig. 8–10; and for the disposition of the intestinal canal in general, see the Plates of *Poli*, loc. cit.

According to *Owen* (Anat. of Clavagella, &c., Pl. XXX. fig. 16, r.), *Clavagella* has a very short and rudimentary caecum.
15 With the exception of *Anomia*, the crystalline stem is wanting in all the Monomya (*Garner*, loc. cit. p. 89). But it exists with many Dimya, as *Pholas*, *Solen*, *Arca*, *Mactra*, *Donax*, *Cardium*, *Tellina*, *Anodonta*, *Unio*, *Mya*, &c.; see *Poli*, loc. cit. Tab. VII. XIII. XIV. XVI. XIX. XX. XXIV. With many of these, there is no caecum and the crystalline stem is situated in the intestine itself. It has always a cylindrical form, and is of a decreased size at its lower end, while at the opposite one it is usually divided in several irregular lobes which project into the cavity of the stomach and appear to close up the orifices of the biliary canals. With the Naiades, where the caecum is wanting, I have found this singular body, which extends from the stomach into the intestine, composed of a cortical and a medullary portion. The first which forms a kind of tube, is homogeneous, transparent, and formed of concentric layers of the consistence of the white of an egg. The second is equally homogeneous and transparent, but is of a more gelatinous nature and contains a quantity of small granules (*Unio*), or batons (*Anodonta*), insoluble in acid, which, at the points where most aggregated, give this organ a whitish color when examined by reflected light. According to *Poli's* description and figure of this organ with *Pholas ductylus*, it has an analogous structure with the other Lamellibranchia (loc. cit. I. p. 47, Tab. VII. fig. 11). As yet nothing positive can be said of the function of this organ. It may be also added that often with some individuals it is looked for in vain, while with others it is very distinct though variable as to its development and the number of layers composing its cortical portion. Hence it seems that it disappears at certain times, to be developed anew.
That of *Anodonta* as figured by *Bojanus* (Isis, 1827, Taf. IX. fig. 9, 10) was undoubtedly in the state of being formed, or disappearing.
1 *Curier* (Sur la Lingule, loc. cit. p. 7, fig. 10, 11, a.), and *Vogt* (loc. cit.) have regarded the glandular mass which, with *Lingula*, opens into the digestive canal, as a solitary organ. But *Owen* (loc. cit.) is opposed to this view and says that all the

connected with the walls of the intestine, almost inseparably, and opens into it through numerous canals.

With the Tunicata, its structure is quite simple, being composed of small, single, or ramified glandular follicles, thickly-set together and covering a large portion of the stomach and intestine. [2]

With the Brachiopoda, there are groups of green follicles removed from the digestive canal but communicating with it by excretory canals. [3]

With the Lamellibranchia, this organ is voluminous and composed of lobes which occupy the upper part of the abdominal cavity. These lobes are made up of distinct Acini formed of brownish-yellow hepatic cells. [4]

The biliary ducts which open into the stomach or the anterior part of the intestine, are always few in number.

CHAPTER VI.

CIRCULATORY SYSTEM.

§ 191.

This system with the Acephala, as well as that of the Mollusca in general, is of a higher grade than that of the Zoophytes and Worms, in having the movement of the blood due always to a contractile central organ, or Heart. This heart is, it is true, very simple in some, but then with others it is so developed as to contain both auricles and ventricles. It receives the blood from the respiratory organs and distributes it over the body, and is therefore an Aortic heart. As to the blood-vessels themselves, the hitherto received opinions have been of late quite seriously objected to; and it appears very probable that all these animals have only arteries and veins,

glandular appendages of the intestine of Brachiopoda are hepatic organs.*

2 The intestinal nucleus of *Salpa* owes its yellowish-brown color to these hepatic organs. But with *Salpa democratica*, and *caerulescens*, it is of a beautiful blue color.

Salpa pinnata, whose straight intestine has already been mentioned, is distinguished also by its liver which is separated from and runs parallel with the intestine; see *Cuvier*, and *Meyen*, loc. cit. This last-mentioned author affirms that he has seen with this species a kind of green gall-bladder (loc. cit. p. 389, Tab. XXVII. fig. 19, m.); but probably he confounded the stomach of the animal with its liver. For the intimate structure of the glandular layer upon the intestine of *Salpa cordiformis*, see *Eschricht*, Over Salperne, p. 27, Tab. III. fig. 20. With the Ascidiae, the liver is a simple glandular layer upon the stomach and intestine in the various species of *Phallusia* and *Diazona*; while with *Cynthia*, it is isolated near the pylorus, and composed of large follicles; see *Savigny*, loc. cit. Pl. XII. fig. 1⁴ (*Diazona*).

3 With *Terebratula*, there are two groups of follicles opening into the stomach; with *Orbicula*, these are replaced by a mass of long hepatic ones; and with *Lingula*, by three principal glandular masses, opening at different points into the intestinal canal; see *Owen*, *Cuvier*, and *Vogt*, loc. cit.

4 *Poli* (loc. cit. Tab. XI. XV. XVI.) has given a good representation of some hepatic lobes with their interanastomosing ducts of several species. See also *Bojanus'* figures of the liver and its ducts of *Anodonta* (Isis, loc. cit. p. 757, Taf. IX.). As to the intimate structure of this organ, I have found with *Cyclas cornea, lacustris*, and *rivicola, Unio pictorum*, and *Tichogonia polymorpha*, short, cylindrical, transparent filaments, a little flexed, but projecting stiffly from the base of the follicles into their cavity. I am yet ignorant as to their function, but have in vain sought for it, with *Unio batava, tumida, Anodonta anatina, cygnea, Mya arenaria, Cardium edule*, and *Mytilus edulis*.

For the intimate structure of the liver of Lamellibranchia, see *H. Meckel* (*Muller's* Arch. 1846, p. 9, Taf. 1.) and *Karsten* (Nov. Act. Nat. Cur. XXI. p. 302, Tab. XX.).

* [§ 190, note 1.] *Frey* and *Leuckart* declare the presence of salivary glands with *Teredo navalis*; see loc. cit. — ED.

which are connected by no capillary net-work except that situated in the respiratory organs. The blood leaving the open ends of the arteries passes into the interstices (*Lacunae*) of the parenchyma of the body; thence it is taken up by the open mouths of the venous radicles.[1]

The Blood is colorless and contains many pale, granular globules, which are indistinctly nucleated.[2]

§ 192.

With *Salpa*, the circulatory system is composed of two main trunks, one upon the dorsal, and the other upon the ventral median line. At the anterior extremity of the body these trunks connect by two arcuate vessels; and at the posterior extremity by a single slightly-dilated canal situated directly in front of the intestinal nucleus. This last-mentioned canal is divided into several chambers by two or three constrictions, and, from its rhythmical contractions, may be regarded as a heart.[1] It is surrounded with a delicate pericardium,[2] and by its pulsations the blood is thrown across the walls of the body in different ways,[3] thus forming extra-vascular currents. But it will here be observed that the heart, thus forcing the blood alternately in one direction and then in another, will regularly change the arterial into a venous current, and *vice versa*.[4]

With the Ascidiae, this system is equally feebly developed. The blood passes for the most part out of the vessels into the lacunae which often consist of ramified canals resembling vessels. The Heart is always present, and is surrounded with a very thin pericardium. It consists of a long canal, which, at both extremities, passes into a vessel which lies loop-like between the vascular sac and the intestine at the lower part of the cavity of the body.[5] Its pulsations quite resemble the peristaltic movements of the

---

1 This effusion of the blood into the parenchyma of the body and its return into the veins without the intervention of capillaries, or in general without walled canals, has been maintained recently, especially by *Milne Edwards* (Observ. et expér. sur la circul. chez les Mollusques, Comp. Rend. XX. 1845, p. 261), and by *Valenciennes* (Nouv. observ. sur la constit. de l'appareil de la circul. chez les Mollusques, Ibid. p. 750). Their observations were not limited to *Salpa* and the Ascidiae, but were extended upon *Ostrea, Pinna, Mactra, Venus, Cardium* and *Solen*. See also Ann. d. Sc. Nat. III. 1845, p. 289, 307, or *Froriep's* neue Not. Nos. 732, 733, 743.

Milne Edwards is about to publish an extended work on the circulation with the Mollusca. He has figured from his beautiful injections the partly lacunal circulatory system of *Pinna*; see Ann. d. Sc. Nat. VIII. 1847, p. 77, Pl. IV.

2 For the blood of *Phallusia, Cynthia*, and *Anodonta*, see *Wagner*, Zur vergleich. Physiol. d. Blutes Hft. I. p. 20, 11. p. 40. The blood-corpuscles of the Naiades have always appeared to me of an irregular form; and they run together when placed in a watch-glass. This is probably due to the fibrin cementing them together. When treated with acetic acid they become separated again, their contour becomes very clear and almost imperceptible, and a hitherto invisible nucleus is seen.

1 See *Cuvier*, loc. cit. p. 19, fig. 2, 9, &c. According to *Meyen* (loc. cit. p. 375, Pl. XXVIII. fig. 1, d.) the heart of *Salpa mucronata* has two constrictions; and, according to *Eschricht*, that of *Salpa cordiformis* is divided into four chambers (loc. cit. p. 26, fig. 8, a).

2 *Meyen* (loc. cit. p. 376) has denied the presence of a pericardium with *Salpa*; but *Cuvier* (loc. cit. p. 10), *Savigny* (loc. cit. p. 127), and *Delle Chiaje* (Descriz. &c. III. p. 43, Tav. LXXVIII.) affirm the contrary.

3 The direction of these blood-currents in the body of *Salpa* is satisfactorily shown by the descriptions and figures of *Quoy* and *Gaimard* (loc. cit.) and especially of *Delle Chiaje* (Descriz. &c.)

Sars (Faun. litt. &c. p. 66), has also observed with *Salpa runcinata*, that the blood beyond the aorta and vena cava, circulates in wall-less passages.

4 This remarkable alteration of the blood-currents which is possible only with a valveless heart, has been observed and described by different observers in a conformable manner. Before the heart changes the direction of its contractions it remains still for a short time, and this slackens the course of the blood-currents in the body a little, before they receive an impulse in the opposite direction; see *Van Hasselt* (Ann. d. Sc. Nat. III. 1824, p. 78). *Eschscholtz* (*Müller's* translation of the annual report of the Swedish Academy upon the progress of Natural History, &c., 1825, p. 94), *Quoy* and *Gaimard* (loc. cit. p. 559, or Isis. 1836, p. 111), and *Delle Chiaje* (Descriz. &c. III. p. 43).

5 For the heart and blood-system of the Ascidiae, see especially, *Milne Edwards* (Sur les Ascidies composées loc. cit. p. 4), who has indicated the presence of the heart in *Phallusia* and *Clavelina*, as well as in *Polyclinum, Botryllus, Didemnum, Pyrosoma*, &c.

intestine; and, as with *Salpa*, the direction of the current is changed so
alternately that the two terminal vessels serve in rotation as an Aorta and
a Vena cava.[6]

The blood not only traverses the lacunae of the intestinal sac, but also
penetrates the walls of the mantle, and even passes into the common support of the compound forms.   In this last case, it circulates in ramified
canals, which, as prolongations of the cavity of the body, extend even into
this portion of the mantle.[7]

With the Brachiopoda, this system is quite remarkable.   The branchial
afferent veins of the mantle do not open into a single heart, but into two
hearts which are situated right and left of the intestinal sac.[8]

These hearts, by pulsation, throw the blood into the intestinal canal,
which ought therefore to be considered as a common visceral sinus.[9]

With the Lamellibranchia, the heart, situated at the posterior extremity
of the back, is divided, usually into three chambers, and surrounded with
a large pericardium.   Two lateral, triangular, thick-walled auricles receive
the blood from the branchiae and send it into a simple muscular ventricle
which is nearly always traversed by the rectum.   Thence the blood passes
into the body by a posterior and an anterior aorta.   Its return into the
two auricles is prevented by valves.[10]   The walls of these aortae disappear
after considerable ramification, and the blood passes into a system of lacunae which extends through the whole body and forms a net-work of sinuses
and anastomosing canals.[11]   The venous blood is received into special

6 This change in the direction of the blood-currents was first noticed by *Lister* (Philos. Trans.
1834, Pt. II. p. 365, or *Wiegmann's* Arch. 1835,
I. p. 309) with *Perophora*, a new genus of the
compound Ascidiae; and *Milne Edwards* has
since confirmed it with *Pyrosoma* (Ann. des Sc.
Nat. XII. 1839, p. 375), and several other Ascidiae
both simple and compound; see his Observ. sur
les Ascidies simples et composées, p. 7.

These inter-alternating peristaltic and anti-peristaltic motions show that the heart of the Ascidiae
is valveless. It is therefore surprising that *Delle
Chiaje* has described it with valves; but this is
not the only point in which he differs from other
observers on this subject, for he describes the heart
of the Ascidiae as bifurcated into two auricles; see
his Mem. &c. loc. cit. III. p. 193, Tav. XLVI. fig.
13, ab. (*Cynthia papillata*), and Descriz. &c. III.
p. 29, Tav. LXXXII. fig. 11, 12 (*Phallusia intestinalis*).

7 This circulation of the blood in the common
Ascidian-stock has been observed by *Lister* (loc.
cit.). *Milne Edwards* has seen also the ascending
and descending currents in the ramified and coecal
prolongations of the peritoneal sac, in *Botryllus*,
Diazona, *Didemnum*, and *Polyclinum*; see *Savigny*, Mém. loc. cit. p. 47; *Delle Chiaje*, Descriz.
&c. III. p. 31, Tav. LXXXIII. fig. 13, 15; and
Milne Edwards, Sur les Ascidies, loc. cit. p. 41,
Pl. VII. fig. I, 1b. 1c. This last-mentioned author
has also observed that, with *Clavelina* (ibid, p.
9. Pl. II.), these canals terminate in caeca which
communicate with the cavity of the body, and are
extended into digitiform prolongations upon the extremity of the peritoneal sac, and herein the blood
moves alternately up and down. The ramified canals which abundantly traverse the mantle of *Phallusia*, are, according to authors, real blood-vessels;
see *Cuvier*, loc. cit. p. 16, Pl. III. fig. 1; *Savigny*,
loc. cit. p. 102, Pl. IX. fig. 1, B., and *Delle Chiaje*,

Descriz. &c. III. p. 33, Tav. LXXXIV. fig. 2.
According to *Kölliker* (Ueber das Vorkommen der
Holzfaser im Thierreich. loc. cit.), these multiramose vessels which come directly from the heart
and whose extremities are penicillated, appear to be
continuous directly beneath the skin with other
vessels returning by the course of these arteries.

8 See *Cuvier*, *Owen*, and *Vogt*, loc. cit.

9 *Owen* was the first to notice this analogy of
the circulation of the Brachiopoda with the extra
vascular one of other Acephala; see his Lettre sur
l'Appareil de la circulation chez les Mollusques de la
Classe des Brachiopodes (Ann. d. Sc. Nat. III. 1845,
p. 315, Pl. IV., or *Froriep's* neue Not. No. 793).

10 For the arrangement of this central part of the
circulatory system, see *Poli*, loc. cit. Tab. IX. fig.
12 (*Unio*); Tab. XIII. fig. 5 (*Solen*); Tab. XXII.
fig. 10 (*Spondylus*); Tab. XXVII. fig. 8, 12 (*Pecten*); Tab. XXIX. fig. 7, 8 (*Ostrea*); Tab. XXXI.
fig. 8, 9 (*Mytilus*), and Tab. XXXVIII. XXXIX.
(*Pinna*). Also *Bojanus*, in the Isis, 1819, p. 42,
Taf. I. II. (*Anodonta*); *Treviranus*, Beobacht aus
d. Zool. u. Physiol. p. 44, fig. 67–69 (*Mytilus* and
Anodonta); and *Garner*, Trans. of the Zool. Soc.
II. p. 90, Pl. XIX. fig. 4 (*Pecten*).

An arrangement quite different from this type is
found with *Arca*, whose two auricles are attached
to the two widely-separated ventricles, and send
out on each side an anterior and posterior aorta,
which meet and join upon the dorsal median line;
see *Poli*, loc. cit. Taf. XXV. fig. 2, 3.*

11 This system of lacunae forms, especially in the
mantle, a beautiful net-work of delicate canals
which, with the Naïades, are visible to the naked
eye. It should not, however, be confounded with
another net-work more difficult to be seen, and
which probably constitutes a system of aquiferous
canals, which is easily seen in the mantle, foot and
other parts of the body by inflation. *Delle Chiaje* has called it *Rete lymphatico-vasculosum*,

* [§ 192, note 10.] See also *Deshayes*, loc. cit. p. 63, 64, &c., Pl. VIII. fig. 1, 2, 3, and *Quatrefages*, loc. cit. p. 47, Pl. I. fig. 7 (*Teredo*). — ED.

lacunae situated at the base of the branchiae, and into which it thence passes.

CHAPTER VII.

RESPIRATORY SYSTEM.

§ 193.

With all the Acephala, the blood, just before returning to the heart, passes through a branchial, or distinctly respiratory organ, which, either simple or multiple in structure, is always hidden in the cavity of the mantle. The renewal of water takes place by special openings of the body, or through the slits of the mantle, which are often prolonged into two respiratory tubes. One of these openings is for the ingress, and the other for the egress of the water, and their currents carry in and out, respectively, food and faeces. In the cavity of the mantle, the water circulates in a definite direction and passes over the branchiae by means of the cilia covering their external surface.

§ 194.

The Branchiae of the Acephala are formed after four different types : —
1. With *Salpa*, there is one only of these organs which stretches, from above downwards and from before backwards, across the cavity of the body. The water enters through an anterior orifice which is usually valvular, and is expelled through a posterior opening by the contractions of the body.[1] The branchia itself, which, near the heart, is bent a little in front at its lower posterior extremity, consists of a narrow band having upon one of its sides numerous transverse, thickly-set folds.[2]. Its remaining portion is flat,[3] or the lateral borders are rolled up like tubes.[4] The branchial vessels are ramified in the interior, communicating, at the superior extremity

---

and has figured it very beautifully ; see his Descriz. &c. Tav. LXXV. fig. 6, and Tab. XC. fig. 1, 2 (Mantle of *Pecten* and *Solen*), Tav. LXXXIX. fig. 11 (foot of a *Mactra*).

The vascular net-work which Poli (loc. cit. Tab. XXXVIII.) has figured in the mantle of a *Pinna*, is probably only one of aquiferous canals.

I shall again (§ 195) allude to this confusion between the blood and aquiferous vessels.*

1 These respiratory motions aid also for the locomotion of *Salpa* ; for, when water escapes by the

posterior orifice, the animal closes the valve of the anterior one, so that the body is thrown forwards. On this account the cavity of the body is often called natatory.

2 See *Cuvier*, and *Savigny*, loc. cit.

3 *Salpa costata*, and *maxima*.

4 *Salpa pinnata, cylindrica, octofora.* When the branchia is contained in a tube it has often been compared to a Trachea ; see *Savigny*, loc. cit. Pl. XXIV.

---

* [§ 192, note 11.] This lacunal system is well-marked with *Teredo* according to *Quatrefages*, who denies that these animals have a proper venous system. The grounds of this conclusion are, that these lacunae are always filled by injecting the

heart, and on the other hand, all the other lacunae and the arterial system beside may be filled by injecting through one lacuna ; see Mémoire, loc. cit. p. 55. — Ed.

of the branchiae with those of the body, and at its opposite one with the heart. Externally, it is covered with large cilia.[5]

2. With the Ascidiae, the walls of the body are, for the most part, lined with a membranous branchial apparatus. In the place of respiratory orifices, there are, what are usually called an oral and an anal tube. By the first of these, the water containing food passes directly into that part of the cavity of the body which contains the branchial apparatus, and which is therefore called the respiratory cavity. By the second, this cavity is emptied of the refuse water containing faeces.[6]

The branchial membrane, which, in some of the simple Ascidiae,[7] forms numerous longitudinal folds extending entirely over the respiratory cavity, presents a trellis-like aspect with rectangular meshes.[8] These meshes which form prominent lines, have often small fleshy papillae,[9] and are always provided on each side with a row of very long cilia which produce regular currents of water.

Two longitudinal sinuses pass off from the base of the respiratory cavity and ascend along its greater and lesser curvature even to the oral tube, where they intercommunicate by a circular canal. These sinuses send numerous transverse vessels into the branchial membrane, where they anastomose vertically and thus form a net-work corresponding to the trellis just mentioned.

From the continual changes in the direction of the blood-currents it is impossible to determine which is the arterial and which the venous of these sinuses.[10]

3. With the Brachiopoda, the internal layer of the mantle serves as a branchia. The internal surface of the halves of this organ is occupied with a system of very apparent blood-canals.

With *Terebratula*, and *Orbicula*, there are four large canals upon the surface corresponding to the imperforate valve, and two upon the other surface. These arise from two hearts, and are subdivided into numerous minute branches. Parallel to these last, are others, smaller, and which appear to communicate with them on the borders of the mantle; perhaps they are the branchial arteries, while the larger canals are veins.[11]

With *Lingula*, the branchial vessels are contained in collar-like projections, giving the inner surface of the mantle a very peculiar aspect.[12]

4. In the cavity of the mantle with the Lamellibranchia, there are two pairs of branchiae, which, as four lamellae, embrace each side of the abdomen, and the foot.[13] The water which bathes them comes in partly through an opening in the mantle, and partly by a particular respiratory orifice upon the border of the abdomen, or by the respiratory tube of the siphon. It passes out through the anal orifice, or by another tube of the siphon.[14]

5 These ciliated organs were first described by *Meyen*, loc. cit. p. 385.

6 With the compound Ascidiae, the arrangement is such that several individuals are disposed in a star-like manner about a cavity in which their anal tubes open.

7 *Cynthia microcosmus, momus*, &c.

8 See the figures of *Savigny*, and *Milne Edwards*, loc. cit.

9 *Phallusia sulcata, monachus, intestinalis*, and *Diazona violacea*; see *Savigny*, loc. cit. Pl. IX.-XII.

10 See *Milne Edwards*, Sur les Ascidies composées, p. 7. *Cuvier*, and *Savigny*, had already partly known this arrangement of the branchial vessels.

11 *Owen*, loc. cit.

12 See *Cuvier*, *Owen*, and *Vogt*, loc. cit.

13 The two external branchial lamellae are usually a little smaller than the two internal; and this difference is well marked with *Cardium*. According to *Valenciennes* (Comp. Rend. XX. p. 1688, XXI. p. 511), there is only a single pair of branchiae with *Lucina jamaicensis*, and *columbella, Cytherea tigerina, Tellina crassa*, and *Solen radiatus*. In this last species, they consist only of two narrow, longitudinal swellings.

14 The ingress and egress of the water through

These four branchial lamellae, whose lower border is free while the other is attached to the viscera, always extend along the abdomen, and not un-' frequently come together above.[15]

Each lamella is formed, essentially, by a widely-projecting cutaneous fold, the two leaves of which are connected by numerous transverse septa, to which correspond externally as many furrows which pass from the base of the branchia to its borders. All these furrows have upon each margin a row of long cilia, which, upon the borders of the branchia, connect with an ordinary ciliated epithelium.[16] The compartments formed by these interleaved septa are also lined with a very delicate ciliated epithelium, and connect with the cavity of the mantle at the base of the branchiae.[17]

With an entire group of this order, the branchial structure is quite different from that just described. Externally, these organs appear like ordinary branchiae, but examined more closely it will be found, that, instead of lamellae, they are composed of numerous thickly-set ribands arranged in rows.[18] These ribands are formed of two lamellae blended together at their extremity. Their circumscribing space is without doubt solely for the lodging of the branchial vessels; for, at their base, there is no orifice analogous to those found in the other species of this order. The blood collects at the base of the branchiae in the longitudinal canal, from which the lateral vessels are given off, at right angles. Thence it passes into the branchiae, traversing a trellis-like net-work quite resembling the analogous one of the Ascidiae.[19] Another series of lateral vessels serves as the branchial veins, pouring the blood into other longitudinal canals, whence it passes into the two auricles of the heart.[20]

the different orifices of the mantle may be clearly seen by observing these animals, when they, at rest, protrude between the valves either their siphon or the borders of the mantle and tinge the surrounding water with coloring matter, which makes the currents quite distinct.

15 With *Unio, Anodonta, Mactra, Cardium, Isocardia, Lutraria,* &c., the four branchial layers are united at their posterior extremity. But with *Pecten, Avicula, Arca, Pectunculus,* and *Pinna,* they are disconnected and extended backwards by two free prolongations.

16 These ciliary movements tend to carry the water, with the internal branchiae, towards their free border ; and with the external, towards their base.

17 These orifices of the branchial compartments are easily seen at the base of these organs ; excepting, however, those belonging to the two external branchiae, which are concealed beneath a kind of canal formed by the mantle. These two canals, closed in front, open behind between the end of the abdomen and the anus into that portion of the cavity of the mantle leading to the anal fissure or tube, and which may be regarded as a Cloaca ; see *Unio, Anodonta, Venus, Cardium, Isocardia, Mactra,* &c. With many, as for instance with *Unio,* and *Anodonta,* the compartments of the external branchiae are much more developed than those of the internal, and their orifices can be closed in a lip-like manner by the vesicular enlargement of the septa which limit them on each side. With *Pinna,* the branchiae are quite different. Their leaves are united by short filaments instead of by septa ; and thus, although not divided into compartments, there are orifices at their base which lead into the interior, and which are situated upon the internal surface of the internal, and upon the external surface of the external branchiae.

18 This pectinated form of the branchiae has been observed by *Baer* (*Meckel's* Arch. 1830, p. 340), with *Mytilus,* and by *Meckel* (Syst. d. vergleich. Anat. VI. p. 60), with *Spondylus, Pecten,* and *Arca ;* see, also, the Règne animal de *Cuvier,* nouv. édit. Mollusques. Pl. LXXIV. fig. 2, a. I have seen similar branchiae with *Pectunculus. Avicula,* and *Lithodomus. Philippi* (*Wiegmann's* Arch. 1835, I. p. 274) has seen them even still more developed with *Solenomya.* This peculiar branchial apparatus with *Mytilus* has been described with much detail by *Sharpey* (Cyclop. of Anat. I. p. 621). I have myself, during the autumn of 1847, completely verified the statements of this author, upon living specimens of the genera *Mytilus, Arca,* and *Pecten,* and would insist here only on a single remarkable fact. Each riband-like branchial filament has, upon both of its surfaces, several cap-like papillae by means of which these filaments are united together in a trellis-like manner. When the branchiae are forcibly distended, the papillae of the filaments are separated from each other. But this separation is not very extended, for there is a cord composed of delicate fibres, between each two papillae and binding them together ; but sometimes, from undue force, this cord is broken in its middle, and then each broken extremity appears as a bundle of moving vibratile cilia which projects from the cavity of the capsule (see *Sharpey,* loc. cit. fig. 305, E. a). The function of this apparatus, which ceases to be visible when the papillae are united together, is yet wholly doubtful.

19 *Unio, Anodonta, Lima, Pinna, Ostrea,* &c. ; see *Treviranus,* Beobacht. aus d. Zool. &c. fig. 62, 63, 65 (*Ostrea* and *Anodonta*) ; and *Poli,* loc. cit. Tab. IX. fig. 17 (*Unio*).

20 For the branchial vessels, see, especially, *Bo-*

§ 195.

It now remains to speak of a particular system of canals traversing in all directions the body of the Lamellibranchia, which as yet has been called the aquiferous system, because it is supposed to serve for an internal respiration like that of the tracheae of insects.[1]   But, in the first place, the existence itself of such a system has been denied, although there are certain facts in its favor.

When one of these animals is suddenly taken from the water, numerous fine jets of water are seen to pass from these organs while the animal is withdrawing its foot and the borders of the mantle within the shell.   From this fact it is evident that these orifices connect with aqueous reservoirs. But these openings are very small and probably are closely contracted, for they cannot be discovered either before or after the jetting out of the water.[2]   Orifices of this kind have as yet been found in a few species only ; such are those in the extremity of the foot of *Solen*,[3] and that singular tube found above the pedunculate anus of *Pinna*.[4]

The aquiferous canals themselves are not very apparent, being seen only after injection.   This last is easily performed by blowing through a small tube inserted under the skin.   There will then be seen a very beautiful network of canals, which, nearly all of the same size, are spread out under almost the whole skin and enter the interior of the body by larger canals. These canals appear to be without walls, and have, in general, the aspect of simple lacunae traversing the parenchyma of different parts of the body.

By some naturalists, this net-work of canals is regarded as a system of lacunae circulating the blood ;[5] but when they are inflated, another net-

janus, Isis, 1819 ; *Treviranus*, Beobacht. &c. p. 44. and the beautiful figures of *Poli*, loc. cit.*

1 *Baer* was the first to call the attention to this aquiferous system with the Naïades (*Froriep's* neue Not. No. 265, 1826, p. 5) after an analogous one had been pointed out with the Gasteropoda by *Delle Chiaje*. Poli, it is true, had recognized it before this, but he had taken them partly for tracheae and partly for lymph or blood-vessels.

2 *Meckel* (Syst. d. vergleich. Anat. VI. p. 64) went certainly too far when he affirmed that these orifices are only accidental fissures.   I have been unable to find the orifices, which, according to *Poli* (loc. cit. Introductio, p. 42, 52), are upon the summit of the cirri of the mantle and lead into a tracheal system.

3 Orifices of this kind have been described and figured by *Delle Chiaje* with *Solen siliqua*, as *Fori aquiferi* (Descriz. &c. III. p. 60, Tav. XC. fig. 1). These pores communicate probably with an aquiferous system which *Treviranus* has seen in the foot of *Solen ensis* (Die Erschein. u. Gesetze

des organisch. Lebens. I. p. 276).   The orifice which *Garner* has figured upon the middle of the foot of *Psammobia* and *Cardium*, and to which he has given the name of *Porus pedalis*, belongs undoubtedly to this system ; see Trans. of the Zool. Soc. II. Pl. XVIII. fig. 2, 13, f.

4 I have easily inflated the reticulated aquiferous canals of this animal by this tube, which, in *Pinna nobilis*, sometimes protrudes far beyond the borders of the mantle, and which *Poli* (loc. cit. II. p. 241, Tab. XXXVI. fig. 3, N. fig. 7, Z. and Tab. XXXVII. fig. 1, S.) has figured as a *Trachea*.

5 See above § 192, note 11.   The vascular net-work which *Poli* (loc. cit. I. p. 8, Tab. IX.) has injected with mercury in the mantle of a *Unio*, and which he regarded as a lymphatic system, belongs probably to the aquiferous system.   The same interpretation ought perhaps to be put upon a sanguineous net-work which he has figured in the mantle of a *Pinna* (loc. cit. Tab. XXXVIII.). *Delle Chiaje* (Descriz. &c. III. Tav. LXXV. fig. 6, Tav. LXXVI. fig. 3, 6, and XC. fig. 1, 2, LXXXIX. fig.

* [§ 194, note 20.] For full details on the branchial vessels of *Teredo*, and beautifully illustrated, see *Deshayes*, loc. cit. p. 69, Pl. VII. and *Quatrefages*, Mémoire, loc. cit. p. 57, Pl. II.   See also *Williams*, On the Structure of the Branchiae and Mechanism of Breathing in the Pholades and other Lamellibranchiate Mollusks, in the Report of the Brit. Assoc. for the Advancem. of Sc. for 1851, p. 82, His first four conclusions are :

"1. That the blood of all lamellibranchiate mollusca is richly corpusculated.

"2. That the branchiae in all species are composed of straight parallel vessels returning upon themselves.

"3. That the heart is systemic and not branchial.

"4. That the parallel vessels of the gills are provided with vibratile cilia disposed in a linear series on either side of the branchial vessel, causing currents, which set in the direction of the current of the blood in the vessels." — ED.

work of much smaller canals is seen expanded between and above them, and which can be only the blood-canals that were already visible before inflation.[b] But the existence with these animals of a double system of lacunae having this interpretation, is attended with many difficulties. For then it must be admitted that one of these systems contains only water, and the other blood; and *it is difficult to understand how two kinds of wall-less canals can traverse the body without passing into each other. But then, on the other hand, if the aquiferous canals are regarded as veins, and the other canals as arteries, how can this be reconciled with the fact that, in this case, the blood system would open externally and the blood escape through the natural orifices, while the water would be mixed with it from passing into the body?[c] At all events, this portion of the organization of these animals still requires a more thorough investigation.

CHAPTER VIII.

ORGANS OF SECRETION.

§ 196.

The relations of the mantle to the secretion of the shell-substance and the byssus-forming organ, have already been spoken of.[a] It now only

---

11) has given very beautiful figures of the aquiferous system of the mantle and foot of *Pecten, Pinna, Solen* and *Mactra*, but has regarded it as a *Rete lymphatico-vasculosum.* Milne Edwards (Compt. Rend. XX. p. 271, or Ann. d. Sc. Nat. III. 1845, p. 300, or *Froriep's* neue Not. No. 733, p. 99), who has seen these canals in *Pinna, Mactra, Ostrea,* &c., regarded them simply as a system of lacunae common to all the Acephala.
6 I have seen it thus, at least with *Unio,* and *Anodonta.*
7 *Delle Chiaje* (Descriz. &c. III. p. 53) thinks that, with the Lamellibranchia, the sanguineous system opens externally through special orifices.

1 See above §§ 174, 179. According to *Deshayes, Teredo* has, at the anterior extremity of the body, a gland concealed between the valves and which communicates with the mouth of the animal. Its product would serve to dissolve the wood in which this animal bores. This glandular apparatus which, according to *Deshayes* exists also with other Teredina which live in calcareous matters, demands a further examination; see Comp. Rend. XXII. p. 38, 300, or *Froriep's* neue Not. XXXVII. p. 324, XXXVIII. p. 103.*

---

* [§ 196, note 1.] The means by which the Teredina penetrate the woody or stony substances in which they live, have received some investigation of late, and I refer here to the subject from its alleged anatomical relations.
According to Hancock (Proceed. Brit. Assoc. for the Advancem. of Sc. 1848, or Ann. of Nat. Hist. 1848, II. p. 225, Pl. VIII. or *Silliman's* Amer. Jour. of Sc. 1849, VII. p. 288), "On a minute examination of the surface of the foot of *Teredo Norvegica* it is found under the microscope to be crowded with minute brilliant points which, on being compressed, consist of comparatively large crystalline bodies imbedded within them. These crystals are numerous and of various sizes and shapes, chiefly five and six sided, but not by any means regularly so. They all agree in having one or more elevated points near the centre. These

bodies are highly refractive, and are for the most part pretty regularly distributed over the whole convex surface of the foot, but are occasionally congregated in masses." This author thinks that this, as also all other boring Mollusks, excavate by means of these parts which rasp down the substance to be removed. See as corroborative of these views, Clark Ann. Nat. Hist. 1850, V. p. 6. But naturalists are not agreed on this point, and however it may be with *Teredo,* yet with *Pholas,* other observers have failed to find these rasping particles in question; see a report on the discussion of *Hancock's* paper in the Athenaeum No. 1086; also *Quatrefages,* Mémoire sur le Genre Taret, Ann. d. Sc. Nat. 1849, XI. p. 33, and History of British Mollusca by *Forbes* and *Hanley,* p. 105.
After all, it would seem that it is most probable that this process is effected by the action of cilia

remains to notice a very remarkable organ found in all the Lamellibranchia, and known as the *Gland of Bojanus.*

This organ, undoubtedly of a renal nature, is always double, and consists of a large long sac with glandular walls, and of a dirty-yellow or dark-green color. It is situated each side of the back between the pericardium and the inferior adductor muscle, and extends usually upon the sides of the abdomen to the base of the branchiae.

Quite often these glands are united upon the median line of the back — their cavities being separated only by a thin septum. They communicate with the cavity of the mantle by two small openings which have swollen borders and are situated sometimes at the upper, and sometimes at the lower end of the sac.[2]

The usually very thin walls of these two sacs have numerous folds or plicae, which form compartments or areolae, all of which are covered with a very delicate ciliated epithelium. The parenchyma of these walls is composed of a very loose tissue, which, upon the least disturbance, separates into small granular cells.[3] Most of these cells contain a blue-black round nucleus, to which is due the more or less deep color of these organs.[4]

2 With *Unio,* and *Anodonta,* these orifices are at the superior extremity of the renal sacs close beside the two genital openings; see *Bojanus,* Isis, 1819, p. 46, Taf. I. fig. 1 ; *Baer,* in *Müller's* Arch. 1830, p. 319, Taf. VII. fig. 1, 2 ; *Pfeiffer,* Naturgesch. deutsch. Land-und Susswasser-Mollusken, Abth. II. Taf. II. fig. 19, b. ; and *Neuwyler,* in the Neue Denkschr. VI. p. 22, Taf. I. II. They lie in the angle formed by the abdomen and the internal branchiae, and concealed beneath the internal leaf of these last. They had already been observed by *Poli* (loc. cit. I. p. 6, Tab. IX. fig. 15, i. i.), who, however, did not recognize their true nature.

With *Pecten,* and *Spondylus,* these renal sacs, which are situated in front of the adductor muscle, have their two orifices at the lower extremity ; see *Garner,* Trans. of the Zool. Soc. loc. cit. Pl. XIX. fig. 2, j. (*Pecten*).

With many, the genital organs open into the urinary ones. This is so according to *Garner* (loc. cit. p. 92), with *Tellina, Cardium, Mactra, Pholas,* and *Mya.* I have very distinctly seen with *Pinna nobilis,* the two orifices common to the kidneys and genital organs. Their borders were swollen, and they were situated upon the anterior surface of the dorsal wall a little in front of the posterior adductor muscle. They opened into a very large sac with thin walls which had no glandular structure except at their lower extremity near the principal adductor muscle ; see *Poli,* loc. cit. Tab. XXXVII. fig. 2, D.

The genital orifices open into the two sacs directly back of these external orifices. With *Mytilus edulis,* the kidneys have a yet more singular arrangement ; their two sacs situated at the base of the branchiae are open their whole length, so

that by spreading apart the branchiae, the compartments and cells of these glands can be distinctly seen ; see *Treviranus.* Beobacht. aus d. Zool. u. Phys. p. 51, fig. 68, b.*

3 It is only recently that the intimate structure of these organs was known. *Neuwyler* was quite mistaken in regarding them as two testicles (loc. cit. p. 25). He speaks of tubes in which he affirms that he has seen spermatic particles, but he gives neither a detailed description nor a figure of one or the other. I have never been able to find anything of this kind in the Lamellibranchia. If the walls of these organs are prepared in any way for microscopic examination, a part of their parenchyma separates into a vesiculo-granular mass, the particles of which have a very lively dancing motion. The motions are due to portions of ciliated epithelium adhering to the cells and granules. It is in this way, probably, that *Neuwyler* has been deceived, taking these moving bodies for spermatic particles.

4 These round nuclei, usually of a deep brown or blue color, can easily be seen in the kidneys of *Unio, Anodonta,* and *Cyclas* ; but with the young individuals their number and size are quite limited, making the kidneys very pale. They resemble, moreover, perfectly the bodies contained in the renal substance of the Gasteropoda (see below). This analogy is particularly striking with *Aspergillum vaginiferum,* whose renal sacs are triangular and situated between the heart and the extremity of the rectum, thus resembling in all respects the kidneys of the Gasteropoda, although *Leuckart* has taken them for the liver (Neue. wirbellose Thiere d. roth. Meeres, loc. cit. p. 46, Taf. XII. fig. 6, g.).

* [§ 196, note 2.] According to *Frey* and *Leuckart,* the bodies of *Bojanus* are absent in *Teredo navalis,* but these observers think the kidneys are present in another part of the body ; see loc. cit. p. 46. — ED.

alone. This would seem inefficient did we not remember their unceasing action ; and this view is the only one which will explain the exact conformation of the excavation to the shape of the body in all its parts. It is the view of *Agassiz,* and others, who have specially examined the subject. I have here thus noticed the matter in a suggestive point of view for microscopical anatomists. — ED.

These nuclei are very solid and ought to be regarded as the secreting bodies. They are sometimes so large as to be visible to the naked eye as inorganic concretions, and, as they contain uric acid, they may well be compared to renal calculi.[5] The walls of these kidneys are surrounded by a distinct net-work which arises from the large venous reservoir in which the afferent blood of the body is accumulated. A small portion of the blood which circulates in the kidneys passes directly to the heart; but the rest is emptied into the pulmonary arteries.[6]

CHAPTER IX.

ORGANS OF GENERATION.

§ 197.

The Acephala throughout, propagate by genital organs. With the Tunicata only, is there also observed multiplication by gemmation.

This occurs with the compound and some of the simple Ascidiae, which remind one of the Zoophytes and more particularly the Polyps, which they resemble from other conditions of the organization. The buds are always developed at the lower extremity of the body, appearing first as small pyriform projections, covered by the general envelope of the mantle, into which the circulation is prolonged.

Gradually, an Ascidian is developed upon the round summit of this projection, while its peduncle is lengthened and somewhat constricted; this continues until the body of the new individual is entirely separated from

5 Similar concretions had already been seen and described with several of the Lamellibranchia by Poli, who has regarded the kidneys as organs for the secretion of the lime of the shell; see his classic work, Introductio, p. 18, also Tom. II. p. 86, Tab. XX. fig. 4, 6, k, fig. 12, 13 (Cytherea chio), p. 143, Tab. XXVI. fig. 11, 12, 13, y. (Pectunculus pilosus), and p. 241, Tab. XXXVII. fig. 5, 6, 3, D (Pinna nobilis).

These concretions were irregular and of a red or yellow color. I have recently found, in several individuals of Pectunculus pilosus, amber-colored concretions, mostly round, of variable size, giving these two organs the appearance of a fish's ovary filled with eggs. Having collected a considerable quantity of these concretions, I sent a part of them to Herro Von Babo of this city, who has favored me with their qualitive analysis. The result was that those with a conchoidal fracture were composed principally of phosphate of lime with a trace of magnesian phosphate, and a small quantity of organic matter which behaved with nitric acid exactly like uric acid. Notwithstanding Bojanus (Isis, 1819, p. 46, 1820, p. 404) has taken much pains to prove that these organs are pulmonary, yet the view that they are kidneys has found most support (Treviranus, in Tiedemann's Zeitsch. f. Phys. I. p. 53, and Carus, Zool. 1834, II. p. 650),

aside from the fact of their containing uric acid (Garner, Trans. of the Zool. Soc. loc. cit. p. 92, and Owen, Lect. on Comp. Anat. &c. p. 284), a point upon which I was not before satisfied.

The chemical composition of these concretions, however, satisfies me that these organs are truly kidneys.

6 This is the mode of circulation of the blood through the kidneys, according to Bojanus, loc. cit. But the opinion of Treviranus is different. According to him all the blood returning from the branchiae traverses the glands of Bojanus before reaching the heart (Beobacht. aus. d. Zool. &c. p. 49). As these organs are not easily found, it will be difficult to determine this point positively by direct observation. It is only by following analogy that Bojanus' opinion can be probable in its essential point, — which is, that if the glands of Bojanus are the analogues of the venous appendages of the Cephalopoda, and of which I am persuaded with Van der Hoeven (Meckel's Arch. 1828, p. 502) is the case, then they connect with the veins which go to the branchiae, and not with the arteries which go from the branchiae to the heart.

The blood-current in the glands of Bojanus, therefore, ought to pass towards the branchiae and not towards the heart.

that of the parent, and the envelope of the mantle alone is common to both.[1]

§ 198.

With the Acephala, the sexes are sometimes separate, sometimes united in one individual. But the genital organs are very fully developed, and, as with the Zoophytes, consist of an ovary and a testicle with an excretory duct; but in none are there copulatory organs, or uterine reservoirs for the eggs.

The eggs are usually spherical, rarely pyriform or elliptical. The pale yellow or reddish vitellus is finely granular, and surrounded with a vitelline membrane and a smooth colorless chorion.

The germinative vesicle has usually two nucleoli cemented together. Often there is a layer of white substance interposed between the chorion and the vitelline membrane.[1]

The sperm is milky, and, at the epoch of procreation, quite full of very active spermatic particles. These always consist of an oblong, oval, or pyriform body, to which is abruptly attached a delicate tail, whose motions are not affected by the water in which these animals live.[2]

With the Acephala of separate sexes, the ovaries and testicles so closely resemble each other, not only as to their form and the arrangement of excretory ducts, but also as to their locality in the body, that they are with difficulty distinguished each from the other, except at the period of procreation.

The copulatory organs being absent, here, as with the Zoophytes, the water is the fecundating medium.

---

1 This multiplication by buds has been observed by *Milne Edwards* with *Botryllus*, *Polyclinum*, *Amaroucium*, *Didemnum*, and *Perophora*. It occurs also, undoubtedly, with other compound Ascidiae, and is the cause of the increase of the Ascidian-stock with the colonies of these animals. With the simple Ascidiae — *Clavelina lepadiformis*, and *producta*, the buds take the form of suckers (*Stolones*), and the new individuals are separated from their parents with the separation of the mantle ; see *Milne Edwards*, Sur les Ascidies composées, loc. cit. p. 41, Pl. III. fig. 2ᵉ. (*Amaroucium proliferum*), Pl. VII. 1, 1ᵇ. 1ᶜ. (*Botrylloides rotifera*), and Pl. II. fig. 1ᵉ. 3 (*Clavelina*), *Eysenhardt* (Nov. Act. Acad. Leop. Carol. XI. p. 263, Tab. XXXVI. fig. 1, &c.), has also observed these stolons upon a simple Ascidian.

1 These eggs have been figured by *Wagner*, Prodromus, &c., p. 7, Tab. I. fig 5 ; *Carus*, Erläuterungstafeln, &c., Hft. V. Taf. 1, fig. 2, and Nov. Act. Acad. Leop. Carol. loc. cit. p. 26, Tab. 1. (*Anodonta* and *Unio*), and by *Milne Edwards*, Sur les Ascid. comp. p. 25, Pl. IV. fig. 1–3 (*Amaroucium*).

2 The spermatic particles of the Acephala have been described and figured by *Wagner*, in *Wiegmann's* Arch. 1835, II. p. 218, Taf. III. fig. 8 (*Cyclas*); *Siebold*, in *Müller's* Arch. 1837, p. 381, Taf. XX. fig. 12–14 (*Unio*, *Anodonta*, *Mytilus*, *Tichogonia*, *Cardium*, *Tellina*, *Mya* and *Cyclas*); *Kölliker*, Beiträge, loc. cit. p. 37 (*Pholas*), and *Krohn*, in *Froriep's* neue Not. No. 356, p. 49, 52 (*Phallusia* and *Salpa*). Those of *Amaroucium* described by *Milne Edwards* (loc. cit. p. 21, Pl. III. fig. 1ᶜ) differ from the usual type in being fusiform and very long, — their tail not being distinct from the body. Those of a *Cynthia* have appeared to me of a similar form, — only the tail was much longer and more delicate. With *Phallusia*, on the contrary, I have distinctly seen them with an oblong body to which is abruptly attached the tail. The sperm of *Polyclinum*, *Botryllus*, *Didemnum*, *Diazona*, and *Phallusia*, contains spermatic particles of a Cercaria-form ; while those of *Salpa* are filamentoid ; see *Kölliker*, Neue schweiz. Denkschr. VIII. p. 43, fig. 30, 49, 55–57.*

---

* [§ 198, note 2.] The spermatic particles of the Acephala throughout, are according to my own observation, of a Cercaria-like form, — that is, having a distinct head to which is attached a more or less delicate tail. Their development, which I have traced in many cases, is in special, daughter-cells as with all other animals. They may, as indeed they often do, assume various groupings afterwards, but the real development appears simple and invariable. The shape of the head of the particle I have found to differ widely, yet in each case to present an uniformity of a zoological value. Sometimes it is perfectly globular (*Polyclinum*), sometimes oval (*Unio*, *Anodonta*), sometimes ovo-globular (*Ostrea*), sometimes oblong (*Ascidia*), sometimes pyriform (*Mytilus*), sometimes conico-pyriform (*Mya*), and sometimes long-conical (*Cyprina*). These forms may seem a refinement more ideal than real, but the exact forms are determined by micrometrical measurements. — Ed.

§ 199.

The genital organs of *Salpa* are yet quite imperfectly known. They cannot be found except at the procreative period, and in a very few species only Ovaries have been discovered.

These consist of two flexuous zigzag cords, situated each side of the median line of the back, between the mantle and peritoneum. Sometimes they are prominent from their violet color.[1] The young are always developed near the nucleus, in a cavity circumscribed by the peritoneum, but it is yet not determined whether it communicates with the ovaries by an oviduct, and whether it has distinct walls, so as to be comparable to an uterus.

As to the Male genital organs, we are yet in almost complete want of reliable researches.

From a single observation, it would appear that there is a testicle concealed in the nucleus, between the coils of the intestine, and communicating, near the anus, with the cavity of the body.[2] But this still leaves it uncertain whether these animals are hermaphrodites or of separate sexes.[3]

The Ascidiae are evidently hermaphrodites, for the male and female organs, varying as to number and position, are found upon one and the same individual between the walls of the muscular and branchial sacs.

With the compound forms, as well as with many of the simple ones, the long, compact and usually yellowish ovarian mass is situated at the base of the cavity of the body. From this there arises a large thin-walled oviduct lined with ciliated epithelium, which ascends along the rectum towards the anal tube and opens into the cloaca through a papilla. Along its side and often beneath it, there is another long mass, which is evidently, from its contents, a testicle. It has a narrow and very tortuous *Vas deferens* filled with sperm, which runs parallel with the oviduct to its very extremity.[4] *Cynthia* presents a remarkable exception in this respect. The gen-

---

1 These two ovaries have been most thoroughly observed with *Salpa pinnata*, see Forskal, Descrip. in itinere orient. observ. p. 13, Tab. XXXV. B. b¹, 4 ; *Cuvier*, loc. cit. p. 12, fig. 1. 2, *b* ; *Chamisso*, loc. cit. p. 6, fig. 1 ; *Delle Chiaje*, Memor. &c. III. Tav. LXV. fig. 8, h. ; *Meyen*, loc. cit. p. 399, Tab. XXVII. fig. 1, 21, f. ; and the Catal. of the Phys. Series, &c., I. Pl. VI. fig. 1–4, p. *Cuvier* (loc. cit. p. 22. fig. 8) has also observed two ovaries with *Salpa cylindrica*. It must also be added here that, according to Forskal and Chamisso, there are two violet ovaries, with *Salpa pinnata* both in a simple and an aggregated form.

2 It was *Krohn* (loc. cit. p. 52) who recognized with *Salpa maxima* a round testicle in the centre of the nucleus. It was composed of numerous seminiferous delicate canals filled with a white seminal liquid, and opening by a short canal into the natatory cavity. This testicle is probably the same organ that *Delle Chiaje* (Descriz. &c. III. Tav. LXXVIII. fig. 4, d.) has described as an ovary. The assertion of *Meyen*, on the other hand (loc. cit. p. 397, Tab. XXVIII. fig. 5–10), that a conical organ which, with *Salpa mucronata*, is situated in front of the cervical ganglion, belongs to the male genital organs, is unfounded and certainly incorrect. But the observation of *Krohn*, on the contrary, gives support to the opinion of *Delle Chiaje* (Mem. &c. III. p. 62, and Descriz. &c. III. p. 42) that this white canal which,

with *Salpa pinnata*, lies along the intestinal canal, is a *Vas deferens*.

3 At all events, the question needs careful examination, whether both the simple and the compound forms of *Salpa* have male organs, or only one of them. In this last case, these animals would have some resemblance to the Aphides. *Sars* (loc. cit. p. 77) having declared that the solitary individuals of *Salpa* are sexless, then the aggregate individuals ought to be considered as representing the perfect state of these animals ; but as yet neither this author, nor *Krohn* (*Froriep's* neue Notiz. XI. p. 151, and Ann. d. sc. Nat. 1846, VI. p. 110) have been able to show the existence of ovaries in these animals. At least these two naturalists pass in silence the violet ovarian striae of *Salpa*, mentioned by other observers.

4 *Cuvier* and *Savigny* have known, and often figured the female organs of the Ascidiae. Those of both sexes have been figured by *Milne Edwards* (Observ. sur les Ascid. comp. p. 21, Pl. III. fig. 1, 2, 11. fig. 1, 3) with *Clavelina*, *Amaroucium*, and *Polyclinum*. The testicle of *Phallusia* and *Rhopalaea* is quite peculiar. It consists of a white multiramose canal widely spread over the hepatic layer of the intestinal canal ; while the ovary always lies in a loop of the intestine ; see *Delle Chiaje*, Memor. III. p. 192, Tav. XLV. fig. 16, i., and Descriz. &c. III. p. 27, Tav. LXXXII. fig. 13, LXXXIV. fig. 1, i. (*Phallusia intestinalis* and

ital organs are situated upon both sides of the body between the branchial membrane and the muscular wall with which they are intimately blended. They form, sometimes several round or angular projections divided into two groups, and sometimes four long crests whose four distinct secretory ducts open, after a short course, into the space included between the branchial membrane and the muscular sac, at a variable distance from the anal tube.[5]

With the Brachiopoda, ovaries only have as yet been found. These surround the liver, and stretch upon both valves of the mantle around the minute branches of the branchial vessels.[6]

With the Lamellibranchia, there are both hermaphrodites and separate sexes. But the last are much the more common; for the first have as yet been confined to *Cyclas*,[7] *Pecten*,[8] and *Clavagella*.[9] The testicles and ovaries lie directly behind each other on each side of the body, between the liver, intestine, and kidney. Their excretory ducts have not yet been satisfactorily made out; all that has been observed, is, that with *Cyclas*, the eggs pass between the lamellae of the base of the external branchiae, and, being here developed, produce sac-like swellings.[10]

With those species which are of separate sexes,[11] the two ovaries or testicles are situated usually in the sub-hepatic region of the abdomen.

mentula). The same arrangement has been observed by *Krohn* (*Froriep's* neue. Not. No. 356, p. 49) with the testicle of *Phallusia*, and by *Philippi* with that of a *Rhopalaea* (*Müller's* Arch. 1843, p. 48, Taf. IV. fig. 9).

5 *Carus* has distinguished, with acuteness, from the ovarian group, as being a testicle, a collection of projecting angular glands lying with *Cynthia microcosmus*, upon the internal surface of the muscular sac (*Meckel's* Arch. II. 1816, p. 577, Tab. II. fig. 1, 2, &c.—and, Nov. Act. Acad. loc. cit. Tab. XXXVII. fig. 1, 2, k. k.). *Savigny* also (loc. cit. p. 92, Pl. VI. fig. 2, 3) has seen with *Cynthia microcosmus*, and *pantex*, two glandular groups with their excretory ducts as the sexual organs, but without determining that one of these was a testicle; while *Cuvier* (loc. cit. p. 28, Pl. I. fig. 3. d. d.) who has observed the testicular group with *Cynthia microcosmus*, did not know what to call it. According to *Delle Chiaje* (Memor. &c. Tav. XLV. fig. 2, h. h.), the genital organs of *Phallusia phusa* consist of numerous glandular projections united into two groups, and having two distinct excretory ducts. With *Cynthia canopus*, the genital glands lie upon four long crests, from the upper extremity of each of which passes off an excretory duct towards the anal tube (*Savigny*, Mém. loc. cit. p. 96, Pl. VIII. fig. 11, 2.). With *Cynthia papillata*, there are two of these crests curved in a loop-like manner. (*Savigny*, loc. cit. p. 92, Pl. VI. fig. 41, 42, or *Delle Chiaje*, Memor. III. p. 191, Tav. XLVI. fig. 1, l. l. and Descriz. &c. III. p. 27, Tav. LXXXII. fig. II, h. h.). From the two extremities of each of these passes off an excretory duct; and in comparing this arrangement with that of *Cynthia canopus*, it might be inferred that this loop-like disposition is due to the fusion of two glandular crests. As to which of these crests are ovaries and which testicles, it would appear from *Krohn* (*Froriep's* neue Not. No. 356, p. 50) that all are ovaries, for he has observed, with a species perhaps identical with *Cynthia canopus*, near the four oviducts, four other excretory ducts not easily seen, and which, as *Vasa deferentia*, arise from the ramified seminiferous tubes spread out over the ovaries.

6 For the ovaries of *Terebratula* and *Orbicula*, see *Owen*, loc. cit. The figure of *Müller* (*Zool.*

danica, I. p. 4, Tab. 5, fig 1, 7) of those of an *Orbicula* is very beautiful; while that of *Poli* is not as good (loc. cit. II. p. 191, Tab. XXX. fig. 19, 20).

7 For the hermaphroditism of this genus see my memoir in *Müller's* Arch. 1837, p. 383.

8 According to *Milne Edwards* (Ann. d. Sc. Nat. XVIII. 1842, p. 322, Pl. X. fig. 1), with *Pecten glaber*, the male gland is situated at the upper, and the female gland at the lower part of the abdomen. The two orifices found at the base of the groove of the byssus belong to the testicles. With another species of *Pecten* which I have examined, I was unable to confirm this hermaphroditism, for I found in the abdomen only either testicles or ovaries alone. Moreover the orifices just alluded to, appear to me to belong to a gland secreting the byssus; see above § 179, note 4.

9 With *Clavagella*, *Krohn* found the testicles beneath the liver, while the ovary surrounded it and the stomach (*Froriep's* neue Not. No. 356, p. 52).

10 These pouches, which, with *Cyclas*, contain but a single egg, have been figured by *Carus* (Erläuterungstafeln, Hft. III. p. 10, Taf. II. fig. 26. 3) after *Jacobson*.

11 *Leeuwenhoek* (Contin. arcan. natur. detec. Lugd. Batav. 1722, Epist. 95, p. 16) had already distinguished male and female individuals with certain Lamellibranchia. Notwithstanding this, he afterwards affirmed that all these Mollusks were exclusively females,—an opinion still entained by some naturalists (see *Deshayes*, in the Cyclop. Anat. I. p. 700, and *Garner*, in the Trans. of the Zool. Soc. II. p. 96). I had, however, several years before, shown that the sexes were separate with *Unio*, *Anodonta*, *Mytilus*, *Trichogonia*, *Cardium*, *Tellina*, and *Mya* (*Müller's* Arch. 1837, p. 380). The fact has been confirmed by *Milne Edwards* (Ann. d. Sc. Nat. XIII. 1840, p. 375) with *Venus*, by *Owen* (Lectures, &c., p. 287) with *Anomia*, and by *Kölliker* (Beiträge, loc. cit. p. 37) with *Pholas*. I can also add to this list, *Arca*, *Pectunculus*, and *Lithodomus*. I have already mentioned (§ 196, note 3) how *Neuwyler*, in taking the ciliary motions for those of spermatic particles, regarded the kidneys of *Anodonta* and *Unio* as the testicles,

They surround the coils of the intestine, and often ascend along the back, covering the liver with their folds. Their excretory ducts are lined with ciliated epithelium, and open each side of the bottom of the abdomen through a fissure with smaller borders, communicating either with the cavity of the mantle close by the renal opening, or with the renal sacs.[12] With those species which have a very small abdomen, these organs are spread out by numerous ramifications into the substance of both halves of the mantle.[13]

The two external branchiae serve, for the most part, the function of an uterus; for the eggs, having escaped from the oviduct, are lodged in their compartments, and, by the aid of the cilia covering the cavity of the mantle, receive the sperm which is introduced in the water for respiration.[14] The quantity of eggs thus accumulated is so great, that with *Anodonta* these organs are extraordinarily enlarged during the development of the young; and on this account the shells of the females of this genus are more convex than those of the males. In this way the sexes of these Naïades can be quickly distinguished from each other by the shell alone.[15]

§ 200.

Most of the Acephala undergo during their development, which always begins by a complete segmentation of the vitellus, a metamorphosis which is quite remarkable in many respects.

Among the Tunicata, the embryology of the Ascidiae is the best known. An oval embryo follows upon the segmented vitellus,[1] and is quickly changed into a Cercaria-like larva.[2] The tail is not formed from a grad-

and therefore considered the Naïades as hermaphrodites.*

12 The genital and urinary openings are contiguous with the Naïades; see above § 196, note 3, and *Neumann*, De Anodontarum et Unionum oviductu. Diss. Regiomont. 1827. This is the same also with *Tichogonia*; see *Van Beneden* Ann. d. Sc. Nat. VII. 1837, p. 128. With *Pinna nobilis*, I have found the genital orifice close behind those of the renal sacs. According to *Garner* (loc. cit. p. 92), a similar arrangement exists with *Tellina, Cardium, Mactra, Pholas, Mya*, and *Pecten*. The two genital orifices which *Valenciennes* (Arch. du Mus. &c. I. Pl. II. fig. 5) and *Delle Chiaje* (Descriz. &c. III. Tav. XC. fig. 2) have seen at the lower end of the abdomen of *Panopaea* and *Solen*, belong probably also to the urinary system.

13 *Mytilus* (*Poli*, loc. cit. II. p. 202, Tab. XXXI. fig. 3), *Anomia, Hiatella, Modiola*, and *Lithodomus* (*Garner*, loc. cit. p. 97). With *Lithodomus dactylus*, I have, however, always found the abdomen filled with testicular or ovarian masses.

14 It is with the Naïades that the branchiae as reservoirs of eggs, are best known; see *Poli*, loc. cit. I. p. 5, Tab. IX. fig. 18; *Pfeiffer*, loc. cit. Abth. II. p. 11, Taf. II. fig. 16–18; *Carus*, Nov. Act. Acad. &c. p. 17, Taf I. fig. 8; and *Neuwyler*, loc. cit. p. 18, Taf. III. fig. 14 (*Unio* and *Anodonta*). While remaining in the compartments of the branchiae the eggs are slightly glued together. With *Unio*, they often escape through

the anal fissures under the form of oval discs shaped like the branchial compartments. For a long time it was inexplicable how the eggs should always pass exclusively into the external branchiae, when the cavities of the internal ones were so much nearer the genital orifices. But *Baer* (*Meckel's* Arch. 1830, p. 313) has shown that their route is circuitous; they glide along the base of the internal branchia to the cloaca, they then ascend by a special canal of the mantle and pass into the external branchiae. This course is the more easily understood from the ciliated structure of these organs. *Will* (*Froriep's* neue Not. No. 620, p. 57) affirms that with *Tellina* the sperm of the males is evacuated in an analogous manner through the anal tube, and being there taken up by the females through their respiratory tube is conducted to the external branchiae.

I have also found embryos within the branchiae of *Teredo navalis*.

15 At present, this difference in the convexity of the valves appears to me to exist only with *Anodonta* (*Wiegmann's* Arch. 1837, 1. p. 415); but *Kirtland* (Ibid. 1836, 1. p. 236) has succeeded very well in distinguishing, by this character, the males and females of *Unio*, of North America.

1 The complete segmentation of the vitellus has been observed by *Milne Edwards* (Sur les Ascid. &c. p. 30, Pl. IV. fig. 1–4) with the eggs of *Amaroucium*.

2 These Cercaria-like foetuses had already been

* [§ 199, note 11.] The hermaphrodite character of the Naïades has recently been urged, and *Kirtland's* marks of the different sexes by the shape of the shell called in question; see article *Zoology* in the Iconographic Encyclopaedia, edited by *Spencer F. Baird*, p. 70. But see *Kirtland's* criticism of

this article in the Proceed. Amer. Assoc. Advancem. of Sc. 5th meeting, Cincinnati, 1851, p. 85. I have examined this subject with some care by the microscope, and have satisfied myself from an analysis of the contents of the organs that the genera in question are of separate sexes. — ED.

ual elongation of the posterior part of the body of the embryo, but is produced by the fusion of a series of globules which result from the vitelline segmentation. These globules lie upon the surface of the embryo and, in their separation from it, assume a tail-like body which is folded in front, and only latterly is extended out behind. With some of the compound forms, it forms also two eye-specks upon the back of the larva.[3] At this period of development the eggs are still in the cloaca, or perhaps have been discharged through the anal tube. Subsequently, the embryos rupture their shell, and then swim freely about by means of their very active tails. Soon after this, the larvae are completely surrounded by a transparent structureless envelope, which ultimately becomes the mantle. They are then fixed by their anterior extremity, — lose their tail and assume their adult form.[4]

With the compound forms, before the larvae have become fixed and deprived of their tails, numerous button-like prolongations arise from the anterior extremity and extend into the mantle; these, after the fixation of the embryo, are changed into as many individuals.[5]

The development of the *Salpae* has yet been incompletely observed, for its earlier conditions have received no attention. But the later ones present very curious phenomena.

In the first place, it is quite remarkable that the two forms of these animals which are always viviparous, produce young wholly dissimilar. The solitary individuals produce others joined together in a chain-like manner,[6] while these last give rise again to the solitary forms. But in neither case do the embryos undergo a metamorphosis. This chain of individuals is usually composed of two rows joined together by several cords and enveloped in a common membranous tube. The individuals at the anterior extremity of this tube are the more developed, — there being a gradation in this respect to the posterior extremity, where they appear only as simple punctiform bodies. This tube usually surrounds also the nucleus of the parent, into the cavity of whose body its anterior extremity often widely

---

observed by *Savigny* (Mém. &c. Pl. XI. fig. 2³, Pl. XXI. fig. 1.) with *Clavelina* and *Botryllus*.

Subsequently they have been described by *Audouin* and *Milne Edwards* (Ann. d. Sc. Nat. XV. 1828, p. 11); *Sars*, Beskrivelser, &c., p. 69, Pl. XII.), and *Dalyell* (Edinb. new Philos. Jour. Jan. 1859, p. 153). Latterly *Milne Edwards* (Sur les Ascidies, &c., loc. cit.) has furnished an exact embryology of these Ascidians, but which has been completed by *Van Beneden* (Mém. sur l'embryog. l'anat. et la physiol. des Ascid. loc. cit.), and by *Kölliker* (Ueber das Vorkommen der Holzfaser im Thierreich loc. cit.).

[Additional note.] The memoirs cited above have since been published, that of *Van Beneden* in Mém. de l'Acad. de Bruxell. XX. 1847, Pl. II. III.; that of *Kölliker* in Ann. d. Sc. Nat. V. 1846, 217, Pl. VII.

3 *Amaroucium* and *Aplidium*, according to *Kölliker*, and as confirmed by *Van Beneden*.

4 For the development of a simple Ascidian, see *Dalyell*, loc. cit.

5 According to *Milne Edwards* (loc. cit. p. 30), these animals use these processes like suckers to fix themselves. But this is contradicted by the observations of *Kölliker*, and *Van Beneden.**

6 This mode of propagation first described by *Chamisso* (loc. cit.), has been doubted by *Eschricht*, who thinks that the young *Salpae* produce solitary foetuses, while those of a more advanced age produce the aggregated form. But, as *Steenstrup* (Ueber den Generationswechsel, p. 36), has justly observed, there is no observation to support this view.

The alternate generation of the *Salpae*, as first described by *Chamisso*, has been confirmed in all particulars by *Sars*, and *Krohn* (loc. cit.). It is however, singular that, according to *Krohn*, the single egg of the aggregate *Salpae* is formed in an ovary, while the whole development of these same animals when solitary occurs from an internal gemmation.

---

* [§ 200, note 5.] The embryology of the Ascidiae has been followed out by *Agassiz* (Proceed. Amer. Soc. for the Advancem. of Sc. 2d meeting, 1849, Cambridge, p. 157), and by *Krohn* (*Muller's* Arch. 1852, p. 312). The observations of *Agassiz* are complete throughout, but unfortunately not yet all

published; the published portion (loc. cit.) refers more particularly to the formation and intimate structure of the egg. Those of *Krohn* are chiefly confirmatory of those of *Edwards* and other observers above mentioned. — ED.

extends, while its opposite end appears attached near the nucleus to the dorsal portion of the walls of the body.[7]

The solitary individuals, which are produced by the chain-like forms, are also developed near the nucleus, and adhere quite singularly to the dorsal wall of the parent by a peduncle resembling an umbilical cord. These pedunculated embryos are always few in number, and have a proper vitellus. Often, however, there is only one.[8] It may be questioned, moreover, if these eggs are not gradually developed with their peduncle at their place of incubation, or whether they become fixed at this place after having been developed in, and separated from the ovary. At least, one might almost think that, after all, this is only an internal gemmation.

With the Lamellibranchia, the Naïades particularly, are those whose embryology has been observed. When the vitellus begins to segment, there are two superficial contiguous vitelline cells that do not participate in this process.[9] These are gradually changed into two three-sided valves, while the remaining portion of the vitellus is transformed into a round embryo covered with cilia, which turns upon itself in the egg — being partly enveloped by the valves.[10] This rotatory movement, however, soon ceases, and the embryo divides itself into halves, each covered by a valve.[11] Each of these valves has a ciliated mouth near the hinge, and a proper intestinal canal.[12]

In the middle of the angle formed by these halves is raised a short, hollow cylinder, — the byssus-forming organ, and out of which projects a very long transparent byssus.[13]

(7) See the figures of *Chamisso*, loc. cit., of *Quoy* and *Gaimard* (Ann. d. Sc. Nat. X. 1825, p. 226, Pl. VIII. fig. 3–6, and Voyage de l'Astrolabe loc. cit.), of *Delle Chiaje* (Descriz. &c. III. p. 42, Tav. LXXVI. fig. 1), and especially those of *Eschricht* (loc. cit. p. 35, Tab. I. II. IV. V.).

(8) See *Chamisso*, loc. cit. fig. 1, D, I, J. (*Salpa pinnata*, with a very developed foetus), fig. 3, F. (*Salpa zonaria* with three button-like foetuses little developed), *Quoy* and *Gaimard*, Isis, 1836, Taf. VI. fig. 12 (*Salpa pinnata*, with a very large foetus), and Ann. d. Sc. Nat. loc. cit. Pl. VIII. fig. 7–9 (*Salpa microstoma*, with four button-like foetuses); *Meyen*, loc. cit. p. 399, Tab. XXVII. fig. 9–16 (*Salpa pinnata*), Tab. XXVIII. fig. 1, 2 (*Salpa mucronata*), Tab. XXIX. fig. 1. h. (*Salpa antarctica*), fig. 2–4 (*Salpa maxima*); *Eschricht*, loc. cit. p. 65. fig. 27, q. 36 (an individual from the chain of *Salpa cordiformis* containing five, isolated, pedunculated foetuses; perhaps here should be mentioned the five pedunculated bodies which he (p. 39, fig. 18, p. 23) has described and figured with *Salpa zonaria*; finally *Delle Chiaje*, Descriz. &c. loc. cit. Tav. LXXVIII. fig. 3 (*Salpa maxima*, with a pedunculated body), and fig. 8, 13 (*Salpa scutigera*, with a developed foetus).

9 These may be easily distinguished with *Unio* and *Anodonta*, a clear nucleus in each segmented division of the vitellus. *Carus* (Nov. Act. Acad. p. 43, Tab. II. fig. 1, 3, 10, 11) has seen the vitelline cells nucleated, but he thought that the eggs, which contained the faceted cells out of which are ultimately formed the valves, were diseased and dead.

10 This rotation of the embryo of mollusks had already excited the astonishment of *Leeuwenhoek* (Continuat. arcan. nat. Epist. 95). Its cause was explained in a very unsatisfactory manner by *Home* (Philos. Trans. 1827, pt. I. p. 39, or in *Heusinger's* Zeitsch. für organische Physik. I. p. 394),

and by *Carus* (loc. cit. p. 27), for they did not know of the existence of cilia.

11 This division of the embryo with valves often lying entirely open, has perhaps led *Rathké* (Schrivter af Naturhist. Selsk. loc. cit. p. 166, Tab. X. fig. 3), and *Jacobson* (Observ. sur le develop. prétendu des oeufs des Moulettes ou Unio et des Anodontes dans leurs branchies, An. d. Sc. Nat. XIV. 1828, p. 22, and *De Blainville's* report upon this work) to regard the young of Naïades for parasites, under the name of *Glochidium parasiticum*. See also the works of *Carus* (loc. cit.), and of *Quatrefages* (Sur la vie interbranchiale des petites Anodontes, Ann. d. Sc. Nat. IV. 1835, p. 283, V. 1836, p. 321, Pl. XII.).

Lovén informs us that the young of *Modiola*, and *Kellia* are formed upon a wholly different type. Their two valves, which are only slightly separated, are overlapped by two lobes (as *Mantle?*) which are everted and provided with very active vibratile cilia; the young swim by means of these lobes (Arch. skandinav. Beitr. zur Naturgesch. Th. I. p. 155, Taf. 1. fig. 9–11). I also have observed with *Teredo navalis*, the embryos swimming freely about by means of a foot-like organ which protrudes between the valves and presents an active ciliary movement.

12 *Quatrefages*, loc. cit. Pl. XII. fig. 20.

13 *Quatrefages* (loc. cit.) has figured with each embryo two byssus-organs out of which projects a double byssus. I have been unable to observe this, and, like *Carus*, have always found a single byssus-organ with a single byssus.

It is quite remarkable that not only the Naïades, but other Lamellibranchia also, have this byssus-organ when quite young. Thus in the young of *Cyclas cornea*, I have distinctly seen a hollow pyriform glandular organ in the foot, and from which projected a long simple byssus.

The embryos of *Kellia* have also a byssus according to *Lovén* (loc. cit.).

Internally, the embryonic halves have three tentacular, stiff points, whose bases are surrounded by collars.[14]

Near the hinge a large muscle passes from one valve to the other; this, from convulsive contractions which occur from time to time, gradually approximates the valves, which are wide open when the young individual escapes from the egg. These valves are trigonal and slightly convex. One of their sides goes to form the hinge, while the two remaining, which are a little arched, unite at an angle opposite. With this angle is articulated a prolongation curved downwards and inwards, and whose convex side has several spines.[15] After their escape from the eggs, these embryos are held together by their entangled byssuses. Subsequently, when the adductor muscle has definitely closed the valves, the embryonic halves are blended together, probably by a new metamorphosis.*

---

14 *Carus*, loc. cit. Tab. IV. fig. 14.

15 See *Rathké*, *Carus*, and *Quatrefages*, loc. cit.

* [End of § 200.] For the embryology of the Acephala with almost a profuseness of detail, see *Loven* (Ofversigt af Kongl. Vetenskaps-Acad. Förhandlingar, 5te Argangen, 1848, Stockholm, 1849, or its transl. into German in *Müller's* Arch. 1848, p. 531, or in *Wiegmann's* Arch. 1849, p. 312). This observer has observed with care the formation of all the organs and their mutual embryonic relations; even the résumé is too lengthy to be here quoted. — ED.

BOOK TENTH.

CEPHALOPHORA.

CLASSIFICATION.

§ 201.

The organization of the animals composing this class is quite dissimilar; and, as in the preceding class there were species which approached the Zoophytes, so here there are those which are scarcely above the Worms. Thus, it is a question whether the genus *Sagitta*, which is placed at the head of this class, is really in its right place, although all the attempts to place it in another group have furnished results no more satisfactory.

Then again, it may be objected that the name *Cephalophora* should have been given to a class composed of the Pteropoda, Heteropoda, and Gasteropoda, since it belongs equally well to *Cephalopoda ;* but I have adopted it for the sole reason of being unable to find a better.[1]

The sub-order Apneusta with its two families Anangia and Angiophora, has been established by *Külliker* in opposition to the other Gasteropoda, which have distinct respiratory organs.[2] This division, composed of small and very interesting species, is the more admissible since it is based upon the anatomical structure of these animals ; and also as the most recent investigations have shown that the term *Phlebenterata* used by *Quatrefages,* is improper.

ORDER I. PTEROPODA.

Animals with natatory organs composed of wing-like or fin-like cutaneous lobes, symmetrically arranged upon the two sides of the body.

---

1 *Meckel* (Syst. der vergleich. Anat.) has already used this word in the same way.

2 *Külliker* has communicated to me a yet unpublished work in which he has described, beside *Flabellina* and *Polycera*, three new genera of the inferior Gasteropoda, under the names of *Acanthina, Lissosoma,* and *Rhodope*

He has placed these, with those which *Quatrefages* has described under the name of *Phlebenterata,* in the division of Apneusta which he divides into two sections : 1. Angiophora, having a heart and rudimentary vascular system. 2. Anangia, without either heart or vessels.

FAMILY: SAGITTINA.

Genus: *Sagitta.*

FAMILY HYALEACEA.

Genera: *Hyalea, Cleodora, Cymbulia, Tiedemannia, Cuvieria, Creseis, Limacina.*

FAMILY: CLIOIDEA.

Genera: *Clio, Pneumodermon, Spongiobranchaea.*

ORDER II. HETEROPODA.

Animals whose locomotion is performed by a carinated natatory apparatus situated under the abdomen and provided often with a sucker.

Genera: *Phyllirrhoë, Pterotrachea, Carinaria, Atlanta.*

ORDER III. GASTEROPODA.

Animals which creep by means of a muscular disc situated under the body.

SUB-ORDER I. APNEUSTA.

Without distinct respiratory organs, and without a shell.

FAMILY: ANANGIA.

Genera: *Rhodope, Pelta, Actaeon, Actaeonia, Lissosoma, Chalidis, Flabellina, Zephyrina, Amphorina.*

FAMILY: ANGIOPHORA.

Genera: *Tergipes, Venilia (Proctonotus), Calliopoea, Eolidina, Aeolis (Eolidia).*

SUB-ORDER II. HETEROBRANCHIA.

The branchiae inserted more or less freely upon various parts of the body. Sometimes there is a very simple patelliform shell.

FAMILY: NUDIBRANCHIA.

Genera: *Scyllaea, Tritonia, Thetis, Doris, Polycera, Plocamophorus.*

FAMILY: INFEROBRANCHIA.

Genera: *Diphyllidia, Phyllidia, Ancylus.*

FAMILY: CYCLOBRANCHIA.

Genera: *Patella, Chiton.*

FAMILY: SCUTIBRANCHIA.

Genera: *Haliotis, Fissurella, Emarginula.*

FAMILY: TECTIBRANCHIA.

Genera: *Gasteropteron, Umbrella, Doridium, Bulla, Bullaea, Aplysia, Notarchus, Dolabella, Pleurobranchus, Pleurobranchaea.*

SUB-ORDER III. TUBICOLAE.

Animals which are enclosed, together with their branchiae, in simple slightly curved or irregularly flexuous tubes.

FAMILY: CIRRIBRANCHIA.

Genus: *Dentalium.*

FAMILY: TUBULIBRANCHIA.

Genera: *Vermetus, Magilus.*

SUB-ORDER IV. PECTINIBRANCHIA.

Branchiae in a special cavity situated at the anterior part of the back. Shell regularly spiral.

FAMILY: SIGARETINA.

Genus: *Sigaretus.*

FAMILY: PURPURIFERA.

Genera: *Buccinum, Harpa, Cassis, Purpura, Eburnea, Terebra.*

FAMILY: CANALIFERA.

Genera: *Murex, Struthiolaria, Tritonium, Turbinella, Fasciolaria.*

FAMILY: ALATA.

Genera: *Strombus, Rostellaria, Pterocera.*

FAMILY: CERITHIACEA.

Genus: *Cerithium.*

FAMILY: VOLUTACEA.

Genera: *Voluta, Oliva, Mitra.*

FAMILY: INVOLUTA.

Genera: *Cypraea, Ovula.*

FAMILY: CONOIDEA.

Genus: *Conus.*

FAMILY: TROCHOIDEA.

Genera: *Scalaria, Turbo, Trochus, Phasianella, Rotella, Littorina, Janthina.*

FAMILY: NERITACEA.

Genera: *Natica, Nerita.*

FAMILY: POTAMOPHILA.

Genera : *Rissoa, Paludina, Ampullaria, Ceratodes, Valvata.*

SUB-ORDER V. PULMONATA.

FAMILY: AMPHIPNEUSTA.

Genus : *Onchidium.*

FAMILY: LYMNAEACEA.

Genera : *Lymnaeus, Planorbis, Amphipeplea, Physa.*

FAMILY: HELICINA.

Genera : *Helix, Caracolla, Succinea, Bulimus, Achatina, Clausilia.*

FAMILY: LIMACINA.

Genera : *Limax, Arion, Testacella, Parmacella.*

FAMILY: AURICULACEA.

Genus : *Auricula.*

FAMILY: OPERCULATA.

Genus : *Cyclostoma.*

BIBLIOGRAPHY.

Cuvier. Mémoires pour servir à l'Histoire et à l'Anatomie des Mollusques. Paris 1817. A collection of monographs upon *Clio, Hyalea, Pneumodermon, Tritonia, Doris, Scyllaea, Aeolis, Glaucus, Thetis, Phyllidia, Pleurobranchus, Aplysia, Bullaea, Bulla, Limax, Helix, Dolabella, Testacella, Parmacella, Onchidium, Lymnaeus, Planorbis, Phasianella, Janthina, Paludina, Turbo, Buccinum, Sigaretus, Haliotis, Fissurella, Emarginula, Patella, Chiton,* and *Carinaria.* Most of these appeared separately in the Ann. du Muséum d'Hist. Naturelle.

Meckel. Beiträge zur vergleichenden Anatomie. Bd. I. Hft. 1, 2, Deutsches Archiv für die Physiologie, Bd. VIII. p. 190, and Archiv für die Anatomie und Physiologie, 1826, p. 13, containing anatomical details upon *Thetis, Doris, Diphyllidia,* &c.

Wohnlich. De Helice pomatia. Wirceburg, 1813.

Leue. De Pleurobranchaea. Diss. Halae, 1813.

Kosse. De Pteropodum ordine. Diss. Halae, 1813.

Feider. De Haliotidum structura. Diss. Halae, 1814.

Stiebel. Lymnaei stagnalis anatome. Diss. Gotting. 1815.

Deshayes. Anatomie du genre Dentale, in the Mém. de la Soc. d'Hist. Nat. de Paris, II. 1825.

Rang. Observations sur le genre Atlante. Ibid. III. 1827, p. 372. An Abstract of both treatises may be found in the Isis. 1832, p. 462, 471.

Treviranus. Ueber die Zeugungstheile und die Fortpflanzung der Mollusken, und über die anatomischen Verwandtschaften des Ancylus fluviatilis, in his Zeitsch. für Physiologie. Bd. I. p. 1, and Bd. IV. p. 192.

Leiblein. Beitrag zu einer Anatomie des Murex brandaris, in *Heusinger's* Zeitschrift für die organische Physik. Bd. I. p. 1, or Ann. d. Sc. Nat. XIV. 1828, p. 177.

Quoy and *Gaimard.* Voyage de la corvette l'Astrolabe sous le commandement de *Dumont Durville.* Zoologie, II. III. 1832; also in abstract in Isis, 1834, p. 283, 1836, p. 31. Contains many observations on the anatomy of the Cephalophora.

Rymer Jones. Cyclopædia of Anatomy and Physiology, II. p. 377. London, 1839. Art. *Gasteropoda.*

Eschricht. Anatomische Untersuchungen über die Clione borealis. Kopenhagen, 1838.

Van Beneden. Exercices zootomiques, Fasc. I. II. Bruxelles, 1839. Monographs on *Amphipeplea* (*Lymnaeus glutinosus*), *Pneumodermon*, *Cymbulia*, *Tiedemannia*, *Hyalea*, *Cleodora*, *Cuvieria*, and *Limacina*, extracted from the Nouv. Mém. de l'Acad. Royale de Bruxelles. That upon *Pneumodermon* is contained in *Müller's* Arch. 1838, p. 296.

Vogt. Bemerkungen über den Bau des Ancylus fluviatilis, in *Müller's* Arch. 1841, p. 25.

Pouchet. Recherches sur l'Anatomie et la Physiologie des Mollusques. Paris, 1842.

Quatrefages. Mémoire sur l'Éolidine paradoxale, and Mémoire sur les Gastéropodes Phlébentérés, in the Ann. d. Sc. Nat. XIX. 1843, p. 274, I. 1844, p. 129.

Nordmann. Versuch einer Monographie des Tergipes Edwardsii. St. Petersburg (1843), from the Mém de l'Acad. impériale de St. Petersburg, IV.; abridged in the Ann. d. Sc. Nat. V. 1846, p. 109.

Krohn. Anatomisch-physiologische Beobachtungen über die Sagitta bipunctata. Hamburg, 1844, also Ann. d. Sc. Nat. III. 1845, p. 102, and Annals of Nat. Hist. XVI. 1845, p. 289.

Hancock and *Embleton.* On the Anatomy of Eolis, in the Annals of Nat. Hist. XV. 1845, p. 1.

Allman. On the Anatomy of Actaeon. Ibid. XVI. 1845, p. 145.

ADDITIONAL BIBLIOGRAPHY.

Rymer Jones. Cyclopædia of Anatomy and Physiology. Art. *Pteropoda.* IV. p. 170.

Wilm. Observationes de Sagitta mare germanicum circa insulam Helgoland incolente. Berol. 1846. [Refers especially to the sexual and nervous systems.]

Middendorff. Beiträge zu einer Malacozoologia Rossica. St. Petersbourg, 1847 (from the Mém. Sciences Nat. Tom. VI.).

——— Mollusken, from the author's Sibirischer Riese II. Th. I. St. Petersbourg, 1851. [These works although chiefly zoological contain many anatomical details.]

Milne Edwards. Observations sur la circulation chez les Mollusques (*Patella, Haliotis, Aplysia, Thetis, Limax, Triton*), in the Ann. d. Sc. Nat. VIII. 1847, p. 37.

Blanchard. Recherches sur l'organisation des Mollusques Gastéropodes de l'ordre des Opisthobranches (Milne Edwards), Tectibranches, Nudibranches, Inférobranches (Cuvier); Ann. d. Sc. Nat. IX. 1848, p. 172, XI. 1849, p. 74.

Alder and *Hancock*. A Monograph of the British Nudibranchiate Mollusca with figures of all the species. Part I. 1844. Part V. 1851. Unfinished. [This fine work contains very valuable anatomical details and includes *Hancock's* contributions upon these animals given also in the Ann. of Nat. Hist.]

———— On the Anatomy of Oithonia, in the Ann. of Nat. Hist. IX. 1852, p. 188.

Hancock and *Embleton*. On the Anatomy of Eolis (Contin.). Ann. Nat. Hist. 1848, I. p. 105, also 1849, III. p. 183.

Hancock. On the olfactory apparatus of the Bullidae. Ibid. IX. 1852, p. 188.

———— On the Anatomy of Doris, in the Philos. Trans. 1852, Part II. p. 1.

Souleyet. Mémoire sur le genre Acteon d'Oken, Journal de Conchologie, 1850, No. I. p. 1.

Leidy. The Terrestrial air-breathing Mollusks of the United States, &c. Described and illustrated by Amos Binney. Edited by Aug. A. Gould, M.D. 2 vol. Boston, 1851. Special Anatomy by Joseph Leidy, M.D. [For beauty of execution *Leidy's* drawings are unsurpassed.]

Leydig. Ueber Paludina vivipara : im Beitrag zur näheren Kenntniss dieses Thieres in embryologischer, anatomischer und histologischer Beziehung; in *Siebold* and *Kölliker's* Zeitsch. II. 1850, p. 125.

———— Anatomische Bemerkungen über Carinaria, Firola und Amphicora. Ibid. III. 1852, p. 325.

Koren and *Damelssen*. Bitrag til Pectinibranchiernes Udviklingshistorie, Bergen, 1851, or its Transl. into the French, in the Ann. d. Sc. Nat. XVIII. 1852, p. 257.

Gegenbaur. Beiträge zur Entwickelungsgeschichte der Landgastropoden, in *Siebold* and *Kölliker's* Zeitsch. III. 1852, p. 371.

———— Bau der Heteropoden und Pteropoden. Ibid. IV. 1853, p. 334.

Moquin-Tandon. Histoire naturelle des Mollusques terrestres et fluviatiles de la France, 1847, avec un Atlas de 25 Planches, &c. Anatomie de Ancyle fluviatile, in the Jour. de Conchol. 1852, No. I. p. 7, No. II. p. 121, and No. IV. p. 337.

Rang and *Souleyet*. Histoire naturelle des Ptéropodes, Paris, 1852.

———————

CHAPTER I.

CUTANEOUS SYSTEM.

§ 202.

The cutaneous envelope of the Cephalophora consists of a dense dermis, of a cellular structure, often containing pigment matter which is free or in cells. It is covered with a delicate ciliated epithelium, which, with the aquatic species, is extended over nearly the whole body, but with those which are terrestrial, is confined to certain spots.[1] With the Gasteropoda, its external surface is striated or tuberculated; on the whole, the skin of these animals closely resembles a mucous membrane and secretes constantly a large quantity of mucus.

It has an extraordinary contractility, due to a muscular layer intimately blended in its texture.

With many species, the skin forms around the neck or back a fold, which is usually circular; the posterior or the upper part of this fold is dilated into a hernial sack containing a portion of the viscera. This portion of the skin is called the *mantle*.[2] With many, it can be wholly withdrawn into the body, and then the orifice of the fold acts as a sphincter.[3]

§ 203.

Very many of the Cephalophora carry upon their back a univalve shell[1] which is formed by the border and external surface of the mantle, and, in a few cases only, in its interior.[2]

The border of the mantle is the tissue most concerned in the formation of the shell. The shell's increase depends upon it, and for this purpose it is always in contact with the orifice.

With the majority of the terrestrial Gasteropoda [3] the completed shell has a lip at its orifice, which, in some aquatic species,[4] is repeated several times at regular intervals during the development. In many Pectinibranchia, the border of the mantle has prolongations, which also secrete lime and produce around the orifice of the shell wing-like or spinous processes.[5] With many species of this sub-order, one of these processes has a kind of canal, called the *Siphon*, which conducts the water into the respiratory cavity. With some, this siphon is contained in an appendix to the orifice of the shell,[6] while, with others, it is protruded through a fissural opening of this last.[7]

With some Gasteropoda, the mantle is folded over a large portion of the external surface of the shell, which it covers with a calcareous substance.[8]

The borders of the mantle have numerous, short, glandular follicles, whose walls are composed of large cells, some of which contain a finely-granular

1 Ciliated epithelium covers the entire surface of the body with *Lymnaeus, Planorbis, Physa, Paludina, Valvata, Terzipes, Flabellina,* and *Polycera.* With the terrestrial Gasteropoda, I have found it only on the surface of the foot, and with *Arion,* upon also the borders of this organ, which is separated from the rest of the body by a longitudinal furrow. I cannot, therefore, support the assertion of *Valentin (Wagner's* Handwört erbuch der Physiol. I. p. 429) that with *Helix* and *Limax,* the whole surface of the body and even the tentacles are covered with this epithelium.

2 With *Limax, Arion,* and some allied genera, this mantle is quite rudimentary, — covering like a shield only a small portion of the back.

3 The genus *Sagitta* differs, in this respect, from all the other Cephalophora. Its skin is without folds, forms a kind of cylinder, and consists of a dense dermis perfectly smooth and non-contractile. At first, its structure appears to be homogeneous, but a more careful examination shows extremely delicate parallel fibres running from before backwards in uninterrupted and apparently varicose rows; they resemble nuclear fibres [of *Henle*].

1 With *Chiton,* exceptionally, the shell is formed of several imbricated pieces so united as to be movable. In some species the organic so much exceeds the calcareous substance, that the shell has a horny aspect, as with *Aplysia, Hyalea,* and *Cleodora.* With *Cymbulia,* the shell is even cartilaginous; that of *Cypraea,* on the contrary, is composed almost exclusively of lime, — the quantity of organic substance being very small.

2 With *Bullaea, Limax,* and *Testacello,* the shell is wholly concealed in the mantle. With *Arion,* the lime secreted in the interior does not unite with the shell, but forms a mass of loosely juxtaposited granulations.

3 With the Auriculacea, and many of the Helicina.

4 *Murex, Harpa, Scalaria.*

5 *Strombus, Pterocera, Murex.*

6 *Cerithium, Murex, Rostellaria, Turbinella, Fasciolaria.*

7 *Harpa, Oliva, Voluta, Buccinum, Dolium, Conus.*

8 *Ovula, Cypraea.*

substance (carbonate of lime) which effervesces with acids,[9] while the others enclose pigment granules.[10] Calcareous cells are also found, but fewer, in the parts of the mantle covered by the shell. These portions of the mantle serve to increase the thickness of the shell, and to repair the loss of substance in places removed from the mantle-borders.

The intimate structure of the shells of these animals is much simpler than that of the Acephala. They are homogeneous throughout, and correspond to the internal layer of the Bivalvia. When the carbonate of lime has been extracted, the remaining organic base consists of a homogeneous membrane having numerous folds varying very much as to form and number, according to the genera.

This organic base is produced by the external surface and border of the mantle, in the form of a mucous liquid containing calcareous and pigment granules, and which, hardening, forms the successive layers of the shell.[11]

Usually there is no epidermis connecting the border of the mantle with the orifice of the shell; it can, therefore, together with the body of the animal, be drawn deeply into the shell. With some Gasteropoda, however, · shell is covered with a kind of epidermis, which has even hair-like processes.[12]

Many of this same order have, upon the post···rsal surface of the foot, a peculiar plate, by which they can tightly close the opening of their shell after having withdrawn their bodies.

This plate, or *operculum*, composed, sometimes of concentric rings, and sometimes of lines spirally rolled together in the same plane, is composed of a calcareous, or a horny substance.[13] In both cases its organic base is lamellated or plicated like that of the shell. The operculum (*Operculum caducum*) with which certain Helicina close their shell at the beginning of winter, is completely structureless, and without rings, spiral lines or lamellae.

Beside these external calcareous products, there are certain species of these animals, which have other deposits of the same nature inside the skin and in various parts of the body, which, in the form of needles, form superposed, reticulated masses.[14]

9 See *H. Meckel*, Ueber die Kalkdrüsen der Gartenschnecke, in *Müller's* Arch. 1846, p. 17.

10 According to *Gray* (Lond. Med. Gaz. pt. V. 1837, 38, vol. I. p. 830), some Gasteropoda have, in the border of their mantle, numerous glands which secrete pigment matter; and the shell will be marked according as this secretion is continuous or irregular.

11 The only solid particles I have been able to find in this mucus are calcareous molecules which disengage gas when dissolved in acids.

In the shells of *Helix, Bulimus, Cyclostoma, Paludina, Neretina,* and *Cypraea*, I have also been unable to find the cellular structure which *Bowerbank* (Ann. of Nat. Hist. No. 68, 1843) affirms exist in those of several Gasteropoda.

12 *Helix hirsuta, hispida, villosa,* and the young of *Paludina vivipara.*

13 The operculum is horny with *Paludina, Conus, Buccinum, Cassis, Murex*; and calcareous with *Nerita, Turbo, Cyclostoma.*

14 With *Paludina vivipara*, there are, between the cutaneous layers, numerous globular calcareous bodies formed of concentric lamellae; and with *Limax*, not only is there a calcareous plate in the mantle, but also a powder of the same nature scattered here and there in other parts of the skin. One white striae which adorn the sides of the neck and foot of *Helix* are composed of short, cylindrical, thickly-set calcareous needles.

According to *Kölliker*, the entire skin of *Polycera* is crowded with analogous, but ramified needles.

Similar, probably, are the concretions which, with *Terxipes*, are found everywhere beneath the skin (*Nordmann*, loc. cit. p. 9, Taf. III. fig. 4 a.), and the calcareous net-work found in the mantle and foot of several species of *Doris* (*Lorén*, Isis, 1842, p. 361, Taf. I. fig. 3).

CHAPTER II.

MUSCULAR SYSTEM AND ORGANS OF LOCOMOTION.

§ 204.

The muscles of the Cephalophora are composed of smooth, primitive bundles, which are easily separated into short oblong fragments, and often have numerous nuclei scattered through their substance.*

The cutaneous muscular system is highly developed; it consists of a muscular layer made up of oblique, longitudinal and transverse fibres, which are not divisible into separate muscles, and are intimately united with the skin.[1] Upon the ventral surface, with the Gasteropoda, this cutaneous layer is very thick and forms a long disc, — the foot. The fibres of this foot, by contraction, produce wrinkles which succeed each other from behind forwards in a wave-like manner; by this means the whole foot glides easily over solid bodies or on the surface of the water.[2] Many Gasteropoda use their foot for a sucker also, and then there are circular, tendinous fibres inwoven between those of the muscle proper.[3]

With the Heteropoda, there is, upon the ventral surface, a laterally compressed process which has numerous muscular fibres. These animals swim upon their back and use this as an organ of locomotion; while a sucker, situated upon its borders, is used, it is said, as an organ for attachment.[4] The Pteropoda, *Thetis*, and *Aplysia*, have, upon certain places of their body, wing-like expansions, which are traversed by numerous muscular fibres, and are used as oars for swimming.[5] The horizontal fins which are found

---

1 Here again the genus *Sagitta* forms an exception. Its muscular fibres are distinctly striated, and its whole muscular system consists of a simple cutaneous layer composed only of longitudinal fibres.

2 The breadth of this foot varies much according to the species. With *Scyllaea*, and *Tritonia*, it forms only a very narrow furrow, with which these animals can embrace marine algae.

3 Thus, with *Patella* and *Haliotis*.

4 See *Forskål*, Icones, &c., Tab. XXXIV. fig. A.; *Delle Chiaje*, Memor. loc. cit. Tav. XLI. fig. 1, and Descriz. loc. cit. Tav. LXIII.–IV.; *Quoy* and *Gaimard*, in the Ann. d. Sc. Nat. XVI. 1829, Pl. II. fig. 4–6, or in Isis, 1833, Taf. VI. (*Pterotrachea* and *Carinaria*); and *Rang*, in Mém. de la Soc. d'Hist. Nat. de Paris, loc. cit. p. 375, Pl. IX. fig. 1, 10, a. d. (*Atlanta*).

5 See *Eschricht*, loc. cit. Tab. I. fig. 5 (*Clio*) and *Van Beneden*, Exercises, &c., Fasc. II. Pl. I. II. (*Cymbulia* and *Tiedemannia*). It is possible that *Thetis* uses as natatory organs, beside its large cephalic fin, the contractile appendages which exist on each side of the back. These last

have had various interpretations as to their nature, from the case with which they are detached (see *Meckel* in his programme; Additamenta ad historiam Molluscorum, Piscium et Amphibiorum, Halae, 1832).

Rudolphi (Synop. Entoz. p. 573), and *Otto* (Nov. Act. Nat. Cur. XI. p. 294, Tab. XLI. fig. 1, a–f.) have taken them for parasites under the names of *Phoenicurus varius*, and *Vertumnus thetidicola*.

Delle Chiaje, who formerly described them under the name of *Planaria ocellata*, has since concurred in the opinion of the last two naturalists; but he suggests that they may be the young of *Thetis* attached to the back of their parents to obtain nourishment; see his Memor. loc. cit. I. p. 59, Tav. II. fig. 9–15, II. p. 265, III. p. 141, Tav. XXXIX. fig. 1, and his Descriz. &c. II. p. 37. Although the real nature of these appendages was made known long ago by *Macri* (Atti della reale academia delle scienze di Napoli. II. 1778, p. 170, Tav. IV.), yet it is only recently that it has been confirmed by *Ferrani* (Isis, 1842, p. 252) and *Krohn* (*Müller's* Arch. 1842, p. 418).

---

* [§ 204.] For histological studies on the muscular tissue of the Cephalophora, see *Lebert* and *Robin* (*Müller's* Arch. 1846, p. 129) and *Leydig* loc. cit. (*Siebold* and *Kölliker's* Zeitsch. II. 1850, p. 191). According to the first-mentioned observers, the intimate composition of this tissue with species they examined (*Mytilus edulis, Buccinum nudatum*, and *Pecten*), is very delicate primitive fibrillae which are either smooth and uniform, or

are finely punctated through their whole length. With *Paludina, Helix, Bulimus, Carocolla*, *Leydig* found the essential element of this tissue to consist of a tube, formed by the fusion of cells linearly arranged; the nuclei of these cells were often visible.

My own observations on *Natica heros* agree with those of *Leydig* — that the essential structure is a fibre and not a fibrilla. — ED.

upon various points of the body of *Sagitta*, differ from the locomotive
organs of the other Pteropoda in being composed wholly of parallel, homogeneous fibres, which decrease in size from the base to the border of this
organ, but which have not the least resemblance to those of muscle.[6]

Some Pteropoda have tentacle-like processes situated in bundles about
the mouth, which have a small sucker at their end; they are, therefore,
probably used as organs of attachment.[7]

§ 205.

Beside this cutaneous muscular system, the cavity of the body contains
isolated muscles which serve different uses. With the turbinated Gasteropoda, a large muscle arises from the columella, and, after dividing into
many parts, is spread over the sides of the body to be inserted into the foot,
and serves as its retractor. Several other muscles of variable size arise
also from the columella, and are distributed, some to the tentacles, and
others to the pharynx and the penis — serving also as retractors of these
organs. With the shell-less Cephalophora, the retractors of these various
organs arise from the inner surface of the mantle, or from the foot.[1]

CHAPTER III.

NERVOUS SYSTEM.

§ 206.

The central part of the nervous system of the Cephalophora consists of a
group of closely approximated ganglia, connected together by several nervous filaments, and which surrounds, like a ring, the base of the pharynx or
the œsophagus. This œsophageal ring may be divided into several portions; one situated above, one below, and one each side of the œsophagus.
The portion lying above consists usually of two very large contiguous
ganglia, which may be called the brain, since they furnish nerves to most of
the organs of sense, — that is, the tactile organs, the eyes, and sometimes
also the organs of hearing.

The portion lying below, varies much in its form and size. It consists,
sometimes of a group of ganglia blended together, or circularly united by
short connecting filaments, and sometimes of a simple transverse cord.
The two lateral portions consist always of two cords connecting the upper
and lower parts just mentioned. The lower portion, which sends nerves
principally to the muscles of the foot and to several viscera, is often asymmetrical. The peripheric nerves are always given off from the ganglia
and never from the connecting cords, of the œsophageal ring.

6 See *Krohn*, lx. cit. p. 6.
7 See *Cuvier*, Mém. loc. cit. p. 8, Pl. 1. B. fig.
8 (*Pneumodermon*); *D'Orbigny*, Voy. dans
l'Amér. mérid., or Isis, 1839, p. 467, Taf. 1. fig.
IX. 1–15 (*Spongiobranchæa* and *Pneumoder-*

mon), and *Eschricht*, loc. cit. p. 8, Tab. II. fig.
12, 13 (*Clio*).
1 For these isolated muscles, see *Cuvier*, Mém.
sur la limace et colimaçon, loc. cit. p. 11, Pl. II.
fig. 2, 3.

§ 207.

The nervous system of the Cephalophora is enveloped by a very distinct fibrous neurolemma containing often various pigments, which, in some species, give it, and especially the ganglia, a well-marked color.[1] The neurolemma enters the ganglia and forms there numerous septa which separate the ganglionic globules into groups. These are very distinct, and although of variable size, always contain a very large nucleus composed of obscure granules in the midst of which are seen usually two to four transparent nucleoli of unequal size.[2]

These ganglionic globules are very often pedunculated,[3] and then their very slender peduncles or processes usually extend far into the nerves which are given off from the ganglion, thus leading one to infer that these globules are the origin or termination of the nervous fibres.[4] Moreover, these primitive fibres which traverse the ganglia, are always situated in that portion of them which is contiguous to the œsophagus or pharynx, while the opposite portion is occupied by the ganglionic globules.

§ 208.

There is a great variety in the form and arrangement of the different parts of the nervous centre, according to the orders and families, as follows.[1]

1. The Heteropoda quite resemble the Lamellibranchia by their widely-separated ganglia connected by very long commissures. At the anterior extremity of their body, and above the œsophagus, there is a cerebral mass which sends backwards two long nervous cords, which, after passing along each side of the intestinal canal, terminate by entering the inferior ganglionic portion (*Ganglion pedale*), situated near the ventral surface. The cerebral portion furnishes nerves to the organs of sense, to the skin, and to the lips, while the posterior portion sends them chiefly to the foot, and to the muscles of the tail.[2]

---

1 These ganglia are orange-colored with *Lymnaeus*, and red with *Planorbis*, *Paludina*, *Hyalea*, *Pleurobranchus*.

2 *Hannover* (Recherch. microscop. sur le Système nerveux, 1844, p. 69, Pl. VIII.) has very well described and figured the ganglion-globules of *Helix* and *Limax*.

3 Judging from *Ehrenberg's* figure (Unerkannt. Struktur &c. Tab. VI. fig. I. 1².) of the pedunculated ganglion-globules of *Arion empiricorum*, he was not aware of the large nuclei which they contained.

4 See *Helmholtz*, De fabr. Syst. nerv. evert. loc. cit. p. 10; *Hannover*, loc. cit. and *Will*, in *Muller's* Arch. 1844, p. 76.*

1 For the descriptions and figures of the nervous system of several Cephalophora, see *Cuvier*, Mém. loc. cit. ; *Garner*, Trans. of the Linn. Soc. XVII. p. 488 ; *Rymer Jones*, Cyclop. of Anat. p. 392, Art. *Gasteropoda* ; Anderson, Ibid. III. p. 605, Art. *Nervous System* ; and *Van Beneden*, Exercises zool. loc. cit.†

2 See *Milne Edwards*, Ann. d. Sc. Nat. XVIII. 1842, p. 326, Pl. XI., and *Delle Chiaje*, Descriz. II.

* [§ 207, note 4.] *Leidy's* results, after very careful dissection, do not accord with these, for he observed none of the nerve-fibres originate or terminate in the ganglionary globules ; see loc. cit. vol. I. p. 243. — ED.

† [§ 208, note 1.] See also *Alder* and *Hancock*, loc. cit. Part. II. Pl. II. fig. 9 (*Dendronotus*), Pl. IV. fig. 16 (*Doto*) ; Part. III. Pl. VIII. fig. 8 (*Eolis*) ; Part. IV. Pl. V. fig. 1, k. (*Scyllaea*) ; Part V. Pl. II. fig. 13 (*Doris*), Pl. XLIII. fig. 10 (*Antiopa*) ; then *Leydig*, Ueber Paludina vivi-

para, &c., loc. cit. p. 152, Taf. XIII. fig. 40, a. β. (*Paludina*); *Blanchard*, Ann. d. Sc. Nat. XI. 1849, p. 78, Pl. III. fig. 1, Pl. IV. fig. 1 (*Janus*) ; *Middendorff*, loc. cit. p. 75, Taf. IX. (*Chiton*); *Leidy*, loc. cit. Pl. I. fig. I. 11-14 (*Limax*), Pl. IV. fig. V. 15-17 (*Vaginulus*), Pl. V. fig. I. 32, 33, 34 (*Bulimus*), Pl. VI. fig. II. 25, Pl. VII. fig. VIII. 20, Pl. IX. fig. IV. 26, 27, Pl. X. fig. IV. 24, 25, 26 (*Helix*) ; Pl. XIII. fig. IV. (*Helicina*), Pl. XIV. fig. IV. Pl. XVI. (*Glandina*). — ED.

The nervous centre of the genus *Sagitta* is in many respects like that of the Heteropoda. A hexagonal cerebral ganglion lies upon the upper surface of the œsophagus; another quite large is situated in the centre of the ventral surface of the trunk. These intercommunicate by two large, very long cords. The cerebral ganglion gives off two pairs of nerves; — of these the anterior are distributed to the base of the oral hooks, and the posterior to the organs of vision; each sends, internally, a filament which passes backwards and joins at the middle of the posterior part of the head with the one from the opposite side, thus forming a loop. The ventral ganglion sends off backwards two considerable, diverging nerves, from whose external surface pass off numerous, delicate, cutaneous filaments. [3]

2. With some Tectibranchia, a simple cerebral ganglion above, and two others, quite widely separated, below, are, all three, connected together by as many cords, thus forming a large œsophageal ring. [4]

3. With many Pteropoda the cerebral ganglia are wanting, while the lower portion is highly developed. This last is composed of two or three pairs of ganglia blended together, and from which passes off a simple commissure embracing the œsophagus. [5]

4. With very many of the Apneusta and Nudibranchia, as also with several other Heterobranchia, the cerebral mass is highly developed, while the remaining part of the œsophageal ring consists of a simple nervous cord. The two or four cerebral ganglia are either connected by transverse commissures, or intimately blended together. [6]

---

p. 99, Tav. LXIII. (*Carinaria*). *Pterotrachea* has a similar disposition. According to *Delle Chiaje* (loc. cit. Tav. LXIII. fig. 14, Tav. LXIV. fig. 11), a short commissure arising from the cerebral ganglionic mass, embraces, in a ring-like manner, the œsophagus of *Carinaria* and *Pterotrachea*; but this is not mentioned by either *Cuvier*, or *Milne Edwards*.

3 See *Krohn*, loc. cit. p. 12, fig. 2, 5, 13.*

4 With *Aplysia*, according to *Cuvier*, Mém. loc. cit. p. 22, Pl. III. IV., and with *Pleurobranchus*, according to *Delle Chiaje*, Memor. loc. cit. Tav. XLI. fig. 8, o. v. v. I have found the œsophageal ring arranged in the same way with *Pleurobranchaea*.

5 This form is found especially in those species where the eyes and tentacles are abortive or entirely wanting; see *Van Benrden*, Exercices zool. Fasc. II. (*Hyalea*, *Tiedemannia*, *Cleodora*, *Cuvieria*, *Limacina*, and *Cymbulia*). Probably to the absence of these organs is due, with the Pteropoda, the often confounding of the dorsal with the ventral surface. It is, moreover, interesting that, among the Gasteropoda, *Chiton*, which is without

eyes and tentacles, has a transversal row of six sub-œsophageal ganglia, but no cerebral ganglia ; see *Cuvier*, *Garner*, and *Rymer Jones*, loc. cit.

For the nervous system of the Pteropoda, see also *Souleyet*, Comp. rend. XVII. No. 14 ; or *Froriep's* neue Not. XXVIII. p. 84. †

6 With *Bullaea*, *Doridium*, and *Phyllidia*, there are two cerebral ganglia united by a more or less long commissure ; while with *Tritonia*, and *Scyllaea*, there are four united by short commissures ; see *Cuvier*, loc. cit. With *Aeolis*, the cerebral mass is also composed of four ganglia transversely arranged (*Delle Chiaje*, Descriz. loc. cit. Tav. LXXXVIII. fig. 12, 16, and *Hancock* and *Embleton*, loc. cit. Pl. V. fig. 16). With *Eolidina*, *Zephyrina*, *Amphorina*, *Pelta*, and *Chalidis*, there are two pairs of fused ganglia which are connected together by a delicate commissure (*Quotrefages*, Ann. d. Sc. Nat. XIX. 1843, p. 233, Pl. XI. fig. 3, 4, 1. 1844, Pl. VI. fig. 1–4). With *Thetis*, and *Doris*, on the other hand, the brain is a single mass, of considerable size, and situated in the neck (*Cuvier*, loc. cit.).‡

---

* [§ 208, note 3.] For the cutaneous nerves and their mode of distribution with *Carinaria*, see *Leydig* (*Siebold* and *Kölliker's* Zeitsch. III. 1851, p. 325). Here, the nerves branch into finer and finer filaments, and finally lose themselves in a terminal net-work ; these terminal branches have frequent ganglionic corpuscles in their course. These corpuscles appear to be developed in the nerve-tube ; see loc. cit. Taf. IX. fig. 5. — Ed.

† [§ 208, note 5.] *Middendorff* (loc. cit. p. 75) has described with *Chiton* a flat and almost perpendicular nervous band situated on the internal sphincter of the mouth, and which he thinks is probably two ganglia cerebralia fused together (Taf. IX. fig. 6, a) ; this band sends off numerous

Nervi labiales to borders of the mouth (Taf. IX. fig. 6, β). — Ed.

‡ [§208, note 6.] *Blanchard* (Ann. d. Sc. Nat. XI. 1849, p. 78) describes the central nervous system of *Janus* (*Eolidia*) as consisting of six medullary masses around the œsophagus, — the cerebral, the cervical and the pedal ganglia ; see his figures, Pl. III. fig. 1, and Pl. IV. fig. 1. For the Cephalic nervous system of the Nudibranchia, see *Alder* and *Hancock*, loc. cit. Part II. Pl. II. fig. 9 (*Dendronotus*, *Doto*), cerebral ganglia, four, and give off ten pairs of nerves ; Part III. Pl. VIII. fig. 3 (*Eolis*), cerebral ganglia, four, and give off twelve pairs of nerves ; Part IV. Pl. V. fig. 13 (*Eumenis*), cerebral ganglia, four, and give

5. With other Apneusta, the œsophageal ring is composed of several contiguous ganglia which closely bind the œsophagus, but have no apparent commissures.[7]

6. The other Gasteropoda, and especially the Pectinibranchia and Pulmonata, have a highly-developed superior and inferior ganglionic mass; the œsophageal ring is formed by these ganglia, being connected on each side by a cord, which is oftener double than single.[8] The superior mass is composed usually of two ganglia which are connected by a transverse commissure, or are contiguous, and sometimes even blended together.[9] The inferior mass also presents many variations. With some species it consists of a circle of distinct ganglia, connected by commissures;[10] while with others, it is composed of a group of ganglia more or less fused together.[11]

§ 209.

With many of the Cephalophora,[1] there is a *Splanchnic nervous system*. This can be divided into a *Plexus splanchnicus anterior* and *posterior*. The first is composed usually of a double *Ganglion pharyngeum inferius*, connected by a transverse commissure, or contiguous, but rarely blended together. They are situated under the œsophagus and connect with the cerebral mass by two filaments; they send off nerves principally to the pharynx, to the œsophagus and the salivary glands; and when the posterior plexus is wanting, they send nerves also to the liver and the genital glands.[2]

---

7 This œsophageal ring is composed of eight ganglia with *Tergipes* (*Nordmann*, loc. cit. p. 35, Tab. II.), but with *Actaeon*, there are only seven, the lower one of which, asymmetrical, sends two very long cords of communication to two large cerebral ganglia, while the two lateral ganglia connect by a short commissure passing under the œsophagus (*Allman*, loc. cit. p. 194, Pl. VII. fig. 1). According to a communication which *Kölliker* has made to me, this ring, with *Flabellina*, has only five ganglia.

8 See *Berthold*, in *Müller's* Arch. 1835, p. 373.

9 There is a transversal commissure between the two cerebral ganglia with *Patella*, *Haliotis*, *Phasianella*, *Janthina*, *Turbo*, *Paludina*, *Lymnaeus*, *Planorbis*, and with many other species having a shell. These two ganglia are contiguous with *Helix*, *Limax*, and *Cypraea*; but they are fused into one with *Buccinum*, *Murex*, *Oliva*, *Harpa*, *Voluta*, and other Pectinibranchia.

10 *Haliotis* has two, and *Patella* four inferior ganglia disposed transversely, which send off from each side a double cord of communication to the brain. With *Ancylus*, *Lymnaeus*, *Planorbis*, *Physa*, *Succinea*, *Bulimus*, the inferior portion is composed of five to seven ganglia, unequal and disposed asymmetrically, and connected together by

off eight pairs of nerves; Part V. Pl. II. fig. 13 (*Doris*), cerebral ganglia, five pairs, and a single ganglion, — the pairs are symmetrically placed with regard to the median line and give off fifteen pairs of nerves; the single or visceral ganglion gives off four nerves which are distributed to the organs of reproduction, to the stomach, to the two hearts, and to the branchiae, and can be traced into ganglia of the sympathetic system belonging to these several organs; Part V. Pl. XLIII. fig. 10 (*Antiopa*), cerebral ganglia, six, and give off ten or eleven pairs of nerves.

commissures; see *Berthold*, loc. cit., and my observations in *Wiegmann's* Arch. 1841, I. p. 153, Taf. VI. fig. 3 (*Lymnaeus stagnalis*). Judging from the figure of *Van Beneden* (Exercices zool. loc. cit. Fasc. I. Mém. sur le Lymnaeus glutinosus, p. 30, Pl. I. fig. 12, and Ann. d. Sc. Nat. VII. 1837, p. 112, Pl. III. B.), of the œsophageal ring of *Lymnaeus*, this genus resembles, in this respect, *Lymnaeus*. With *Pneumodermon violaceum* (*Van Beneden*, loc. cit. p. 45, Pl. I. fig. 2), and with *Clio* (*Eschricht*, loc. cit. p. 6, Tab. III. fig. 28), the lower portion appears also to consist of a circle of ganglia.

11 *Helix*, *Limax*, *Arion*. With *Limax* (*Pouchet*, Recherch. loc. cit. p. 8), there remains in the middle of the fused ganglia only a small opening, which, with several species of *Helix*, entirely disappears.*

1 See *Brandt*, Ueber der Mundmagennerven der Evertebraten, loc. cit. p. 43.

2 The two ganglia of the *Plexus splanchnicus* or *Sympathicus anterior*, which is situated more or less in front of the inferior portion of the œsophageal ring, have, together with their corresponding nervous filaments, already been regarded by *Cuvier* as a sympathetic system, with several Gasteropoda; see his Mém. sur le Genre Aply-

With all these genera, the œsophageal ring is formed by lateral commissures which unite with the sub-œsophageal ganglia which are sometimes two (*Dendronotus*, *Doto*, *Eumenis*), sometimes four (*Eolis*, *Doris*, *Antiopa*). — Ed.

* [§ 208, note 11.] The nervous system of the terrestrial Gasteropoda has been most carefully described and beautifully figured by *Leidy* (loc. cit.). The details are so full that I can only indicate the work. — Ed.

The posterior plexus is composed of a single mass, rarely of two separate ganglia. It is situated under the digestive canal or between its coils, and from it pass off nerves to the intestine, the liver, and the genital glands, beside two cords of communication with the lower portion of the œsophageal ring.[3]

CHAPTER IV.

ORGANS OF SENSE.

§ 210.

The Tactile organs of the Cephalophora consist of two to four contractile tentacles situated upon the head, or the anterior part of the back.[1] They receive nerves of considerable size from the cerebral mass, which have sometimes a ganglionic enlargement in the extremity of the tentacle.[2] With some Gasteropoda these tentacles are hollow and button-like at their extremity, and can be inverted like the finger of a glove.[3] But with the

---

sin, p. 23, Pl. IV. fig. 1, c.; sur la Lymnée, p. 9, Pl. I. fig. 11, u.; sur l'Onchidie, p. 14, Pl. I. fig. 6, o. *Brandt* (Med. Zool. 11. p. 328, Tab. XXXIV. fig. 11, 13) has described it with *Helix pomatia*; *Van Beneden* (loc. cit.), with *Amphipeplea*, and *Treviranus* (Beobacht. aus. d. Zoot. und Physiol. p. 42, Taf. IX. fig. 60), with *Limax*. See also the researches of *Schlemm* (Dissert. de hepate ac bile crustaceorum et molluscorum quorundam, Berol. 1844, p. 22, Tab. I. fig. 2, 3), upon the hepatic nerves of Gasteropoda. *Delle Chiaje* also, has seen this plexus with *Doridium* and *Pleurobranchus* (Memor. II. p. 123, Tav. X. fig. 7, p. and III. p. 153, Tav. XLI. fig. 8, p.). According to *Garner* (loc. cit.), there is a double *Ganglion pharyngeum inferius* with, also, *Scyllaea, Doris*, and *Eolis*. With *Patella*, on the contrary, he found this anterior *Plexus splanchnicus* composed of three ganglia, two upon the sides, and the third median and a little behind.

According to *Van Beneden* (Exercices zoot. Fasc. I. p. 30, Pl. I. fig. 12, e.), there is a similar disposition with *Amphipeplea*. With the Heteropoda, this plexus is highly developed, composed of two ganglia, from which pass off long cords of communication to the cerebral mass; see *Milne Edwards*, Ann. d. Sc. Nat. XVIII. p. 327, Pl. XI. fig. 1, s. x. and fig. 2, e. f.; and *Delle Chiaje*, Descriz. &c. loc. cit. Tav. LXIII. fig. 14, l. and LXIV. fig. 11, d. (*Carinaria* and *Pterotrachea*). The Pteropoda also have this plexus; but its two ganglia are more or less intimately fused, and as the cerebral mass is here replaced by a simple collar, it does not connect with it, but with the inferior portion of the œsophageal ring; see *Van Beneden*, Exercices zoot. Fasc. II. p. 11, et seq. Pl. I fig. 9, 10, 11. fig. 8, 10, III. fig. 6, 9, and V. fig.

13 (*Cymbulia, Tiedemannia, Hyalea* and *Limacina*.)*

3 The *Plexus splanchnicus posterior* with its two long cords of communication is quite apparent with *Aplysia* (see Cuvier, loc. cit. p. 23, Pl. IV. fig. 1, R.). *Delle Chiaje* (Memor. Tav. V. fig. 1, m, X. fig. 7, o. and XLI. fig. 8, y. y.), has observed it with *Doridium*, and *Pleurobranchus*, beside the genus just mentioned, and in *Pleurobranchus*, he found it composed of two entirely separated ganglia. *Van Beneden* (Exerc. zoot. Fasc. I. p. 46, Pl. I. fig. 3–5) has found it composed of only a single ganglion with *Pneumodermon*. *Milne Edwards* (loc. cit. p. 329, Pl. XI. fig. 1, u. v. 6), has observed in the visceral sac of *Carinaria*, first, a double *Ganglion abdominale*, which receives two long cords of communication from the cerebral mass and from the *Ganglion pedale*, and then a *Ganglion anale*, communicating with the two abdominal ganglia.

1 There are most usually two tentacles. But with *Limax, Arion, Helix, Achatina, Clausilia*, and other Helicina, there are four. They are wholly wanting with *Sagitta, Cleodora, Cuvieria, Hyalea, Pterotrachea, Lissosoma, Rhodope, Phyllidia*, and *Dentalium*.

2 This swelling exists not only with the inferior and eyeless tentacles, but also the superior ones bearing eyes, with several Limacina and Helicina. However, no ganglionic globules are seen in it, and only a finely-granular substance lies interposed between the primitive fibres.

3 With the Limacina and Helicina, these organs are invested by a muscle which arises upon the columella or upon the internal surface of the mantle, and is inserted at the extremity of each tentacle.

---

* [§ 209, note 2.] For the splanchnic nervous system of the terrestrial Gasteropoda, see *Leidy*, loc. cit. Pl. XIII. fig. IV. 2 (*Helicina*), Pl. XIV. fig. IV. 3 (*Glandina*), and Pl. XVI. fig. II. 2 (*Helix*).

See also for that of the Nudibranchia, *Alder*

and *Hancock*, loc. cit. (*Eolis, Doris, Antiopa*, &c.).

According to *Middendorff* (loc. cit. p. 76), *Chiton* has a complex splanchnic nervous system which is widely distributed over the digestive organs and their auxiliary glands. — ED.

majority of the Cephalophora, they are solid and usually conical, and sometimes are replaced by two groove-like, cutaneous processes, which, from contractions of their muscular fibres, can be shortened, but not inverted.[4]

Beside these tentacles, many Cephalophora have also as tactile parts, organs, which consist of two contractile lobes situated on each side of the cutaneous fold which rests over the mouth like a second lip.[5] The prehensile organs about the mouth of certain Pteropoda, and the contractile filaments and processes on the border of the mantle of other Cephalophora, are also used, probably, as tactile parts.[6]

§ 211.

The organs of Hearing, which as yet have been found in all the orders of these animals, are, as in the Acephala, of a very low order. Like them also they consist only of two simple round auditive capsules whose transparent, solid walls contain sometimes a single, sometimes several otolites, suspended in a clear liquid, and which are composed of carbonate of lime.[1]   When

4 With the Pectinibranchia, there are usually two conical tentacles ; more rarely are there four as with *Amphorina, Eolidina, Flabellina,* and *Aeolis.* Cutaneous furrow-like prolongations are observed with *Notarchus, Dolabella, Pleurobranchus, Pleurobranchaea,* and *Aplysia.* With *Doris, Tritonia,* and *Scyllaea,* the two conical tentacles can be withdrawn into particular tubular excavations of the mantle.*

5 *Flabellina, Aeolis, Doris, Phyllidia, Doridium, Aplysia, Pleurobranchus, Pleurobranchaea, Dolabella, Ampullaria, Ceratodes.* These cutaneous lobes are often so large, that one is disposed to include them among the real tentacles.

6 I refer here to the tentacle-like organs by which *Clio, Pneumodermon,* and *Spongiobranchaea* fix themselves upon marine bodies (§ 204), to the filaments of the anterior lobes of the mantle of *Thetis, Plocamophorus,* and *Tritonia thetidea,* and to the prolongations of the lateral border of the same organ with *Haliotis, Doris fimbriata,* and *Cypraea erosa.*

1 *Eudoux* and *Souleyet* (Institut. 1838, No. 255, p. 376, or *Froriep's* neue Not. No. 174, 1838, p. 312,) were the first to notice the auditive organ with the Cephalophora. They found with *Pterotrachea, Carinaria, Pneumodermon,* and *Phyllirhoë,* as also *Gaudichaud* with *Atlanta,* that the auditive capsules are small round semi-transparent bodies attached by a peduncle upon the cerebral mass. *Laurent* (Append. aux recherch. sur les organes auditifs des Mollusques, in the Ann. franc. et étrang. d'Anat. et de Physiol. Mai, 1839, p. 118, fig. 1–16) has described these organs with their crystalline contents a little more fully, for, beside the figures of *Eudoux* and *Souleyet* relative to *Hyalea. Cleodora,* and *Creseis,* he has added others concerning *Limax* and *Helix.* Since then these organs with their otolites of different Heteropoda, Pteropoda, and Gasteropoda have been described by *Krohn* (*Müller's* Arch. 1839, p. 335, or *Froriep's* neue Not. XIV. 1840, p. 310, XVIII. 1841, p. 310). In another series of the terrestrial and fresh-water Gasteropoda, I

have attempted to show the analogy of these organs with the auditive organs of the embryos of fishes (*Wiegmann's* Arch. 1841, I. p. 148, Taf. IV. or Ann. d. Sc. Nat. XIX. 1843, p. 193, Pl. II. B.). *Kölliker* (Ueber das Gehörorgan der Mollusken, in *Froriep's* neue Not. XXV. 1843, p. 133) also has described them with many marine Heteropoda, and Gasteropoda, so that they may be said to exist in all the Cephalophora which have been subjected to dissection. The following are the genera in which they have been observed. Among the Pteropoda : *Cymbulia, Tiedemannia, Hyalea, Creseis, Pneumodermon, Limacina* ; Heteropoda : *Carinaria, Pterotrachea, Phyllirhoë, Atlanta* ; Gasteropoda : *Rhodope, Flabellina, Lissosoma, Amphorina, Pelta, Chalidis, Zephyrina, Actaeon, Actaeonia, Acolis, Venilia, Tergipes, Doris, Polycera, Tritonia, Thetis, Diphyllidia, Ancylus, Doridium, Aplysia, Gasteropteron, Umbrella, Notarchus, Pleurobranchus, Pleurobranchaea, Paludina, Lymnaeus, Planorbis, Physa, Bulimus, Clausilia, Succinea, Helix, Arion,* and *Limax.* It is remarkable that the auditive organs are developed so early, for they may be distinguished while the embryo is still in the egg. From the account of *Pouchet* (Ann. d. Sc. Nat. X. 1838. p. 64), it appears that he saw the otolites in motion in an embryo of a *Lymnaeus,* but without knowing their nature. *Loven* also, who saw the two capsules in the young *Eolis* (Kongl. Vetensk. Acad. Handl. 1839, p. 227, or Isis, 1842, p. 560, Taf. I. fig. 1, o.) did not know what to think of them. *Van Beneden* (Ann. d. Sc. Nat. XV. 1841, p. 127, Pl. I. fig. 13, 15, 17, d.) mistook them in the embryos of *Limax* and *Aplysia,* for nervous ganglia ; while *Allman* (loc. cit. p. 153, Pl. VII. fig. 10–12, d.) regarded them as eyes in the embryos of *Actaeon. Sars* (*Wiegmann's* Arch. 1845. I. p. 8, Taf. I. fig. 7–11) and *Nordmann* (loc. cit. p. 44, 87, Taf. IV. V.), on the other hand, very correctly recognized them as organs of hearing in the embryos of *Doris, Tritonia, Tergipes, Buccinum, Littorina, Cerithium,*

* [§ 210, note 4.] *Hancock* and *Embleton* (loc. cit.) regard these tentacles as olfactory organs, a view which is sustained by their special anatomy, by their special and comparative relations. *Moquin-Tandon* also (Bibl. Univer de Genève. Nov. 1851, p. 247) regards this sense as located in the end of the tentacles, with the Gasteropoda

See also *Hancock* (Ann. Nat. Hist. 1852, IX. p. 188) on this apparatus with the Bullidae. In these, no proper tentacles exist, as is well known, but this author shows that here the head-lobe, which is the result of the fusion of tentacles, is the seat of this sense. — ED.

single this otolite is spherical and crystalline, but when multiple they are fusiform, a little compressed, and usually very numerous, there being with some Gasteropoda, thirty or forty, and even eighty in each capsule.[2]

The movements of these bodies are even more marked with the Cephalophora than with the Acephala; and the balancing and rotation of each, producing a kind of trembling of their whole mass which occupies the centre of the capsule, is a wonderful spectacle. It has been recently discovered that these motions are due to very small cilia upon the internal surface of the capsule.[3]

The situation of these two auditive capsules varies according to the orders, families, and genera. With several Heteropoda, and Apneusta, they lie a little under the skin, behind the eyes, and are connected with the cerebral mass by a longer or shorter auditive nerve.[4] In some Nudibranchia, they lie upon the cerebral mass itself, contiguous with the posterior part of the eyes.[5] With the other Cephalophora, they are situated at the lower side of the body, and usually touch the inferior portion of the œsophageal ring. In only a very few of the genera are the two auditory nerves separated and distinct from each other.[6]

Phasianella, and *Rissoa*. I have myself seen them quite early in the embryos of *Vermetus*.*

2 There is a single otolite only with the Heteropoda, the Tubulibranchia and several of the Apneusta ; see *Delle Chiaje*, Descriz. &c. II. p. 100, Tav. LXIII. fig. 5, 6 (*Carinaria*), and *Quatrefages*, Ann. d. Sc. Nat. I. 1844, p. 160, Pl. VI. fig. 8–10 (*Actaeon*, *Pelta*, *Chalidis*). According to *Krohn's* and my own observations, there are groups of small fusiform otolites with some Pteropoda, and very many of the Gasteropoda, as *Cymbulia*, *Hyalea*, *Doris*, *Tritonia*, *Thetis*, *Aeolis*, *Venilia*, *Pleurobranchaea*, *Paludina*, *Planorbis*, *Lymnaeus*, *Helix*, *Limax*, and many others. It is not rare to find among these fusiform otolites, others composed of two or four calcareous corpuscles. Those of a spherical or spindle shape divide, from pressure, into four to eight fragments in the direction of cruciform lines which may often be seen before division. According to the observations of *Laurent*, *Krohn*, and myself, in the centre of these bodies, a single otolite is first developed, in the capsules which are to contain several, and others are added as the embryo increases ; see *Frey*, in *Froriep's* neue Not. XXXVII. No. 801, p. 132, and *Wiegmann's* Arch. 1845, I. p. 217. Taf. IX.†

3 *A priori*, it might have been inferred that these motions are due to cilia, for the otolites never come in contact with the sides of the capsule, but always remain at a little distance from it, and when there are several, they are grouped in the centre ; indeed when one has strayed from this central position it is always quickly returned. *Wagner* (Lehrbuch der Physiol. ed. II. 1843, p. 463) positively affirms that he has seen cilia in these capsules. They have been very distinctly seen by *Külliker* also (loc. cit.) with *Tritonia*, *Thetis*, *Pleurobranchaea*, *Diphyllidia*, *Hyalea*, *Lissosoma*, and *Rhodope*.

4 See *Laurent*, loc. cit. fig. 1–6, and *Quatrefages*, Ann. d. Sc. Nat. I. loc. cit. Pl. IV. VI. According to *Delle Chiaje* (Descriz. &c. loc. cit. Tav. LXIII. fig. 3, d. 14, f.), and *Milne Edwards* (Ann. d. Sc. Nat. XVIII. 1842, Pl. XI. fig. 1, z. fig. 3, h.), the auditive nerves are very long with *Carinaria*. With many Cephalophora which are transparent, the auditive organs may be perceived by the naked eye, through the skin, as two white spots.‡

5 *Doris*, *Thetis*, *Tritonia*, *Aeolis* (Krohn, loc. cit.), and *Tergipes* (Nordmann, loc. cit. p. 44, Tab. II.).

6 According to *Krohn* (loc. cit. No. 394, p. 311) the two auditive capsules of *Pleurobranchaea*, and *Paludina* receive distinct auditive nerves from the inferior portion of the œsophageal ring. He has observed the same with *Cymbulia*, and *Hyalea* (loc. cit. No. 306, p. 311) ; but *Van Beneden* (Ex-

---

* [§ 211, note 1.] See also *Alder* and *Hancock* loc. cit. Part II. Pl. II. fig. 11 (*Dendronotus*), Pl. IV. fig. 18 (*Doto*); Part. III. Pl. VIII. fig. 4, 5, 6 (*Aeolis*); Part V. Pl. II. fig. 15 (*Doris*); then *Leydig*, Ueber Paludina vivipara, &c., loc. cit. p. 139, 155, Taf. XI. fig. 12, k. Taf. XIII. fig. 14–24, 49, R. (*Paludina*); and *Leidy*, loc. cit. p. 246, Pl. IX. fig. VII.–IX. (*Helix*), Pl. XIII. fig. IV. 4 (*Helicina*). *Leydig* has furnished valuable contributions in the development of this organ ; in *Paludina*, it appears, prior to the nervous system with which it is connected, as an almost solid body with a very small, round, central cavity ; with the growth of the organ, this cavity increases, and finally the whole becomes a capsular organ in which are developed otolites. — ED.

† [§ 211, note 2.] See, for the auditory apparatus of *Aeolis*, *Hancock* and *Embleton*, Ann. Nat. Hist. 1849, III. p. 196. The otolites which have hitherto been regarded as calcareous, they found not to be materially affected by long treatment with acetic acid. — ED.

‡ [§ 211, note 4.] See also *Leydig* (Anat. Bemerk. üb. Carinaria, Firola und Amphicora, in *Siebold* and *Külliker's* Zeitsch. III. 1851, p. 325). His Taf. IX. fig. 4 (*Carinaria*) gives a very clear idea of the structure and relations of the auditory capsules with these animals. His observations upon the cause of the movements of the otolites are confirmatory of those of *Milne Edwards* with *Firola* ; see L'Instit. Jour. univ. des Soc. sav. XIII. 1845, p. 43. — ED.

§ 212.

The organs of Vision are absent with only a very few genera of the Cephalophora.[1] They are never more than two in number, and their size, compared to that of the body, is usually small; they are smallest with some Heterobranchia, and the largest with the Pectinibranchia.[2]

* The eyes consist usually of two round bulbs concealed under the skin; this last is colorless at this point, and lies over them like a thin lamella. Each bulb is limited outwardly by a tissue resembling a *Sclerotica*, but beneath the skin, this tissue is more convex than elsewhere, and thus forms a kind of *Cornea*.[3] The sclerotica is lined by a dark pigment layer, or *Choroidea*, which, near the corner, ends in a free border, forming thus a *Pupilla*. With some Gasteropoda, the pupillary border has a very thick pigment layer which serves, perhaps, as an *Iris*.[4]

The internal surface of the choroïdea is covered by a whitish pellicle which undoubtedly is a *Retina*, for the optic nerve enters the sclerotica at a point opposite the cornea.[5] The cavity of the eye-bulb is filled with a gelatinous, vitreous body, which, in front, envelops a spherical crystalline lens.[6]

The Optic nerve arises from the cerebral ganglia, and runs along, for a longer or shorter distance, in company with the tentacular nerve of the same side.[7]

---

erc. zoot. Fasc. II. p. 13, Pl. I. fig. 8, f. 9, c. 10, Pl. V. fig. 13, x.) affirms that with the first of these genera, and with *Tiedemannia*, and *Limacina*, the auditive vesicles lie directly upon the two principal inferior ganglia; this agrees with *Delle Chiaje's* description of these organs with *Cymbulia*; see his *Descriz. &c.* I. p. 94, Tav. XXXII. fig. 2, i. *Eschricht* (loc. cit. p. 6, Tab. III. fig. 28, s.) has figured, with a *Clio*, two ganglia with short peduncles, situated close by the two anterior ganglia of the œsophageal ring. These, I infer, are only the auditive capsules receiving two short auditive nerves.

With those Gasteropoda whose inferior œsophageal ganglia are arranged in a circle, as, for examples, with *Lymnaeus*, *Planorbis*, *Physa*, *Succinea*, *Bulimus*, *Ancylus*, these capsules lie upon the posterior surface of the two large anterior ganglia. But when, on the other hand, these ganglia are approximated, or even fused into one common mass, as with *Helix*, these capsules lie upon the inferior surface of this mass, and especially upon the nodules corresponding to two large anterior ganglia.

1 *Phyllirrhoë*, *Diphyllidia*, *Chiton*, *Dentalium*, and the Pteropoda with the exception of *Sagitta* and *Clio*, are blind. In many of the Pteropoda, the auditive appear to have been taken for the ocular organs.

2 *Swammerdamm* (Bibel der Nat. p. 47, Tab. IV. fig. 5–8) made out very correctly the structure of the eyes of *Helix*. The later works of *Stiebel* (*Meckel's* Deutsch. Arch. 1819, p. 206, Tab. V.), *Huschke* (Beitr. zur Physiol. u. Naturgesch. 1824, p. 57, Taf. III. fig. 8), and of *De Blainville* (De l'Organisat. des Animaux, 1823, p. 445), upon the eyes of *Helix*, *Paludina*, and *Voluta*, have been much improved by those of *Muller* (*Meckel's* Arch.

1829, p. 208, Taf. VI. fig. 4–8, and Ann. d. Sc. Nat. XXII. 1831, p. 7, Pl. III. IV., or in the Isis, 1835, p. 347, Taf. VII.), and *Krohn* (*Muller's* Arch. 1837, p. 479, 1839, p. 352, Taf. X. fig. 6–8) upon the eyes of *Helix*, *Murex*, *Paludina*, and *Pterotrachea*.

3 The eyes of the Heteropoda present a remarkable exception; their very convex cornea is surrounded by a collar of skin; the ocular bulb is very long, and, at its base, the sclerotica spreads out interiorly and posteriorly, forming a round prominence; see Milne Edwards, Ann. d. Sc. Nat. XVIII. 1842, Pl. XI. fig. 1, c. (*Carinaria*), and especially the description of *Krohn* (loc. cit. 1839) of the eye of *Pterotrachea*. The ocular bulb of *Clio* is also very long, but has no prominence; see *Eschricht*, loc. cit. p. 7, Tab. III. fig. 29. Those of *Actaeon* are long and pyriform; see *Quatrefages*, Ann. d. Sc. Nat. I. 1844, Pl. VI. fig. 5, and *Allman*, loc. cit. Pl. VII. fig. 2.

4 A dark iris is distinctly seen with *Paludina* and *Murex*. That of *Strombus* is very brilliant and multicolored, according to *Quoy* and *Gaimard*; see, Voy. de l'Astrolabe, Zool. III. p. 56, Mollusques, Pl. I. LI. I am not yet certain whether or not the iris of these Gasteropoda is susceptible of movements of contraction and dilatation. It may be well to add that the choroïdea of the Heteropoda has several spots of its surface free from pigment.*

5 *Krohn* (loc. cit. 1837, p. 482) affirms that he has seen this white layer with a *Paludina*.

6 The existence of a distinct vitreous body was known to *Swammerdamm*, and has been confirmed by *Krohn* (loc. cit. 1837).

7 According to *Krohn* (loc. cit. 1839), the two optic nerves of *Paludina*, *Murex*, *Aplysia*, Cy-

---

* [§ 212, note 4.] For the visual organs of *Paludina*, with histological details, and especially confirmatory of *Krohn's* observations, see *Leydig*, loc. cit., *Siebold* and *Kölliker's* Zeitsch. II. 1850,

p. 159, Taf. XII. fig. 25, Taf. XIII. fig. 26–28. See also this same author in *Siebold* and *Kölliker's* Zeitsch. 1851, III. p. 327 (*Carinaria*).— ED.

There is, moreover, a series of Cephalophora with which the eyes are much more simple and often nearly abortive. Such is the case with *Sagitta*, and many of the Apneusta and Heterobranchia.[8] Here, the eyes are not always nicely limited by a sclerotica, but the light-refracting bodies lie surrounded in a mass of pigment granules, and situated more or less distant from the external surface of the cervical region. The cornea is absent, and often also the optic nerve, in which case, the eyes lie directly upon the cerebral mass.[9]

The most complete eyes are nearly always connected with the tentacles, although their position varies quite widely.[10] Very often they are situated at the base of the external surface of these organs.[11] With many Pectinibranchia, they are more or less elevated upon the outer side of the tentacle on a protuberance or on a support which exceeds the extremity of the tentacle in length and size.[12] With many Pulmonata, these organs are situated upon the very extremity of the tentacles, and are upon the posterior pair, when these last are four in number.[13]

---

praea, *Rostellaria*, *Buccinum*, and *Littorina*, arise from the cerebral ganglia by an origin which is distinct from that of the tentacular nerves. I have been able to confirm this for *Helix*, *Limax*, *Caracolla*. According to *Muller* (Ann. d. Sc. Nat. loc. cit. p. 12, Pl. III. fig. 5), the optic nerve is only a special branch of the end of the tentacular nerve.

8 The two pretty simple eyes of *Sagitta*, forming two prominences on the top of the head, are spherical, and rest directly upon the ganglionic enlargement of the optic nerve , see *Krohn*, loc. cit. p. 13, fig. 5, 14.

9 According to *Quatrefages* (loc. cit. I. p. 158, Pl. V I. fig. 6, 7), the eyes of *Pelta*, and *Chalidis*, have, instead of a choroidea, a mass of pigment containing neither a sclerotica nor a cornea. According to the observations of *Nordmann*, and *Külliker*, the eyes of *Tergipes* and *Polycera* are without optic nerves, and lie directly in contact with the cerebral ganglia. With *Doris*, *Glaucus*, *Thetis*, *Aeolis*, *Doridium*, *Aplysia*, *Bulla*, *Bul-*

lnea, &c., these organs are comparatively small and appear more or less distinct through the skin, sometimes in front of, and sometimes behind, the tentacles.*

10 This is so with various Heteropoda, all the Pulmonata, Pectinibranchia, and with some of the Heterobranchia ; see *Loven*, loc. cit. and Isis, 1842, p. 364.

11 The eyes are situated at the base of the tentacles on a small prominence, with *Carinaria*, *Atlanta*, *Vermetus*, with the Lymnaeacea, the Operculata, *Patella*, *Emarginula*, *Fissurella*, *Sigaretus*, *Paludina*, *Littorina*. A kind of peduncle replaces this prominence with *Haliotis*, *Navicella*, *Phasianella*, *Trochus*, *Ceratodes*, *Ampullaria*.

12 The prominences supporting the eyes are situated on the outer side of the tentacles with *Buccinum*, *Harpa*, *Dolium*, *Cypraea*, *Murex*, *Oliva*, *Turbo* ; and at a variable distance from the extremity which they sometimes surpass in breadth and length, as, for example, with *Strombus*.

13 Amphipneusta, Helicina, and Limacina.†

* [§ 212, note 9.] For the eyes and their intimate structure with the Nudibranchia, see *Alder* and *Hancock*, loc. cit. (*Dendronotus*, *Doto*, *Aeolis*, *Scyllaea*, *Eumenis*, *Doris*, *Antiopa*) ; with all these, the optic nerves were distinct, and the eye itself was furnished with a well-rounded, black pigment-cup, often a spherical crystalline lens (*Doris*, *Aeolis*, *Antiopa*), with an arched cornea in front, and the whole enveloped by a transparent membranous capsule. — ED.

† [§ 212, note 13.] See, in this connection, *Lespès* (Recherches sur l'oeil des Mollusques Gastéropodes terrestres et fluviatiles de France, Thesis. Toulouse, 1851). His conclusions are :

"1. All the terrestrial and fluviatile Gasteropoda have eyes ;

"2. These organs present, as to their position, three different types :

"(1.) The eye at the extremity of the tentacle (*Helix*) ;

"(2.) The eye at the internal base of the tentacle (*Limnaea*) ;

"(3.) The eye at the external base of the tentacle (*Cyclostoma*).

"3. These organs present also three types as to their organization :

"(1.) The lenticular crystalline lens, the vitreous humor fluid, non-adherent (*Helix*) ;

"(2.) The lenticular crystalline lens, the vitreous humor thick and united to this last ;

"(3.) The crystalline lens thick and slightly convex, the vitreous humor viscous and slightly adherent to the lens." — ED.

CHAPTER V.

DIGESTIVE APPARATUS.

§ 213.

The highly-developed digestive organs of the Cephalophora always commence at the anterior extremity of the body, with a round, oral orifice, which is surrounded with tumid lips, but rarely has special prehensile organs.[1] These lips are quite contractile, and can evert and invert the mouth; with many species, they can be prolonged into a cylindrical proboscis.[2] The walls of the oral cavity are very muscular, and, with the majority of the species, form a round and often very large pharynx. The epithelium of this cavity is frequently developed into collars or callosities which serve as masticatory organs. With some Gasteropoda, this apparatus is composed of two horny, lamelliform jaws, which have a truncate, convex, internal border, and move upon each other in a lateral manner.[3] These jaws are situated, sometimes directly behind the oral orifice, and sometimes at the base of the pharynx.

Many other Gasteropoda have only an upper jaw enchased in the roof of the oral cavity, and which is easily seen from its deep-brown color. It consists of a transverse, semilunar, horny plate, upon whose anterior surface are several vertical crests, which terminate upon the free border by as many tooth-like processes.[4]

Nearly all the Cephalophora have a longer or shorter fleshy mass, adhering to the base of the pharynx, and which is sometimes grooved longitudinally; it is quite comparable to a Tongue. Sometimes it is very large and contained in a membranous sheath at the base of the pharynx. It is always armed with horny, denticulated spines and plates, which are very delicate, and arranged in quite elegant, longitudinal and transverse rows. The

---

1 Such are the tentacular appendages which have a sucker, of Pteropoda (*Clio, Spongiobranchaea,* and *Pneumodermon*), already mentioned above (§ 204).

2 There is a retractile proboscis with *Pneumodermon, Spongiobranchaea, Pterotrachea, Thetis, Buccinum, Dolium, Cypraea, Murex, Conus, Voluta,* and many other Pectinibranchia.

3 The external borders of these jaws are easily perceived between the lips, as with *Scyllaea* (Cuvier, Mém. loc. cit. fig. 6, a. 6, b.), with *Tritonia* (Savigny, Descript. de l'Egypte, Hist. Nat. II. Pl. II. fig. 1 –1⁹), and *Delle Chiaje,* Descriz. loc. cit. Tav. XLII. fig. 1), and with *Diphyllidia* and *Bulla.* They are found also directly behind the lips with *Fenilia, Aeolis, Amphorina,* and *Tergipes* (Alder, Hancock and Embleton, Ann. of Nat. Hist. XIII. p. 162, Pl. II. fig. 3, 4, XV. p. 4, Pl. II.; also *Quatrefages,* Ann. d. Sc. Nat. I. p.

147, Pl. V. fig. 5, and *Nordmann,* loc. cit. p. 12, Tab. I. fig. 7). With *Dentalium,* on the contrary, the jaws are situated at the base of the oral cavity (*Deshayes,* loc. cit. p. 333, Pl. XV. fig. 11, b. b. 15, 16, or in the Isis, 1832, p. 463, Taf. VI. fig. 15, 19, 20).

4 This upper jaw is particularly developed with the Limacina and Helicina; see *Cuvier,* Mém. loc. cit. Sur la Limace, &c., Pl. II. fig. 4 (*Limax*); *Troschel,* in *Wiegmann's* Arch. 1836, I. p. 257, Taf. IX. fig. 3–9 (*Arion, Limax, Helix, Clausilia,* and *Succinea*), and *Erdl,* in *Mor. Wagner's* Reisen in der Regentsch. Algier. III. p. 268, Tab. XIII. XIV. With *Lymnaeus,* and *Planorbis,* there are, beside, two small lateral jaws; these exist also with *Valvata,* and *Paludina,* where the upper jaw is wanting. With *Zephyrina,* there are also three jaws at the base of the pharynx; see *Quatrefages,* loc. cit. I. p. 132, Pl. V. fig. 1.

* [§ 213, note 4.] For many details upon the oral organs of the Helicina, of an anatomical as well as a zoological import, see *Troschel* (Ueber

die Mundtheile einiger Helicien, in *Wiegmann's* Arch. 1849, p. 225). — ED.

21

points of these spines turn backwards, and thus the retractile tongue can serve as an organ of ingestion, and as such is used with much address.[5]

§ 214.

The intestinal canal has often longitudinal folds and a ciliated epithelium extending from the œsophagus to the rectum, and even into the hepatic ducts.[1] It is usually two or three times the length of the body, and has therefore several convolutions, which, with the species which have a shell, are contained in its spiral cavity.

It commences at the base of the pharynx by an Œsophagus, of variable length, which is sometimes dilated at its posterior extremity into a kind of crop.[2] The stomach, which, from constrictions,[3] is often divided into several portions, consists sometimes of a simple dilatation with thin walls,[4] and at other times of a nicely-defined cavity whose walls are thick and fleshy,[5] and provided, sometimes, with thick epithelium, and even, in certain cases, with plates and horny teeth.[6] The cardiac and pyloric ori-

---

5 See the description and figures of *Troschel* (loc. cit. Taf. IX. X.) of the tongue of our terrestrial and fresh-water Gasteropoda, and also of *Amphipeplea* (Ibid. 1839, I. p. 182, Taf. V. fig. 8). For that of the marine Gasteropoda, see principally *Quoy* and *Gaimard* (loc. cit.), also *Poli*, Testacea Siciline, &c., I. p. 5, Tab. III. fig. 9 (*Chiton*), *Savigny*, Descrip. de l'Égypte, Hist. Nat. II. Pl. II. fig. 2⁴-2¹⁵, III. fig. 5⁷, 5⁹ (*Aplysia* and *Chiton*), *Rang*, Hist. Nat. des Aplysiens, Pl. XX. fig. 9–13 (*Aplysia*), *Delle Chinje*, Memor. &c. Tav. XV. fig. 7–10 (*Carinaria*), and *Eschricht*, loc. cit. p. 10, Tab. III. fig. 20–23 (*Clio*).

The tongue is very long with most of the Apneusta; see *Quatrefages*, loc. cit. I. Pl. IV. V. (*Actaeon* and *Amphorina*), *Alder*, *Hancock* and *Embleton*, Ann. of Nat. Hist. XIII. Pl. II. fig. 5–6, XV. Pl. I. II. (*Venilia* and *Aeolis*), *Altman*, Ibid. XVI. Pl. VI. VII. fig. 5 (*Actaeon*), and *Nordmann*, loc. cit. Tab. I. fig. 7–10 (*Tergipes*). With *Patella*, this organ nearly exceeds the body in length, and bends loop-like, near its posterior extremity (*Cuvier*, Mém. loc. cit. Pl. II.). With *Trochus pagodus*, it is seven times longer than the body (*Quoy* and *Gaimard*, loc. cit., and Isis. 1836, p. 69, Taf. IV. fig. 3). With *Pleurobranchaea*, there are spines, not only on the tongue, but on a considerable portion of the lateral walls of the oral cavity. To the same category belong the spines which *Eschricht* (loc. cit. p. 9) found upon the pharynx of a *Clio*, and described as lateral teeth. This apparatus with *Pneumodermon* is quite remarkable — being composed of two tongues which are contained in two cœcal sheaths (*Van Beneden*, Exer. zool. loc. cit. Fasc. I. p. 47, Pl. II. fig. 2). With *Pterotrachea*, the tongue consists only of a simple transversal row of curved spines. The circle of hooks surrounding the mouth of *Sagitta* may also be regarded as an abortive tongue (*Krohn*, loc. cit. p. 7, fig. 3–6), for they are exactly like the lingual spines of *Pterotrachea* (*Delle Chinje*, Mem. loc. cit. Tav. LXIX. fig. 1).

Lebert has given a very detailed description of the parts of the mouth and the tongue of *Patella*, *Buccinum*, *Doris*, *Haliotis*, *Paludina*, and *Limax*; see *Müller's* Arch. 1846, p. 435, Taf. XII. -XIV.

1 The intestine is lined with cilia, with *Patella*, *Buccinum* (*Sharpey*, Cyclop. of Anat. 1. p. 620), *Lymnaeus stagnalis*, *Paludina vivipara*, and *Helix cellularis* (*Purkinje* and *Valentin*, De

Phaenom. motus vibrat. loc. cit. p. 48), and with the Apneusta (*Quatrefages*, Ann. d. Sc. Nat. I. p. 166).

I have also seen the ciliary motions, with *Lymnaeus*, *Planorbis*, and *Clausilia*; but not with *Limax*, *Arion*, and *Helix*. *Valentin* may, therefore, be mistaken in affirming (*Wagner's* Handwörterbuch d. Physiol. I. p. 492) that ciliated epithelium exists generally in the intestine of the Gasteropoda.

A ciliary movement has also been observed in the intestine of *Sagitta* by *Krohn* (loc. cit. p. 8), and by *Wilms* (Observ. de Sagitta, Diss. Berolini, 1846, p. 12).

2 The œsophagus is very long with *Buccinum*, *Paludina*, *Lymnaeus*, and *Planorbis*; but very short with *Thetis*, *Haliotis*, *Testacella*, *Helix*, and *Limax*. It has a kind of crop close upon the stomach with *Cymbulia*, *Onchidium*, *Lymnaeus*, and *Planorbis*, while with *Buccinum*, and *Voluta*, a long, crop-like caecum arises from the upper portion of the stomach near the œsophagus.

3 *Aplysia*, *Dolabella*, *Notarchus*, *Ancylus*, *Pleurobranchus*, and *Onchidium*; see *Cuvier*, Mém. loc. cit. I am unable to say anything upon the crystalline, gelatinous stem, which, according to *Cuvier* (Edinb. new Philos. Jour. VII. 1825, p. 225, and Isis. 1832, p. 815) is found in all the species of *Strombus* and some of *Trochus* and *Murex*, and is contained in an internally projecting appendix of the stomach.

4 *Cypraea*, *Cassis*, *Murex*, *Testacella*, *Limax*, *Helix*, &c.

5 *Lymnaeus*, *Planorbis*, *Thetis*.

6 There are three horny lamellae in the stomach of *Bullaea* (*Cuvier*, loc. cit. fig. 11), and of certain species of *Pleurobranchus* (*Meckel*, Beitr. &c. I. Hft. 1, p. 31, Tab. V. fig. 36, 37); four in that of *Cymbulia*, *Tiedemannia*, *Hyalea*, and *Limacina* (*Van Beneden*, Exerc. zool. loc. cit. Fasc II.).

That of *Pella* has four denticulated horny plates (*Quatrefages*, loc. cit. I. p. 153, Pl. IV. fig. 5, V. fig. 7), as is also true of *Lissosoma*, according to *Kölliker*. With *Scyllaea* (*Cuvier*, loc. cit. fig. 6, d.), and with *Tritonia* (*Meckel*, Syst. d. vergleich. Anat. IV. p. 188), there is a complete row of lamellae with sharp edges. With *Dentalium* also, the entrance of the stomach has a very complicated dental apparatus (*Deshayes*, loc. cit. p. 333, Pl. XV. fig. 13, or Isis. 1832, p. 463, Taf. VI. fig. 17). But *Aplysia*, of all the Cephalophora, is best provided for in this respect, for here the second muscular stomach is lined with a triple row of

fices are usually situated opposite each other; but in some they are so approximated that the stomach has the form of a caecum.[7]

The intestine, having made more or less numerous [8] convolutions, seldom forms a rectum, but opens, usually, close by the respiratory orifice on the right side of the anterior end of the body, and rarely at the posterior extremity.[9] With the Pectinibranchia, the rectum often projects widely into the cavity of the mantle, as a longer or shorter prolongation upon whose extremity the anus is situated.

Sagitta and the Apneusta present wide differences from this just-described type of structure. With the first, the mouth opens into a short œsophagus which passes directly, without any stomachal dilatation, into the intestine; this last runs straight backwards, and, curving downwards, terminates in the anus situated on the median line of the ventral surface at the posterior end of the body.[10] With the Apneusta, on the other hand, there is a stomach with several and often highly-ramified caecal appendages — which, in some species, extend even into the dorsal appendages. A short rectum follows directly upon the stomach, and ends in an anus, often difficult to be found, and situated at the anterior part of the right side of the body.[11]

cartilaginous lamellae, and the third has, beside, numerous horny hooks which point forwards (Cuvier, loc. cit. Pl. III.).

7 Murex, Voluta, Sigaretus, Phyllidia, Diphyllidia, and many species of Doris and Carinaria.

8 The intestine is very short and slightly tortuous with Clio, Carinaria, Thetis, Tritonia, Diphyllidia, Pleurobranchaea, Buccinum, Murex, and Janthina. With the other Cephalophora, it has usually many convolutions, which are quite numerous especially with Haliotis, Patella, and Chiton (Cuvier, loc. cit. Pl. I.–III., and Poli, loc. cit. Tab. III. fig. 6).

9 With the Pectinibranchia, and most of the Pulmonata, whose anus is near the respiratory orifice, the position of the first is determined by that of the last, and therefore is most usually upon the right, and rarely upon the left side. This is the case also with nearly all the other Gasteropoda.

With Patella, it is situated directly back of the head; with Tritonia, Scyllaea, and Thetis, a little further back; and even still more behind with Diphyllidia, Dolabella, Notarchus, and Pleurobranchaea. In this last genus it is above the branchia, while in Pleurobranchus, and Aplysia, it is behind this organ. With Chiton, Phyllidia, Doridium, Bullaea, Testacella, and Onchidium, it is at the very posterior end of the body. With Doris, and Polycera, it is somewhat elevated on the side of the back and surrounded by branchiae.

With Haliotis, it is anterior and on the left side; and with Sigaretus, Fissurella, and Emarginula, it is even in front of the oral cavity.

Its position is varied with the Heteropoda and Pteropoda. With Carinaria, and Pterotrachea, it is situated at the base of the intestinal sac, — with Atlanta, upon a prolongation of the right side of the neck; with Phyllirhoë, upon the middle of the right side; with Pneumodermon, directly behind the right pinion; with Tiedemannia, at the middle of the abdomen; with Hyalaea, at the same point but a little at the left; and with Cymbulia and Limacina, in the respiratory cavity. See, for these various positions, the works especially of Cuvier, Meckel, and Van Beneden.

10 Krohn, loc. cit. p. 8.

11 For the intestinal canal of the Apneusta, see Milne Edwards, Ann. d. Sc. Nat. XVIII. 1842, p.

330, Pl. X. fig. 2 (Calliopaea), also Quatrefages, Alder, Hancock and Embleton, Allman, and Nordmann, loc. cit. According to a communication from Kölliker, that of Rhodope is the most simple; it consists only of a caecum which extends even to the posterior extremity of the body, and near the cardia sends off a short caecum which passes along the left side of the œsophagus to the pharynx, and upon the right side of the other end of the body terminates in a short rectum. With Actaeon, according to Souleyet (Compt. rend. XX. 1845, p. 94), the intestine, after forming a stomachal dilatation, bends, first forwards, then backwards, opening on the right side of the neck. But the descriptions and figures of this animal by Quatrefages (Ann. d. Sc. Nat. I. p. 141, Pl. IV. fig. 2, V. fig. 4) and Allman (loc. cit. p. 148, Pl. VI.), are remarkably contradictory to these statements of Souleyet. According to these authors, the stomach is followed by a short rectum opening upon the right side of the neck, which is attended by two superior and two inferior intestinal tubes which send numerous ramified appendages into the parenchyma of the body. With Chalidis, the œsophagus is followed by four caeca, the two shortest of which extend in front, and the others behind. With Pelta, there is a large intestinal tube having many short coecal appendages, situated in the middle of the body. With Aeolis, Flabellina, Tergipes, which have only a single intestinal tube closed posteriorly, and with Zephyrina, Amphorina, and Calliopaea, which have two such, the caeca from this canal extend even into the dorsal appendages. With Eolidina, which has three intestinal tubes intercommunicating by numerous transversal anastomoses, these last give rise to the caeca of the dorsal appendages. Quatrefages (Ann. d. Sc. Nat. XIX. p. 285, Pl. XI. fig. 2, c.), who at first declared that the median tube of these animals opened by an anus at the posterior end of the body, has since (Compt. rend. XIX. p. 811) rectified this error; for here, as also with Actaeon, Aeolis, Tergipes, and Rhodope, the anus is anterior and on the right side.

A similar correction will perhaps be made with Venilia, whose stomach, according to Alder and Hancock (Ann. of Nat. Hist. XIII. p. 163, Pl. II. fig. 7), not only sends many ramified caeca into the lateral appendages of the body, but also is followed by a rectum, opening, they say, at the posterior portion of the back.

§ 215.

With those Cephalophora which are nourished by solid food, and which often have, therefore, masticatory organs, there are, almost without exception, highly-developed Salivary organs. These are usually composed of two lobular yellow glands surrounding the œsophagus or stomach, and which have in front two excretory ducts which are lined with ciliated epithelium.[1]

These ducts pass, in company with the œsophagus, through the œsophageal ring, and, extending over the base of the pharynx, end in the oral cavity on each side of the tongue. With some species, these glands consist of two very long tubes.[2] Some Gasteropoda have two pairs of these organs, one of which opens at the anterior part of the mouth.[3] In a few cases only these organs appear to be wholly wanting.[4]

The Biliary organs are always present; and their glandular follicles contain hepatic cells filled with a brownish-yellow substance.[5] Most commonly, the liver is large and distinctly separated from the digestive canal; and it is with a few genera only that it is more or less blended with it.

1. This last is the case with some Pteropoda, and Apneusta, whose intestinal walls, as with the Worms, are partly composed of the hepatic substance, or furnished with numerous small follicles which open into the intestinal cavity.[6]

---

1 *Helix, Limax, Onchidium, Haliotis, Pleurobranchus,* and the Pectinibranchia. For the internal structure of these glands, see *Müller*, De Gland. secern. struct. p. 54, Tab. XVII.*

2 *Clio, Aplysia, Thetis, Lissosoma, Tergipes,* and many species of *Doris.*

3 *Janthina, Flabellina, Actaeon,* and *Atlanta.* With some Gasteropoda, as for instance, with *Rhodope,* and *Eolidina,* there are only two salivary glands; these open in front into the oral cavity and so far from the œsophagus that they appear to correspond to the anterior pair of those species in which there are four. †

4 *Sagitta, Cymbulia, Tiedemannia, Dentalium,* and *Chiton.*

5 For the internal structure of the liver of the Gasteropoda, see *Müller,* De Gland. secern. &c. p. 71, Tab. X.; *Schlemm,* De hepate ac bile Crustac. et Mollusc. quorundam, loc. cit. p. 19, Tab. I. II.; *Karsten,* Nov. Act. Acad. Nat. Cur. XXI. p. 304, Tab. XXI.; and *H. Meckel, Müller's* Arch. 1846, p. 9, Tab. I. ‡

6 With *Sagitta,* the hepatic substance appears

to be blended with the intestinal walls (*Krohn,* loc. cit. p. 8). This is distinctly so with *Venilia, Aeolis, Eolidina, Amphorina,* and *Zephyrina,* and is especially seen upon the coecal ends of the branches of the intestinal canal which terminate partly in the dorsal appendages, and partly in the parenchyma of the body; see *Quatrefages,* loc. cit. XIX. p. 289, Pl. XI. fig. 5, I. Pl. IV. V.; *Alder, Hancock* and *Embleton,* Ann. of Nat. Hist. XIII. p. 163, Pl. II. fig. 9, XV. p. 80, Pl. IV. According to *Nordmann* (loc. cit. p. 20, Tab. II. III. fig. 3), the liver is isolated with *Tergipes*; but as the organ here described appears to open externally by a special duct, it resembles an urinary organ (see below, § 223). With *Pneumodermon,* and *Clio,* the stomach is lined with a layer of small hepatic follicles (*Cuvier,* loc. cit. p. 8, fig. 7, p.; and *Eschricht,* loc. cit. p. 11).

According to a communication from *Kölliker,* the intestine of *Rhodope* also has numerous follicles of this kind, which are pyriform and filled with cells having yellow nuclei.

---

* [§ 215, note 1.] See also *Leidy* (loc. cit.) for the salivary glands and their intimate structure, of *Limax, Helix, Tebennophorus, Vaginula, Succinea, Glandina.*

† [§ 215, note 3.] With *Paludina,* the salivary glands are highly developed and two in number. They are situated on the upper and posterior side of the pharynx, behind the brain; their excretory ducts pass under the cerebral commissure, forwards, and perforate the upper wall of the pharynx. In structure they consist of ramose cæca, made up essentially of cylindrical epithelium situated on a basement membrane; see *Leydig,* loc. cit. p. 165, Taf. XII. fig. 10, a, b.

For further details on these organs with the Nudibranchiate Cephalophora in general, see

Alder and *Hancock,* loc. cit. Part II. Pl. IV. fig. 1, f. (*Doto*); Part III. Pl. VII. fig. 6, a. (*Acolis*); Part IV. Pl. V. fig. 1, c. (*Scyllaea*); Part V. Pl. II. fig. 1, h. (*Doris*). — ED.

‡ [§ 215, note 5.] See also *Leydig,* Ueber Paludina vivipara, &c., loc. cit. p. 143, 166 (*Paludina*); he gives its development and its adult structure. It is developed from cells as an appendage to the alimentary canal; and its structure, when complete, is follicular, as above described. See furthermore, for the liver of the terrestrial Gasteropoda, *Leidy,* loc. cit. Of its internal structure, he says: "The lobuli of the liver are composed of the rounded commencement of the biliary ducts, and are lined with polygonal cells." — ED.

2. With the other Cephalophora, the liver is wholly isolated, nearly always asymmetrical,[7] and often divided into several lobes of a yellowish-brown or brownish-green color; often, also, it wholly envelops the intestinal convolutions. The biliary canals, which arise from the hepatic lobes, form usually, two, three, or more excretory ducts, which empty the bile into the stomach or intestine, rarely into the œsophagus.[8]

CHAPTER VI.

CIRCULATORY SYSTEM.

§ 216.

For a long time it was erroneously supposed that the circulatory system of the Cephalophora was completely closed. But the heart or central portion of this system, is developed in an inverse ratio to the imperfect peripheric part which is without a capillary net-work. This incompleteness is often so great that, in many genera, the arteries are wanting and the veins more or less wholly absent. The circulation is, therefore, extravascular for a longer or shorter course, and passes into cavities (*Lacunae*) situated in the parenchyma of the body.[1]

The blood is colorless, often opalescent, and always very poor in corpuscles. These last are also colorless and consist of smooth cells, with a granular, indistinct nucleus.[2]

---

7 With *Dentalium*, there are two symmetrical livers, one on each side of the intestinal canal ; see *Deshayes*, loc. cit. Pl. XV. fig. 11, or Isis. Taf. VI. fig. 15, m. m. With *Diphyllidia*, also, there are two livers, one on each side of the stomach into which they open by several transverse excretory canals ; see *Meckel's* Arch. 1826, p. 15, Taf. I. fig. 11.

8 For the external form of the liver, consult *Cuvier*, loc. cit. The hepatic ducts open, near the pyloric orifice, with *Limax, Helix, Testacella, Doridium,* and *Dentalium ;* into the intestine, with *Haliotis, Vermetus, Pleurobranchus, Diphyllidia, Doris, Planorbis,* and *Lymnaeus ;* into the third stomach, with *Aplysia, Dolabella,* and *Notarchus;* while with *Onchidium,* two of the ducts open into the œsophagus, and the third into the first stomach.*

1 The tenacity with which the opinion was entertained that there is a completely-closed vascular system with the Mollusca, is shown in the fact that *Cuvier* (Règne anim. I. p. 50), after having seen,

with *Aplysia*, the veins communicate distinctly with the cavity of the body by special orifices, still persisted in the old view, — regarding this as an exception ; see Mém. loc. cit. p. 13. It is only latterly that the circulation of the blood through the lacunae and interstices of the body, has been shown to be the rule, by *Pouchet* (Recherches loc. cit. p. 13), *Milne Edwards* and *Valenciennes* (Compt. rend. XX. 1845, p. 261, 750, or *Froriep's* neue Not. XXXIV. p. 81, 257).

Milne Edwards, in his memoir already cited upon the circulation of *Patella, Haliotis, Helix, Aplysia, Thetis,* and *Triton,* has abundantly shown that the vascular system of the Cephalophora is also incomplete, and that the aorta terminates in a large lacunal sinus containing the brain, the muscles, and the retracted tongue, and which forms also a part of the visceral cavity ; see Ann. d. Sc. Nat. VIII. 1847, p. 37, Pl. I.–III., or *Schleiden* and *Froriep's* Not. V. p. 1, fig. 1–4.

2 For the blood of the Gasteropoda, see *Carus*,

* [§ 215, note 8.] For the details of the hepatic structure with the Nudibranchia, see *Alder* and *Hancock*, loc. cit. Part II. Pl. II. fig. 2, e. hh., and fig. 3 (*Dendronotus*); Part III. Pl. VIII. fig. 9 (*Aeolis*); Part IV. Pl. V. fig. 1, g. g. g. (*Scyllaea*), and fig. 8, g g. (*Eumenis*); Part V. Pl. I. fig. 2, d. d., and Pl 11. fig. 1, f. (*Doris*).

For the liver of *Chiton*, see *Middendorf* (Beiträge zur einen Malacozoologia rossica, St. Petersburg, 1847, p. 63, Taf. V. fig. 2, l.). Its ducts open into the alimentary canal near the stomach. — ED.

21*

§ 217.

The Heart is wanting in only a few genera of the Cephalophora.[1] Almost always it has a pericardium,[2] and is divided into a simple, very muscular ventricle, and a thin-walled auricle which is equally simple, rarely double.[3]

The arterial blood passes from the respiratory organs into the auricle, thence into the ventricle, from which it is forced through a very short aorta over the body. These two chambers of the heart are usually pyriform, and are joined together at their large extremity by a constriction in which is sometimes situated a valve, which prevents the return of the blood into the auricle.[4]

The position of the heart usually depends upon that of the respiratory organs. It is generally situated at the base of the branchiae, or in the bottom of the pulmonary cavity. It is most often found, therefore, upon the right side of the body.[5]

Von den äusser. Lebensbeding. d. weiss-und Kaltblütigen Thiere. p. 72; *Ehrenberg*, Unerkannte Struct. loc. cit. Tab. VI. fig. I. 1, II. I (*Arion* and *Paludina*); and *Erdl*, De Helicis algirae vasis sanguiferis. Diss. Monach. 1840, p. 10.

With *Planorbis*, the blood is red. With the Cephalophora in general, there is only a very small quantity of fibrine, at least there is only a trace in the blood of *Helix*; it forms a kind of a web, scarcely visible, uniting the globules into masses and rows. The nuclei of these blood-globules become very distinct by the addition of acetic acid.*

1 *Forbes* (Instit. 1843, p. 358), and *Darwin* (Ann. of Nat. Hist. XIII. p. 3), have been unable to find a heart with *Sagitta*; although *D'Orbigny* (Voy. dans l'Amer. mér., or Isis. 1839, p. 501) affirms that he has seen the movements of this organ in this enigmatical animal, and *Darwin* (loc. cit. p. 6) has perceived a pulsating organ at the anterior extremity of the embryos. The heart is wanting, according to *Quatrefages* (loc. cit. I.), in *Zephyrina*, *Actaeon*, and *Amphorina*; and according to *Kölliker*, in *Flabellina*, *Rhodope*, and *Lissosoma*. However, *Souleyet* (Comp. Rend. XX.1845, p.73) contradicts, very positively, the assertions of *Quatrefages*, and assigns a heart to all the Apneusta. The difficulties in the study of these animals, from their non-transparency, are undoubtedly the cause of many of these contradictory statements. One should not, also, conclude as to the organization of the adults from the development of the embryos; for it is very singular that the embryos of *Actaeon* are completely developed without a heart (*Vogt*, Comp. Rend. XXI. No. 14, XXII. No. 9, or *Froriep's* neue Not. No. 795,

and 820), while with the other Gasteropoda the heart appears very early in the embryonic development. According to *Nordmann* (loc. cit. p. 93), the embryos of *Tergipes*, which has a heart, are developed as those of *Actaeon*, thus showing that the absence of the organ in these last is only a delay of its appearance.

Wilms (loc. cit. p. 11) has been also equally unable to find a heart with *Sagitta*.

2 The pericardium is apparently wanting with the Apneusta.

3 *Chiton*, *Haliotis*, *Fissurella*, and *Emarginula* have two lateral auricles; the last three of these Scutibranchia resemble moreover the Lamellibranchia in their heart being traversed by the rectum; see *Cuvier*, loc. cit., and *Meckel*, Syst. d. vergleich. Anat. V. p. 115.

4 See *Cuvier*, loc. cit. Pl. 1. fig. 2–4, II. fig. 1; *Carus*, Erläuterungstafeln IIft. VI. Taf. II. fig. 6 (*Helix*); and *Van Beneden*, Exerc. zool. loc. cit. Pl. III. fig. 11 (*Hyalea*).

Nordmann (loc. cit. p. 26, Tab. III. fig. 4) found with *Tergipes*, the auriculo-ventricular valvular apparatus replaced by a very movable valve situated between the ventricle and the bulb of the aorta. With *Limax*, and *Arion*, the valves are wholly wanting (*Treviranus*, Beobacht. aus d. zool. u. Phys. p. 40).

5 The heart is situated on the right side of the back, with most of the Tectibranchia, with the dextral Pectinibranchiae, and Pulmonata, and with all the Limacina; while it is on the opposite side with *Ancylus*, *Haliotis*, and all the sinistral Gasteropoda. That of *Carinaria*, *Clio*, *Hyalea*, and *Cleodora*, is upon the dorsal median line, a little to the left.†

---

* [§ 216, note 2.] *Leydig* (loc. cit.) describes the blood of *Paludina* as containing two forms of corpuscles; one, round, which became granular nucleated cells after the action of acetic acid; the other provided on one side with processes which disappeared upon the action of acid; see loc. cit. p. 170, Taf. XII. fig. 47, 48. In this connection, see also for the blood-corpuscles of the Gasteropoda (*Buccinum magnum*) and their development, *Wharton Jones*, Philos. Trans. 1846, Part II. p. 96, Pl. II. fig. 1–7, of the Gasteropoda division. *Jones* also mentions the stellate form of the corpuscle (fig. 4). It would appear to me that this peculiarity is, after all, only a crenulation due

to an exosmotic passage of the cell-contents — leaving the cell-membrane thus deeply wrinkled, as may often be observed also with the blood of vertebrates. — Ed.

† [§ 217, note 5.] With *Firola*, and *Atlanta*, the heart is situated near the posterior extremity of the body; its auricle and ventricle are composed of interlaced, striated muscular fibres; both the auriculo-ventricular and the aorto-ventricular orifices are valvular; see *Huxley*, Ann. d. Sc. Nat. XIV. 1850, p. 193.

See also, for the heart its positions and connections, with the Lymnaeacea, *De St. Simon*, Jour de Conchol. 1852, II. p. 113. — Ed.

It lies upon the median line, and its ventricle and aorta are directed forwards, in those genera whose respiratory organs are symmetrical, or wholly wanting.[6] With many other Cephalophora, they have also the same direction, without, however, being situated on the median line; but in the turbinated genera, the apex of the ventricle and the aorta are directed backwards.

§ 218.

The vascular system of the Cephalophora consists almost solely of arteries with their branches, of large venous canals receiving the blood from the cavity of the body, and of the lacunae in the parenchyma which return it to the respiratory organs.

With *Sagitta*,[1] and some Apneusta[2] there are no traces of blood-vessels; and, as with the Nematodes, the nutritive liquid transudes directly through the intestinal canal into the cavity of the body.

In another series of the Apneusta, there are rudiments of arteries and veins, in the form of a short aorta, which passes in front from the ventricle, and has a bifurcated extremity, — and two vena cava even shorter, which open each side of the posterior end of the auricle.[3]

With the other Cephalophora, the aorta divides, after a short course, into two principal arteries, the anterior of which passes through the œsophageal ring, and, sending branches to the cephalic organs, finally ramifies in the fleshy walls of the body; but the other, posterior, ramifies over the organs in the intestinal sac. These ramifications, which sometimes form a beautiful vascular net-work, never pass into a capillary system which

---

6 The heart is situated on the median line of the back in *Dentalium, Tritonia, Scyllaea, Thetis, Phyllidia, Fissurella,* and *Emarginula*; upon the posterior part of the body, with *Doris,* and *Chiton,* as is also the case with *Onchidium* which is remarkable in other respects. It is singular that with *Patella,* whose respiratory apparatus is symmetrically disposed, the position of the heart is in front and on the right side ; see *Meckel,* Syst. d. vergleich. Anat. V. p. 119, and Arch. für Anat. u. Phys. 1826, p. 19. Several of the Apneusta have the heart on the dorsal median line, as for examples, *Tergipes (Nordmann,* loc. cit. p. 24, Tab. II. T., Tab. III. fig. 4), *Eolidina (Quatrefages,* loc. cit. XIX. p. 288, Pl. XI. fig. 3), *Aeolis (Hancock* and *Embleton,* loc. cit. Pl. V. fig. 16), and *Actaeon (Allman,* loc. cit. p. 149, Pl. V. fig. 4).

1 *Krohn,* loc. cit. p. 8.

Notwithstanding the absence of a heart and a vascular system with *Sagitta, Wilms* (loc. cit. p. 12), has found in the visceral cavity of these animals regular blood-currents, due, probably, to ciliated organs.

2 *Flabellina, Lissosoma,* and *Rhodope,* according to *Kölliker ; Zephyrina,* and *Amphorina,* according to *Quatrefages.*

3 Such a rudimentary vascular system situated in the anterior part of the back, has been seen by *Nordmann* (loc. cit. p. 24), with *Tergipes,* by

Quatrefages (loc. cit. p. 288), with *Eolidina,* and by *Van Beneden* (Instit. No. 627, or *Froriep's* neue Not. No. 797, p. 68), with *Aeolis.*

Allman, judging from one of his figures (loc. cit. Pl. V. fig. 4, c.), has found it also with *Actaeon. Nordmann* has observed, that in spite of this imperfection of the blood-vessels, the blood effused into the cavity of the body circulates regularly, so that with *Tergipes,* the whole body, including the appendages, is traversed by arterial and venous currents which can be traced even to the two venae cavae which arise from open mouths. This circulation is quite like that of insects ; except that here, the blood of the Apneusta continues a longer course in the arteries. For *Nordmann,* with *Tergipes,* and *Quatrefages,* with *Eolidina,* have been able to trace on each side of the body an anterior and posterior branch of the aorta. *Quatrefages,* however, commits an error at the outset concerning this simple circulation of the Apneusta, in declaring that with these Gasteropoda the ramified intestinal canal serves also the function of a vascular system ; this has induced him to give the name *Phlebenterata* to an entire group of these animals. In the controversy between him and *Souleyet* on this subject (Comp. Rend. XIX. XX.), and which threatens to be interminable, this last has gone too far in asserting that, not only with the Apneusta, but even with all the Gasteropoda, there is a completely-closed vascular system.*

* [§ 218, note 3.] See, for detailed remarks against the doctrine of Phlebenterism with the Eollididae, *Hancock* and *Embleton* (loc. cit. 1849). They have shown here the existence of a pretty highly-developed vascular system commencing in a well-formed heart which consists of a ventricle and an auricle, and enclosed in a pericardium. — ED.

opens into the veins, but gradually disappear,[4] so that it is probable that
the blood is effused from their open extremities into the interstices of the
parenchyma of the viscera, as well as into the cavity of the body; and is
thence taken up through numerous orifices on the inner surface of this last,
and conducted to the respiratory organs through the wall-less venous canals
which are hollowed in the muscular substance of the envelope of the
body.[5]

CHAPTER VII.

RESPIRATORY SYSTEM.

§ 219.

The respiratory organs are absent with only a few of the Cephalophora;
namely : with *Sagitta*, the Apneusta, and with some of the Pteropoda and
Heteropoda.[1] With these, therefore, it may be inferred that the respira-

---

4 *Erdl* (De Helicis algirae, &c., loc. cit.) has, indeed, figured venous net-works on the digestive
apparatus of an *Helix* (see also its copy in *Carus'*
Erläuterungstafeln, Hf. VI. Tab. 11. fig. 5) ; but I
regard these as of an arterial nature, and this so
much the more, since *Erdl*, in his dissertation, has
nowhere shown a direct communication between
the arteries and veins. The absence of a capillary
net-work and of venous radicles, is quite apparent
with *Arion*, in which the posterior artery forms
beautiful ramifications of a white color upon the intestine and liver. If the larger branches of this
artery are examined, their muscular walls will be
distinctly seen to be internally lined with a granulated layer composed of carbonate of lime and which
gives the color just mentioned.
If also the smaller branches are examined, their
muscular walls will be found to have gradually disappeared so that the blood circulates inside of the
granular layer only; and this last in its turn will also
be found to have disappeared leaving no trace of
capillaries or venous radicles. For the details of
the arterial system of the Cephalophora, see the
Mémoires of *Van Beneden*, loc. cit. (Pteropoda) ;
Milne Edwards, Ann. d. Sc. Nat. XVIII. 1824,
p. 325, Pl. XI. fig. 1 (*Carinaria*), and *Cuvier*,
Meckel, and *Delle Chiaje*, loc. cit. (Gasteropoda).
5 Although *Cuvier* in 1803 (Ann. du Mus. d'Hist.
Nat. II. p. 299, Pl. II. fig. 1, 3) perceived, on the inner surface of the envelope of the body, the orifices
of the venous canals, which as a net-work traverse
the fleshy walls of *Aplysia* even to the base of the
branchiae, and although this was confirmed by
Treviranus (Biologie, IV. p. 248) and *Delle
Chiaje* (Memor. &c. I. p. 63), yet it is only lately
that the opinion has been recognized that this
might be so with all the Cephalophora, for the observation upon *Aplysia* remained thus long isolated. But now, facts of this kind are so numerous as
not to be based upon exceptional observations. It
should be understood, however, that the absence of
capillaries and of venous radicles, as well as the
presence of numerous orifices opening into the venous canals, are the rule with all the Cephalophora

which have respiratory organs. These orifices may
be easily seen, especially by asphyxing species of
Limax and *Arion*, — by which experiment, will be
appreciated the correctness of *Delle Chiaje's* figure
of *Arion* which was engraved in 1830 (Memor. loc.
cit. Tav. CIX. fig. 16 without text, and Descriz.
loc. cit. II. 1841, p. 10, Tav. XXXVII. fig. 16, the
same plate with text), with the exception that there
are orifices on their ramifications as well as on the
two principal canals. *Pouchet* (loc. cit. p. 19,
has named these *Orifices absorbants*, and his observations were also made on *Arion; but Milne
Edwards* and *Valenciennes* (Compt.Rend. loc. cit.)
have demonstrated this structure with *Aplysia,
Doris, Polycera, Scyllaea, Patella, Chiton,
Haliotis, Notarchus,Umbrella, Pleurobranchus,
Dolabella, Buccinum, Tritonium, Turbo, Ampullaria, Onchidium, Helix, &c.,* and therefore
with the Nudibranchia, Cyclobranchia, Scutibranchia, Tectibranchia, Pectinibranchia, and Pulmonata. I must here repeat that these venous canals
are only lacunae excavated in the muscular
walls of the body, and are without proper walls,
as *Meckel* (Syst. d. vergleich. Anat. V. p. 128) has
pretended is the case with those of *Aplysia.* To be
convinced of their wall-less structure it is only necessary to examine microscopically a longitudinally
incised *Arion.* They will be found composed wholly
of muscular fibres interlaced in every direction, and
some of which surround, sphincter-like, the venous
orifices, thus showing that these last are not closed
by valves, but by the contraction of these fibres.
Souleyet himself could not deny this wall-less structure in the veins of the Gasteropoda, although it is
in contradiction with his statements against Phle-
benterismus. He declares (Compt. Rend. XX. p.
81, note 3) "que le système veineux des Mollusques n'est pas toujours formé par des vaisseux dis-
tincts, mais qu'il, se compose en grande partie
de ces canaux creusés dans l'épaisseur ou dans
l'interstice des organes." See also below § 216,
note 1.
1 Respiratory organs appear to be wholly absent in *Sagitta,* and *Phyllirrhoë.*

tion is cutaneous, which, with the Apneusta, is probably favored by ciliated epithelium.[2] With some of these species, there is an aquiferous system which also serves, perhaps, for respiration.[3]

I. Branchiae.

§ 220.

With nearly all the Cephalophora, excepting the Pulmonata, there is a Branchial apparatus; this is usually very contractile, and always covered with very lively cilia.[1] It is composed either of lamellae, or of filaments arranged in rows or in bundles, or of plumose or pectinate ramified prolongations. With some, the branchiae are situated, uncovered, on the back or on the sides of the body; with others, they are more or less covered by the mantle; but with the majority, they are contained in a special cavity of this last.

This Branchial cavity communicates externally by the *Siphon*, which is simply a canaliculated, contractile prolongation of the mantle itself.[2]

1. With the Pteropoda, the respiratory organs are very unequally developed. In some genera, they appear wholly wanting, while in others, there is a spacious branchial cavity containing one or two groups of fringed lamellae from which pass out as many veins towards the auricle of the heart.[3]

2. With most of the Heteropoda, there is, upon the median line of the posterior part of the back, a pectinate or plumose branchial apparatus, which connects with the heart by a short vein.[4]

3. This apparatus is most variable as to form and situation with the Gasteropoda, and the different groups of this class are founded upon its modifications. The Cirribranchia have a bundle of small filaments on each side of the neck.[5] The Nudibranchia have on each side of the back, in one or more rows, or in a circle upon the middle of the posterior part

---

2 The opinion that the dorsal and lateral appendages of *Aeolis, Eolidina, Fenilia, Zephyrina, Amphorina, Flabellina, Calliopaea,* and *Tergipes,* are branchiae, is untenable, since it has been shown that they contain prolongations of the digestive canal.

3 For the aquiferous system of *Actaeon,* and *Fenilia,* see below, § 222.

1 For the ciliated organs of the branchiae of Gasteropoda, see *Sharpey,* Cyclop. Anat. &c. 1. p. 619.

2 For the branchial apparatus of the Cephalophora, I must refer principally to the works of *Cuvier* (Mémoires, &c.), *Savigny* (Descript. de l'Égypte, loc. cit. Il. Pl. I.–III.), *Meckel* (Beiträge zur vergleich. Anat., and Syst. d. vergleich. Anat., loc. cit.), *Quoy* and *Gaimard* (Voyage de l'Astrolabe, or Isis, loc. cit.), and *Delle Chiaje* (Mem. and Descriz. loc. cit.).

3 With *Clio,* one does not know what to think of the form and position of their respiratory organs, since that *Eschricht* (loc. cit. p. 5, 16) has shown that the vascular net-works observed by *Cuvier* upon the two fins of these animals (Mém. loc. cit. p. 5), and which have been taken for branchial vessels, are only muscular fibres. *Van Beneden* also, could find no respiratory organs with *Limacina* and *Cuvieria.* Moreover, more accurate observations are required to determine whether or not the four-rayed cutaneous appendage of the poste-

rior extremity of *Pneumodermon,* and the circular cutaneous lobe in the same locality with *Spongiobranchaea,* are really branchiae; see *Cuvier,* Mém. loc. cit. p. 7, Pl. B. fig 1–6, g. ; *Van Beneden,* loc. cit. p. 49, Pl. I. fig. 1, d. (*Pneumodermon*); and *D'Orbigny,* Isis, 1839, p. 497, Taf. I. fig. IX. 1–3, 11, 12 (*Spongiobranchaea*). On the other hand, *Van Beneden* (loc. cit. p. 17, 40, Pl. 1. fig. 2, 12, III. 1, 5, 6) has distinctly seen branchiae and branchial veins in *Hyalea, Cymbulia* and *Cleodora.* In the first of these genera, there lie in a very large respiratory cavity situated on the back of the intestinal sac, numerous branchial lamellae arranged in an arcuate manner, and bound together by a branchial vein. In the other two genera, the cavity of the mantle has, on each side, a fan-shaped branchia. See also *Delle Chiaje,* Descriz. &c. I. p. 59, Tav. XXXIV. fig. 9, 11.

4 With *Atlanta,* the single branchia is simple, pectinated, and always concealed in the interior of their cell (*Rang,* loc. cit. p. 378, Pl. IX. fig. 12, or Isis, loc. cit. p. 473, Taf. VII. fig. 12). With *Carinaria,* and *Pterotrachea,* the branchia is also simple, but very developed and demi pinnate, and in the first of these genera it projects outside the shell (*Delle Chiaje,* Mem. loc. cit. Tav. XIV. XV. LXIX., and Descriz. loc. cit. Tav. LXIII.–IV.).

5 *Dentalium,* according to *Deshayes,* loc. cit. p. 334, Pl. XV. fig. 12, or Isis, loc. cit. p. 464. Taf. VI. fig. 16.

of the body, numerous fasciculated plumose, or dendritic branchiae.[6] With the Cyclobranchia, and some of the Inferobranchia, the lamelliform branchiae are situated on the furrow which separates the border of the mantle from the foot,[7] under the form of a continuous cord, or of two lateral rows.

With the Scutibranchia, the two pectinal rows, which are wholly concealed in the cavity of the mantle, have, nevertheless, a certain symmetry [8] which is wholly absent with the other Gasteropoda. Thus, all the Tectibranchia have only a single lamellate or pinnate branchia situated on the right side, rarely on the left, and which is more or less covered and sometimes wholly concealed by a fold of the mantle.[9] The Pectinibranchia and Tubulibranchia have a pinnate or pectinate branchia, contained in a cavity which is situated upon the anterior portion of the back and often provided with a siphon on its left side.[10]

With many Nudibranchia, the returning blood from the branchiae is emptied by several veins into the simple auricle of the heart, which (the heart) is situated upon the middle line of the back.[11] With only a few Gasteropoda, as also with the Cirribranchia, Cyclobranchia, and Scutibranchia, the branchial veins are united into two trunks which open into the simple or double auricle.[12] With the other Gasteropoda, which have an uneven, lateral branchia,[13] the blood passes from this last, through a short, simple, venous trunk, to the heart situated near its base.

II. Lungs.

§ 221.

The pulmonary cavity, formed in the mantle of the Pulmonata, is situated

---

6 With *Scyllaea*, there are, on the back, two pairs of cutaneous lobes, between and which are numerous branchial vessels. With *Glaucus*, there are, upon the sides of the body, three pairs of prolongations which have long, digitiform branchial filaments. With *Thetis*, the back is surrounded by a double row of semi-pinnate branchiae ; while with *Tritonia*, there is on each of its sides a single row of multiramose branchial tufts. With *Doris*, and *Polycera*, there are twenty to twenty-five more or less ramified branchiae, arranged circularly around the arms, and capable, from contraction, of being withdrawn into the mantle.

7 The branchial lamellae form a complete circle with *Patella*, *Chiton*, and *Phyllidia*, and two lateral rows with *Diphyllidia*.

8 With *Fissurella*, and *Emarginula*, there is a row of branchiae on each side of the cavity of the mantle, while, with *Haliotis*, there are two rows on the left side.

9 With *Umbrella*, *Pleurobranchaea*, and *Pleurobranchus*, this branchia, situated on the right side and half exposed, is easily seen. On the same side also is situated the branchial lamella often deeply concealed between the folds of the mantle, of *Gasteropteron*, *Aplysia*, *Bullaea*, *Notarchus*, &c. But with *Doridium*, the branchia is on the left side and quite behind.

Ancylus, which differs from the other Inferobranchia by its simple branchia, has, moreover, this peculiarity, that this organ has the form of a simple cutaneous enlargement on the left side, con-

cealed under a fold of the mantle (*Treviranus*, loc. cit. p. 192, Taf. XVII. fig. 1, 2, d., or *Vogt*, loc. cit. p. 28, Taf. II. fig. 1–3, p.).

10 With *Valvata*, there is a single pinnate branchia which projects out of a cavity in which it is contained (*Gruithuisen*, Nov. Act. Acad. Nat. Cur. X. p. 441, Tab. XXXVIII. fig. 2, 3, 5, 12). The branchia is simple and pectinate with *Vermetus* (*Philippi*, Enumer. Mollusc. Sicil. I. p. 169, Tab. IX. fig. 24), *Rostella*, and *Struthiolaria*. It is bi-pectinate with *Turbo*, and *Janthina*, and tri-pectinate with *Paludina*. With many Pectinibranchia, as, for examples, with *Harpa*, *Cassis*, *Conus*, *Buccinum*, *Terebra*, *Murex*, *Voluta*, *Oliva*, &c., there is, beside a very large unipectinate branchia, another organ of this kind which is smaller and bi-pectinate. The epithelium, which covers not only the branchia, but also the walls of the respiratory cavity, plays an important part in the renewal of the water in the branchial cavity, which takes place through its opening, or by the siphon of these Gasteropoda which is situated upon the neck usually a little to the left side.*

11 *Scyllaea*, *Thetis*, *Doris*.

12 *Tritonia*, *Dentalium*, *Patella*, *Chiton*, *Haliotis*, *Fissurella* and *Emarginula*. Among the Inferobranchia, *Phyllidia* should also be cited here. But with *Diphyllidia*, on the contrary, the veins appear to pass each into the auricle of the heart.

13 The Tubulibranchia and Pectinibranchia.

---

* [§ 220, note 10.] See also *Leydig*, loc. cit. for the branchiae and their intimate structure, with *Paludina*. According to this observer, they

are unipectinate with *Paludina vivipara*, and not tri-pectinate, as above mentioned of this genus in general. — ED.

at the anterior part of the back, rarely at the posterior part.[1] Its orifice, which can be closed by a kind of sphincter, is upon the right side; it is upon the left with those species only which have sinistral shells, and in one genus alone, it is upon the median line at the posterior extremity of the body.[2] The pulmonary cavity is triangular with those species which have a shell, and round with those which are without it.[3] Its interior is lined with a raised vascular net-work which, with the aquatic species, is covered with a ciliated epithelium.[4] With the naked Gasteropoda, this net-work forms a uniformly-meshed trellis;[5] while with the others, there may here be usually seen several large pulmonary veins, which, in passing towards the middle principal vein, are spread over the borders of the respiratory cavity, frequently anastomose with each other, and receive several other veins of a dendritic form. The principal vein opens, at last, into the auricle of the heart at the posterior corner of the pulmonary cavity.[6]

Carefully examined, these veins will be found to be wall-less canals directly surrounded by the transverse and longitudinal fibres of the mantle, so that, apparently, they are only a continuation of the venous canals of the walls of the body.

III. Aquiferous System.

§ 222.

The existence of aquiferous vessels and reservoirs, with the Cephalophora, is not yet satisfactorily settled. However, it appears that here, as with the Acephala, there is an aquiferous system with wall-less canals, of which some are singly ramified, while others form an anastomotic net-work, but all accompany the venous canals and open upon the surface of the body, — presenting an arrangement analogous to the trachean system of insects.

With some Apneusta, the existence of this system, which may have the function of an internal respiratory apparatus, can scarcely be doubted;

1 The respiratory cavity is situated in the middle of the back with *Parmacella*, and wholly behind with *Testacella*, and *Onchidium*.
2 *Onchidium*. Whether or not the contractile, ramified excrescences at the posterior part of the back of this amphibious mollusk, of which *Ehrenberg* has counted more than twenty, serve really as branchiae as this naturalist asserts (Symb. physic. animal. evertebr. Mollusca), cannot be determined except from a most exact analysis of these organs. *Troschel (Wiegmann's Arch.* 1845, I. p. 197, Taf. VIII.) has shown with more certainty that *Ampullaria* is amphibious, for he found a pulmonary above the branchial cavity communicating with this last, and lined with blood-vessels.
3 With *Limax*, and *Arion*, the respiratory cavity has an annular form, its centre being occupied by the heart and kidney.
4 I have found ciliated epithelium in the pulmo-

nary cavity of the Lymnaeacea, but not in that of *Helix* or *Arion*.
5 *Onchidium, Limax*, &c.; see Cuvier, Mém. loc. cit. Pl. II. fig. 8–10 (*Arion*).
6 See Cuvier, Ibid. Pl. I. fig. 2–4, and *Treviranus*, Beobacht. aus. d. Zool. u. Physiol. Tab. VIII. fig. 57, 58 (*Helix pomatia*). In the vascular net-work which *Erdl* (De Helicis algirae, &c., fig. 6, copied in *Carus*, Erläuterungstafeln, Taf II. fig. 10) has figured with many details, all the vascular trunks do not run towards the principal vein, but with some their large extremity is directed towards the border of the lungs.
This disposition, however, does not exist in nature. The pulmonary vessels of this species are arranged like those of *Helix pomatia*, which is also confirmed by *Van Beneden's* figure of it; see his Anat. de l'Helix algir t, in the Ann. d. Sc. Nat. V. 1836, Pl. X. fig. 3, f.*

* [§ 221, note 6.] See, for the respiratory organs of the terrestrial Gasteropoda, *Leidy*, loc. cit. p. 235. — ED.

for, upon the back and directly behind the heart, there is a reservoir filled with water, from which ramifying canals pass off in all directions.[1]

The older observations upon these aquiferous canals of the Pteropoda, Heteropoda, and Gasteropoda, have been but indifferently increased by more recent labors. With these Cephalophora, the substance of the envelope of the body is permeated by a beautiful net-work of wall-less canals, which are filled with water, it is supposed, through several orifices upon the surface of the body.[2] It is, nevertheless, far from being settled that these canals belong to an aquiferous system, for the existence of their external orifices is doubtful, and it may be urged that they are only a continuation of the venous system.[3] At all events, this question demands further researches based upon facts observed with the Acephala and Cephalophora.

CHAPTER VIII.

ORGANS OF SECRETION.

I. Urinary Organs.

§ 223.

With most of the Cephalophora, the Urinary apparatus consists of an uneven, lamellate gland, which is usually situated near the branchial or princi-

---

1 According to *Souleyet* (Compt. Rend. XIX. p. 360, XX. p. 93), there is, with *Actaeon*, an aquiferous system, arising from a reservoir of water situated behind the heart, and which he has called *Poche pulmonaire*, which is spread through the whole of the body. *Vogt*, as he has written me, has distinctly seen this system with a canal opening on the right side behind the anus. *Allman* (loc. cit. p. 148, Pl. V. fig. 4, a. a. b.) has also observed it in the same species, but he took it for a blood system. The canal, which, with *Venilia*, opens at the posterior part of the back, and which has been taken by *Alder* and *Hancock* (loc. cit. XIII. Pl. II. fig. 1, 7, b.) for the rectum with its anus, belongs also, perhaps, to an aquiferous system, as well as the orifice figured by *Delle Chiaje* (Descriz. loc. cit. Tav. LXXXVIII. fig. 2, d.) in the same region, with *Aeolis cristata* (*Venilia ?*).

2 *Delle Chiaje* is as yet the only naturalist who has published quite detailed researches upon the aquiferous canals of the Cephalophora indicated in the text. In an earlier work, he has described them with *Doris, Thetis, Aplysia, Pleurobranchus, Pleurobranchaea, Bulla, Doridium, Diphyllidia, Turbo, Trochus, Nerita, Conus, Cypraea, Voluta, Buccinum, Murex, Cerithium, Rostellaria, Haliotis,* and *Patella,* as canals which traverse the foot, opening, for the most part, on its borders by numerous orifices (see his Descrizione di un nuovo apparato di canali acquosi scoperto negli animali invertebrati marini, in his Memor. &c. II. p. 259, Tav. XVII. fig. 10–15). Since then, he has described this system, which, he says, is wanting with the aquatic Pulmonata, as a beautiful, subcutaneous net-work. He

has named it *Apparato idro-pneumatico* or *Sistema linfatico-venoso ;* see his Descriz. I. p. 83, &c., Tav. XXXII. XXXIV. XL. &c. (*Cymbulia, Hyalea, Carinaria, Pterotrachea, Doris, Tritonia, Thetis, Pleurobranchaea, Diphyllidia, Doridium, Gasteropteron, Aplysia, Bulla, Sigaretus,* and *Janthina*). With *Cymbulia,* and *Gasteropteron,* this aquiferous canal communicates with a large sinus from which passes off a long afferent canal which projects from the surface of the body (see *Delle Chiaje,* Descriz. loc. cit. Tav. XXXII. fig. 1, 2, g. LV. fig. 2, b. f. 4. c. a.).

3 *Meckel* (Syst. d. vergleich. Anat. VI. p. 72) positively denies the existence of an aquiferous system and its external orifices. But he maintains that the marine Cephalophora can absorb and reject simply by their skin, considerable quantities of water, without the need of special orifices. *Milne Edwards* (Compt. Rend. XX. p. 271, or *Froriep's* neue Not. No. 733, p. 98) declares that this apparatus, such as described by *Delle Chiaje,* belongs to the venous system. He also denies the existence of external orifices, explaining the ingress and egress of water which has been observed with these animals, as due to endosmose and exosmose. *Van Beneden,* also (Ann. d. Sc. Nat. IV. 1835, p. 250), says that he is convinced that with *Aplysia* the so-called aquiferous canals are only a dependence of the venous system. On the other hand, he is inclined to admit that, with *Aplysia,* and *Carinaria,* &c., there are small orifices by means of which these animals can mix water with their blood (Compt. Rend. XX. p. 520, and l'Institut. No. 627, or *Froriep's* neue Not. No 72?, p. 4, and No. 797, p. 65).

pal pulmonary vein. Its excretory duct accompanies the rectum and often opens near the anus.[1]

The kidney is nearly always of a dirty yellow, or reddish color, of a lamellated structure, and its surface is wholly without vibratile organs. It is surrounded by a sac-like envelope which is continuous with the internally ciliated, excretory duct. Each renal lamella is composed of thickly-set, delicate cells loosely bound together. In their transparent liquid floats an obscure nucleus which, by direct light, appears brown or violet.

These nuclei, which are round and embossed, have a very dense crystalline structure, and are undoubtedly a product of the renal secretion.[2] Certainly they contain the uric acid which is found when the whole gland is chemically analyzed.[3] The ramified canals upon the membranous envelope of the kidneys, return, probably, the blood into the respiratory organs. But in the gland itself no blood-vessel has been observed.[4]

With *Sagitta*, and the other Pteropoda, nothing like a renal organ has yet been found. With the Heteropoda, and Apneusta, on the contrary, there are vestiges of certain organs which further researches may show to be of a urinary nature.[5]

With the Pectinibranchia, the kidney is replaced by a gland which is situated behind the branchia, between the heart and liver, and which, in some marine species, secretes the purple liquid.

It is composed of several ramified lamellae, and opens by a large orifice, or by a duct of variable length which accompanies the rectum, at the base of the branchial cavity.[6] With the other branchiated Gasteropoda, the existence of this gland is yet doubtful, although with most of them, and

1 This is the gland which, with the Gasteropoda, has been considered by the older anatomists such as *Swammerdamm*, *Poli*, and *Blumenbach*, as an organ secreting the calcareous salts, and by *Cuvier* as a muciparous gland.

2 This gland corresponds, consequently, as to its position and intimate structure, to the bodies of *Bajanus*, which, with the Lamellibranchia, have been considered as kidneys; excepting that they have no ciliated organs. For the intimate structure of the kidneys of Gasteropoda, see *H. Meckel*, in *Muller's* Arch. 1846, p. 15, Taf. 1.

3 *Jacobson* (Jour. de Physique, XCI. p. 318, or *Meckel's* Arch. VI. 1820, p. 570) was the first who showed the presence of uric acid in this gland, with *Helix pomatia*, and *nemoralis*, *Limax niger*, *Lymnaeus stagnalis*, and *Planorbis cornea*. But, some time previous, *Döllinger* and *Wahulich* (Diss. de Helice pomatia, Wirceb. 1813, p. 23) had regarded this organ as a kidney. The presence of uric acid can be easily shown in the dried kidneys of *Helix pomatia* and *Paludina vivipara*, for when treated with nitric acid and ammonia, a considerable quantity of murexid is disengaged.

4 According to *Treviranus* (Beobacht. aus. d. Anat. u. Physiol. p. 39), with *Helix* and *Arion*, a portion of the blood of the lungs, instead of going to the heart, passes into the kidneys, and thence enters the great pulmonary vein. But it must be very difficult to show the course of this liquid in the interior of the kidneys.

5 The spongy substance mentioned by *Delle Chiaje* (Descriz. II. p. 96, Tav. LXIII. fig. 3, s.) as existing near the heart and at the base of the branchiae, with *Carinaria*, is undoubtedly a urinary gland. The long, yellow ciliated body, but without excretory organs, which *Nordmann* (loc. cit. p. 24, Taf. II. Q.) observed with *Tergipes*, between the stomach, liver, heart and rectum, is also, perhaps, a kidney; at all events, as such cannot be regarded another and neighboring body, larger, lobulated and of a yellowish color, having apparently an excretory canal opening externally, and which already has been mentioned as being an hepatic gland. Perhaps a like interpretation should also be put upon the yellow bodies observed by *Quatrefages* in the posterior part of the body of *Zephyrina*, *Actaeon*, and *Amphorina* (Ann. d. Sc. Nat. I. p. 136, Pl. IV. fig. 1–3).

6 With *Tritonium*, and *Murex*, this gland opens by a large orifice into the cavity of the mantle; see *Eysenhardt* (*Meckel's* Deutsch. Arch. VIII. p. 210, Taf. III. fig. 4, r.), and *Leiblein* (*Heusinger's* Zeitsch. für d. Organ. Phys. I. p. 4, Taf. I. h. i., or Ann. d. Sc. Nat. XIV. 1828, p. 179, Pl. X. h. i.). A similar urinary gland has been described with *Janthina*, by *Delle Chiaje* (Descriz. II. p. 108, Tav. LXVII. fig. 3, e., LXVIII. fig. 14, f. l.), as an accessory respiratory cavity. With *Paludina*, this gland has a quite long excretory duct; see *Cuvier*, Mém. loc. cit. fig. 3, l. 7, p. q. The kidney has, moreover, been described by *Cuvier* (loc. cit.), and *Quoy* and *Gaimard* (Voy. de l'Astrolabe Zool. II. or, Isis, 1834, p. 285, 1836, p. 31) under the names of *Muciparous gland, Organ of the purple*, and *Depurating organ*, with *Phasianella*, *Turbo*, *Buccinum*, *Mitra*, *Oliva*, *Cypraea*, *Harpa*, *Dolium*, *Cassis*, *Purpura*, *Fusus*, *Auricula*, &c.*

* [§ 223, note 6.] For the renal organs with *Paludina*, see *Leydig*, Ueber Paludina vivipara, &c., loc. cit. p. 180, Taf. XIII. fig. 49, O. — ED.

22

especially with the Nudibranchia, and Tectibranchia, there is a glandular apparatus which may perhaps be of this nature.[7]

With the terrestrial and aquatic Pulmonata, the lamellated kidney is quite distinct. In the species having a shell, it is of riband-like, or triangular form, and situated beside the heart and the large pulmonary vein. Its excretory duct arises from the anterior extremity and passes, first, backwards to the rectum, near the posterior corner of the gland, then turns and runs forwards terminating, finally, in the respiratory cavity near the anus.[8] With the Limacina, on the contrary, the kidney surrounds the pericardium like an annular collar, and its excretory duct opens near the respiratory orifice.[9]

II. Organs of peculiar Secretions.

§ 224.

Mention has already been made of the parts of the mantle which secrete the calcareous substance,[1] and further on, I shall speak of the different glandular appendages attached to the genital organs.[2]

As to the other organs of particular secretions which are less common, I will mention the following :

1. With those Apneusta which have cutaneous appendages, there is, in the dorsal and lateral lobes, a follicle whose excretory orifice opens at the extremity of the lobe, and which secretes a granular mucous substance, and peculiar corpuscles which resemble the nettling organs of certain Zoophytes.[3]

---

7 With *Doris*, there is found between the lobes of the liver a gland, which sends off backwards a long excretory duct which opens externally close by the anus and has sometimes near its extremity, a vesicular dilatation. This gland, formerly taken for a liver, is probably a urinary organ ; see *Cuvier*, loc. cit. p. 16, Pl. I. II. ; *Meckel*, Beitr. zur vergleich. Anat. 1. Hft. 2, p. 9, Taf. VI. fig. 3, I. and *Delle Chiaje*, Descriz. II. p. 25, Tav. XLI. fig. 12, u. y. C. fig. 21.

The orifice found with *Thetis*, directly behind the anus in the dorsal region, is also in communication with a gland which may be regarded as a kidney ; see *Cuvier*, loc. cit. fig. 1, c. and *Delle Chiaje*, Descriz. II. p. 35, Tav. XLVII. fig. 1, q., XLIX. fig. 3. *Delle Chiaje* (Ibid. Tav. XLII. fig. 1, 3) has seen with *Tritonia*, a similar gland opening into the rectum ; and with *Gasteropteron* (Ibid. p. 56, Tav. LIV. a.), another situated between the base of the branchiae and the heart.

The large triangular glandular mass, which, with *Aplysia*, is situated in the cutaneous fold enveloping the shell, and lies in the space between the heart, the base of the branchiae and the anus, se-

cretes a large quantity of a red liquid ; this also is probably a kidney ; see *Cuvier*, loc. cit. p. 11, Pl. II. fig. 1, C. D. E. fig. 3, B. C. D., and *Delle Chiaje*, Memor. II. p. 55, Tav. II. fig. 2, r. t. 5, 6. With *Vermetus*, and *Magilus*, there is an analogous gland behind the branchiae. However, this renal apparatus of the branchiferous Gasteropoda demands a more careful investigation in both an histological and a chemical point of view.*

8 See the figures of the kidney of *Helix* and *Lymnaeus* in *Cuvier*, loc. cit., and in *Treviranus* Beobacht. &c. Tab. VIII. fig. 58 ; see also *Paasch*, in *Wiegmann's* Arch. 1843, 1. p. 78, and, De Gasteropodum nonnullorum hermaphroditicorum, systema. genit. et uropoëtico, Diss. Berol. 1842.

9 See *Cuvier*, loc. cit. Pl. II. fig. 8–10, und *Treviranus*, Beobacht. Tab. IX. fig. 59 (*Arion*), and *Paasch*, loc. cit. p. 82.]

1 See § 203.

2 See below, Chapter IX.

3 These glandular follicles which, from spontaneous contraction can empty their contents, communicate, according to *Quatrefages* (Ann. d. Sc. Nat. XIX. p. 287, 291, Pl. XI. fig. 5, 6), with *Eolidina*,

---

* [§ 223, note 7.] See, in reference to this gland with *Doris*, *Alder* and *Hancock*, loc. cit. Part V. Pl. II. fig. 1, g. g.

For the renal organs of *Chiton*, see *Middendorff*, loc. cit. p. 72, Taf. VI. fig. 1, N. and Taf. VII. fig. 5, N. They consist of a velvet-looking substance which stretches on each side of the body, over the tendinous mass of the ventral muscles, and join together horse-shoe-like on the anterior border of the posterior diaphragm. Their intimate

structure consists of arborescent digitations from a central canal. — ED.

† [§ 223, note 9.] For the renal organ and its intimate structure with the terrestrial Gasteropoda, see *Leidy*, loc. cit. p. 239. See also for the different varieties of this organ with this order, *De St. Simon* (Jour. de Conchol. 1851, No. IV. p. 342), who speaks of it as *La Glande praecordiale*. — ED.

2. The genus *Aplysia* has an apparatus of particular secretion, consisting of a group of pyriform follicles situated under the branchia, inside of the skin. Its excretory orifice is behind the female genital opening, and its secretion is a whitish liquid with attributive corrosive qualities.[4]

3. Many of the Pectinibranchia, and Tubulibranchia, have upon the upper wall of the cavity of the mantle, a row of folds which secrete an extraordinary quantity of viscous mucus which is not excreted through any particular duct.[5]

4. With several terrestrial Gasteropoda, the median line of the foot is occupied by a straight canal lined with ciliated epithelium, which ends in a large orifice situated under the mouth. On each side of this canal, are rows of follicles that secrete a granular mucus which, passing into its cavity, is excreted externally, probably by means of cilia.[6]

CHAPTER IX.

ORGANS OF GENERATION.

§ 225.

The Cephalophora propagate solely by means of male and female genital

---

by a narrow canal, with the prolongations of the digestive cavity which enter into the dorsal appendages, and their contained liquid is subjected to a process of respiration. But *Nordmann* (loc. cit. p. 33, Tab. II. R. R.) has been unable to find any such communication between these two organs, with *Tergipes*, and he has distinctly seen the granular mucus which is expelled from the follicles from contraction, escape through an orifice on the extremity of each dorsal appendage. With *Æolis*, according to *Hancock* and *Embleton* (loc. cit. p. 80, Pl. IV. V.), the product of these follicles is quite interesting. It contains elliptical vesicles which immediately burst when put in water, exposing a transparent cylinder, out of which a filament, sometimes of a spiral form, is projected as swift as lightning. They compared these bodies to spermatic particles; but to me, they appear exactly like the nettling organs of *Actinia*. *Hancock* and *Embleton* have also seen and figured with *Æolis*, a canal of communication between these follicles and the prolongations of the digestive canal, but it may be questioned if this was not an artificial formation produced by compression of these organs during the examination.*

4 See *Cuvier*, loc. cit. p. 4, fig. 2, Σ.; *Delle Chiaje*, Memor. II. p. 56, Tav. II. fig. 2, O. fig. 3; and *Rang*, Hist. Nat. des Aplysies, p. 25.

5 These muciparous organs described by *Cuvier*, with *Buccinum* (Mém. loc. cit. p. 5, fig. 3, f.) as

Feuillets muqueux, are also found with *Murex* (*Eysenhardt*, in *Meckel's* Deutsch. Arch. VIII. p. 215, Taf. III. m. m.), *Terebra*, *Turbo*, *Voluta*, *Cypraea*, *Harpa*, *Dolium*, *Cassis*, *Tritonium*, &c. (*Quoy* and *Gaimard*, Voy. de l'Astrolabe, loc. cit., or 1-is. 1836, p. 35, Taf. II. fig. 6, q. Taf. III. fig. 10, X. 18, m.). *Carus* (Museum Senckenberg. II. 197, Taf. XII. fig. 8, h.) has seen similar mucous folds with *Magilus*. With *Vermetus*, on the contrary, I have found only a single, but a very considerable, longitudinal fold which runs along by the side of the rectum and covers the excretory duct of the genital organs.

6 This muciparous apparatus of *Bulimus*, *Helix*, *Limax*, and *Arion*, was announced in 1829, by *Kleeberg*, at the Congress of Naturalists at Heidelberg (Isis, 1830, p. 574); but it had not escaped the observation of *Delle Chiaje* with many Helicina and Limacina (Descriz. II. p. 10, Tav. XXXVII. fig. 17, x.). It is therefore surprising that it remained thus long unknown to other naturalists. The assertion of *Kleeberg*, that with *Limax*, and *Arion*, the mucous canal communicates with the venous system, I have been unable to confirm by observations upon *Arion*. *Leydig* declares that this mucous canal with the terrestrial Gasteropoda is the seat of the sense of smell; see *Schleiden* and *Froriep's* Notiz. IV. p. 24, or Ann. of Nat. Hist. XX. p. 210.

* [§ 224, note 3.] See for further description, together with figures of these peculiar bodies containing a spiral thread, *Alder* and *Hancock*, loc. cit. Part III. Pl. VIII. fig. 14 (*Æolis*); they correct their former view (mentioned above) and admit, what I think is not in the least doubtful, that they

are analogous to the nettling organs of the Polyps. *Agassiz* has carefully observed them, and they have all the characteristics of a true lasso-cell; to this I may add my own observations upon other Mollusca. — ED.

organs. These are either combined in one individual, or the sexes are separate. In most species there are copulatory organs. The genital organs have several uneven divisions, which, when fully developed, are arranged as follows: A *Tuba Fallopii* passes from the ovary into the uterine sac, at whose base is an organ which secretes albumen, while at the point where it is continuous with the vagina, there is a *Receptaculum seminis*. The male genital organs consist of a testicle, a *Vas deferens*, and a *Ductus ejaculatorius* which opens into a retractile penis. With the hermaphroditic species, these two kinds of genital organs are more or less blended together, — the testicle with the ovary, and the *Vas deferens* with the *Tuba Fallopii*; very often also the vagina is united with the *Ductus ejaculatorius*, forming a cloaca into which open several particular secreting organs. These different male and female organs are usually lined internally with ciliated epithelium.

The eggs of these animals have, at their escape from the ovary, a round and sometimes an elliptical form, and are composed of a thin chorion enclosing a finely-granular vitellus of variable color, which contains a germinative vesicle and dot.[1] The sperm is white and opalescent, and quite crowded with very active spermatic particles. These last are either of the form of *Cercaria*, or consist of a very long filiform body, one extremity of which is incrassated and often of a spiral form. The trembling, undulatory movements of these particles cease when placed in water, with those species which have copulatory organs; they become twisted into loop-like forms and are finally rigid and motionless.[2]

1 See *Carus*, Erläuterungstafeln Hft. V. Taf. II. fig. 4, a³. (*Limax*), and in *Muller's* Arch. 1835, p. 491, Taf. XII. fig. 2 (*Helix pomatia*); *Wagner*, in *Wiegmann's* Arch. 1835, I. p. 368, and Prodromus, loc. cit. p. 7, Tab. I. fig. 6, 7 (*Helix* and *Buccinum*); and *Allman*, loc. cit. p. 152, Pl. VII. fig. 7 (*Actaeon*).

2 *Wagner* and *Erdl* (*Froriep's* neue Notiz. No. 249, p. 98) have found with *Chiton*, *Patella*, and *Haliotis*, spermatic particles of a Cercarian-form, that is with a long body to which is abruptly attached a hair-like tail. I have seen a similar form with *Vermetus gigas* and *triqueter*. Those of *Trochus* also have this form, according to *Kölliker* (Beitr. loc. cit. p. 28), but the middle of their body has a slight constriction. They are hair-like, and taper at both extremities with *Turbo*, *Buccinum*, *Purpura* (*Kölliker*, loc. cit. p. 25, Taf. I. fig. 5), and *Sagitta* (*Krohn*, loc. cit. p. 10, fig. 12). With other marine Gasteropoda, as for example, *Carinaria*, these particles are hair-like, but with one of their extremities slightly incrassated (*Milne Edwards*, Ann. d. Sc. Nat. XVIII. p. 324, Pl. XI. fig. 7); and with *Doris*, *Tergipes*, and *Paludina*, this thickened extremity has a spiral form (*Kölliker*, Beitr. loc. cit. p. 35, Taf. I. fig. 6; *Nordmann*, loc. cit. p. 52, Taf. III. fig. 8, 9, and my observations in *Muller's* Arch. 1836, p. 240, Taf. X.). With the pulmonate Gasteropoda, the spermatic particles have only a short incrassated extremity of a spiral form; see my observations loc. cit. 1836, p. 45, Taf. II.; *Paasch*, in *Wiegmann's* Arch. 1843, I. p. 71, Taf. V., and *Dujardin*, Observ. au Microscope, Atlas, Pl. III.

The development of these spermatic particles takes place in two large cells (Mother-cells), in which are formed others (Daughter-cells) which are changed into the spermatic particles. But the cell-membrane of the mother-cell, disappears quite

early and its contents are condensed into a solid nucleus around which are grouped the daughter-cells, ultimately forming a bundle of spermatozoon. See, beside these observations of *Kölliker*, *Nordmann*, and *Paasch*, loc. cit., those of *H. Meckel*, in *Muller's* Arch. 1844, p. 483, Taf. XIV. fig. 9-13, and the more recent researches of *Kölliker*, in the Neue Denkschrift. d. allgem. schweizer. Gesellsch. f. d. gesammt. Naturwissensch. VIII. 1846, p. 4, Taf. I. fig. 1-10 (*Helix pomatia*).

The presence of two kinds of spermatic particles in the sperm of *Paludina vivipara*, is a very remarkable fact; see my observations in *Muller's* Arch. 1836, p. 245, Taf. X.

Beside the hair-like spermatic particles already mentioned, there are long cylindrical bodies, from one of the extremities of which project many delicate filaments having very lively motions. These have been described by *Ehrenberg* (Symbol. physic. Anim. evertebrat. Dec. I. Phytozoa entozoa, Appendix) as parasites under the name *Phacelura paludinae*.

Paasch (*Wiegmann's* Arch. 1843, p. 99, Taf. V. fig. 8), on the other hand, regards them as bundles of spermatic particles of the normal form, and *Kölliker* (Beitr. loc. cit. p. 65, and Neue Denkschr. loc. cit. p. 41) considers them only as two forms of the same kind of spermatic particle: the second he regards as elongated mother-cells containing many ordinary spermatic particles.

For my part, I do not know how to explain this fact, and I would willingly place the second form in the category of Spermatophora; but against this opinion, as against that of *Kölliker*, and *Paasch*, it can be urged that, with the second form, the extremities are never thickened or spiral, as is true of the first, and that both forms are simultaneously developed in the testicle.*

* [§ 225, note 2.] My own observations on the spermatic particles of the Cephalophora and their

development, correspond closely with the above account. Their development in special c***ls I have

§ 226.

Among all the hermaphrodite Cephalophora, the genus *Sagitta* stands wholly alone, in having all parts of its genital apparatus double.

The ovaries consist of two straight, non-ciliated tubes situated at the posterior extremity of the cavity of the body; these open externally by an arcuate oviduct, situated upon the back directly over the median line of each of the posterior lateral fins.

The two internally ciliated testicles fill the caudal cavity, which is divided into two chambers by a longitudinal septum. They send backwards two short deferent canals, which open in front of the caudal fin, by two tumid orifices, but are without copulatory organs.[1]

§ 227.

As for the other hermaphroditic Cephalophora, to which belong the Pteropoda and a majority of the Gasteropoda, the genital organs of the Nudibranchia, Inferobranchia, Tectibranchia, and Pulmonata, have been the most thoroughly investigated. But the different divisions of these organs have been interpreted in a manner so varied and contradictory, that one can almost despair of having any positive knowledge of their relations.[1]

1 See *Krohn*, loc. cit. p. 9, fig. 2, 7-9. The ciliated epithelium which covers the male genital organs of *Sagitta*, from the posterior extremity to the genital orifice, produces a general up-and-down movement of the sperm in the testicle, a phenomenon which *Darwin* has compared to the motions of the sap in *Chara* (Ann. of Nat. Hist. XIII. p. 3, Pl. I. fig. 1, or *Froriep's* neue Notiz. No. 633, p. 3, fig. 62, and Ann. d. Sc. Nat. I. 1844, p. 362, Pl. XV. B). For the genital organs of *Sagitta*, see also the researches of *Wilms* (loc. cit. p. 12).

1 It has been quite difficult to reconcile the fact that, with these animals, the testicle and the ovary are united in a single body, — the Hermaphrodite gland. *Cuvier*, whose opinion has been followed by *Meckel*, and *Carus*, in their different publica-tions, regarded this gland, with the Pulmonata, as an ovary, and the albumen-secreting organ, as a testicle. *Treviranus* (Zeitsch. für Physiol. I. p. 3, V. p. 140) was of the opposite opinion; he considered the hermaphrodite gland as a testicle and the other as an ovary. This view has been adopted by *Prevost* (Mém. d. l. Soc. Phys. de Genève, V. p. 119, and Ann. d. Sc. Nat. XXX. p. 33, 43), and by *Paasch* (Diss. loc. cit. and *Wiegmann's* Arch. 1843, I. p. 71, 1845, I. p. 34). In England, *Rymer Jones* adopts the view of *Cu-vier*, and *Owen* that of *Treviranus*. *Wohnlich* (loc. cit. p. 32) names as ovary, the albumen gland ; and as testicle, the half-canal which runs along the uterus ; but he is in doubt as to the function of the hermaphrodite gland. *Erdl*, who

observed exactly like *Kölliker*. I have found these particles throughout this group, generally, to consist of a delicate thread, one end of which is more or less incrassated and twisted in a cork-screw manner (*Æolis*, *Physa*, *Lymnaeus*, *Natica*, *Helix*, *Limax*, &c.) ; in some, however, the form is remarka-bly different. Thus, with *Buccinum*, it consists of a thread with a terminal third somewhat incrassated, but which terminates in a delicate filament. This, as will be seen above, *Kölliker* has also noticed.

As to the remarkable statements made above upon two kinds of spermatic particles with *Palu-dina vivipara*, they deserve our especial attention. My own investigations have led me to regard it as a law in Spermatology, that each animal had only one kind of spermatic particle, the shape and size of which in that animal, are invariably the same ; this point I have regarded as so well established that I have proposed the basis of an animal classifica-tion from spermatological data. I was therefore surprised to find an observer like *Leydig* who has recently gone over the ground, according his views with those of *Siebold* and others above-mentioned (see Beitrag, loc. cit. in *Siebold* and *Kölliker's* Zeitsch. II. 1850, p. 125, Taf. XIII. fig. 31-43). *Leydig*, however, has watched their formation from cells ; and here I may remark as being evi-dence against their being spermatic particles, that, according to him, they are produced by the meta-morphosis of an entire nucleolated cell, and not, as is the grand law with spermatic particles, from a cell-nucleus. From this and from the above-men-tioned reasons, based upon analogy, I cannot ad-mit that these peculiar bodies are true spermatic particles. *Leydig's* observations on their develop-ment of course render invalid the hypothesis of *Gratiolet* that they are modified spermatic parti-cles, having undergone changes, like those of the Helicina, in the *Vesicula copulatrix*; see Jour. de Conchol. No. II. 1850, p. 116, and No. III. p. 236, Pl. IX. fig. 3-7. — ED.

It is only lately, that, from microscopical analyses of the contents of these parts, this point has been made clear. A peculiarity which distinguishes principally the Pteropoda, Apneusta, Nudibranchia, Inferobranchia, Tectibranchia, and Pulmonata, is the existence of a hermaphrodite gland. An exact knowledge of the structural relations of this gland has been the means of reconciling the hitherto confused opinions upon the genital organs of the Cephalophora.

This gland, which is nearly always buried in the substance of the liver, is composed of digitiform or botryoidal ramose caeca, bound together in groups of variable size forming a lobulated organ. Upon each caecum is an external sac, producing eggs, and an internal one, folded in the first, producing sperm. The walls of these two invaginated follicles are usually in direct contact, and are not separated from each other except at the points where there are eggs which push the ovarian sac outwards and the testicular one inwards.[2]

From these sacs pass off excretory canals, which, also, are invaginated, and terminate in two principal ducts, the external of which is the *Tuba Fallopii*, and the internal the *Vas deferens* which is usually tortuous.[3]

attributes to this last the function of an ovary (Beitr. zur Anat. d. Helicinen, loc. cit.), has expressed no positive opinion as to the function of the albumen gland. *Steenstrup* (Undersögelser over Hermaphroditismens Tilvaerelse i Naturen, 1845, p. 76, Tab. II.) has expressed a very singular opinion on the subject of the genital organs of the Pulmonata. He regards the Gasteropoda as of separate sexes with which the different parts of the genital apparatus are double, and that only one side is developed, the other remaining atrophied as in female birds. According to this, the hermaphrodite gland would represent the active ovary, in the individuals which *Steenstrup* regards as females, and the albumen-gland would be the ovary on the other side imperfectly developed. The uterine canal would belong to the active side, the *Vas deferens* would be the abortive uterus on the other side, and the penis as an abortive analogous vesicle would correspond to the pedunculated vesicle of the active side.

In the other individuals of the same species regarded by *Steenstrup* as males, the hermaphrodite gland would be the active testicle, and the albumen-gland, the same organ on the other side, abortive; the uterus would be the developed *Vas deferens*, and the proper *Vas deferens* the undeveloped organ on the other side. The pedunculated vesicle would have the same signification as with the female individuals, and the penis would be this vesicle imperfectly developed.*

2 After *R. Wagner* (Wiegmann's Arch. 1836, I. p. 370) had found in various Pulmonata, eggs and spermatic particles at the same time in one and the same genital gland, and I myself had expressed my conviction (Ibid. 1837, I. p. 51) that with these Gasteropoda the ovary and testicle were united in a single organ, *H. Meckel* was the first who de-

scribed exactly the structure of this hermaphrodite gland (Müller's Arch. 1844, p. 483, Taf. XIV. XV).

It is, therefore, astonishing that *Steenstrup* (Undersögelser, &c., p. 76, Tab. II. fig. 3, 4), who knew the researches of *Meckel*, and who, judging from his figures, saw distinctly the line of separation between the ovarian and testicular follicles, has determined two fragments of this gland taken from different individuals of *Helix pomatia*, as being one an ovary, and the other a testicle. In this last-mentioned fragment, he has called spermatic cells not only those really such of the internal follicle, but also the eggs contained in the external follicle; while in the first-mentioned fragment, or the so-called ovary, he has named as eggs not only the real eggs but also the internal spermatic cells. The spermatic particles, which he also saw at the same time, would, according to him, be brought out by coition.

3 Under the Pteropoda, *Kölliker* (Denkschrift. &c. VIII. p. 39) has found the hermaphrodite gland with *Hyalea*. From this, the organs described by *Cuvier*, *Eschricht*, and *Van Beneden*, (loc. cit.), as ovaries and oviducts with *Clio*, *Cymbulia*, *Cleodora*, *Cuvieria*, *Limacina*, &c., may be regarded as an hermaphrodite gland, and as invaginated excretory canals. Under the Apneusta, this gland has been seen by *Kölliker*, with *Aeolis*, *Lissosoma*, and *Flabellina*. It exists also with *Actaeon*, judging from the description of *Allmann* (loc. cit. p. 152, Pl. VI. VII. fig. 8) of its voluminous and multiramose ovaries, in which, he says there are observed, beside the projecting sacs filled with eggs, others smaller filled with a granular substance. The first are very probably ovarian, and the others testicular follicles. *Tergipes*, also, has a similar ramified ovary; but it was incorrectly interpreted by *Nordmann* (loc.

* [§ 227, note 1.] This structure — a hermaphrodite gland — is not mentioned by *Alder* and *Hancock* in their anatomical details of the Nudibranchia; see loc. cit. With those genera (*Aeolis*, *Doris*, &c.) with which they have given in special detail the generative organs, this combination of the two sexual organs is not spoken of.

See especially *Hancock* and *Embleton's* Anatomy of *Aeolis*, Ann. Nat. Hist. I 1848, p. 93,

where the androgynous apparatus is minutely described.

These authors affirm that although self-impregnation is, perhaps, possible, yet there is usually a congress of two individuals, and therefore a reciprocal copulation.

See also upon this point,—the real relations of the hermaphrodite gland, *Gratiolet*, Jour. de Conchol. 1850, No. 11. p. 116. — Ed.

The testicular follicle and the deferent canal are lined with ciliated epithelium, which, however, is wanting in the ovarian follicle.

The disposition of the various parts of the genital apparatus varies very much according to the families and genera of these hermaphroditic Cephalophora. The two invaginated excretory ducts of the hermaphrodite gland either pass to the base of the uterus, or the *Vas deferens* leaves the *Tuba Fallopii* a little way from it, and passes in a tortuous course to the penis.[4] In the first case, the *Vas deferens* leaves the Fallopian tube at the point where it enters the uterus, and continues its course on the sides of this organ, but as a semi-canal open upon its inner surface.[5]

In some genera, this semi-canal continues on along the vagina to the genital cloaca,[6] while in others, it becomes a complete canal upon leaving the uterus, and passes, after a longer or shorter course, into the penis.[7] The *Vas deferens* has, at different points of its course, glandular or vesicular appendages, which sometimes contain sperm. These may be compared, on the one hand, to an *Epididymis* or *Vesicula seminalis*, and, on the other, to a *Glandula prostata*.[8]

At the base of the uterus there is an Albumen-gland which is usually tongue-shaped, and sometimes very long, being rolled up and bound together by a cellular tissue so as to have a round form. The walls of this gland are composed wholly of cells filled with drops of albumen which is undoubtedly used to envelop the eggs as they pass into the uterus.[9]

cit. p. 64, Tab. II. III. fig. 5, O. S.). He regarded the testicular follicles containing spermatic particles in various degrees of development, as so many *Receptacula seminis*; and he attempted to sustain this view, in erroneously supposing that the spermatic particles could be produced in the pouches of fecundation. As to the Heterobranchia, *H. Meckel* (loc. cit.) has found this gland with the Nudibranchia (*Doris, Tritonia,* and *Thetis*), the Inferobranchia (*Diphyllidia*), and the Tectibranchia (*Aplysia, Bullaea, Doridium, Umbrella, Pleurobranchaea, Gasteropteron*). *Kölliker* (Denkschrift. loc. cit. p. 40), beside confirming the existence of this gland in the marine Gasteropoda just mentioned, has also added the genera *Notarchus* and *Pleurobranchus*. In the separated lobules of the ovary of a *Tritonia* figured by *Sars* (*Wiegmann's* Arch. 1840, I. p. 197, Taf. V. fig. c.), there can be easily recognized the hermaphrodite gland, such as has been represented with this animal by *H. Meckel* (loc. cit. Taf. XV. fig. 14). According to *Kölliker* (Rhodope nuovo gen. die Gasterop., in the Giornale dell' Inst. Lomb. di Scienze, &c., XVI. Milano, 1847, fig. 2), the testicular and ovarian follicles are grouped separately, the one above and the others below.

4 This last case is found with *Thetis, Doris,* and *Pleurobranchaea* (*H. Meckel*, loc. cit. Taf. XV. fig. 1, 2, 5). The *Vas deferens* pursues a similar course with the Apneusta, — at least with *Flabellina*, and *Rhodope*, according to the communication from *Kölliker*; and, from *Allman's* (loc. cit.) figure of that of *Actaeon*, it may be concluded that it there leaves the tube very high up and passes to the penis.

5 *Prevost* was the first to point out this half-canal (Mém. de Genève, &c., V. p. 123, Pl. I. fig. 12, 11. fig. 3, and Ann. d. Sc. Nat. XXX.).

* [§ 227, note 8.] This *Prostate* has been well developed by *Leidy* (loc. cit.) with the terrestrial Gasteropoda. He regards it as belonging to the male apparatus, notwithstanding its close connection with the ovary, since, in *Vaginulus*, it emp-

6 *Aplysia*, and perhaps also, *Bullaea, Doridium*, &c. (see *H. Meckel*, loc. cit. Taf. XV. fig. 7). With the Pteropoda, also, the *Vas deferens* does not leave the female canal, until it reaches the genital cloaca.

7 The Pulmonata.

8 With *Helix pomatia, Aplysia camelus, Tritonia ascanii,* and *Diphyllidia lineata*, there is a *Vesicula seminalis* at the point where the *Vas deferens* unites with the base of the uterus (*H. Meckel*, loc. cit. Taf. XIV. fig. 8, d. XV. fig. 7, d. 12, c. 16, c). It yet remains to be determined whether the dilatation upon the common excretory duct of the hermaphrodite gland with the Pteropoda, belongs to the *Vas deferens* or to the Fallopian tube. In the first case it would represent an epididymis or a seminal vesicle; in the second, perhaps an uterus; see *Eschricht*, loc. cit. Tab. III. fig. 25, r.* (*Clio*), and *Van Beneden*, Exerc. zoot. loc. cit. Pl. I. et. seq. (*Cymbulia, Hyalea*, &c.). This last-mentioned naturalist has simply, though erroneously, called this dilatation a testicle. A glandular mass, similar to a prostate, surrounds the deferent canal shortly after it leaves the oviduct, with *Thetis, Pleurobranchaea* (*H. Meckel*, loc. cit. Taf. XV. fig. 1, h. 5, f.), *Lymnaeus stagnalis* (*Treviranus*, Zeitsch. f. Physiol. I. Tab. III. fig. 14, d, or *Paasch*, in *Wiegmann's* Arch. 1843, 1. Taf. V. fig. 7, i.), *Bulimus radiatus*, and *Physa fontinalis* (*Paasch*, Ibid. 1845, I. Taf. V. fig. 12 nl. 13, i.).

According to *Leuckart* (Zur. Morphol. u. Anat. d. Geschlechtsorg., 1847, p. 128), the hermaphrodite gland of the Gasteropoda has a common excretory duct, and the eggs reach it by passing through the walls of the testicular follicles.*

9 This albumen-gland which formerly has been taken, sometimes for a testicle, and sometimes for an ovary (see above), has lately been designated

ties solely into the *Vas deferens*. It is composed of tortuous, tubular, simple follicles, lined with short, thick, pyramidal epithelia, which are densely granular, and contain a round, nucleolated nucleus. — ED.

The Uterus is very often a long, large canal, with transversely plicated glandular walls; it is distinctly separated from the ensuing vagina,[10] but often, also, it is only a simple dilatation of the oviduct,[11] which is sometimes insensibly continuous with the vagina.[12] This last communicates usually with the excretory duct of a pyriform vesicle, which, as a *Receptaculum seminis*, is filled with fresh sperm directly after the epoch of procreation.[13] This vesicle has, moreover, sometimes a lateral, caecal deverticulum.[14]

as a muciparous or an uterine gland. It is tongue-shaped with nearly all the Pulmonata (see the figures of *Cuvier*, *Treviranus*, *Erdl*, *Paasch*, &c. loc. cit.). It is a round, glandular body with *Thetis*, *Tritonia*, *Umbrella*, and *Gasteropteron* (*H. Meckel*, loc. cit. Taf. XV. fig. 1, 12, 15, 17). According to *Kölliker*, there is with *Rhodope*, and *Lissosoma*, a similar gland annexed to the uterus; and without hesitation I should pronounce as of the same nature, the glandular body which *Allman* (loc. cit. Pl. VI. 7.) has regarded as a testicle with *Actæon*. With *Doris*, *Aplysia*, and *Diphyllidia*, it is a twisted knotted tube (*H. Meckel*, loc. cit. Taf. XV. fig. 2, 7, 16).*

10 Such is the case with the Pulmonata (see the figures of *Cuvier*, *Treviranus*, *Erdl*, *Paasch*, &c.). Undoubtedly the glandular walls of this uterus secrete the calcareous crystals which incrust the eggs of many Helicina (see *Turpin*, Analyse microscop. de l'oeuf du limaçon, in the Ann. d. Sc. Nat. XXV. 1832, p. 426, Pl. XV.), or which supply the gelatinous substance enveloping in the form of a cylinder or a disc the eggs of the Lymnaeacea (*Pfeiffer*, Naturg. deutsch. Land-und Susswässer. Mollusken. Abth. I. Taf. VII. VIII.).

11 With the Pteropoda, the common excretory duct of the hermaphrodite gland, before passing into the vagina, has one or two dilatations, the inferior of which corresponds perhaps to an uterus (*Van Beneden*, Exerc. zoot. loc. cit. Pl. III. fig. 18, e. IV. A. fig. 6, d. and B. fig. 4, d. *Hyalea*, *Cleodora*, and *Cuvieria*). With *Clio*, *Cymbulia*, and *Limacina*, it is not yet determined whether the dilatation which is here found belongs to the deferent canal or to the oviduct, and therefore the name of uterus cannot be given to it.

12 The uterus is short and is directly continuous with the vagina with the Nudibranchia, Infero-branchia, Tectibranchia (*H. Meckel*, loc. cit. Taf. XV.), and perhaps also with the Apneusta. I am yet undecided if in this undeveloped uterus are formed the envelopes which, in the form of a riband, a cord, or a capsule, surround the eggs of the Nudibranchia, the Tectibranchia, and the Apneusta. Thus with *Aplysia*, *Doris*, *Tritonia*, &c., their spawn has the form of a riband or cord; and with *Glaucus*, and *Actæon*, it is wound in a spiral manner about various objects; while with *Tergipes*, it is attached to marine plants under the form of kidney-shaped capsules with short pedun-cles. With *Tritonia*, *Aeolis*, and *Aplysia*, there is observed the remarkable fact that there are several vitelluses each surrounded by an albuminous layer, in one and the same envelope; see *Sars*, in *Wiegmann's* Arch. 1837, I. p. 402, 1840, I. p. 196, Taf. V.-VII.; *Van Beneden*, Ann. d. Sc. Nat. XV. 1841, p. 123, Pl. I.; and *Loven*, in Isis, 1842, p. 359.

13 This *Receptaculum seminis* was formerly designated under the name of pedunculated vesi-cle; although *Treviranus* regarded it as an urinary bladder, and, with *Arion*, erroneously assigned to it a communication with the kidney (Zeitsch. f. Phys. I. p. 10). However, there can now be no further doubt as to its nature, for if its contents are examined shortly after coition, they will easily be found to consist of fresh sperm containing fully-developed, active, spermatic particles. Later than this, when the eggs have been deposited for a time, the sperm will be found to have lost its freshness, and to have changed into a viscous granular substance of a reddish or brown color, containing sometimes traces of dead, rigid spermatic particles. The resemblance of this matter then to excrement is, without doubt, the reason why this organ has been compared to an urinary bladder, or con-founded with the sac for purple (kidney) of other Cephalophora.

With the Pteropoda, this organ is a pyriform vesicle with a short peduncle, — at least with *Clio* (*Eschricht*, loc. cit. Tab. III. fig. 23, s.), *Cymbulia*, and *Limacina* (*Van Beneden*, Exer. zoot. loc. cit. Pl. I. fig. 17, d. V. fig. 12, A., where this organ is figured as a sac for purple). The Ap-neusta, also, have a sac for fecundation; at least I can give no other name to a long-pedunculated, pyriform vesicle which *Nordmann* (loc. cit. p. 40, Tab. II. L., III. fig. 5, b. d.) has described as a testicle with *Tergipes*; and so much the more as he always found perfect spermatic particles, and not developing seminal cells.

The pedunculated vesicle with its semi-liquid contents, which *Allman* (loc. cit. p. 152, Pl. VI. 3.) has observed with *Actæon*, is also a *Receptaculum seminis*.

According to *Kölliker*, this organ exists also with *Flabellina*, and *Rhodope*, as a pedunculated vesicle communicating with the lower extremity of the vagina. The excretory duct of this organ is short with *Thetis* (*Cuvier*, loc. cit. fig. 7, c. ; *Delle Chiaje*, Descriz. loc. cit. Tav. XLVII. fig. 1, s.), *Aplysia* (*Cuvier*, loc. cit. Pl. IV. 7 ; *Delle Chiaje*, Memor. loc. cit. Tav. IV. fig. 1, p.), and *Pleurobranchaea* (*H. Meckel*, loc. cit. Taf. XV. fig. 5, n. fig. 1, q. 7. o.). It is longer with *Scyl-laea*, *Bulla*, *Bullaea* (*Cuvier*, loc. cit. fig. 5, l. fig. 10, i.), *Doridium*, *Tritonia*, *Umbrella*, *Diphyl-lidia* (*H. Meckel*, loc. cit. Taf. XV.), and *Notar-chus* (*Delle Chiaje*, Descriz. loc. cit. Tav. LXIV. fig. 5, n.). In the Pulmonata, this peduncle is very long with *Helix*, and *Clausilia*; less so, with *Lymnaeus*, *Planorbis*, *Bulimus*, and *Physa*, and pretty short with *Limax*, *Arion*, and *Suc-cinea* (see the figures of *Cuvier*, *Wohnlich*, *Treviranus*, *Erdl*, and *Paasch*, loc. cit.).†

14 This deverticulum exists with many Helicina.

De St. Simon (Observations sur l'organe de la Glaire des Gastéropodes terrestres et fluviatiles, in the Jour. de Conchol. 1853, p. 1); this author is very minute in his details on the color, form and size of this organ, with these animals. — ED.

† [§ 227, note 13.] This receptacle is the organ called *genital bladder* by *Leidy*, and which he

* [§ 227, note 9.] For the muciparous appa-ratus with the Nudibranchiata, see *Alder* and *Hancock*, loc. cit. Part II. Pl. IV. fig. 15, (*Doto*); Part III. Pl. VIII. fig. 2, g. g. (*Eolis*); Part IV. Pl. V. fig. 8, i. (*Eumenis*); Part V. Pl. II. fig. 7, h. h. '*Doris*); also *Hancock*, Ann. Nat. Hist. VIII. 1851, p. 34, Pl. III. fig. 6, g. (*Antiopa*). See also

Underneath the point of insertion of this vesicle upon the vagina, are various glandular appendages which open into this last or into the genital cloaca. But as yet their function is unknown. With the Pteropoda, and Heterobranchia, there is a single appendage only, consisting of a simple tube.[15] To this same category belongs, also, the dart-sac, — a very remarkable cylindrical organ opening into the genital cloaca.[16] Its walls are quite thick, and on each side of its base is a group of more or less numerous cæca.[17] At the bottom of this sac is a conical papilla which secretes a calcareous concretion of the form of a lance-head with the point downwards, — the Dart. This is projected during copulation, and often remains sticking in the skin near the genital opening. Its loss is subsequently replaced by the secretion of another in the same place.[18]

The male copulatory organs consist of a more or less long, projecting Penis, which, when at rest, is either retracted freely between the other viscera of the cavity of the body, or enveloped wholly or in part in a proper sheath (*Præputium*).

This penis consists, nearly always, of a hollow fleshy cylinder, which is usually closed at its posterior extremity, and has, behind, a long flagelli-

It is very long with *Bulimus radiatus, Helix arbustorum, lactea,* and *vermiculata ;* very short, on the other hand, with *Helix pomatia, nemoralis,* and *candidissima.* It is entirely wanting with *Helix fruticum, strigella,* and *rhodostoma.* With *Helix algira,* it communicates directly with the seminal sac (see the figures of *Erd!,* and *Paasch*). With *Doris,* the *Receptaculum seminis* has a peculiar structure ; it is kidney-shaped, and from its concavity arises a very large excretory duct, arcuate, and opening into the genital cloaca, which has not only a cæcal appendage, but also a short canal that communicates with the base of the uterus (*H. Meckel,* loc. cit. p. 496, Taf, XV. fig. 2). Further research must determine if the canal which *Nordmann* (loc. cit. p. 50, Tab. III. fig. 5, d.) has observed upon the seminal sac of *Tergipes* without being able to trace it to its extremity, is a simple deverticulum, or a canal communicating with the female genital organs.

15 An analogous appendix, of a round form, has been described with *Cymbulia,* and *Limacina,* as a prostate by *Van Beneden* (Rech. zoot. loc. cit. Pl. I. fig. 17, e. V. fig. 12, B.), and as a testicle by *Eschricht,* with *Clio* (loc. cit. Tab. III. fig. 25, 26). There is a long glandular appendix upon the genital cloaca with *Doridium, Pleurobranchaea,* and *Diphyllidia* (*H. Meckel,* loc. cit. Taf. XV.). As yet the function of this gland is only hypothetical. Perhaps it furnishes the viscid substance enveloping the eggs during their deposition, or it may be a copulatory pouch (*Bursa copulatrix*). But it is quite probable that the penis enters the peduncle of the *Receptaculum seminis* during copulation, for with most Cephalophora the penis and the peduncle are of the same length.

16 The dart-sac, which is more or less long, is found with many species of *Helix.* It is absent with *Helix algira, candidissima, cellaria,* and *verticillus.* It is double with *Helix ericetorum,*

and with *Helix strigella,* is replaced by two very long cæca ; see *Wohnlich, Erdl,* and *Paasch,* loc. cit.

17 Two considerable groups of dichotomously ramified cæca are found with *Helix pomatia, adspersa, austriaca, lactea, naticoides,* and *vermiculata,* while with *Helix umbrosa, strigella,* and *striata,* there are only four cæca on each side. With *Helix incarnata,* and *nemoralis,* there are three, and two only with *Helix lapicida, arbustorum,* and *personata ;* see *Cuvier, Wohnlich, Erdl, Paasch,* loc. cit. and *Wagner,* Icon. zoot. Tab. XXX. fig. 11, 12. As to the use of these glandular tubes, I would suggest the view that they secrete a coagulable substance, which, during the coition, envelops the sperm like a spermatophore to conduct it into the seminal sac. Indeed, I am much inclined to regard as the debris of a spermatophore the thin horny bodies of a peculiar aspect, which, with *Helix hortensis, arbustorum,* and *nemoralis,* often project out of the genital cloaca after fecundation, and which, when they have left it, are rolled in a spiral form at both extremities. When carefully examined they will be found composed of several layers of coagulated albumen, and to be involved in the peduncle of the *Receptaculum seminis ;* see *Huschke,* in *Meckel's* Arch. 1826, p. 620, Taf. VII. fig. 9, and *Carus,* in *Muller's* Arch. 1835, p. 495, Taf, XII. fig. 4–7.

18 The dart is hollow and of the form of a cylindrical stylet with *Helix ericetorum,* and *striata ;* but with *Helix pomatia, hortensis,* and *adspersa,* four, sharp denticulated edges, extending its whole length give it a very elegant form ; see *Prevost,* in Mém. de Genève, loc. cit. V. p. 121, Pl. I. fig. 7, and *Carus,* in *Muller's* Arch. 1835, p. 494, Taf. XII. fig. 9, 12.

It is probably an excitatory organ, for the snails reciprocally prick each other before coition.*

* [§ 227, note 18.] For the relations of the dart-sac with the American Helices, see *Leidy,* loc. cit. He adds, "The dart-sac and multifid vesicles, so common in European species, are very rare in American species." The dart-sac has been found in only four species. — ED.

has so well figured. He found its contents to be spermatic particles, but, in regard to its being a seminal receptacle, he remarks : "This, however, cannot be considered wholly as its use ; for it secretes a mucoid matter which may probably facilitate the passage of the ova through the vagina and cloaca," p. 234. — ED.

form prolongation (*Flagellum*). In many of the genera of the Gasteropoda, the *Vas deferens* is inserted upon the penis near its base, or at the posterior end of its cavity.[19] The penis has also inserted into it many small retractor muscles which arise from the walls of the envelope of the body, or on the columella.[20]

The external orifices of these hermaphroditic genital organs are usually on the right side, and present the following relations : 1. The vagina and penis open into a common genital cloaca which communicates externally upon the sides of the anterior part of the body.[21] 2. The two orifices are situated side by side, — that of the penis directly in front of that of the vagina.[22] 3. The orifices are quite removed from each other, and then the penis, which is usually concealed beneath the testicle of the right side, communicates with the genital cloaca situated behind, by a groove which runs along the sides of the body.[23] This groove is lined with ciliated epithelium, and, without doubt, conducts the semen from the genital cloaca to the penis, during copulation.

§ 228.

The Cephalophora with which the sexes are separate, may be divided into two sections, in one of which, the copulatory organs are wanting, while in the other, they are highly developed.

1. To the first section, belong the Cyclobranchia, the Scutibranchia, and also, probably, the Tubulibranchia and Cirribranchia; with all of which, the genital glands are easily seen at the epoch of procreation, from the presence of sperm or of eggs.[1]

19 The penis is short, and of a compact form with the Pteropoda ; see the figures of that of *Cymbulia, Tiedemannia, Hyalea, Cleodora, Cuvieria,* and *Limacina,* in *Van Beneden,* Exerc. zoot. loc. cit. *Clio,* however, forms an exception in this respect, its penis being long and flexuous (*Eschricht,* loc. cit. Tab. III. fig. 24). With the Apneusta, this organ is pretty long, spiral-form, and concealed in a pyriform sac, and the *Vas deferens* is inserted at its base ; see *Allman,* loc. cit. Pl. VI. t. (*Actaeon*), and *Nordmann,* loc. cit. Tab. III. fig. 5, p. q. r. (*Tergipes*). There is a similar disposition with *Thetis, Tritonia, Doris,* and *Pleurobranchaea* (*H. Meckel,* loc. cit. Taf. XV.). In the last-mentioned genus, it is distinguished for its extraordinary length. That of *Arion, Limax, Succinea, Lymnaeus, Planorbis, Physa, Clausilia, Helix cellaria,* and *fruticum,* is thick, very short, and unites either abruptly or gradually with the *Vas deferens;* while that of *Bulimus,* and most species of *Helix,* ends posteriorly in a long lash which projects freely into the cavity of the body, and upon which is inserted the deferent canal at a variable distance from the extremity (*Wohnlich, Treviranus, Erdl, Paasch,* loc. cit.). With *Onchidium, Bullaea,* and *Gasteropteron,* the penis has a very long, flexuous lash, which, with *Aplysia,* and *Pleurobranchus,* is shorter, but never in connection with the *Vas deferens* (*Cuvier,* and *H. Meckel,* loc. cit.).

20 These retractor muscles are inserted at the posterior extremity of the penis with *Arion, Limax,* and *Planorbis;* and more in front and on the sides with *Lymnaeus,* and *Helix* (*Wohnlich, Erdl,* and *Paasch,* loc. cit.).

21 Such a common genital orifice is found with *Helix, Limax, Arion, Succinea, Bulimus,* and *Clausilia,* on the right side of the neck behind the tentacles ; it is situated further behind, but always

on the right side, with *Æolis, Tergipes, Scyllaea, Doris, Tritonia, Thetis, Pleurobranchus, Pleurobranchaea,* and *Diphyllidia.*

22 With *Planorbis,* and *Physa,* the male and female orifices are situated on the left side of the neck behind the tentacle ; with *Flabellina, Rhodope, Cleodora,* and *Cuviera,* a little further behind on the right side.

23 With most of the Pteropoda (*Clio, Cymbulia, Tiedemannia, Hyalea,* and *Limacina*), the orifice of the penis is in the neck, and that of the genital cloaca a little further behind on the right side. With *Actaeon,* and *Lissosoma,* the two orifices are also on the right side, but even more widely separated from each other. With *Gasteropteron, Bulla, Bullaea,* and *Aplysia,* the genital cloaca is very far behind, while the penis is under the right tentacle. With *Doridium,* the cloaca is also quite in the rear, but upon the left side, and consequently the penis is under the tentacle of the same side. But with *Onchidium,* these orifices are the widest apart, — the cloaca opening close by the anus, and the penis under the right tentacle. The furrow passing from the cloaca to the penis is found with all these Gasteropoda, and it is very probable that it will be found also with all the other Cephalophora, whose penis is entirely removed from the other male genital organs. The furrow which *Van Beneden* (Exerc. zoot. Fasc. II. p. 46) observed with a *Hyalea,* between the two genital orifices, shows that there is such a communication with the Pteropoda also.

1 The separation of the sexes with *Chiton, Patella,* and *Haliotis,* was first shown by R. *Wagner* and *Erdl* (*Froriep's* neue Notiz. No. 249, 1839, p. 102). It has been confirmed with *Patella,* by *Milne Edwards* (Ann. d. Sc. Nat. XIII. 1840, p. 370), and by *Robin* and *Lebert* (Ibid. V. 1846, p. 191). With many individuals of *Vermetus gigas,* I have

In the genus *Chiton*, the male and female genital gland is long and lobulated; it lies over the other viscera, and, from each side of its posterior extremity, passes out a short excretory duct which opens upon the border of the mantle.[2]

With *Patella*, and *Haliotis*, this gland is covered by the liver, and its single duct passes in front and opens near the anus, at the right with the first of these genera, and at the left with the second.[3]

2. In the second section, there is a protractile penis with various Heteropoda, all the Pectinibranchia,[4] and operculate Pulmonata.

The Ovary, or the testicle, always lies concealed at the base of the visceral sac between the liver, and its excretory duct, as *Tuba Fallopii* or *Vas deferens*, passes on to and accompanies the rectum during the remainder of its course.

The oviduct opens near and often a little behind the anus, and, with the Heteropoda, has frequently several glandular appendages; [5] while, with the Gasteropoda, the portion accompanying the rectum is dilated into a kind of uterine tube which has glandular walls.[6] From the walls of this tube are secreted, without doubt, the often very regular envelopes with which the eggs of many Pectinibranchia are surrounded.[7] In this last-mentioned order, there has as yet very rarely been found an albumen-gland or a receptacle of the sperm which communicates with the uterus.[8]

found, in the posterior region of the body, and in the greenish-brown liver, a yellowish-brown glandular body, containing active spermatic particles, and very large caudate cells enclosing undeveloped spermatic particles, from which passed off a long excretory duct opening near the anus, without the appearance of any penis. This apparatus is undoubtedly a male genital organ.

The other individuals, in which I could find no spermatic particles, were the females. The details by *Ruppell* (Mém. d. l. Soc. d'Hist. Nat. à Strasbourg, I. p. 3, fig. 4), and by *Carus* (Museum Senckenberg, II. p. 199, Taf. XII. fig. 8) upon the genital organs of *Magilus antiquus*, render probable the separation of its sexes also ; but it is doubtful if the males have a penis, as *Ruppell* says, for it is difficult to comprehend how copulation can take place with this animal which lives buried in the coralla of the Madreporina, any more than with the *Vermetus* which are fixed upon stones. But *Carus* declares that he has seen, instead of a penis, an indistinct papilla on the neck of *Magilus*. The ovary, which, according to *Deshayes* (loc. cit. p. 334, Pl. XV. fig. 8, f., or Isis, 1832, p. 469, Taf. VI. fig. 12, f.), fills almost entirely the cavity of the body with *Dentalium*, will probably, after more careful research, prove, with many individuals, to be a testicle.

2 See *Cuvier*, Mém. loc. cit. p. 24, Pl. III. fig. 10, 13, or Isis, 1819, p. 734, Taf. XI. fig. 10, 13.

3 See *Cuvier*, Mém. loc. cit. p. 12–18, Pl. II. fig. 11, e. 14, 15, or Isis, 1819, p. 728, 731, Taf. XI. fig. 11, e. 14, 15.

4 The genus *Littorina* is the only one which contains hermaphrodite species ; here the voluminous penis, having a longitudinal furrow, projects under the right tentacle (*Quoy* and *Gaimard*, Voy. de l'Astrolabe, Zool., or Isis, 1834, p. 209).

5 The genital organs of the Heteropoda are yet little known, and what has been said in the text relates only to *Carinaria*. Among the two to four deep-colored appendages of the vagina of *Carinaria mediterranea*, may be especially distinguished a spiral tube containing internally transverse glandular folds (see *Delle Chiaje*, Memoir. II. p. 208, Tav. XV. fig. 5, 6, and Descriz. II. p. 97.). These appendages, the existence of which I have verified

with individuals preserved in alcohol, must be more carefully studied before it can be decided if they are the analogue of an uterus, seminal sac, &c.

6 See *Cuvier*, Mém. loc. cit. fig. 2, 3, h. ; *Treviranus*, Zeitsch. f. Physiol. I. p. 32, Taf. IV. fig. 21 ; *Paasch*, in *Wiegmann's* Arch. 1843, I. p. 100, Taf. V. fig. 8 (*Paludina vivipara*) and *Leiblein*, in *Heusinger's* Zeitsch. I. p. 32, Taf. I. fig. 6 (*Murex*). *Quoy* and *Gaimard* have furnished many facts on this point (loc. cit., or Isis, 1834, 1836). With *Strombus lambis*, they have described a furrow which arises from the female genital orifice, and passes along the right side of the foot.

7 These envelopes or capsules filled with eggs are cylindrical, pyriform, infundibuliform, and sometimes pedunculated. They are attached singly or in groups to objects, and sometimes are aggregated in considerably-sized masses around a common axis. Often they open by a special fissure, which, in some species, has a particular operculum ; see *Lund*, in Ann. d. Sc. Nat. I. 1834, p. 84, Pl. VI., or *Froriep's* Not. No. 881, 882, and *D'Orbigny*, in Ann. d. Sc. Nat. XVII. 1842, p. 117. Such a mass, arranged around an axis, in which the eggs of *Janthina* are deposited, and which is carried about with them a long time attached to their foot, was long regarded as an enigmatical body under the name of *Spuma cartilaginea*, and, by some naturalists, has been even considered as a modified operculum of the shell ; see *Lund*, loc. cit. fig. 23 ; *Lesson*, in the Voy. de la Coquille. Zool. II., or Isis, 1833, p. 134, Taf. I. fig. 1 ; and *Delle Chiaje*, Descriz. II. p. 108, Tav. LXVII. fig. 1, 2.

8 With *Paludina vivipara*, there is an albumen-gland beneath the last convolution of the intestine (See *Treviranus*, loc. cit. p. 31, Taf. IV. fig. 21, u., and my observations in *Muller's* Arch. 1836, p. 245). In this same species, the bottom of the uterus communicates by a large orifice with a sessile *Receptaculum seminis* in which I have always found, after copulation, numerous active spermatic particles (*Muller's* Arch. 1836, p. 244). This sac for fecundation appears to be absent with all the other Pectinibranchia, and *Berkeley* (Zool. Jour. II. 1829, p. 278, or Isis, 1830, p. 1264) could not find it with the females of *Cyclostoma*.

The course of the seminal duct, and that of the oviduct also, is the same as that of the uterus, until it reaches the extremity of the rectum, when it passes into the penis which always projects from the right side of the body. With the Heteropoda, the penis is often bifid, but then the seminal canal does not traverse except one of its divisions.[9]

With the Gasteropoda, the penis is either very long,[10] tongue-shaped and often flexuous,[11] or short and lanceolate.[12] It projects under and usually behind the right tentacle, and extends upon the side of the body — rarely being in a wholly retracted state, but is capable of being easily folded under the border of the mantle.

With some genera, its extremity has a small hook.[13] With several Pectinibranchia, the seminal canal terminates behind the anus, and then takes the form of a furrow, which communicates with the base of the penis, extending even to its extremity, either as an external, or an internal semicanal.[14]

§ 229.

The development of the Cephalophora has, as yet, scarcely been observed except with the Gasteropoda, and in particular with the Apneusta, the Heterobranchia, and the Pulmonata.[1]

All observations concur as to the fact that the vitellus undergoes a regular and complete segmentation,† after which, there appears an usually long, round embryo, one of the poles of which is indented and covered with

9 The penis is double and on the right side at the base of the visceral sac, with *Carinaria* and *Pterotrachea* (*Milne Edwards*, Ann. d. Sc. Nat. XIII. 1840, p. 195, XVIII. p. 323, Pl. X. fig. 3). *Quoy* and *Gaimard* (Voy. de l'Astrolabe, Mollusq. Pl. XXVIII. fig. 10, or Isis, 1834, Taf. III. fig. 10) have figured a long bifid penis with *Phyllirrhoë amboinensis*; and so, if with the other Heteropoda the penis is not retractile, as appears to be the case with *Carinaria*, according to *Milne Edwards*, this species would be a male, while *Phyllirrhoë bucephalus*, figured by *Péron* (Ann. du Muséum XV. fig. 1, or *Kosse*, or Pteropodum ordine. Diss. fig. 1), apparently without a penis, would be a female, although *D'Orbigny* (Voy. dans l'Amér. mér., or Isis, 1839, p. 519,) regards this genus as hermaphrodite. With *Atlanta*, there is a simple, pointed penis on the right side of the neck directly near the arms; but as *Rang* (Mém. loc. cit. p. 378, Pl. IX. or Isis, 1832, Taf. VII.) has found this penis with all the individuals he has examined, it may be questioned if the sexes are really separate with this Heteropod.

The internal genital organs of *Atlanta*, and *Phyllirrhoë*, should be thoroughly studied for the elucidation of this point.*

* [§ 228, note 9.] See *Gegenbauer* (*Siebold* and *Kölliker's* Zeitsch. IV. p. 233), who has described some follicular penis-glands with *Littorina*, and which serve some purpose in the copulatory act. — ED.

† [§ 229.] Upon the vitellus of various Gasteropoda, there appears at the time of its segmentation, a small round, colorless body, resembling a vesicle. This was first mentioned by *Pouchet* (Ann. d. Sc. Nat. 1838, X. p. 63) and has since attracted the attention of *Van Beneden* in his embryology of *Aplysia depilans* (Ann. d. Sc. Nat. 1841, XV. p. 126). Quite recently, the subject has

10 For the male genital organs of the Pectinibranchia, see especially the works of *Cuvier*, and of *Quoy* and *Gaimard*, loc. cit.
11 *Buccinum, Murex, Dolium, Harpa, Ampullaria, Mitra, Littorina, Strombus, Cyclostoma*.
12 *Janthina, Eburnea, Conus*, &c.
13 *Cassis, Dolium, Buccinum, Strombus, Sigaretus,* and *Paludina*. With *Paludina vivipara*, the penis is, moreover, so united to the right tentacle, that this last appears to be a detached prolongation from the inferior surface of its apex (*Treviranus*, loc. cit. Taf. IV. fig. 18).
14 With *Dolium, Harpa, Ampullaria, Tritonium, Strombus,* &c., this semi-canal extends even to the end of the penis (*Quoy* and *Gaimard*, loc. cit.); while with *Murex*, it ceases at the base of this organ (*Leiblein* in *Heusinger's* Zeitsch. I. p. 31, Taf. I.).

1 With the Cephalophora, the embryonic development does not generally begin until after the eggs have been deposited. A few only of the Gasteropoda, and among them *Paludina vivipara*, and *Clausilia ventricosa* (*Held*, Isis, 1834, p. 1001), are viviparous.

been brought up by Fred. *Müller* (Zur Kenntniss des Furchungsprocesses im Schneckeneie, in *Wiegmann's* Arch. 1848, p. 1) who ascribes to it a great importance in the primitive developmental changes of the ovum, and has called it the directive vesicle (*Vesicula directrix*, or Richtungsbläschen). But the special importance of this body seems not yet well made out, and it may be questioned if it is not rather a secondary formation, than a primitive, directive organ. See *H. Rathké* (*Wiegmann's* Arch. 1848, p. 157) and *Gegenbauer* (*Siebold* and *Kölliker's* Zeitsch. III. 1852, p. 373). — ED.

a delicate ciliated epithelium. By the means of these cilia, the embryo rotates upon its axis for a long time. From this period, the aquatic differ widely from the pulmonate Gasteropoda. With the Apneusta, and the Heterobranchia, the two lobules produced by the indentation just indicated, enlarge and change into round pinions (*Vela*), upon whose borders very long cilia are gradually developed. A third eminence is developed between these two pinions, and, ultimately, changed into the foot.

Although the ciliated epithelium is always most widely spread around these two pinions, which should be regarded as situated on the anterior extremity of the body, yet there is formed a thin shell upon the posterior extremity of the embryo, whether this last belongs to a conchiferous species or not. At the same time, there appears upon the dorsal part of the foot, an operculum corresponding as to size with the opening of the shell.

Among the internal organs, the two auditive capsules appear first; and when these have become quite distinct, the eyes are seen. Following these, are developed the tentacles, the border of the mantle, and the mouth which appears between the two pinions. At the same time, the stomach, the intestine, and the liver, individually appear in the interior. At this epoch, the young leave the egg and swim freely about by means of the long cilia which are situated on their extended and rigid pinions.[2] Subsequently these pinions disappear, or are changed into two tentacular prominences situated on each side of the mouth.[3] At the same time, also, the naked Gasteropoda lose their shell and operculum. From the isolated facts hitherto published upon the embryology of other branchiferous Gasteropoda, it may be concluded that they experience a similar metamorphosis, only the shell of the embryo, at this time, usually presents some convolutions.[4]

In the development of the operculate Pulmonata, there is no analogous metamorphosis.[5] The embryo lengthens a little when it begins to rotate

2 The embryology of the Apneusta, and the Heterobranchia owes its progress principally to the following works : Sars, in *Wiegmann's* Arch. 1837, I. p. 402; 1840, I. p. 196, Taf. V.–VII. 1845, I. p. 4, Taf. I. fig. 7–11 (*Tritonia, Doris, Aplysia,* and *Aeolis*); *Lovén,* in the Kongl. Vetensk. Akad. Handl. 1839, p. 227, or Isis, 1842, p. 360, Taf. I. (*Aeolis*); *Van Beneden,* Ann. d. Sc. Nat. XV. 1841, p. 123, Pl. I. (*Aplysia*); *Nordmann,* loc. cit. p. 71, Taf. IV. V. (*Tergipes*); *Allman,* loc. cit. p. 152, Pl. VII. fig. 10–12; *Vogt,* Compt. Rend. XXI. 1845, No. 14, XXII. No. 9, or Froriep's neue Not. No. 795, 820 (*Actaeon*); and *Reid,* Ann. of Nat. Hist. XVII. 1846, p. 377, Pl. X. (*Doris* and *Polycera*). *Vogt* has since published his entire Memoir on the development of *Actacon viridis* in the Ann. d. Sc. Nat. 1846, p. 5, Pl. I.–IV. ; see also *Schleiden,* and Froriep's Not. II. p. 77, fig. 1–12.

3 These remains of the two pinions are easily seen with *Tergipes. Aeolis, Doris, Tritonia, Aplysia,* and other Heterobranchia. The ciliated lobes on the head of *Thetis,* are only these pinions persisting in an embryonic form; see *Lovén,* loc. cit.

4 According to *Lovén* (loc. cit. or Isis, 1842, p. 366, Taf. I. fig. 22), the young of *Rissoa* have a very large pinion. *Nordmann* (loc. cit. p. 98) has confirmed this, and found an analogous one with *Littorina,* and *Phasianella.* The small Mollusks with a pinion and a turbinated shell, of which Sars (Beskrivel. loc. cit. p. 77, fig. 38, 39) has formed the genus *Cirropteron,* have since been found by himself to be young individuals of *Turbo, Trochus,* or *Nerita ;* this accords with Grant's observations (Edinb. new Philos. Jour. No. 13, 1827) upon *Tur-*

bo, Nerita, Buccinum, and *Purpura.* Judging from *Carus'* figure (Nov. Act. Acad. Nat. Cur. XIII. 1827, p. 767, Tab. XXXIV. fig. 2) of the embryo of *Paludina vivipara,* it also has at this age a pinion.

This remark is also applicable to the young animals found by *Lund* (Ann. d. Sc. Nat. I. 1834, Pl. VI. fig. 9–14) in the egg-capsules of a *Murex* (?) and a *Natica* (?). I have found, in the pyriform ovigerous capsules adhering to the orifice of the shell of *Vermetus,* young with highly-developed pinions having long cilia, and with a regularly convoluted shell, such as has been described by *Philippi* (*Wiegmann's* Arch. 1839, I. p. 128, Taf. IV. fig. 8). *Lovén* has observed similar embryos swimming with two pinions, with the Heterobranchia of the genera *Elysia, Bulla, Bullaea,* and with the Pectinibranchia of the genera *Lacuna, Cerithium,* and *Eulima ;* see Arch. Skandinav. Beitr. &c. I. 1845, p. 154, Taf. I. fig. 1–8.

5 The development of the Pulmonata which have a shell, has often fixed the attention of naturalists. See *Stiebel,* loc. cit. p. 38, Tab. II. and in *Meckel's* Arch. deutsch. I. p. 423, II. p. 557, Taf. VI. ; *Hugi,* Isis, 1823, p. 213 ; *Carus,* Von den äusseren Lebensbed. loc. cit. p. 60, Taf. I. ; *Prevost,* Ann. d. Sc. Nat. XXX. 1833, p. 40 (*Lymnaeus*) ; *Pfeiffer,* Naturg. deutsch. Land-und Süsswasser-Mollusk. Abth. III. p. 70, Taf. I. (*Helix*) ; *Quatrefages,* Ann. d. Sc. Nat. II. 1834, p. 107, Pl. XI. B. (*Lymnaeus* and *Planorbis*), Jacquemin, Ibid. V. 1836, p. 117, 119, and in the Nov. Act. Acad. &c. XVIII. 1838, p. 636, Tab. XLIX. L. (*Planorbis*) ; *Dumortier,* Nouv. Mém. de l'Acad. Roy. de Bruxelles, X. 1837, Pl. I.–IV. and Ann. d. Sc. Nat.

upon itself; its posterior extremity soon assumes a spiral form and is covered with an alveolate wrapper, upon which gradually appear the convolutions of the shell, without there being formed, at the same time, an operculum. During this period, the eyes, tentacles, border of the mantle, and the foot, appear at the anterior extremity; and, in the interior, the auditive capsules, the intestinal canal, the liver and heart are gradually developed. Here, therefore, the development of the cephalic pinions, which characterize the embryos of the Branchiata, is also incomplete.

The development of the naked Pulmonata is quite different.[6] When the round embryos begin to rotate, two crests appear side by side, upon the previously divided vitellus; one of these is changed into the shield and into the respiratory and circulatory organs situated beneath, while the other goes to form the foot. At its anterior extremity, appear the eyes, tentacles, and lips; and at the posterior extremity, a peculiar contractile vesicle is formed. This vesicle presses its contents towards the vitelline substance which is still contained in a kind of vitelline sac projecting anteriorly between the two crests, and which, also, becomes contractile.[7] By this arrangement there is an interchange of the contents of the vitelline sac and the caudal vesicle, due to their alternate contractions. Subsequently, the liver and digestive canal are formed out of the vitelline substance between the two crests. The vitelline sac and caudal vesicle are in this way considerably diminished, and, at last, wholly disappear.

The development of *Sagitta*, as far as yet known, differs essentially from that of the Gasteropoda,[8] in that its embryo is not formed at the expense

VIII. 1837, p. 129, Pl. III. IV; *Pouchet*, Ann. d. Sc. Nat. X. 1838, p. 83 (*Lymnaeus*); and *Rathké*, *Froriep's* neue Not. XXIV. 1842, p. 161 (*Lymnaeus, Planorbis*, and *Helix*).

6 See *Laurent*, in the Ann. d. Sc. Nat. IV. 1835, p. 248 (*Limax* and *Arion*); *Van Beneden*, and *Windischmann*, in the Bull. de l'Acad. roy. de Bruxelles, V. No. 5, p. 286, Ann. d. Sc. Nat. IX. 1838, p. 306, and in *Muller's* Arch. 1841, p. 176, Taf. VII. VIII. (*Limax*).*

7 This contractility shows itself quite early in the

vitellus of the Limacina, for *Dujardin* (Ann. d. Sc. Nat. VII.1837, p. 374, or, Observ. au Microsc. Atlas, 1842, Pl. V. fig. 10, 11) has seen in the eggs of *Limax cinereus* soon after their deposition, singular vitelline movements exactly resembling the alternate protrusions and retractions of the parenchyma of *Amoeba*.

8 See the observations of *Darwin*, in the Ann. of Nat. Hist. XIII. p. 4, or Ann. d. Sc. Nat. I. 1844, p. 363.†

* [§ 229, note 6.] See also *O. Schmidt* (Ueber die Entwickelung von Limax agrestis, in *Muller's* Arch. 1851, p. 278) who differs in many points from *Van Beneden* and *Windischmann*, as to the histological development of some of the organs. See, furthermore, *Gegenbaur, Siebold* and *Kölliker's* Zeitsch. III. 1852, p. 371. — ED.

† [§ 229, note 8.] See, for some of the more recent contributions to the embryology of the Cephalophora, *Koren* and *Danielssen* (Bidrag til Pectinibranchiernes Udvicklings histoire, Bergen, 1851, or its Translation into French in the Ann. d. Sc. Nat. XVIII. 1852, p. 257, and XIX. 1853, p. 89), and *Gegenbaur* (Beiträge zur Entwickelungsgeschichte der Landgastropoden, in *Siebold* and *Kölliker's* Zeitsch. III. 1852, p. 371.) These works are quite complete as far as they go, and that of *Gegenbaur*, especially, has full details upon the formation of all the organs and their mutual embryological relations. No just résumé can be given in the proscribed limits of my notes.

I cannot here well omit at least an allusion to that

most remarkable episode in the embryology of the Mollusca, the development of certain Mollusks in Holothurioidea. The facts of the case were discovered and announced by *J. Muller* (Verhandl. der Akad. zu Berlin, 1851, p. 628 (October 23), and Nachtrag, p. 679 (Nov. 13), or *in extenso* in *Muller's* Arch. 1852, p. 1), and they are indeed so wonderful that it is well they were first brought out by so reliable a physiologist and embryologist.

The main facts, briefly stated, are as follows: In certain individuals of *Synapta digitata* there are found from one to three sac-like bodies in the cavity of the body, and attached by their superior extremity to the head, and by the lower end to the intestine; but this connection of the sac with the abdominal and other organs, is one of simple contiguity and not of very direct communication. The upper portion of the sac is of a yellow, and the lower of a green color; the lower portion, moreover, is intussuscepted, with a blind end, like an inverted finger of a glove. It is in this sac-like organ that are developed true Mollusks; in the upper or more ca-

of the entire surface of the vitellus, but surrounds the last in a ring-like manner, and is gradually detached by its cephalic and caudal extremities.

pacious portion are found both male, (testes), and female (ovarium) organs in the shape of sacs, which are not attached in any way to the main molluskigerous sac. These genital organs bear no resemblance whatever to ordinary testes or ovaria, except in their products, which are identical. When the ovarium is perfectly developed, it and its capsule burst and discharge the ova which are then contained in the main molluskigerous sac; after this, fifteen to twenty ova become invested with a common capsule, though their fecundation takes place previous to this investment. Upon this succeeds their development.

The sperm-capsules vary from four to eighteen in number, and lie perfectly free in the main sac, not far from the ovary. The spermatic particles are set free by the bursting of these capsules, and they resemble those of the Gasteropod Mollusks in shape and form.

The development in the egg here proceeds exactly as with the Mollusca (e. g. *Actaeon*, according to *Vogt*), and finally it assumes pretty definite characters indicating rather its relation to the Pectinibranchia. Of its zoölogical character as a Mollusk there can, therefore, be no doubt, and the whole story in a word is, that a true Mollusk is developed within a *Synapta*, not by gemmation, but by means of the normal sexual products which occur under otherwise amorphous and anomalous parts and conditions. It should, moreover, be remarked that the connection of this molluskigerous sac is not special or direct with the *Synapta*, but this last appears to serve as a kind of nest in which the Mollusk carries out its ulterior and remarkable changes.

Such being the facts, the question now arises, What interpretation shall be given these phenomena? The distinct sexual mode of reproduction would seem to remove these phenomena from the category of the so-called alternation of Generation, or gemmiparity as we now understand it. Then again, the doctrine of " heterogeneous generation " as suggested by *Muller*, does not seem to me admissible, beside being particularly unsound, — for if an animal can produce, by true sexual generation, an offspring zoölogically dissimilar to itself, zoölogists may well look about for the stability of their science. If I may be allowed an opinion or rather a view on a subject on which I have made no observations, I would say that an approximate solution of this enigma seems obtained by admitting the possibility of new and hitherto unknown parasitic conditions in the life of the Mollusk in question.

Why may not this Mollusk undergo a form of retrograde metamorphosis during which its life is parasitic and very peculiarly connected with the life of another and wholly different animal? Or again, why may not the phenomena observed be the final conditions of certain low modes of life which are connected with points in the economy of these animals that we do not yet understand? I throw out these remarks in a suggestive way. If we refer for a moment to the historical relations of the Cestodes, it will be perceived that there was a time when the conditions of their life were equally if not more obscure. *Siebold*, however, has shown that here, although the path taken by Nature is circuitous and intricate, yet, after all, no new features of a heterogeneous nature are introduced, and that all required for the observer was care and patience. It does not seem to me any more improbable that this Mollusk should have entered in some of its stages the body of the *Synapta*, since the anomalous undeveloped forms of many Helminthes pursue a similar course. Let the naturalist also bear in mind the remarkable phenomena of the Hectocotyli. In the *Nachtrag* to this first account before the Berlin Academy, but more especially in a subsequent and more complete account (Ueber die Erzeugung von Schnecken in Holothurien, in *Müller's* Arch. 1852, p. 1) lately given, *Müller* discusses still further these facts. After some remarks upon the importance of a careful study of the embryology of this curious form, he says : " I do not give up the hope that we may yet determine at least the genus of this Mollusk ; and I found this hope mainly upon the very characteristic form of the spermatic particles, beside the other features above mentioned. . . . . The spermatic particles of *Natica* and its allies are yet unknown. . . . . . In studies bearing upon this matter, one should particularly bear in mind the terminal enlargement of the spermatic particles, which up to this time has been observed in no Gasteropod, but which with the spermatic particles of the Mollusk in question is never wanting." Although for some time familiar with the details of the spermatic particles of the Gasteropod Mollusks, yet I have very recently reëxamined the spermatic particles of *Natica* (*N. heros*) with reference to this point. They resemble closely those of the pulmonary Gasteropoda (*Helix*, for instance), and consist of a well-defined cork screw head to which is attached a very delicate tail ; they agree, therefore, in general with the form given by *Müller* of the Mollusk in question. — Ed.

BOOK ELEVENTH.

CEPHALOPODA.

CLASSIFICATION.

§ 230.

The Cephalopoda present, in their organization both internal and external, so many peculiarities which distinguish them from all the other Mollusca, that it is necessary to consider them in a class by themselves, although their genera are not numerous.

It is, moreover, necessary to state why we here regard the different forms of *Hectocotylus* which hitherto have been considered as parasites of these animals, as the males of certain Octopoda.[1] The researches of *Kölliker* have led us to make this change. This naturalist founds his opinion upon the following convincing reasons:[2] The specimens of *Hectocotylus* have branchiae, and a heart with arteries and veins, and they cannot, therefore, be regarded as Helminthes. On the other hand, they have, in common with the Cephalopoda, the contractile chromatophoric cells of the skin, and the same kind of spermatic particles and suckers; and the muscular substance of their body is arranged exactly like that of the arms of the Cephalopoda. All of them are males, and the Cephalopoda, with which they are connected, are all females; finally, the embryos found in the eggs of certain Octopoda exactly resemble them. Whoever has had the opportunity of examining the species yet known, viz: *Hectocotylus argonautae, octopodis,* and *tremoctopo-*

---

1 At present there are known two or three species of these singular beings resembling the torn-off arms of the Octopoda, and which live in the cavity of the mantle of certain Octopoda, attached by the means of suckers. *Hectocotylus argonautae* was first described quite imperfectly by *Delle Chiaje* (Memor. &c. II. p. 225, Tav. XVI. fig. 1, 2, and Isis, 1832, Taf. X. fig. 12, a. b.) under the name of *Trichocephalus acetabularis.* Another description by *Costa* (Ann. de Sc. Nat. XVI. 1841, p. 181, Pl. XIII. fig. 2, a.–c.) has not added much to our knowledge of the real nature of this animal. Another species, *Hectocotylus octopodis,* established by *Cuvier* (Ann. d. Sc. Nat.

XVIII. 1829, p. 147, Pl. XI. A. fig. 1 5, or *Froriep's* Not. XXVII. 1830, p. 6, fig. 16–19, or Isis, 1832, p. 559, Taf. IX. fig. I–5) should be found in the cavity of the mantle of *Octopus granulatus* (*Lamarck*). It is probably identical with *Octopus tuberculatus* of *Delle Chiaje* (*Octopus Verany, Wagner*), which lives in the Mediterranean Sea, and perhaps, also, with *Tremoctopus violaceus.* If this last is not so, there is then a third species of *Hectocotylus,* viz: the male of *Tremoctopus violaceus.*

2 See *Kölliker,* On the *Hectocotylus* of *Tremoctopus violaceus,* and *Argonauta argo,* in the Ann. of Nat. Hist. XVI. 1845, p. 414.

dis, as well as the females on which they are found, will admit the correctness of the preceding statement, and, also, must have perceived the very remarkable abortiveness of the males of *Argonauta* and *Tremoctopus*.[3]

FAMILY: NAUTILINA.

Genera: *Nautilus, Spirula.*

FAMILY: OCTOPODA.

Genera: *Argonauta, Tremoctopus, Octopus, Eledone.*

FAMILY: LOLIGINA.

Genera: *Sepia, Loligo, Onychoteuthis, Sepioteuthis, Ommastrephes, Loligopsis (Perothis), Cranchia, Rossia, Sepiola.*

BIBLIOGRAPHY.

Swammerdamm. Zergliederung der spanischen Seekatze (Sepiae mariss, in the Bibel der Natur. Leipzig, 1752, p. 346, Taf. L.–LII.

Needham. An Account of some new Microscopical Discoveries. London, 1745, or the same in French. Leide, 1747. It contains the anatomy of *Loligo vulgaris.*

Cuvier. Mémoire sur les Cephalopodes et leur Anatomie, in the Mém. sur les Mollusques, loc. cit. Pl. I.–IV.

Brandt. Medizin. Zool. II. p. 298, Taf XXXI.–II.

Owen. Memoir on the pearly Nautilus. London, 1832 (in the Isis, 1835, p. 1, or in the Ann. d. Sc. Nat. 1833, XXVIII. p. 87). Also description of some new and rare Cephalopoda, in the Trans. of the Zool. Soc. II. 1841, p. 103; and the Art. *Cephalopoda,* in the Cyclopaedia of Anatomy, I. p. 517; also the twenty-third of the Hunterian Lectures, on Cephalopods with chambered shells, 1843.

Grant. On the structure and characters of Loligopsis, in the Trans. of the Zool. Soc. I. 1835, p. 21, Pl. II.; also on the Anatomy of the Sepiola vulgaris, Ibid. p. 77, Pl. XI. (in Isis, 1836, p. 389, Taf. X. fig. 4–10).

Rathké. Perothis, ein neues Genus der Cephalopoden, in the Mém. présentés à l'Acad. impér. de St. Petersberg, II. 1835, p. 149, Taf. I. II.

Van Beneden. Mém. sur l'Argonaute, in the Nouv. Mém. de l'Acad. Roy. des Sc. et belles-lettres de Bruxelles, XI. 1838, or in the Exer. zoot. Fasc. I. 1839, Pl. I.–V.

Férussac and *D'Orbigny.* Hist. nat. gén. et particulière des Mollusques. Cephalopodes acétabulifères. Paris, 1834.

Valenciennes. Nouv. Recherch. sur le Nautile flambé, in the Arch. du Muséum d'hist. nat. II. 1841, Pl. VIII.–XI.

Kölliker. Entwickelungsgeschichte der Cephalopoden. Zürich, 1844.

---

3 I am indebted to *Kölliker* for the examination of these individuals of *Hectocotylus tremoctopodis,* of which I found two in one and the same cavity of the mantle of a female. Although they have been preserved in alcohol a long time, yet I was able to assure myself of the correctness of many of *Kölliker's* statements, and thereby to be convinced of the real nature of these animals.*

* [§ 230, note 3.] See my note below under § 261, note 6. — ED.

ADDITIONAL BIBLIOGRAPHY.

Reinhardt and *Prasch.* Om Sciadephorus Mülleri en Nudersögelse. Kjöbenhavn, 1846, mit 5 Tafeln.

Prasch. Nogle nye Cephalopoder, beskrevne og anatomisk undersögte. (From the Mem. of the Danish Academy 5th ser. I.) Kjöbenhavn, 1847.

Milne Edwards. Règne animal, éd. illustr. Mollusques, Pl. 1ᵃ. 1ᵇ. 1ᶜ. 1ᵈ. 1ᵉ. 1ᶠ. (Octopus). — ED.

CHAPTER I.

INTERNAL SKELETON.

§ 231.

The Cephalopoda have many cartilages, which, serving as points of insertion for muscles, and surrounding the nervous centres, may therefore be regarded as the rudiments of an internal skeleton.

Their texture is essentially the same as that of the true cartilages of the vertebrata. There is a homogeneous, usually yellowish base, having the aspect of ground glass, in which are scattered numerous dark-colored molecules. This base contains, moreover, the proper cartilage cavities, which enclose a mass of granules, and each a more or less distinct nucleus. These cavities are more or less numerous, and are often partitioned each into two by a thin septum.

§ 232.

These rudiments of an internal skeleton may be divided into the cephalic, dorsal, articular, branchial, and pinnate cartilages.[1]

1. The Cephalic cartilage is concave in front and convex behind. It is perforated in the centre by the œsophagus and by two lateral conchoidal prolongations. At its upper part there is a deep excavation for the reception of the brain; and, at the inferior part, an enlargement containing the auditive organs. It is, moreover, traversed by canals of different sizes for the passage of nerves. The two lateral prolongations cover, by their anterior and concave surface, the ocular bulbs, and are thus the analogues of a kind of orbits. With *Loligo*, and *Sepia*, there are, beside, two lanceolate, cartilaginous lamellae, which join with the anterior and inferior parts of the cartilage, covering the ocular bulb in front, and thus completing the orbit. *Nautilus* differs very much from the other Cephalopoda in this respect. The lateral prolongations are wanting, and the body, which is incomplete above, is much developed below, and has two prolongations extending in front in a forked manner and concealing the auditory organs.[2]

---

[1] For the different cartilages, see *Schultze*, in *Meckel's* deutsch. Arch. IV. p. 334, Taf. IV. fig. 1, A–G; *Spix*, Cephalogenesis, p. 33, Taf. V. fig. 15–17; *Meckel*, Syst. d. vergleich. Anat. II. Abth. I. p. 125; *Brandt*, Medizin Zool. II. p. 303, Tab. XXXII.; *Owen*, Cyclop. Anat. and Phys. I. p. 524, fig 212, A–D; *Wagner*, Icon. zool. Tab.

XXIX.; and *Van Beneden*, loc. cit. Pl. I. (*Argonauta*).

[2] See *Owen*, On the Nautilus, p. 16, Pl. VIII. fig. 1, or Isis. 1835, p. 14, or Ann. d. Sc. Nat. XXVIII. p. 102, Pl. IV. fig. 1, and *Valenciennes*, loc. cit. p. 271, Pl. IX. fig. 4–6.

2. The Dorsal cartilages are found only with *Sepia* and *Loligo*. They are two in number; the inferior is situated in the neck, and the superior in the mantle at the anterior extremity of the internal shell. The cervical cartilage of *Loligo* is very long, rhomboid, and pretty massive; while that of *Sepia* consists of a thin, semilunar plate, with the cavity directed backwards. In both genera, its median line has a longitudinal groove; and in both also, the superior cartilage is only a thin semi-lunar lamella, both extremities of which are extended backwards by a long prolongation.

3. Those are called Articular cartilages, which, with *Argonauta*, and the Loligina, are found on each side of the base of the funnel, in the form of long cupels whose cavities receive, when the mantle is closed, the two cartilaginous prominences of its (the mantle's) internal surface.[3]

4. The Brachial cartilage is found only with *Sepia*. It is a narrow plate, transversely situated directly in front of the superior border of the cephalic cartilage. It has, anteriorly, three short apophyses for the support of the base of the arms. With the Loligina, there is found in the mantle, at the base of the lateral fins of the body, two other narrow lamellae, — the Fin-cartilages; these serve as points of insertion of the muscles of the fins, and extend more or less along the sides of the body, taking the form of the fins.[4]

CHAPTER II.

CUTANEOUS ENVELOPE.

§ 233.

The Cephalopoda are distinguished from the other Mollusca by a wholly peculiar structure of their skin. The skin is easily detached from the subjacent muscular layer, to which it is united by a loose cellular tissue, the fibres of which are interlaced in every direction. The extremely thin epithelium of the skin is lamellated, but never ciliated, with the adult individuals. The Corium is composed of a contractile fibrous tissue, in the meshes of which are contained the remarkable contractile Chromatophoric cells.[1] These consist of flattened, contractile cavities surrounded by a very

---

3 See the figures of *Férussac*, loc. cit. (*Sepia, Sepiola,* and *Argonauta*). With *Argonauta*, the two projections of the mantle are round tubercles, while with *Loligo, Onychoteuthis,* and *Sepiola,* they are two very long longitudinal ridges, to which corresponds a groove-like excavation in the two oppositely situated cartilages of the funnel.

By means of these articular cartilages, together with the two dorsal, when present, the collar-like border of the mantle is exactly fitted about the neck of the Cephalopoda.

4 The cartilages of the fins are very long with *Sepia* (*Schultze*, loc. cit. fig. C. D., and *Owen*,

Cyclop. &c. fig. 212, D. D.), but very short with *Sepiola.*

1 For the chromatic cells, see *San Giovanni,* in the Giornale enciclopedico di Napoli Ann. XIII. No. 9, or *Frorierp's* Not. V. 1823, p. 215, or Ann. d. Sc. Nat. XVI. 1829, p. 308; *Frenaze*, Observ. sur la mobilité des taches que l'on remarque sur la peau des Calmars, &c., Paris, 1823; *Delle Chiaje,* Memor. &c. IV. 1829, p. 63, and Descriz. I. 1841, p. 14; *Wagner,* Isis, 1833, p. 159, in *Wiegmann's* Arch. 1841, I. p. 35, and Icon. zool. Tab. XXIX. fig. 8–13, and *Harless,* in *Wiegmann's* Arch. 1846, I. p. 34, Taf. I.*

---

* [§ 233, note 1.] I have made some careful observations with the microscope upon the chromatic relations of the skin of the Cephalopoda, selecting for

my subject the common Squid (*Loligo illecebrosa*). My results differ somewhat from those of *Harless* above-mentioned. I found only one kind of pig-

delicate elastic membrane; when contracted, their form is round, but it becomes dentate on dilatation. The pigment granules, which they enclose, are always of the same color in each cell, and produce the red, the yellowish-brown, the blue, or violet spots, whose extent and shade vary, according as the cells are contracted or dilated.[2] Usually, adjacent cells have very different colors, and to their alternate contractions and dilatations in groups, are due those magnificent chromatic changes which have long made celebrated the skin of the Cephalopoda.[3]

These contractions, and consequently these chromatic changes, are under the influence of the nervous system. This is the reason of their decrease or disappearance, or their reappearance and increased brilliancy, in certain places, when the neighboring or even the distant skin is irritated. Moreover, the fibres of the corium preserve their contractility after having been detached, so that the chromatic changes may be observed on portions of the skin that have been removed.

§ 234.

Behind the neck of the Cephalopoda, the skin forms a large sac-like mantle, which completely envelopes the trunk, but is adherent only upon the back. Its anterior border is free, and can embrace, like a sphincter, the neck and posterior part of the head. Under the throat, the skin is prolonged in the form of a *funnel*, the free apex of which extends in front, while the broad base communicates with the cavity of the mantle, and is

---

2 The movements of the chromatic cells are not directly due to the cell-membrane, but to the contractile fibres of the dermis which are united in them, and which, upon contraction, pull at their point of insertion, thus producing the ragged aspect of these cells when expanded. They return to their round form when the fibres are relaxed, from the elasticity of the cell-walls (see *Kölliker*, Entwick. d. Cephalopden, p. 71, and *Harless*, loc. cit.). When these cells are dilated, the pigment granules are often removed from the centre to the periphery of the cell, thus forming a central, colorless transparent spot, which has been regarded by *Wagner* (loc. cit.) as the nucleus of the cell.

3 These so highly characteristic chromatic cells of the Cephalopoda, are found also in the skin of *Hectocotylus*, and are, therefore, one of the data for determining the nature of these animals, which have hitherto been regarded as trematode parasites. *Delle Chiaje* and *Costa* (loc. cit.) have represented these cells in a colored figure of *Hectocotylus argonautae*. I have, also, distinctly seen them with individuals of *Hectocotylus tremoctopodis* preserved in alcohol.

As *Grube* (Aktinien, Echinodermen und Würmer des Adriat. und Mittel-Meeres, p. 49, fig. 2) has observed these same chromatic cells in the skin

of a genus of parasites, which he has called *Polyporus chamaeleon*, it is certain that this animal, found on the branchiae of a marine fish, is only a torn off arm of one of the Loligina. The presence of these cells in the skin of *Nautilus* seems proved, for *Rumph* (Amboinische Raritäten-Kammer von Schnecken und Muscheln, p. 7) expressly declares of this animal which he saw living, that "its upper portion is reddish or bright brown with some black spots, which as with the cuttle-fish, become faded."

The fragment of the Mollusk, which *Quoy* and *Gaimard* found at the Celebes islands, and which they thought to belong to *Nautilus pompilius* (Ann. d. Sc. Nat. XX. 1830, p. 470, Pl. XIV. A. or Isis, 1834, p. 1146, Taf. XV. A. B.) deserves our attention in various ways. If it really belonged to a Cephalopod, it should have the chromatic cells, a point which may yet, perhaps, be determined from the preserved specimen at Paris.

In the colored figure which these naturalists have given of it, the skin is dotted with red,—a presumption in favor of the existence of these cells.

But, indeed, is it not possible that this animal, from its resemblance to the Hectocotyli, is not a mutilated one, but the male of *Nautilus pompilius*, abortive as to its form and size?

ment, deposited, as he has so well described, in the chromatophoric contractile sacs. The splendid changeable colors of the surface appeared to be due, not to the pigment spots alone, but to the intervening tissue; and the surface color over the pigment spots is subject to the same variations. Thus, a bistre-brown spot will sometimes appear blue, then green, &c. These facts may be tested by placing a small portion of the skin on a plate of

glass, and introducing a little water under it, the evaporation of which, by changing the surface conditions, generally produces a variety of colors.

The chromatic appearances of these animals appeared to me, therefore, as due full as much to surface phenomena as to pigment, and I have failed to detect different layers of pigment as described by *Owen*; see *Burnett* Proceed. Bost. Soc. Nat. Hist. IV. p. 252. — ED.

covered by its anterior border.[1] The sea-water, which enters into the interior of the mantle, passes, with its various contents, into this funnel, and is thence expelled through its anterior orifice.

Many Cephalopoda have, on the dorsal wall of this organ directly behind the anterior orifice, a tongue-shaped valve, which prevents the reflux of the water.[2]

With the Loligina, the sides of the trunk have variously-shaped cutaneous lobes, which these animals use as fins.[3] The Octopoda, on the other hand, swim by rowing with their arms, which are bound together at their base by a kind of natatory membrane, whose extremities have, each, a broader or narrower cutaneous dilatation.[4]

§ 235.

The mantle of many of the Cephalopoda secretes a shell, which may be either external or internal.

1. An external shell is found with *Argonauta* and the Nautilina. That of the Paper-Nautilus is very thin and flexible, — and, in its composition, the organic base predominates above the calcareous matter, which consists of thickly-set, small, round masses. The substance of the shell, which, with *Argonauta*, is nowhere attached to the animal it encloses, is secreted principally by the two large cutaneous lobes of the two median dorsal arms, which lie upon the external surface of the shell. On this account, the structure of the two surfaces of these lobes is different; — the external surface is quite smooth and has many chromatic cells; while the internal has scarce any of these last, but is covered with numerous reticulated, projecting lines, which become the more prominent when the lobes are contracted, and between which, cell-like depressions are formed.[1]

With the Nautilina, the shell has a very complicated structure; its walls are composed of two distinct layers, clearly separate, the internal of which has a beautiful mother-of-pearl aspect. The cavity of the shell is divided, even to the last spiral turns, by numerous transverse septa, which are all perforated. With *Nautilus*,[2] a tube traverses the septa, while with

1 With *Nautilus*, the funnel is composed of two pretty large, cutaneous lobes, placed upon both sides of the throat, and reciprocally covering each other on the ventral surface in a cornet-like manner; see *Owen*, On the Nautilus, p. 10, Pl. I. or Isis. p. 10, or Ann. d. Sc. Nat. loc. cit. p. 93, Pl. I. III. and *Valenciennes* loc. cit. p. 263, Pl. X. fig. 1.

2 This is so with *Sepia*, *Sepiola*, *Loligo*, *Sepioteuthis*, *Onychoteuthis*, and *Nautilus*. For this last, see *Owen*, loc. cit. Pl. II. fig. 2, e., and *Valenciennes*, loc. cit. Pl. XI. fig. 4, λ. I have sought for it in vain with *Argonauta*, *Eledone*, and *Tremoctopus*. It is also wanting with *Loligopsis* and *Cranchia*; with *Octopus*, there exists in its place, that is, on the ventral surface of the funnel, a transverse ridge.

3 With *Sepia*, and *Sepioteuthis*, both sides of the body are bordered their entire length with a cutaneous lobe. With *Loligo*, and *Onychoteuthis*, the two fins are triangular and inserted on the posterior extremity of the body; they are round and short with *Sepiola*, *Loligopsis*, and *Cranchia*; in the first of these genera, they are situated on the middle of the sides of the body, and in the last two, upon its extremity.

4 These interbranchial natatory membranes exist with *Octopus*, *Eledone*, and *Tremoctopus*; they are particularly developed in this last genus, between the two pairs of dorsal arms.

In this same genus, as also with *Argonauta*, the two dorsal arms are terminated by a very large cutaneous lobe, and are used not only as locomotive organs, but also for keeping the shell in place by being applied on its external surface; see *Ferussac*, loc. cit. *Argonauta*, Pl. 1. fig. 5, 6, Pl. VI. fig. 2, and in the Mém. de la Soc. d'Hist. Nat. de Paris, II. 1825, p. 160, Pl. VI. fig. 2, or Isis, 1832, p. 460, Taf. V. fig. 2; *Rang*, Docum. pour servir à l'Hist. nat. des Céphalopodes, in the Magaz. de Zool. 1837, Livr. IV. p. 13, Pl. LXXXVI.—LXXXVIII., or Ann. d. Sc. Nat. VII. 1837, p. 176; and *Delle Chiaje*, Descriz. loc. cit. Tav. VII. fig. 1, 2.

1 It has been attested by several observers, that these two cutaneous lobes furnish the substance of the shell, and that, also, with which the animal repairs accidental lesions; see *Rang*, Magaz. de Zool. loc. cit.; *Jeanette Power*, in the Atti dell' Acad. di Scienz. Nat. di Catania, XII. 1839, or Isis. 1845 p. 606, or in *Wiegmann's* Arch. 1845, I. p. 369; and, Further experiments and observ. on the Argonauta Argo, in the Reports of the Brit. Assoc. 1844, Notices and Communic. p. 74. For the non-parasitism of the animal, see, moreover, *Van Beneden*, loc. cit. p. 4, and *Ferussac*, loc. cit. p. 114.

2 *De Blainville*, in the Nouv. Ann. du Museum d'Hist. Nat. III. 1834, p. 3, Pl. I. II.

Spirula,[3] an analogous calcareous tube extends close upon the inner surface of the shell from one septum to another. The animal, whose trunk occupies only the first chamber, is loosely attached to it by the cartilaginous border of its mantle.

With *Nautilus*, this border has a lobe which extends along the back of the animal, surrounding the spiral portion of the shell.[4] With all the Nautilina, there is another prolongation in the form of a membranous tube, or *Sipho*, which arises from the posterior part of the body, — traverses the orifices or calcareous tubes of the septa, and penetrates even into the last chambers of the shell. These chambers are lined with a thin membrane, and have no external communication except through the Siphon.

2. With the Loligina, an internal shell lies free in the dorsal portion of the mantle. In most genera, it is composed of a homogeneous, horny substance, of a yellowish-brown color, and has a form like a feather (*Calamus*), or the head of a lance. At one of its extremities is an attenuated stem, and two delicate lateral winglets of variable length.[5] With *Sepia*, this shell differs very much from that of the other Loligina. Its two surfaces are covered by very distinct calcareous layers, which have erroneously given it the name of *Os sepiae*.[6] As a whole, it is tongue-shaped; its two surfaces are convex and its borders are sharp. Behind, the lateral borders become thinner and are slightly bent toward the ventral surface; and a short conical point projects from the middle of the posterior border. The horny substance is reduced to a thin sheet, situated between the calcareous layers, but its borders usually extend out beyond those of these last. The calcareous layer of the dorsal surface is very thin, but quite solid, and its surface in front, is granulated and striated; that of the ventral surface, on the other hand, is very thick, especially in the middle, and its very loose tissue contains numerous quite thin, porous lamellae, which, superposed almost horizontally, alternate regularly with layers of small, transversely-striated, dichotomous, vertical prisms.[7] This ventral layer is truncated obliquely from its middle backwards, and the horizontal layers may easily be counted upon its truncated surface.[8]

3 *De Blainville*, Ibid. p. 18, Pl. I. fig. 6, A–F.
4 *Owen*, and *Valenciennes*, loc. cit.
5 See *Wagner*, Icon. zoot. Tab. XXIX. fig. 32 (*Loligo*), and *Férussac*, loc. cit. (*Loligo, Loligopsis, Onychoteuthis, Sepiola,* and *Sepioteuthis*).

I cannot here omit speaking of the remains of an antediluvian animal, which, under the name of *Aptychus*, has much engaged the attention of palæontologists, and, up to the present time, been the object of discussion.

Some have regarded it as the operculum of an Ammonite or of another Mollusk (*Rupped*, Abbild. und Beschreib. einig. Versteiner. von Solenhofen, 1829, and *Voltz*, in the Neuen Jahrbuch für Mineralogie, &c., 1837, p. 304, 432); others as a shell of a bivalve (*H. von Meyer*, in the Nov. Act. Acad. Nat. Cur. XV. pt. II. p. 125 and in the Jahrbuch f. Mineral. &c. 1831, p. 391); and others, finally, as an internal shell of one of the Cephalopoda (*Coquand*, in the Bull. de la Soc. Géol. de France, XII. 1840–41, p. 376).

This last opinion is undoubtedly the correct one. As for myself, I am able to perceive in the different species of *Aptychus* only shells whose shaft is abortive, and the wings excessively developed. I was therefore quite surprised to hear my colleague, *Alexander Braun*, express himself in a conversation, that, "after all, the animal called *Aptychus*

might well have been the male of certain Ammonites." If the relations of the Hectocotyli to certain Octopoda are borne in mind, the idea of *Braun*, that there have existed Ammonites, the males of which are quite different in form from the females, certainly merits much consideration. For the males of these animals were, perhaps, abortive like those of *Argonauta* and *Tremoctopus*, and obliged, therefore, to shelter themselves in the mantle of their females, and this would explain why it is that the specimens of *Aptychus* are so often found at the base of the first chamber of Ammonites.

Judging from the form of the shell, the bodies of these animals must have been very large. There will be an additional analogy in favor of this view, if it is proved that the large and flattened animal found by *Quoy* and *Gaimard* is really the male of a *Nautilus* (§ 233, note 3).

6 The error of *Spix* (Cephalogenesis, loc. cit. p. 33) in comparing it to a rudimentary vertebral column, is still wider.

7 According to *Kölliker* (Entwickel. loc. cit. p. 72, Taf. V. fig. 45, 46) these calcareous prisms begin to be formed in the embryo.

8 A very detailed description of this *Os sepiae* has been given by *Cuvier* (Mém. loc. cit. p. 46), *Brandt* (Mediz. Zool. II. p. 301, Taf. XXXI. fig. 3, 6), *Wagner* (Icon. zoot. Tab. XXIX. fig. 34), and *Férussac* (loc. cit.).

Although it must be supposed that the calcareous matter of this shell is secreted by the internal surface of the dorsal cavity, yet the thin fibrous membrane which lines this last, is without a glandular structure.

CHAPTER III.

MUSCULAR SYSTEM AND ORGANS OF LOCOMOTION.

§ 236.

The muscular system of the Cephalopoda is highly developed. Its primitive fibres are smooth, but are not so diversely interlaced as with the other Mollusca. These fibres are usually parallel, and the fasciculi which they form, are of equal thickness. When isolated, they often show a zigzag tendency, which, probably, belongs also to their state of contraction. The fasciculi are very compactly bound together in one direction by a cellular tissue, and, in this manner, form clearly-defined, long, flat muscles.

§ 237.

The mantle of the Cephalopoda has a very distinct layer of circular fibres.[1] From the internal surface of the sac which it forms, arise, in the dorsal region, two pairs of large cylindrical muscles. One pair of these passes in front and is extended into the walls of the base of the funnel; the other pair extends to the posterior part of the neck, and is inserted partly into the cephalic cartilage, and partly at the base of the arms. The other muscles, which are thinner, arise from the sides of the cervical cartilage, and are inserted upon the funnel.[2] By means of a part of this muscular apparatus, these animals can vigorously contract the cavity of the mantle and the funnel, and, by tightly embracing the neck and base of the funnel with the border of the mantle, can eject, through the orifice of this last, the liquids contained in the cavity of the body. With many species, these muscular contractions serve, also, as a means of a backward locomotion in the water.

§ 238.

The principal locomotive organs of the Cephalopoda are the arms fixed upon the cephalic cartilage; they serve also as prehensile organs. Each of these consists of a tubular axis composed of a dense cellular tissue, of muscular fibres radiating towards the surface, between which are inter-

---

1 This muscular layer is incomplete with *Sepia* — being wanting in the dorsal portion of the mantle.
2 A very full description of the muscles of the trunk and the head of Cephalopoda may be found in *Curier*, loc. cit. p. 9; *Brandt*, loc. cit, p. 303, and especially in *Delle Chiaje*, Memor. IV. p. 72, and Descriz. &c., I. p. 21.

posed others which are longitudinal, and, lastly, of a layer of circular fibres directly beneath the skin.[1]

Over the entire length of the internal surface of these arms, are suckers, arranged in a single, double, or multiple row.[2]   But with the Loligina, these suckers occur in groups of variable extent only at the extremity of the ninth and tenth arms (*tentacular arms*).   These suckers are moved by muscular fasciculi which pass from the arms and are spread upon the former in a ray-like manner, and which form, moreover, with the Loligina, a peduncle.

With *Tremoctopus*, the suckers are cylindrical and very simple, while with the other Octopoda, their opening is closed by a membrane perforated centrally by an orifice which can be closed by a papilla that projects from the base of the sucker.[3]   These organs are applied to objects at the moment when the papilla is withdrawn and removed to the base of the sucker. With the Loligina, the lateral walls of the suckers are very thin, extensible, and have upon their borders a horny and denticulated ring; in this ring the fleshy base of the sucker adjusts itself in an urceolate form, and upon withdrawal, produces a vacuum.   With *Loligopsis*, and *Onychoteuthis*, many of the suckers on the two tentacular arms are imperfect, but, on the other hand, some of the teeth of their horny border are disproportionably developed, or the whole is changed into a strong claw.[4]

The arm-like processes about the mouth of *Nautilus* differ very much from the preceding.[5]   They have no trace of suctorial organs, and are composed of thirty-eight prismatic filaments, which are a little flattened and transversely curled.

Each of these filaments is surrounded, at its base, by a contractile sheath into which it can be wholly withdrawn.[6]   The whole fasciculus is, moreover, enveloped in a common sheath, which, upon the back, is flattened so as to resemble the foot of the Gasteropoda, and like it, probably, may serve for creeping.[7]

The portion of the skin, which extends as a kind of Natatory membrane between the arms of many Cephalopoda, contains a very loose net-work of longitudinal and transverse muscular fibres.[8]

The fins of the Loligina, have, on the contrary, large muscles composed of parallel, contiguous fasciculi which arise from the cartilages of these organs.

1 This axis is usually of a prismatic form; consequently if an arm is cut transversely the section presents a quadrilateral or rhomboidal spot in its centre; see *Savigny*, in the Descript. de l'Égypte, Hist. Nat. Pl. I. fig. I. w.; *Owen*, in the Cyclop. I. p. 528, fig. 214, c, and *Ferussac*, loc. cit. *Octopus*, Pl. II. fig. 3, and Pl. XV. fig. 11, b. A similar section of the body of an *Hectocotylus* presents a like aspect.

2 The suckers form a single row with *Eledone*; a double one with the other Octopoda, and with most of the Loligina; but the rows are multiple with *Sepia*.

3 Although the double row of suckers on the body of *Hectocotylus* does not diminish towards the anterior extremity, yet, in other respects, it so closely resembles the suckers of *Argonauta* and *Tremoctopus*, that it may well be asked why this single fact was not sufficient to discover to the older observers the real nature of this pretended parasite.

4 *Férussac*, loc. cit. *Loligopsis*, Pl. IV. and *Onychoteuthis*, Pl. VI. VIII. &c.

5 *Owen*, and *Valenciennes*, loc. cit.
6 The internal structure of these filaments agrees pretty closely with that of the arms of other Cephalopoda. *Owen*, on Cephalopods with chambered shells, loc. cit. p. 8, fig. 131, Cyclop. loc. cit. p. 526, fig. 213, and Annals of Nat. Hist. XII. p. 305.

7 *Owen*, and *Valenciennes*, loc. cit.
8 The large cutaneous lobes of the median dorsal arms of *Tremoctopus* and *Argonauta*, have a similar structure. In this last genus, they are not used as ours, nor as sails, but are thrown back upon the shell to keep it in place (§ 255, note 4); they move in the water, moreover, like the other Cephalopoda, by the contractions of the mantle and the funnel (*Rang*, Magaz. d. Zool. 1837, p. 22. Pl. LXXXVII.).   It is therefore astonishing that *Jeanette Power* (*Wiegmann's* Arch. 1845, I. p. 373) should have revived the old fable that these animals raise these two large arms above the surface of the sea to be used as sails.

With *Nautilus*, there are two large, particular muscles, which arise from the under surface of the cephalic cartilage, and extend, divergingly, backwards; they serve, by means of a horny plate, to fix the animal to the internal border of the shell.[9]

CHAPTER IV.

NERVOUS SYSTEM.

§ 239.

The nervous system of the Cephalopoda attains a very high degree of development. Its central portion, especially, quite resembles the brain of the Vertebrata, in the extraordinary increase of its ganglionic substance, and by the presence of a cartilaginous cavity containing it, comparable to a cranium. This cavity is incomplete, it is true, but at its anterior part where the cartilaginous substance is wanting, it is closed by a tendinous cellular tissue which takes the place of a *Dura mater*.

The brain itself, which is far from filling the cavity of the cephalic cartilage, is enveloped by a fibrous membrane, which sends off sheaths to the nerves which leave the brain and traverse, in different places, the cephalic cartilage. The cavities remaining between the brain and this cartilage are filled with a fat-like liquid.

The primitive nerve-fibres are straight, finely granulated, and bound together into fasciculi of variable size by a very distinct neurolemma.[1] The long and oval corpuscles which are often found in abundance between them, belong probably to the neurolemma.

§ 240.

The central mass of the nervous system, with the Cephalopoda, forms also an œsophageal ring, which consists of a superior and an inferior ganglionic mass connected by lateral commissures. The superior portion is small and sends some delicate nerves to the parts of the mouth. The inferior portion on the contrary, is very large, and extends along the sides of the œsophagus in order to be directly continuous with the broad commissures. The olfactory, and the two optic nerves arise from the lateral portions of this ganglion, while the auditory nerves have their origin from its inferior surface.

From its anterior border pass off four or five pairs of large nerves to the arms, and, also, others to the muscles of the head. From its posterior border arise small nerves for the funnel, and also two large trunks for the

9 *Owen*, On the Nautilus, p. 17, Pl. IV. fig. 2, k., or Isis, p. 15, or Ann. d. Sc. Nat. p. 103, Pl. II. fig. 3, k., and *Valenciennes*, loc. cit. p. 268, Pl. XI. fig. 4, P.

1 *Kölliker*, Entwickelung. d. Cephalop. p. 79. According to the researches of *Lebert* and *Robin*

(*Müller's* Arch. 1846, p. 128), the histological composition of the ganglia with the Cephalopoda is very remarkable. They have here found very large ganglionic globules, of even one-twenty-fifth of an inch in diameter, and containing, each, several nuclei.

.back of the mantle.[1] With *Sepia*, this inferior portion has several swellings; from the two anterior of these, which are the largest, arise the nerves of the arms; while the two lateral posterior send off the two optic nerves.[2] With *Nautilus*, this same portion is divided into an anterior and a posterior transverse band,[3] which may be compared, to a certain extent, to the semi-circle of ganglia upon the inferior surface of the œsophagus with certain Gasteropoda.

§ 241.

Among the Peripheric nerves, those of the arms and mantle should be specially mentioned.

The Brachial nerves enter into the axial canals of the arms at the base of these last, and extend even to their extremity after intercommunicating, each, by a transverse anastomosis with the two neighboring nerves.[1] In their course through this canal, they give off numerous filaments to the muscular substance of the arms and to the suckers. With the Octopoda, these nerves are composed of two parallel cords, each one of which has, alternately right and left, ganglionic enlargements.[2]

The two Pallial nerves, which are easily seen from their size, pass, at first, between the cervical muscles, and, having reached the internal surface of the back of the mantle, terminate in two very large ganglia (*Ganglion stellatum*) from the external border of which pass off numerous nervous filaments, which enter, ray-like, the fleshy portion of the mantle.[3] With those Loligina, which have fins, the pallial nerves, before terminating in the star-like ganglia, send off a large branch, which, at a short distance from its origin, is joined by another large branch from the pallial ganglion, and is then distributed to the muscles of the fin.[4] With the long-bodied species of this family, this nerve pursues a long course by the side of the me-

1 The nervous system of the Cephalopoda has been carefully described by *Cuvier*, Mém. p. 34, Pl. I. fig. 4 (*Octopus*); *Brandt*, Mediz. Zool. p. 308, Taf. XXXII. fig. 23 (*Sepia*); *Owen*, and *Valenciennes*, loc. cit. (*Nautilus*); and by *Van Beneden*, loc. cit. (*Argonauta*); see, moreover, the figures given by Owen, of that of *Sepia* (On the Nautilus, Pl. VII. fig. 3, or Isis, 1835, Taf. IV. 7, fig. 3, or in the Ann. d. Sc. Nat. XXVIII. Pl. III. fig. 5, and Cyclop. I. p. 549, fig. 232), and those of *Loligo*, *Sepia*, and *Octopus*, in *Delle Chiaje*, Memor. &c. Tav. XCV. C.–CII., and Descriz. Tav. XXV. XXIX.–XXXI.*

2 *Brandt*, loc. cit.

3 *Owen*, On the Nautilus, p. 36, Pl. VII. fig. 1, or, Isis, 1835, p. 30, Taf. IV. 7, fig. 1, or, Ann. d. Sc. Nat. XXVIII. p. 134, Pl. III. fig. 4, and *Valenciennes*, loc. cit. p. 287, Pl. VIII. fig. 2–4.

1 *Cuvier*, Mém. p. 36, Pl. I. fig. 4, (*Octopus*); *Delle Chiaje*, loc. cit. Tav. CII. (29), C. (31), (*Octopus* and *Sepia*); *Férussac*, loc. cit. Pl. 15, fig. 1, and *Van Beneden*, loc. cit. p. 15, Pl. II. fig. 2, and Pl. IV. (*Argonauta*).†

2 *Van Beneden*, loc. cit. p. 14, Pl. II. fig. 3–5, Pl. III. fig. 4, and Pl. IV. (*Argonauta*). I have found the same organization with *Octopus*, and *Tremoctopus*. In this last genus, the ganglia may, from their reddish color, be very clearly separated from the white nervous substance. Both the smooth and the nodulated cords send off nerve-filaments, but with the last, they arise exclusively from the ganglionic swellings.

I have been unable to decide if the smooth cords send off filaments only to the muscles, and the nodulated ones to the suckers ; or if the first contain only motory fibres, and the second sensitive fibres. I should add that in the axis of *Hectocotylus tremoctopodis* I have also found a highly-developed, nodulated trunk, the number of swellings of which corresponded with that of the suckers.

3 See the figures of *Van Beneden*, *Delle Chiaje*, *Brandt*, loc. cit., and of *Owen*, in the Cyclop. I. fig. 232 (*Argonauta*, *Octopus*, *Loligo*, and *Sepia*).

4 See the figures of *Delle Chiaje*, and *Owen*, loc. cit. (*Loligo* and *Sepia*).

* [§ 240, note 1.] See especially the excellent illustrations of *Milne Edwards*, Règne anim, loc. cit. Pl. 19. See, for a very detailed description of this system with *Ommastrephes*, Hancock (Ann. Nat. Hist. X. 1852, p. 1), who has sought to point out the homologies of the Cephalopoda with the Mollusca. — Ed.

† [§ 241, note 1.] See also the illustrations of *Milne Edwards*, Règne animal, loc. cit. Pl. I. fig. 3, f. f. (*Argonauta*). — Ed.

dian line of the body until it reaches the base of the large fin-muscles, situated at the extremity of the body.[5]

With *Nautilus*, numerous filaments arise from the posterior ganglionic band, and, without forming a ganglion, are distributed to the two muscles of the shell. From their origin, they may be regarded as the representatives of the pallial nerves of the other Cephalopoda.[6]

Another pair of nerves corresponding to the Pneumogastric nerves of the Vertebrata, arises from the middle of the inferior cerebral mass, between the two pallial nerves, descends along the neck behind the funnel, the posterior wall of which it pierces, and thence passes under the peritoneum; here it sends several nerves to the ink-sac, and then ramifies upon the heart, the large vascular trunks, the branchial hearts, and the branchiae. Both of these nerves have ganglia, here and there, in their net-works,[7] and these net-works communicate probably with the *Plexus splanchnicus posterior*.

§ 242.

The Splanchnic nervous system is particularly developed with the Cephalopoda. It may be divided into an anterior and a posterior plexus.[1]

The *Plexus splanchnicus anterior* consists of a *Ganglion pharyngeum inferius*, situated under the œsophagus sending filaments forwards to the parts of the mouth, and backwards to the œsophagus, and connecting at the same time with the inferior cerebral mass by two commissures.[2]

With the Loligina, there is, beside, opposite this ganglion, a *Ganglion pharyngeum superius*, which, also, sends several filaments to the parts of the mouth, gives off two filaments to the inferior œsophageal ganglion, and appears to connect, likewise, with the superior cerebral mass.[3]

The *Plexus splanchnicus posterior* is characterized by a large *Ganglion gastricum* lying upon the stomach. From this, filaments pass off in different directions to the other viscera, and it receives two filaments of communication, which, after having arisen from the inferior pharyngeal ganglion, accompany the œsophagus through the œsophageal ring.[4]

---

5 *Delle Chiaje*, loc. cit. Tav. XCV. (25) and CI. (30), (*Loligo*). The two parallel nerves which, with *Loligopsis*, extend backwards along the inferior dorsal surface of the mantle belong also to this class of nerves. *Grant* (loc. cit. p. 21, Pl. II. fig. 5, 6), has compared them to the spinal marrow of Vertebrata. With *Onychoteuthis*, I have also seen the two nerves of the fins running along the internal surface of the mantle, while with *Loligo*, as *Delle Chiaje* has indicated in his Tav. CI. (30), they afterwards pass into the muscular layer of the mantle and continue their course between it and the skin, sending off filaments to the two muscles of the fins.

6 *Owen*, On the Nautilus, p. 38, Pl. VII. fig. 1, No. 13, or, Isis, p. 32, Taf. IV. 7, fig. I, or, Ann. d. Sc. Nat. p. 137, Pl. III. fig. 4, No. 13.

7 These two nerves, analogous to the *Par vagum*, have been observed by all the Anatomists of these animals; see *Cuvier* Mém. p. 36, Pl. I. fig. 4, *u*. (*Octopus*); *Brandt*, loc. cit. Tab. XXXII. fig. 3, g. and fig. 23, k.; *Owen*, Cyclop. loc. cit. I. fig.

232, c. (*Sepia*); *Van Beneden*, loc. cit. p. 19, Pl. I. fig. 7, h. Pl. III. fig. 5, k. and Pl. IV. r. (*Argonauta*); *Owen*, On the Nautilus, Pl. VII. fig. 1, No. 15, or, Isis, Taf. IV. 7, fig. I, or Ann. d. Sc. Nat. Pl. III. fig. 4, No. 16 (*Nautilus*); and *Delle Chiaje*, loc. cit. Tav. XCV. (25), C. (31), and CII. (29), (*Loligo, Sepia*, and *Octopus*).

1 For the sympathetic nervous system see *Brandt*, Ueber die Mundmagennerven der Evertebraten, loc. cit. p. 40.

2 *Brandt*, Mediz. Zool. II. p. 309, Taf. XXXII. fig. 23, 3; *Owen*, Cyclop. loc. cit. fig. 232 (*Sepia*); *Van Beneden*, loc. cit. p. 16, Pl. II. fig. 6 (*Argonauta*); *Delle Chiaje*, loc. cit. Tav. XCV. C.–CII. (25, 29–31), (*Loligo, Sepia*, and *Octopus*).

3 *Brandt*, *Owen*, and *Delle Chiaje*, loc. cit. (*Sepia* and *Loligo*).

4 *Van Beneden*, loc. cit. Pl. III. fig. 1–5 and Pl. IV. (*Argonauta*); *Brandt*, loc. cit. Taf. XXXII. fig. 3, 20; and *Delle Chiaje*, loc. cit. Tav. C. (31), and CII. (29), (*Sepia* and *Loligo*).

CHAPTER V.

ORGANS OF SENSE.

§ 243.

The sense of Touch is well developed with the Cephalopoda, and is situated in the whole cutaneous envelope, in the fringed labial membranes, and, especially, in the arms.[1] *Nautilus* is particularly rich in tactile organs, which are situated on the head; and this animal has, beside the thirty-eight tentacular arms, two external, and two median, large, labial prolongations, placed about the mouth, the border of which has twelve small, curled filaments, whose internal structure quite resembles that of the arms.

The nerves of the filaments of the two external of these prolongations have an origin common with those of the arms, arising, consequently, from the front border of the anterior cerebral band. Those of the filaments of the median prolongations arise from the same band (but nearer the median line), by two common roots which, before dividing, have a flat ganglion.[2] This animal has, also, four other curled tentacles, which can be retracted in a sheath, two in front of, and two behind, the eyes. These tentacles receive a special tactile nerve, which has its origin by the side of the optic nerve.[3]

§ 244.

With the Cephalopoda, the fleshy point of the tongue is undoubtedly a Gustatory organ. It is concealed in the anterior angle of the lower jaw, and its rounding surface is covered with numerous soft villosities, which very probably serve as gustatory papillae.[1]

§ 245.

The Olfactory organs of the Cephalopoda are situated in the neighborhood of the eyes, and consist, each, of a cavity with tumid borders, or of a cutaneous fossa which has an opening, and, sometimes, at the bottom, a whitish papilla. The nerves of these organs arise from the optic ganglion of the œsophageal ring, near the optic nerves. At first, they are closely united with these last, enter the orbit with them, and extend along its posterior wall, thence to the olfactory papillae, to which they are distributed in a ray-like manner.[1]

---

1 Touch appears the only sense developed with *Hectocotylus*. If *Costa's* figure (Ann. d. Sc. Nat. XVI. Pl. XIII. fig. 2, c. f.) is exact, *Hectocotylus argonautae* has a special tentacle-like tactile organ on the anterior extremity of the body.
2 *Owen*, On the Nautilus, Pl. IV. Pl. VII. fig. 1, or Isis, 1835, Taf. III. IV., or Ann. d. Sc. Nat. XXVIII. Pl. II. fig. 1, Pl. III. fig. 4.
3 *Owen* and *Valenciennes*, loc. cit. Pl. VIII. fig. 2, i. and Pl. IX. fig. I, l.
1 This organization appears to have eluded the observation of most naturalists. I have seen it very distinctly, not only with the Loligina, but also with the Octopoda. *Owen* (On the Nautilus, p. 23, Pl. VIII. fig. 7, or, Isis, p. 20, Taf. II. or, Ann. d. Sc. Nat. p. 113, Pl. IV. fig. 7, and Cyclop. I. p. 551, fig. 236,) and *Valenciennes* (loc. cit. p. 280, Pl. X. fig. 3, 4,), only, have represented with *Nau-

tilus* this part of the tongue as having all the characteristics of a gustatory organ. With *Sepia*, the soft papillae have already been figured by Savigny (Descript. de l'Égypte, loc. cit. Pl. 1. fig. 4, 5, and in *Férussac*, loc. cit. *Sepia*, Pl. IV. fig. 2, 3).
1 The cavities here mentioned were for a long time regarded as the external auditory passages, and the cutaneous folds surrounding them as a Pavilion (*Férussac*, loc. cit.), until *Külliker* (*Froriep's* neue Notiz. XXVI. 1843, p. 166, and, Entwickel. d. Cephalopoden, p. 107) discovered a special nerve, and declared, with reason, that the whole was an olfactory organ. The Cephalopoda being poor in vibratile organs, it is quite desirable to ascertain if these olfactory organs are ciliated, for they are so in fishes with which ciliated epithelium is likewise feebly developed.

With *Nautilus*, the two olfactory papillae are situated, directly beneath the eyes, in a cavity which is surmounted by a wart-like swelling.[2] With the Octopoda, the olfactory organs are concealed, behind the eyes, in the angle of insertion of the mouth upon the occiput. With *Argonauta*, and *Tremoctopus*, they consist of two naked papillae; and with *Octopus*, and *Eledone*, of two membranous cavities.[3]

With the Loligina, these organs are situated behind and a little below the eyes, and consist of fossae having narrow apertures; but they are easily seen from the elongated or round cutaneous swelling with which they are surrounded.[4]

§ 246.

The Auditory organs of the Cephalopoda are situated in the lower middle portion of the cephalic cartilage, where they form two more or less large, round cavities, separated by a cartilaginous septum, and without any external communication.[1] With the Octopoda, the internal walls of these cavities are smooth;[2] but with the Loligina they have many tubercles or papillae, which are sometimes quite prominent.[3] This portion of these organs may best be compared to the osseous *Labyrinthus* of the Vertebrata. These cavities are filled with a liquid substance, and contain, also, each, a small pyriform sac — membranous labyrinth — adhering to the cartilaginous labyrinth at the point where the auditory nerve enters it, and upon which this nerve is spread out. This sac contains a single, white, irregular otolite of a crystalline texture.[4]

2 *Valenciennes* (loc. cit. p. 290, Pl. VIII. fig. 2, h. Pl. IX. fig. 1, h. x., and fig. 3) in 1841, and consequently before *Kölliker*, described these organs as olfactory with *Nautilus*. He found not only the nerve which goes to the olfactory papilla, but also an orifice at the base of this last, leading into a cavity lined with a mucous membrane which had two regular rows of folds. *Owen* (On the Cephalopods with chambered shells, p. 11) has regarded these papillae, which he appears to have completely overlooked in his earlier memoir, as short hollow tentacules. On the other hand, he regards as the olfactory organs a row of twenty membranous lamellae arranged longitudinally at the entrance of the mouth between the two internal labial prolongations (On the Nautilus, p. 41, Pl. IV. 1., Pl. VII. fig. 1, g. fig. 2, or Isis, p. 34, Taf. III. IV., or Ann. d. Sc. Nat. p. 141, Pl. II. fig. 1, l., Pl. III. fig. 4, g. fig. 6); but it would appear to me that these lamellae are tactile lobules, for they receive numerous nerve-filaments from the ganglia of the nerves of the internal labial prolongations (*Owen*, loc. cit.).

3 With *Argonauta*, and *Tremoctopus*, these olfactory nerves have a ganglion lying on the optic nerve (*Kölliker*, Entwickel. d. Ceph. p. 168); this was seen by *Van Beneden* (loc. cit. p. 13, Pl. I. fig. 5, 6, k.), but not explained. The olfactory cavities of *Octopus* did not, indeed, escape the notice of *Rapp* (Naturwiss. Abhandl., von einer Gesellsch. in Würtemberg, 1826, p. 69), and of *Delle Chiaje* (Descriz. &c. Tav. VI. fig. 1. k. and Tav. XVIII. fig. 1, y), but they did not in the least suspect their nature.

4 According to *Owen's* account accompanied with a figure (On the Nautilus, pl. VII. fig. 3, No. 9, or Isis, 1835, Taf. IV., or Ann. d. Sc. Nat. XXVIII. Pl. III. fig. 5, No. 9, and Cyclop. I. p. 549, fig. 232, k), the olfactory nerves of *Sepia* and *Loligo* appear to arise from a special ganglion situated near the *Ganglion opticum*. The entrance, with its tumid borders, of the olfactory cavities, has often been figured with the Loligina, by *Férussac* (loc. cit. *Sepia*, Pl. XVII. fig. 2. c. Pl. XVIII. fig. 3, b. Pl. XXVII. fig. 1, 6; *Loligo*, Pl. XX. fig. 7, Pl. XXIII. fig. 5, 17, Pl. XXIV. fig. 2, 14; *Sepioteuthis*, Pl. VI. fig. 2, b.; *Sepiola*, Pl. III. fig. 5, 15 b.).

1 It has already been seen (§ 245) that the olfactory organs of the Loligina have been taken by some naturalists for an external ear.

A very remarkable organ — a flexuous canal lined with ciliated epithelium, has been seen by *Kölliker* (Entwick. d. Ceph. p. 105, fig. 60–63), but, only with the embryos of *Sepia* and *Loligo*; departing from the auditive vesicles, it ran in front without opening either upon the surface of the body, or into the œsophagus, so that it could have been neither an external auditory duct, nor a *Tuba Eustachii*.

2 See *Scarpa*, Anat. disquis. de auditu et olfactu, p. 3, Tab. IV. fig. 11 (*Octopus*); *Delle Chiaje*, Descriz. &c. Tav. XIV. fig. 1, d.; and *Van Beneden*, loc. cit. Pl. I. fig. 3 (*Argonauta*).

3 See *Brandt*, Mediz. Zool. p. 309, Taf. XXXII. fig. 14; *Wagner*, Icon. zool. Tab. XXIX. fig. 37–39; *Owen*, Cyclop. I. p 554, fig. 235, and Transact. of the Zool. Soc. II. Pl. XXI. fig. 17; and *Delle Chiaje*, Descriz. &c. I. p. 68, Tav. XVI. fig. 12, 21 (*Sepia* and *Loligo*). This last-mentioned author has compared some of these cartilaginous prominences to the Ossicula of the ear; but to me they appear to represent rather the first traces of semicircular canals, which, with the embryos of fishes, appear to consist, likewise, of simple prominences on the internal surface of the auditive vesicle.

4 These otolites are composed mostly of carbonate of lime, and vary considerably in their forms. With the Octopoda, they resemble, more or less, a

The auditory organs of *Nautilus* are somewhat different. They are widely separated from each other, and situated in the prolongations of the cephalic cartilage which extend in front; they consist of a very long, narrow labyrinthian cavity containing a homogeneous, thick liquid without otolites.[5]

§ 247.

The Eyes of the Cephalopoda are very highly developed and disproportionately large.[1] Although resembling very much those of the Vertebrata, yet they differ from them in many respects.[2] With the Octopoda, and Loligina, each eye has an ocular Bulb and a Capsule.

The capsule is formed by the cartilaginous orbit, and by a fibrous membrane attached to the borders of this last, and is blended externally with the cutaneous envelope. This envelope, in the form of a circular swelling, covers the eye, and, being thin and transparent, takes the place of a Cornea, — a part which, properly, does not exist with the Cephalopoda.[3]

The circular swelling often has, above and below, a semilunar fold of skin containing muscular fibres, which, upon contraction, cover the convexity of the eye like an upper and under lid. The ocular bulb, contained in this capsule, is round and a little flattened in front; and, as it is not adherent to its capsule in front nor upon the sides, there is a free space, which, from the absence of a cornea, would coincide with the anterior chamber of the eye.[4] In most cases, this space contains a transparent liquid, and is lined by a serous membrane covering not only the posterior surface of the anterior part of the capsule, but also the anterior surface of the bulb. It is remarkable that this same space, which contains in part the anterior chamber, communicates, externally, by a circular orifice which, with the Octopoda, is covered by the upper lid, and with the Loligina, is situated upon the anterior border of the cutaneous fold which takes the place of the cornea. Internally, this space can be closed by a kind of fold

disc, concave on one side, and very convex and sometimes even conical, on the other *Scarpa*, loc. cit. Tab. IV. fig. 9, and *Weber*, De Aure et Auditu, p. 11, Tab. II. fig, 8, (*Octopus*); also *Delle Chiaje*, Memor. Tav. LVIII., and Deseriz. &c. Tav. XII. fig. 15, 19, 23, 24, (*Octopus* and *Eledone*). With *Octopus*, they have a crystalline structure; but with *Eledone*, where they are very flat and colored brown on one of their surfaces, they consist of a soft limeless substance, — often rendering in vain the search for them in specimens long preserved in alcohol.

The irregular otolites of the Loligina, which appear bristling with points and serratures (*Scarpa*, loc. cit. Tab. IV. fig. 8, and *Delle Chiaje*, loc. cit. Tav. LVIII. (12) fig. 13, 14, 16, 25, 26, (*Sepia* and *Loligo*),), have the aspect, under the microscope, of an aggregation of very fine, acute prisms, the points of which turn inwards (*Carus*, Lehrb. d. vergleich Zoot. I. p. 358).

5 See *Valenciennes*, loc. cit. p. 291, Pl. VIII. fig. 2, No. 3, and Pl. IX. fig. 4, 5, *a*. He has seen it supplied with nerves coming directly from the brain. *Owen* (On the Cephalop. with chambered shells, p. 10) took them for venous sinuses, and could not admit that they were auditive organs, since they contained no otolites. But it might be argued that these otolites are limeless like those of *Eledone*, and may, therefore, dissolve and entirely disappear after death.

1 The largest eyes are found with the Loligina; the smallest with the Octopoda.

2 For the structure of the eyes of Cephalopoda, see, beside *Cuvier*, Mém. p. 37, Pl. II. fig. 5, and Pl. III, fig. 7, and *Owen*, Cyclop. I. p. 551, fig. 234, —*Massalien*, Descript. oculorum Scombri Thynni and Sepiæ, Diss. Berol, 1815, p. 10; *Soemmerring*, De Oculorum hominis animaliumque sectione horizontali, p. 76, Tab. III.; *De Blainville*, Princ. d'Anat. comp. p. 441; *Mayer*, Analekt. f. vergleich. Anat. IIft. I. p. 52; *Krohn*, Nov. Act. Acad. Leop. Carol. XVII. Pt. 1. p. 339, Tab. XXVI., and XIX. Pt. II. p. 43; *Wharton Jones*, Lond. and Edinb. Philos. Mag. 1836, Jan'y, or *Froriep's* Notiz. XLVIII. p. 2, fig. 1-3; *Delle Chiaje*, Descriz. &c. I. p. 70, Tav. XIX. and XXIX. also Osservaz. Anatom. su l'occhio umano 1838, Tav. IX. fig. 1–11; *Valentin's* ideal section of an eye of a Cephalopod, in *Wagner's* Icon. zoot. Tab. XXIX. fig. 42; and *John Power*, Dublin Jour. of Med. Science, XXII. 1843, p. 350.

3 *Krohn*, *Valentin*, and others, admit the existence of a particular horny substance situated between the cutaneous layers of the anterior part of the ocular capsule.

4 *Treviranus* (Vermischte Schrift. III. p. 154; says he has observed a thin, transparent, but solid membrane, placed directly in front of the lens, and continuous with the conjunctiva (Argentea), thus forming a completely-closed anterior chamber; but this statement requires confirmation.

or Pupil. The serous membrane just mentioned, which is spread over the ocular bulb even to the papillary border of the iris, contains a particular pigment of a silvery lustre, called the *Argentea*, and comparable to a *Conjunctiva*.[5]

With *Onychoteuthis, Loligopsis,* and allied genera, the anterior wall of the ocular capsule is entirely wanting, and as there is also no cornea, the crystalline lens is in direct contact with the surrounding medium (the water of the sea). In the first of these genera, the free border of the capsule has, in front, a deep fissure corresponding, perhaps, to a lachrymal canal.[6]

The *Iris* is formed from the argentea, which is covered on its posterior surface by a black *Uvea*, while its anterior surface often has chromatic cells.

The pupil is usually of a transverse, or semilunar, rarely of a circular form, and is capable of being completely closed.[7] Under the Argentea extends a thin cartilaginous tunic — *Sclerotica* — which, behind, circumscribes the ocular bulb, and, in front, penetrates a certain distance into the iris. It furnishes points of insertion for the muscles of the eye, and is cribriform behind for the passage of numerous filaments of the optic nerve.

The cavity of the bulb is filled with a transparent, watery liquid which takes the place of the vitreous body, and is contained in a very thin *Hyaloïdea.*

The Crystalline lens is spherical, and lodged in a deep depression of the vitreous body. It is of a brownish color, and its anterior surface projects through the pupil, so that the posterior chamber of the eye is only a small circular space. As with the Vertebrata, this organ is composed of numerous concentric layers, but has the remarkable peculiarity of being divisible into halves, the anterior of which is less convex than the posterior, but both are exactly joined together; the borders of these halves are quite bevelled, but are kept in place by the Ciliary body which arises from the sclerotica and iris. One part of this ciliary body embraces the borders of the lens, while the other penetrates between its halves as a thin, transparent septum.[8]

5 Zoolomists are not agreed upon the interpretation to be put on this membrane. *Krohn,* and *Owen* (loc. cit.), who regard the anterior part of the ocular capsule as a cornea covered by a conjunctiva, consider the cavity found behind it as a large anterior chamber, filled with a *Humor aqueus. Cuvier, Wharton Jones* (loc. cit.), and *J. Muller* (in his Arch. 1836, Jahresb. p. 91), regard the capsular cavity with its serous membrane, as a closed conjunctival sac ; so that the transparent convexity of the capsule is not a cornea, but a continuous closed eyelid. Moreover, as there are often found two rudimentary eyelids in the eyes of Cephalopoda (*Mayer,* Analekt. f. vergleich. Anat. Hft. I. p. 52, Taf. IV. fig. 6–11), this transparent convexity may be regarded as a third lid or a nictitating membrane adherent throughout except at the point of the opening. Many anatomists, and especially *Cuvier,* and *Owen,* have not noticed this opening of the ocular capsule. But *De Blainville* (Princip. d'Anat. comp. I. p. 444, and Dict. d. Sc. Nat. XLVIII. p. 262) mentions it with *Loligo, Octopus,* and *Sepia* ; and *Wagner* (Analekt. &c. p. 55) has described it carefully.

In the large work of *Férussac,* it is often figured under the name of *Orifice lacrymal* ; see *Loligo,* Pl. XX. fig. 7, Pl. XXIII. fig. 5, a, 17,

* [§ 247, note 8.] The microscopic structure of this lens corresponds also with that of the Vertebrata — that is, composed of delicate tubes or fibres.

Pl. XXIV. fig. 2, d, 14 ; and *Sepiola,* Pl. III. fig. 5, 15. a, Pl. VI. fig. 2. a, Pl. IV. fig. 10. a.

6 On account of this singular organization, *D'Orbigny* (in *Férussac,* loc. cit. Introduct. p. 15) has separated, under the name *Oigopsidés,* the genera mentioned in the text from the other Loligina which he calls *Myopsidés.* The segment of the border of the capsule, and which is wanting with *Loligopsis,* is spoken of as a *Sinus lacrymalis* in *Férussac,* loc. cit. *Onychoteuthis,* Pl. III. fig. 1, Pl. III. fig. 2, Pl. XII. fig. 4, 13, Pl. XIV. fig. 1 ; *Ommastrephes,* Pl. I. fig. 15, Pl. II. fig. 3, 11.

7 The pupil is not circular except with *Onychoteuthis, Ommastrephes,* and *Loligopsis.*

The upper papillary border, usually convex with the other Cephalopoda, is often prolonged as a *Velum* or *Operculum pupillare.* With *Sepia,* it is often bilobed after death ; see the figures of *Férussac,* loc. cit. and *Delle Chiaje,* Osservaz. anat. loc. cit. Tav. IX. fig. 1, 2, 3.

8 For the lens and the ciliary body, see *Huschke,* Comment. de pectine in oculo Avium, 1827, p. 9, fig. 11, and *Delle Chiaje,* Descriz. &c. Tav. V. fig. 18, and Tav. XIX. fig. 6–8. Although *Mayer* (Analekt. loc. cit. p. 54) declares that this lens has a capsule, yet I am undecided on this point, for the other anatomists are silent.*

These fibres however are more than twice as small as those of any of the Vertebrata I have examined. — ED.

The Optic nerves enter the posterior part of the orbit through a kind of *Foramen opticum*, after which they swell into a large kidney-shaped ganglion in which a portion of the nerve-fibres are completely interlaced with those from the opposite side.[9] Leaving this *Ganglion opticum*, the nerve divides into numerous filaments which traverse the cribriform sclerotica, and then unite with the other elements of the *Retina*. The external layer of the retina is composed of these filaments; beneath it, is a pigment layer of a reddish-brown color, and pierced by numerous fibres given off rectangularly from the external layer. The internal layer is composed of granules, among which the fibres of the optic nerve probably terminate.[10] The external layer is continuous as a thin membrane upon the ciliary body, and even upon the septum of lens.[11]

The two optic ganglia are enveloped by a peculiar white substance composed of fat-cells, which, perhaps, serves only as a fat-cushion.[12] The eye is moved by several straight and oblique muscles, which arise from the cartilaginous portion of the orbit, and are inserted, usually, upon the middle of the bulb.

But with *Nautilus*, the eyes differ in many respects from those of the other Cephalopoda. They are supported upon a muscular stalk and project from the head; while with the other Cephalopoda, excepting *Loligopsis*, they are sunken deeply in the head.[13] From the rudimentary lower lid a narrow furrow passes over the anterior surface of the eye even to the small, circular pupil. As yet, neither cornea nor lens has here been found.[14]

---

9 For the interlacement of the nerve-fibres in the *Ganglion opticum*, — see, especially, *Wharton Jones*, and *John Power*, loc. cit.

10 The intimate, very complicated structure of the retina, has been described principally by *Treviranus* (loc. cit. p. 155), *Wharton Jones* (loc. cit.), and *Paccini* (Nuove ricerche microscop. sulla tessitura intima della retina nell' Uomo, nei Vertebrati, nei Cefalopodi e negli Insetti. Bologna, 1845, p. 55, fig. 13, 14). The mysterious phenomenon, that, according to the older anatomists, the surface of the retina exposed to the light is covered with a pigment-layer, rests only on an imperfect knowledge of the structure of this organ, as has been shown by *Wharton Jones* (loc. cit.), and *Valentin* (Repert. f. Anat. II. 1837, p. 109).

11 See *Krohn*, and *Wharton Jones*, loc. cit.

12 *Mayer* (Analekt. &c. p. 53) regards this substance as a semi-adipose gland with several excretory ducts,— a kind of lachrymal gland the product of which is poured into the conjunctival sac; but *Kölliker* (Entwick. d. Ceph. p. 103) could find nothing glandular in its structure.

13 For the pedunculated eyes of *Loligopsis*, see *Rathké*, in the Mém. d. St. Petersburg, loc. cit. Pl. I., and *Férussac*, loc. cit.

14 The eyes of *Nautilus* having been studied for a long time upon dead specimens, it may be suspected that the exceptionable peculiarities observed by *Owen*, and *Valenciennes*, are referable to the want of fresh specimens. It is, at first, singular that *Owen* (On the Nautilus, p. 39, Pl. I. v. w., or Isis, p. 32, Taf. I. 1, fig. 1, v. w., or Ann. des Sc. Nat. p. 139, Pl. I. fig. 1, v. w.) speaks of a ridge, and *Valenciennes* (loc. cit. p. 289, Pl. IX. fig. 1, No. 3) of a furrow, running from the border of the lower lid to the pupil. As the cornea is wanting, it might almost be supposed, from examining *Valenciennes'* figure (Pl. VIII. fig. 2, P.), that *Nautilus* belonged to the *Oigopsides* of *D'Orbigny*, except, that with this animal, instead of a complete absence of the anterior part of the ocular capsule, there exists only a fissure, regarded by one of the authors in question as a ridge, and by the other as a furrow. The lens, not perceived by either *Owen*, or *Valenciennes*, escaped perhaps through this fissure, after having been detached by maceration.

As for the pigment layer, spoken of by *Owen*, as situated upon the concave surface of the retina of *Nautilus*, his problem will be explained, from researches upon fresh specimens in the same way, as with the other Cephalopoda.

CHAPTER VI.

DIGESTIVE APPARATUS.

§ 248.

The mouth of the Cephalopoda[1] is always surrounded by the arms, (which serve partly as prehensile organs), and by a circular fleshy lip which is fringed or denticulate on its free border. It is, moreover, covered externally by a thin cutaneous fold having a crucial opening. With the Loligina, there is, beside, a third external lip, arising as a cutaneous fold from the base of the arms; it has an heptagonal, rarely an octagonal, opening, from the angles of which project longer or shorter tentacular prolongations.[2] With *Nautilus*, this lip is extraordinarily developed, — having four considerable prolongations provided with long tentacles.[3]

Behind these lips is a round pharynx, very fleshy, and armed with two blackish-brown, horny jaws, which move against each other vertically.

Upon each of these jaws are two large lateral branches which join at an acute angle, thus forming a hooked point. The edges of these jaws being very sharp, the whole has the form of a reversed parrot's-beak, for, the edges of the lower jaw project far beyond those of the upper.[4] The pharynx is enveloped by a very complicated muscular apparatus, which arises in part from the cephalic cartilage, and moves the jaws as well as serves in producing the protraction and retraction of the pharynx.[5]

Between the two branches of the lower jaw is a Tongue, which is fleshy, and resembles a long swelling adherent to the floor of the oral cavity. Upon its anterior extremity are soft gustatory papillae, and over the rest of its surface there are horny lamellae arranged in regular longitudinal rows, and golden-yellow spines which point backwards.[6] Its posterior extremity is often folded over, thereby forming a kind of cavity, the opening of which is directed backwards, and continuous with a semi-canal leading into the œsophagus.

§ 249.

The intestinal canal of the Cephalopoda is wholly without ciliated epi-

1 I have been unable to find in *Hectocotylus tremoctopodis*, the orifice which *Cuvier* (Ann. d. Sc. Nat. loc. cit. p. 151, fig. 1, 3, 4, f., or Isis, 1832, p. 560, Taf. IX., or *Froriep's* Notiz. loc. cit. p. 8, fig. 16, 18, 19, f) has regarded as a mouth with *Hectocotylus octopodis*; and as *Kölliker* (loc. cit.) says nothing about a digestive apparatus with these animals, I suspect that it is wanting here, nutrition taking place by cutaneous absorption while these bodies are in the mantles of their females.

2 See *Férussac*, loc. cit. the figures for *Sepia, Loligo, Sepioteuthis, Onychoteuthis,* and *Ommastrephes.*
3 See § 243.
4 *Cuvier*, Mém. p. 25, Pl. III. fig. 6; *Savigny*, Descript. de l'Égypte, loc. cit. Pl. I.; *Delle Chiaje*, loc. cit. Tav. LX. (10) fig. 9; *Wagner*, Icon. zoot. Tab XXIX. fig. 18; and the numerous figures

given by *Férussac*, loc. cit. According to *Owen* (On the Nautilus, p. 20, Pl. VIII. or Isis, p. 18, Taf. I. or Ann. d. Sc. Nat. p. 10.9, Pl. IV.).with *Nautilus*, the extremities of the jaws are covered with a bluish-white calcareous substance, and the border of the lower jaw is denticulated; but *Valenciennes* (loc. cit. p. 279, Pl. XI. fig. 1, 2) has not confirmed these observations.

5 For this muscular apparatus, see *Cuvier*, loc. cit. Pl. III. fig. 3-5, and Anat. comp. V. p. 9 (*Octopus*); and *Owen*, loc. cit. (*Nautilus*).
6 *Needham*, Nouv. Decouv., loc. cit. p. 28, Pl. III. fig. 1; *Brandt*, loc. cit. p. 305, XXXII. fig. 6-10; *Savigny*, loc. cit. Pl. I.; *Férussac*, loc. cit. *Octopus*, Pl. III. *Argonauta*, Pl. IV. *Sepia*, Pl. IV.; *Owen*, On the Nautilus, p. 22, Pl. VIII. fig. 6, 7, or Isis, p. 19, Taf. II. or Ann. d. Sc. Nat p. 113, Pl. IV.; and *Valenciennes*, loc. cit. p. 280, Pl. X. fig. 3, 4.

thelium. It begins behind the pharynx by a straight, long, very narrow œsophagus, whose internal surface is longitudinally plicated. After leaving the annular opening of the cephalic cartilage, it enters the peritoneal cavity, which is highly developed and divided by constrictions into several chambers. With the Loligina, the œsophagus is of uniform calibre throughout to the stomach;[1] but with the Octopoda, it is abruptly dilated, upon leaving the cephalic cartilage, into a kind of crop, which extends to the stomach.[2] With *Nautilus*, also, it is dilated, but gradually, into a very large crop, which communicates with the stomach by a narrow, short canal.[3]

The Stomach invariably consists of a sac lined with a very solid epithelium, which is plicated longitudinally; the Cardia and Pylorus are situated close to each other at its upper portion.[4] As soon as the intestine has left the pylorus, it forms a Caecum which has glandular, plicated walls, and, with many genera, is more or less elongated and spirally convoluted.[5] The rest of the intestine is short, rarely flexuous, and extends from the peritoneal sac to the base of the funnel,[6] where it terminates in a small anal prolongation, the borders of which are often fringed; sometimes it has two lateral tongue-shaped valves, placed opposite each other, and by which the anal opening can be closed.[7]

§ 250.

The Salivary organs of the Cephalopoda are highly developed, and consist of a superior and an inferior pair, the former of which is sometimes, but the latter very rarely, wanting. The superior pair consists of two glandular lobes situated at the posterior extremity of the pharynx, which open by short excretory ducts behind the root of the tongue.[1] The inferior pair lies on each side of the œsophagus at the upper portion of the peritoneal sac, directly behind the cephalic cartilage. These organs, usually of a dull-white color, are composed of numerous inter-

1 *Sepia, Loligo, Onychoteuthis, Loligopsis, Sepiola,* &c.

2 *Cuvier*, Mém. Pl. IV. fig. 1, 2, b.; *Wagner*, Icon. zool. Tab. XXIX. fig. 14 (*Octopus*); *Van Beneden*, loc. cit. Pl. III. fig. 3, d. (*Argonauta*); *Férussac*, loc. cit. *Octopus*, Pl. XIII. fig. 9, 10, *Argonauta*, Pl. I⁵. fig. 1, 2; and *Delle Chiaje*, Descriz. Tav. XV. fig. 3 (*Tremoctopus*).

3 *Owen*, On the Nautilus, Pl. IV. or Isis, Taf. III., or Ann. d. Sc. Nat. Pl. II. fig. 1.

4 See the figures in *Cuvier, Brandt, Férussac, Owen,* &c. The stomach of *Octopus* and *Eledone*, from its muscular walls, and its almost horny epithelium, resembles very much the gizzard of birds.

5 This caecum, regarded as a second stomach by many zoötomists, corresponds, probably, to the pyloric appendages of fishes. With *Nautilus*, it is a round sac, the internal surface of which has longitudinal folds, so that its cavity has a lamellated appearance (*Owen*, On the Nautilus, p. 25, Pl. IV. y. and Pl. VIII. fig. 8, f., or Isis, Taf. II. III., or Ann. d. Sc. Nat. Pl. II. fig. I, y. Pl. IV. fig. 8, f.). With *Loligopsis*, and *Sepiola*, this round sac is lined internally with spiral folds (*Grant*, Transact. loc. cit. p. 25, Pl. II. fig. 7, g. and p. 8], Pl. XI. fig. 7, 8, c.). With *Sepia*, and various Octopoda, it is oblong, and lined internally with transverse spiral folds supported by a kind of mesentery; — see *Van Beneden*, loc. cit. Pl.

III. (*Argonauta*); *Delle Chiaje*, Descriz. Tav. XIII. XV. XVIII. (*Tremoctopus, Sepia,* and *Loligo*); *Cuvier*, Mém. Pl. IV. fig. 1, 2, f.; *Wagner*, loc. cit. fig. 14, f. (*Octopus*); *Home*, Lect. on Comp. Anat. Pl. LXXXIII. (*Loligo sagittata*); and *Férussac*, loc. cit. But, in this respect, *Loligo vulgaris* forms an exception; its caecum is straight, oblong, and its thin walls are without internal plicae; see *Meckel*, Syst. d. vergleich. Anat. IV. p. 193, and *Delle Chiaje*, Descriz. &c. Tav. XVI. fig. 5, s.

6 The intestine is straight with *Argonauta, Loligo, Sepia, Sepiola,* and other Loligina; but it is flexuous with *Octopus, Eledone,* and *Nautilus.*

7 *Owen* (Transact. of the Zool. Soc. II. Pl. XXI. fig. 16) has found two lateral valves projecting into the anal cavity with *Sepioteuthis.* I have seen two similar with a *Tremoctopus, Rathke* (Mém. de St. Pétersburg, loc. cit. p. 160 Pl. II.) has found them replaced, with *Loligopsis,* by two tentaculiform prolongations.

1 *Cuvier*, Mém. p. 27, Pl. III. fig. 3, e. (*Octopus*); *Férussac*, loc. cit. *Octopus*, Pl. XII. fig. 6, n. Pl. XIII. fig. 9, n.; *Owen*, Cyclop. I. p. 532, fig. 218, i. (*Onychoteuthis*). With *Nautilus, Owen* found no lower, and only the traces of the upper glands (On the Nautilus, p. 23, Pl. VIII. fig. 7, g., or Isis, p. 29, Taf. II., or Ann. d. Sc. Nat. p. 114, Pl. IV. fig. 7, g.).

anastomosing glandular tubes, forming, sometimes, several lobes, and sometimes, a single triangular mass with a smooth exterior.

In their passage in front, the two excretory ducts converge and form, under the œsophagus, a common canal which traverses the pharynx and terminates in the mouth near the root of the tongue.[2]

The Liver is generally of a reddish-yellow color, and is rarely lobulated. Usually, it is a compact glandular mass capsulated by a fold of the péritoneum.[3]   With the Octopoda, it is a large, smooth, ovoid gland,[4] while with the other Cephalopoda, with a few exceptions,[5] it is divided into two or four portions symmetrically surrounding the œsophagus.[6]   The bile, when this organ is single or double, is excreted by two ducts arising from the inferior extremity of the organ ; but when this organ is quadruple, as with *Nautilus*, and *Loligopsis*, each division has a special excretory duct, and all these ducts soon unite into a common *Ductus choledochus*, which, after a short course, opens upon the sides of the cœcum.[7]

As a Pancreatic gland may, certainly, with reason, be regarded the pale-yellow, short, ramified glandular tubes, which, with many species, are appended to the hepatic ducts with which they communicate by many orifices.[8]

---

2 For the intimate structure of these glands which appear to be wanting with *Nautilus*, and *Loligopsis*, see *J. Muller*, De Gland. struct. p. 54, Tab. V. fig. 9.   They are lobulated with *Loligo*, and consist only of a small compact body with *Octopus*, *Eledone*, *Sepia*, &c ; see *Cuvier*, Mém. Pl. III. fig. 2, 3 ; *Wagner*, Icon. zoot. Tab. XXIX. fig. 14, k.; *Brandt*, loc. cit. Taf. XXXII. fig. 3, 5 ; *Férussac*, loc cit. *Octopus*, Pl. XII. XIII. Their surface is granulated with *Sepiola*, according to *Delle Chiaje*, Descriz. Tav. XXVI. fig. 14, L., and *Grant*, Trans. &c. Pl. XI. fig. 8, g.

3 For the intimate structure of the liver, see *Muller*, De Gland. struct. p. 71 (*Octopus*), and *Rathké*, loc. cit. p. 137 (*Loligopsis*).

4 See *Cuvier*, *Wagner*, and *Férussac*, loc. cit.

5 With *Onychoteuthis Banksii*, the liver is a single, very oblong mass ; see *Owen*, in the Cyclop. I. p. 537.

6 With *Nautilus*, the liver is divided into four large portions, each composed of numerous lobes embracing on each side the crop-like œsophagus ; see *Owen*, On the Nautilus, p. 26, Pl. IV. z., or Isis, p. 22, Taf. III., or Ann. d. Sc. Nat. p. 117, Pl. II. fig. 1. z.   With *Loligopsis guttata*, the four hepatic divisions are, according to *Grant* (Trans. &c. p. 25, Pl. II. fig. 4, c. and 7, a.), deeply concealed in the cavity of the body ; while with *Loligo Eschscholtzii*, and *dubia*, it is a single mass, according to *Rathké* (Mém. de St. Pétersb. loc. cit. p. 137, 170. Pl. II.).   With *Sepia*, *Loligo*, *Sepiola*, &c., this organ is divided into long halves, smooth externally, and extending from the neck along the dorsal median line, their length depending on that of the animal ; see *Brandt*, loc. cit. Taf XXXII. fig. 3, p. (*Sepia*), and *Grant*, loc. cit. Pl XI. fig. 7, 8. f. (*Sepiola*).

7 See *Cuvier*, Mém. p. 30, Pl. IV. fig. 2, 4, n. n.; *Férussac*, loc. cit. *Octopus*, Pl. XIV. fig. 5, 6, *Argonauta*, Pl. 1⁵. fig. 2, d.; *Owen*, On the Nautilus, Pl. VIII. fig. 8, h., or Isis, Taf. II., or Ann. d. Sc. Nat. Pl. IV. fig. 8 h.; and *Grant*, Trans. of the Zool. Soc. I. Pl. II. fig. 7, b. Pl. XI. fig. 7, g. (*Loligopsis* and *Sepiola*).

8 This structure and arrangement of the glandular appendages of the hepatic ducts which were noticed and regarded as a pancreas by *Hunter* (The Catal. of the Physiol Ser. I. p. 229, No. 775) with *Sepia*, remind one very much of what is found in fishes, where, according to *Stannius* investigations, the pyloric appendages communicate with the *Ductus choledochus* (see *Brockmann* (*Stannius*) De Pancreate piscium, Diss. Rostoch. 1846).*   According to *Delle Chiaje* (Descriz. I. p. 32, Tav. XIII. XVIII.), these bodies exist not only with *Octopus*, *Eledone*, *Tremoctopus*, and *Argonauta*, but also with *Sepia*, *Loligo*, and *Sepiola*.   *Grant* (The Edinb. Philos. Jour. XIII. 1825, p. 197) has described them with *Loligo sagitta*, and *Owen* sought in vain for them with *Nautilus*, but found them highly developed with *Sepiola*, *Onychoteuthis*, *Sepioteuthis*, and *Rossia* (Cyclop. I. p. 537). See also *Grant*, I. Trans. of the Zool. Soc. I. Pl. II. fig. 7, c., Pl. XI. fig. 7, 8, 13 (*Loligopsis* and *Sepiola*).

In the species of *Loligopsis* examined by *Rathké* (loc. cit. p. 160, Pl. II.) the *Ductus choledochus* was dilated into a round sinus at the point where the pancreatic tubes opened into it.

* Note. These recent researches modify essentially what *Stannius* has said in the second volume of his work, upon the *Appendices py'orica* and the *Pancreas* of fishes, — organs not in the least identical.

CHAPTER VII

CIRCULATORY SYSTEM.

§ 251.

The circulatory system of the Cephalopoda does not appear more highly developed than that of the other Mollusca.[1] However, this subject is still deficient in creditable observations, and especially in those relating to the absence of completely-closed vessels.

The blood is usually colorless, or of a green-hue, or violet-hue color, and contains, proportionably, numerous round corpuscles enclosing many granules most of which are colorless, but with a few, scattered here and there, of a violet hue.[2]

§ 252.

The Central organ of the circulation consists, with all the Cephalopoda, of a simple ventricle, situated in the centre of the cavity of the body, and surrounded with a pericardium. It is round, or oblong,[1] and serves as an aortic heart.

With *Nautilus* (Tetrabranchiata), this organ receives, on each side, two branchial veins; while with the Dibranchiata there is one vein only, and the heart sends off a superior and an inferior aortic trunk.[2] The mouth of the veins and the origin of the arteries are furnished with valves.[3] The Ascending aorta first sends two branches to the mantle, then gives off branches to the liver, to the upper portion of the digestive canal, to the inferior salivary glands, and to the funnel. Behind the cephalic cartilage it bifurcates, forming a ring embracing the upper extremity of the oesophagus, and from which arise two arteries for the ocular bulbs,[4] eight or ten for the arms, and many small branches for the parts of the mouth.[5] The Descending aorta furnishes branches to

1 See *Milne Edwards*, and *Valenciennes*, Compt. rend. XX. 1845, p. 261, 750, or *Froriep's* neue Notiz. XXXIV. p. 84, 258; also *Milne Edwards*, Ann. d. Sc. Nat. III. 1845, p. 341. This last author has also described (Ann. d. Sc. Nat. VIII. p. 53), the circulatory system of the Ldigina, which is interrupted by a large sinus; but he makes no mention of the aquiferous system. As of late there is increasing evidence for the opinion, that, with various invertebrate animals, the blood-system communicates externally at certain points on the body, and can therefore receive water into its interior, it is now important to investigate the direct or indirect relations between this and the aquiferous system which is so widely spread through the Mollusks, the Worms and the Zoophytes. It may be that this aquiferous system, if it really communicates with the blood system, corresponds to a lymphatic apparatus, although it seems hardly reasonable to suppose that canals, which carry a portion of the nutritive fluids, should open externally.*

2 *Wagner*, Zur vergleich. Physiol. d. Blutes. Hft. I. p. 19; and *Delle Chiaje*, Descriz. I. p. 57.

1 The form of the heart depends upon that of the posterior part of the body; it is large in the genera with a short body, and elongated in those of a long body. According to *Kölliker* (Ann. of Nat. Hist. XVI. p. 414), *Hectocotylus* has also a heart communicating with arteries and veins, but he says nothing of its locality.

2 See *Owen*, On the Nautilus, Pl. VI. fig. 1, or Isis. Taf. IV. or Ann. d. Sc. Nat. XXVIII. Pl. III. fig. 2; *Brandt*, loc. cit. Taf. XXXII. fig. 22; The Catal. of the Physiol. Ser. II. Pl. XXII. (*Sepia*); and *Van Beneden*, loc. cit. Pl. III. fig. 5 (*Argonauta*).

Often the two branchial arteries are widely dilated before entering the heart, and these dilatations may be regarded as auricles.

3 See *Cuvier*, Mém. p. 22, Pl. II. fig. 4 (*Octopus*); and *Owen*, Cyclop. I. p. 541, fig. 227 (*Onychoteuthis*).

4 For the distribution of the ophthalmic arteries see *Krohn*, Nov. Act. Nat. Cur. XIX. pt. II. p. 47.

5 *Delle Chiaje*, loc. cit. Tav. LXXXVIII. XC. XCII. XCIV. (or 20, 28, 22, 24) has represented in detail the arterial system of *Octopus vulgaris*, *Sepia officinalis*, *Loligo vulgaris* and *sagittata*.

* [§ 251, note 1.] For *Milne Edwards'* beautiful figures see Règne anim. loc. cit. Pl. 1b, 1c (*Octopus*).— ED.

the stomach, the small intestine, the rectum, the branchiae, and the genital organs ; the artery of these last, however, sometimes arises directly from the heart.

Nothing positive can now be said as to the terminal relations of these arteries ; — that is, whether they are directly continuous with the venous radicles by means of a capillary system with proper walls, or whether they terminate by orifices so that the blood is effused immediately into the parenchyma of the body.[6]

The Venous system begins in the different parts of the body by numerous small vessels, of which we are still ignorant whether they are continuous with the terminal arterioles, or whether they commence by themselves with proper orifices. Their radicles unite and form longer branches which finally open into a large Sinus. One of these sinuses, which is of a circular form, surrounds the upper extremity of the œsophagus, and receives the veins coming from the eyes, the arms,[7] and the parts of the mouth. From this sinus arises another, of an oblong form, which, since it extends into the cavity of the body and receives the different veins from the viscera, may be called a *Vena cava superior*. In the centre of the body it divides into two large venae cavae which extend on each side to the base of the branchiae[8] and terminate in the two so-called branchial hearts.[9] These two veins receive, also, two trunks, which bring the blood from the mantle and are often dilated into two large sinuses.[10]

The distinct, but often very thin walls of the venous sinuses, are sometimes so intimately blended with the adjacent organs, that these sinuses may be easily taken for wall-less lacunae.[11]

6 *Milne Edwards* and *Valenciennes* (loc. cit.) throw no light on these questions. It is moreover singular that in the numerous and often very detailed figures of *Delle Chiaje* of the vascular system of Cephalopoda, he has nowhere represented in the least a capillary net-work between the arteries and veins ; while *Kölliker* (Entwick. der Cephal. p. 81), declares that he has seen numerous capillary vessels in the embryos of *Sepia*.

7 All the arms of the Cephalopoda have two venous trunks. *Lebert* and *Robin* (*Muller's* Arch. 1846, p. 130) have observed, in the venous system *Sepia officinalis*, a valve preventing the reflux of the blood towards the head.

8 With *Nautilus*, this sinus is divided into four venae cavae (*Owen*, loc. cit).

9 The so-called Branchial hearts of the dibranchiate Cephalopoda have no muscular fibres, but have a very glandular aspect, and are in close relation with the urinary organs ; see below, § 255.

10 *Delle Chiaje*, loc. cit. Tav. LXXXVII. LXXXIX. XCI. XCIII. (17, 27, 21, 23), has also figured with many details the venous system of *Octopus, Sepia,* and *Loligo*.

11 It is, therefore, difficult to decide if the large cavities which *Milne Edwards* (Ann. d. Sc. Nat. III. loc. cit. Pl. XIII.–XVI.) has injected, were dilated veins or simple lacunae. In this last case, the venous system would communicate directly with

the cavity of the body, and there are many circumstances in favor of this view. It is, therefore, to be regretted that *Milne Edwards* did not, in his researches, pay more attention to the aquiferous system which is spread through the whole body of the Cephalopoda, and thus, for the present at least, prevent the objection, that these aqueous reservoirs should be confounded with the venous sinuses. The lymphatic reservoirs which, according to *Erdl* (*Wiegmann's* Arch. 1845, I. p. 163) surround, and can be injected by means of the arteries, are also perhaps, venous sinuses. An observation of *Owen* (On the Nautilus, p. 27, Pl. VI. fig. 1, No. 1 , or Isis, p. 24. Taf. IV. or Ann. d. Sc. Nat. p. 121, Pl. 1.1. fig. 2, No. 1¹), and of *Valenciennes* (loc. cit. p. 287), that the large superior vena cava communicates with the abdominal cavity by numerous orifices, is of much importance. For, in this way, this vein must be regarded as a large blood-reservoir, conducting, very probably, the nutritive fluid, after its transudation through the intestinal canal, into the general blood current.

The pericardium of the Cephalopoda sustains, perhaps, analogous relations to the blood-system, for, with *Nautilus*, it is said to communicate with the abdominal cavity, and with the principal vena cava, with the other Cephalopoda ; see the concluding paragraph of note 1, § 251.

CHAPTER VIII.

RESPIRATORY ORGANS.

$ 253.

All the Cephalopoda respire by means of Branchiae. These are situated in the cavity of the mantle, separated from the other viscera, and outside of the peritoneum. *Hectocotylus* forms the only exception in this respect, — its branchiae being free, and placed along the sides of the anterior half of the body under the form of numerous oblong, thin, thickly-set lamellae.[1] *Nautilus* has, on each side, two branchiae, while the other Cephalopoda have only one.

These organs have a more or less oblong, pyramidal form, and are attached, at one of their borders, to the external surface of the mantle by a thin cutaneous fold, — leaving their extremity to extend freely in front. The adherent edge is bordered by the trunk of the branchial artery, and by a large glandular band,[2] while the free border is occupied, from its base to the top, by the principal branchial vein. With *Nautilus*, and the Loligina, there are, between these vessels, numerous, triangular, branchial lamellae lying upon each other, and plicated upon both surfaces. But with the Octopoda, these lamellae are replaced by arches, which, on each side, pass from one vascular trunk to another, and have, upon their convex edge, a multi-plicated membranous band.[3] The branchial vessels extend from the branchial artery to the branchial vein through the lamellae and the branchial arches;[4] and in this passage, the venous is changed to arterial blood.[5] As there is no ciliated epithelium on the surface of the branchiae, the water is renewed exclusively by the rhythmical respiratory movements.[6] It enters, from both sides of the funnel, into the interior of the mantle when its borders are open, and is ejected through the funnel by the contractions of the mantle when its borders are closed.[7]

---

1 I have found such with *Hectocotylus tremoctopodis*; according to *Kölliker* (loc. cit.) that of *Argonauta* has also branchiae.

2 This glandular body has been regarded by *Cuvier* (Mém. p. 20, Pl. II. fig. 3, Pl. III. fig. 1, A.) and other zootomists as a muscular stripe; while *Mayer* (Analekten, &c., p. 56, Taf. V. fig. 1, No. 14), from its cellulo-vascular texture, has taken it for a spleen. I have been unable to find in it any muscular fibres, but only numerous cells, and I am of the opinion that this enigmatical organ holds some special relations with the venous system.

3 See *Owen*, On the Nautilus, p. 30, Pl. VI. fig. 1, 2, or Isis, p. 26, Taf. IV., or Ann. d. Sc. Nat. p. 124, Pl. III. fig. 2, 3, and *Valenciennes*, loc. cit. p. 281, Pl. IX. X.; The Catalog. of the Physiol. Ser. II. Pl. XXI. XXII. (*Sepia*); *Treviranus*, Beobacht. aus. d. Zoot. u. Physiol. p. 37, Taf. VIII. fig. 52–54; *Grant*, Transact. of the Zool. Soc. I. Pl. II. XI. (*Loligopsis* and *Sepiola*); *Cuvier*, Mém. p. 20, Pl. II. III.; *Delle Chiaje*, Descriz. Tav. XIX. fig. 1–5 (*Octopus*); and *Férussac*, loc. cit.

4 The number of the branchial lamellae and arches varies very much. With *Nautilus*, each branchia is composed of a double row of forty-eight lamellae; the long-bodied Loligina have a double row also, composed of sixty to ninety lamellae. With *Sepia*, there are thirty pairs; and with the Octopoda, the number of branchial arches is still less; there are only fifteen pairs with *Argonauta*, and twelve alone with *Octopus*, and *Eledone*.

5 For the distribution of the blood-vessels in the branchiae of *Sepia*, see *Tilesius*, De Respirat. Sepiae officinalis, Tab. I. II.

6 That, with the Cephalopoda, which are in general so poor in cilia, there should be no ciliated epithelium on the branchiae, is so remarkable a fact, that I have had it confirmed from fresh specimens by my friend *H. Koch* at Trieste, although *Sharpey* (Cyclop. I. p. 619) had already spoken of it.

7 For these respiratory movements, see *Gravenhorst*, Tergestina, p. 1, and *Wagner*, in the Isis, 1833, p. 159.

§ 254.

The existence of an Aquiferous system with the Cephalopoda cannot be doubted.[1] It occupies the entire trunk of these animals, and terminates by two orifices between which lies the excretory duct of the ink-sac, and which are often situated upon a small tubular eminence of the peritoneum. Each of these orifices leads into a spacious, thin-walled cavity (lateral cell),[2] situated near the pericardium. It contains the two venae cavae with their appendages, and communicates, by orifices and canals, with other aquiferous cells surrounding the various viscera, — such as the stomach and the caecum, as well as with the two so-called branchial hearts. These cells send a canal to the special genital glands.[3]

With *Nautilus,* there are, on each side, in the abdominal peritoneum, three orifices, through which the water of the cavity of the mantle enters into the lateral cavities.[4] There is another system of aquiferous canals under the skin of the head and neck. It consists of several large reservoirs which extend somewhat deeply between the organs of this portion of the body. These reservoirs communicate externally by orifices situated upon different points of the head.[5]

CHAPTER IX.

ORGANS OF SECRETION.

I. Urinary Organs.

§ 255.

The Urinary organs of the Cephalopoda, which have hitherto been much doubted, are particular appendages of the Venae cavae. With all the species having two branchiae, the two Venae cavae, formed by the division of the great median sinus, and which extend obliquely through the two lat-

---

1 For this aquiferous system, see *D'Orbigny,* in *Férussac,* loc. cit. Introduct. p. 20, Ouvertures, aquifères, and *Delle Chiaje,* Descriz. I. p. 53, Apparato-acquoso o idro-pneumatico. Both of these naturalists have included in this system the lachrymal openings and the space circumscribed by the ocular capsules.

2 See *Swammerdamm,* loc. cit. p. 354, Taf. LI. fig. 1, q. q, and Taf. LII. fig. 10, g. g ; *Brandt,* Media. zool. II. p. 309, Taf. XXXII. fig. 1, 24, l. i (*Sepia*); *Cuvier,* Mém. p. 15, Pl. I. fig. 1 r. r, and *Mayer,* Analekt. &c. p. 51, Taf. V. fig. 1. t. u. (*Octopus*); *Savigny,* loc. cit. Pl. I. fig. 12, 31 : g. g (*Octopus* and *Sepia*), and *Férussac,* loc. cit. (*Octopus*), Pl. XII. fig. 1. Pl. XIII. fig. 2, Pl. XIV. fig. 1, f. f r. r. See also *Krohn,* in *Müller's* Arch. 1839, p. 353.

3 *Delle Chiaje,* Descriz. Tav. XV. fig. 1. q. (*Tremoctopus*).

4 *Owen,* On the Nautilus, p. 32, or Isis, p. 27, or Ann. d. Sc. Nat. p. 127, and *Valenciennes,* loc. cit. p. 285, Pl X. fig. 1, 2.

5 With *Tremoctopus violaceus,* there are four very distinct *Foramina aquifera.* Two of these are situated at the superior part of the head behind the base of the superior arms, and the other two on the sides of the funnel (*Delle Chiaje,* loc. cit. Tav. LXXI. (11) fig. 10, p., *Férussac,* loc. cit. p. 92, *Octopus,* Pl. XVIII. XIX. fig. 1). With *Octopus tuberculatus,* there are only these last two openings (*Delle Chiaje,* loc. cit. Tav. LV. (3), fig. 1, d. d. ; *Wagner,* in *Heusinger's* Zeitsch. f. d. organ. Physik. III. p. 227, Taf. XII. fig. 1, and *Férussac.* loc. cit. p. 88, *Octopus,* Pl. VI.3 fig. 2).

It is the same also with *Ommastrephes todarus* (*Férussac* loc. cit. *Ommastrephes,* Pl. II. fig. 3, 10). With *Octopus indicus,* there are eight small aquiferous orifices between the arms near the mouth (*Férussac,* loc. cit. p. 25, *Octopus,* Pl. XXVI. fig. 1). According to *D'Orbigny,* there are only six, in the same situation with *Sepia, Loligo, Onychoteuthis,* &c.

eral aquiferous cells to the base of the branchiae, have, exteriorly, variously ramified, glandular tufts which project into the aquiferous cells.[1] Sometimes similar appendages are found also upon the principal veins which open, in these cells, into the venae cavae.[2] With *Nautilus*, which has on each side in the peritoneum four venae cavae, each of these last extends between two cells each of which receives a part of the glandular appendages with which it is provided.[3]

These organs, for a long time known as the Spongy bodies, can now be regarded positively as kidneys; for, by chemical analysis, it has been proved that they secrete uric acid.[4] Careful examination of them has shown that their parenchyma consists of a tissue of contractile fibres,[5] among which are spread branches coming from the venae cavae.

This parenchyma is surrounded by a structureless membrane, covered with several layers of nucleated granular cells. The urine is secreted from the external surface of this cellular layer; it is of a dirty-yellow color, and escapes immediately into the peritoneal cavities, and thence is discharged externally through their orifices, which may, therefore, be taken for urethral canals. These spongy appendages of the veins ought, therefore, to be regarded as everted glandular follicles, the secreting cells being situated externally and the blood-vessels within.[6] Not unfrequently, the reddish crystals formed in the urine, completely incrust these glands, giving them their peculiar color.[7]

The so-called Branchial hearts of the Dibranchiata contain no trace of muscular fibres, and appear to be in some way connected with the urinary organs. They are round, hollow, thick-walled, and lie upon the course of the venae cavae between the last renal masses of the branchiae, so that the blood of these veins passes into their cavities and bathes their spongy walls.[8] Their color is violet with the Octopoda, and pale-yellow with the

1 *Cuvier*, Mém. p. 18, Pl. II. fig. 1, 3, Pl. III. fig. 1, x. x; *Wagner*, Icon. zool. Tab. XXIX. fig. 14, q. q. 16; *Delle Chiaje*, loc. cit. Tav. LXXXVII. XCI. XCIII. XCIX. (17, 21, 23, 19); *Carus*, Erläuterungstafeln, Hft. VI. Tab. II. fig. 15, 17; *Mayer*, Analekten, Taf. V. fig. 1, s. s. (*Octopus*); *Grant*, Transact. of the Zool. Soc. I. Pl. II. fig. 8, a. b, Pl. XI. fig. 9, b. b. (*Loligopsis* and *Sepiola*); *Van Beneden*, loc. cit. Pl. III. fig. 5, f. f. (*Argonauta*).*
2 *Krohn*, in *Muller's* Arch. 1839, p. 355, and *Brandt*, loc. cit. Taf. XXXII. fig. 2. x.
3 *Owen*, On the Nautilus, p. 31, Pl. V. No. 6, Pl. VI. fig. 1, No. 6, or Isis. p. 26, Taf. III. IV., or Ann. d. Sc. Nat. p. 126, Pl. III. fig. 1, 2; and *Valenciennes*, loc. cit. p. 286, Pl. X. fig. 2, 7.
4 These appendages have been successively regarded as absorbent vessels, a rudimental portal system, a spleen, accessory branchiae, blood-reservoirs, genital organs, &c. *Mayer* (Analekt. &c. loc. cit. p. 54) was the first to regard them as urinary organs, but this view was not commonly received. The two peritoneal cavities containing these organs, were also taken by him for urinary bladders, and their orifices as urethrae. The same function has also been attributed to these organs by *Savi* (Atti della terza riunione degli scienziati

tenuta nel Firenze, 1841, p. 396, or Isis, 1843, p. 417).
At my request, *E. Harless*, while at Trieste, subjected these organs to a chemical analysis, and, as he obtained from their contents purpurate of Ammonia, there can be no doubt that they are really kidneys.
5 The contractility observed in these appendages is due, without doubt, to this fibrous tissue (*Krohn*, in *Froriep's* neue Notiz. XI. 1839, p. 214, and *Erdl*, in *Wiegmann's* Arch. 1843, p. 162).
6 I am indebted for this remarkable histological fact to a recent communication from *Harless*.†
7 I have often found in the kidneys of the *Sepia officinalis* groups of rhomboidal crystals of a crimson red color. *Krohn* (*Froriep's* neue Notiz. XI. p. 215) has found them constantly with *Sepia*, but has sought in vain for them with *Octopus*, and *Loligo vulgaris*.
8 The so-called branchial hearts, which are wanting with *Nautilus*, and which, it is supposed, pour the blood into the branchiae with the other Cephalopoda, are surrounded by a smooth peritoneal envelope, and have, internally, a cavernous aspect; see *Cuvier*, Mém, Pl. II. fig. 3, No. 9 (*Octopus*); *Carus*, Erläuterungstaf. Hft. VI. 1843, Tab. II. fig. 18 ([*Sepia*). With the Loligina, a constriction sit-

* [§ 255, note 1.] See also *Milne Edwards*, Règne anim. loc. cit. Pl. 1*. r. (*Octopus*). — ED.
† [§ 255, note 6.] For this communication in full with figures of *Harless*, see *Wiegmann's* Arch.

1847, p. 1, Taf. I. His chemical as well as his histological results, can leave little doubt as to the Renal nature of these organs. — ED.

Loligina. Their walls are composed of a dense web of cells, which, with the Octopoda, contain round, violet nuclei, of a crystalline texture and resembling entirely those found in the renal cells of the Gasteropoda.[9]

II. Organs of Special Secretions.

§ 256.

The Ink-sac is an organ generally common with the Cephalopoda. It is usually pyriform, situated upon the median line of the abdomen, and often enveloped with a peritoneal layer of silvery lustre.[1] Its apex points forwards and upwards, towards the funnel. The walls of its generally small cavity are cavernous,[2] and secrete the well-known black pigment, which, through contractions, passes into the funnel, and is then expelled, mixed with the water of the sea, which is passing out of the body. The excretory duct of this sac runs along the rectum, and terminates just behind the anus, or opens into the rectum.[3]

As an organ, also, of special secretion, ought to be regarded the completely-closed chambers found in the shell of the Nautilina; for it is said that their walls, like those of the natatory bladder of fishes, secrete a gas.[4]

CHAPTER X.

ORGANS OF GENERATION.

§ 257.

The Genital organs of the Cephalopoda are always distributed upon two individuals, and present very remarkable peculiarities.

uated at the inferior or lateral portion, separates these bodies into two unequal divisions ; see *Brandt*, loc. cit. Taf. XXXII. fig. 22, q. r. ; The Catal. of the Physiol. Ser. II. Pl. XXII. f. x. (*Sepia*); *Delle Chiaje*, loc. cit. Tav. XCI. XCIII. XCV. XCVI. (21, 23, 25, 26), (*Loligo* and *Sepiola*).

9 *Erdl* (*Carus*, Erläuterungstaf. Hft. VI. p. 7) has published an observation on the glandular nature of these bodies, and the resemblance of their parenchyma with that of the kidneys of *Helix*, all of which I have been able to fully confirm. Nevertheless, these organs demand further chemical and histological investigation.

1 This organ, which is entirely wanting with *Nautilus*, and *Hectocotylus*, has an elongated form with the long-bodied species, and is large with those whose body is short ; see *Wagner*, Icon. zoot. Tab. XXIX. fig. 20, h. i. (*Octopus*); *Ferussac*, loc. cit. *Argonauta*, Pl. I. fig. 2, 3 ; *Brandt*, loc. cit. Tab. XXXII. fig. 1, 24, o. (*Sepia*). That of *Sepiola* is quite remarkable in having an extraordinary development in certain individuals, at particular seasons, without reference to the sexes. There are then found two long bodies adherent by a constriction to the sides of the otherwise simple ink-sac. These lateral bodies are black, composed of a glandular tissue continuous with that of the sac,

and surrounded by a muscular layer. *Peters* (*Müller's* Arch. 1842, p. 329, Taf. XVI. fig. 1, b. b. 3–10) has seen it contract regularly. With the specimens preserved in alcohol the ventral surface appears colorless. *Grant* appears to have been unaware of this increased development at certain times with *Sepiola*, for he has attributed to them in general a trilobed ink-sac (Transact. of the Zool. Soc. I. p. 82). In the *Sepiola* figured by *Delle Chiaje* (Descriz. Taf. XI. fig. 4, i.), the two lateral lobes in question are very distinct.

2 *Delle Chiaje*, Descriz. I. p. 74, Tav. XIII. fig. 1, 2. Tav. XVIII. fig. 4 (*Loligo*, *Octopus* and *Eledone*).

3 This last case obtains with the Loligina.

4 *Owen* (On the Nautilus, p. 47, or Isis, p. 39) has, it is true, left undecided the question whether these chambers are filled with gas or with liquid ; but, according to *Frolik* (Ann. of Nat. Hist. XII. p. 174) the chambers of *Nautilus pompilius* really contain a gas composed chiefly of nitrogen without any trace of carbonic acid. These chambers resemble, in many respects, the swimming bladders of fishes, and, like them, serve, perhaps, to facilitate the ascension and descension of these animals in the water.

The Eggs, at their escape from the ovary, are oval, and have a yellow, or rose-colored vitellus, containing a germinative vesicle and dot.

The vitelline membrane has transverse and longitudinal folds on its internal surface, which extend into the vitellus, — giving the eggs a reticulated aspect.[1]

The Spermatic particles are very active, of a cercarian, or a simply capillary form, and, as a whole, give the sperm a white color. Those of a cercarian form are proper to the Loligina, and consist of a cylindrical body to which is rather abruptly attached a small and pretty long tail.[2] Those of a capillary form are found with the O......oda, not only with the males of *Octopus* and *Eledone*, but also with *Licetocanycus*.[3]

§ 258.

The Ovary, always simple, is situated, at the base of the sac of the mantle, in a solid envelope (ovarian capsule) of a round or oblong form, and derived from the peritoneum. At its circumscribed point the proper ovary commences as a multi-lobulated body filling its cavity.[1] The eggs, which are developed in the parenchyma of these lobes, appear first as round prominences; they gradually increase, and, finally, are attached to the ovary only by a small peduncle. At this epoch the ovary furnishes them with a thin envelope (egg-capsule), through which, in the mature eggs, the reticulated folds of the vitelline membrane can be seen.[2] When the eggs are fully matured, their capsules burst, and they fall into the ovarian capsule, after which, their proper capsules fade and finally disappear.[3] From the ovarian capsule they pass into the oviduct through an infundibuliform opening; but, beside this opening, some Octopoda have also two others belonging to an aquiferous canal, and which, perhaps, play an important part in the fecundation of the eggs.[4] The oviduct is simple, or double, and extends directly in front opening at the base of the funnel near the rectum. With *Argonauta*, and *Tremoctopus*, alone, the two female genital orifices are situated,

1 See *Kölliker*, Entwickelungsgesch. &c. p. 1, 9, Taf. I. fig. 9-12. These longitudinal and transverse folds have been observed in the eggs of *Sepia* and *Sepiola*; the longitudinal only are found in those of *Argonauta*, *Tremoctopus*, *Octopus*, *Eledone*, &c.

2 See my Beiträge z. Naturgeschichte z. wirbell. Thiere, in the Neuesten Schrift. d. naturforsch. Gesellsch. in Danzig. III. 1839, Hft. II. p. 54, Taf. II. fig. 47 (*Loligo*); *Milne Edwards*, Ann. d. Sc. Nat. XVIII. 1842, p. 337, Pl. XII. fig. 6, Pl. XIII. fig. 7 (*Loligo* and *Sepia*), and *Peters*, in *Muller's* Arch. 1842, p. 334, Taf. XVI. fig. 14 (*Sepiola*).

It is easy to observe the development of the spermatic particles in the testicles. According to my observations, the daughter-cells in the mother-cells, are developed into as many spermatic particles the tails of which rupture one of the thin sides of the mother-cell.

3 *Milne Edwards* (loc. cit. Pl. XIII. fig. 11, Pl. XIV. fig. 5), has erroneously figured the spermatic particles of *Octopus* and *Eledone* with a very large body; for it is only a small button-like enlargement; see *Valentin*, Report. 1837, p. 140, and *Philippi*, in *Muller's* Arch. 1839, p. 398, Taf. XV. fig. 11. This last has represented the tail of that of *Eledone* too short.

I have found the spermatic particles of *Hectocotylus tremoctopodis* to be exactly like those of *Eledone*.

1 *Cuvier*, Mém. p. 31, Pl. IV. fig. 6, a. b., *Van Beneden*, loc. cit. Pl. V. fig. 2, a.; *Delle Chiaje*, Descriz. Tav. XIV.-XVI., and *Grant*, Trans. of the Zool. Soc. I. Pl. II. fig. 9 (*Octopus*, *Argonauta*, *Eledone*, *Loligo*, and *Loligopsis*).

2 *Delle Chiaje*, Descriz. Tav. XV. fig. 15, and *Kölliker*, Entwick. &c. Taf. I. fig. 9 (*Sepia*): *Carus*, Erläuterungstaf. Hft. V. Taf. II. fig. 9 (*Eledone*).

3 With *Sepia*, the dehiscence produces a rent with irregular borders (*Kölliker*, loc. cit. p. 13); with *Rossia*, and *Sepiola*, a simple round opening, which, with *Nautilus*, according to a figure of *Owen's*, has crenulate borders, and, according to *Delle Chiaje*, is regularly denticulated with *Eledone*; see *Grant*, Transact. loc. cit. I. p. 84, Pl. XI. fig. 12, and *Owen*, Ibid. II. Pl. XXI. fig. 18, also, On the Nautilus, p. 42, Pl. VIII. fig. 9, c. c., or Isis, p. 35, Taf. III. or Ann. d. Sc. Nat. p. 142, Pl. IV. fig. 9, c. c., and *Delle Chiaje*, loc. cit. Tav. LV. (3) fig. 15.

4 These two aquiferous canals form a communication between the ovarian capsules and the aquiferous cells surrounding the branchial hearts (§ 250). They are found with *Octopus*, *Eledone*, and *Tremoctopus*; see *Krohn*, in *Muller's* Arch. 1839, p. 357; *Kölliker*, Entwick. &c. p. 11, and *Delle Chiaje*, Descriz. Tav. XV. fig. 1, q. (*Tremoctopus*).

wide apart, in the region of the base of the branchiae.[5]　When there is
only a single oviduct, it terminates always on the left side.[6]　With most
of the Octopoda, the oviducts, at near the middle of their course, traverse
a round glandular body, the internal surface of which is longitudinally plicated ; from this point to their extremity, they are covered with analogous
glandular folds.[7] · This glandular body is absent with the Loligina, but the
walls of their oviduct become thick and glandular before terminating.[8] It
is very probable that this glandular apparatus furnishes the materials of the
various envelopes of the eggs after their escape from the ovary.

There is another peculiar glandular apparatus (Nidamental glands) having no direct connection with the genital organs, which consists of two hollow, pyriform, whitish bodies, situated upon the ink-sac of the female Loligina. These bodies have a lamellated structure and their obtuse extremity
extends forwards ; they open near the genital orifice.[9]　Sometimes,
directly in front of these glands, there is another gland, simple or double
(accessory nidamental gland), of a reddish color, lobulated posteriorly, composed of cœca, but apparently without any excretory duct.[10]　This whole
glandular apparatus secretes, perhaps, a substance with which the eggs are
coated as they pass from the oviduct, and which serves to glue them to
foreign bodies.

The deposited eggs (spawn) are always surrounded with envelopes and
prolongations of various forms, by which they are bound together and
attached to submarine bodies.　Thus, those of *Sepia* are enclosed, each, in a
black, oval capsule, composed of several horny layers, which is prolonged
at one of its extremities into a short, cleft peduncle, by which the eggs are
attached, singly or in groups, to marine plants ;[11] but those of the Loligina
are united by a colorless gelatinous substance into a chaplet, and are
enclosed, moreover, each, in a special capsule one of the extremities of which
has a small peduncle ; thus arranged, they form large masses floating free

5 With *Octopus*, *Eledone*, *Tremoctopus*, and
Loligo sagittata, there are two oviducts ; see
Cuvier, Mém. Pl. I. fig. 1, q. q. Pl. II. fig. 1, r. r.;
Mayer, Analekten, Taf. V. fig. 1, i. i. f f.; *Ferus-
sac*, loc. cit. *Octopus*, Pl. XV. fig. 2, l. l.; *Carus*,
Erläuterungstaf. IIft. V. Taf. II. fig. 7, h. h.;
Wagner, Icon. zool. Tab. XXIX. fig. 20, m. m.;
Owen, Trans. of the Zool. Soc. II. p. 121, and
Cyclop. I. p. 553. With *Argonauta*, these two
oviducts are very long and flexuous ; see *Delle
Chiaje*, Descriz. Tav. XIV. fig. 1, z. n.; and *Van
Beneden*, loc. cit. Pl. V. fig. 1, 2; *Ferussac*, loc.
cit. *Argonauta*, Pl. I. 4 fig. 2, s. s.*

6 Such is the case with *Nautilus pompilius*
(*Owen*, loc. cit.), *Loligo vulgaris* (*Carus*, Erläuterungstaf. IIft. V. Taf. II. fig. 10, m. l.), *Sepia
officinalis*, *Sepioteuthis*, *Rossia*, &c. According
to *Rathké* (Mém. d. St. Petersburg, loc. cit. p. 161,
Pl. II. fig. 10. p. q.), the simple oviduct of *Loligopsis* passes directly to the posterior part of the
body and terminates at the ventral median line
between the two fins.

It is, however, desirable that this remarkable
exception to the general rule should be confirmed
by other anatomists, for *Grant* (loc. cit.) is wholly
silent upon the course of the oviduct in the females
of *Loligopsis* which he examined.

7 With *Octopus*, *Eledone*, and *Tremoctopus*,
each oviduct has such a glandular enlargement,
but it is entirely wanting with *Argonauta* ; see
Cuvier, Mém. p. 32, Pl. IV. fig. 6, g.; *Ferus-

sac*, loc. cit. *Octopus*, Pl. XV. fig. 9, 10.; *Mayer*,
Analekten, Taf. V. fig. 1, g. h. (*Octopus*) ; *Delle
Chiaje*, Descriz. Tav. XV. fig. 1, n. Tav. XVI. fig.
6 ; and *Wagner*, Icon. zool. Tab. XXIX. fig. 20, n.
n. (*Tremoctopus* and *Eledone*).

8 This is so with *Loligo*, *Sepia*, *Sepioteuthis*,
Sepiola, &c.; see *Owen*, Trans. &c. II. p 121,
Pl. XXI. fig. 18. e. (*Rossia*). With *Nautilus*, the
very short oviduct has glandular walls in its whole
extent (*Owen*, loc. cit.).

9 See *Swammerdamon*,Bib. d Natur. p. 354, Taf.
LII. fig. 10, g. g.; *Brandt*, loc. cit. p 310, Taf.
XXXII. fig. 25, k. l ; fig. 28–31 (*Sepia*) ; *Delle
Chiaje*, Mém. IV. p. 102, and Descriz. I. p. 37,
Corpi adiposi, Tav. LVIII. (12), fig. 10, a , 11, c.;
Peters, in *Muller's* Arch. 1842, p. 335, Taf. XVI.
fig. 6, f. f. (*Sepiola*) ; *Owen*, Trans. of the Zool.
Soc. II. Pl. XXI. fig. 18, g. g. (*Rossia*)

10 With *Sepia*, and *Sepiola*, this gland is single,
and divided by deep fissures into three lobes (see
the figures cited in the preceding note, and *Owen*,
Trans. &c. loc. cit. Pl. XXI. fig. 19, 20); with
Loligo, and *Rossia*, it is double, and each is divided into two lobes (*Owen*, Ibid. Pl. XXI. fig. 18.
b. b.).

11 *Cuvier*, Nouv. Ann. du Mus. d'Hist. Nat. I.
1832, p. 153, Pl. VIII. fig. 1–4 ; *Carus*, Erläuterungstaf. IIft. III. Taf II. fig. 16 ; *Owen* (Cyclop.
loc. cit. p. 560, fig. 244), and *Kölliker*, Entwickel.,
&c. p. 14.

* [§ 258, note 5.]　See also *Milne Edwards*, Règne anim. loc. cit. Pl. 1b. 1c. n. (*Octopus*). — ED.

in the sea.[12]   With those of *Argonauta* and *Tremoctopus*, the envelope
is composed of a solid, homogeneous, colorless substance, and at their pointed
extremity there is a small filament; these filaments being entangled together, the eggs form large botryoidal masses.   *Argonauta* attaches these
bunches to the convex portion of its shell;[13] but *Tremoctopus*, with which
these masses are in chaplets, forms them into a staff-like structure, by
means of a tissue of leathery consistence, secreted for this purpose.[14]   With
the other Cephalopoda, the eggs are arranged in tubes or in fillets.[15]

§ 259.

The simple, round, or oblong, whitish Testicle, is situated, in most
species, at the bottom of the cavity of the mantle.   It is surrounded by a
capsule derived from the peritoneum but adherent to it only at one point.   It
is composed of numerous ramified cylinders, converging from the periphery
towards the centre, which is occupied by a narrow irregular cavity.

The sperm is formed in the intervals of the cylinders and thence passes
into the deferent canal, which, at its passage from the testicular capsule, is
narrow and very flexuous, but suddenly dilates at its upper extremity, —
where its walls are thick and glandular, and have a longitudinal fold on
their internal surface.

This glandular portion of the *Vas deferens* receives the orifice of an
equally flexuous caecum, which is probably an organ of secretion; and
terminates, finally, in a large sac with muscular but thin walls which are
plicated longitudinally.   This sac, known as the *Bursa Needhamii*, is
followed by a fleshy tube (*Ductus ejaculatorius*), which extends directly in
front and projects, as a short penis, into the cavity of the mantle near the
rectum.[1]

With the Loligina, and with *Octopus*, and *Eledone*, the sperm is not freely
evacuated, but is enclosed in very complicated organs (*Spermatophores*),
which, at the epoch of procreation, accumulate in abundance in the *Bursa
Needhamii*, and are large enough to be seen with the naked eye.   They are
always cylindrical and consist of a homogeneous, colorless, solid tube, round
at the anterior extremity, but at the posterior end, somewhat constricted and
then dilated into a kind of sphere.   Each of these Spermatophores contains
two kinds of organs: a very thin sac filled with spermatic particles, and an
apparatus to project these particles outwards.

The Sperm-sac always contains fully-developed spermatic particles
bundled together: it nearly entirely fills the cavity of the tube, through

12 These chaplets are long with *Loligo vulgaris*,
and short with *Sepioteuthis*; see *Burdach*, De
quibusd. anim. marin. p. 155, Tab. XII.; *Ferussac*, loc. cit. *Loligo*, Pl. X. fig. 1, 1¹, and *Kölliker*,
loc. cit. p. 14.

13 *Rang*, Magaz. d. Zool. 1837, V. Taf. LXXXVII.
LXXXVIII.; *Férussac*, loc. cit. *Argonauta*, Pl.
I.J.

14 This body has a peculiar structure, and undoubtedly, is made by the animal itself.   It is
composed of numerous, superposed, very distinct
layers of a granular, probably coagulated substance; — forming a kind of staff or baton by which
the eggs are bound together in groups; for, according to *Kölliker* (loc. cit. p. 14), it (*Tremoctopus*)
carries the entire mass attached to the suckers of
one of its arms.

15 See *Quoy* and *Gaimard*, Ann. d. Sc. Nat.
XX. 1830, p. 472, Pl. XIV. B., and *Ferussac*,
loc. cit. *Octopus*, Pl. XXVIII. fig. 3.

1 For the male genital organs of *Octopus*, see
Cuvier, Mém. loc. cit. p. 32, Pl. IV. fig. 5; he
regards the upper glandular portion of the deferent
canal as a *Vesicula seminalis*, and the coecum
appended to it as a *Prostata*.   See, also, *Delle
Chiaje*, Descriz. Tav. VI. fig. 2, Tav. XI. fig. 2, 3,
Tav. XII. fig. 28 (*Octopus, Sepia* and *Loligo*);
Wagner, Icon. zoot. Tab. XXIX. fig. 22 (*Octopus*);
Peters, in *Müller's* Arch. 1842, p. 332, Taf. XVI.
fig. 2, 3 (*Sepiola*), and especially the beautiful
figure of those of *Sepia* by *Milne Edwards*, in
the Ann. d. Sc. Nat. XVIII. 1842, p. 344, Pl. XV.

which it is seen of a milk-white color. Its posterior extremity is attached, by a short, small ligament, to a kind of piston which forms the anterior portion of the projectile apparatus.

This piston is a solid, cylindrical body, continuous behind with a spiral ligament which is contained in a thin sheath extending to the posterior extremity of the tube, in a fold of which it terminates.[2]

The Spermatophores are evidently formed in the upper glandular portion of the deferent canal, where droplets of sperm are often seen arranged in rows, and, at first, appear surrounded by simple, colorless envelopes; these, as they advance in the *Vas deferens*, gradually resemble more and more the perfect Spermatophores.

Those found in the *Bursa Needhamii* are always regularly arranged, and sometimes form, lengthwise, several superposed layers. Their anterior extremities always point forwards, and not unfrequently their posterior ends are bound together by long, flattened, interlaced filaments. These Spermatophores are in the highest degree hygroscopic: they absorb liquids very quickly, and then their posterior extremity bursts, allowing the escape of the compressed spiral ligament together with its sheath, and the piston, which draws with it the sperm-sac to which it is attached.[3]

The projection of the seminal sacs occurs, most probably, at the moment when, during coition, the Spermatophores pass from the penis of the male into the sac of the mouth of the female. A true intromission of the penis into the female genital opening, appears impossible with these animals, so that coition consists only in a simple juxtaposition of the genital organs.[4]
The fecundation of the eggs should occur very early — while the eggs are

2 *Needham* (An account of some new Microscopical Discoveries, London, 1745, or Nouv. découv. faites avec le Microsc. Leyde, 1747, Pl. III. IV.), was the first who described accurately the Spermatophores of *Loligo vulgaris*. With those of the Loligina, the posterior extremity is enlarged, with one or two constrictions, and contains the spiral ligament with its sheath; the piston also, from its deep-brown color, is easily seen; see *Krohn*, in *Froriep's* neue Notiz. XII. 1839, p. 17, fig. 20 (*Sepia*); *Milne Edwards*, Ann. d. Sc. Nat. XVIII. 1842, p. 335, Pl. XII. fig. 1-5; XIII. fig. 1-6 (*Loligo* and *Sepia*); *Peters*, in *Muller's* Arch. 1842, p. 334, Taf. XVI. fig. 11 (*Sepiola*). With those of Octopus, and *Eledone*, the posterior enlargement is very slight, and often, at this point, the envelope is entirely involuted; the seminal sac, moreover, is remarkable from its spiral form; see *Milne Edwards*, loc. cit. p. 338, Pl. XIII. fig. 8-10, XIV. fig. 1-6 (*Octopus* and *Eledone*); *Philippi*, in *Muller's* Arch. 1839, p. 301, Taf. XV. fig. 1-6 (*Eledone*); this last author has erroneously taken the spiral turns of the ligament for hooks pointing backwards.

3 *Redi* (De Animalculis vivis quæ in corporibus animalium vivorum reperiuntur, Lugd. Batav. 1723, p. 252, Tab. II. fig. 2), was the first who saw these Spermatophores; but he took them for worms. *Swammerdamm*, on the other hand (Bib. d. Natur, p. 353, Taf. LII. fig. 6, 7), and especially *Needham* (loc. cit.), had a correct idea of their nature, for they regarded the white substance they contained as sperm, and the Spermatophores themselves as a kind of cases or machines. But this did not prevent the later anatomists from regarding them as parasites. Thus *Delle Chiaje* described those of *Octopus* and *Sepia* under the names of *Monostomum octopodis*, and *Scolex dibothrius* (Mem. IV. p. 53, Tav. LV. fig. 8, 14, 9, 9.d). Even latterly, this naturalist has not relinquished this opinion, for he

has figured anew, as an Entozoa, and even as an *Echinorhynchus*, the uncoiled Spermatophores of *Loligo* (Descriz. III. 1841, p. 138, Tav. XI. fig. 12, 13). *Wagner*, also, formerly regarded those of *Sepia* as containing an *Echinorhynchus*, and has figured as such the piston with the spiral ligament in a rudimentary state (Lehrb. d. vergleich. Anat. 1835, p. 312, and, *Muller's* Arch. 1836, p. 230, Taf. IX. fig. B. C.).

Carus went even still further, and described the Spermatophores as gigantic spermatic animals under the name of *Needhamia expulsoria*; and regarded the parts they contained, such as the sperm-sac, the piston, the spiral ligament, &c., as a colon, a small intestine, a stomach, a crop, and an œsophagus (Nov. Act. Acad. Nat. Cur. XIX. I. 1839, p. 3, Tab. I., and Erläuterungstaf. loc. cit. IIft. V. 1840, p. 4, Taf. I. fig. 10). It was not until 1839, a time when several naturalists were convinced of the presence of spermatic particles in the sperm-sac, that the true nature of these bodies was fully understood; see *Philippi*, in *Muller's* Arch. 1839, p. 301; *Krohn*, in *Froriep's* neue Notiz. XII. 1839, p. 17; *Siebold*, Beiträge z. Naturgeschichte d. wirbell. Thiere, 1839, p. 51; *Peters*, in *Muller's* Arch. 1840, p. 98, and *Milne Edwards*, Ann. d. Sc. Nat. XIII. 1840, p. 193. *Leuckart* (Zool. Bruchstücke, IIft. II. 1841, p. 93) has given the history and criticism of the opinions relating to the sperm machines of *Needham*.

4 *Aristotle* (Hist. Animal. lib. V. cap. 6) had already declared that the Cephalopoda copulate by a kind of embrace. From the observations of *Lebert* and *Robin* (loc. cit. p. 135, and Ann. d. Sc. Nat. IV. 1845, p. 95, Pl. IX. fig. 5, 6), it would appear that the males do not deposit the spermatic particles further in than the cavity of the mantle of the females; for they observed, with a female *Loligo*, numerous Spermatophores glued to the internal surface of this cavity, near the oviduct.

still at the bottom of the female genital organs; for, later, the action of the sperm would be obstructed by their solid envelopes. It must, therefore, be supposed that the sperm is carried from the cavity of the mantle into the ovarian capsule, either by means of the anti-peristaltic movements of the oviduct, or by the aid of the aquiferous system.[5]

§ 260.

With the individuals hitherto known as the *Hectocotyli*, the genital organs occupy a space disproportionably large to the size of the body. The round, smooth enlargement on their posterior extremity is a genital capsule, with thin walls, and containing the sperm and the copulatory organs.[1] The sperm forms a long, moniliform, clustered string, composed of thick oval bundles of spermatic particles, regularly bound together by fasciculi of hair-like spermatic particles.[2] In this clustered string are included, also, the *Ductus ejaculatorius* and the very long and retractile penis. With *Tremoctopus violaceus*, this penis sometimes projects between the fifth of the posterior pair of suckers, as a small cylindrical, folded prolongation.[3]

§ 261.

The Development of the Cephalopoda is almost without analogy, and, from the remotest times, has excited the curiosity of Naturalists; but it is only very recently that it has been correctly understood and followed from its first stages.[1]

After the disappearance of the germinative vesicle, the vitellus divides; but this segmentation is only partial. Usually, at the acute extremity of the vitellus, where the germinative vesicle is found, there appears a small elevation from the vitelline mass, divided into halves by a furrow. Each

5 According to *Kölliker* (Entwickel. &c. p. 11), the eggs are fecundated while yet contained in the ovarian capsule.

1 See *Hectocotylus octopodis*, in Ann. d. Sc. Nat. loc. cit. fig. 1–3, b.

2 I have so observed it with the males of *Tremoctopus violaceus*; and *Dujardin*, also (Hist. Nat. d. Helminth. p. 482), has observed a smooth cord composed of capillary spermatic particles with *Hectocotylus octopodis*, Cuv.

I am uncertain as to the origin of this cord, but, in the specimens preserved in alcohol, which I have examined, it appeared probable that it was primarily contained in the deferent or in the testicular canal.

3 The *Ductus ejaculatorius* of *Tremoctopus violaceus*, which is gradually continuous into the penis, begins by a well-marked, clavate thickening, projecting into the genital capsule, and apparently perforated at its upper extremity, at which point, perhaps, the sperm enters. Near the end of the penis, this canal has, over a considerable extent, small horny tubercles, and it is probable that this portion can be everted, thus allowing the possibility of an intimate union with the female organs. I am unable to say whether this is so with the other *Hectocotyli*. The penis of *Hectocotylus argonautae*, according to a figure of *Delle Chiaje* (loc. cit. Tav. XVI. fig. 1, a), and *Costa* (loc. cit. Pl. XIII. fig. 2s, e), projects from the posterior extremity of the body; but it may be, that with the specimens examined by these naturalists, this organ

had become free from an accidental rent of the genital capsule.

1 *Aristotle* (Hist. Animal. lib. V. cap. 16, 4), and in the last century, *Cavolini* (Abhandl. über die Erzeugung der Fische u. d. Krebse, 1792, p. 54) had already declared that, with *Sepia*, the vitelline sac is situated on the head of the embryo, and, as it were, hanging from the mouth; but it is only latterly that this statement has been thoroughly verified; see *Froriep*, Das Thierreich, Abth. V. 1806, p. 24, fig. 8–10; *Carus*, Erläuterungstaf. Hft. III. 1831, p. 10, Taf. II. fig. 16–30; *Cuvier*, Sur les œufs de Sèiche, in the Nouv. Ann. du Mus. I. 1832. p. 153, Pl. VIII. fig. 6–14, also in abstract in Ann. d. Sc. Nat. XXVI. 1832, p. 63, or *Froriep's* Notiz. XXXIV. p. 193; *Coldstream*, On the fœtus of *Sepia officinalis*, in the Lond. and Edinb. philos. Magaz. Oct. 1833, or *Froriep's* Notiz. XXXIX. p. 6; *Dugès*, Note sur le développement de l'embryon chez les Mollusques céphalopodes, in Ann. d. Sc. Nat. VIII. 1837, p. 107, Pl. V., or *Froriep's* neue Notiz. VII. p. 200, fig. 3–9.; *D'Orbigny*, in *Férussac*, loc. cit. *Loligo*, Pl. X. fig. 3–6; *Van Beneden*, Recherches sur l'embryogénie des Sépioles, in the Nouv. Mém. de l'Acad. de Bruxelles, XIV. 1841, Pl. I.; *Delle Chiaje*, Descriz. I. p. 38, Tav. VI. fig. 6, 7 (*Sepia*), Tav. XIV. fig. 14–21 (*Argonauta*), and Tav. XXIX. fig. 2–5 (*Sepiola*). But the first phases of their development remained unobserved, until *Kölliker* in 1844, filled this deficiency by his masterly work — Entwickelungsgeschichte der Cephalopoden.

of these halves is also divided, and so on, forming four, eight, &c., segments, each resembling a more and more acute triangle, with a converging apex, while its base is directly continuous with the remaining vitellus. After a certain number of segments have been formed by these longitudinal divisions, transverse furrows are seen separating the apices of the segments, and forming, at first, in the centre of the eminence, a ring composed of eight to sixteen portions; but finally, from a further segmentation in both directions, these furrows become a mass of increasingly smaller and more numerous parts.[2]

This portion of the vitellus, which, during this time, has also been developed at its periphery, is changed into a blastoderma composed of two layers. Upon this blastodermal membrane several folds appear, which are the first traces of the future embryo, viz. first, a median, uneven fold or rudiment of the mantle, and then two others, lateral, which ultimately form the eyes. Between these three folds are placed two others which become the two lateral halves of the funnel. Subsequently, the folds of the branchiae and arms appear; and among these last the two belonging to the ventral surface are first seen. Still later, the folds of the eyes and arms, and their surrounding parts, become more and more prominent upon the vitelline mass, thus forming the cephalic portion of the embryo. From this last, opposite the vitellus, the fold of the mantle is sketched as the future posterior portion of the body. The general form of the animal may, therefore, be recognized very early, although the cephalic portion quite exceeds that of the mantle.

At the posterior or dorsal surface of the cephalic portion, the mouth appears, first as a semilunar depression, and the internal layer of the blastoderma gradually extends from the border of this portion over the whole vitelline mass, producing, finally, a true vitelline sac. The external surface of this is covered with ciliated epithelium which gradually spreads over the other parts of the embryo, such as the lobes of the head, the arms, the eyes, and the mantle, while that of the branchiae, and the funnel, the halves of which have then united, is never ciliated.[3]

Of the Cartilages, the articular and cephalic are the first developed. The internal shell,[4] the nervous system, the heart with the vascular and respiratory systems, the digestive canal and its appendages, and the ink-sac, are formed successively, and may be easily seen at the termination of the embryonic life, when, also, are found some chromatic cells.

As to the vitelline sac, it should be remarked that it never communicates with the intestinal canal, as has hitherto been supposed.[5] The cephalic portion, which always extends upon this sac, embraces a part of it, so that it is divided by a constriction into an internal and external portion, the former of which extends even into the cavity of the mantle. The constricted portion is gradually elongated, and finally becomes a long, very narrow canal, extending from the cephalic extremity to the side of the

2 See *Kölliker*, loc. cit. p. 17, Taf. I.

3 With *Loligo*, the vitelline sac and its ciliated epithelium are formed quite early; while with *Sepia*, they do not appear until the embryo and its different organs have become quite large. With the first, the embryos have rotatory movements, but with the second this is not the case (*Kölliker*, loc. cit. p. 51).

4 With *Argonauta*, the external shell is formed while the embryo after its escape from the egg, is still persistent in the spawn inside the shell of its parent; see *Power*, in *Wiegmann's* Arch. 1845, I.

p. 379, and *Maravigno*, Ann. d. Sc. Nat. VII. 1837, p. 174.

5 Most of the earlier anatomists were led into error from the tenuity of the canal of communication, and the difficulty of its examination. They supposed that the external vitelline sac communicated with the oesophagus and stomach, by this canal; see *Carus*, loc. cit. Taf. II. fig. 27 (*Loligo*): *Cuvier*, loc. cit. Pl. VIII. fig. 9; *Dugès*, loc. cit. Pl. V. fig. 3 (*Sepia*) and *Van Beneden*, loc. cit. Pl. I. fig. 13 (*Sepiola*). *Kölliker* was the first to view it correctly (loc. cit. p. 56, Taf. IV.).

mouth, and producing a communication between the internal and external sacs.

The vitellus of the internal sac gradually disappears, and is replaced through this canal of communication by that of the external sac.   During the successive development of the organs contained in the cavity of the mantle, the internal vitelline sac is divided into lobes which are finally broken up and absorbed; while the intestinal canal, the remaining organ of the embryo, is developed, quite independently, from the vitelline mass.

As to the development of the males (*Hectocotyli*) of *Argonauta* and *Tremoctopus*, nothing is yet known except of its last period.   It has been observed, however, that, during the time they are in the egg, they have their proper form which is so remarkably different from that of the females.[6]

6 Had the fact that the *Hectocotyli* exist in the egg with their proper form, among the eggs of females of certain species, been properly observed, the true relations of these supposed parasites to the animals in which they live, would have been known long ago.   A passage of *Maravigno*, first properly interpreted by *Kölliker* (Ann. of Nat. Hist. loc. cit. p. 414) shows clearly that this Italian naturalist, in his researches connected with *Argonauta*, was in error only as to the eggs which contained the male individuals.

He says thus : " Mais encore que le petit poulpe, au sortir de l'oeuf, ne ressemble pas entièrement à ce qu'il sera par la suite ; c'est alors une sorte de petit ver (vermicello) pourvu de deux rangées de ventouses dans la longueur, avec un appendice filiforme à une extrémité, et un petit renflement vers l'autre, où il paraît que sont les organes de la digestion." — (Ann. d. Sc. Nat. VII. 1837, p. 173.) The mode of development of these males differs undoubtedly very much from that of the females.*

* [§ 261, note 6.]   The subject of the Hectocotylus to which such frequent mention has been made in these pages, is one that has elicited a good deal of attention of late years, but, now, happily, seems pretty definitely settled.   Chief among these investigators are *H. Müller*, and *Vérany* and *Vogt*. They have pretty clearly shown the non-independent character of these forms.   The details of these researches cannot here be given ; it may be remarked, however, that these observers have all studied these forms upon living specimens on the coast.   It has been shown that the *Argonautae* on which these Hectocotyli are found, have a highly-developed testicle, the situation and structure of which correspond to those of the common Cephalopoda, and which communicates with the Hectocotylus.

In conclusion, I may quote H. Müller's own words : " It is then proved that the Hectocotylus is formed on a male *Argonauta*, and is nothing but an arm metamorphosed in a very irregular manner.   This arm, or the Hectocotylus, is detached when it has been filled with the sperm which is formed in a true testicle of the *Argonauta* itself, and it then plays an apparently independent life.   In this condition, it meets the female *Argonautae* which, by a true copulation, it impregnates, as I have observed with the Hectocotylus of a *Tremoctopus*, and it resembles in this, as also by its movements, by a kind of circulation, and by the long duration of its life after detachment, a true male animal."

For the literature of this subject, see, beside the writings referred to in the above pages, *Kölliker*, Transact. Linn. Soc. London, XX. 1846, p. 9. Pl. I. ; Bericht von der zootomischen Anstalt zu Würtzburg, Leipzig, 1849, p. 67, Taf. I. II ; *Power* (*Madame*) Mollusques Méditerranéens, 1re partie Gênes, 1847-51, p. 34, 126, Pl. XLI. ; *Vérany* and *Vogt*, Ann. d. Sc. Nat. XVII. 1852, p. 146, Pl. VI. -IX. ; *H. Müller*, Ann. d. Sc. Nat. 1852, XVI. p. 132 ; also, *in extenso*, in *Siebold* and *Kölliker's* Zeitsch. IV. p. 1, Taf. I. and p. 346 ; and *Siebold*, in Ibid. IV. p. 122. — ED.

WITHIN a short time, the class Crustacea has received a contribution of so valuable a character that I cannot omit to mention it specially in a note. I refer to the large and comprehensive work of *Dana*, published this year (1853). This work, aside from its high zoological value, includes anatomical details and the discussion of principles in animal morphology, of great importance to the student of this interesting yet difficult class of animals.

It will be found that constant reference has been made to the anatomical details, in my notes; but the doctrines advanced as to the morphological structure of these animals, more than equally important, could be here given only in a separate form. I have been the more induced to include them here, from the fact that the work in question will have a very limited circulation, comparatively, and can be accessible only to a few. With these views, I have solicited *Professor Dana* to put his particular principles into a condensed form for this work, and he has kindly favored me with the following account:

The several types of structure among Crustacea are distinguished, primarily, by the different degrees of centralization or *cephalization* in the species, which degrees of cephalization are exhibited in the form of the body, and position, number, form or length of the appendages. The higher cephalization is seen in the larger number of organs that are pressed into the service of the senses and mouth; in the closely-crowded position and small size of these organs; in the little elongation of the antennae; and in the obsolescence of the abdomen and absence of abdominal appendages. Thus, in the Brachyura, nine segments and their pairs of appendages, out of the fourteen cephalothoracic, belong to the senses and mouth; they are all small, and gathered into a short space; the antennae are exceedingly small, excepting the basal joint which is the seat of sense; the abdomen in the males is small and without appendages. In the Maioids, the highest Brachyura, the head is very narrow, with the anterior antennae *longitudinal*, and the base of the outer antennae soldered *without suture* to the shell. The concentration is here most complete. The widening of the front in the Cancroids shows a relaxation of the concentration,

26

as do also other characteristics; the loosening of the outer maxillipeds in the true Cancers, and most swimming Crustacea and Corystoids, is another step in this relaxation; the elongation of the antennae in the Corystoids and Anomoura is another step; the loosening of the abdomen from the ventral surface of the cephalothorax; its becoming loosely inflexed or even extended; its taking appendages — are among the other steps seen in the Anomoura; the outer maxillipeds becoming pediform, and then the next pair pediform also, showing a tendency to a passage from the mouth-series to the foot-series, are other steps downward, observed in the Macrura; and the elongated abdomen with its regular series of organs as well as the elongated antennae, the union without fossettes, and eyes without sockets, all exhibit the relaxation of centralization that marks the Macrura.

A further degradation is seen in the obsolescence of some of the pairs of feet and abdominal appendages, as in the Mysis group; and the same principle is exemplified in the Brachyura, where the posterior cephalothoracic legs become small or rudimentary, or swimming legs.

There are, hence, two methods by which the passage of Crustacea from the higher to the lower grades takes place:

1. A diminution of the centralization leading to an enlargement of the circumference or sphere of growth at the expense of concentration, as in the elongation of the antennae, a transfer of the maxillipeds to the foot-series, and the elongation of the abdomen and abdominal appendages.

2. A diminution of force as compared with the size of the structure, leading to an abbreviation or obsolescence of some of the circumferential organs, as the posterior or cephalothoracic legs, or anterior antennae, or the abdominal appendages (if such appendages belong to the type embracing the species).

The Macrura, Anomoura, and Brachyura are alike in having normally nine cephalic annuli (out of the fourteen cephalothoracic), and but five foot-annuli. The Mysis and Squilla groups are in the same category. There are species that show a tendency to a transfer of the posterior mouth-annuli or appendages to the foot-series, but it is only a tendency. These together constitute the *First type* among Crustacea.

In the *Second type*, there are seven cephalic annuli and pairs of appendages, and seven foot-annuli or pairs of feet; such are the Isopoda, Anisopoda and Amphipoda.

In the *Third type*, there are normally six (or five) cephalic annuli, out of the whole normal number, fourteen, — the eight (or nine) posterior annuli belonging to the foot-series, part of which (the three posterior pairs and often more) are usually obsolete. Moreover, the abdomen, by the second law of degradation, mentioned above, is without appendages — such are the Entomostraca.

In the *Fourth type*, there are six (or five) cephalothoracic annuli, as in the Entomostraca, with which group they might be associated. But other peculiarities lead to a separation, and the species referred to are the Cirripedia.

In the *Fifth type*, there are five (or four) cephalothoracic annuli, out of the whole normal number fourteen; in other words, the mouth never includes more than a single pair of maxillae with the mandibles. Moreover, by the second law of degradation, all the jointed cephalothoracic appendages are wanting. These are the Rotatoria.

The following table presents a view of the number of cephalic annuli in these Types, and also the mean size:

	Typical No. of cephalic annuli.	Mean normal length in lines.
Type I. Decapoda or Podophthalmia. — Sub-type I. Brachyura,	9	24 (and breadth 24).
Sub-type II. Macrura,		36 (and breadth 6).
Type II. Tetradecapoda,	7	6
Type III. Entomostraca,	6–5	1
Type IV. Cirripedia,	6–5	1
Type V. Rotatoria,	5–4	1-9

See pp. 1406 and 1407 (loc. cit.), for observations on mean size in the Entomostraca and Cirripedia, where an important principle is brought out, and where, also, some explanations are furnished which make the statement given above of the mean size, intelligible. — Ed.

BOOK TWELFTH.

CRUSTACEA.

CLASSIFICATION.

§ 262.

In the Classification of the Crustacea, the remark of *Erichson*[1] should be adduced, that, with these animals, the external locomotive organs are not limited, as with the other Arthropoda, to the anterior part of the body, but may exist on all its segments, and often with a shape so changed, that, they become foot-jaws, or anal-feet, or oars. If the Crustacea are examined from this point of view, it will not appear surprising that the Myriapoda are classed among them; for they do not properly belong either to the Arachnoidae or to the Insecta.

ORDER I. CIRRIPEDIA.

FAMILY: BALANODEA.

Genera: *Balanus, Chthamalus, Coronula, Tubicinella.*

FAMILY: LEPADEA.

Genera: *Otion, Cineras, Lepas, Pollicipes.*

ORDER II. SIPHONOSTOMA.

FAMILY: PENELLINA.

Genera: *Penella, Peniculus, Lernaeocera, Lernaea.*

FAMILY: LERNAEODEA.

Genera: *Achtheres, Tracheliastes, Brachiella, Lernaeopoda, Anchorella, Chondracanthus.*

FAMILY: ERGASILINA.

Genera: *Dichelestium, Lamproglena, Ergasilus, Nicothoë.*

---

1 *Erichson,* Entomographien, IIft. J. p. 12.

FAMILY : CALIGINA.

Genera: *Caligus, Pandarus, Trebius, Dinematura, Euryphorus, Phyllophora.*

FAMILY : ARGULINA.

Genus : *Argulus.*

ORDER III. LOPHYROPODA.

Genera: *Cyclopsina, Cyclops, Anomalocera, Calanus, Peltidium, Hersilia, Polyphemus, Daphnia, Evadne, Lynceus, Cypris.*

ORDER IV. PHYLLOPODA.

Genera: *Lymnadia, Isaura (Estheria), Apus, Branchipus, Artemia, Chirocephalus.*

ORDER V. POECILOPODA.

Genus: *Limulus.*

ORDER VI. LAEMODIPODA.

Genera: *Cyamus, Caprella, Leptomera, Aegina.*

ORDER VII. ISOPODA.

FAMILY : BOPYRINA.

Genera : *Bopyrus, Phryxus, Jone, Cepon.*

FAMILY : CYMOTHOIDEA.

Genera : *Cymothoa, Aega, Nerocila, Anilocra, Serolis.*

FAMILY : SPHAEROMATODA.

Genera : *Sphaeroma, Cymodocea, Nesea, Amphoroïdea.*

FAMILY : IDOTHEOÏDEA.

Genus : *Idothea.*

FAMILY : ASELLINA.

Genera : *Lygia, Janira, Asellus, Lygidium, Porcellio, Oniscus, Armadillidium, Tylos.*

ORDER VIII. AMPHIPODA.

Genera : *Vibilia, Hyperia (Hiella), Phronima, Iphimedia, Amphithoë, Talitrus, Gammarus.*

26*

ORDER IX. STOMAPODA.

Genera: *Phyllosoma, Amphion, Mysis, Leucifer, Cynthia, Thysanopoda, Alima, Squilla, Squillerichthus.*

ORDER X. DECAPODA.

SUB-ORDER I. MACRURA.

Genera: *Penaeus, Pasiphaea, Alpheus, Caridina, Hippolyte, Palaemon, Aristeus, Gebia, Callianassa, Crangon, Nephrops, Astacus, Homarus Palinurus, Scyllarus, Galathea.*

SUB-ORDER II. ANOMURA.

Genera: *Pagurus, Porcellana, Remipes, Ranina, Homola, Lithodes, Dromia, Dorippe.*

SUB-ORDER III. BRACHYURA.

Genera: *Lupea, Portunus, Eriphia, Carpilius, Cancer, Maia, Leucippa, Hyas, Pisa, Stenorhynchus, Mithrax, Camposcia, Ilia, Grapsus, Ocypoda, Uca, Gecarcinus, Thelphusa.*

ORDER XI. MYRIAPODA.

SUB-ORDER I. CHILOGNATHA.

Genera: *Glomeris, Blaniulus, Platyulus, Polydesmus, Spirobolus, Julus.*

SUB-ORDER II. CHILOPODA.

Genera: *Cryptops, Geophilus, Scolopendra, Lithobius, Scutigera.*

BIBLIOGRAPHY.

Swammerdamm. Von der Zergliederung einer Krebsschnecke, in the Bibel der Natur, 1752, p. 84.

Schäffer. Der fischförmige Kiefenfuss, 1754. Die geschwänzten und ungeschwänzten zackigen Wasserflöhe, 1755. Der krebsartige Kiefenfuss, 1756.

Rösel. Der Flusskrebs; in his Insekten — Belustigungen, Th. III. 1755, p. 307.

O. F. Müller. Entomostraca, 1785.

Cavolini. Abhandlung über die Erzeugung der Fische und der Krebse, 1792.

Ramdohr. Beiträge zur Naturgeschichte einiger deutschen Monokulus-Arten; in his Micrographischen Beiträgen zur Entomologie und Helminthologie, 1805.

Jurine. Mémoire sur l'Argule foliacé; in the Ann. du Mus. d'Hist. Nat. VII. 1806, p. 431. Histoire des Monocles, 1820.

Prevost. Mémoire sur le Chirocephale ; an appendix to *Jurine s* Histoire des Monocles, p. 201.

Cuvier. Mémoire sur les animaux des Anatifes et les Balanes ; in the Mém. du Mus. d'Hist. Nat. II. 1815, p. 85, also in his Mémoires sur les Mollusques, loc. cit.

Treviranus. Abhandlungen über den inneren Bau der ungeflügelten Insekten ; in his Vermischten Schriften anatomischen und physiologischen Inhalts. Bd. I. II. 1816-17.

Geveke. De Cancri Astaci quibusdam partibus. 1817.

Suckow. Anatomisch-physiologische Untersuchungen der Insekten und Krustenthiere. 1818.

Straus. Mémoire sur les Daphnia ; in the Mém. du Mus. d'Hist. Nat. V. 1819, p. 380. Mémoire sur le Cypris, Ibid. VII. 1821, p. 33. Mémoire sur les Hiella, nouveau genre de Crustacés Amphipodes, Ibid. XVIII. 1829, p. 51. Ueber Estheria dahalacensis, eine neue Gattung aus der Familie der Daphniden ; in the Museum Senckenbergianum. Bd. II. 1837, p. 117.

Brongniart. Mémoire sur le Limnadia ; in the Mém. du Mus. d'Hist. Nat. VI. 1820, p. 83.

Savigny. Description de l'Égypte, Hist. Nat. Crustacés. 1820-30.

Rathké. Anatomie der Idothea Entomon ; in the Neuesten Schriften der naturforschenden Gesellschaft in Danzig. Bd. I. 1820, p. 109. Zur Fauna der Krym. 1836. De Bopyro et Nereïde, commentationes anatomico-physiologicae duae, 1837. Bemerkungen über den Bau des Dichelesthium Sturionis und der Lernaeopoda stellata ; in the Nov. Act. Nat. Cur. XIX. 1839, p. 127. Beiträge zur Fauna Norwegens, Ibid. XX. 1843, p. 3.

Léon Dufour. Recherches anatomiques sur le Lithobius forficatus et la Scutigera lineata ; in the Ann. d. Sc. Nat. II. 1824, p. 81.

Desmarest. Considerations générales sur la Classe des Crustacés, 1825.

J. Müller. Zur Anatomie der Scolopendra morsitans ; in the Isis, 1829, p. 549.

John Thompson. Zoological Researches and Illustrations, or Natural History of nondescript or imperfectly known animals. I. Pt. 1, 1831-34.

Nordmann. Mikrographische Beiträge zur Naturgeschichte der wirbellosen Thiere, IIft. II. 1832.

Zenker. De Gammari Pulicis historia naturali atque sanguinis circuitu commentatio, 1832.

Kutorga. Scolopendrae morsitantis anatome, 1834.

Roussel de Vauzème. Sur le Cyamus Ceti, in the Ann. d. Sc. Nat. I. 1834, p. 239.

Burmeister. Beiträge zur Naturgeschichte du Rankenfüsser, 1834. Beschreibung einiger neuen oder weniger bekannten Schmarotzerkrebse ; in the Nov. Act. Nat. Cur. XIX. 1835, p. 271.

Martin St. Ange. Mémoire sur l'organisation des Cirripèdes, 1835, also, in the Mém. présentés à l'Acad. Roy. d. Sc. de l'Inst. de France, VI. 1835, p. 513.

Kollar. Beiträge zur Kenntniss der lernäenartigen Crustaceen ; in the Ann. des Wiener Mus. d. Naturgeschichte, Bd. I. 1835, p. 79.

Milne Edwards. Histoire naturelle des Crustacés, 1834-40 ; also, article *Crustacea*, in Cyclop. of Anat. I. p. 750.

John Coldstream. *Cirrhopoda*, in the Cyclop. of Anat. I. p. 683.

Kröyer. Ueber die Schmarotzerkrebse, in his Naturhistorisk Tidskrift. Bd. I. II. 1836–37, also in the Isis, 1840, p. 702, 1841, p. 187. Ueber den Bopyrus abdominalis, Ibid. III. or in Isis, 1841, p. 693, or in Ann. d. Sc. Nat. XVII. 1842, p. 142. Monografisk Fremstilling af Slaegten Hyppolytes Nordiske Arter. 1842.

Brandt. Beiträge zur Kenntniss des inneren Baues von Glomeris marginata; in Müller's Archiv., 1837, p. 320. Recueil de Mémoires relatifs à l'ordre des Insectes Myriapodes, 1841; Extracted from the Bull. scientifique publ. par. l'Acad. Impér. d. Sc. de St. Pétersburg, V.—IX.

Van der Hoeven. Recherches sur l'histoire naturelle et l'anatomie des Limules, 1838.

Lovén. Evadne Nordmanni, ein bisher unbekanntes Entomostracon; in Wiegmann's Arch. 1838, Bd. I. p. 143.

Joly. Histoire d'un petit Crustacé (Artemia salina); in the Ann. d. Sc. Nat. XIII. 1840, p. 225. Recherches zoologiques, anatomiques et physiologiques sur l'Isaura cycladoïdes, Ibid. XVII. 1842, p. 293.

Zaddach. De Apodis cancriformis anatome et historia evolutionis, 1841.

Newport. On the organs of reproduction and the development of Myriapoda; in the Philosoph. Transact. 1842, Pt. II. p. 99, also in *Froriep's* neuen Notiz. XXI. p. 161. On the Structure and Development of the nervous and circulatory systems, and on the existence of a complete circulation of the blood in vessels in the Myriapoda and the Macrourous Arachnida, Ibid. 1843, Pt. I. p. 243; also in *Froriep's* neuen Notiz. XXVIII. p. 177, or in Ann. d. Sc. Nat. I. 1844, p. 58, or in Ann. of Nat. Hist. XII. p. 223.

Rymer Jones. Myriapoda, in Cyclop. of Anat. III. 1842, p. 544.

Lereboullet. Mémoire sur la Ligidie; in the Ann. d. Sc. Nat. XX. 1843, p. 103.

Vogt. Beiträge zur Naturgeschichte der schweizerischen Crustaceen (Argulus und Cyclopsina), in the Nouv. Mém. de la Soc. helvétique, VII. 1843.

Frey. De Mysidis flexuosae anatome commentatio, 1846.

ADDITIONAL BIBLIOGRAPHY.

J. D. Dana and *E. C. Herrick.* Description of the Argulus Catostomi. Amer. Jour. of Sc. XXXI. p. 297.

J. D. Dana and *C. Pickering.* Description of the Caligus Americanus. Ibid. XXXIV. p. 235.

J. D. Dana. On the Eyes of Sapphirina, Corycaeus, &c. Proceed. Acad. Nat. Sc. Philad. 1845, II. p. 185, and Amer. Jour. of Sc. 2nd Ser. IX. p. 133.

———— Report on the Crustacea of the U. S. Exploring Expedition, &c. &c., part I. 1852, 4to. pp. 688, part II. 1853, pp. 689–1618. With a folio Atlas of 96 Plates.

Leydig. Ueber Argulus foliaceus; ein Beitrag zur Anatomie, Histologie und Entwickelungsgeschichte dieses Thieres; in *Siebold* and *Kölliker's* Zeitsch. II. 1850, p. 323.

———— Ueber Artemia salina und Branchipus stagnalia; Beitrag zur anatomischen Kenntniss dieser Thiere, in Ibid. III. 1851, p. 280.

Baird. The Natural History of the British Entomostraca. Published by Ray Soc. London, 1850. In this work are embodied the results of his earlier researches published in the Trans. of the Berwick Nat. Club, &c.

Darwin. A monograph of the sub-class Cirripedia with figures of all the species. Published by Ray Soc. London, 1851. This work is rich in anatomical details, many of which are new.

Van Beneden. Recherches sur quelques Crustacés inferieurs, in the Ann. d. Sc. Nat. XVI. p. 71. — Ed.

CHAPTER I.

EXTERNAL ENVELOPE AND CUTANEOUS SKELETON.

§ 263.

The External envelope of the Crustacea is more or less solid, and has the form of a multi-articulated, cutaneous skeleton, sometimes of a leathery or horny consistence, but generally consists of a hard, calcareous shell.

It has, consequently, no contractility, and participates in the movements of the body only in a passive manner, that is, by the interarticular soft skin, and by the antennal and foot-like processes.

In this cutaneous skeleton, whether it is leathery, horny, or calcareous, there is a peculiar organic substance as its base. This substance, which is found in the cutaneous skeleton of other Arthropoda also, has received the name of *Chitine.* It resembles cellulose of plants in its insolubility in caustic potass, but differs essentially from it in containing nitrogen.[1]

§ 264.

Nothing in general can be said as to the Histological composition of this cutaneous skeleton of the Crustacea, for it differs widely not only in the various orders and families, but even in the different parts of the body of the same species.[1] Whether hard or soft, it is usually composed of

---

1 This Chitine which was formerly taken for a horny substance, was first discovered with the Insecta by *Odier* (Mém. d. l. Soc. d'Hist. Nat. de Paris, I. 1823, p. 29). Latterly, it has been carefully investigated by *C. Schmidt* (Zur vergleich. Physiol. d. wirbellos. Thiere, 1845, p. 32), who found, moreover, that the cutaneous skeleton of Crustacea has the same composition as that of Insecta.

* [§ 264, note 1.] Of the results of *Lavalle's* observations it may be well to add his concluding remarks; after a minute description of each portion of the tegumentary apparatus, he concludes: " I shall here only observe further, that my investigations seem to be in complete opposition to the theories which make the shell of the Crustacea analogous (homologous) to the scaly epidermis of Serpents and Lizards. I see no analogy (homology) between the shedding of the shell of the Crustacea, — which divests them of organs des-

1 We are indebted to *Valentin* for the researches, few as they are, which have hitherto been made on the internal structure of the skeleton of Crustacea; see his Repertor. f. Anat. u. Physiol. I. 1836, p. 122.

Lavalle is about to publish microscopical researches on the structure of the cutaneous skeleton of the Decapoda; see Ann. d. Sc. Nat. 1847, p. 352.*

tined to give the body its form and volume, to serve as points of attachment to the locomotor muscles, to furnish the instruments of prehension and mastication; organs placed not only on the surface of the body, but often immersed in the midst of soft parts, and in which we find an organization such as I have described, — and the periodical shedding observed in reptiles of a thin epidermis, without consistency, completely unorganized and incapable of fulfilling any of the uses to which the shell is destined. My re-

numerous very thin layers, made up of very fine, interlaced fibres. However, sometimes this fibrous texture is scarcely distinguishable, and often the lamellae are perfectly homogeneous. Frequently, also, these lamellae are traversed, either in a parallel or perpendicular direction, by canals, which are often so small that, seen under the microscope by reflected light, they appear only as lines or black points.[2]   In some species, this skeleton has a distinct cell-structure ; for the skin, here and there, has the aspect of a net-work composed of numerous round, or polyhedral meshes.

This net-work is, undoubtedly, the result of the fusion of the walls of numerous cells lying on the same plane. In the calcareous shells, the carbonate and phosphate of lime so intimately combined with the chitine, that their particles, as such, cannot be distinguished.[3]   In those portions of the skin which serve a respiratory function, the calcareous matter is always wanting.

The pigments are due to very fine granules which exist either as such in the cutaneous lamellae, or are so thoroughly fused in these last, that they are indistinguishable. In some cases, these granules are contained in polyhedral cells which form a simple layer under the transparent skin ; in others, radiating pigment cells, isolated, or reticulated, are seen through the colorless skin. The red, green, or blue color of many of the lower Crustacea, is due to oil-globules in the interior of the body, which are seen through the transparent integument.[4]

The tubercles, points, bristles, single or bifid hairs, which are usually hollow and exist on the surface or borders of different parts of the cutaneous skeleton, are always mere prolongations or simple excrescences of the integument, and contain its characteristic substance, — Chitine.

With Crustacea, as also with the other Arthropoda, the cutaneous envelope, whatever may be its tenuity — as for instance on the respiratory organs, is never covered with ciliated epithelium.[5]   This absence of vibratile organs is due, probably, to the presence of chitine.

The internal surface of this envelope is usually lined with a peculiar, thin, fibrous membrane, analogous to an internal periosteum. In the moulting process, which is common to all Crustacea, it plays an important part, for it probably secretes, in layers, the materials for the new envelope.

§ 265.

Beside the cutaneous skeleton, there is, with the Cirripedia, an envelope, entirely resembling the mantle and the valves of the Acephala.

2 *Astacus, Apus, Julus,* and *Glomeris.* But in the last two of these, the cutaneous canals are pretty large and consequently have not the appearance of black lines.

3 According to *Valentin* (loc. cit. p. 124), the cutaneous canals of the *Astacus fluviatilis* are filled with carbonate of lime, a point which I have not had the opportunity to confirm.

4 *Cyclops, Cyclopsina,* and other Entomostraca.

5 *Templeton* (Trans. of the Entomol. Soc. I. p. 195, Pl. XXI. fig. 9, a. b.) has observed with *Calanus arietis* (an animalcule allied to *Cyclopsina castor*), two bristles at the extremity of each of the long antennae, and which, he says, are provided with a row of vibratile cilia. But this observation does not invalidate what I have remarked in the text, for how often have ciliary phenomena been observed on organs which really have no such appendages. I doubt if this observation of *Templeton* will be confirmed by other observers.

searches have convinced me of the vitality of the shell, at least in the first period of its existence ; and in reference to this, I am fully of Cuvier's opinion, when he said, in his 'Anatomie Comparée,' 'The envelope of the Crustacea is at first, soft, sensible, and even furnished with vessels, but a quantity of calcareous mole soon collects there, hardens it, and obstructs the pores and vessels ;'" see loc. cit. p. 376, also Comp. rend. 1847, XXIV. p. 12. — ED.

The body of these animals, as well as its articulated appendages, are enclosed in a cutaneous skeleton containing the chitine ; and, moreover, is enveloped in a peculiar mantle having, externally, calcareous plates which vary in number and are so united together as to be movable in some species, and fixed in others. With the Lepadea, the mantle is prolonged into a kind of siphon.* Not only this mantle, but also its ligaments uniting the movable pieces of the shell and the siphon, are composed of a lamellated tissue analogous to that of the proper cutaneous skeleton, and like it also, contain chitine. It is covered with a thin layer of dark-colored pigment cells.

But the valves of the Cirripedia differ essentially from the calcareous shell of the other Crustacea. In the first place, they have no participation in the moulting, to which the cutaneous skeleton and the mantle are regularly subjected ;[1] then again, their structure and chemical composition resemble that of many of the Bivalvia.[2]   The valves of the Balanoidea form the only exception in this respect. They are traversed, in part, by numerous parallel tubes, dilated at their inferior or their external portion, which pursue a vertical course in the vertical valves, but are radiated in the horizontal plate. These tubes, which are wanting in the movable opercula of these shells and in the transversely-striated valves which, in the genus *Balanus*, are intercalated between the longitudinally-striated ones, are often laterally compressed, and their interior has imperfect longitudinal septa, or is even divided into several chambers by transverse partitions.[3] The horizontal plate which forms the base of the shell, is perforated centrally, and hollowed on its under surface, with the genus *Coronula*. This cavity is divided, by numerous vertical and symmetrically-arranged septa, into compartments filled with a fibrous substance.[4]   With *Tubicinella*, this plate is entirely wanting, and is replaced by a fibrous substance. This fibrous matter, by which *Coronula* and *Tubicinella* are fixed firmly to foreign bodies, is comparable to the pedicle of the Lepadea, which has become internal and overgrown by the shell.

The increase of the shells of Cirripedia follows the same laws as that with the bivalve or multivalve mollusca, judging from the course of the lines of growth which they present.

§ 266.

The form and number of the different segments of the cutaneous skeleton, which are sometimes extraordinarily developed, and sometimes equally

---

1 *Thompson* (Zool. Research. &c. p. 79, Pl. X. fig. 1), has observed with *Balanus pusillus*, that the Cirripedia, like the other Crustacea, cast off their entire skin at certain seasons. I have myself often seen this animal deprived of its skin with all the appendages, and even the mantle which lines its shell. In captivity, these little animals repeat this process at irregular and often very short intervals, as in twelve, eight, and even five days.

* [§ 265.] With the *Anatifae*, the siphon or pedicle corresponds to a pair of antennae in the young ; the animal attaches itself by the sucker-like disc terminating these organs, before the metamorphosis commences, and in a group of these animals all the different stages may be observed,

2 See *Schmidt*, loc. cit. p. 60.

3 See *Poli*, loc. cit. Tab. IV. fig. 6–10 ; *Rapp*, in *Wiegmann's* Arch. 1841, I. p. 168 ; and *Coldstream*, in the Cyclop. of Anat. loc. cit. p. 685.

4 For *Coronula diademe* and *balaenaris*, see *Chemnitz*, Neues Conchylien-Cabin. VIII. p. 319, Taf. XCIX. fig. 844, 846 ; *Lamarck*, Ann. du Mus. d. Hist. Nat. I. p. 461, Pl. XXX. fig. 3, and *Burmeister*, Beiträge, &c., p. 34, Taf. I. fig. 2, 3.

from the pair of distinct antennae to the fixed simple pedicle ; see *Dana*, Notice of some Genera of Cyclopacea, *Silliman's* Jour. Vol. I. 2nd Ser. p. 225, note, also Rep. on Crustacea, Ex. Exped. of the U. S. p. 1393. — Ed.

abortive and fused several together, — serve, in descriptive zoology, to characterize orders, sub-orders, families, and genera; consequently they need not
be mentioned here.[1]

With many Crustacea, the internal surface of the skeleton in widely
different parts of the body, has prolongations and processes of the most
manifold form; some of these serve as points for the insertion of muscles
and tendons, and others as partitions separating and shielding particular
organs.

CHAPTER II.

MUSCULAR SYSTEM AND ORGANS OF LOCOMOTION.

§ 267.

The voluntary muscles of Crustacea are composed exclusively of transversely-striated fibres, and are, moreover, perfectly colorless.[1]

They are always inserted upon the interior of the skeleton, either
directly, or by means of its prolongations. These last are often very long,
resembling tendons; from which, however, they differ in their intimate
structure and chemical composition. They are composed of straight, parallel, flattened fibres, and show their direct relations with the cutaneous
skeleton by containing chitine.

The isolated muscles have usually a riband-like form,* and are especially
accumulated in those regions of the body displaying great power or
extensive movements. There are, therefore, for their reception, cavities or
large canals in certain parts of the skeleton. Generally, the flexors are
upon the ventral, and the extensors on the dorsal surface of the body.
The first are always larger and more powerful than the second. Usually,
the muscles pass from one segment to the adjacent one, and by this
arrangement, the interarticular movement between the segments is produced. Their course is longitudinal, but, especially where there are
several superposed layers, they assume also an oblique and crucial direction.[2] Rarely are transverse muscles observed.[3]

The muscular system in general is very unequally developed in the
various orders of Crustacea. It is most complicated when the number of

---

1 See *Savigny*, Mém. &c. part I. and *Erichson*,
Entomograph. Hft. 1, 1840, p. 1, Taf. II.
1 For the muscles of *Astacus*, see *Will*, in *Muller's* Arch. 1843, p. 358.*
2 In the tail of many Decapoda, and in the abdominal segments of Myriapoda.

3 With the Myriapoda, the transverse muscles
pass off right and left from the ventral median line
to the sides of the abdominal segments. With the
Lernaeodea, and Ergasilina, there are, under the
skin, transverse as well as longitudinal muscles.

* [§ 267, note 1.] For researches on the intimate composition of the muscular tissue with Crustacea (*Argulus, Artemia, Branchipus*), see *Leydig*, loc. cit. *Siebold* and *Kölliker's* Zeitsch. II. p.
327, III. p. 301). The fibres of muscles can often
be easily separated into fibrillae which appear to
consist of piles of discs exactly as with some of the

other Arthropoda, and especially the Insecta, under
which, this point will be specially treated. — ED.

* [§ 267.] For a circular muscle quite extraordinary in *Caligus*, see *Dana*, Descript. of a species
of Caligus, Amer. Jour. of Sc. XXXIV. p. 247, Pl.
IV. fig. 7, r. — ED.

the segments of the body is greatest;[4] and most simple when these segments are atrophied or blended together.[5]

§ 268.

The locomotive organs of the Crustacea are, in general, very numerous; for, often all the segments, from the head to the extremity of the tail, that is, the three corresponding to the thoracic segments of insects, and those of the posterior part of the body, have, each, a pair of articulated appendages. In the order Myriapoda, the Chilognatha have two pairs of legs on each segment of the body.[1] The form of these organs may be most variously modified, and even so much so that their function is entirely changed.[2] But those of the first five segments of the abdomen are most constant in their form; although they change their function, being sometimes ambulatory legs, sometimes prehensile organs, and sometimes oars. When prehensile organs, their last joint is armed with a very hooked, sharp claw; when oars, this same joint becomes a plate bordered with stiff bristles or bifid hairs.

The locomotive organs of the three thoracic segments are usually pressed towards the mouth and changed into foot-jaws, which serve either as masticatory, or as tactile and prehensile organs. The appendages of the posterior part of the body may have even yet wider variations. They may be changed into false or abdominal feet serving sometimes as oars, as fins, or as respiratory organs; and, in the act of generation, they may play the part, some, of copulatory organs, and others, as porters of the eggs.

When they are ambulatory, or when prehensile organs, these appendages may be divided into six pieces, viz.: The *Coxa*, the *Trochanter*, the *Femur*, the *Tibia*, the *Metatarsus*, and the *Tarsus* the extremity of which, with the ambulatory feet, is often prolonged into a short, stiff claw. When they serve as natatory organs, the separate joints are more or less flattened and spread out. When used as prehensile organs, they are either monodactyle — the entire tarsus being transformed into a strongly-curved hook which can be applied against the metatarsus, — or they are didactyle or like pincers, — the metatarsus being thickened or increased in a hand-like manner, and prolonged into an immovable process (*Index*), against which the tarsus (*Pollex*) can be applied in a finger-like manner.

From these metamorphoses and the complete abortion of these appendages, the various forms of Crustacea may be reduced to a few principal types, as follows:

---

4 The muscular system is highly developed with Decapoda, Stomapoda, Amphipoda, Isopoda, Myriapoda, Poecilopoda and Phyllopoda; see *Gervke*, De Cancri astaci quibusd. partib. p. 7, fig. 1–7; *Suckow*, Anat. physiol. Untersuch. loc. cit. p. 04, Taf. IX. X. (*Astacus fluviatilis*); *Milne Edwards*, Hist. Nat. d. Crust. I. p. 155, Pl. XIII. (*Homarus marinus*); *Kutorga*, Scolopendr. morsit. Anat. p. 12, Tab. 11. fig. 1, 2; *Van der Hoeven*, Recherch. sur l'hist. nat. et l'anat. d. Limulus, p. 24, Pl. III.; *Zaddach*, De Apolis canceriformis Anat. p. 4, Tab. I. III.

5 The abortion of the muscular system is often so extensive in the lower parasitic Crustacea, that, beside the few muscles belonging to the tactile and locomotive organs, there are found only some longitudinal and transverse fibres under the skin; see *Nordmann*, Microgr. Beiträge, Hft. 2, p. 6, Taf. I. V. VII. (*Lamproglena, Actheres* and

Trachcliastes), *Rathké*, in the Nov. Act. Nat. Cur. XIX. p. 141, Tab. XVII. fig. 2. 3 (*Dicheles-tium*); *Pickering* and *Dana*, in the Isis, 1841, Taf. IV. (*Caligus*).

1 This anomaly, in which the three segments back of the head do not participate, is due, perhaps, to the segments of the body being always fused in twos.

2 In the interpretation of the movable appendages, I have relied for the most part on the principles of *Erichson* (Entomograph. loc. cit.), for they appear most consistent and unconstrained. In the instances where, at first sight, they appear unwarranted, they may be very well explained by recourse to the phenomena of development of Crustacea; and by this means, here, especially, where the metamorphoses occur gradually and continuously, may be found the solution of many obscure questions in morphology.

1. With the Myriapoda, these appendages are ambulatory and have the same form with all the segments of the body ; and only with the Chilopoda the anterior and middle pairs of the first segment corresponding to a thorax, are changed into tactile organs.

2. With the Isopoda, Laemodipoda, and Amphipoda, the first thoracic pair are tactile organs. With the Amphipoda, the second and third thoracic pairs are changed into prehensile organs armed with a claw. The five anterior abdominal pairs are ambulatory and unchanged, with the Isopoda, and Amphipoda. But the remaining posterior pairs are transformed, with the first of these orders, into lamelliform respiratory organs ; and the second, into short, very movable appendages, terminated, each, by a double uni- or multi-articulate cirrus, which serve sometimes as oars, sometimes as gyratory organs.

3. With the Decapoda, the thorax is entirely abortive, and its three pairs are changed into oral and tactile organs ; while the first pair, belonging to the anterior abdominal segments, is usually transformed into a forficulate prehensile organ. The four succeeding pairs are simply ambulatory organs. But the appendages of the posterior part of the abdomen are reduced to tendril-like processes, which play a part in the act of generation.

With the Squillina, the three thoracic, and the first two abdominal pairs have the form of prehensile organs, while the three succeeding pairs retain their character of ambulatory organs, and those of the remaining posterior segments are changed into lamelliform fins.

4. In the section of the inferior Crustacea, designated usually under the name of Entomostraca, the head and thorax are fused into a single part called Cephalothorax, and the mouth is situated so far behind, that the first pair of feet is in front of it. The locomotive apparatus here consists usually of oars or prehensile organs. With the Poecilopoda, the first three pairs of appendages are forficulate, as, also, are the three pairs of jaws. With the Phyllopoda, and Lophyropoda, the first two pairs of feet resemble antennae ; of these sometimes the first, as well also as the second, which are usually branched, serve as oars ; [3] here, also, the often very numerous, anterior abdominal appendages are used usually as fins, while the posterior ones are scarcely at all developed.

5. With the Cirripedia, the first thoracic pair is transformed in a remarkable manner. With the Lepadea, they are changed into a soft foot ; and with the Balanodea, into a shell. [4] The remaining six pairs are multiarticulate cirrate organs, and the abdomen is prolonged into a tail free from appendages. The three anterior pairs of these cirrate organs are the shorter, and have a tactile function ; while the three posterior are used as gyratory organs.

6. With the Siphonostoma, the mouth is even still further behind, and

3 With *Cyclops, Cyclopsina,* and *Cypris,* it is the first pair of feet that is transformed into oarlike organs, but it is the second with *Apus, Limnadia, Daphnia,* and *Polyphemus.*

With the Branchiopoda, the disposition is quite different. The anterior pair is changed into two slightly movable appendages, hook-like or digitiform, and rolled spirally. With the embryos and the young animals, these organs are used clearly as oars ; see *Jurine,* Hist. d. Monocles, loc. cit. Pl. XX. fig. 9, and Pl. XXI. fig. 1, 2 (*Chirocephalus*), and *Joly,* in the Ann. d. Sc. Nat. XIII. Pl. VII. (*Artemia*).*

4 See *Thompson,* Zool. Research. Pl. IX. fig. 3 (*Balanus*), and *Burmeister,* Beitr., &c., Taf. I. fig. 3–5 (*Lepas*).

* [§ 268, note 3.] The first pair of feet is, generally, the second pair of antennae. For a full discussion of this point, see *Dana,* Report on Crustac. &c. p. 1031.— ED.

the number of appendages much less; so that the three and only pairs, corresponding to the thoracic, are in front of the mouth. With the Caligina, and the Ergasilina, the thoracic appendages are prehensile organs, while those of the abdomen are changed into rudimentary oars. With the genus *Argulus* alone, the first abdominal pair has the form of suckers,* the remaining ones being fin-like as usual. With the Lernaeodea, the abdominal appendages are entirely wanting, and there are only a few anterior prehensile ones, two of which, in some genera, are prolonged arm-like, and united, at their extremity, into a button-like, suctorial organ.[5] Sometimes these arms are wanting, there being present only the suctorial organ.[6] With the Penellina, the locomotive organs are reduced to non-articulated rudiments; or even these may be wanting, and then the cephalic extremity of the unsegmented body has stiff, forked, horny, processes, by means of which these parasites enter the parenchyma of other animals.[7]

§ 269.

Certain Crustacea have, moreover, a special locomotive apparatus. With *Cypridina*, the body is shielded with a bivalve shell, the halves of which move on a kind of hinge. Upon their internal surface are inserted muscular fibres, arising from the back of the animal, which act like the adductor muscles of the bivalve Acephala.

With the Cirripedia, there is a considerable transverse adductor muscle, which, with the Balanodea, and Lepadea, is situated in the anterior or cephalic angle of the fissure of the mantle, which is nearly always closed by an operculum.[1] In this same angle, the body, with all the Cirripedia, is in connection with the mantle, partly by its cutaneous envelope, which, at this point, is folded in so as to line the cavity of the mantle, and partly by various muscles. These muscles arise from the anterior extremity of the body, which is inverted within the cavity of the mantle, and from both the ventral (or upper), and from the dorsal (or lower) surface of the animal.

When those of the upper or abdominal surface are contracted, and, at the same time, the adductor muscle of the valves is relaxed, the animal comes out through the fissure of the mantle; but it is withdrawn into the mantle-cavity when those of the lower or dorsal surface are contracted.[2]

---

5 *Tracheliastes, Achtheres, Brachiella.*
6 *Anchorella.*
7 *Lernaea, Lernaeocera.*
1 *Poli,* loc. cit. Tab. IV. fig. 3 J.; *Cuvier,* Mém. &c. p. 5, fig. 2, 7 e. 11 A, and *Martin St. Ange,* Mém. &c. p. 15, Pl. II. fig. 18, S.

2 *Poli,* loc. cit. Tab. IV. fig. 13, y. z. 17; *Cuvier,* loc. cit. p. 5, fig. 18 b. ix, and *Martin St. Ange,* loc. cit. p. 14, Pl. II. fig. 17, 19, J.

* [§ 268.] For a very complete description of these sucker-like organs, with excellent figures, see *Dana,* Amer. Jour. Sc. XXXI. 1837, p. 297, and Rep. on Crustac. loc. cit. p. 13, 18. — ED.

CHAPTER III.

NERVOUS SYSTEM.

§ 270.

The Nervous system of the Crustacea, is developed in different degrees according to the various orders.[1]

Its central mass consists of an abdominal cord, connecting, usually, with the cerebral ganglia by an œsophageal ring. With the long-bodied species, this abdominal cord is composed of numerous ganglia, arranged in successive pairs from before backwards, and connected together by longitudinal commissures. But when the cutaneous skeleton is shortened by a diminution or a fusion of the segments, the ganglionic chain is lessened in a like manner by a coalescence or a disappearance of several of its ganglia.

With the Macrura, the Stomapoda, the Amphipoda, and Isopoda, the abdominal cord consists of ten to thirteen pairs of unequal ganglia, situated, usually, on the median line, and shielded by septa given off from the internal surface of the thoracic and abdominal segments of the skeleton.

The size of these ganglia is in direct ratio with the development of the segments and their appendages, to which they belong. Those of the thorax, — the anterior abdominal ones, as well as the last caudal one, are consequently very large, for they send filaments to the various chelate, prehensile, ambulatory, and natatory appendages, and to the caudal lamellæ, which are usually highly developed. With the Myriapoda, the abdominal cord is remarkable for the great number of its ganglia, which are of equal size. Quite often, the ganglia of the same pair are fused into a single mass ; in which case, the two interganglionic commissures are more or less approximated or even blended together. With some species, a portion of the abdominal ganglia are so closely approximated, successively, that the interganglionic commissures are wholly wanting. With the Brachyura, the whole abdominal cord is concentrated into a large central mass.

The peripheric nerves arise from the ganglia, rarely from the interganglionic commissures. The cerebral mass, which is situated above or in front of the œsophagus, is composed of a pair of considerable ganglia, more or less fused together. The nerves sent off from these, go principally to the organs of sense ; and in the inferior Crustacea, where these last are wanting, the cerebral mass is absent also. In such case, there are usually wanting likewise the two cerebral commissures, which are given off from the anterior thoracic ganglion, and surround the œsophagus.*

---

[1] *Audouin* and *Milne Edwards* (Ann. d. Sc. Nat. XIV. 1828, p. 77, Pl. II.–VI.) have given a general review of the disposition of the nervous system in the different orders of Crustacea ; see also *Milne Edwards*, Hist. Nat. d. Crustac. I. p. 126, Pl. XI., and his article *Crustacea* in the Cyclop. of Anat. loc. cit. p. 762.

* [End of § 270.] It is regretted that no example, illustrative of this last statement, is given, for certainly none is now recollected where the grand typical structure is not present, — in other words, where the œsophagus does not pierce the cerebral nervous system at some point. In many of the inferior Crustacea, such as *Caligus*, and some of not all of the *Cyclops* tribe, the cephalic, thoracic, and abdominal ganglia, are fused into a single mass through the anterior part of which the œsophagus passes ; see *Dana*, loc. cit. Caligus. Amer. Jour. Sc. XXXIV. p. 250. — Ed.

§ 271.

The intimate structure of the nervous system in many of the orders of Crustacea, can be made out without difficulty, by dissection and the microscope;[1] for its elements are not as liable to change as in the other classes of the Invertebrata already described.

In many species, there may be observed in the nerves surrounded by a delicate fibrous neurilemma, the primitive nerve-fibres so large that their double contour is easily seen; but these gradually assume a varicose aspect.[2] In the ganglia, the ganglionic globules may be easily seen, as very large, round, and sometimes pyriform cells, having each a disproportionately large nucleolated nucleus.[3] As to the course and arrangement of the nerve-fibres in the interior of the abdominal ganglia, two kinds of these fibres may sometimes be distinguished: the first pass uninterruptedly through all the ganglia successively, and thus contribute to the formation of the longitudinal commissures; but the second pass round among the ganglionic globules, and emerge laterally from the ganglion to form the peripheric nerve.[4]

§ 272.

From a more particular examination of the arrangement of the nervous system in the different orders of Crustacea, the following remarkable facts have been noticed.[1]

With the Macrura, where this system is most highly developed,[2] the abdominal cord is composed of twelve pairs of ganglia, generally blended

1 See *Helmholtz*, De fabric. syst. nerv. evertebrat. loc. cit. p. 17.*

2 *Ehrenberg*, Unerkannte Struct. &c. p. 56, Tab. VI. fig. 3–5 (*Homarus marinus*, *Astacus fluviatilis* and *Palaemon squilla*). The varicose enlargements are represented too regular in some of these figures. See, also, *Hannover*, Recherch. &c. p. 68, Tab. VI. fig. 76, e. c.

3 *Hannover*, loc. cit. p. 67, fig. 75, 76 a. (*Astacus fluviatilis*), and *Valentin*, in the Nov. Act. Acad. Nat. Cur. XVIII. p. 210, Tab. IX. fig. 72–85. This last author declares that he has observed, in the abdominal ganglia of the common crawfish, the ganglionic globules divided symmetrically into two groups, right and left; and in the caudal ganglia of the same species, that he has seen two double groups, two anterior and two posterior.

4 *Valentin* (loc. cit. p. 211) has seen these two kinds of primitive fibres in the abdominal cord of the common crawfish. We are indebted to this observer, *Newport* for very complete observations on the disposition of the nervous fibres in the abdominal cord of Myriapoda, and illustrated by numerous figures; see Philos. Transact. 1843, p. 243, Pl. XI., or in abstract in *Froriep's* neue Notiz. XXVIII. p. 177, or in the Ann. d. Sc. Nat. I. 1844, p. 58, or Annals of Nat. Hist. XII. p. 223. According to this observer, four fasciculi of primitive nerve-fibres may be observed in the ventral cord of the Myriapoda. An upper and a lower, extending longitudinally, contain the one, motor, and the other, sensitive fibres. A third is composed of transverse fibres which pass from one side of the ganglion to the other; and

the fourth extends from one ganglion to the next succeeding, by the side of the longitudinal commissures. To these last, *Newport* has given the name of *fibres of reinforcement*. Each peripheric nerve given off from the abdominal cord, contains fibres from all of these four fasciculi. The associate and reflex motions between the feet of the same pair, are due to the transverse fibres, and the sympathy between the posterior and anterior feet is referable to the fibres of reinforcement.

1 If, in proceeding from the higher to the lower species in the description of the nervous system, I have deviated from the plan hitherto pursued, it is because, with the Crustacea, this system, notwithstanding the various forms of the body, is found upon one and the same type, which is not true in any of the preceding classes, as, for instance, in the Acephala. This type is especially apparent during the young age of these animals, and does not change except from their ulterior metamorphosis, when, often some portions of the nervous system disappear; on this account, this last will be best understood when studied in its primitive state, or from the more perfect forms it presents in the higher Crustacea.

2 For the nervous system of the macrourous Crustacea, see *Audouin* and *Milne Edwards*, loc. cit. (*Homarus*, *Palaemon*, and *Palinurus*); *Suckow*, loc. cit. p. 61, Taf. XI. fig. 7 (*Astacus*); *Brandt*, Medizin. Zool. I. p. 64, Taf. IX. fig. 1, and especially *Newport*, Philos. Trans. 1834, p. 406, Pl. XVII. fig. 40–42 (*Homarus*).

* [§ 271, note 1.] See, also, for histological details on this system, *Leydig*, loc. cit. *Siebold* and *Köliker's* Zeitsch. II. p. 328, III. p. 291 (*Argulus*, *Artemia*, *Branchipus*). — Ed.

27*

together laterally, on the median line. Of these, the first six pairs, belonging to the thoracic and to the anterior abdominal segments, are the larger, and send off nerves principally to the foot-jaws, to the prehensile, and to the ambulatory organs. The two longitudinal commissures between the anterior abdominal ganglia, are separate; but those between the posterior ganglia are, on the contrary, blended into a single cord. In some species, these commissures are wholly wanting between the anterior ganglia.[3] The brain consists of a single transverse ganglion; from its front and sides pass off several nerves for the antennae, the olfactory organs, the eyes, and the auditory organs; while, from behind, it sends off the two long cords which surround the œsophagus. These last give branches on their course to the organs of mastication, and interanastomose behind the œsophagus, just before reaching the first thoracic ganglion, by a transverse filament.[4]

With the Stomapoda, the nervous system is composed of a cerebral ganglion, and of about ten abdominal ganglia; of these, with the Squillina, the last six belong to the tail, while the remaining four, anterior, send nerves to the thorax, and to the first three abdominal segments. The size of the first, which sends nerves to the prehensile feet, is due to its being composed of several ganglia fused together.[5] With the Mysina, the five or six largest ganglia belong to the thorax and to the anterior part of the abdomen, and are connected together by short, double commissures.[6]

The genus *Phyllosoma* has two extraordinarily long and very small œsophageal cords extending from the brain to the abdominal cord. The thoracic portion of this last is composed of three pairs of ganglia, blended almost into a single mass; these are succeeded by six pairs of large abdominal ganglia, arranged in two longitudinal rows, and interconnected by six very short, transverse filaments. In the short tail, there are, moreover, six pairs of ganglia, smaller and laterally contiguous, but connected successively by very small longitudinal filaments.[7]

With the Anomoura, which resemble the Brachyura in the abortion of the post-abdomen, the structure of the nervous system confirms this affinity. With *Pagurus*, the anterior portion of the abdominal cord consists only of three ganglia, which send nerves to the foot-jaws, to the cheliform, and to the partially abortive ambulatory feet. But the posterior part of this cord consists of two cords which arise from the third abdominal ganglion, and unite, just in front of the arms, in the fourth and last ganglion.[8] With the genus *Homola*, the five pairs of the anterior abdominal ganglia are fused into a single mass which is perforated through its centre. From the posterior border of this mass a simple nervous cord as rudiment of the posterior part of the ventral cord, passes off to the wholly abortive post-abdomen.[9]

With the Brachyura, the nervous system has only two central masses, one cerebral, the other abdominal. The first sends nerves, as in the other

3 *Palinurus*, and *Palaemon* (Audouin and Milne Edwards, loc. cit.). In these two Crustacea, there is only a small fissure in the centre of the principal ganglionic mass, after the fusion of the ganglia.
4 This transverse filament is absent neither with *Palaemon*, *Palinurus*, nor with *Homarus*, and *Astacus*. It was overlooked by Suckow in the crawfish, although distinctly seen by Brandt; see his Medizin. Zool. loc. cit., and his Bemerkungen über die Mundmagen-nerven, loc. cit. Tab. 1. fig. 1, 2 E., or Ann. d. Sc. Nat. V. 1856, Pl. IV.

5 Cuvier, Leçons &c. III. 1845, p. 330, and Delle Chiaje, Descriz. &c. Tav. LXXXVI. fig. 5.
6 Frey, De Mysidis flexuosae anat. p. 9.
7 Audouin and Milne Edwards, Ann. d. Sc. Nat. loc. cit. p. 81, Pl. III.
8 Cuvier, Leçons, &c., loc. cit. p. 329; and Owen, Lect. on Comp. Anat. p. 170.
9 Milne Edwards, Hist. Nat. d. Crust. Pl. XI. fig. 9.

Decapoda, to the organs of sense; the second is large, round or oval, and situated in the centre of the thorax, — it is sometimes perforated centrally,[10] and supplies all the nerves of the trunk, beside sending off the two œsophageal cords. These cords are connected by the transverse commissure already mentioned, and give off filaments to the organs of mastication. From the posterior extremity of the cord, there arises a simple nervous trunk, free from ganglia, and extending along the median line to the very extremity of the tail.[11]

Among the Amphipoda, the Gammarina have a brain scarcely larger than the first of the abdominal ganglia ; these last, twelve in number, are connected by double commissures, and the posterior ones belonging to the segments of the body which have false feet, are always smaller than the others.[12] With the large-headed Hyperina, the two cerebral ganglia are considerably larger than the abdominal ones, which are ten in number, and of unequal size. Their commissures are contiguous, and the first ganglion, which is the largest, is probably the result of the fusion of two pairs.[13]

With the Isopoda, the abdominal cord, which is connected with the cerebral ganglia by two short œsophageal cords, is composed of seven pairs of ganglia, situated in the thoracic and anterior abdominal segments, and connected together, successively, by double commissures. In some genera, the posterior ganglia send off radiating nerves to the partially abortive and partially fused terminal segments.[14] In others, these seven pairs are succeeded by five or six pairs of others, smaller, and which, with *Idothea*, are connected together by double commissures;[15] but with the genera *Cymothoa*,[16] *Aega*[17] and *Lygidium*,[18] are contiguous. With many Isopoda, the peripheric nerves are given off, not only from the ganglia, but also from their longitudinal commissures, and the posterior ones are distributed to the dorsal region of the animal.[19]

With the Laemodipoda, the abdominal cord is composed of eight pairs of ganglia, of which the first two are situated in the cephalic segment, one behind the other, and send off nerves to the organs of mastication, and to the first pair of feet, — thus corresponding to the result of the fusion of the first two thoracic segments with the head. The other pairs are connected by very distinct double commissures, which, between the last two pairs, are quite short, thus bringing the last three pairs almost together in the third terminal segment of the body.[20]

With the Myriapoda the ganglia of the abdominal cord are very numerous, and nearly all of the same size. The brain consists, usually, of a

10 *Maia squinado.*

11 *Audouin* and *Milne Edwards,* Ann. d. Sc. Nat. loc. cit. p. 91, Pl. VI., and *Milne Edwards,* Hist. Nat. d. Crust. I. p. 141, Pl. XI. fig. 5, 10 (*Maia squinato* and *Cancer maenas*).

12 *Audouin* and *Milne Edwards,* loc. cit. p. 79, Pl. II. fig. 1, and *Milne Edwards,* loc. cit. p. 129. Pl. XI. fig. 1 (*Palitrus*).

13 *Straus,* Mém. sur les Hiella, in the Mém. du Mus. d'Hist. Nat. XVIII. 1829, p. 60, Pl. IV. fig. 16 (*Hyperia*).

14 *Treviranus,* Verm. Schrift. I. p. 63, Tab. IX. fig. 53 (*Porcellio scaber*); *Brandt,* Medizin. Zool. II. p. 75, Tab. XV. fig. 23 (*Oniscus murarius*), and *Rathké,* De Bopyro et Nereide, p. 14, Tab. III. fig. 4 (*Bopyrus squillarum*).

15 *Rathké,* Danzig. Schrift. loc. cit. p. 127, Tab. IV. fig. 2 (*Idothea entomon*).

16 *Audouin* and *Milne Edwards,* loc. cit. p. 83, Pl. II. fig. 2, and *Milne Edwards,* loc. cit. Pl. XI. fig. 2.

17 *Rathké,* Nov. Act. Nat. Cur. XX. Pt. I. p. 33, Tab. VI. fig. 15.

18 *Lereboullet,* Ann. d. Sc. Nat. XX. p. 124, Pl. V. fig. 24.

19 *Porcellio, Oniscus, Armadillidium, Idothea.*

20 *Treviranus,* Vermisch. Schrift. II. p. 8, Taf. I. fig. 5, and *Roussel de Vauzème,* Ann. d. Sc. Nat. I. p. 253. Pl. IX. fig. 19 (*Cyamus*).

According to *Frey* and *Leuckart* (Beitr. p. 102), the ventral cord of *Caprella* somewhat resembles that of *Cyamus.*

distinct right and left half, upon each of which is a kind of *Ganglion opticum*, of a size proportionate to that of the development of the eyes. With the Chilopoda, the abdominal ganglia are widely separated from each other, but connected by double commissures which are closely approximated, and in some cases, fused together as a ventral cord.[21]   With *Lithobius*, and *Scutigera*, there are sixteen pairs of these ganglia ; with *Scolopendra*, twenty-two, and with *Geophilus*, fifty to one hundred and forty.   Of these ganglia, the first pair, belonging to the two anterior feet, which are changed into prehensile or tactile organs, are much the largest. The size of the others corresponds, for the most part, to the development of the feet.[22]

Of the Chilognatha, the genus *Polydesmus*, the long feet of which are widely separated, is allied to the preceding section of Myriapoda.   Above each two pairs of feet, the abdominal cord is enlarged into two successive ganglia, and the medullary mass between them corresponds to a simple longitudinal commissure.[23]   With the other Chilognatha, of which the pairs of feet are close together, the longitudinal commissures are wholly wanting, so that the ganglia, of a number corresponding to that of the pairs of feet, form a moniliform cord ; and in some Julidae, the constrictions of this last are entirely effaced.[24]

The disposition of the nervous system of *Limulus* is remarkable.   Its principal mass surrounds the mouth like a ring.   From the anterior portion of this, corresponding to a brain, pass off nerves in front, among which the two optic nerves are conspicuous for their length ; while its posterior arc, which surrounds the œsophagus, has three transverse commissures succeeding each other.   From the lateral portions of this ring, pass off six pairs of large nerves for as many pairs of prehensile feet.   From its posterior border arises a large trunk composed of two bands which extend backwards along the median line of the abdomen, furnishing nerves to the fin-like and gill-like appendages, and then separating, terminate in the tail in a ganglion from which are given off many filaments to the neighboring parts, and a very long one which enters the caudal spine.[25]

Of the nervous system of the Phyllopoda, that of the genus *Apus* is the best known.[26]   The brain consists of a flattened, quadrilateral body, from the superior angles of which arise the optic nerves, while from the posterior angles pass off the two long, œsophageal commissures.   These last, before reaching the thoracic ganglia, are connected by a transverse commissure. Upon the thoracic succeed numerous abdominal ganglia, those of each pair of which, as in the first, are wide apart, but they gradually approximate posteriorly, and at last are fused into a single mass.

The two thoracic ganglia, as well also as the anterior abdominal pairs,

---

21 *Geophilus.*

22 *Treviranus*, Verm. Schrift. II. p. 31, Taf. VII. fig. 2, 5 (*Lithobius* and *Geophilus*) ; *Kutorga*, loc. cit. p. 15, Tab. II. fig. 2, Tab. III. fig. 1, 2 (*Scolopendra morsitans*) ; but especially *Newport*, Philos. Trans. 1834, p. 408, Pl. XVII. fig. 43-48 (*Scolopendra*), and 1843, p. 257, Pl. XI. fig. 11-13 (*Geophilus*).

23 *Newport*, loc. cit. 1843, p. 252, Pl. XI. fig. 6, 10, or *Owen*, Lect. &c. p. 200, fig. 99.

24 There are only six pairs of ganglia with *Glomeris* ; see *Brandt*, in *Müller's* Arch. 1837, p. 324, Taf. XII. fig. 6. But these ganglia are very numerous with *Julus* ; see *Treviranus*,

Verm. Schrift. II. p. 16, Taf. IX., and *Newport*, Phil. Trans. 1843, p. 247, Pl. XI. fig. 1. The number of ganglia increases in general with the increase of the body together with that of the number of the segments and feet.

25 *Van der Hoeven*, Recherch. &c. p. 21, Pl. III. fig. 2, 3.

26 This system has been described by *Gaede* (*Wiedemann's* Zool. Magaz. I. Stück 1, p. 91, Taf. I. fig. 1), and by *Berthold* (Isis, 1830, p. 690, Taf. VII. fig. 4). But for the most careful researches on this subject we are indebted to *Zaddach* (loc. cit. p. 35, Tab. III.)

are connected together by double, transverse commissures, which, posteriorly, become single, and, finally, wholly disappear. The longitudinal commissures are disposed in a like manner; they are double and wide apart in front, but, posteriorly, approximate and are proportionably shortened, until they fuse together, and then entirely disappear, — the cord terminating in a simple moniliform band which ends above the last pair of feet. The other abdominal segments which have no feet, receive their nerves from two long cords which arise from the twenty-fourth and twenty-fifth abdominal ganglia and accompany the intestinal canal to the last segment of the tail, where they end in a ganglionic enlargement from which are given off several short filaments, beside a long nerve to the two caudal bristles. In the other Phyllopoda, the nervous system is observed with difficulty, probably from its tenuity; and, as yet, only a single flattened cephalic ganglion has been found.[27] With the very small Lophyropoda, these difficulties are even greater, for here there has been observed a multi-constricted, nervous mass, situated in front of the œsophagus, which may be regarded as a cerebral ganglion, since it sends off, in front, several filaments to the tactile and ocular organs; and behind, two cords which surround the œsophagus, and join, perhaps, in an abdominal ganglion.[28]

Among the Siphonostoma, with *Argulus*, as with the Lophyropoda, the nervous centre is reduced to a cerebral mass situated above the proboscis, — and composed of three ganglia arranged triangularly.[29] With the other parasitic Crustacea, of which the head and organs of sense have gradually disappeared, the cerebral ganglion always becomes correspondingly less apparent, while the abdominal cord is the more distinct. This is so with the genus *Chondracanthus*, which has a cerebral ganglion, and in the few segments of the body, several widely separated (laterally) ganglia connected together by longitudinal, double commissures.[30] With *Diche-*

---

27 *Brongniart*, loc. cit. p. 87, Pl. XIII. fig. 2, 3, a. (*Limnadia*), and *Joly*, loc. cit. p. 310, Pl. V. fig. 5, k. and Pl. VIII. fig. 21, a. (*Isaura*). This last naturalist has been unable to find a cerebral ganglion with *Artemia* (loc. cit. p. 242).*

28 An analogous brain, divided by constrictions into three ganglia placed in a row, has been figured by *Schäffer* (Die zackigen Wasserflöhe, loc.

* [§ 272, note 27.] The investigations of *Leydig* (loc. cit. *Siebold* and *Kölliker's* Zeitsch. III. p. 200) have shown that, with at least *Artemia* and *Branchipus* of the Phyllopoda, the nervous system is well developed. This system seems, for the most part, to have escaped the observation of former investigators from want of manipulation; *Leydig* has described it with detail, and divides it, as usual, into a central and a peripheric portion. The first consists of the brain which sends off nerves to the organs of sense (eyes, antennae, &c.) and connects, by two commissural cords which embrace the œsophagus, with the ventral cord. This cord is composed of eleven (*Branchipus*), or twelve (*Artemia*) ganglia, which are connected, successively, by two longitudinal commissures, and, laterally, each, by a double, transverse commissure. Each of these ganglia sends off, from its outer border, three nerves which are distributed to the abdominal organs and appendages, and to the skin. — ED.

cit. p. 39. Tab. II. fig. II. 1, 2, 3), by *Straus* (loc. cit. p. 396, Pl. XXIX. fig. 6, b. d. e. (*Daphnia*)), and by *Lovén* (loc. cit. p. 151, Taf. V. fig. 5, d. (*Evadne*)).

29 *Jurine*, Ann. du Mus. VII. p. 447, Pl. XXVI. fig. 11, and *Vogt*, loc. cit. p. 14, fig. 1, L, 11.†

30 *Rathke*, Nov. Act. Nat. Cur. XX. p. 125.

† [§ 272, note 29.] The recent researches of *Leydig* (loc. cit. *Siebold* and *Kölliker's* Zeitsch. II. p. 328) have extended our knowledge of the nervous system with these lower Crustacea. In *Argulus*, this observer found the central nervous system to consist of a cerebral portion and a ventral cord. The first, or brain, is composed of two parts — one anterior and club-shaped, the other, beneath the first, pyriform and much the larger. This portion connects, by two commissures which embrace the œsophagus, with the ventral cord. This cord is composed of six ganglia. He observed the following distribution of the peripheric portion of the nervous system. From the brain arise the optic nerves, and behind these, two pairs of nerves for the antennae; of the ventral ganglia, the first, third and sixth give off nerves to the appendages of the body and its internal organs. *Leydig* found no trace of a splanchnic system with these animals. — ED.

lestium, the cerebral ganglion is entirely wanting, but, in its stead, there is a conspicuous thoracic ganglion under the œsophagus, from which passes off an equal number of nerves in front and behind, and which is succeeded, posteriorly, by a large nervous trunk as the abdominal cord. This cord has ganglionic enlargements in the three anterior abdominal segments, and finally divides into two branches which extend to the very extremity of the tail.[31] With *Achtheres,* and *Peniculus,* the nervous centres consist only of two trunks lying on the lower surface of the abdomen, each side of the intestinal canal.[32]

With the Cirripedia, which are headless, the nervous centre consists of two parallel abdominal trunks, which, in their course, form six to seven ganglionic enlargements from which pass off, laterally, nerves to the cirri. The two anterior ganglia are connected by a nerve which stretches arcuately over the œsophagus, and sends filaments to the organs of mastication, so that a brain proper is wanting. The last two pairs of ganglia are blended into a single mass, which sends nerves to the cirri, and two filamens into the long tail.[33]

§ 273.

The Vegetative nervous system is distinctly developed with many Crustacea. It consists of a single or of a double Splanchnic nerve.

With the Decapoda, and Squillina, a single splanchnic nerve arises from the posterior border of the brain, — passes over the stomach, at the same time enlarging into one or two ganglia, distributes its branches to the walls of this organ, and, finally, enters the liver right and left.

This nerve is reinforced by two filaments which, conjointly with the nerves of the masticatory organs, are given off from the ganglionic enlargements of the two œsophageal commissures, and, before entering the splanchnic nerve, send off filaments directly to the lateral walls of the stomach.[1]

With the Oniscidae, there are two splanchnic nerves. On each side of the small stomach are two ganglia which connect with the brain by a short filament, and send off, posteriorly, small branches to the walls of the stomach.[2]

With the Myriapoda, there are also two systems of splanchnic nerves.

31 *Rathké,* Ibid. XIX. p. 150, Tab. XVII. fig. 3, 4.

32 *Nordmann,* Microgr. Beitr., Hft. 2, p. 72, 103, Taf. V. fig. 7, J., 6.

33 *Cuvier,* Mém. loc. cit. p. 11, fig. 11, and *Martin St. Ange,* loc. cit. p 18, Pl. 11. fig. 8 (*Lepas*); also *Wyman* in Silliman's Amer. Jour. XXXIX. 1840. p. 182 (*Otion*).*

1 We are indebted to *Brandt* for very complete contributions on the sympathetic system of the Decapoda; see his Bemerk. über d. Mundmagennder Eingeweid-nerven der Evertebr. loc. cit. p. 7, Tab. I. fig. 1-3 (*Astacus* and *Squilla*), (also in the Ann. d. Sc. Nat. V. 1836, p. 87, Pl. IV. and in the Mediz. Zool. II. Taf. XI. fig. 1, i.); see, also,

Krohn, Isis, 1834, p. 52?, Taf. XII. fig. 1-4, and *Schlemm,* De hepat- ac bile Crustace arum et Molluse arum, loc. cit. p. 16, Tab. I. fig. 2, Tab. II. fig. 13 (*Astacus fluviatilis*). *Suckow* (l c. cit. p. 62, Tab. XI. fig. 7, g.) in the Crawfish, and *Newport* (Philos. Trans. 1834, Pl. XVII. fig. 40, f.) in the lobster, have observed only a single splanchnic nerve, which they have regarded as a cardiac nerve. *Audouin* and *Milne Edwards,* on the other hand, have described and figured with both the Macrura and the Brachyura, double splanchnic nerves, but the single one was entirely overlooked.

2 *Brandt,* Bemerk. &c. p. 14, and Medizin. Zool. II. p. 75, Taf. XV. fig. 27, c.

* [§ 272, note 33.] Subsequent researches have shown that with some at least of the Cirripedia, there is a proper brain furnishing nerves to the organs of sense; see *Darwin,* Monogr. of the sub-class Cirripedia, &c., p. 48 (*Lepas*). Even in the description referred to above, of *Otion,* by *Wyman* in *Silliman's* Jour., a brain is really spoken of. — Ed.

The single stomato-gastric system consists of two short trunks which extend from the brain in front, send several small filaments to the parts of the mouth, and finally pass in front of the brain, — ending in a small ganglion. From this last, arises a single nerve, which passes under the brain and extends along the œsophagus to the stomach, being, in its course, sometimes enlarged like a ganglion. The double splanchnic system, on the other hand, is composed of a double row of ganglia accompanying the œsophagus, and connected, partly by the posterior border of the brain and the single nerve, and partly by nervous branches. The filaments given off from these ganglia are distributed not only to the œsophagus, but also to the salivary glands.[3]

In the genus *Limulus*, there is observed, as a single splanchnic nerve, only one nerve, having a ganglion and situated on the heart;[4] while, with *Apus*, the splanchnic nervous system is highly developed. The two œsophageal commissures furnish, as with the Decapoda, two nerves which, shortly after their origin, are connected by a transverse commissure. On the œsophagus, they are blended into a single nerve, and send to this canal numerous filaments.[5] With the other inferior Crustacea, no splanchnic nerves have as yet been observed.

CHAPTER IV.

ORGANS OF SENSE.

§ 274.

The sense of Touch is highly developed with Crustacea. Its seat is in the multi-articulate antennae, situated on the head, or cephalic extremity, which always contain large nerves arising directly from the brain. Often, the masticatory organs have one or several pairs of tactile appendages; and, not unfrequently, several pairs of the feet neighboring the mouth, are changed into tentacular, tactile organs, which play an important part in the choice and prehension of food.[1]

§ 275.

As Olfactory organs, with the Crustacea, may undoubtedly be regarded the two shallow excavations which, with the Macrura, and with *Pagurus*, are situated in the basal joint of the two median antennae. Each of these cavities communicates, externally, by a fissure-like opening, placed on the upper surface of the joint, and usually fringed with fine bristles. Inter-

---

3 *Brandt.* Bemerk. &c. p. 34, Taf. III. fig. 6–9, and in *Müller's* Arch. 1837, Taf. XII. fig. 7 (*Scolopendra, Spirobolus,* and *Glomeris*); also *Newport,* Philos. Trans. 1843, p. 246, Pl. XI. fig. 1, 2 (*Julus*). *Treviranus* had already seen something of a single sympathetic system with

Julus; see his Verm. Schrift. II. p. 47, Taf. IV. g.
4 *Van der Hoeven,* loc. cit. p. 23.
5 *Zaddach,* loc. cit. p 36, Tab. III. fig. 5.
1 The various differences of form of these tactile organs belong rather to Zoölogy.

nally, these organs arc lined by a soft membrane, which contains a nerve
arising from the brain in common with the internal antennal nerve. [1]

§ 276.

Organs of Hearing, with the Crustacea, have as yet been observed only
with the Decapoda. [1] With these Crustacea, there is a hollow conical process, perforated at its obtuse apex, on the lower surface of the basal joint of
the external antennae. Its opening is always closed by a kind of Tympanitic membrane, in the centre of which there is usually a fissure. [2] Behind
this conical process, and in the cephalothorax, there is a large, thin-walled
sac, filled with a clear liquid; this is prolonged by a kind of neck into the
process, and has, undoubtedly, the function of a *Labyrinthus*, [3] for, a
special nerve, arising from the lateral parts of the brain, in common with
the external antennal nerve, is spread upon its walls. [4]

The base of this labyrinth is in connection with a singular glandular
organ, of a usually greenish color, but whose nature is yet undetermined. [5]

1 These olfactory organs were first described and
considered as such by *Rosenthal* (*Reil's* Arch. X.
1811, p. 433, Taf. VIII. fig. 1-4) with the craw-fish
and lobster. *Treviranus* (Biologie, VI. 1822, p.
303) has subsequently confirmed these observations
with the lobster. See, for this same animal, *Milne
Edwards*, Hist. Nat. d. Crust. Pl. XII. fig. 1.
These organs have been found latterly, also, by
Farre, with *Palinurus* and *Pagurus*, (Philos.
Trans. 1843, p. 233, Pl. IX. X. and Ann. of Nat.
Hist. XII. p. 220). I have myself observed them
with *Palaemon*, *Nephrops*, and *Maia*. It is difficult to understand how *Farre* could have taken
these cavities for organs of hearing into which
grains of sand, entering by accident, would serve
as otolites.

1 Although special auditive organs have not yet
been observed with the other Crustacea, yet it
cannot be denied that they are sensible to sounds.
At least, the observations of *Coldstream* (Cyclop.
of Anat. I. p. 688) show that the Cirripedia have
a very acute sense of hearing, for they appear
cognizant of the slightest sound, and quickly close
the shell.

2 This cylindrical protuberance, with its tympanitic membrane, is easily seen in the basilar article mentioned, with *Homarus*, *Astacus*, *Nephrops*, *Palinurus*, and other Macrura; — see
Scarpa, Anatom. disquis. de auditu et olfactu.
p. 2, Tab. IV. fig. 4, a. b. ; *Weber*, De aure animal.
aquatil. p. 8, 106, Tab. 1. fig. 1, No. 1, and *Milne
Edwards*, Hist. Nat. d. Crust. Pl. XII. fig. 11. o,
(*Astacus*). This protuberance is long and cylindrical with *Pagurus striatus*, and *Homola Cuvieri*.
With the Maiina, whose antennal articles are
large and immovable, the auditive organs are
slightly protuberant, and situated near the mouth.
See *Savigny*, Descrip. de l'Egypte, loc. cit. Pl.
VI. fig. 4.: and 6.2 a. e (*Maia* and *Stenorhynchus*), and *Milne Edwards*, loc. cit. I. p. 264. Pl.
III. fig. 2, e. Pl. XV. fig. 2, 10, 16 (*Maia*, *Mithrax*, *Leucippa*, and *Camposcia*).
With *Scyllarus latus*, whose antennae are very
large and fixed at their base, the large but flat
auditive cylinders are very short and near together
on the borders of the mouth (*Savigny* loc. cit. Pl.
VIII. fig. 1.: a. c.). With *Scyllarus arctus*, I

have found the same concealed in the semilunar depressions which are underneath the mouth. With
Maia, these cylinders are obliquely truncated, and
are articulated with the large and basilar articles
of the antennae. They can be depressed towards
the inner side, and then righted as a kind of external auditive conch, and for this purpose the internal surface of the cylinder has a pair of muscles which are inserted on an internal, stirrup-like
process ; see *Cavolini*, Abhandl. über die Erzeugung d. Fische and der Krebse, p. 133, and *Milne
Edwards*, Hist. Nat. d. Crust. I. p. 124, Pl. XII.
fig. 10, c. l. m., and fig. 11, or in Cyclop. of Anat. loc.
cit. p.768, fig. 397, 398. Further researches are necessary, before the opinion of *Souleyet* (*Froriep's*
neue Notiz. XXVIII. p. 84) can be admitted, that
a small, round, glittering body which, with *Leucifer*,
is situated at the base of the internal antennae, is
an an litive organ.*
3 Formerly, the attention had been called only to
the portion of this labyrinth which is concealed in
the auditive cylinder (*Scarpa*, loc. cit. Tab. IV.
fig. 6, and *Weber*, loc. cit. Tab. 1. fig. 2). It is
only lately that it has been shown that this small
auditive vesicle belongs to a very large ampulla situated at its base ; see *Brandt*, Mediz. Zool. II. p.
64, Taf. XI. fig. 13, a. a., and *Neuwyler*, Anatom.
Untersuch. über den Flusskrebs, in the Verhandl.
d. schweizer. Naturf. Gesellsch. bei ihrer Versam-
mel. zu Zurich, 1841, p. 176.
4 *Scarpa*, loc. cit. Tab. IV. fig. 5, g. g., and *Weber*, loc. cit. Tab. I. fig. 2, No. 7; *Brandt*, and
Neuwyler, loc. cit. ; *Farre*, Philos. Trans. 1843,
Pl. IX. fig. 10, e. e.
5 This glandular body which appears to be present with the Brachyura also, is situated, with the
Astacina, behind the base of the external antennae,
concealed in the lower portion of the shell, and
covered, in part, by the membranous labyrinth ; see
Rosvel, loc. cit. p. 322, Tab. LVIII. fig. 9, c.;
Suckow, loc. cit. p. 55, Taf. IX. fig. 2, a.; *Brandt*
Mediz. Zool. p. 64, Taf. XI. fig. 8, k. (*Astacus*),
Milne Edwards, Hist. Nat. d. Crust. Pl. XII.
fig. 9, a. 10, g. (*Astacus* and *Maia*). *Neuwyler*
has given the green glands of the craw-fish a special examination (loc. cit.). He found that they
consisted of an intestinal tube communicating
with the membranous labyrinth. At first he

* [§ 276, note 2.] The organ of hearing in *Leucifer* first noticed by *Souleyet*, has since been
studied by *Huxley* (Ann. of Nat. Hist. 1851, p.
304) who appears to have clearly made out the
structure which resembles the ordinary form of

auditory apparatus in the Mollusca. See also
Schöller (*Wiegmann's* Arch. 1840, p. 363) upon
this organ with *Acanthocereus rigidus* ; finally,
Darwin, loc. cit. Cirripedia, p. 53. — ED.

§ 277.

The sense of Sight is present quite universally with Crustacea.[1]

The Cirripedia, the Penellina, and the Lernaeodea, alone, are without it; and even here this deficiency occurs only during the last phases of their retrograde metamorphosis, when these animals remain fixed to foreign bodies.[2]

There is, moreover, in the other orders, here and there a genus which contains blind individuals. Such is the case with the females of certain parasitic Isopoda,[3] and with some subterranean Myriapoda.[4]

The eyes of Crustacea present very various grades of development. The lowest of these is seen in the so-called Simple-eyes. With these, there is observed a convex cornea, and, behind it, a round, light-refracting body. This lens is surrounded by a layer of black, brown, red, or blue pigment, which, at its most convex point, is perforated by an optic nerve. The young individuals of the Cirripedia, the Penellina, and the Lernaeodea, have an eye of this kind in the middle of their forehead, but which gradually disappears in the course of their metamorphosis.[5] Certain Ergasilina, as likewise the Lophyropoda and Phyllopoda, have, also, at their escape from the egg, a simple eye, which, with the Ergasilina, and certain Lophyropoda,

thought them comparable to a cochlea, but he was unable to find any nerve going to them, and has relinquished this idea, — doubting that these organs, and the ampulla mentioned, are really auditive organs. Farre (loc. cit.) has gone further ; he has taken these bodies for olfactory organs, and has endeavoured to show, as already mentioned, the organs of smell to be real organs of hearing. It is true that, in the organs of hearing, no otolites are found ; but the principal parts exist, such as a *Cavum tympani*, at the entrance of which is extended a tympanitic membrane and an auditive vesicle, upon which is spread a nerve.

The view of *Frey* (De Mysidis Anat. p. 13), then, is inadmissible ; he regards the seat of hearing, with *Mysis*, as the two internal caudal valves, where he has observed a cavity containing a radiated body, the nucleus of which has a crystalline structure, and which he regards as an otolite. But, aside from the singular structure of this body, he does not mention its having any special nerves.

It is, moreover, unnecessary to seek, with those Crustacea whose antennae are highly developed, the auditive organs anywhere but on the head ; for, at the base of these antennae, as, for example, with the Amphipoda, there are several other hollow processes which, in part, have been regarded as palpi, but which, upon more careful examination, will undoubtedly be found to be, some auditive, and others olfactory organs.

Frey and *Leuckart* (Beitr. p. 114, Taf. II. fig. 18) have, from the first, described in more detail the organs of the caudal valves of *Mysis*, as proper auditory organs ; but, aside from the two so-called otolites which, contrary to all analogy, are provided with stiff bristles, the correctness of this interpretation is always open to question, for these authors have been unable to perceive any

nerve destined for these so-called auditive capsules, with *Mysis*.

1 For the eyes of Crustacea, see especially *J. Müller*, Zur vergleich. Physiol. d. Gesichtsinnes, p. 307 ; or its abstract in Ann. d. Sc. Nat. XVII. 1829, p. 225, or his later researches on the eyes of the Insecta and Crustacea, in *Meckel's* Arch. 1829, p. 38, and in *Tiedemann's* Zeitsch. f. Physiol. IV. p. 97.

2 The adult Cirripedia, notwithstanding the absence of eyes, are very sensitive to light. This I have observed with individuals of *Balanus pusillus*, which I had captive several weeks at Danzig. These animals, when undisturbed, came out of their shell, and executed the usual motions of their cirri, but they withdrew as quick as lightning into the shell, when, from passing my hand over the vessel, I shaded them. *Coldstream* (Cyclop. loc. cit. p. 688) has made similar observations."

3 The females of *Ione*, *Phryxus*, and *Bopyrus*.†

4 For example, with *Polydesmus*, *Blanulus*, *Cryptops*, and *Geophilus*.

5 For example with *Achtheres*, *Tracheliastes*, *Lernaeocera* (*Nordmann*, loc. cit. p. 80, &c., Tab. IV. fig. 5, Tab. VI. fig. 5, 6). The Cirripedia have, at their escape from the egg, a single, black eye, according to *Thompson* ; see the Philos. Trans. 1835, p. 355, *Owen*, Lectures, &c., p. 161, fig. 88 ; and *Goodsir*, Edinb. new Philos. Jour. 1843, No. 69, p. 97, Pl. III. fig. 8, and Pl. IV. fig. 13-17 (*Lepas* and *Balanus*); but with the embryos of *Balanus pusillus*, I have found this eye of a red color. The reason why *Burmeister* (Beitr. &c. p. 15, Taf. I. fig. 2) could perceive no eye with the young of *Lepas*, is, as he himself has remarked, because they had been effaced by the alcohol in which the specimens examined had long been preserved.

---

* [§ 277, note 2] Recent investigations have disclosed the existence of eyes with the Cirripedia. *Leidy* (Proceed. Acad. Sc. Phil. IV. 1848, No. 1) discovered them with *Balanus*, and this discovery has led to the confident and successful search of them in other genera. With *Lepas*, according to *Darwin* (Monograph, &c., loc. cit. p. 49), there are two closely-approximated eyes, forming a double eye, situated at the extremity of two optic nerves which proceed each from an ophthalmic ganglion. These ganglia are situated on two nervous cords which arise from the supra oesophageal ganglia. — ED.

† [§ 277, note 3.] Quite remarkable among the blind Crustacea is the *Astacus pellucidus* Telk. from the Mammoth Cave, Kentucky. — ED.

remains as a visual organ during the whole life;[6] while, with other Lophyropoda, and with the Phyllopoda, it either entirely disappears,[7] or remains in a condition apparently rudimentary, by the side of the other eyes, which are subsequently formed.[8] With certain Ergasilina,[9] and some Lophyropoda,[10] with the Caligina,[11] and the males of some parasitic Isopoda,[12] there are two permanent eyes, right and left, on the vertex of the head. The Poecilopoda, also, have, beside their compound eyes, two simple ones, contiguous on the middle of the forehead.[13] These simple eyes are also sometimes the more numerous, and are then situated on each side of the head, in fours, sixes or eights, in a single or double row, constituting the *Oculi seriati*, as is observed with some Myriapoda;[14] or they are collected in a thick group of twenty to forty, constituting the *Oculi gregati*, as is the case with other Myriapoda, and with the Isopoda.[15] Each of these eyes has a separate branch of the optic nerve; this nerve, therefore, divides as many times as there are eyes.

Another form of eyes which is pretty common among Crustacea, but which has many modifications, has received the name of Compound Unfaceted Eyes.

These organs are composed of a common cornea, covering numerous simple eyes, closely set against each other. They are found in their simplest form, with the Cirripedia at a certain epoch of development, with the Argulina, the Laemodipoda, and certain Lophyropoda, Phyllopoda and Amphipoda. Here, directly under the cornea, are a greater or less

---

6 *Lamproglena, Ergasilus* (*Nordmann,* loc. cit. Tab. II. fig. J, 7), *Cyclops, Cyclopsina, Cypris,* &c.

7 With *Limnadia,* and *Isaura,* it is replaced by a compound eye; see *Joly,* Ann. d. Sc. Nat. XVII. Pl. IX. fig. 39-41.

8 This is so with the adult individuals of *Apus* and the Branchiopoda, where the simple embryonic eye persists in an atrophied condition between the two faceted eyes; see *Schäffer,* Der krebsartige Kiefenfuss, Taf. II. fig. 1 c., and Taf. V. fig. 3-5; also, *Zaddach,* loc. cit. p. 48, Tab. II. fig. 18-22, C. and Tab. IV. (*Apus*); *Prevost,* in *Jurine's* Hist. d. Monocles, Pl. XX.-XXI. (*Chirocephalus*), and *Joly,* loc. cit. XIII. Pl. VII. (*Artemia*). The black spot observed front of the compound eye, with *Lynceus,* and certain species of *Daphnia,* is certainly only the remains of the simple eye; see *Müller,* Entomostr. Tab. IX.-XI., and *Jurine,* Hist. d. Monocles, Pl. XV.-XVI.
But, with this simple rudimentary eye, should not be confounded the problematical vesiculiform organ which is found behind the compound eyes of certain Phyllopoda and Lophyropoda. With *Apus,* this organ contains a nucleus, divided into four parts (*Schäffer,* loc. cit. Taf. II. fig. 1, b., or *Zaddach,* loc. cit. p. 48, Taf. II. fig. 10. P., 25). The vesiculiform body which, with *Limnadia,* stretches from the inner surface of the head, behind the eye, towards the forehead (*Brongniart,* loc. cit. p. 83, Pl. XIII. fig. 6), may serve, according to *Straus,* to fix the animal to foreign bodies; (see Mus. Senckenb. II. p. 126, or *Férussac,* Bull.

* [§ 277, note 8.] With *Artemia* and *Branchipus, Leydig* (loc. cit. *Siebold* and *Külliker's* Zeitsch. III. p. 295) has found very highly-developed eyes. In structure they correspond to the compound faceted eyes described below. In regard to the pigment-spots found on the head of these animals, and regarded as of a visual character by *Joly* and others, this observer considers

d. Sc. Nat. XXII. 1830, p. 333). With *Evadne,* there is found at the same place, behind the large eye, a circular muscle, which also, perhaps, is for the attachment of the animal.*

9 *Nicothoë* (*Rathké,* Nov. Act. Nat. Cur. XX. p. 102, Tab. V. fig. 1, 8, 10).

10 *Hersilia, Peltidium,* &c. (*Philippi,* in *Wiegmann's* Arch. 1839, I. p. 128, Taf. IV. fig. 9, 13, or *Milne Edwards,* Hist. d. Crust. Pl. XXXVII.).

11 *Pandarus, Caligus, Trebius, Dinematura,* &c. (*Milne Edwards,* Hist. d. Crust. Pl. XXXVIII., and *Kröyer,* in the Naturhist. Tidskr. I., or in the Isis, 1841, p. 188, Taf. I.).

12 *Phryxus* and *Bopyrus* (*Rathké,* Nov. Act. Nat. Cur. XX. p. 44, Tab. I. fig. 13, in Tab. II. fig. 3, and, De Bopyro et Nereide, Tab. I. fig. 2).

13 See *Van der Hoeven,* Recherches, &c., 23, Pl. III. fig. 5, a. a., 6, C.

14 With *Platyulus,* there are, on each side, six eyes, arranged in two rows. *Scolopendra* has four, while with *Glomeris,* there are eight, which form a simple arcuate row on each side; see *Müller,* in *Meckel's* Arch. 1823, p. 40, Tab. III. fig. 3, 4, also *Kutorga,* loc. cit. p. 17, Tab. III. fig. 3, 4 (*Scolopendra*), and *Brandt,* Mediz. Zool. II. p. 90, Taf. XV. fig. 43 (*Glomeris*).

15 *Treviranus,* Vermisch. Schrift. II. p. 32, Taf. VII. fig. 1 (*Lithobius*) and *Müller,* in *Meckel's* Arch. 1823, p. 43 (*Julus*); see also *Treviranus* loc. cit. I. p. 64, Taf. IX. fig. 54 (*Porcellio*), *Müller,* loc. cit. p. 42, Taf. III. fig. 5, 6 (*Cymothoa*), and *Lerebouillet,* Ann. d. Sc. Nat. loc. cit. p. 107, Pl. IV. fig. 2, 2′ (*Lygidium*).

them as mere accumulations of pigment granules, having no special function whatever. This naturalist alludes, also, to the problematical body above mentioned. He did not observe it with *Artemia,* but it was present with *Branchipus,* and larger in the larval than in the adult conditions. He hesitates to express an opinion as to its nature. — ED.

number of round, pyriform, or cuneiform lenses, the pointed posterior extremity of which is surrounded by a pigment matter of usually a deep brown or black color, while the rounded anterior extremity is always widely protuberant. The optic nerve, before reaching this pigment, divides into as many branches as there are lenses.

With *Argulus*,[16] *Cyamus*,[17] and with the Amphipoda,[18] there are always two considerably flattened eyes ; while with *Daphnia*, *Lynceus*, *Polyphemus*, *Evadne*,[19] (the Lophyropoda) and also with the young bivalve Cirripedia,[20] there is, on the other hand, only a single ocular bulb, spheroidal, and the result of the fusion of   s ; it receives, therefore, the two optic nerves which are separate l from each on or by the median line of the body. With *Limnadia*, and *Artemia*, of the Phyllopoda, this fusion is less complete, for, upon close examination, the line of separation may be seen.[21] With many Daphnioïdae, this cyclopean eye has several muscles, corresponding to the recti muscles of the Vertebrata, which give the eye a movement of rotation about its centre.[22]

With some Crustacea belonging to the orders Amphipoda, Phyllopoda, and Poecilopoda, the compound eyes are so modified, that, beneath the cornea which is simple, there is another cornea that is faceted. Each of these facets consists of a depression, in which fits the truncated extremity of an oblong, conical lens ; and the opposite extremity of this lens is surrounded by pigments, and connected with a filament of the optic nerve.[23]

A second modification of these compound eyes is also observed with some Amphipoda and Phyllopoda. Here, the cornea is likewise double, but between the faceted one and the conical lenses, are interposed peculiar lenses of an oval form.[24]

The third form of eyes observed with Crustacea has received the name of Compound Faceted Eyes. These are found in the genus *Scutigera*, and in the higher groups of Crustacea, namely : the Stomapoda, and Decapoda, with which the eyes are situated at the extremity of two peduncles, or, what is more rare, at a point below their extremity.[25] These peduncles are movably inserted on the anterior border of the cephalothorax, and are

---

16 *Jurine*, loc. cit. p. 446, Pl. XXVI. fig. 13, and *Muller*, in *Tiedemann's* Zeitsch. f. Physiol. IV. p. 97, Taf. VI. fig. 5, 6 (*Argulus foliaceus*).*

17 *Roussel de Vauzème*, loc. cit. p. 242, Pl. VIII. fig. 5.

18 *Muller*, in *Meckel's* Arch. loc. cit. p. 57, Taf. III. fig. 16, 17 (*Gammarus*).

19 The lenses are pyriform with *Daphnia* (*Straus*, loc. cit. p. 397, Pl. XXIX. fig. 6, 7), cuneiform with *Polyphemus* and *Evadne* (*Jurine*, Hist. d. Monocles, Pl. XV. fig. 1–3, and *Lovén*, loc. cit. p. 143, Pl. V.).

20 It is very remarkable that the Cirripedia, after the disappearance of the simple eye, which, during the embryonic state, is situated on the front, acquire another, compound but equally transitory. This last is situated at the lower border of the cephalic extremity, directly in front of the mouth, during the period when these animals are contained between two shells, and swim like a *Cypris*. It is pedunculated, and has the same structure as that of *Daphnia* ; see *Thompson*, Zool. Research.

* [§ 277, note 16.] For the intimate structure with many details, of the eyes of *Argulus*, see *Leydig* (loc. cit. *Siebold* and *Kölliker's* Zeitsch. II. p. 331, Taf. XX. fig. 1) ; they are not immovable as *Jurine* has described. — ED.

loc. cit. p. 77, Pl. IX. fig. 3, 4 ; and *Burmeister*, Beitr. p. 17, Taf. I. fig. 3–5.

21 See *Branzniart*, loc. cit. p. 85, Pl. XIII. fig. 3, 4 (*Limnadia*), and *Joly* loc. cit. p. 309, Pl. VII. fig. 3, Pl. VIII. fig. 24, 26 (*Isaura*). In this last-mentioned Crustacean the eyes contain ovoid lenses.

22 *Daphnia* and *Evadne* ; see *Jurine* and *Lovén*, loc. cit.

23 This modification is found with *Amphithoë*, *Apus*, and *Limulus* ; see *Milne Edwards*, Hist. d. Crust. I. p. 116 ; *Zaddach*, loc. cit. p. 45, Tab. II. fig. 18–24, and *Van der Hoeven*, loc. cit. p. 23, Taf. III. fig. 6, A. B.

24 *Hyperia* (*Milne Edwards*, Hist. d. Crust. III. p. 74, and Ann. d. Sc. Nat. XX. 1850, p. 388, and *Muller*, in his Arch. 1836, p. 102), and *Branchipus* (*Burmeister*, in *Muller's* Arch. 1835, p. 525, Taf. XIII. fig. 1–4). The lenses of this last-mentioned Phyllopod are situated in the cup-like cavities of the cones, so that this kind of eyes which, moreover, are pedunculated, form the transition to the faceted ones.*

25 With some species of *Ocypoda*.

* [§ 277, note 24.] The peculiarity in the structure of the eye of *Branchipus*, as above mentioned by *Burmeister*, *Leydig* (loc. cit. *Siebold* and *Kölliker's* Zeitsch. III. p. 295), was unable to verify with *Branchipus stagnalis* — ED.

usually concealed in special fossae. The tetragonal or hexagonal facets of the cornea are always very numerous; [26] — behind each of them, is a conical, or prismatic lens, the round extremity of which is fitted into a transparent conical fossa, corresponding to a vitreous body; while the conical extremity of these bodies is received into a kind of calyx, formed by the filaments of the optic nerve. Each of these filaments, together with its calyx, is surrounded by pigment matter in a sheath-like manner. [27] *

CHAPTER V.

DIGESTIVE APPARATUS.

§ 278.

The opening of the digestive apparatus with the Crustacea is usually situated directly in front of the first pair of feet, which, as foot-jaws, grasping or prehensile organs, are used for the seizing, the tasting, and the bearing to the mouth of food. [1] With many species, there are, as auxiliary organs for this purpose, the oar-like, the post-abdominal, and branchial feet, the movements of which not only produce currents of water necessary for respiration, but also direct towards the mouth a great quantity of nutritive matter. [2]

The mouth is generally situated underneath and somewhat removed from the anterior border of the head. It is covered with a soft upper lip,

26 The facets are tetragonal with *Astacus*, *Homarus*, *Palinurus*, *Galathea*, *Scyllarus*, *Palaemon*, *Pasiphaea*, and *Penaeus*; hexagonal with *Scutigera*, *Squilla*, *Phyllosoma*, *Pagurus*, *Calianassa*, *Maia*, *Campilius*, *Portunus*, and *Ilia*; see *Milne Edwards*, Hist. de Crust. l. p. 117, Pl. XII. and *Will*, Beitr. z. Anat. der zusammengeset. Augen, &c., p. 7. fig. 3, c.

27 *Will*, loc. cit. p. 12, fig. 3, 4 ; see also *Suckow*, loc. cit. Taf. X. fig. 19, 20 ; *Soemmering*, De oculor. sect. horizont. p. 75, Tab. III., and *Milne Edwards*, Hist d. Crust. Pl. XII. fig. 8 (*Astacus*).

1 See above, § 268.

2 These acts may be distinctly seen with the Phyllopoda, the Lophyropoda, and the Cirripedia. These last use principally their long, posterior, cirrus-like feet, which they unroll and roll up alternately, maintaining regular currents in the water. During these movements, the three pairs of anterior and shorter feet seize, with much address, the particles of food borne against them by the current. Often the oar-like feet with the Daphnioidae become dirty in this act, and are glued together by particles of food which have been ejected from the mouth. But these animals easily relieve themselves by curving in front their spinous tail and combing out the oar-like feet, which are themselves ciliated and bristled.

* [§ 277, end.] There is another form of eye observed by *Dana* (Report on Crust. loc. cit. p. 1020) with *Corycaeus* and *Sapphirina*, and of so remarkable a character that I quote his description : "A pair of simple eyes, consisting of an internal prolate lens, situated at the extremity of a vermiform mass of pigment, and of a large, oblate, lens-shaped cornea. The cornea is connected intimately with the exterior shell of the front or the under side of the head, and the two corneae are like spectacles adapted to the near sighted lenses within ; their size is extraordinary, being often one-third of the greatest breadth of the body in *Corycaeus*. The lens and the cornea are often very distant from each other, being separated by a long clear space. The external surface of the cornea is spherical ; but the inner is conoideo-spherical, or parabolic. The texture is firm, and when dissected it breaks or cuts like a crystalline lens. The true lens is always prolate, with a regular contour, excepting behind, where it is partly penetrated by the pigment. The pigment is slender, vermiform, of a deep color, either red or blue, but at its anterior extremity usually lighter, and often orange or yellow." — ED.

beneath which is a pair of strong upper jaws (*Mandibulae*), which move laterally by means of large muscles arising from the internal surface of the cephalic and dorsal parts of the skeleton; the internal border of these jaws is hard and often denticulated. With the higher Crustacea, these mandibles have a tactile organ (*Palpus*).[3] . Behind these mandibles are two pairs of lower jaws (*Maxillae*), which are weaker, softer, and deficient in palpi. They are composed of several pieces, except with the Myriapoda, where they are fused into a kind of lower lip. Between the two mandibles and the first pair of maxillae, there is a soft, tongue-shaped, and sometimes bifid process, which, also, may be regarded as an under-lip.[4]

With many of the lower Crustacea, the parts about the mouth are variously modified, whereby they lose their peculiarities as masticatory organs. Thus with the Pœcilopoda, the mouth is simple, infundibuliform, and jawless, — the mandibles and maxillae being changed into cheliform legs.[5]

With the parasitic Crustacea, the organs of the mouth are changed into parts for Suction. The two lips are prolonged into a kind of proboscis, and the masticatory organs become more and more indistinct and finally disappear entirely. This is best observed with the Caligina where the mouth has the form of a beak pointing backwards, and the upper and lower lips are joined together forming a long tube which contains the two very long, horny, denticulated mandibles, while at its base are two palpiform, rudimentary maxillae.[6] With *Argulus*, the oral parts form a suctorial apparatus even more complete. This is a very long proboscis, pointing forwards, and out of which the two mandibles project in the form of two small pointed stylets, while the maxillae are entirely wanting.[7] With the Lernaeodea, and Penellina, on the other hand, the proboscis is short, and contains two short mandibles, which are denticulated and hooked; and on its outside are two palpiform, rudimentary maxillae.[8] But the oral parts are most abortive with the Ergasilina and Bopyrina. Here, the upper and under lips are blended together into a short proboscis without mandibles,

*

---

3 With the Decapoda, Stomapoda, Amphipoda, and the majority of Isopoda. With the Chilopoda, these palpi exist only in a very rudimentary condition ; and they are entirely wanting with *Idothea*, the Chilognatha, and the other lower Crustacea. For the parts of the mouth of Crustacea, see the descriptions and figures contained in the works of *Savigny, Milne Edwards,* and *Erichson ;* also the various monographic works upon the Decapoda, Isopoda, Myriapoda, Phyllopoda, Lophyropoda, and Cirripedia, by *Suckow, Brandt, Rathké, Treviranus, Zaddach, Jurine, Lovén, Burmeister, Martin St. Ange,* &c.

4 *Astacus, Palaemon, Palinurus, Squilla.*

5 See *Van der Hoeven,* loc. cit. p. 16, Pl. II. fig. 1, A. (*Limulus*).

6 *Milne Edwards,* Sur l'organisation de la bouche chez les Crustacés suceurs, in the Ann. d. Sc. Nat. XXVIII. 1833, p. 78, Pl. VIII. ; and especially *Burmeister,* in the Nov. Act. Nat. Cur. XVII. p. 278, Tab. XXIII.-XXV.

7 *Jurine,* loc. cit. p. 440, Pl. XXVI. fig. 3–7, 16 ; *Vogt,* loc. cit. p. 7, fig. 5.*

8 *Nordmann,* Microgr. Beitr. loc. cit. Taf. V.-IX., and *Kollar,* loc. cit. Taf. IX. X. (*Actheres,* *Branchiella, Chondracanthus, Tracheliastes,* and *Basanistes*) ; also *Burmeister,* loc. cit. p. 310, Tab. XXIV. A. (*Lernaeocera*).

* [§ 278, note 7.] According to *Dana,* the proboscis here mentioned is simply a spicula without any mouth-opening or mandibular appendages ; the true mouth is posterior to this and has a trunk-form, with the buccal orifice on the under surface, as in some genera of the Caligoidea, and provided with regular mandibles ; see Amer. Jour. of Sc. 1837, XXXI. p. 299, also Rep. on Crust. loc. cit. p. 1322. This has since been verified by *Vogt* (Beiträge, &c., p. 7), and by *Leydig* (loc. cit. Sie-bold and *Kölliker's* Zeitsch. II. p. 332, Taf. XLX.

fig. 2. b.). This last-mentioned author thinks that the spicula in question is a poison weapon ; a view which is rendered probable from the fact that it has glands connected with it, as in the stings of insects, which glands have hitherto been considered salivary ; but they do not open into the mouth like ordinary salivary glands. Moreover, *Leydig* quotes the observation of *Jurine* that tadpoles pierced by this organ of *Argulus,* seemed poisoned and soon died. — ED.

and the palpiform maxillae, with only a few exceptions are wholly want-ing.[9]

§ 279.

The Intestinal Canal with nearly all the Crustacea, traverses the body without convolutions on the median line,[1] and the anus is situated at the extremity of the tail.[2] Its walls are composed of three to four different layers, of which the outer, answering to a peritoneal envelope, consists of a dense fibrous membrane.

The internal layer consists of a structureless, transparent epithelium, always non-ciliated. In the anterior portion of the intestine, which is often dilated into a kind of stomach, as also in the rectum, this epithelium is quite dense and is directly continuous with the external skin, and like it also, contains chitine; it is moreover, cast off, at the moulting, with the skin to which it remains attached, partly by the mouth, and partly by the anus.[3] Between this epithelium and the peritoneal envelope, there is a granulo-vesicular, mucous layer, surrounded by smooth, simple, and in-terlaced, muscular fibres.

With the higher Crustacea, alone, the digestive canal consists always of a very short œsophagus, a stomach, an intestine and rectum. With the lower Crustacea, it is only a simple tube of the same calibre throughout, except near the anus where it is sometimes constricted by the accession of a muscular layer. With the Siphonostoma,* and many of the Lophyro-poda and Phyllopoda, it is straight throughout;[4] but with the Daph-

9 *Nordmann*, loc. cit. Taf. I.–III. (*Lamproglena* and *Ergasilus*); *Rathké*, De Bopyro &c. p. 4, Tab. I. and Nov. Act. Nat. Cur. XX. p. 42, 103, Tab. II. V. (*Nicothoë* and *Phryxus*); also *Kroyer*, Isis, 1841, p. 343, Taf. V. fig. 7. c. (*Nico-thoë*). There is an exception in this respect with *Dichelestium*. Its proboscis is prolonged into a kind of beak surrounded by numerous movable processes, of which one pair of denticulated stylets concealed in a fold of the proboscis corresponds perhaps to mandibles, while another pair may perhaps be regarded as maxillae; see *Rathké* Nov. Act. Nat. Cur. XIX. p. 136, Tab. XVII. fig. 12–14, and *Milne Edwards*, Hist. d. Crust. Pl. XXXIX. fig. 4, a–c. or Cyclop. of Anat. loc. cit. p. 773, fig. 412–415.

1 *Glomeris* and *Lynceus* form here an excep-tion. With the first, the intestine has one curve in front and another behind (*Brandt*, in *Muller's*

Arch. 1837, p. 322, Taf. XII. fig. 2); with *Lyn-ceus*, it has one or two spiral turns (*Muller*, En-tomostr. Tab. IX. X., and *Jurine*, Hist. d. Mono-cles Pl. XV. XVI.).

2 The Cirripedia form an exception to this rule; their anus is situated between the last pair of cirri and the base of the tail; see *Cuvier*, Mém. loc. cit. fig. 7, k., and *Martin St. Ange*, loc. cit. Pl. II. fig. 4, 5, &c., h.

3 See *Schmidt*, Zur vergleich. Physiol. p. 30.

4 For the straight intestine of the Penellina, Lernaeodea, and Ergasilina, see *Nordmann*, loc. cit. Taf. I.–X., also *Burmeister*, Nov. Act. Nat. Cur. XVII. p. 311, Tab. XXIV. A. fig. 1. (*Ler-naeocera*); *Rathké*, Ibid. XIX. p. 156, Tab. XVII. fig. 2 (*Dichelestium*); *Jurine* Hist. d. Monocl. Pl. I.–VII. (*Cyclops* and *Cyclopsina*); *Prevost*, Ibid. Pl. XX.–XXII. (*Chirocephalus*); *Joly*, loc. cit. Pl. VII. VIII. (*Artemia*). †

* [§ 279.] The alimentary canal of the Cali-goidea, according to *Dana* (Report. Crust. loc. cit. p. 1337), is divided into four very distinct parts,—an œsophagus, small and slender; a stomach broad and heart-shaped; an intestine, marked by light constrictions, and a rectum provided with power-ful muscles. The œsophagus has a valve at its entrance into the stomach, and thereby regurgita-tion of the food is prevented.

See, also, for the digestive canal of *Argulus*, with its histology, *Leydig* (loc. cit. *Siebold* and *Kölliker's* Zeitsch. II. p. 332, Taf. XIX. fig. 2). — ED.

† [§ 279, note 4.] For details upon the struc-ture of the digestive canal of *Artemia* and *Bran-chipus*, see *Leydig* (loc. cit. *Siebold* and *Kölli-*

ker's Zeitsch. III. p. 283). This observer divides it into three distinct portions: Œsophagus, Stomach, and Intestine. The stomach is composed histologi-cally of four tunics; 1. A Muscular, made up of circular and longitudinal muscles; 2. A Homoge-neous, serving as a support for this organ; 3. A Cellular; and lastly 4, A Homogeneous, which ap-pears to be merely a continuation in words of the external Chitine layer. The intimate structure of the intestine is quite the same as that of the stom-ach, but the elementary particles of the muscles composing its muscular tunic, are spindle-shaped, giving this tissue here a structure quite peculiar, and unlike anything found elsewhere; see loc. cit. Taf. VIII. fig. 6, 10. — ED.

nioïdae, and Apodidae, on the contrary, its anterior extremity mounts towards the dorsal surface of the head, and then curves backwards to the mouth.[5]

With the other Crustacea, there is, more or less distant from the œsophagus, a stomach, formed by a pyloric constriction of the intestine. This stomach is small with the Cirripedia, Laemodipoda, Isopoda, and Amphipoda ; [6] but is pretty long with the Myriapoda.[7] In many of the Isopoda and Laemodipoda, the stomachic epithelium has stiff cilia, or presents a cartilaginous, or horny aspect, thus constituting a stomachic support and dental apparatus,[8] which is also observed in the somewhat larger stomach of the Pœcilopoda and Stomapoda.[9] But this structure of the stomach is most prominent with the Decapoda. Remarkable for its size and form, it consists of two portions; one, anterior, vesiculiform, communicating with the œsophagus, the other pyloric, pyramidal, and with the apex pointing backwards. The internal tunic of the stomach is composed of chitine and covered with stiff bristles, or sometimes with groups of very singular hairs of a forficulate form. Moreover, its callous and cartilaginous portions form, in the pyloric region, a remarkable support, on which are three solid movable pieces. One of these pieces is a single tooth placed in the middle of the posterior wall of the stomach; while the other two, longer and somewhat crenulated, are situated on the sides opposite each other. Several muscles, arising from the internal surface of the cephalothorax, are inserted on this stomach, and it is very probable that, by these, the animal can voluntarily bring the three pieces together, making them serve as internal masticatory organs.[10]

5 For the arcuate intestine of *Daphnia*, *Lynceus*, and *Polyphemus*, see the figures given by Jurine, Hist. d. Monocl. ; *Straus*, loc. cit. Pl. XXIX. (*Daphnia*) ; *Brongniart*, loc. cit. Pl. XIII. (*Limnadia*) ; *Straus*, Mus. Senckenb. loc. cit. p. 112, Taf. VII. fig. 12, and *Joly*, loc. cit. Pl. VII. fig. 5 (*Isaura*). With *Cypris*, there is a kind of stomach on the curved digestive canal (*Straus*, loc. cit. p. 50, Pl. I. fig. 10).

6 For the intestine and stomach of the Cirripedia, see the writings of *Cuvier*, *Burmeister*, and *Martin St. Ange*; also *Roussel de Vauzème*, loc. cit. Pl. VIII. fig. 12, 18 (*Cyamus*) ; *Brandt*, Mediz. Zool. II. Tab. XV. fig. 39 (*Oniscus*) ; *Lereboullet*, loc. cit. p. 126, Pl. V. fig. 25 (*Lygidium*), and *Rathké*, loc. cit. Taf. IV. fig. 19 (*Idothea*).

7 See *Ramdohr*, Abhandl. üb. d. Verdauungsw. d. Insek. p. 148, Taf. XV. fig. 1 ; *Trevirunus*, Verm. Schrift. II. p. 23, 43, Taf. V. fig. 4, Taf. VIII. fig. 6 (*Lithobius* and *Julus*) ; *L. Dufour*, loc. cit. p. 84, 95, Pl. V. fig. 1, 4 (*Lithobius* and *Scutigera*) ; *Kutorga*, loc. cit. p. 5, Tab. I. fig. 2 (*Scolopendra*) ; and *Brandt*, in *Müller's* Arch. loc. cit. Taf. XII. fig. 2 (*Glomeris*).*

8 The stomach of *Oniscus* contains a cartilaginous support of a peculiar form (*Brandt*, Mediz. Zool. II. p. 74, Taf. XV. fig. 41, 42). That of the stomach of *Idothea entomon* is composed of several solid pieces (*Rathké*, loc. cit. p. 119, Taf. IV. fig. 20, 21). With *Lygidium*, the epithelium is supported by several horny pieces, and provided with numerous stiff bristles (*Lereboullet*, loc. cit. p. 127, Pl. V. fig. 26–30). Finally, with *Cyamus*,

there are in the cardiac region of the stomach two lateral horny tridentate folds (*Roussel de Vauzème*, loc. cit. p. 251, Pl. VIII. fig. 13, 14).

9 With *Limulus*, the œsophagus extends in front and opens into a very muscular backwardly-curved stomach the epithelium of which has fifteen longitudinal rows of horny teeth (*Van der Hoeven*, loc. cit. p. 17, Pl. II. fig. 3, B.). With *Squilla*, the stomach is pyramidal, and has, at its pyloric region, horny plates and very regular rows of hairs (*Duvernoy*, in *Cuvier's* Leçons d'Anat. Comp. V. p. 231). With *Mysis*, also, the epithelium of the pyriform stomach is supported by several solid lamellae composed of chitine and covered with bristles mixed with hairs (*Frey*, loc. cit. p. 16).

10 The stomach of the Crawfish is the one best known ; see the descriptions and figures given by *Roesel*, *Suckow*, *Brandt*, loc. cit., and *Milne Edwards*, Hist. d. Crust. I. p. 67, Pl. IV. The intimate structure of this stomach and its internal appendages have been carefully studied by *Valentin* (Repertorium, I. p. 115, Taf. I. fig. 15–21) and by *Oesterlen* (*Müller's* Arch. 1840, p. 387, Taf. XII.).

The teeth and bristles here observed, are found also with the three divisions of the Decapoda. I have seen them with *Homarus*, *Palinurus*, *Galathea*, *Pagurus*, *Cancer*, *Maia*, *Lupea*, &c. With *Crangon* and *Palaemon*, I found the dental lamellae wanting but the epithelium was hairy. With *Caridina*, according to *Joly* (loc. cit. p. 73, Pl. III. fig. 27), hairs of this kind are inserted on the band-like condensations of the stomach.

* [§ 279, note 7.] For the alimentary canal of *Julus* in all its details, see *Leidy*, A Flora and

Fauna within living Animals, in Smithsonian Contributions to Knowledge, V. 1853. — ED.

§ 280.

A large portion of the Crustacea have glandular appendages to the digestive canal. But it is only a few of these organs to which can be attributed the function of Salivary Glands. Two such of a lobular form, are found in the Cirripedia on the stomach, and pour their secretion into the anterior part of this organ.[1] But with the Myriapoda, these organs are very distinct. There are two or more on each side of the œsophagus and stomach, and their rather long, excretory ducts open into the oral cavity.[2] With all the other Crustacea, these organs are wholly wanting.*

The Liver, which exists sometimes as a glandular layer enveloping the digestive canal, and sometimes as a separate organ, is composed of greenish, or of yellowish-brown tubes of variable size, the walls of which are formed by numerous granular cells, between which are interposed fat-vesicles.[3] With most of the lower Crustacea, with the Siphonostoma, the Lophyropoda, the Phyllopoda, and Myriapoda, the liver is not isolated from the digestive tube, but the follicles of its glandular layer are somewhat protuberant on the external surface of this tube, and open on its internal surface, each probably by a separate orifice.[4] With *Argulus*,

1 See *Cuvier*, Mém. loc. cit. p. 10, fig. 9, u. u., 11. d. (*Lepas*); *Burmeister*, loc. cit. p. 42, Tab. 11. fig. 13, 14, c. (*Coronula*); *Karsten*, Nov. Act. Nat. Cur. XXI. Tab. XX. fig. 1, d. (*Balanus*).

2 With *Lithobius*, and *Scutigera*, there are two compact salivary glands which extend from the head into the first segments of the body (*L. Dufour*, loc. cit. p. 83, 95, Pl. 1. V.). *Treviranus* (Verm. Schrift. II. p. 25, Taf. V. fig. 4, q. q.) regarded them as a mass of fat. The botryoidal glands, which open by several excretory ducts into the oral cavity, have been observed by *Gaede* (*Wiedemann's* zool. Mag. I. p. 107, Taf. 1. fig. 7, i l.), by *Muller* (Isis, 1829, Taf. II. fig. 5), and by *Kutorga* (loc. cit. p. 4, Tab. I. fig. 4), in the anterior extremity of the body of *Scolopendra*. With *Glomeris*, there are only two short, slightly flexuous glandular tubes situated in the lower portion of the head and opening into the mouth (*Brandt*, in *Muller's* Arch. 1837, p. 323, Taf. XII. fig. 3). With other Myriapoda, these organs quite resemble those of the Insects. Thus, with *Geophilus*, there are two flexuous tubes situated pretty far behind the head, and from which pass off two very long, small excretory ducts along the œsophagus to the mouth (*Treviranus*, loc. cit. p. 37, Taf. VII. fig.

3). With *Julus*, the salivary organs are even longer and form, with the urinary canals, a very complicated net-work about the stomach, and from which pass off, according to *Treviranus* (loc. cit. p. 44, Taf. VIII. fig. 6), three excretory ducts to the mouth. But *Ramdohr* (Abhandl. &c. p. 149, Taf. XV. fig. 1, g. g.) has figured only two simple salivary canals with *Julus*, and this number has been verified by *Burmeister* (Isis, 1834, p. 136). I have seen these two canals with *Julus sabulosus* anastomose in an arcuate manner at their posterior extremity.*

3 For the intimate structure of the biliary tubes, see *Schlemm*, De hepate ac bile Crust., loc. cit. p. 14, Tab. II. fig. 12 (*Astacus*), and *Karsten*, Nov. Act. Nat. Cur. XXI. p. 295, Tab. XVIII. -XX. (*Oniscus*, *Astacus*, and *Balanus*).

4 An hepatic layer of this kind may be observed with the Penellina, Lernaeoidea, Ergasilina (*Nordmann*, loc. cit. Taf. 1.-X.), and with *Artemia* (*Joly*, loc. cit. p. 239, Pl. VIII. fig. 4). The numerous cœca, which, according to *Rathke* (Nov. Act. Nat. Cur. XX. p. 122, Tab. V. fig. 15), belong to the entire digestive canal of *Chondracanthus*, are perhaps formed by an hepatic substance. With the Chilognatha, and Chilopoda, I have found the

* [§ 280, note 2.] For the salivary glands of *Julus* see *Leidy* (A Flora and Fauna within Living Animals, p. 17, Pl. VII. fig. 21, a. b. In Smithsonian Contributions to Knowledge, V. 1853). Beside the long tubular glands mentioned by the authors above, he has described two others which are placed on each side of the œsophagus and are pyriform, conglomerate, and cellular in structure.

Wright (Ann. Nat. Hist. 1848, p. 140) has also made observations on the glands of *Geophilus* which open into the head; he has shown them to be veneniferous, for a single excretory duct passes off from the anterior part of this gland and termi-

nates, on each side, in a canal of the jaw or mandible, as in the Arachnidae. — ED.

† [§ 280, *Dana* (Report, &c., loc. cit. p. 1339) speaks of several small glands about the mouth, and communicating with it by ducts, with the Caligoidea, and which are probably of a salivary nature. These organs in *Argulus* have been carefully examined by *Leydig* (Ueber Argulus, &c., loc. cit. p. 333, Taf. XIX. fig. 2, a.), and especially as to their relations to the spicula (see § 278, note 6) of these animals. *Leydig* thinks they may as well be regarded poisonous as salivary glands. See my note under § 278, note 7. — ED.

Daphnia, and *Apus*, alone, the anterior extremity of the intestinal canal has several single or ramose cacca, the walls of which appear to serve principally as hepatic organs.[5] With the Cirripedia, similar caeca exist on the stomach,[6] and form, evidently, the transition to the hepatic organs of the other Crustacea,—that is, to an isolated liver with special, though short excretory canals. Such an isolated liver occurs with the Laemodipoda, Isopoda, and Amphipoda, and consists of long varicose caeca arising from the base of the stomach, and accompanying the intestine a considerable distance.[7]

With the Poecilopoda, Stomapoda, and Bopyrina, the hepatic organs are inserted at various points along the digestive canal.[8] Finally, with the Decapoda, the liver consists of two glandular masses composed of more or less ramose caeca loosely bound together. Each of these glands, which sometimes occupies only the sides of the cephalothorax, but sometimes, also,

largest portion of the alimentary canal dotted with small, yellowish-brown follicles, which I can only regard as hepatic organs, although other Zootomists consider them as Malpighian canals (see § 287). L. *Dufour* (loc. cit. p. 96, Pl. V. fig. 4, B.) has found these follicles in the stomach of *Scutigera*, but did not regard them as hepatic. The numerous large cells, which, according to *Serres* (Ann. du Mus. d'Hist. Nat. XX. p. 259), cover the external tunic of the intestine of *Lithobius*, are certainly only follicles of this kind.*

5 With *Daphnia*, there are two lateral, backwardly-curved caeca, which ascend from the anterior extremity of the digestive canal towards the dorsal surface of the head; see *Schäffer*, loc. cit. p. 41, Taf. II. fig. 2, k. k.; *Straus*, loc. cit. p. 401, Pl. XXIX. fig. 6, s. o. s., and *Jurine*, Hist. d. Monocl. Pl. IX. X. fig. 7, XI.–XIII. With *Branchipus*, and *Artemia* (*Joly*, loc. cit.), the anterior extremity of the digestive tube has also two short caeca which, with the glandular tunic with which the remainder of the intestine is covered, should be regarded as a liver. With *Argulus*, the stomach has two multiramose caeca, which lie in the parenchyma of the body (*Jurine*, loc. cit. p. 441, Pl. XXVI. fig. 1–3, 9, or *Vogt*, loc. cit. p. 8, fig. 1, 9). With *Apus*, these caeca are given off from the anterior extremity of the digestive canal and do not extend beyond the anterior border of the cephalothorax (*Schäffer*, loc. cit. p. 70, Taf. V. fig. 15, a. a.). According to *Zaddach* (loc. cit. p. 8, Tab. I. fig. 10–13, and Tab. IV.), these caeca contain numerous glandular follicles.†

6 Beside the figures of these stomachic appendages in the works of *Cuvier*, *Burmeister*, and *Martin St. Ange*, see also particularly those which *Karsten* (Nov. Act. Nat. Cur. XXI. p. 301, Tab. XX. fig. 1–4) has given of the hepatic organs surrounding the pylorus with *Balanus*.

7 With *Cyamus*, there are two long hepatic canals which wind over the digestive canal (*Roussel de Vauzème*, loc. cit. p. 252, Pl. IX. fig. 19). The two stomachic appendages of *Idothea* which *Rathké* (loc. cit. p. 121) has taken for adipose bodies, belong to the hepatic apparatus which here, according to my observations (*Muller's* Arch.

1837, p. 435) consists of three pairs of yellow varicose tubes. With *Oniscus*, *Porcellio*, *Asellus*, and *Lygidium*, there are four very long varicose hepatic tubes which open right and left into the pylorus (*Treviranus*, Verm. Schrift. I. p. 57, Taf. VII. fig. 38, Taf. IX. fig. 50, Taf. XI. fig. 64; *Brandt*, Mediz. Zool. II. p. 75, Taf. XV. fig. 39; *Lereboullet*, loc. cit. p. 130, Pl. V. fig. 25; *Karsten*, loc. cit. p. 296, Tab. XXVII. fig. 1). *Treviranus*, who did not observe the excretory ducts of these glands regarded them as masses of fat, while *Ramdohr* (Abhandl. üb. d. Verdauungsw. &c. p. 204, Taf. XXVIII. fig. 5), who, probably by mistake, has figured with *Porcellio* three similar appendages, has taken them for salivary organs. There are three pairs of hepatic canals with *Cymothoa* (*Meckel*, Syst. d. vergleich. Anat. IV. p. 154), *Aega* (*Rathke*, Nov. Act. Nat. Cur. XX. p. 30, Tab. VI. fig. 16, d. d. 18), and *Lygia* (*Milne Edwards*, Hist. Nat. d. Crust. Pl. IV. fig. 3). I cannot now decide whether *Iliella* has really only one varicose hepatic tube, or whether the others were overlooked by *Straus* (loc. cit. p. 59, Pl. IV. fig. 15). With *Gammarus*, and the other Amphipoda, I have found two pairs of long hepatic tubes.

Frey and *Leuckart* (Beitr. p. 101) have found with *Caprella*, as with *Cyamus*, two simple hepatic caeca.

8 With *Limulus*, there are four groups of interlaced caecal canals situated in both sides of the cephalothorax. The bile is poured into the anterior portion of the intestine by four distinct excretory ducts, which are widely separated from each other (*Van der Hoeven*, loc. cit. p. 18, Pl. II. fig. 1, 5, 8). With *Squilla*, *Bopyrus*, and *Phryxus*, the digestive canal has ramose or varicose hepatic caeca on both sides, at irregular intervals, of its whole length (*Muller*, De Gland. Struct. p. 70, Tab. IX.; *Duvernoy*, Ann. d. Sc. Nat. VI. 1836, p. 243, Pl. XV. fig. 1 (*Squilla*); and *Rathke*, De Bopyro et Nereïde, p. 9, Tab. I. fig. 7, and Nov. Act. Nat. Cur. XX. p. 47, (*Bopyrus* and *Phryxus*.) The genus *Mysis* (*Frey*, loc. cit. p. 19) on the contrary, which has eight hepatic canals opening right and left into the base of the stomach, resembles again the Amphipoda and Isopoda.

* [§ 280, note 4.] For the liver of *Julus*, see *Leidy*, loc. cit. He says, "At the termination of the proventriculus, there open two biliary tubes, and from it, surrounding the commencement of the ventriculus, is suspended a broad, white, opaque, reticulated band, apparently composed like the rete adiposa of insects." — Ed.

† [§ 280, note 5.] The hepatic nature of these appendages with *Artemia* and *Argulus* is denied

by *Leydig* (Ueber Argulus, &c., and Ueber Artemia, &c., loc. cit. *Siebold* and *Kölliker's* Zeitsch. II. p. 334, and III. p. 286) on histological grounds; it is most probable however that they serve as a liver, since *Will* (*Muller's* Arch. 1848, p. 506) has shown, by chemical analysis, the hepatic nature of analogous caecal tubes with *Daphnia* and *Cyclops*. — Ed.

reaches even to the tail, pour their secretion, by a short duct, into the digestive canal on both sides close behind the pylorus.[9]

§ 281.

With many Crustacea, the digestive canal is surrounded with fat-cells, the contents of which are often of a beautiful orange or blue color. These cells either consist of a few scattered globules,[1] or are disposed in lobes of various forms.[2]   This tissue is undoubtedly analogous to the *Corpus adiposum*, so common in insects.

The fat which these cells contain, plays a part, probably, in digestion and assimilation ; for with these animals the excess of nutriment is deposited as fat to be used in times of need, as, for example, during the act of moulting. This explains why the quantity found is so variable, or even may be entirely wanting.

CHAPTER VI.

CIRCULATORY SYSTEM.

§ 282.

Although the blood of Crustacea traverses the body by a very regular circulation, yet, as with all the Arthropoda, the vascular system is here quite imperfect, the blood-currents not always being contained in proper canals. But a central, propelling organ is very rarely absent, and consists of a heart, sometimes round and vesiculiform, sometimes long and tubular. With the higher Crustacea, it is the point of departure of an arterial system which, with the lower orders, gradually becomes abortive, and at last entirely disappears. The more or less long arteries do not terminate peripherically in a capillary net-work, but the blood is freely effused into the

---

9 For the liver of the common crawfish, which is large but contained in the cephalothorax, see the descriptions and figures of *Roesel, Suckow, Geecke, Brandt,* and *Schlemm*, also those of *Muller* (De Gland. Struct. p. 63). This last mentioned author found the liver conformable with that of many of the other Macrura and Brachyura. *Milne Edwards* (Hist. d. Crust. Pl. IV, fig. 5) has found, with *M tea*, a hepatic mass very remarkable in being symmetrically divided into several lobes. With *Pagurus*, there is, on each side of the pylorus, a long biliary vessel, which extends along the intestine to the extremity of the tail, and into which numerous lateral follicles empty their product ; see *Swammerdamm*, loc. cit. p. 86, Taf. XI. fig. 4, 5 ; *Muller*, De Gland. &c. p. 70, Tab. VIII. fig. 12,

13 ; and *Delle Chiaje*, Descriz. &c. Tav. LXXXVI. fig. 6.[4]

1 These fat-globules, of an orange color, are often found scattered about in *Cyclops, Daphnia* and *Gammarus*.

2 Such lobes and of a blue color are found with *Branchipus* on the sides of the digestive canal. Other whitish adipose masses form a kind of net-work around the intestinal canal of *Lernava, Lernaeocera* and *Lamprogirna* (*Rathké*, Nov. Act. Nat. Cur. XX. p. 123, and *Nordmann*, loc. cit. p. 6, 125, 132, Taf. I. fig. 4, Taf. VI. fig. 4). This last observer has regarded this reticulated mass as a liver. With the Myriapoda, these adipose masses are large, lobulated, and occupy quite a space in the visceral cavity.

* [§ 281, note 9.] For the intimate structure of the liver of Crustacea, as elucidated by the microscope, see *Leidy,* Amer. Jour. Med. Sc. 1848, XV. p. 1. — *Ed.*

lacunae which lie between the different visceral organs and appendages of the body. But, notwithstanding the absence of vascular walls in these interstices, the blood moves in determinate directions, until, after a course of variable length, it is returned to the heart. During their course, the blood-currents are often taken up by particular reservoirs, which, as venous sinuses, may be regarded as forming the rudiments of the venous system. In this manner, notwithstanding the imperfection of this vascular apparatus, all the organs constantly receive fresh blood, which is nowhere stagnant ; also, the arterial may be clearly distinguished from the venous currents, even when the arterial walls are wanting.

The Blood, itself, is either colorless, or of a faint red or violet hue. These colors belong to the blood-liquid, and not to the contained globules, which are few and always colorless. These globules are round, oval, or pyriform ; their surface is rough, and they contain fine granules, and, often a very large nucleus.[1]

§ 283.

The Heart of the Crustacea is always situated in the axis of the body, directly under the shell, at the anterior part of the back, and is often attached to the internal surface of the skeleton by muscular fibres.

Usually, its walls are thin and composed of scattered muscular fibres interlaced in various ways. By the contraction of these fibres the blood is propelled from behind forwards through the arterial orifices, — those of the veins being closed at the same time by valves.

The number of these different orifices, and the form and divisions of the Heart, have the following modifications :

1. With many of the lower Crustacea, especially with the Siphonostoma, and the Lophyropoda, the heart is a simple, thin-walled sac, of either a spheroidal or an elongated form, but invariably with only two orifices, — a posterior or venous, and an anterior or arterial.[1]

---

1 For the blood of Crustacea, see *Wagner*, Zur vergleich. Physiol. d. Blutes, p. 21. It is pale red with the craw-fish ; I have found it deep red with *Apus*, and violet with *Gammarus*. That of *Palinurus* is also pale red, according to *Lund* and *Schultz* (Isis, 1830, p. 1225). See also *Nordmann*, loc. cit. p. 73 (*Achtheres*) ; *Joly*, loc. cit. p. 238 (*Artemia*) ; *Zenker*, loc. cit. p. 20 (*Gammarus*) ; *Frey*, loc. cit. p. 21 (*Mysis*), and *Carus*, Von d. äusseren Lebensbeding. d. weiss-und Kaltblütigen Thiere, p. 80.

1 The heart is round or ovoid, and its pulsations quite frequent, with *Daphnia, Lynceus, Polyphemus*, and *Evadne*, where it is situated at the anterior part of the back, and very easily seen (see the figures of it given by *Straus, Jurine*, and *Loven*, loc. cit.). According to *Nordmann* (loc. cit. p. 11), there is also a round heart in the cephalothorax of *Ergasilus*. *Jurine* (Hist. d. Monocl. p.

57. Pl. V. fig. 4) thinks he has observed a distinct auricle underneath the heart of *Cyclops* ; but for my own part I have been unable to see it. As to a second or ventral heart, situated under the dorsal heart, which, according to *Perty* (Isis, 1832, p. 725), is found with *Daphnia*, I have been as unable as *Wagner* (Vergl. Anat. 1834, p. 166) to find it. With *Argulus*, the heart is long and situated under the dorsal shell, as *Vogt* (loc. cit. p. 9, Taf. I. fig. 1, 10, M.) has shown, contrary to the opinion of *Jurine* (loc. cit. p. 457, Pl. XXVI.). With *Achtheres, Dichelestium, Chondracanthus*, the heart consists of a long cylindrical tube (*Nordmann*, loc. cit. p. 73, and *Rathké*, Nov. Act Nat. Cur. XIX. p. 153, and XX. p. 125). The anterior and posterior valvular system which *Pickering* and *Dana* (Isis, 1840, p. 206) have seen with *Caligus*, lead us to think that here also there is a heart between these valves.*

---

* [§ 283, note 1.] With *Caligus*, the circulation is wholly lacunal, and appears to consist of broad irregular streams, passing through the spaces left by the internal organs, — there being in no part distinct vessels. A single centre of circulation, or a *heart*, can scarcely be said to here exist, but there are two points in the median line where there is a valvular action, and which perhaps perform the functions of this organ ; see *Dana*, Caligus, &c., Amer. Jour. Sc. XXXIV. p. 257, Pl. III. fig. 6, a. 6, b.

A corresponding structure has been found with *Argulus*, by *Leydig* (loc. cit. *Siebold* and *Kölliker's* Zeitsch. II. p. 235, Taf. XIX. fig. 3), who has given, moreover, many histological details upon the circulatory system of these animals. — ED.

2. With the other Crustacea, excepting the Myriapoda, the heart has, likewise, the form of a short simple sac, or that of a simple tube. In both cases, it is perforated by very numerous arterial and venous orifices. During the systole, the blood is propelled through the arterial orifices leading, nearly always, into vessels of the same nature ; at the same time, the venous orifices are closed by valves, which open, however, during the diastole, to allow the ingress of the blood into the heart. With the Decapoda, the heart is vesiculiform, situated in the middle of the cephalothorax, and its projecting corners often give it a star-like aspect. This heart has arteries passing off in front, behind, and below, and the returning venous blood enters it through venous orifices on its upper lateral portion.[2] With the Poecilopoda, Isopoda, Amphipoda, and probably, also, with the Laemodipoda, and Cirripedia, the tubular heart, occupying a large portion of the anterior and middle regions of the back, sends off arteries before, behind, and laterally, and receives the venous blood through lateral venous orifices.[3] This organ is most highly developed with the Stomapoda, where it occupies nearly the whole length of the body like a tube ;[4] but with the

2 There are, usually, in the polygonal heart of the Decapoda, three anterior arterial orifices, two below and one behind. These open distinctly into as many main arteries ; see *Swammerdamm*, loc. cit. p. 87, Taf. XI. fig. 8 (*Pagurus*); *Roesel*, loc. cit. p. 58, Taf. IX. fig. 14, and *Suckow*, loc. cit. p. 58, Taf. IX. fig. 1, Taf. XI. fig. 2–4 (*Astacus*); *Audouin* and *Milne Edwards*, Ann. d. Sc. Nat. XI. 1827, p. 353, 363, Pl. XXIV. XXVIII. fig. 1 ; and *Milne Edwards*, Hist. d. Crust. Pl. V. VII. (*Maia* and *Homarus*), and *Cyclop.* of Anat. loc. cit. p. 775, fig. 418 (*Cancer*). Not so easily seen are the six venous orifices which always are only valvular fissures, chiefly because they do not open into veins. According to *Lund*, and *A. W. F. Schultz* (Isis, 1825. p. 504, Taf. III. fig. 2–4 ; Ibid. 1829, p. 1299, (*Homarus*), and 1830, p. 1226, with the figure of p. 1228, (*Maia*)). the heart of the macrourous Decapoda has two upper, two lower, and two lateral venous orifices, while that of the Brachyura have only four upper and two lateral. *Krohn* (Isis, 1834, p. 524, Taf. XII. fig. 1–3), has confirmed this observation with the crawfish. *Suckow*, however (loc. cit. p. 58, Taf. XI. fig. 2, a. a.), did not perceive in this species only the two upper orifices, while *Audouin* and *Milne Edwards* (Ann. d. Sc. Nat. loc. cit. p. 357, 364, Pl. XXVI. fig. 3, N.nn) have not observed in the heart of *Homarus* and *Maia* only the two lateral orifices. This last naturalist (Hist. d. Crust. L. p. 94. Pl. V. VI.) refuses to admit the description of the heart of the Decapoda given by *Lund*, and brings to his support (*Cyclop.* loc. cit. p. 777) *Hunter's* preparations of the lobster ; but, judging from the beautiful figures of them given by *Owen* (Catal. of the Physiol. Ser. II. Pl. XV. h. h. Pl. XVI. fig. 2, d. and especially fig. 1, f. f. f.) these are just the preparations to support the view of *Lund*, *Schultz*, and *Krohn*. I, at least, have perceived distinctly the upper, lower and lateral venous orifices, as " the three orifices of the veins passing into the heart, f. f. f." See also the description of *Owen* of the heart of the lobster in his Lectur. on Comp. Anat. p. 179, fig. 91.

3 For the heart of *Limulus*, see *Straus*, Consid. gén. sur l' anat. comp. des anim. articulés, p. 346, and especially *Van der Hoeven*, loc. cit. p. 18, Pl. II. fig. 9. Beside the anterior and posterior arterial orifices, there are, with these Crustacea, seven others belonging to the seven pairs of lateral arteries, and on the dorsal portion of the organ, an equal number of valvular openings belonging to the venous system. With the

Isopoda, the tubular heart is continuous with an anterior and a posterior aorta ; it receives only three to five pairs of lateral vessels which have been regarded sometimes as arterial and sometimes as venous ; see *Treviranus*, Verm. Schrift. I. p. 58, 65, Taf. VIII. fig. 46, and Taf. IX. fig. 55 (*Porcellio* and *Armadillidium*); *Brandt*, Med. Zool. II. p. 75, Taf. XV. fig. 38 (*Porcellio*); *Leeeboullet*, loc. cit. p. 131, Pl. V. fig. 35 (*Lygidium*); *Rathké*, in the Neuest. Danzig. Schrift. I. p. 122 (*Idothea*), and Nov. Act. Nat. Cur. XX. p. 31 (*Asga*). It is, however, very probable that these orifices are arterial, for they open into vessels, and, moreover, the venous orifices are found, as with *Limulus*, on the dorsal surface of the organ. For the Amphipoda, *Gammarus pulex* may be cited as a type, and of which the heart as a cylindrical vessel occupies the axis of the anterior segments of the body. In this animal may be very easily seen how the blood, with the diastole, enters the heart through the several dorsal venous orifices, and how, with the systole, it is thrown forwards, backwards and laterally through the arterial openings. We have not yet complete researches on the heart of the Cirripedia ; but since *Martin St. Ange* (loc. cit. p. 18) states that these animals have a dorsal vessel with lateral trunks, it may be concluded that their heart is like that of the Amphipoda, Isopoda, &c. As to the Laemodipoda, we have only the imperfect details given by *Treviranus* (Verm. Schrift. II. p. 8), and *Roussel de Vauzème* (loc. cit. p. 254), according to which there is, with *Cyamus*, only a simple tube opening before and behind ; and we are therefore unable to say whether this heart is formed after the first or second type indicated in the text.

4 With *Mysis*, the heart consists, according to *Frey* (loc. cit. p. 21), of a dorsal vessel extending from the cephalothorax into the back part of the body ; but the blood enters it only through a posterior *Ostium venosum*, and passes out into the body through an anterior *Ostium arteriosum*. If this organization is confirmed, *Mysis* will differ remarkably in this respect from the Isopoda, Amphipoda, &c., but especially from another Stomapode genus, — *Squilla* ; for in this last, the heart with its anterior, posterior and lateral orifices, reaches its greatest development, occupying the entire abdominal cavity except the cephalothorax, and sends off laterally fourteen to seventeen pairs of arteries, beside being perforated on its upper portion by various pairs of venous orifices ; see *Davernoy*, Ann. d. Sc. Nat. VIII. 1837 p. 42, Pl.

Phyllopoda, it is less elongated and has numerous constrictions, thereby resembling the following type.[5]

3. This type, the third, is found with the Myriapoda, and considerably resembles that of the so-called Dorsal Vessel of the Insecta. With the Chilognatha, and Chilopoda, it consists of a more or less articulated tube, occupying the whole dorsal line of the body. It is divided by constrictions and imperfect muscular septa into chambers, nearly as numerous as the segments of the body. Each chamber is attached, as with the Insecta, right and left to the internal surface of the segments of the body, by triangular muscles. The Diastole is produced chiefly through these muscles. At its anterior extremity, this dorsal vessel passes through an *Ostium arteriosum* into an aorta, while, from the posterior extremity of each of these chambers are given off two lateral arteries. The returning blood enters the heart through the two venous orifices on the dorsal surface of each compartment. The Systole consists of an undulating action from behind forwards, and the blood is thereby propelled partly from one chamber to the next forward, and partly into the lateral arteries.[6]

§ 284.

The Circulation outside of the heart, with the Crustacea, has very varied relations, as has already been mentioned. With the lower Crustacea, with the Siphonostoma, the Lophyropoda, and the Phyllopoda, the blood forms regular currents in the intervisceral lacunae and interstices, but there is no trace of vascular walls. The aortic current, shortly after leaving the heart, divides into a right and left portion, which, also, sub-divide, enter the appendages of the cephalic extremity, then turn and run along the abdominal surface of the body — furnishing, in their course, several lateral, looplike currents, which enter the locomotive organs, then turn again towards the posterior extremity of the back, where they enter the heart.[1]

II. fig. 1, and especially *Audouin* and *Milne Edwards*, Ibid. XI. 1827, p. 376, Pl. XXXII. These last naturalists have very distinctly represented the dorsal venous orifices of the heart just mentioned.

5 With *Branchipus*, *Artemia*, *Isaura*, and *Apus*, the heart which has several constrictions and whose venous orifices are very apparent, occupies the entire dorsal median line excepting in the caudal extremity ; see *Joly*, Ann. d. Sc. Nat. XIII. p. 239, Pl. VIII. fig. 4, j., XVII. p. 307, Pl. IX. fig. 43, r. ; also *Krohn*, *Frariep's* neu: Not. XLIX. p. 305, fig. 1, 2 ; and *Zaddach*, loc. cit. p. 17, Tab. I. fig. 17, C., Tab. II. fig. 4–14.*

6 Although *Treviranus* (Verm. Schrift. II. p. 31, Taf. VI. fig. 6), and *Kutorga* (loc. cit. p. 18) have, indeed, furnished some communications on the heart of *Lithobius* and *Scolopendra*, yet we are really indebted for what is known of the structure of this organ with the Myriapoda to the excellent researches of *Newport* ; see Philos. Trans. XXIII. p. 272, Pl. XIII. fig. 18–22 (*Scolopendra*),

and fig. 25 (*Scutigera*). According to these investigations, the interventricular septa are scarcely developed with the Chilognatha, although very much so with the Chilopoda.

1 An extra vascular circulation has been observed with the Lernaeodea, by *Nordmann* (loc. cit. p. 73, 98), and with the Caligina, by *Pickering* and *Dana* (Isis, 1840, p. 205, 1841, Taf. IV.). *Jurine* (loc. cit. p. 437, Pl. XXVI. fig. 8), and, with more exactness, *Vogt* (loc. cit. p. 9, Taf. I. fig. 10), have described the circulation with *Argulus*. For that of *Daphnia*, see *Gruithuisen*, N ov. Act. Nat. Cur. XIV. p. 403, Tab. XXIV. fig. 6 ; *Perty*, Isis, 1832, p. 725, and *Ehrenberg*, Abhandl. d. Berl. Akad. 1835, p. 189, note. *Zaddach* (loc. cit. p. 23, Tab. I. fig. 17) has represented in much detail that of *Apus*. In order to be convinced of the entire want of vascular walls with the lower Crustacea, there is perhaps no species which will serve better than *Argulus foliaceus* whose body is wholly flattened and transparent throughout.†

---

* [§ 283, note 5.] See also *Leydig*, loc. cit. *Siebold* und *Kölliker's* Zeitsch. III. p. 257 (*Artemia* and *Branchipus*). — ED.

† [§ 284, note 1.] This statement of the complete absence of true vessels in *Argulus*, is confirmed by the researches of *Leydig* (loc. cit.

Siebold and *Kölliker's* Zeitsch. II. p. 337) upon this same species. His schema of the circulation with these animals is as follows : "The blood is thrown from the heart into the interstitial lacunae of the organs ; thereupon it is collected in the posterior portion of the heart ; a portion of it

With the other Crustacea, with which the heart is unarticulated, the blood passes from this organ into arterial canals; but the walls of these last sooner or later entirely disappear, so that here also the blood circulates at liberty between the interstices of the body. The regular arterial currents thus formed finally bend about and become those of the venous system. With the Isopoda, and the Amphipoda, perhaps, also, with the Poecilopoda, and Laemodipoda, the anterior, posterior, and lateral arterial trunks disappear after a very short course.[2]

With the Stomapoda, and Decapoda, the arterial system is pretty well developed, and can be traced even to its ultimate ramifications. With the first, the heart, at its anterior extremity, sends off a simple, pretty long aorta, which ramifies to the eyes and tentacles; while from its sides, pass off numerous arteries for the segments of the body and their appendages, and, posteriorly, a branch which extends to the very extremity of the tail.[3]

With the Decapoda, on the other hand, the heart has three anterior aortae, of which the middle one goes, almost unbranched, to the eyes, while the two lateral, belonging to the antennae, give off, in their course, branches to the cephalo-thoracic organs. The two hepatic organs, alone, have special arteries, which arise directly from the lower surface of the heart. Behind, there is a posterior aorta which, immediately after its origin, divides into a dorsal and an abdominal branch. The first of these, either simple as with the Macrura, or bifurcated as with the Brachyura, extends even to the end of the tail, sending off branches right and left. The second passes below, and is distributed principally to the feet, the pincers, the foot-jaws, and the maxillae.[4]

2 According to the researches of Treviranus (Verm. Schrift, I. p. 78) upon Asellus, and of Zenker (loc. cit. p. 21) upon Gammarus, the arterial system is very rudimentary with the Isopoda, and Amphipoda. This may be easily proved by an examination of allied species. It may be asked, however, if the blood-currents of these Crustacea are not enveloped in vascular walls so delicate as to escape observation; but with proper care one may be satisfied that no such walls exist. From muscular contractions or the bending of the articulations, the current of the blood is often stopped, and then the blood-globules evade the obstacle by passing at any point directly from the arterial into the venous current.

Goodsir (Edinb. new Philos. Jour. July, 1842, p. 184) was certainly deceived when he affirmed that he had observed the blood of Caprella circulating in arterial and venous vessels.

The absence of vascular walls with Caprella, already observed by Wiegmann (Arch. 1839, I. p. 111), has been confirmed by Frey and Leuckart (loc. cit. p. 104, Taf. II. fig. 19, 20), and, according to them, the circulation here is analogous to that of the Amphipoda.*

3 This disposition of the arterial system has been observed by Audouin, Milne Edwards, and Duvernoy (Ann. d. Sc. Nat. XI. 1827, p. 377, Pl. XXXII. and VIII. 1837, p. 33, Pl. II. fig. 1), with Squilla, while Mysis appears from its circulatory organs to be allied to the Isopoda and Amphipoda; see Thompson, Zool. Research. loc. cit. 1. p. 13, and Frey, loc. cit. p. 13.

4 The arterial system of Maia and Homarus has been described with many details in the so-often-quoted memoir of Audouin and Milne Edwards (Ann. d. Sc. Nat. XI. 1827, p. 352, Pl. XXIV.-XXIX.). Lund, also (Isis, 1825, p. 393, Taf. III. fig. 1), has very well described the arteries of the lobster. But especially should be noticed the excellent preparations of Hunter of the arterial system of this same animal (Catal. of the Physiol. Ser. II. Pl. XV.-XVIII.). For this system with the crawfish, see Brandt, Med. Zool. loc. cit. p. 63, Taf. XI. fig. 2; and for that of Cancer pagurus, Milne Edwards, in the Cyclop. loc. cit. p. 775, fig. 418.

enters this organ without passing to the branchiae, but the other portion traverses the gills and afterwards returns to the heart." — Ed.

* [§ 284, note 2.] In a private letter Agassiz has communicated some interesting facts on the circulation of Caprella. He says "Caprella has a tubular, dorsal vessel with lateral valves, exactly like the larvae of Insecta, — the blood is emptied, in front, into the main cavity of the body, moves backwards along the lower part of that cavity without being enclosed in vascular walls, and returns to the dorsal vessel through the lateral valves. The circulation was traced in a living animal into which a solution of a small quantity of carmine had been injected." — Ed.

With all the Crustacea, the venous currents gradually converge from the lower part of the body into various intercommunicating sinuses, situated, some upon the median line, and others at the base of the feet.[5] From these sinuses the blood proceeds to the branchiae, and thence into the dorsal sinus the walls of which are thin and uncontractile, and within which the heart is entirely enclosed. This dorsal sinus is filled during the systole, and the arterialized blood which it contains is absorbed during the diastole through the venous orifices of the heart, without any aid on the part of the walls of the sinus.[6]

With the Myriapoda, also, the arterial system is highly developed. Not only are there numerous arteries arising from the sides of the heart, which ramify in the segments of the body, but also, beside an anterior dorsal aorta, two other considerable arteries which embrace the œsophagus, then bend below and unite to form, on the abdominal cord, a Supra-spinal artery. This artery gives off numerous lateral branches, which accompany the principal nerves, and terminate, at last, in ramuscules.[7]

But what distinguish the Myriapoda from the higher Crustacea, are the venous currents, which, equally extra-vascular, do not run towards the respiratory organs, but pass directly into the dorsal sinus, and thence are absorbed into the chambers of the heart through the venous orifices.[8]

---

5 Of the absence of vessels around the venous currents one may easily be convinced from an examination of small Amphipoda and Isopoda. This absence exists also with the higher Crustacea; see *Duvernoy*, Ann. d. Sc. Nat. VIII, 1837, p. 34, or in *Cuvier*, Leçons d. Anat. Comp. VI. p. 404 (*Squilla*). I am quite of the opinion of *Lund* and *Schultz* (Isis, 1830, p. 1225), who have combated the opinion of *Audouin* and *Milne Edwards* and have described the venous system of the Decapoda as having proper walls (Ann. d. Sc. Nat. Pl. XXVI.-XXXI.). But *Milne Edwards*, who, at this time, advocates with so much zeal the wall-less condition of the circulating currents with Mollusca, appears, moreover, to entertain the opinion of a similar circulation with the Decapoda; at least, such would be inferred from what he has said upon the circulation in general of Crustacea; see Hist. d. Crust. I. p. 101, and Cyclop. loc. cit. p. 777.

6 According to *Audouin* and *Milne Edwards* (loc. cit. Pl. XXVI. fig. 3), the returning blood from the branchiae enters the heart direct through inter-anastomosing *vasa branchio-cardiaca*. But this statement has been reasonably doubted by various observers, for these naturalists had overlooked the sinus which envelopes the heart of the higher Crustacea, and receives, first of all, the branchial blood; see *Straus*, Considér. &c. p. 345; *Lund* and *Schultz*, Isis, 1830, p. 1226; and *Krohn*, ibid. 1834, p. 522. This dorsal sinus has been

compared sometimes to an auricle, sometimes to a pericardium; but, strictly speaking, neither of these comparisons is correct.

7 The division of the anterior dorsal aorta, with *Scolopendra*, was first noticed by *Straus* (Considér. &c. p. 347). More detailed researches on the arterial system of the same have been published by *Kutorga* (loc. cit. p. 18, Tab. III.), and *Lord* (Med. Gaz. part VI. vol. I. 1837, p. 892), who were chiefly occupied with the supra-spinal artery. But of all the observers, *Newport* (Philos. Trans. 1843, p. 274, Pl. III. XIV.) has worked out the arterial system of the Myriapoda in the most complete and masterly manner. His researches have shown that this system is least developed with the Julidae, and rises gradually through the Glomeridae and Geophilidae — reaching its highest grade of structure with the Scolopendridae. *Kutorga* has entirely mistaken the nature of the heart in regarding it as a vena cava, and the supra-spinal artery as an aorta. *Gaede*, also (Zool. Magaz. 1. p. 108, Taf. I. fig. 7, g. f.), is quite in error as to the vascular system of *Scolopendra*; for he has evidently seen the three vessels arising from the anterior extremity of the heart, namely : the dorsal aorta and the two vessels which, uniting, form the supra-spinal artery; but he has taken them for nerves.

8 *Newport*, who has so well observed the circulatory system of the Myriapoda, says nothing of veins, and describes the dorsal sinus as a pericardium.

CHAPTER VII.

RESPIRATORY SYSTEM.

§ 285.

The majority of Crustacea respire by Branchiae; but among the lower orders, there are many which have no trace of respiratory organs, while the Myriapoda respire by aeriferous tracheae.

With most Siphonostoma, Lophyropoda, and many Stomapoda, there are no particular respiratory organs, the respiration being, therefore, cutaneous; and with some species of these orders, the water is renewed by the oarlike action of some of the locomotive organs.[1]

The Branchiae of Crustacea are sometimes lamelliform, sometimes cylindrical, and often appear either distinct and separate, or consist of compound serrated organs, branched in various ways, on which the branchial lamellae are disposed in a regular row, and the branchial tubes united in larger and smaller tufts. But these lamellate or tubular branchiae are invested with a membrane so thin that it widely differs from those of the other regions of the body. It is never ciliated, and is usually without fringes, bristles, &c. The interior of these organs presents only a few parenchymatous points, and, whatever may be their form, they are always traversed by numerous canals and large interanastomosing lacunae, which are wholly without proper walls, and are filled by the arterial and venous currents.[2]

The branchiae are often in connection with their neighboring appendages. These last consist of multi-articulate lashes or cirri, or of scales, or large plates, and serve either as gyratory organs, or as opercula shielding the respiratory organs; sometimes, indeed, they perform both of these functions at the same time. Nearly always these organs are fringed with long, stiff, and often pinnate bristles.[3]

---

1 The branchiae are wanting with the Penellina, Lernaeodes, Ergasilina, and with some Caligina. With *Daphnia*, *Lynceus*, and some other allied Lophyropoda, the small oar-like feet concealed under the belly are probably designed for the agitation of the water, while the two feet projecting in front of the body, and which are larger and usually branched, are the principal swimming organs. Indeed, even when these animals are at rest, these organs are seen in perpetual motion — thus causing in the cavity of the shell a continual current of fresh water; this supports the observation of *Ehrenberg* (in his third Beitr. loc. cit. p. 189, note) that, with these Entomostraca, the internal surface of the valves performs the function of branchiae. The active, hairy, clavate corpuscles inserted on the base of the first pair of feet with *Cyclopsina castor*, and which have been usually regarded as posterior antennae (*Muller*, Entomostr. p. 106, Tab. XVI. fig. 5, 6, c., or *Jurine*, Hist. d. Mon. cl. p. 52, Pl. IV. fig. 1, Pl. V. fig. 1. b. Pl. VI. fig. 13, a.), are nothing but organs for the agitation of the water. With *Cypris*, only, are there perhaps special branchial organs. These little animals have at the base of the posterior pair of jaws two semilunar, pectinated plates, curved upwards, having completely the aspect of branchiae; see *Ramdohr*, Beitr. loc. cit. p. 15, Taf. IV. fig. 5, B. and fig. 8, L.; also *Straus*, loc. cit. p. 49, Pl. I. fig. 4, v. and fig. 8, c., or *Baird*, in the Magaz. of Zool.

and Bot. I. p. 520, Pl. XVI. fig. 8. These organs appear to have been wholly misapprehended by *Treviranus* (Verm. Schrift. II. p. 59, Taf. IX. fig. 5). With *Mysis*, *Leucifer*, and *Amphion*, there are no traces of branchiae, while with the other allied Stomapoda, such as *Alima* and *Phyllosoma*, they sometimes exist in a rudimentary form. As branchiae, have been regarded, also, the articulated processes of the cloven feet of *Mysis* and some other Stomapoda; but, certainly, they are organs for swimming or for the agitation of the water, and their organization has nothing in common with that of branchiae.

2 It is owing to this small quantity of parenchyma in the lamelliform branchiae and to the numerous lacunae filled with blood that, when the circulation in these organs is arrested, the two lamellae of which they are composed, separate from each other, and the whole branchia, swollen from accumulated blood, has the form of an ampulla. The blood then changes its natural color. This pathological state may be easily seen with individuals of *Asellus*, *Gammarus*, and *Apus*, when allowed to be a long time dying. These ampullae are violet, with *Gammarus*; and of a beautiful red, with *Apus*; see my note upon the ampullae of *Apus cancriformis*, in the Isis, 1831, p. 429.

3 Hairy and bristled appendages of this nature are often taken for branchiae. These organs are not only surrounded by a thick skin which of it-

The branchiae are usually inserted at the base of the anterior true feet, or the posterior false feet, floating freely in the water; or, they are contained in a special respiratory cavity, into which water is admitted through various ways.

§ 286.

The principal differences observed with the Crustacea in the disposition and structure of their branchiae, are the following:

1. Many genera of the Caligina and Argulina, have upon various parts of their body, such as the back, the abdomen, and the tail, several thin, simple, naked lamellae, which may be regarded as branchiae. [1]

2. The feebly-developed branchiae of the Lepadea consist of cylindrical or lanceolate processes inserted at the base of some of the cirrate feet, and curved towards the back of the animal, so that they are always concealed in the cavity of the shell. But the water is renewed upon their surface by the regular movements of the long posterior feet. [2] With the Balanodea, the branchiae have left the body of the animal, and are developed on the internal surface of the mantle as more or less numerous soft folds or lamellae. [3]

3. With the Laemodipoda, and some Stomapoda, the branchial apparatus is reduced to a few vesicular or cylindrical, sometimes wholly rudiment-

---

self would render them unfit for the respiratory function, but they are not traversed except by feeble blood-currents which do not enter the bristles or hairs. From the complete absence of ciliated epithelium, the vortex-producing organs are of much importance. On this account, many species with which these organs are wanting, use their feet for this purpose.

1 *Euryphorus* has four such pedunculated branchial lamellae on the dorsal surface of the two grand segments of the body. These are what *Milne Edwards* (Hist. Nat. d. Crust. III. p. 462, Pl. XXXIX. fig. 1) has called *Appendices elytroides*. With *Dinematura*, the last pair of feet is changed into two naked, deeply-tissured branchial lamellae (*Kröyer*, Isis, 1841, p. 279, Taf. I. fig. 5, 1.). With *Phyllophora*, the branchial apparatus is still more fully developed, for each foot of the last four pairs terminates with two ovoid, glabrous, branchial lamellae (*Milne Edwards*, loc. cit. III. p. 471. Pl. XXXVIII. fig. 14). The two thin, lanceolate caudal lamellae of *Argulus*, through which pass strong blood-currents interrupted only by some islets of substance, are certainly respiratory organs to which the oar-like organs which *Jurine* (loc. cit. p. 442) has erroneously considered as branchiae, serve as vortex-producing organs. I am yet undetermined if the respiration is performed by these lamellae alone, or in part by the lateral portions of the dorsal shield through which pass numerous blood-currents. But in any case, I cannot believe

it possible, as has *Vogt* (loc. cit. p. 11), that these lateral portions are the only respiratory organs of these animals.*

2 In the various species of *Lepas*, there are from two to five arcuate branchiae which hang from each side at the base of the first pair of cirri. With *Cineras*, beside the six branchiae, there is one, very short, upon the back of the animal, at the base of the third, fourth, and fifth pairs of feet ; while with *Otion*, there is a seventh pair inserted on the second pair of feet ; see *Mertens* in *Müller's* Arch. 1835, p. 532; *Wagner*, Lehrb. d. vergleich. Anat. p. 200 ; *Cuvier*, Mém. loc. cit. p. 6, fig. 2, 5, o. p.; *Burmeister*, Beitr. &c. p. 31, Pl. I. fig. 14, c. c., and *Martin St. Ange*, Mém. loc. cit. Pl. II. fig. 17, 19, K. K. (*Lepas*).

3 These branchial lamellae are extraordinarily developed with *Coronula diadema* (*Burmeister*, Beitr. &c. p. 38, Taf. II. fig. 10, a. a.). They are few in number with *Balanus* (*Cuvier*, Mém. loc. cit. p. 14, fig. 18, c. c.). It is true that *Burmeister* subsequently (Handb. d. Naturgeschicht. p. 551) did not regard as branchiae but rather as ovarian sacs, these organs which as to form and position correspond somewhat to the branchiae of certain Branchiopoda (*Lingula*). But even if they do serve at the same time as receptacles of the eggs, this would be no reason for refusing to the folds of the mantle of the Balanodea the function of a respiratory organ, for, with other lower animals, as for example with the Lamellibranchia, the branchiae serve as receptacles for the eggs.

---

* [§ 286 note 1.] *Leydig* (loc. cit. *Siebold* and *Kölliker's* Zeitsch. II. p. 337) has carefully examined the intimate structure of these caudal appendages with *Argulus*; they are composed, 1st, of simple glands such as are found under the skin over the whole body ; 2nd, of a rich muscular net-work;

and 3rd, of a lacunal net-work (*Lückennetz*). The glands and the muscles constitute what is described above as the islets of substance. *Leydig* denies that these lamellae, thus composed, have, peculiarly, a gill function. — ED.

ary appendages, which hang freely from the base of some of the feet, or are inserted isolatedly on the sides of the body.[4]

4. The Phyllopoda have, at the base of each of their numerous swimming feet, an ovoid or lanceolate branchial lamella, pointing forwards. It is quickly distinguished by its thin, glabrous covering, in opposition to that of the other divisions of feet, which are bristled.[5]

5. With the Amphipoda, the rapacious and ambulatory feet, excepting the first and last pairs, are those only which are provided with respiratory organs. These last consist of oval or round glabrous lamellae, situated internally at the base of the five middle feet. They receive, constantly, fresh water by the movements of the three anterior pairs of post-abdominal feet, which act as gyratory organs.[6]

4 With *Phyllosoma*, there is, at the base of the anterior feet, a small, ovoid, pedunculated appendage, which may perhaps be regarded as a rudimentary branchia ; see *Milne Edwards*, Hist. Nat. d. Crust. II. p. 474, Pl. XXVIII. fig. 15, a. It is remarkable that, with another Stomapode genus, *Squilla*, there are at the base of the ten rapacious feet similar pedunculated appendages of the form of oval lamellae (*Milne Edwards*, Hist. Nat. d. Crust. II. p. 512, Pl. XXVI. fig. 15, Pl. XXVII. fig. 13, 14, b.). These, also, would be regarded as rudimentary branchiae, did not these Crustacea have distinct branchial organs (see below). With *Atina*, the oval feet have sometimes very rudimentary branchiae in the form of simple vesicles or ramified processes (*Milne Edwards*, loc. cit. II. p. 506). With *Caprella*, and *Aegina*, the first two posterior abdominal segments have, upon the sides, a simple, very soft, pyriform branchia ; while with *Leptomera*, there is a vesicle of the same nature at the base of the six feet of the first three posterior abdominal segments ; see *Müller*, Zool. Danic. Tab. LVI. fig. 5, and Tab. Cl. fig. 2 ; *Templeton*, Transact. of the Entomol. Soc. I. p. 193, Pl. XXI. fig. 7, f. ; and *Kröyer*, Naturhist. Tidskr. IV. p. 490, Pl. VI.-VIII. With *Cyamus*, the respiratory organs are even more developed. They consist of four long, simple cylinders inserted on the sides of the first two posterior abdominal segments, and projecting over the back ; see *Treviranus*, Verm. Schrift. II. p. 9, Taf. I. fig. 1-3, and Beobacht. aus. d. Zoot. u. Physiol. p. 32, Taf. VII. fig. 48—50 ; also *Kröyer*, loc. cit. IV. p. 474, Pl. V. fig. 70-76 ; and *Roussel de Vauzème*, loc. cit. p. 248, Pl. VIII. ; according to this last mentioned author, *Cyamus ovalis* has four double, branchial cylinders. These branchiae of *Cyamus* have often been taken for metamorphosed feet, but it is only necessary to examine them in their earliest condition in order to be convinced that they are special organs (*Milne Edwards*, Ann. d. Sc. Nat. III. 1835, p. 329, Pl. XIV. fig. 14). At this epoch they are as pyriform as those inserted on the side of the feet of *Leptomera*. The passage to *Squilla*, whose branchiae are more highly developed, is made by *Cynthia*. Each anal foot has here a bifurcated branchia, the two cylindrical divisions of which are curved towards each other ; see *Milne Edwards*, Hist. Nat. d. Crust. II. p. 462, Pl. X. fig. 5.

5 The delicate branchial lamellae usually assume after death the form of vesicles, from being filled with blood, a phenomenon already mentioned (§ 285, note 2). But, formerly, they were taken for special organs whose function was unknown, and *Berthold* (Isis, 1830, p. 643) has regarded those of *Apus* as male genital organs ; while the remaining pilose divisions of swimming feet were, according to him, respiratory organs.

These branchiae are easily perceived with *Apus* after death, and from the form which they then assume, they have long been known as the problematical red sacs (*Schaeffer*, loc. cit. Tab. II. III. VI. ; *Zaddach*, loc. cit. p. 14, Tab. II. fig. 13, B. Tab. XIV.). In 1830 (Isis, p. 429) I gave the correct interpretation of these organs which, like the swimming feet of these animals, diminish in size from before backwards ; but, already, before m·, *Laschge* (Naturforsch. Stück. XIX. p. 68, Taf. III. fig. 6, 7, 10) had recognized their nature. With *Limnadia*, and *Isaura*, the branchiae are very long and of a brown-red color, but are wanting on the last swimming feet ; see *Brongniart*, loc. cit. p. 86, Pl. XIII. fig. 7, 8 ; *Straus*, Mus. Senckenb. loc. cit. p. 124, Taf. VII. fig. 13, 14, r., 15, k. ; *Joly*, loc. cit. p. 299, Pl. VII. fig. 2, 6, 7, f. and Pl. VIII. fig. 8, f. With *Chirocephalus*, *Branchipus*, and *Artemia*, they have a more oval form and exist on all the swimming feet. *Ruthké* (Zur Fauna der Krym. p. 108, Taf. VI. fig. 14, 19-21), has figured, probably from dead individuals, those of *Artemia* as vesicular bodies. In the figures of *Jurine* (Hist. d. Monocl. Pl. XXI. XXII.), made for the memoir of *Prévost*, the branchiae of *Chirocephalus* are not seen at first, but with a little attention may be discovered. *Gaede* (*Wiedemann's* zool. Magaz. I. p. 88), *Berthold* (Isis, 1830, p. 689, Taf. VII. fig. 1), and *Zaddach* (loc. cit. p. 11, Tab. I. fig. 17, Tab. II. fig. 10) have regarded the large dorsal shield of *Apus* as a respiratory organ, since its lateral halves are traversed by blood-currents running close to each other (*Schaeffer*, loc. cit. p. 72, Tab. I. fig. 5, b. b.), and thence passing directly towards the anterior extremity of the heart.

Indeed, from the vascularity and delicateness of the under surface of this shield, one would be quite disposed to attribute to these parts a participation in the respiratory act.

6 For a long time, the multi-articulated, bristly, anal feet of these small Crustacea were regarded as branchiae, for the true branchiae are quite concealed under the internal surface of the anterior feet. Even after the attention had been directed to these organs, their form was often misapprehended ; for when these animals are a long time dying, their branchiae are changed, from congestion, into ampullae. With the Amphipoda, it is easy to distinguish the branchial lamellae from the incubatory lamellae at their side, for the borders of these last are bristled ; see *Straus*, loc. cit. p. 57, Pl. IV. fig. 10, 11, h. (*Hiella*) ; *Zenker*, loc. cit. p. 8 (*Gammarus*) ; *Milne Edwards*, Ann. d. Sc. Nat. XX. 1830, p. 357, Pl. X. fig. 7, Pl. XI. fig. 1, also Ibid. III. 1835, Pl. XIV. fig. 9, and Hist. d. Crust. III. p. 6, Pl. II. fig. 15, c., Pl. XXX. fig. 1, 13, 16 (*Gammarus Phronima*, *Vibilia*, *Hyperia*). According to *Savigny's* figure (Descrip. de l'Égypte, loc. cit. Pl.

6. With the Isopoda, the five pairs of post-abdominal feet are nearly always concerned exclusively in the function of respiration. The two multi-articulate cirri of each of those feet, have been changed into plates, which, pointing backwards, are imbricated and applied against the under surface of the last caudal segment, which is usually very large.[7] The form of these plates is sometimes lanceolate, sometimes discoidal or rhomboidal, and they often differ widely in the different sexes of even the same species. Upon the same foot, the external or anterior plate is usually leathery and bristled on its external border ; while the internal or posterior plate is covered with a very thin envelope, and is usually entirely glabrous. This last, therefore, should be regarded as the proper branchia, of which the first is only the operculum, serving, also, often as a gyratory organ. The first case is observed with the terrestrial Isopoda, where the branchial opercula are fixed, rhomboidal, slightly concave, and completely cover the branchial lamellae preserving them from desiccation.[8]

With most of the aquatic Isopoda, on the other hand, this apparatus is in perpetual motion, and the branchiae are often of the same form and size as the operculate plates. The opercula of the first pair are so large that they extend beyond all the rest.[9] With the Idotheoïdae, the operculate apparatus has an entirely peculiar structure. The two feet of the last caudal segment are developed into two valves which move laterally like the two folds of a door, and can open and close the branchial cavity, which is provided with five pairs of double plates.[10] The branchial apparatus of the Bopyrina differs, in many respects, from that of the other Isopoda. With some species, it is reduced to four or five pairs of simple, superposed plates, without any accessory organ ; while with others, there are four to six branchiae which, as more or less deeply fissured cordiform plates, or as long and

XI. fig. 4[2], 4[3].) of *Amphithoe filosa*, this animal has, beside the ten round branchial lamellae, a sixth and rudimentary pair on the two posterior feet.*

7 For the respiratory organs of the Isopoda, see especially *Duvernoy* and *Lereboullet*, Ann. d. Sc. Nat. XV. 1841, p. 177, Pl. VI.

8 With the terrestrial Isopoda, the branchial apparatus is somewhat abortive, for true branchiae are wanting beneath the two anterior pairs of opercula, and those back of the three posterior pairs are very small and delicate ; see *Treviranus*, Verm. Schrift. 1. p. 62, Taf. VI. VIII. IX. (*Porcellio*); *Savigny*, Descript. de l'Égypte, loc. cit. Pl. XII. fig. 7 (*Lygia*), and Pl. XIII. (*Tylos*, *Porcellio* and *Armadillidium*); *Brandt*, Mediz. Zool. II. Taf. XV. fig. 35–37 (*Porcellio*), and *Lereboullet*, loc. cit. p. 118, Pl. IV. fig. 17, Pl. V. fig. 18–22 (*Lygidium*). This abortion of the branchiae is compensated with some Oniscidae by the existence of lung-like organs. (See below, § 287.)

9 *Asellus* has two very large, common, anterior

branchial opercula ; but the branchial apparatus, moreover, is composed of only three pairs of plates on each side (*Treviranus*, Verm. Schrift. I. p. 75, Taf. X. XII.), while with *Sphaeroma*, *Cymothoa*, and allied genera, there are five pairs on each side (*Savigny*, loc. cit. Pl. XI. XII.).

With some species of *Sphaeroma*, *Cymodocea*, *Nesea*, and *Amphoroidea*, the branchial plates of the last two pairs of branchiae, have numerous transverse plicae, which connect these Sphaeromatoba with the Poecilopoda (*Duvernoy* and *Lereboullet*, loc. cit. p. 215, Pl. VI. fig. 15–23, and *Milne Edwards*, Hist. d. Crust. III. p. 223, Pl. XXXII. fig. 9). With *Serolis*, the branchial structure is quite different, the fourth and fifth pairs of feet being changed into broad branchial plates (*Milne Edwards*, Arch. du Mus. d' Hist. Nat. II. p. 21, Pl. II. fig. 1–6).

10 See *Rathke*, loc. cit. p. 115, Taf. IV. and *Milne Edwards*, Hist. d. Crust. Pl. X. fig. 6, 7 (*Idothea*).

* [§ 286, note 6.] *Leydig* (loc. cit. *Siebold* and *Kölliker's* Zeitsch. III. p. 289) does not admit that the red pouches, above-mentioned with *Apus*, are of a respiratory character, at least with *Artemia* and *Branchipus*, where he has examined their histological composition. In this connection it may be mentioned that this observer has found on each

natatory foot of *Branchipus*, a peculiar and new structure. This is a roundish, dark-orange-colored, pedunculated body, situated on the under side of the leg near the coxal joint. This body is composed of large nucleated cells which contain a yellowish liquid. The use of this structure is unknown. — Ed.

sometimes branched tubes, project considerably beyond the lateral borders of the posterior segments of the body.[11]

7. The Pœcilopoda hold a place between the Isopoda and the Decapoda, their branchiae being, as in the first, inserted on the abdominal feet, and, as with many of the second, composed of numerous plates. With *Limulus*, the five posterior abdominal feet, which are inserted on the second dorsal segment, and changed, as well as the first pair of abdominal feet, into very large plates, have upon their posterior surface numerous semi-oval, branchial plates lying upon each other. The first pair of feet appears to play, also, at the same time, the part of an operculate apparatus.[12]

8. The Stomapoda, with which the respiratory apparatus is most highly developed, have numerous branchial filaments disposed pectinately on a long stalk, and float freely in the water.

The Squillina have a similar branchial tuft on the anterior surface of the external plate of each of the ten swimming feet, which are only the ten post-abdominal feet of the posterior part of the body, transformed.[13]

With *Thysanopoda*, only, these branchial tufts are inserted at the base of the anterior abdominal feet.[14]

9. With the Decapoda, all the branchiae are joined together at the base of the anterior abdominal feet and of some of the foot-jaws; but at the same time they are contained in a special branchial cavity, which is covered by the lateral parts of the cephalothorax. Each of these two cavities communicates externally by two fissures. One of these is situated at the under surface of the body between the lower border of the cephalothorax and the base of the feet; through it the water enters the branchial cavity. The other is upon both sides of the masticatory organs, and through it the water is ejected. In this last, which is sometimes prolonged into a semi-canal,[15] are several multi-articulate cirri and lamellae, which belong to the second and third pairs of foot-jaws.[16] Their continual motion produces a regular current of water from the branchial cavity outwards.[17] As to the number of branchiae, there are wide differences in the various families of this order. There may be six, seven, fourteen, eighteen or even twenty-one in the same respiratory cavity. When numerous, there are usually two or three fixed on the four posterior foot-jaws, three or four on

11 Both sexes of *Bopyrus squillae* have five pairs of small branchial plates lying over each other like scales (*Rathké*, De Bopyro, &c., p. 7, Tab. i.). This is probably true also of the males of *Phryxus hippolytes* (*Rathké*, Nov. Act. Nat. Cur. XX. p. 48). The females of this same species and of *Phryxus paguri* have four pairs of cordate, and nearly double plates, which stand off laterally a little from the posterior part of the body; see *Rathké*, Ibid. p. 46, 50, Tab. II.; *Kröyer*, Naturhist. Tidskr. III. p. 102, Pl. I. II., or in Isis, 1841, p. 6,d, 707, Taf. II. Tab. I, and Taf. III. Tab. 2, or in Ann. d. Sc. Nat. XVII. 1842, p. 142, Pl. VI. With *Cepon*, the branchial apparatus is highly developed in that, beside the five pairs of lanceolate and pretty long plates which project from the sides of the tail with the males, the five abdominal and the last caudal segment, have six pairs of long, narrow diverging lamellae with pectinated borders. *Duvernoy* (Ann. d. Sc. Nat. XV. 1841, p. 126, Pl. IV. fig. 1-11), has described these twelve appendages as the principal branchiae of *Cepon*, while to me, they appear to be accessory, and are, perhaps, vortex-producing organs — the result of a metamorphosis of the anal feet. With *Ione*, all the abdominal segments have a pair of long branchial tubes pointing backwards, and with the females of this same genus, the five anterior pairs are branched on one side. In this sex, also, the organization of the Amphipoda appears to be repeated, for, from the base of the anterior feet hangs a long riband-like band (branchia?). See *Milne Edwards*, Hist. d. Crust. III. p. 279, Pl. XXXIII. fig. 14, 15.

12 See *Van der Hoeven*, loc. cit. p. 19, Pl. I. fig. 10, Pl. II. fig. 1, 11-15; and *Duvernoy*, Ann. d. Sc. Nat. XV. 1841, p. 10. Pl. III.

13 *Squilla* and *Squillerichthus*; see *Trevira-nus*, Beobacht. aus d. Zool. u. Physiol. p. 22, Taf. VI. fig. 36-39; and *Milne Edwards*, Hist. d. Crust. Pl. X. fig. 4, Pl. XXVII. fig. 7.

14 *Milne Edwards*, Ibid. Pl. X. fig. 3, Pl. XXVI. fig. 6, and Ann. d. Sc. Nat. XIX. 1830, p. 453, Pl. XIX.

15 With many Brachyura.

16 See *Suckow*, loc. cit. Taf. X. fig. 1, p. q., fig. 2, p. r., fig. 3, d. s. c. (*Astacus*); *Milne Edwards*, Hist. d. Crust. Pl. III. fig. 3-10, i. j. (*Maia*).

17 For this mechanism of the respiratory organs of the Decapoda, see *Milne Edwards*, Ann. d. Sc. Nat. XI. 1839, p. 126, Pl. III. IV.

each of the four anterior pairs of feet, and one only on the last pair. With these Crustacea, moreover, the organs have no connection with the movable basal joint of the feet; but, on the other hand, most of them are inserted on the base itself of the respiratory cavity above this joint. Many Macrura, which have numerous branchiae, are those exclusively which have one of these organs inserted on the coxa of the feet.[18] As to their structure, these organs vary also very much. Usually, they have the form of a long, acutely-pointed pyramid with a solidly-attached base, the axis of which is formed in its whole length by a shaft traversed by an arterial and venous canal, and covered by numerous thin lamellae or cylindrical filaments, the size of which decreases gradually towards their apex.[19]

§ 287.

Many terrestrial Isopoda have a branchial apparatus, the organization of which is entirely peculiar, and distinctly indicates a pulmonary respiration. With *Porcellio*, and *Armadillidium*, there are four white spots on the two anterior pairs of the branchial opercula. These spots communicate with as many cavities which ramify like vessels. They are situated between the two plates of these four opercula, and are filled with air. At the base of each of these opercula there is a narrow opening through which, when these cavities are compressed, the air will escape, and then the white spots disappear. By these means, these animals are undoubtedly in

18 The branchiae are fewest with the Brachyura, and Caridoidae; among these last, *Crangon* and *Alpheus* have only six in each respiratory cavity, and *Palaemon* and *Hippolyte* seven. *Uca*, also, has only six on each side, while with the majority of Brachyura, namely, *Portunus*, *Grapsus*, *Thelphusa*, *Gecarcinus*, *Pisa*, *Maia*, *Cancer*, &c., there are eighteen in all, the two anterior pairs of which are usually only feebly developed and belong to the two pairs of posterior foot-jaws, while the others are in general (*Maia*, *Cancer*, *Lupea*, &c.) so aggregated at the anterior part of the bottom of the respiratory cavity, that the space corresponding to the last two pairs of feet appears gill-less. The majority of the Brachyura have fourteen branchiae on each side, and these organs are even more numerous with various Macrura. Thus, I have counted eighteen with *Astacus*, *Homarus*, and *Palinurus*; two of which, with *Palinurus*, and *Astacus*, are in connection with the middle, and three with the posterior foot-jaw; while with *Homarus*, this last has also three branchiae, but the second foot-jaw has only a rudimentary one. As to the other branchiae, there is, in these three genera, a branchia inserted on the coxa of the four anterior feet. Above each of these same feet are other branchiae disposed, in couples, with *Astacus*, and in threes above the fourth foot, with *Homarus*, and above the second, third and fourth, with *Palinurus*; while above the last foot that is gill-less, there is only a single branchia. With *Nephrops*, there are twenty branchiae on each side, and with *Scyllarus*, twenty-one. See, for the number and disposition of these organs with the Decapoda, *Duvernoy*, in *Cuvier's Leçons d'Anat. Comp.* VII. p. 393.
19 The various forms of the branchiae of the Decapoda may be reduced to two types. The first, the less common, exists with many Macrura, for example with *Scyllarus*, *Palinurus*, *Gebia*, and *Homarus*. The shafts of their branchial arches

support numerous cylinders set together in a brush-like manner. In the figures which *Audouin* and *Milne Edwards* (Ann. d. Sc. Nat. XI. 1827, Pl. XXIX. fig. 1, Pl. XXX. fig. 2, Pl. XXXI.) have given of the branchiae of *Homarus*, this structure may be easily seen. With *Astacus*, the cylinders are much less numerous, and disposed only on two of the sides of the branchial shaft, — giving it a pinnate aspect; and those which are inserted on the coxae are terminated by a thin, multiplicate lamelliform dilation, which has completely the structure of a branchial lamella (*Suckow*, loc. cit. p 54, Taf. X. fig. 1, 2, 25, 26, Taf. XI. fig. 5, 6; *Brandt*, Medic. Zool. II. Taf. XI. fig. 23). With *Homarus*, and *Palinurus*, also, the coxae have an analogous plate inserted close by the side of the coxal branchia; but it is of a leathery consistence and covered with numerous hairs, so that it cannot participate in the function of respiration, but is probably only a septum to separate the different groups of branchiae. *Aristeus*, which has sixteen branchiae on each side, differs widely from the other Macrura in having its penniform branchiae composed of a shaft from which pass off right and left numerous curled filaments whose convex border is covered by tufts of very delicate thick-set branchial cylinders (*Duvernoy*, Ann. d. Sc. Nat. XV. 1841, p. 104, Pl. V.). The second type is formed by these branchiae to the shafts of which adhere at right angles numerous thin sometimes rhomboidal, sometimes spheroidal lamellae, contiguous, and decreasing in size towards the apex of the shaft. This type occurs especially with the Brachyura, the Anomura, and with *Galithea* of the Macrura; also of the Caridoidae, with *Palaemon*, *Hippolyte*, *Alpheus*, *Penaeus*, *Crangon*, &c.; see *Audouin* and *Milne Edwards*, Ann. d. Sc. Nat. XI. 1827, Pl. XXVI. and XI. 1830, Pl. III. fig. 1, Pl. IV. fig. 1, 4 (*Maia*, *Ranina*, *Palaemon*); also *Kröyer*, loc. cit. Tab. I.–V. (*Hippolyte*), and *Joly*, loc. cit. p. 71, Pl. III. fig. 24 (*Caridina*).

a condition to respire atmospheric air. [1]   But with *Tylos*, this pulmonary apparatus is still more highly developed; for, under the four pairs of opercula, there are, instead of simple branchial plates, oblong appendages on which is a transverse row of aeriferous sacs having a kind of stigma on their under surface. [2]

All the Myriapoda respire by true tracheae. Their blood does not require, therefore, special organs to receive the influence of the air, for this last is carried into every part of the body.

The stigmata for the ingress and egress of the air, are easily seen with the Chilopoda, for they are usually surrounded with a ring of brown chitine, and situated on each side of the body between the base of the feet and the dorsal shields; they are not found, however, above all the feet, for the segments which have them alternate more or less regularly with those that are without them. [3]   With the Chilognatha, the very small stigmata are on the ventral surface. They are situated on the anterior border of the ventral plates, from the posterior border of which arise the feet. [4]   The intimate structure of these tracheae, which are usually brown, is exactly like that of those of insects. [5]   Among the Chilognatha, the Julidae are noticeable for the very simple character of their tracheal apparatus. Each stigma leads into a tuft of tracheae from which arise air-canals which neither ramify nor anastomose but gradually become smaller and smaller and surround the various organs. [6]   With the Glomerina, on the contrary, the tracheae, which arise from the stigmata by two trunks, are branched, but do not anastomose with the neighboring branches. [7]   Those of the Chilopoda most closely resemble those of the Insecta, — being very ramose, and their large trunks intercommunicating at their origin by longitudinal and transverse anastomoses, so that each stigma can introduce air into the entire tracheal system. [8]

1 According to *Duvernoy* and *Lereboullet* (loc. cit. p. 231, Pl. VI. fig. 14), these cavities secrete a liquid for the moistening of the branchiae. See upon this subject, my observations in *Müller's* Arch. 1842, p. 141, note 2.

2 See *Savigny*, Descript. de l'Égypte, loc. cit. Pl. XIII. fig. 1.⁶—1.⁸ ; but especially *Milne Edwards*, Institut. 1839, p. 152, and Hist. d. Crust. III. p. 187, and his figures in the Iconograph. du Règne anim. Crust. Pl. LXX.

3 With *Lithobius*, there is a stigma above the first, third, fifth, eighth, tenth, twelfth, and fourteenth pairs of feet (*Treviranus*, Verm. Schrift. II. p. 29, Taf. IV. fig. 7, Taf. VI. fig. 5). With *Scolopendra*, the stigmata have a similar disposition (*Kutorga*, loc. cit. p. 14).

4 See *Savi*, Isis, 1823, p. 219, Taf. II. fig. 9, a. a., and *Burmeister*, Ibid. 1834, p. 134, Taf. I. fig. 2,

a. a. (*Julus*). These stigmata with *Julus* were entirely overlooked by *Treviranus*. He had regarded as such the orifices of a row of glands which are situated on the sides of the segments of the body (Verm. Schrift. II. p. 42, Taf. VIII. fig. 4, 8. 8.).

5 The characteristic spiral filament of the Insecta is also not wanting here : see *Kutorga*, loc. cit. p. 14, Tab. II. fig. 8.

6 *Straus*, Considérat. &c. p. 307, and *Burmeister*, loc. cit. Taf. I. fig. 3 (*Julus*).

7 *Brandt*, in *Müller's* Arch. loc. cit. p. 323, Taf. XII. fig. 4, 5 (*Glomeris*).

8 *Straus*, loc. cit. p. 307, and Traité d'Anat. comp. II. p. 161 ; *Treviranus*, Verm. Schrift. II. p. 30, Taf. VI. fig. 6 (*Lithobius*), and *Müller*, Isis, 1829, p. 551, Taf. II. fig. 1.

CHAPTER VIII.

ORGANS OF SECRETION.

I. Urinary Organs.

§ 288.

As yet, Urinary organs have not been observed with the Crustacea except in the Myriapoda.   Here, as with the Insecta, they consist of long, small, brownish vessels, caecal, and describing many convolutions about the stomach and intestine.   These Malpighian vessels, as they have been termed, open into the digestive canal at the boundary between the stomach and intestine, and secrete as certainly as do those of the Insecta, uric acid.[1]   With the Chilopoda, there is usually one on each side of the pylorus; but with the Chilognatha, there are two, which open, however, into the intestinal canal by a common orifice.[2]

It is now undetermined whether these organs exist also in the other families of Crustacea.   But with some Decapoda, there are certain caecal vessels which are imperfectly known.   They open into the intestine at various points between the pylorus and rectum, and a more complete examination may, perhaps, show them to be of a urinary nature.[3]

II. Organs of Special Secretions.

§ 289.

The Astacina have a very remarkable secretion commonly known as Crabs-eyes.   These are a kind of calculi composed of carbonate of lime

1 For the Malpighian vessels, which were for a long time regarded as biliary canals, see further under the anatomy of the Insecta.

2 Ramdohr, Abhandl. über d. Verdauungsw. &c. p. 149, Taf. XV. fig. 1 (Julus); Treviranus, Verm. Schrift. loc. cit. p. 24, 44, Taf. V. fig. 4, Taf. VIII. fig. 6 (Lithobius and Julus), and L. Dufour, Ann. d. Sc. Nat. loc. cit. p. 86, 96, Pl. V. fig. 1, 4 (Lithobius and Scutigera). Scutigera differs from the other Chilopoda in having two pairs of urinary canals.
See also Kutorga, loc. cit. p. 6, Tab. I. fig. 2, and Müller, Isis, 1829, p. 550, Taf. II. fig. 5 (Scolopendra); finally Brandt, in Müller's Arch. loc. cit. p. 322, Taf. XII. fig. 2 (Glomeris).

3 Swammerdamm (loc. cit. p. 87, Taf. XI. fig. 3) had already figured, with Pagurus, a pretty long coecum opening at the posterior extremity of the intestine. With Maïa squinado, there are three such pretty long, of which two are inserted on each side of the pylorus, and the third a little further behind (Milne Edwards, Hist. Nat. d. Crust. I. p. 76, Pl. IV. fig. 1, m. n.). Lund (Isis, 1829, p. 1302) has also seen two glandular canals rolled up in a knot, which open each side of the pylorus, while a third entered the rectum. This last, according to Cuvier (Leçons d'Anat. Comp. III. p.

678) is very common with the Macrura, Brachyura, and Anomura, and notably with Astacus fluviatilis, Homarus marinus, Cancer pagurus, Portunus puber, and Cancer maenas. But although Milne Edwards admits the same also (loc. cit. I. p. 76), yet it does not appear to be agreed upon, for Meckel (Syst. d. vergleich. Anat. IV. p. 161) contradicts Cuvier in this respect, and declares that he has never found this caecum either with the Crabs or with Astacus, Scyllarus, and Palinurus, but only with Pagurus, Penaeus, and Palaemon. Duvernoy, also (Leçons d'Anat. Comp. V. p. 228), has not observed it in the Macrura just cited, nor with Galathea squamifera, and Palaemon serratus, although he perceived it with Portunus puber directly behind the pylorus, and with Cancer pagurus, near the rectum. Like Milne Edwards (Hist. d. Crust. I. p. 115, Pl. X. fig. 2, j. (Maïa)), I must leave undetermined the point whether or not, this glandular mass which, with the Decapoda, is concealed under the floor of the respiratory cavities in the bottom of the cephalothorax, and which opens externally by an excretory canal between this same cephalothorax and the first abdominal segment, — is really a urinary organ.

and formed in the two lateral pouches of the stomach of the *Astacus fluriatilis*.[1] As they are not observed during the whole year, but only just before the moulting, and as, when this process occurs, they pass from the cast-off stomach into the cavity of the new one, it may be inferred that they are in some way connected with the act of ecdysis, and that if the lateral pouches of the stomach secrete from the blood the excess of calcareous salts, it is in order that these last may be subsequently used for the formation of the new shell.[2]

The caustic, brown fluid, which most Myriapoda, when touched, emit from a row of orifices situated on the sides of the segments of the body (*Foramina repugnatoria*), and which exhales an odor like that of chlorine, is secreted by small, pyriform, glandular follicles, situated immediately beneath the skin. Its use is, perhaps, for the lubrication of the articulations of the segments of the body.[3]

In the following chapter will be mentioned many other glandular organs connected with the genital functions.

CHAPTER IX.

ORGANS OF GENERATION.

§290.

The Crustacea reproduce by Male and Female Organs, situated in different individuals, and have, for the most part, copulatory organs.

Nevertheless, the Cirripedia form an exception in this respect, being hermaphrodites; while, on the other side, many Entomostraca differ from the general rule, in their species being almost exclusively females, which produce, during many successive generations, individuals exclusively of the female sex, and only at long intervals, those of the male sex.[1]   There is,

---

1 See *Suckow*, loc. cit. p. 53, Taf. X. fig. 10, 11, e. This author is mistaken in supposing that these green glandular bodies mentioned in connection with the organs of hearing (§ 276), secrete the "Crabs-eyes." See also *Brandt*, Medic. Zool. II. p. 63, Taf. XI. fig. 8, 9, c.

2 For the nature of these crabs-eyes, see the researches of *Baer* (*Muller's* Arch. 1834, p. 510) and *Oesterlen*, Ibid. 1840, p. 432.

3 *Treviranus* (Verm. Schrift. II. p. 42, Taf. VIII. fig. 4, f. f. fig. 5, d. e.) has regarded these organs as respiratory with *Julus*, while *Savi* (Isis, 1823, p. 21), Taf. II. fig. 1, 13, 14, a. b.), and *Burmeister* (Ibid. 1834, p. 136, Taf. I. fig. 1, a. a.) have well perceived that they are cutaneous glands. According to *Waga* (Revue Zool. 1839, N. 3, p. 76, or in *Wiegmann's* Arch. 1840, II. p. 350), *Polydesmus*, *Platyulus*, and *Geophilus electricus*, have, upon the sides of the body, analogous glands, out of which this last Myriapod emits a

luminous liquid. *Brandt* (Recueil, &c., p. 154, 157) has observed, with *Glomeris*, that these follicles are situated, in pairs, on the dorsal surface of each segment of the body.*

1 This is so with the Daphnioidae, Cyprinoidae, and Apodidae. In the second of these groups, the males are so rare, that these Entomostraca have been taken for hermaphrodites, and *Straus* (loc. cit. p. 52, Pl. 1. fig. 15) has said that if this was really the case, he regarded as testicles two long cylindrical problematical bodies which he had observed with all the females of *Cypris*. In the genus *Apus*, no individuals which can with certainty be regarded as males, have been found. *Berthold* (Isis, 1830, p. 645) has taken the red ampullae found with these Crustacea for testicles; but, as I have already remarked, these ampullae are only branchial lamellae filled with blood during the dying of the animal. (See § 286, note 5.)

* [§ 289, note 3.] These odoriferous glands have been successfully studied by *Leidy* (Proc. Acad. Sc. Philad. 1849, IV. p. 235) with *Julus*. Here, they consist of a globular body or sac, with an elongated conical neck, and resemble in form a florence flask with the mouth drawn to a point. This sac is composed of a basement membrane lined with a single layer of secreting cells. The neck of the glands has muscular bands. — ED.

probably, some relation between this remarkable mode of generation and the fact that some females lay two kinds of eggs, one of which is developed spontaneously, that is, without the influence of sperm, while the other requires to be fecundated.[2]

The structure and disposition of the genital organs is so different in the various divisions of Crustacea, that it is difficult to make any general statement about them. Usually, there is a complete duplication of these organs, internal and external, with both sexes. With the females, there is nearly always, right and left, a longer or shorter, rarely-branched, ovarian tube. This is succeeded by a narrow oviduct, usually long, and often flexuous. This last continues into a large vagina, which opens at very different points on the ventral surface, sometimes quite in front, sometimes near the middle, or at the posterior extremity. It is rare that this vagina has a *Receptaculum seminis;* but, more commonly, the females have special glandular canals annexed to the genital orifice. The product of these last is a viscous mucus, which hardens in water, and serves to envelop the eggs, and to glue them together. The eggs, thus bound together in chaplets or clusters, remain glued to the parts neighboring the genital orifice, or to the post-abdominal feet, and are borne about by the females, until the embryos are fully developed. With other females, where these organs are wanting, they are replaced by a special pouch (*Marsupium*) situated, usually, at the inferior surface of the thorax. In this pouch the eggs are deposited and remain until their embryos are completely developed.

With the males, the internal genital organs are disposed in a similar manner, and often have the same form as those of the females. A careful examination is, therefore, necessary, to perceive their distinctive character. Moreover, they open, also, at the most varied points of the body. In many species, there are, near the genital orifices, copulatory organs in the form of stylets, or canaliculi, which serve to transfer the sperm into the female organs. With others, the antennae, or some of the feet, are provided with a kind of hook, or pincers, with which they seize and retain the females during copulation. Sometimes the internal organs of the left communicate, by anastomoses, with those of the right side, or, in the place of two lateral genital openings, there is only one, situated on the median line. With many species, the genital organs, internal and external, are simple, and placed in the axis of the body; but it is rarely observed, that the oviducts and deferent canals are single where the ovaries or the testicles are double, or that there are two genital openings for single internal organs.

The Eggs of the Crustacea are usually of either a lively green, yellow, or violet color. They are always spherical, and composed of a dense chorion, containing a vitellus, which surrounds a germinative vesicle, with one or more nuclei. The vitellus is composed of numerous oil-globules, which are held together by a clear, albuminous liquid, and give the egg its peculiar color.[3]

The sperm is white and sometimes opalescent. The spermatic particles are of very varied and remarkable forms. Nearly always, they are

2 This phenomenon is undoubtedly analogous to that of the alternation of generation, which is so general with the other lower animals.
3 For the eggs of Crustacea, see *Rathké,* De Animal. Crust. generat., 1844, and his Bemerkungen in *Froriep's* neue Notiz. XXIV. 1842, p. 181; *Erdl,* Entwickelung d. Hummereies, p. 13; and especially *Wagner,* Prodromus, &c., p. 8, Tab. I. fig. 12–17.

stiff and motionless, and may be arranged under the following principal types.

1. With the Cyclopidae, and Chilognatha, the spermatic particles, which are developed in cells, retain their cell-form to their perfect state, without any trace of processes or appendages.[4]

2. With the Decapoda, they are likewise nearly always of a granular or cell form, but have small, filiform, sac-like processes; sometimes they are divided into two portions by a constriction.[5]

3. With the Mysina, Amphipoda, and Isopoda, they have the form of very long threads, pointed at both extremities, or with a cylindrical incrassation at one of them. They are motionless, and, upon the addition of water, do not roll up in a loop-like manner.[6]

4. With the Cirripedia, and Chilopoda, they are capillary, very lively, and, from contact with water, become entangled, forming loops and rings.[7]

---

4 With *Cyclopsina castor*, the spermatic particles are small, finely-granular, oval corpuscles (see my Beitr. zur Naturg. d. wirbellosen Th. p. 41, Taf. 11. fig. 41-43 c., or Ann. d. Sc. Nat. XIV. 1840, p. 30, Pl. V. B.). As to the other Entomostraca, we have not yet sufficient data to say anything in general. But the form observed with *Cyclopsina castor* cannot be regarded as a typical one with these animals, for *Wagner* (*Wiegmann's* Arch. 1836, I. p. 369) has observed large, filiform, flexuous, spermatic particles with *Cypris*. I, myself, have found those of *Daphnia rectirostris* to consist of a long, semi-circular body, which became motionless and disappeared by bursting on the addition of water. *Stein* (*Muller's* Arch. 1842, p. 263, Taf. XIV. fig. 37, 40) has rightly figured those of *Glomeris* as fusiform cells; but he was less exact with those of *Julus* and *Polydesmus*, in describing them as small transparent vesicles (Ibid. fig. 36, 39.) With *Julus subulosus*, they look exactly like very short cylinders containing a very distinct, round nucleus. With *Julus hispidus*, they are of the same form, but are not nucleated; while with *Julus terrestris*, they are conical and nucleated (see my notice in *Muller's* Arch. 1843, p. 13). Those of the Siphonostoma are, also, of a celloid form; see *Frey* and *Leuckart*, Beitr. loc. cit. p. 135, Taf. II. fig. 21 (*Caligus*).

5 *Henle* (*Muller's* Arch. 1835, p. 603, Taf. XIV. fig. 12) and myself (Ibid. 1836, p. 26, Taf. III. fig. 23, 24) first called attention to the singular form of the spermatic particles of the common craw fish; but, subsequently, *Kölliker* (Beitr. &c. 1841, p. 7, Taf. II. III. and in the Schweizerisch. Denkschrift. t. d. gesammt. Naturw. VIII. 1846, p. 26, Taf. II.) has shown that those of the most diverse species of Brachyura, Anomura, and Macrura, are motionless, *radiated cells*, one part of which is separated by a constriction, and prolonged sometimes into a kind of peduncle; the rays are often only three or four in number, and the cells themselves are sometimes conical or cylindrical. The most simple of these spermatic particles are observed with *Crangon vulgaris*, and *Palaemon squilla*; and, according to my own observations, consist only of flattened cells with a short pointed process.

6 Simple capilliform spermatic particles are found with *Mysis*, *Oniscus*, *Porcellio*, *Idothea*, and *Gammarus*; see my researches in *Muller's* Arch. 1836, p. 27, Taf. III. fig. 19, 20, and Ibid. 1837, p. 433; also *Kölliker*, Beitr. loc. cit. p. 15. This last naturalist (Beitr. &c. p. 14, Taf. III. fig. 28, 29) has stated that the long capillary, but motionless spermatic particles of *Iphimedia obesa* and *Hyperia medusarum* are terminated by a cylindrical and slightly flexuous incrassation. I have found those of *Asellus aquaticus* to be similar, but the cylindrical extremities were straight.

7 The spermatic particles of the Cirripedia, which are simply capillary and very active when fully developed, have been observed by me (*Muller's* Arch. 1836, p. 29), with *Balanus pusillus*, and by *Kölliker* (Beitr. p. 16, Taf. III. fig. 30, and Schweiz. Denks. loc. cit. p. 33) with many other species of *Balanus*, and with *Chthamalus*, *Lepas*, and *Pollicipes*. Those of *Lithobius* and *Geophilus* present a remarkable aspect from their extreme activity, and may well be recommended for study from their size (*Stein*, in *Muller's* Arch. 1842, p. 250, Taf. XIII. XIV. fig. 19-33). *Treviranus* (Verm. Schrift. II. p. 26, Taf. VI. fig. 2, 3) has taken those of *Scolopendra*, which are bound together in a long white cord, for a Helminth.*

---

* [§ 290, note 7.] The spermatic particles of the Crustacea are the most remarkable of any in the whole animal kingdom. The strange, bizarre forms, here observed, have led to singular views as to their development and character. The recent researches of *Kölliker* have done much to clear up this intricate subject, and these researches have been continued by *Wagner* and *Leuckart* (Art. *Semen*, Cyclop. Anat. and Physiol.). The most singular of these particles, as is well known, are those belonging to the higher forms of this class. The development and nature of these I have recently studied, and with results quite different from those of the authors just mentioned. My observations were made on those of *Pagurus*, *Pilumnus* and *Astacus*. Here, the development occurs in special cells like that of those of other animals, and the particle, however singular its form, is the transformed nucleus of these cells. The spine-like processes lie reverted on the body of the particle when this last is in the special cell, but become erect and prominent when the particle escapes. The body of the particle, therefore, is solid, and not hollow and nucleated, as has been supposed (see *Kölliker* and his

With very many Crustacea, the sperm, at its emission, is contained in capsules (*Spermatophores*).

I. *Hermaphrodite Crustacea.*

§ 291.

With the Cirripedia, the male and female genital organs are quite removed from each other. With the Lepadea, the ovaries are lodged in the upper extremity of the peduncle and in the midst of the spongy substance, filling its cavity.[1] They consist of ramified caeca, while with the Balanodea, the ovarian follicles are situated between the lamellae of the mantle.[2] With the Lepadea, the canal which extends from the lower extremity of the shell into the peduncle and communicates by a narrow opening with the cavity of the mantle, may properly be regarded as an oviduct.[3] But new researches are required to show by what means the eggs of the Balanodea reach this same cavity, for, as is the case with the Lepadea, they remain there until the embryos are fully formed. These eggs, of a blue or yellow color, are always intimately glued together, and form, after the laying, a large sheet or layer which, with the Balanodea, is applied to the internal surface of the mantle, and often retained there by the branchial lamellae ;[4] while with the Lepadea, it covers, bonnet-like, the rounded portion of the body.

The Testicles are composed of numerous ramified follicles spread out between the skin and the two sides of the digestive canal. They join from the right and left into two very long and tubular *Vasa deferentia* which accompany, serpentinely, the alimentary canal to the anus, and then blend together, forming a more narrow *Ductus ejaculatorius*. As this last traverses the whole tail and opens at its extremity, it has been usually regard-

---

1 *Burmeister* (Beitr. p. 46), and *Wagner* (Müller's Arch. 1834, p. 469, Taf. VIII. fig. 10), were the first to notice the ovarian follicles of the foot of the Lepadea, the first with *Otion*, the second with *Lepas*. *Martin St. Ange* (loc. cit. p. 20, Pl. 1. fig. 10, 11) has verified this fact with this last genus. I have found them also in the foot of *Cineras*, and I will remark that in the remaining spongy substance of this foot, there are other round nucleated bodies which appear to be solid concretions, and should not be confounded with the germs.

2 The ovaries of the Balanodea are more difficult of study than those of the Lepadea, probably because they are scattered in the walls of the mantle, and consequently scarcely visible, especially when empty. It is undoubtedly on this account that *Poli* (Testac. utriusq. Sicil. &c. 1. p. 19, 28, Tab. IV. fig. 13, x. x. Tab. V. fig. 13, 15) has taken for ovaries, with a *Balanus*, the testicular follicles, although he distinctly saw and has figured the ovarian follicles in another species of this same genus.

3 This canal, regarded as an oviduct by *Wagner* (loc. cit.), had already been mentioned by *Cuvier* (Mém. loc. cit. p. 4, fig. 4).

4 With *Balanus*, the layers of eggs form usually two large discs (*Poli*, loc. cit. Tab. IV. fig. 18, c. c.).

figures). The nuclear appearance is due, sometimes to a depression in the body (as with *Astacus*), sometimes to a plastic membrane lying about one of the spinous processes (as with *Pilumnus*), all made prominent by the refraction of the light ; see my researches in the Proceed. Boston Soc. Nat. Hist. IV. p. 258.

In regard to the spermatic particles of the Entomostraca, I have examined those of *Cypris*, *Cyclops*, and *Daphnia*. They are developed, as usual, in special cells — are exceedingly minute, and in form closely resemble those of the Araneae ; consisting of an arcuate rod to which is attached a short but very delicate tail. My results, therefore, do not agree with those above-mentioned.

The whole subject of the spermatic particles of the Crustacea is sadly deficient in well-authenticated observations, and particles and cell-like forms are constantly described as spermatic particles, which, according to all the laws of Spermatology as yet known, cannot be such. It should be remembered that the spermatic particle is never a cell, but is the metamorphosed nucleus of a cell ; it is, therefore, always a more or less solid corpuscle (whatever be its form, &c.), and to which, moreover, there may be attached one or more appendages — Ed.

ed as a Penis.[5] The length and mobility of this tail is such that it can be used, with the Cirripedia, as a copulatory organ, and, being brought in contact with the orifice of the oviducts, which is situated on the mantle, self-impregnation may thus take place.*

II. Female Crustacea.

§ 292.

The female genital organs of the Siphonostoma and Lophyropoda consist nearly always of two long and sometimes flexuous ovarian pouches, situated on both sides of the digestive canal. The oviducts pass backwards and terminate on both sides of the end of the body by separate orifices, or by a single genital opening on the median line. With those species whose body has a tail, these orifices are not situated like the anus, at its extremity, but at its base. Very often, they serve as the outlets of the excretory ducts of two caecal organs which secrete a viscous substance by which the eggs are glued together in clusters or chaplets.[1] Not unfrequently, there are hook-like or capsular appendages near the genital openings, for the retention of these clusters.[2] But these are wanting with the Daphnioïdae, there being in their place an incubating cavity, situated between the

---

5 *Cuvier* (Mém. loc. cit. p. 9, fig. 8) has taken, with *Lepas*, the testicles for the ovaries, and the *Vasa deferentia* for the testicles. This error could not be rectified until the discovery of the true ovaries (see *Burmeister*, Beitr. loc. cit. p. 33, Taf. II. fig. 16 ; *Wagner*, in *Muller's* Arch. loc. cit. p. 463, Taf. VIII. fig. 8 ; and *Martin St. Ange*, loc. cit. p. 21, Pl. II. *Lepas*). It is more singular to see *Goodsir* (Edinb. New Philos. Jour. 1843, July, p. 88, Pl. III. IV. or Ann. d. Sc. Nat. I. 1844, p. 107, Pl. XV. C. or *Froriep's* neue Notiz. No. 651, 1844, p. 193), endeavor to confuse this question by declaring the hermaphrodite animals of *Balanus* to be females which carry, in the cavity of their mantle, dwarfish and abortively-formed males. Very probably those so-called males are parasitic Crustacea, as *Kölliker* (Schweiz. Denks. loc. cit. p. 33) has supposed.

1 The female genital organs are completely double with the Penellina, the Lernaeodea, the Ergasilina and the Caligina ; see *Nordmann*, loc. cit. p. 6, Taf. I. fig. 4, Taf. V. fig. 7, Taf. VI. fig. 10 (*Lamproglena*, *Achtheres* and *Peniculus*); *Goodsir*,

Edinb. New Philos. Jour. July, 1842, p. 178, or Ann. d. Sc. Nat. XVIII. 1842, p. 181 ; *Kröyer*, Naturh. Tidskr. I. Pl. VI. or Isis, 1841, p. 194, Taf. I. Tab. VI. fig. 4, C. (*Caligus*); and *Rathké*, Nov. Act. Nat. Cur. XIX. p. 145, Tab. XVII. fig. 2 (*Dichelestium*). This last-mentioned author was the first to notice the organs which, in various Crustacea, secrete the viscous matter mentioned in the text. He found them highly developed with the *Nicothoe* (loc. cit. XX. p. 106), where they extend, with the ovaries, even into the wing-like appendages of this parasite. The ovarian follicles of *Chondracanthus*, which are multiramose, differ very much from the usual form (*Rathké*, Ibid. XX. p. 123, Tab. V. fig. 18). With the Cyclopidae, there is only a single genital opening, although the ovaries and the organs secreting the viscous matter, are double. But with *Argulus*, the female genital organs are the most simple, consisting only of a single ovarian tube, opening at the base of the tail (*Jurine*, loc. cit. p. 448, Pl. XXVI. fig. 3).†

2 *Nordmann*, loc. cit. p. 8, Taf. II. fig. 6 (*Ergasilus*).

---

* [§ 291, end.] That the Cirripedia are not universally hermaphrodites, was first discovered by *Goodsir* (El. New Phil. Jour. XXXV. p. 88), upon *Balanus balanoides*. The male is very small, and it is not strange that it before eluded observation.

Darwin has made some researches, lately, with a similar result in some respects. Exceptions to the rule were found by this naturalist in the genera *Ibla* and *Scalpellum*. With *Ibla*, the males lie within the sac of the female, and have an elongated body with a pedicle below. He has also observed that, with these genera, there are both females and hermaphrodites ; and in some hermaphrodites, males have been observed so similar in general

character to those of *Ibla*, that he considers them to be true males of the species with which they are connected. Being thus supernumeraries, he has termed them *complemental* males. As spermatic particles were distinctly observed in them, their male nature is clear, but it would not appear equally clear that they really belong to the genus and species with which they are connected. Facts so singular require further research. See *Darwin*, Monogr. &c. loc. cit. p. 207, 231. — ED.

† [§ 292, note 1.] For further details on the female genital organs of *Argulus*, with illustrations, see *Leydig*, loc. cit. in Siebold and *Kölliker's* Zeitsch. II. p. 339 Taf. XIX. fig. 5, a. Taf. XX. fig. 8, 10. — ED.

shell and the posterior part of the back.[3]   The females of *Daphnia* have, beside those eggs which are rapidly developed in this cavity, another kind known as the hibernating eggs, and in which no germinative vesicle is observed.   They are always found in couples in a thickened, saddle-like portion of the shell of the animal, which is often of a black color, and separated from the shell by a kind of moulting.   Thus enveloped in a bivalved capsule, they are protected against the severities of the winter.[4]

Among the Phyllopoda, the Apodidae are distinguished for the very large, multiramose, ovarian follicles which border the two oviducts on every side ; these last are straight and large, and situated on the side of the digestive canal.   With *Apus*, each of them sends off, at about its middle, a short, excretory canal, to the eleventh pair of feet, in which there are two alveolate receptacles with covers for the reception of the eggs.[5]   With the Branchiopoda, the ovaries consist of two straight cocca, situated in the tail, on each side of the intestine.   Their upper extremity, beneath the last pair of feet, passes into an elongated receptacle.   These two receptacles, separated from each other only by a thin partition, have a narrow outlet at their posterior extremity, and form, under the base of the tail, a kind of oblong tumor, into which the hard and granular eggs are constantly cast from the contractions of special muscular bands.[6]   There are, moreover, on the sides of the body above the last pair of feet, two oblique horny plates, which the males, during copulation, seize with their cephalic pincers.

With the Poecilopoda, the cephalothorax contains two ramified ovaries,

---

3 See *Straus*, Mém. sur les Daphnia, loc. cit. p. 413, Pl. XXIX. and *Jurine*, Hist. d. Monocl. Pl. VIII.-XVI. The genera *Argulus* and *Cypris* differ, moreover, from the other Entomostraca, in that they do not bear about their eggs after laying, but deposit them on foreign bodies ; see *Jurine*, Mém. sur l'Argule, loc. cit. p. 451, and *Straus*, Mém. sur les Cypris, loc. cit. p. 54.

4 The formation of the saddle, which is intimately connected with the deposition of the hibernating eggs, has been called by *Jurine* the *Matadie de la selle*. But it has been before observed by *Muller* (Entomostr. p. 84, Tab. XI. fig. 9-11, Tab. XII. fig. 5), and by *Ramdohr* (loc. cit. p. 28). See, also, *Straus*, loc. cit. p. 415, Pl. XXIX. fig. 16, 17, and *Jurine*, Hist. d. Monocl. p. 120, Pl. XI. fig. 1, 4.*

5 See *Schaeffer*, Der krebsartige Kiefenfuss, p. 79, Taf. IV. fig. 2-7, and *Zaddach*, loc. cit. p. 51,

Tab. I. (*Apus*). With *Limnadia*, and *Isaura*, a special receptacle is wanting ; the eggs are attached to the feet probably by the aid of their hairy external envelope ; see *Brongniart*, loc. cit. p. 88 ; *Straus*, Mus. Senckenb. loc. cit. Taf. VII. fig. 16, and *Joly*, loc. cit. p. 308, Pl. IX. A.†

6 See *Prevost*, in *Jurine*'s Hist. d. Monocles, p. 228, Pl. XX. fig. 1, 10 (*Chirocephalus*). This author erroneously declares, moreover (loc. cit. p. 207), that the females of this animal have, also, at the end of the tail, openings into which the sperm is received during coition. See also *Joly* (loc. cit. p. 240, Pl. VII. fig. 12, Pl. VIII. fig. 4), who regards the receptacles of the eggs, with *Artemia*, as the ovaries. The eggs with a solid, granular shell, of *Branchipus*, have been pretty distinctly figured by *Schaeffer* (Der fischförm. Kiefenfuss, fig. 14).

---

* [§ 292, note 4.] Recent investigations upon the economy and development of the Crustacea indicate that the phenomena above-mentioned, of the reproduction by means of a second kind of eggs (so-called), is far from being limited to a few of these animals. Indeed, it is probable that all or most of the Entomostraca reproduce by this mode. As mentioned on a preceding page, these phenomena do not appear to me to belong to true oviparous reproduction, but must be considered as a kind of internal gemmiparity. The so-called winter-eggs are, therefore, not eggs, but buds (*gemmae*) — a view which is borne out by their composition, — there being no germinative vesicle and dot. This subject will be discussed with some detail below (note under §355) when speaking of the Aphides — animals with which I have traced these phenomena with some care. For many interesting details on this

subject, see *Liévin*, Die Branchiopoden der Danziger Gegend, 1848, p. 11. et seq.; *Baird*, Brit. Entomostr. &c. loc. cit. passim ; *Zenker*, Physiologische Bemerkungen über die Daphnoidae, in *Muller*'s Arch. 1851, p. 112 ; *Leydig*, Ueber Artemia salina und Branchipus stagnalis, in *Siebold* and *Kölliker*'s Zeitsch. III. 1851, p. 297. — Ed.

† [§ 292, note 5.] For the female genital organs of *Artemia* and *Branchipus*, see *Leydig* (loc. cit. *Siebold* and *Kölliker*'s Zeitsch. III. p. 300). *Joly*, it would appear, did not observe the ovaries of *Artemia*, but has described the egg-capsules as such. The real ovaries here consist of sacs or pouches, lying near the dorsal surface of the abdomen, and extending to the second abdominal ring ; these ovarian sacs pass into a vesiculiform dilatation, which has non-muscular walls, and corresponds to a uterus. — Ed.

the large oviducts of which open at the base of the first pair of feet, near the median line of the body.[7]

With the Laemodipoda, Isopoda, Amphipoda, and Mysina, there are two simple ovarian tubes wound about the digestive canal; these oviducts are sometimes terminal, sometimes lateral. The two vulvae are usually situated on the internal side of the fifth pair of feet.[8] The eggs, after laying, are always deposited in an incubating pouch, situated beneath the anterior extremity of the body, and the walls of which are formed in part by from two to five pairs of imbricated, and often concavo-convex lamellae.[9] These last are generally bristled on their borders, and are chiefly developed at the epoch of procreation, after which they disappear.[10]

The ovaries of the Squillina differ remarkably from those of the other higher Crustacea. They consist of numerous, ramified lobes, filling the lateral portions of the posterior abdominal segments, and the digitations of which extend even into the last and flattened caudal segment. All these divisions of each ovary join in a large, long tube, which surrounds the digestive canal. The portion of the ovary contained in the three segments to which are attached the ambulatory feet, sends towards the ventral surface, three branches, which join, upon the median line beneath the abdominal cord, with those of the opposite side, and form, in the middle of each of these three segments, a round sinus. These sinuses are connected by longitudinal anastomoses, and the anterior one is prolonged into a common papillary vulva, situated in the middle of the first abdominal segment beneath a horny process.[11]

---

7 *Van der Hoeven*, loc. cit. p. 21, Pl. II. fig. 15, Pl. III. fig. 1 (*Limulus*).

8 There are two ovarian tubes, each continuous posteriorly into a short oviduct with *Cyamus* (*Roussel de Vauzème*, loc. cit. p. 253, Pl. IX. fig. 19), with *Aega* (*Rathké*, Nov. Act. Nat. Cur. XX. p. 32, Tab. VI. fig. 17), and with *Mysis* (*Frey*, loc. cit. p. 25). The two oviducts meet in a common vulva in front of the anus with *Bopyrus* and *Phryxus* (*Rathké*, De Bopyro, &c., p. 19, Tab. I. fig. 7, and Nov. Act. Nat. Cur. XX. p. 47). With the Asellina, the ovarian tubes are coecal at both of their extremities; the oviducts pass off laterally and open in the articulation of the fifth and sixth segment of the body (*Brandt*, Mediz. Zool. II. p. 76, Taf. XV. fig. 32). I have observed an analogous structure in the genital organs of *Idothea* (*Muller's* Arch. 1837, p. 434). With *Caprella*, the ovarian tubes are coecal in the same way, but they interanastomose by two pairs of short, transversal oviducts (*Goodsir*, Edinb. New Philos. Jour. July, 1842, p. 184, Pl. III. fig. 2). This author adds, contrary to all analogy, that these oviducts terminate in two vulvae situated one behind the other on the middle of the belly.

9 The incubating sac of *Cyamus* and *Caprella* is composed of four lamellae situated back of the branchiae upon the two footless segments of the body (*Roussel de Vauzème*, loc. cit. p. 249, Pl. VIII. fig. 3, and *Goodsir*, loc. cit. p. 165, Pl. III. fig. 3, 10). With *Mysis*, this cavity contains also only four lamellae covered with stiff bristles and attached to the coxae of the last two pairs of feet (*Muller*, Zool. Danic. Tab. LXVI. fig. 1, 2; *Milne Edwards*, Hist. d. Crust. Pl. XXVI. fig. 8, d.; and *Rathké*, in *Wiegmann's* Arch. 1839, 1. p. 199). With *Nerocila*, there are also four large lamellae arising from the coxae of the sixth and seventh pairs of feet. With *Idothea*, the Asellina and the Gammarina, on the other hand, the five anterior segments of the body have as many

pairs of ventral lamellae of this kind. With *Gammarus*, the borders of these ten lamellae are covered with long bristles (*Zenker*, loc. cit. p. 8, fig. N. b.). With *Cymothoa*, the coxae of the first six pairs of feet have a semilunar lamella (*Milne Edwards*, Ann. d. Sc. Nat. III. 1835, Pl. XIV. fig. 2, and Cyclop. loc. cit. p. 784. fig. 436). The same is true of *Anilocra*, judging from *Savigny's* figure (loc. cit. Crust. Pl. XI. fig. 10-). With *Bopyrus*, and *Phryxus*, the incubating sac contains six pairs of lamellae which, in the first of these genera, are not wholly superposed (*Rathké*, De Bopyro, &c., p. 6, Tab. I. fig. 5, and Nov. Act. Nat. Cur. XX. p. 44, Tab. II. fig. 12). The sixth or anterior pair of these lamellae is wanting with *Cepon* (*Duvernoy*, Ann. d. Sc. Nat. XV. 120, Pl. IV. fig. 2), but with the Bopyrina in question, is singularly attached to the head. According to *Treviranus* (Verm. Schrift. 1. p. 61, Taf. IX. fig. 52), there are at the bottom of this sac with Oniscidae, four short conical processes which secrete a yellowish fluid; but *Brandt* (loc. cit. II. p. 72, Taf. XII. fig. 2, Taf. XV. fig. 33) was unable to find them, while *Rathké* (loc. cit.) has been led to regard as secreting organs also, two filaments which, with *Mysis*, are attached to the ventral wall, and stretch into the incubating sac.

10 This origin and disappearance of the incubating lamellae I have seen very distinctly with *Idothea entomon* (*Muller's* Arch. 1837, p. 435). The females of *Cyamus* which *Muller* (Zool. Danic. Tab. CXIX. fig. 16), and *Treviranus* (Verm. Schrift. II. Taf. I. fig. 2) have figured, appear to have been individuals whose incubating sac was not then fully developed.

11 The ovaries of *Squilla* are so intimately blended in part with the liver, that they may be easily confounded with that organ. It is on this account that *Duvernoy's* figure (Ann. d. Sc. Nat. VI. 1836, p. 248, Pl. XV. and VIII. 1837, p. 42, Pl. II.) of this organ with this same animal, is not

With the Brachyura, the cephalothorax contains four long ovarian tubes, two anterior, and two posterior. The first wind outwardly over the liver, and are anastomosed by a short transverse canal; while the second are straight, lie close to each other, and cover the anterior part of the intestine. The anterior and posterior tubes of each side unite in a short vagina, and, at their point of junction, open into a pyriform sac, which has been regarded by some as a *Bursa copulatrix*, and by others as a gland secreting the viscous substance which envelops the eggs, but which, upon a more careful examination of its contents, will be found to be a *Receptaculum seminis*.[12] The two vaginae open near the ventral median line in the segment which bears the third pair of feet.[13] With the other Decapoda, — the Anomura and the Macrura, — these sacs, just mentioned, are wanting, while the ovaries themselves are disposed, in general, like those of the Brachyura.[14] But the genera *Pagurus* and *Astacus*, alone, form an exception in this respect. In the first, the two ovaries with their oviducts lie concealed principally beneath the dorsal surface of the tail; while in the second, they are aggregated in a trilobed mass in the pyloric region, from which pass off two short oviducts.[15] The female genital openings are situated, with the Anomura, as with nearly all the Macrura, in the coxal joints of the third pair of feet.[16]

With the females of all the Decapoda, the feet of all the caudal segments are highly developed and very hairy. They serve to support the eggs which are glued together in clusters by a viscid substance which hardens in water; these clusters are attached to the bristles or hairs of these feet. But with the Brachyura, and Anomura, these eggs have an additional protection in the tail, which is folded against the body.[17]

The Chilognatha have only a single long and large ovarian tube, provided with two short oviducts which are narrower, and open externally at two squamous bodies situated on the under surface of the third segment of the body. These two bodies contain two short caeca one of which is dilated at its base into a vesicle, and each pair opens by a common orifice in the vulva. They represent a *Receptaculum seminis*.[18] With the

perfectly clear. This naturalist has, moreover, regarded a large part of the ovaries as venous sinuses, and the white eggs which they contain as coagulated blood. In order to have a general idea of the disposition of the female genital organs with these animals, it is only necessary to cast a glance over *Delle Chiaje's* figure (Descriz. &c. Tav. LXXXVI. fig. 4, b. g. g.); it is true that he has represented them as testicles, but they are perfectly exhibited, with the exception, however, of the anterior portion.

12 See *Cavolini*, loc. cit. p. 138, Taf. II. fig. 3 (*Grapsus*); *Milne Edwards*, Hist. d. Crust. I. p. 170, Pl. XII. fig. 12, and *Cyclop*. loc. cit. p. 784, fig. 434; *Carus*, Erläuterungstaf. loc. cit. Heft. V. p. 7, Taf. III. fig. 7, and *Erdl*, Entwickel. d. Hummereics, p. 11 (*Maia*).

13 *Cavolini*, loc. cit. Taf. II. fig. 2, a. (*Grapsus*); *Milne Edwards*, Hist. d. Crust. Pl. III. fig. 4, i.; and *Carus*, loc. cit. Taf. III. fig. 8, h. (*Maia*).

14 *Milne Edwards*, Hist. d. Crust. I. p. 171, and *Duvernoy*, in *Cuvier's* Leçons, &c., loc. cit. VIII. p. 340.

15 The internal female genital organs of the crawfish are represented in *Roesel*, loc. cit. Taf. LX. fig. 24, 25; in *Suckow*, loc. cit. Taf. X. fig. 16; and in *Brandt* and *Ratzeburg's* Mediz. Zool. II. Taf. XI. fig. 15.

16 The two vulvae of the crawfish may be seen in the figures already cited. For those of the Anomura, which, except with *Pagurus*, are covered by the tail curved in front; see *Milne Edwards*, Hist. d. Crust. III. p. 172, Pl. XXI. fig. 8, 18 (*Dromia* and *Remipes*), and Arch. du Mus. II. Pl. XXVI. fig. 1, c. (*Lithodes*). But the Caridoidae form an exception in this respect, — their female genital openings being situated in the same places as those of the males, that is on the external side of the coxae of the posterior feet; see *Kroyer*, loc. cit. p. 27, fig. 54, A. f. and fig. 97, B. g. (*Hippolyte*).

17 With *Pagurus*, the anal feet are developed only on one side of the tail.

18 Many erroneous opinions have been entertained by Zoötomists on the subject of the female genital organs of the Chilognatha. Thus, *Treviranus* (Verm. Schrift. II. p. 45) with *Julus*, and *Brandt* (*Muller's* Arch. 1837, p. 325, Taf. XII. fig. 8) with *Glomeris*, think they have observed double ovaries, as is also true of *Stein* (*Muller's* Arch. 1842, p. 246, 248); but *Newport* (Phil. Trans. 1842, p. 99, or in *Froriep's* neue Not. XXI. p. 161; see, also, *Rymer Jones*, Cyclop. loc. cit. p. 552, fig. 315, 316) has noticed only a simple ovarian tube with *Julus*, which I have been able to confirm; and as for *Glomeris*, *Brandt* (Recueil, loc. cit. p. 157) has himself recently perceived the same.

Chilopoda, also, the ovary is a single long tube, but extends from before backwards and terminates by a short oviduct in the last segment of the body. The *Receptaculum seminis* consists, here, of two ovoid capsules, sessile or pedunculated, and inserted upon the sides of the extremity of the oviduct. Into this last, moreover, just before its termination, long excretory ducts enter from the two to four *Glandulae sebaceae*, which furnish probably the viscous coating of the eggs. [19]

III. Male Crustacea.

§ 293.

The males of the Siphonostoma often differ very much from the females, not only as to their external form, but also in their smaller size, — their development being arrested at a very early period. On this account, some are still unknown, and the organization of others is not understood. [1]

With the Caligina, however, they have received more attention, for in size they are scarcely smaller than the females. Their posterior abdominal segment, which, usually, is not as large as that of the other sex, has, at its extremity, two genital openings, side by side. No testicles or excretory canals have yet been observed, but it may be inferred that their external form and their disposition are analogous to those of the oviducts. [2]

But with *Dichelestium*, the male organs are better known. The two spheroidal testicles, and the somewhat tortuous *Vasa deferentia* of these

There has been the same misapprehension and changing of opinion on the subject of the position of the external genital openings. According to *Treviranus*, and *Brandt* (loc. cit.), they should be situated, with *Julus* and *Glomeris*, at the posterior extremity of the body ; but *Latreille* (Hist. Nat. d. Fourmis, 1802, p. 385) had before indicated their true position with *Polydesmus*, and *Savi* (Isis, 1823, p. 217) has confirmed this with *Julus*. *Brandt* (Recueil, loc. cit. p. 154) has since rectified his error in respect to *Glomeris*. But *Stein* has treated with most detail the subject of the genital openings with *Julus* and *Glomeris*, as well as their seminal receptacles (*Müller's* Arch. 1842, p. 246, Taf. XII. fig. 12, and Taf. XIII. fig. 15. See, also, my observations, Ibid. 1843, p. 9).

19 For the female genital organs of *Lithobius* and *Scutigera*, see *L. Dufour* (loc. cit. p. 89, Pl. V. fig. 1, 4) who regards the two stalkless *Receptacula seminis* of *Lithobius* as a reservoir of the four *Glandulae sebaceae*, and with *Scutigera*, as the *Glandes sébacées* themselves. I cannot now say whether these last organs are wanting in *Scutigera*, or whether they escaped the attention of this naturalist. *Treviranus* (Verm. Schrift. II. p. 28, Taf. V. fig. 8) has very well observed the simple ovary with its appendages of *Lithobius*; but he did not recognize the use of these last. *Kutorga* (loc. cit. p. 8, Tab. I. fig. 5) has not been more fortunate with the female genital organs of

Scolopendra. Those of *Scolopendra morsitans* represented by *Müller* (Isis, 1829, p. 550, Taf. II. fig. 5) are probably the male organs. *Stein* (loc. cit. p. 239, Taf. XII. fig. 2, 8) has described very accurately these organs with *Lithobius*, and *Geophilus*. This last has two long-pedunculated seminal receptacles, and only two very long *Glandulae sebaceae.*

1 *Nordmann* (loc. cit. p. 76, &c., Taf. V. VIII. IX. X.) was the first to discover some of these small male Crustacea which are nearly always attached to their females in the neighborhood of the genital openings. He observed them with *Achtheres*, *Brachiella*, *Chondracanthus*, and *Anchorella*. But with an individual of the first of these genera only, he found in the posterior part of the body, four round masses, which perhaps may be the internal genital organs. *Burmeister* (Nov. Act. Nat. Cur. XVII. p. 320) refuses to recognize these small males for the above-mentioned Siphonostoma, while *Kröyer* (Natur. Tidskr. I. Pl. III. or Isis, 1840, p. 710, Taf. I. Tab. III.) sustains the opinion of *Nordmann* with cogent arguments, and has described and figured several of these males belonging to *Lernaeopoda* and *Lernaea*. See also the description of *Chondracanthus* published by *Rathké*, Nov. Act. Nat. Cur. XX. p. 126, Tab. V. fig. 13.

2 See *Kröyer*, Naturh. Tidskr. I. Pl. VI. or Isis, 1841, p. 194, Taf. I. Tab. VI.*

---

* [§ 293, note 2.] *Dana* (*Caligus*, loc. cit. Amer. Jour. Sc. XXXII. p. 261, also, Report. Crust. &c. p. 1344) has observed, with *Caligus*, a well-formed male apparatus. Here, the testicle (and the ovary, also, is the same) consists of a large pyriform body of an internal glandular appearance, and continuous into a duct extending the whole length of the thorax into the abdomen where it passes into the seminal organs. Described more particularly, the testicles are rather larger than the buccal mass, and are situated just anterior to the stomach, in part beneath the base of the prehensile legs, and the spine of the preceding pair. — Ed.

Crustacea completely resemble, as to form and situation, the ovaries and oviducts of the females, except that the deferent canals are dilated, before their termination, into two seminal vesicles.[3]

With *Argulus*, the males have, at the base of their last pair of feet, a hook which is used in copulation. But as to their internal organs, there are, as yet, no credible observations.[4]

With the Cyclopidae, the male organs consist of a single pyriform testicle, the *Vas deferens* of which curves, first forwards, then backwards, and opens at the base of the tail on the median line. In the lower end of this canal, a homogeneous, cylindrical envelope is formed around the sperm, — a real spermatophore, which has a narrow neck, and which the males glue to the vulva of the females.[5] For effecting this last, the males have one or even both of their antennae incrassated at their base, and provided with a special article near their extremity, which gives these organs a forficulate character.[6] When the male, by the aid of these antennae, has embraced the abdomen of the female, he bends the posterior part of his body forwards, and seizes hold of the female a second time with the forficulate foot of the second pair, at the same time grasping, with the other and digitiform foot, the spermatophore as it is escaping from the genital opening, and attaches it to the vulva.[7]

As yet we possess only quite incomplete observations upon the males of Daphnioïdae, Cypridoïdae, and Apodidae, which are found only at certain seasons of the year.[8] With the species yet observed, the testicles

3 *Rathké*, Nov. Act. Nat. Cur. XIX. p. 149, Tab. XVII. fig. 17. I do not know how it is with the males of the other Ergasilina, for as yet we know only the females of these animals.

4 *Jurine*, who was the first to notice these copulatory organs of the male *Argulus*, says he perceived at the base of the penultimate pair of feet a vesicular swelling containing, he thinks, a fecundating liquid (Ann. du Mus. loc. cit. p. 448, Pl. XXVIII. fig. 1, 21).*

5 For the formation of these spermatophores with *Cyclopsina castor*, and *minutus*, see my Beitr. zur Naturg. d. wirbellosen Thiere, p. 36, Taf. II. fig. 41-44, or Ann. d. Sc. Nat. XIV. 1840, p. 26, Pl. V. B. I have shown how their contents are thrust out in passing the neck by the action of a peculiar substance which swells when in contact with water.

6 The two antennae are thus endowed, with *Cyclops quadricornis, Cyclopsina minutus*, and *alpestris*; while this organization obtains with one antenna, only, with *Cyclopsina castor*, and *Anomalocera Patersonii*; see the figures of *Muller*, Entomostraca; and *Jurine*, Hist. d. Monocles; also *Vogt*, Schweiz. Denksch. loc. cit. p. 18, Taf. II.; and *Templeton*, Trans. of the Entomol. Soc. II. p. 36, Pl. V. fig. 1, 5. The asymmetrical posterior pair of feet has been figured by *Jurine*, loc. cit. p. 61, Pl. IV. fig. 2, Pl. VI. fig. 11 (*Cyclop-*

sina castor), and by *Templeton*, loc. cit. p. 37, Pl. V. fig. 1, 18 (*Anomalocera*).

7 These spermatophores, the true signification of which was unknown until lately, are found, often in the numbers of four to six, upon the same female, after several coitions occurring at different intervals; see *Muller*, loc. cit. Tab. XVI. fig. 5, 6, and *Jurine*, loc. cit. Pl. IV. fig. 6 (*Cyclopsina castor*); also *Ramdohr*, loc. cit. Taf. III. fig. 6, 9, and *Jurine*, loc. cit. Pl. VII. fig. 2, 14 (*Cyclopsina minutus*). The spermatophores of this last have the form of a curved horn, and become, after a time, of a brown color. With *Cyclops quadricornis*, the sperm does not appear to contain spermatophores at the moment of its evacuation.

8 The males of *Polyphemus, Limnadia*, and *Apus*, have not yet been observed. It is said, it is true, that *Kollar* (Isis, 1834, p. 680) has discovered those of *Apus cancriformis*; but as yet nothing definite has been learned about the matter. At all events, the description given by *Zaddach* (loc. cit. p. 53, Taf. I. fig. 15, 16, and Taf. III. fig. 1, Pl.) of the male genital organs of these Crustacea, is unsatisfactory, for at the point, where, according to this naturalist, are found the two male genital orifices surrounded by short spines, that is, on the dorsal surface of the last segment of the body, are found, with all the females also, similar orifices. It is therefore probable that the ramose testicles which

* [§ 293, note 4.] For the male genital organs of *Argulus*, see *Leydig* (loc. cit. *Siebold* and *Kölliker's* Zeitsch. II. p. 341). The testicles consist of two pouch-like organs, situated, one in each caudal fin; they send off, each, a vas deferens which terminates in a seminal vesicle; from this last pass off two deferent ducts which end in the common genital orifice. Just before reaching this orifice, each of these ducts is joined by another

coming from an accessory gland, which is pouch-like, and stretches back of the seminal vesicle. As auxiliary copulatory organs may be regarded a hook situated on the anterior border of each of the last pair of feet, and a nodule or papilla in the posterior border of the penultimate pair, corresponding, oppositely, with the hook. These hooks were taken by *Jurine* for penises, and the papillae for seminal capsules. — ED.

consist of two spheroidal bodies which open externally, by two deferent canals, in front of the tail.[9] The copulatory organs are attached to the anterior feet, and consist of hooks and long bristles, by which these animals adhere to the under surface of the thorax of the females.[10] With the Branchiopoda, the male genital organs have a very remarkable organization. The testicles consist of two long, straight, caecal tubes, stretching the whole length of the tail. From the upper and dilated extremity of each passes off, inwards and backwards, an excretory canal. These canals, shortly after their origin, dilate into a seminal vesicle, and then pursue their course between two longitudinal ridges which run backwards from the base of the tail. At the posterior extremity of these ridges, they open near a process covered with short spines. For the seizure and retention of the females for copulation, the two anterior cheliform feet are provided with antler-like hooks, and, also, at their base, with two peculiar, sometimes digitiform processes, curved above the front.[11]

With the Poecilopoda, the testicles co   ` t of ramified canals situated in the cephalothorax, which terminate at that same point on the first pair of post-abdominal feet where are situ   the genital openings with the females, in two short, perforated, penis-like organs.[12]

With the Laemodipoda, Isopoda, and Amphipoda, the testicles consist of two caeca situated by the side of the digestive canal, and continuous, posteriorly, into two more or less flexuous deferent canals upon the sides of

Zaddach thinks he has observed with a small number of individuals which had been preserved in alcohol a long time, are only ovaries, the characteristics of which have been effaced by the spirit. As to *Cypris*, all we know about their males is that their spermatic particles, according to *Wagner* (loc. cit.), are disproportionately large, and that *Ledermüller* (Microscop. Gemüths-und Augen-Ergötzung, p. 141, Taf. LXXIII. fig. d.) thinks he has seen them in copulation. *Baird*, also (Magaz. of Zool. and Bot. I. p. 522), has often seen two individuals of *Cypris* together, but was not sure that they were copulating.*

9 *Lovén*, in *Wiegmann's* Arch. p. 160, Taf. V. fig. 13 (*Evadne*).

10 With the males of *Daphnia*, there is a hook together with a small long lash on the two anterior pairs of feet situated close under the head. The first pair of feet situated on the beak in front of the mouth, is very long and provided with two small pointed hooks ; while, with the females, these feet

have the form of two short, obtuse antennae (*Muller*, Entomostr. p. 87, Tab. XII. fig. 6 ; *Ramdohr*, loc. cit. p. 25, Taf. VII. ; *Straus*, Mém. du Mus. V. p. 419, Pl. XXIX. fig. 18, 19 ; and *Jurine*, Hist. d. Monocles, p. 105, Pl. XI. fig. 5–8). With the males of *Evadne*, only the feet of the first abdominal pair are provided each with a hook and some pretty long bristles on their last two articles (*Lovén*, loc. cit. p. 157, Taf. V. fig. 11). With *Isaura*, on the other hand, the first two pairs of abdominal feet are armed at their extremity with stout nails (*Straus*, Mus. Senckenb. II. p. 123, Taf. VII. fig. 4, 13 ; and *Joly*, loc. cit. p. 298, Pl. VII. fig. 2, 6).

11 *Schaeffer*, Der fischförm. Kiefenf. fig. 3–11 ; and *Muller*, Zool. danic. Tab. XLVIII. (*Branchipus*). The frontal digitiform processes are especially developed with *Chirocephalus* ; see *Prevost*, in *Jurine's* Hist. d. Monocl. p. 202, Pl. XXII.†

12 *Van der Hoeven*, loc. cit. p. 20, Pl. II. fig. 14, 18 (*Limulus*).

* [§ 293, note 8.] For the genital organs of *Cypris*, see *Zenker* (*Muller's* Arch. 1850, p. 191). They closely resemble those of *Cyclops*. He has also described the spermatophores (Taf. V. fig. 6) ; they are probably the very large spermatic particles seen by *Wagner* as mentioned above. These observations I have recently confirmed. *Wagner* and *Leuckart* (Cyclop. Anat. and Physiol. Art. *Semen*, p. 496, note) must, therefore, be mistaken, when they assert the hermaphroditic nature of *Cypris*, and say, "We beg to direct the attention to the simultaneous appearance of egg together with the spermatozoa in the same individual ; and therefore to the hermaphroditic condition of the genitals in *Cypris*." It is probable that they observed only females, and if what they called such

were really spermatic particles, the time of observation must have been soon after copulation. — Ed.

† [§ 293, note 11.] For the details of the male genital organs of *Artemia* and *Branchipus*, see *Leydig* (loc. cit. Siebold und Kölliker's Zeitsch. III. p. 297). With these Phyllopods, these organs consist of testes, vasa deferentia, and penises ; all of which are double and symmetrical. The testes consist, each, of an oblong pouch which is directly continuous into its vas deferens ; and this last passes into its penis. The two penises are situated at the base of the abdomen, and point, long and naked, backwards. Besides these parts, there is an external organ, style-like, used in copulation (loc. cit. Taf. VIII. fig. 4, n.). — Ed.

which, with the Idotheoïdae, and Asellina, are two pairs of similar seminal tubes. The two *Vasa deferentia* converge towards the posterior portion of the body, where they pass into a double, or a single excretory canal, which usually commences directly in front of the first pair of post-abdominal feet, on the median line of the body.[13] With the Isopoda, this excretory canal opens into a short, backwardly-curved penis, upon which are two long processes (secondary penises) inserted on the internal border of the second pair of feet.[14]

With the Stomapoda, the testicles consist of more or less ramified, glandular lobes, from which pass off, laterally, two *Vasa deferentia* which terminate in two hollow penises projecting at the base of the last pair of feet.[15]

With the Brachyura, and the short-tailed Anomura, the two testicles consist of a net-work of very small semeniferous canals, occupying the lateral portions of the cephalothorax, which gradually increase in size until they pass into the long *Vasa deferentia*. These last form numerous convolutions, and are finally continuous into two larger *Ductus ejaculatorii*.[16] With the male Paguridae, the testicles are contained, like the ovaries, in the tail. They consist of two large tubes which rapidly contract into a *Vas deferens*, which is straight, at first, but afterwards spiral. This then becomes larger and is gradually continuous into a *Ductus ejaculatorius*.[17]

With some Macrura, the cephalothorax contains two anterior and two posterior testicular tubes, a portion of the last being extended even into the tail ; while the first are connected, by a transverse anastomosis, behind the middle of the body. The two posterior join with the two anterior in the posterior extremity of the cephalothorax, and form on each side, a short, narrow, deferent canal, which terminates in a larger *Ductus ejaculatorius*.[18]

---

13 With *Cyamus*, whose caudal extremity is atrophied, the orifices of the two excretory ducts are situated directly in front of the arms on two, side by side papilliform penises (*Roussel de l'auzéme*, loc. cit. p. 252, Pl. VIII. fig. 7, 15). With *Aega*, the two testicular tubes are curved S-like on the sides of the œsophagus. Their deferent canals are dilated at the posterior extremity each into a seminal vesicle of the same S-like form. They open through two approximated papillae situated on the under surface of the last foot-bearing abdominal segment (*Rathké*, Nov. Act. Nat. Cur. XX. p. 32, Tab. VI. fig. 16).

The three testicles which are found on each side of the thorax with *Idothea*, *Lygia*, *Lygidium*, *Asellus*, *Porcellio*, *Oniscus*, &c., are very attenuated in front, but behind, are enlarged into a kind of bulb before passing into the *Vas deferens*. *Carolini* (loc. cit. p. 155) has already carefully described these with *Lygia oceanica*. See, moreover, *Milne Edwards*, Hist. d. Crust. Pl. XII. fig. 13 (*Lygia*); *Brandt*, Medīz. Zool. II. p. 76. Taf. XV. fig. 31 (*Oniscus*), and *Lereboullet*, loc. cit. p. 132, Pl. V. fig. 134 (*Lygidium*).

14 The copulatory organs of the Asellina have been described and figured by *Brandt* (loc. cit. p. 73 and Taf. XV. fig. N. V. Z.). *Treviranus*, also (Verm. Schrift. I. p. 59, 74, Taf. VIII. fig. 48, 49, Taf. XII. fig. 65–67), has well represented them with *Porcellio*, and *Asellus*, although he entirely overlooked the six seminal tubes of these Crustacea. The penis, and its auxiliary stalks, which, with the Isopoda are always concealed in the midst of the branchial lamellae, have been figured by *Degeer* (Abhandl. zur Geschichte d. Insekt. VII. p. 191, Taf. XXXII. fig. 6, 20), and by *Rathke* (loc. cit. p. 125,

Taf. IV. fig. 16, 17, f. h. 25) with *Idothea entomon ;* but this last author is quite mistaken about the internal genital organs, having confounded the male with the female (loc. cit. p. 123, fig. 22). I have already corrected this error in *Muller's* Arch. 1837, p. 434. *Savigny*, also (Descript. de l'Égypte, Crust. Pl. XII. XIII.), has given beautiful figures of the copulatory organs of *Sphaeroma*, *Lygia*, *Idothea*, *Tylos*, and *Oniscus*. The secondary or auxiliary penises have been represented by *Lereboullet*, loc. cit. p. 120, Pl. V. fig. 19 (*Lygidium*), and by *Milne Edwards*, Arch. d. Mus. II. p. 21, Pl. II. fig. 3.* b.¹. (*Serolis*), and Ann. d. Sc. Nat. XV. 1841, Pl. VI. fig. 4 (*Lygia*).

15 As to both form and position, the multilobular testicles of *Squilla* almost exactly resemble the ovaries. But their lateral lobes are not blended together at the anterior extremity of the body, and the two deferent canals are given off laterally (*Delle Chiaje*, Descriz. &c. Tav. LXXXVI. fig. 4). The two penises of these Crustacea have been correctly figured in *Desmaret's* Considérat. &c. Pl. XLII. n. n.

For the male organs of *Mysis*, of which the testicles are composed of only a few lobes, see *Frey*, loc. cit. p. 26.

16 *Cavolini*, loc. cit. p. 144, and *Milne Edwards*, Hist. d. Crust. I. p. 166, and *Cyclop*, loc. cit. p. 783, fig. 418 (*Cancer pagurus*).

17 *Swammerdamm*, loc. cit. p. 86, Taf. XI. fig. 6, and *Delle Chiaje*, Descriz. &c. Tav. LXXXVI. fig. 6.

18 *Milne Edwards*, Hist. d. Crust. Pl. XII. fig. 15 (*Homarus*), and *Delle Chiaje*, loc. cit. Tav. LXXXVII. fig. 6 (*Scyllarus*).

With other Macrura, the testicles consist only of a trilobed glandular mass covering the pyloric portion of the stomach, and from which pass off two long, very flexuous *Vasa deferentia*, which are dilated, near their extremity, into a nearly straight *Ductus ejaculatorius.*[19] The excretory ducts of the sperm are very distinct with the Decapoda, when filled with this fluid, from their chalk-white color. With many species, the sperm, as it approaches the end of these ducts, is divided into portions, around which capsules or spermatophores are developed.

These last are usually pyriform, and connected together by a common band.[20] The external genital organs of the male Decapoda are quite varied, although these excretory ducts almost invariably open on the coxal joint of the last pair of feet.[21] With the Paguridae, and Macrura, the male genital orifices are surrounded by a soft sphincter, without any trace of a penis, but out of which the *Ductus ejaculatorius* is perhaps protruded during copulation.[22] But with the Brachyura and short-tailed Anomura, on the contrary, there are two longer or shorter tubular penises, always covered by the tail, which is pressed against the belly.[23] With very many Decapoda, the two feet of the first caudal segment are transformed into pedicellated processes (secondary penises), the extremity of which is sometimes grooved. With some short-tailed Anomura, the feet of the second post-abdominal pair take part also in the act of copulation, and, for this purpose, are prolonged into stalk-like organs.[24]

Among the Myriapoda, the Glomerina have two testicular tubes extending into the abdomen and composed of numerous vesicles partially blended together. They unite in the thorax into a common *Vas deferens*. With the Julidae, the testicles have a similar structure, but the vesicles open separately into the external side of the two *Vasa deferentia*, which are close together, and are connected, in a ladder-like manner, by numerous trans-

19 With *Astacus*; see *Roesel*, loc. cit. Taf. LVIII. fig. 9, and Taf. LX. fig. 23; *Suckow*, loc. cit. Taf. X. fig. 15; *Brandt*, Mediz. Zool. II. Taf. XI. fig. 14; *Milne Edwards*, Hist. d. Crust. Pl. XII. fig. 14; and *Carus*, Erläuterungstaf. Heft. V. Taf. III. fig. 9.

20 These spermatophores, first made known by *Kölliker*, are bound together, with *Galathea*, by ramified pedicles; and with *Pagurus*, by simple filaments; see *Kölliker*, Beitr. zur Kenntniss d. Geschlechtsv. &c. p. 9, fig. 21, 22, also, Schweiz. Denksch. VIII. p. 52, fig. 32–35. See, also, the description which I have given of the spermatophores of *Pagurus Bernhardus*, in Muller's Arch. 1842, p. 136, note 1. But one must be careful not to take, in the testicles of the Decapoda, the mother-cells in which are developed the radiating cells for the spermatophores.

21 The land crabs make an exception in this respect, their male genital orifices being situated on the last segment of the body; see *Milne Edwards*, Hist. Nat. d. Crust. I. p. 168, Pl. XVIII. fig. 6 (*Gecarcinus*).

22 For *Astacus*, see the figures cited above; for *Palinurus*, *Milne Edwards*, Hist. d. Crust. Pl. XXIII. fig. 2; and for *Hippolyte*, Kröyer, loc. cit. p. 27, fig. 54, B. f.

23 There are two very short, and soft penises with *Maia*, *Pisa*, *Cancer*, *Grapsus*, *Lupea*, *Gecarcinus*, *Porcellana*, *Homola*, &c. They are long, hard, and point forwards with *Dromia*.

24 The canaliculated, secondary penises may be very easily seen on the first caudal segment of *Homarus*, *Nephrops* and *Astacus*; see *Roesel*, loc. cit. Taf. LVI.; and *Carus*, Erläuterungstaf. Heft.

V. Taf. III. fig. 12 (*Astacus*). In this last genus, these organs are slightly spiral at their extremity. These penises are long, secondary, and concealed under the tail with the male Brachyura and Anomura, with which the majority of the other anal feet are wanting; see *Milne Edwards*, Hist. d. Crust. I. p. 169, Pl. III. fig. 6, 15, 16 (*Maia*), in this genus the two pairs of anal feet are rudimentary. This abortion is observed, also, with *Grapsus*, *Cancer*, *Lupea*, *Ocypoda*, *Porcellana*, &c. See the beautiful figures of *Savigny*, in Descript. de l'Égypte, Crust. Pl. II.-VII., and *Carolini*, loc. cit. Taf. II. fig. 10 (*Grapsus*). With *Dromia*, the two feet of the second caudal segment have the form of two long spines. With *Homola*, the same feet are equally pedicellated, but terminate with a kind of sucker, and, therefore, are undoubtedly auxiliary in the act of copulation. No auxiliary organs have been found with *Galathea*, *Palinurus*, and *Scyllarus*; but in the last two of these genera the feet of the first caudal segment are wholly wanting. With the Caridoidae, the copulatory organs are usually absent, and the first pair of anal feet does not differ from the others; with *Crangon*, only, have I found the internal prolongation of these feet highly developed and glabrous; while with the posterior feet, it is very small, and, like the external one, very hairy. According to *Joly* (loc. cit. p. 43, Pl. III. fig. 20), it is somewhat similar with *Caridina*. Kröyer (loc. cit. p. 27, Pl. II. fig. 54, B. g.) has observed, with *Hippolyte*, between the feet of the fourth pair, two short hooked appendages which may be regarded as secondary penises.

verse anastomoses. In front, the testicular vesicles are lost in these canals, which finally diverge from each other in an arcuate manner, as is also true of the *Vas deferens* of the Glomerina. In this manner, these canals, as two *Ductus ejaculatorii*, extend to a triangular scale situated under the third thoracic segment, and terminate at the lower angles of this scale in two short, conical, penis-like protuberances.[24]

With the Chilopoda, the male organs are very complicated and formed upon a wholly different type. Their orifices are always situated at the posterior extremity of the abdomen. With some species, there is only a single, long, testicular tube into which pass two lateral, also very long, coecal tubes (*Epididymes ?*). At their point of junction, arise two short *Vasa deferentia*, which terminate in a common, short, campanulate penis. Other Chilopoda have two to three varicose testicular tubes which anastomose, loop-like, at both of their extremities, and terminate, at last, in a longer or shorter *Vas deferens*, which bifurcates in its course, but its branches come together again in a short penis. With all the Chilopoda, the common genital orifice is connected with the short excretory ducts of two to four oblong accessory glands, the nature of which is yet unknown.[25]

§ 294.

The Development of the Crustacea occurs, as with all Arthropoda, according to a special type.[1]

After the disappearance of the germinative vesicle, a partial segmentation occurs upon a given point of the surface of the vitellus. By this process, a transparent, finely-granular, proligerous disc is formed.[2] The borders of this disc gradually extend over, and finally cover the surface of the vitellus. It is then changed into a proligerous vesicle enclosing the remainder of the vitellus.

At the pole of the egg where the proligerous disc is first formed, are de-

24 For the male organs of the Chilognatha, see *Newport*, Philos. Trans. 1842, loc. cit. p. 99 ; *Rymer Jones*, Cyclop. III. p. 551, fig. 314 ; and *Stein*, in *Müller's* Arch. 1842, p. 246, Taf. XII.–XIV. (*Julus, Polydesmus,* and *Glomeris*). The two testicles of *Glomeris* were formerly described as ovaries by *Brandt* ; see his Beitr. loc. cit. p. 325, Taf. XII. fig. 8 ; but he has rectified this in his Recueil, loc. cit. p. 157. For the copulatory organs of the Julidae, may be cited, also, the researches of *Latreille*, and *Savi* (loc. cit.).

25 *Lithobius* has only a single testicular tube with two epididymes and four accessory glands (*Treviranus*, Verm. Schrift. II. p. 25, Taf. V. fig. 7 ; *L. Dufour*, loc. cit. p. 87, Pl. V. fig. 2, 3, and *Stein*, loc. cit. p. 240, Taf. XII. fig. 1). *Geophilus* has three interanastomosing, varicose testicles, and two accessory glands (*Stein*, loc. cit. p. 243, Taf. XII. fig. 7). Judging from *Müller's* figure (loc. cit. Taf. II. fig. 5), *Scolopendra morsitans* has also two anastomosing varicose testicles. But this point is made somewhat uncertain from the researches of *Kutorga* (loc. cit. p. 10, Tab. II. fig. 4–6), who has shown positively the existence of four accessory glands with this animal. *L. Dufour's* figures (loc. cit. p. 97, Pl. V. fig. 5) of the male organs of *Scutigera* indicate here a very different organization. There are two testicular tubes which unite loop-like at the anterior extremity and then send off a long very flexuous canal which has two

pedunculated vesicles (*Vesiculae seminales*). The posterior extremity of these testicles is continuous into two *Vasa deferentia* which become dilated into as many *Ductus ejaculatorii*. Perhaps this abnormal organization of these animals in this respect, will be reduced from further researches to the type of the *Scolopendra*.

1 The embryology of Crustacea has been brought out, especially by the numerous and exact researches of *Rathké* ; see his Untersuch. über d. Bild. u. Entwickel. d. Flusskrebses, 1829, then his notes in *Burdach's* Physiol. II. 1837, p. 250 ; his Abhandl. zur Bild. u. Entwickel. d. Mensch. u. d. Thiere, 1833 ; his Mittheilung. über d. Entwickel. d. Decapoden, in *Müller's* Arch. 1836, p. 187, or in *Wiegmann's* Arch. 1840, I. p. 241 ; and in the Neuest. Schrift. d. Danzig. naturf. Gesellsch. III. Heft. IV. 1842, p. 23 ; then, Zur Morphol., Reisebemerk. aus Taurien, 1837 ; his Beobacht. u. Betracht. über d. Entwickel. d. Mysis vulgaris, in *Froriep's* neue Not. XXIV. 1842, p. 181 ; and finally his Comment. de Animal. Crust. generat. 1844. See, also, *Erdl*, Entwickel. d. Hummercies, 1843 ; and *Joly*, Sur le développ. des Caridina, in the Ann. d. Sc. Nat. XIX. loc. cit. p. 57, Pl. IV.

2 *Cancer maenas* forms perhaps the only exception in this respect. Here, the segmentation appears to be complete ; see *Rathké*, in *Froriep's* neue Not. loc. cit. p. 182 ; and *Erdl*, loc. cit. p. 27.

veloped the ventral portion together with the abdominal cord of the future embryo; while, at the opposite pole, where the borders of the disc meet, the dorsal portion of the animal appears. Quite early, the blastoderma can be seen composed of an external or serous, and of an internal or mucous layer. This last, after having enveloped the entire vitellus, is changed gradually into the alimentary canal. The hepatic organs are only deverticuli of this last, while the antennae, the oral apparatus, the feet, and the branchiae, are developed from the serous layer.

The embryos, thus formed, differ considerably, and their form is often so dissimilar from that of the adult animal, that, during their ulterior development, there is a real metamorphosis, which takes place by more or less numerous stages coincident with the act of moulting.

An embryonic type quite general among the lower Crustacea, that is, the Cirripedia, Siphonostoma, Lœmyropoda and Phyllopoda, is that which was first observed with *Cyclops*. There is here a long series of metamorphoses. The monocle-like larvæ have an ovoid, unarticulated body, usually provided with a single, simple eye, and two or three pairs of oar-like appendages covered with long hairs.[3]

With some Brachyura, there is an equally well-marked metamorphosis; for, in leaving the egg, they have a long tail and two very large eyes; but with the first moulting they acquire two enormous, spur-like apophyses, one on the front, and the other on the back.[4]

---

3 It is remarkable that the young Cirripedia, which are hexapod, have the characteristics of the larvae of Monocles; see *Thompson*, Zool. Research. loc. cit. p. 69, Pl. IX. (*Balanus*); *Burmeister*, Beitr. loc. cit. p. 12, Taf. I. (*Lepas*); *Goodsir*, Edinb. New Philos. Jour. No. 69, July, 1843, p. 97, Pl. III. IV., or Isis, 1844, p. 901, Taf. I. fig. 8, 11-17 (*Balanus*). The larvae of these Crustacea, before becoming fixed in order to undergo their metamorphoses, change into a bivalve animal resembling *Cypris*. Among the Siphonostoma, the monocle-like embryos are very general. *Nordmann* (loc. cit. p. 11, &c., Taf. II.-VII.) has recognized larvae of this kind, some with three (*Ergasilus* and *Lernaeocera*), and others with only two (*Achtheres* and *Tracheliastes*) pairs of feet. According to *Kollar* (loc. cit. p. 87, Taf. X. fig. 10), the embryos of *Basanistes* are monocle-like and have six feet, as are also those of *Lernaeopoda* described by *Rathké* (Zur Morphol. &c. p. 34, Taf. I.). *Goodsir* (loc. cit. No. 65, July, 1842, p. 178, Pl. III. fig. 19-23) has observed embryos with four feet in the eggs of *Caligus*. The larvae of *Nicothoë* (*Rathké*, Nov. Act. Nat. Cur. XX. p. 109, Tab. V. fig. 8-10) and of *Argulus* (*Muller*, Entomostr. p. 122, Tab. XX. fig. 2, and *Jurine*, loc. cit. p. 453, Pl. XXVI. fig. 4) form an exception in this respect, for when they leave the egg they have two simple eyes, an articulated body, and more than three pairs of feet. Those of the Cyclopidae which have six feet have been known a long time. But *Muller* (Entomostr. p. 39, Taf. I. II.) formerly divided them under the names of *Nauplius* and *Amymone*. See *Degeer* Abhandl. &c. VII. p. 181, Taf. XXX. (*Cyclops*); *Ramdohr*, loc. cit. p. 5, &c., Taf. I. III.; but especially *Jurine*, Hist. d. Monocl. p. 15, &c., Pl. I.-VII. (*Cyclops* and *Cyclopsina*). The young Daphnioïdae and Cypridoïdae, on the contrary, resemble the adults on their escape from the egg. The simple eye is evidently the result of a very early fusion of two eyes; see *Jurine*, loc. cit. p. 113, Pl. VIII. IX. (*Daphnia* and *Cypris*); *Rathké*, Abhandl. z. Bildungs u. Entwickelungsgesch. &c. p. 85 (*Daphnia* and *Lynceus*); *Baird*, Magaz. of Zool. and Bot. I. p. 522, and II. Pl. V. fig. 12 (*Cypris*); finally *Lovén*, loc. cit. p. 161, Taf. V. fig. 12 (*Evadne*). Of the Phyllopoda, the monocle embryos of the Apodidae have two pairs of feet, while those of the Branchiopoda have three; see *Scharffer*, Der krebsartige Kiefenf. p. 118, Taf. I. fig. 3; and *Zadduch*, loc. cit. p. 55, Tab. IV. fig. 1-3 (*Apus*); *Joly*, loc. cit. p. 321, Pl. IX. fig. 39 (*Isaura*); *Prevost*, in *Jurine's* Hist. d. Monocl. p. 214, Pl. XX. fig. 9 (*Chirocephalus*); and *Joly*, loc. cit. p. 257, Pl. VII. fig. 4 (*Artemia*).[4]

4 These embryos with such singular forms, have, hitherto, been figured as separate genera under the

---

* [§ 294, note 3.] For many highly-interesting details on the economy of the Entomostraca, see *Baird* (British Entomostr. &c. loc. cit. passim). These details with their corresponding figures will render clear many obscure economical points alluded to above. For the embryology of *Argulus*, *Artemia* and *Branchipus*, see *Leydig*, loc. cit. *Siebold* and *Kölliker's* Zeitschrift, II. p. 344, and III. p. 304. The descriptions of this observer are quite rich in details upon the successive appear-ances of the different organs. *Argulus* is quite well developed when hatched, its muscles are transversely striated and the locomotive organs well formed. *Artemia* has, at this period, two antennae, two pairs of feet on the head, and the red pigment spots on the forehead, but these last have as yet no light-refracting body. The muscles are still without striae, and even here and there are filled with vitelline globules. The heart and blood-circulation are still unformed. — Ed.

The young of the Paguridae and Macrura differ more or less from the adult animals.[5]   But this difference is less with the Poecilopoda, Laemodipoda, Stomapoda, Isopoda, and Amphipoda.[6]   Finally, with the Myriapoda, the metamorphosis is limited to the increase of the number of the segments of the body, and of the feet.[7]

names *Megalops, Monolepis* and *Zoëa* (*Milne Edwards*, Hist. d. Crust. II. p. 260, 431), until *Thompson* perceived their true nature; see his Zool. Research. &c. Pl. I. and his Memoir on the double Metamorphosis in the Decapodous Crustacea, in the Philos. Trans. 1835, pt. II. p. 539; see also the Edinb. New Philos. Jour. No. 20, p. 221, and the Entomol. Magaz. No. 14, p. 370. Although these observations have been confirmed from different sides, yet they did not, at first, receive full assent, especially on account of the authority of *Rathké* (*Müller's* Arch. 1836, p. 187), who opposed them.   *Templeton* (Trans. of the Entomol. Soc. II. p. 115, Pl. XII.) and *Westwood* (Philos. Trans. 1835, pt. II. p. 311, Pl. IV.) refuse to give up the genus *Zoëa*; but since *Du Cane* (Ann. of Nat. Hist. III. 1839, p. 438, Pl. XI. or *Froriep's* neue Notiz. XIII. p. 5, fig. 10–13), has verified, with *Cancer maenas*, the observations of *Thompson*, and *Rathké* himself (*Wiegmann's* Arch. 1840, I. p. 246, and Neuest. Danzig. Schrift. loc. cit. p. 39, Taf. IV.) has seen the embryos of *Hyas* under the form of a *Zoëa*, this wonderful metamorphosis of the Brachyura can no longer be doubted.   See also *Steenstrup*, in the Oversigt over det kgl. danske Videnskabernes Selskabs Forhandlinger, 1840, p. 15, or *Müller's* Arch. 1841, p. 218 (*Hyas*), and *Goodsir*, Edinb. New Philos. Jour. No. 65, 1842, p. 181, Pl. III. fig. 16–18 (*Cancer maenas*).

5 The embryos of *Pagurus* which have a frontal spine, were also, before the discovery of *Thompson*, taken for species of *Zoëa*; see *Philippi*, in *Wiegmann's* Arch. 1840, I. p. 184, Taf. III. fig. 7, 8; also *Rathké*, Ibid. p. 242, and, Danzig. Schrift. loc. cit. p. 29, Taf. III.; *Steenstrup*, loc. cit.; and *Goodsir*, loc. cit. No. 65, p. 182, Pl. III. fig. 12–14.   The difference in form between the embryos and the adults is less marked with *Astacus*, *Homarus* and other Macrura; see *Rathké*, Entwick. d. Flusskr.) and in the Danzig. Schrift. loc. cit. p. 24, Taf. II. (*Homarus*); *Du Cane*, Ann. of Nat. Hist. II. 1839, p. 178, Pl. VI. VII. or *Froriep's* neue Notiz. XIII. p. 3, fig. 4–9 (*Palaemon* and *Crangon*); *Kröyer*, Monogr. loc. cit. p. 37, Pl. VI. (*Hippolyte* and *Homarus*); *Joly*, Ann. d. Sc. Nat. XIX. loc. cit. Pl. IV. (*Caridina*), and *Erdl*, loc. cit. p. 18, Taf. III. IV. (*Homarus*).

6 According to *Milne Edwards* (Instit. 1838, No. 258, p. 397), a cephalothorax and abdomen may already be distinguished with the hatching embryos of *Limulus*. But the abdomen has only three pairs of appendages and its long spine is wholly wanting.   This naturalist, also, has figured the embryo of *Cyamus* which closely resembles the adult (Ann. d. Sc. Nat. III. 1835, p. 328, Pl. XIV. fig. 14).

For the embryos of the Isopoda and Amphipoda, see *Rathké*, Abhandl. loc. cit.; Ann. d. Sc. Nat. II. 1834, p. 139, Pl. XI.; Zur Morphol. &c. 41, Taf. II. III. (*Bopyrus, Idothea, Janira, Lygia* and *Amphithoë*); Nov. Act. Nat. Cur. XX. p. 49, Tab. I. (*Phryxus*); also *Milne Edwards*, Ann. d. Sc. Nat. III. 1835, p. 323, Pl. XIV. (*Cymothoa, Anilocra, Phronima* and *Amphithoë*); finally *Rathké*, in *Wiegmann's* Arch. 1839, loc. cit. Taf. VI. (*Mysis*).

7 See *Gervais*, Ann. d. l. Soc. Entomol. de France, 1837, and Institut. 1839, p. 22; *Waga*, Rev. Zool. 1839, No. 3. p. 76, or *Wiegmann's* Arch. 1840, II. p. 351; and especially *Newport*, Philos. Transact. 1842, part II. p. 99, and Cyclop. loc. cit. III. p. 355, fig. 317–326, also in *Froriep's* neue Notiz. XXI. p. 161.

BOOK THIRTEENTH.

ARACHNOIDAE.

CLASSIFICATION.

§ 295.

The Arachnoidae, which are organized after very different types, have always four pairs of feet. The Tardigrada form no exception in this respect; and although it may appear singular to find them placed in this class, yet this seems their most proper place; only they should be placed at the head, for they form the transition of the Arachnoidae to the Annelides, exactly as do the Cirripedia from the Crustacea to the Acephala.*

The Arachnoidae are usually defined as Arthropoda wanting the antennae; this, however, is incorrect, for these organs are not wanting, strictly speaking, but take the place of the mandibles, which are absent, as will be shown hereafter.

ORDER I.

Cephalothorax multi-articulate. Special respiratory organs wanting.

SUB-ORDER I. TARDIGRADA.

Legs rudimentary. Abdomen wanting.

Genera : *Milnesium, Macrobiotus, Emydium.*

SUB-ORDER II. PYCNOGONIDAE.

Legs very much developed. Abdomen rudimentary.

Genera : *Nymphon, Ammothea, Pallene, Phoxichilidium, Pariboea, Endeis, Phoxichilus, Pycnogonum.*

---

* [§ 295.] For a detail of the data which fully justify this position of the Tardigrada, see *Kauf-mann*, Ueber die Entwickelung und systematische Stellung der Tardigraden, in *Siebold* and *Kölliker's* Zeitsch III. 1851, p. 220. — ED.

ORDER II.

Cephalothorax unarticulated, or biarticulated. Respiratory organs consisting of tracheae.

SUB-ORDER III. ACARINA.

Abdomen unarticulated and fused with the cephalothorax. Palpi simple.

FAMILY: ACAREA.

Genera: *Demodex, Sarcoptes, Glycyphagus, Tyroglyphus, Melichares, Dermaleichus, Acarus, Pteroptus.*

FAMILY: HYDRACHNEA.

Genera: *Limnochares, Arrenurus, Eylaïs, Diplodontus, Hydrachna, Atax.*

FAMILY: ORIBATEA.

Genera: *Hoplophora, Oribates, Zetes, Pelops, Damaeus.*

FAMILY: GAMASEA.

Genera: *Dermanyssus, Uropoda, Gamasus, Argas.*

FAMILY: IXODEA.

Genus: *Ixodes.*

FAMILY: BDELLEA.

Genera: *Bdella, Molgus.*

FAMILY: TROMBIDINA.

Genera: *Erythraeus, Trombidium, Smaridia, Tetranychus, Rhyncholophus, Rhaphygnathus, Penthaleus.*

SUB-ORDER IV. OPILIONINA.

Abdomen articulated, but indistinctly separated from the cephalothorax. Palpi simple.

Genera: *Phalangium, Gonyleptes, Eusarcus.*

SUB-ORDER V. PSEUDOSCORPII.

Abdomen articulated, but indistinctly separated from the cephalothorax. Palpi forficulate.

Genera: *Obisium, Chelifer.*

SUB-ORDER VI. SOLPUGIDAE.

Abdomen articulated, distinctly separated from the cephalothorax. Palpi simple.

Genus: *Galeodes.*

31*

ORDER III.

Abdomen and cephalothorax unarticulated, distinct from each other. Respiratory organs consisting of tracheae and lungs.

SUB-ORDER VII. ARANEAE.

Genera : *Mygale, Thomisus, Uptiotes, Lycosa, Dolomedes, Salticus, Segestria, Dysdera, Scytodes, Clubiona, Drassus, Argyroneta, Clotho, Agelena, Lachesis, Tegenaria, Micryphantes, Theridion, Linyphia, Epeira, Tetragnathus.*

ORDER IV.

Abdomen articulated. Cephalothorax unarticulated. Respiratory organs consisting only of lungs.

SUB-ORDER VIII. PHRYNIDAE.

Abdomen distinct from the cephalothorax. Cheliceres unguiculate.

Genera : *Thelyphonus, Phrynus.*

SUB-ORDER IX. SCORPIONIDAE.

Abdomen indistinctly separated from the cephalothorax. Cheliceres forficulate.

Genera : *Scorpio, Buthus, Androctonus.*

BIBLIOGRAPHY.

Leeuwenhoek. Continuatio arcanorum naturae, 1719, Epist. 138, p. 312.

Roesel. Insekten-Belustigungen, Thl. IV. 1761, p. 241.

Degeer. Abhandlungen zur Geschichte der Insekten, VII. 1783.

Hermann. Mémoire aptérologique, 1804.

J. F. Meckel. Beiträge zur vergleichenden Anatomie, I. Hft. 2, 1809, p. 105.

Treviranus. Ueber den inneren Bau der Arachniden, 1812. Vermischte Schrift. anat. u. physiol. Inhalts. I. 1816. Ueber d. Bau d. Nigua (*Acarus americanus*), in his Zeitsch. f. Physiol. IV. 1831, p. 185.

Serres. Mémoires du Mus. d'hist. Nat. V. 1819, p. 86.

Léon Dufour. Ann. génér. d. Sc. physiq. d. Bruxelles, IV.–VI.

J. Müller. Beiträge zur Anatomie des Scorpions, in *Meckel's* Arch. f. Anat. 1828, p. 29.

Lyonet. Anat. de différentes espèc. d'Insectes, in the Mém. du Mus. d'hist. Nat. XVIII. 1829, p. 282, 377.

Brandt. Medizin. Zool. II. 1833, p. 87. Recherch. sur l'anat. d. Araignées, in the Ann. d. Sc. Nat. XIII. 1840, p. 180.

Savigny. Descript. de l'Égypte. Hist. Nat. Arachnides, Pl. I.–IX.

Audouin. Lettre contenant des recherches sur quelques Araignées

parasites, in the Ann. d. Sc. Nat. XXV. 1832, p. 401. Cyclop. of Anat.
I. 1836, p. 196, Art. *Arachnida.*

Dugès. Recherches sur l'ordre des Acariens, in the Ann. d. Sc. Nat.
I. 1834, p. 5, and II. p. 18; also, Sur les Arancides, Ibid. VI. 1836, p.
159.

Walckenaër. Hist. Nat. d. Insectes aptères, I.-III. 1837-44.

Doyère. Sur les Tardigrades, in the Ann. d. Sc. Nat. XIV. 1840, p.
269.

Van der Hoeven. Bijdragen tot de kennis van het geslacht Phrynus,
in the Tijdschrift voor natuurlijke Geschiedenis en Physiologie, IX. 1842,
p. 68.

Grube. Einige Resultate aus Untersuchungen über d. Anat. d. Arancï-
den, in *Müller's* Arch. 1842, p. 296.

Menge. Ueber d. Lebensweise d. Arachniden, in the Neuest. Schrift. d.
naturf. Gesellsch. in Danzig. IV. Hft. 1, 1843, p. 1.

Tulk. Upon the anatomy of Phalangium opilio, in the Ann. of Nat.
Hist, XII. 1843, p. 153, or in *Froriep's* neue Notiz. XXX. 1844, p. 97.

Dujardin. Sur les Acariens, in the Comp. rend. XIX. 1844, p. 1158,
and Mém. sur les Acariens, in the Ann. d. Sc. Nat. III. 1845, p. 5.

Quatrefages. Mém. sur l'organis. d. Pycnogonides, in the Ann. d. Sc.
Nat. IV. 1845, p. 69.

Blanchard. Observat. sur l'organis. d'un type de la classe des Arachnides, le genre Galéode, in the Comp. rend. XXI. 1845, p. 1383.

Wasmann. Beiträge zur Anat. d. Spinnen, in the Abhandl. d. naturwissensch. Vereins in Hamburg, 1846.

ADDITIONAL BIBLIOGRAPHY.

Wilson. Researches into the Structure and Development of a newly-
discovered parasitic animalcule of the Human Skin, — the Entozoon folliculorum, in the Philos. Trans. 1844, p. 305.

Wittich. Dissertatio sistens. observ. quaed. de Arancarum ex ovo
evolut. Halis, 1845.

—— Die Entstehung des Arachnideneies im Eierstocke; die ersten
Vorgänge im demselben nach seinem Verlassen des Mutterkörpers; in
Muller's Arch. Hft. 2, 1849, p. 113.

Leuckart. Ueber den Bau und die Bedeutung der sog. Lungen bei den
Arachniden, in *Siebold* and *Kölliker's* Zeitsch. 1849, I. p. 246.

J. V. Carus. Ueber die Entwickelung des Spinneneies; in *Siebold* and
Kölliker's Zeitsch. II. 1849, p. 97.

Blanchard. De l'Appareil circulatoire et des Organes de la respiration
dans les Arachnides, in the Ann. d. Sc. Nat. XII. 1849, p. 317.

Dufour. Observations sur l'Anatomie du Scorpion, in the Ann. d. Sc.
Nat. XV. 1851, p. 249.

Kaufmann. Ueber die Entwickelung und systematische Stellung der
Tardigraden, in *Siebold* and *Kölliker's* Zeitsch. III. 1851, p. 220. — ED.

CHAPTER I.

EXTERNAL ENVELOPE AND CUTANEOUS SKELETON.

§ 296.

The external envelope of the Arachnoidae is usually soft, or coriaceous, rarely horny;[1] but in no instance does it possess a proper contractility. In place of this, however, it is extensible in the highest degree with many species. This extensibility is seen especially with those species which are accustomed to long fasts, having only an occasional opportunity to fill their digestive canal with food consisting of the animal juices.[2]

The envelope is composed here, as with all the Arthropoda, chiefly of chitine.[3] To this last are undoubtedly due its solidity and indestructibility, which may be observed with the small and delicate Acarina and Tardigrada, not only when it is in a fresh state, but even after it has been cast off by a kind of moulting.[4]

§ 297.

With most Arachnoidae, the cutaneous envelope may be separated into two tunics; an external and an internal. The first is the more solid and thick, and, in the cephalothorax and the extremities, has often a cellular structure. Upon the abdomen of the Araneae and Acarina, it presents peculiar, waving markings which, as concentric rings, surround the base of the hairs;[1] but it is difficult to determine if they are due to delicate plicae, or the effect of the intimate structure of the skin. With *Ixodes*, only, these prominent lines appear, unmistakably, as folds of the epidermis, for they completely disappear when these animals are gorged with food.

The epidermis is often provided with papillae, clavate excrescences, spines, bristles, simple or plumose hairs, and even, sometimes, with scales.[2] These various cutaneous formations, which are usually hollow, either occupy only certain points, or are extended over the whole surface of the body, giving it a velvety or a furry aspect.

The internal tunic of the skin consists of a thin, always colorless membrane, finely granular or fibrillated, which is perforated at those points where there are hair-like or other formations of the epidermis.[3] Directly beneath this membrane, which, undoubtedly, reproduces the epidermis after

---

1 For example, with the Scorpionidae and Phrynidae. The cutaneous envelope is hardest and most fragile with the Oribatea, where it breaks like glass from the lightest pressure.

2 For example, with *Ixodes*, and *Argas*, as also with the parasitic larvae of certain Hydrachnea and Trombidina, known under the names of *Achlysia* and *Leptus.*

3 *Lassaigne*, Compt. rend. XVI. 1843, No. 19, or *Froriep's* neue Not. XXVII. p. 8, and *Schmidt*, Zur vergleich. Physiol. p. 47.

4 This solidity of the skin with the Tardigrada, is one evidence that these animals are more properly classed with the Arachnoidae, instead of with the worms whose skin contains no chitine and is, therefore, quickly dissolved in caustic potass. See the analyses of the skin of the earth-worm by *Las-*

saigne (loc. cit.), and of that of *Ascaris*, *Mecklia*, *Sabella*, *Hermione* and *Nephtys*, made by *Loewig* and *Kölliker* (Ann. d. Sc. Nat. V. 1846, p. 198).

1 For example, with *Epeira*, *Segestria*, *Thomisus*, *Argyroneta*, *Salticus*, *Sarcoptes*, &c.

2 Plumose hairs are very often found with the Araneae; and I have found lanceolate scales with *Salticus*, and clavate excrescences with the Trombidina; see *Hermann*, loc. cit. Pl. III. fig. O—Y.

3 I am unable to say whether the internal membrane is prolonged at these points into the hollow excrescences of the skin, or whether the appearances alluded to are not produced artificially when the outer is separated from the inner layer of the skin.

moulting, is a layer of colored vesicles and granules, which can be seen through the skin, giving it the often very beautiful colors which are observed in many species.

The various divisions of the cutaneous skeleton, of which the number is quite limited, have been so thoroughly studied in zoology, that they may well be passed over here without notice.

The Cephalothorax sends off from its inner surface, especially with the Opilionina, and Araneae, various processes, which serve, as with the Crustacea, as points for the insertion of muscles, and as septa between certain organs. With the Araneae, they form, at the bottom of the cephalothorax, a solid horizontal plate, — a kind of internal skeleton, which, before and behind, is attached to the sternum by two tendinous ligaments. This plate is deeply indentated on its anterior border, and furnishes points of insertion for the muscles of the extremities, as well as for several other parts.[4]

CHAPTER II.

MUSCULAR SYSTEM AND ORGANS OF LOCOMOTION.

§ 298.

The voluntary muscles of the Arachnoidae are of a dirty-yellow color, and, like those of the Crustacea, are distinctly striated transversely.[1] Their general disposition agrees, also, with those of Crustacea.[2]

The principal muscular masses are found in the cephalothorax, for here arise, not only the muscles of the parts of the mouth, but also those of the first article of the tactile organs and legs. With those species having an unarticulated abdomen, the muscles of this part of the body differ from those of Crustacea. For, directly beneath the skin, is a thin layer, composed of numerous short, riband-like fibres, interlaced in various directions, and frequently anastomosing with each other.[3] Moreover, with many species, there are, on both the dorsal and the ventral surface of the abdomen, depressions of the skin, from which pass off small muscular bands, which penetrate into the interior of the abdomen, and pass among the viscera.

With the Araneae there is, generally, on each side of the ventral median

---

4 This plate, already recognized by *Lyonet* (loc. cit. p. 405, Pl. XXI. fig. 26), and by *Treviranus* (Bau d. Arach. Taf. 11. fig. 24), has been described more exactly by *Wasmann* (loc. cit. p. 2, fig. 2–4). A similar, but rudimentary plate, exists, perhaps, with *Phalangium*, and, as it lies under the ventral cord, the muscles have the appearance of rising from this last ; see *Tulk*, loc. cit. p. 325, or in *Froriep's* neue Notiz. XXX. p. 136.

1 The Tardigrada form an exception in this respect, their muscles being smooth ; see *Doyère*, loc. cit. p. 336.

2 For the disposition and arrangement of the muscular system of Scorpionidae and Araneae, see

Meckel, Syst. d. vergleich. Anat. III. p. 47 ; and for the muscles of *Phalangium* and *Myzale*, see *Tulk*, and *Wasmann*, loc. cit. The very complicated muscular system of the Tardigrada is quite apparent from the transparency of these animals ; see *Doyère*, loc. cit. p. 335, Pl. XVII.–XIX.

3 This cutaneous layer, already observed by *Treviranus* (Verm. Schrift. 1. p. 9, Taf. 1. fig. 3, a. n.), and by *Brandt* (Mediz. Zool. 1. p. 88, Taf. XV. fig. 8, a. a., or Ann. d. Sc. Nat. XIII. p. 180, Pl. IV. fig. 1, a. a.), with *Epeira*, has been confirmed by *Tulk* (loc. cit. p. 154) with *Phalangium*, and described in more detail by *Wasmann* (loc. cit. p. 8, fig. 7, 8) with *Mygale*.

line, a tendinous ligament, on which are inserted several of these muscular bands.[1] It is very probable that these animals can, by this apparatus, compress their abdomen in various directions.

§ 299.

The locomotive organs of the Arachnoidae are situated exclusively on the cephalothorax. They consist of only four pairs of legs, of which the first may, perhaps, be regarded as the posterior pair of metamorphosed maxillae.[1]

Some Mites, only, when young, have six feet, and the young of the Pycnogonidae have, also, only four. With *Phrynus*, and *Thelyphonus*, the first pair considerably resembles two multi-articulated tactile organs; but with *Galeodes*, these same organs have wholly the appearance of legs, excepting they are without claws. With *Mygale*, the maxillae of the first pair have the form of feet, and their extremity is not only unguiculated, but also provided with a tarsus. The other Arachnoidae have usually nails on all their feet, and, with some, each foot may have four nails.[2] With many Araneae, the nails have, on their convex side, a pectinated appendage.[3]

As to the types of the articulations of the legs, they are usually as follows; first, a movable *Coxa*; then a short *Trochanter*; then a longer, stiff *Femur*; then a *Tibia*, divided by an articulation into two unequal parts; and, finally, a *Tarsus*, composed of a long and a short article. With the Phrynidae, not only are the first and antenniform pair of feet already mentioned, different from this type, but the three other pairs have a great number of articles, each tarsus having four. But the Phalangidae differ the most,—the tarsi of all the feet having an extraordinary number of articles. On the other hand, among the lower Arachnoidae, and especially with the Acarina and Tardigrada, there are species with which the seven articles just mentioned cannot be easily distinguished, for the articulations are less in number, or wholly indistinct. With many of these species, some of the pairs of legs, or even all, are reduced to real foot-stumps.[4] Numerous parasitic Acarina have, between the nails, a small organ (*Arolium*), by which, as with a sucker, they can attach themselves to foreign bodies.[5] These organs are most developed with *Sarcoptes* and allied genera, which are without nails, for they here consist of a long, pedunculated disc upon all, or only upon some of the feet.[6] With the aquatic Hydrachnea, the swimming feet have no other peculiarities than that one of their sides is thickly pilose.

4 For these muscles, the cutaneous insertions of which, with the Araneae, with *Chelifer* and *Phalangium*, have been taken by *Treviranus* (Bau d. Arach. p. 23, Taf. II. fig. 17–19, Taf. III. fig. 28, and Verm. Schrift. I. p. 18, 33, Taf. II.) for the stigmata, see *Brandt*, Mediz. Zool. loc. cit. p. 88, Taf. XV. fig. 8, c. e., and Ann. d. Sc. Nat. loc. cit., and *Wasmann*, loc. cit. p. 3. fig. 1, 6, 24.

1 See *Dugès*, Ann. d. Sc. Nat. 1. p. 7, and *Erichson*, Entomogr. Heft. I. p. 7.

2 Most usually there are two nails to each foot; but *Phalangium*, *Hoplophora*, and *Damaeus*, have only one; while *Segestria*, *Lachesis*, and *Clotho*, as well as *Demodex*, *Pteops*, *Zetes*, and *Oribates*, have three, and *Eusydium* and *Macrobiotus* have even four.

3 See the figures of *Savigny*, loc. cit.

4 The articulations are few and indistinct with all the eight legs of *Tyroglyphus* and *Glyciphagus*, but with the anterior legs, only, with *Sarcoptes*. The posterior legs of this last genus, and all of them with the Tardigrada, and with *Demodex folliculorum*, are only simple stumps.

5 For example, with *Ixodes*, *Argas*, *Dermanyssus*, *Pteroptus*, &c.

6 With *Sarcoptes ovis* and *cati*, this arolium is absent with the penultimate pair of legs; and with *Sarcoptes equi*, with the last pair. With *Sarcoptes cynotis*, *rupicaprae*, and *scabiei*, it is wanting with the last two pairs. With *Sarcoptes hippopodes*, *Glyciphagus prunorum*, and *Melichares agilis*, all the legs have long pedunculated organs of this kind; see *Hering*, Die Krätzmilben der Thiere, in the Nov. Act. Nat. Cur XVIII. part II. Tab. XLIII.–XLV.

CHAPTER III.

NERVOUS SYSTEM.

§ 300.

The grades of development of the Nervous System with the Arachnoidae are very different, being connected with the divisions of the cutaneous skeleton. For, when these last disappear, those of the nervous system belonging to them, and often the ventral cord, are concentrated, as with the brachyurous Decapoda, into a single ganglionic mass, occupying the ventral portion of the cephalothorax; while, if the body is multi-articulate, this system resembles that of the macrurous Decapoda. In both cases, with only a few exceptions, there is a cerebral ganglion situated above the œsophagus, and connected with the ventral cord by two short commissures surrounding this canal. From this ganglion pass off nerves to the eyes, and the maxillary palpi or so-called mandibles; while the first pair of maxillae, changed into tactile organs, receive their nerves from the anterior extremity of the ventral cord.

The intimate structure of the nervous system, with the Arachnoidae, consists of primitive fibres much finer, and ganglionic globules much smaller, than those of Crustacea.[1] As to the direction and disposition of these fibres, those of the Scorpionidae almost exactly resemble those of the Myriapoda.[2]

§ 301.

The nervous system is most simple in its organization with the Acarina. In those species where, as yet, it has been found,[1] it consists only of a simple abdominal ganglion, from which pass off, from all sides, the peripheric nerves; and, upon the upper surface of which, is detached a simple transverse band, under which the œsophagus passes.

With the Tardigrada, this system is a little more developed, although the brain is still wanting.[2] It consists of four ganglia, corresponding to the four segments of the body, and connected together by double longitudinal commissures. Between each of the ganglia, the commissures are connected by a transverse filament. The nerves which proceed from the ganglia belong to the muscles; but the first ganglion sends, moreover, in front, four larger trunks, which are the nerves of sense, and are distributed to the eyes and palpi.[3]

---

1 *Hannover*, loc. cit. p. 71, Pl. VI. fig. 83, 84.
2 See § 271, and *Newport*, Philos. Trans. 1843, loc. cit.
1 With many small Acarina, particularly *Sarcoptes* and *Demodex*, no traces of a nervous system have been found, notwithstanding the most careful researches; but this is not surprising, considering the minuteness of these animals.
2 *Treviranus* (Verm. Schrift. 1. p. 47, fig. 32) has investigated the nervous system of *Trombidium*, and the results he obtained have been

confirmed with this genus and with *Limnochares*, by *Dujardin* (Ann. d. Sc. Nat? III. p. 19). Subsequently, *Treviranus* (Zeitsch. f. Physiol. loc. cit. p. 189, Taf. XVI. fig. 7. c.) has also confirmed, with *Ixodes*, this passage of the œsophagus through the principal ganglionic mass. With *Trombidium*, whose ganglion is somewhat reddish, the cerebral commissure is quite distinct.
3 See *Doyère*, loc. cit. p. 343, Pl. XVII. (*Milnesium*).

The ventral chain of the Pycnogonidae is composed likewise of four ganglia, but these, which send off, each, a nerve from its side to the corresponding foot, are contiguous, and the first connects with the ovoid cerebral ganglion by two lateral commissures.[4]

With the Araneae, the central portion of the nervous system consists of a large sub-œsophageal ganglion, and another, smaller and above the œsophagus. They are separated from each other only by a narrow fissure through which the œsophagus passes. The super-œsophageal ganglion, which is somewhat emarginated in front, corresponds to the brain, and sends off nerves to the eyes and cheliceres. The sub-œsophageal ganglion, situated in the middle of the cephalothorax, sends off, on each side, four larger processes, from which arise the nerves of the feet. Its anterior border supplies, moreover, the nerves of the two palpi; and from its posterior margin pass off two nerves for the abdominal viscera.[5]

The nervous system of *Galeodes*,[6] *Phrynus*, and *Thelyphonus*,[7] has a like disposition.

The central mass of the nervous system of the Phalangidae begins by two conical, contiguous cerebral ganglia, which connect with a sub-œsophageal, fused ganglion, by two short lateral commissures. This ventral ganglion is composed of a transverse portion, which is situated in the centre of the cephalothorax, and of two lateral portions which consist, each, of an anterior or larger, and a posterior or smaller lobe. These lobes send off nerves to the eight legs, and in front, others to the palpi; while from the posterior border of the transverse portion pass off several nerves to the viscera of the abdomen.[8]

With the Scorpionidae, the nervous system is very highly developed. The brain, which is not large, is composed of two spheroidal, super-œsophageal ganglia fused together. Above, and in front, they send off nerves to the eyes and the cheliceres; and below, they connect with the first ventral ganglion by two short, large filaments, which embrace the œsophagus. The first ventral ganglion is pretty large, being the result, probably, of the fusion of several ganglia. It is situated in the middle of the cephalothorax, and sends nerves to the palpi and to the eight legs. In the rest of the body there are three ventral ganglia, smaller, and followed by four others situated in the tail. All these ganglia are connected by double, longitudinal commissures, and the posterior seven give off, from each side, two nerves; while from the last ganglion arise also two others, which, passing backwards, soon unite and extend to the very extremity of the tail, sending off nerves right and left.[9]

---

4 *Quatrefages*, loc. cit., 77, Pl. I. fig. 1ᴬ 2ᵃ; also Pl. II. fig. 2, 3 (*Ammothea* and *Phorichilus*).

5 *Treviranus*, Ueber d. inn. Bau d. Arach. p. 44, Taf. V. fig. 45, and Zeitsch. f. Physiol. IV. p. 94, Taf. VI. fig. 4 ; *Lyonet*, loc. cit. p. 405, Pl. XXI. fig. 22 ; *Brandt*, Mediz. Zool. II. p. 90, Taf. XV. fig. 3, 4, or Ann. d. Sc. Nat. XIII. p. 184, Pl. IV. fig. 4. *Dugès*, Ibid. VI. p. 174 ; *Grube*, loc. cit. p. 392, and finally *Owen*, Lectures, &c., p. 255, fig. 109. This last author has represented, in a very instructive manner, the nervous system of a *Mygale* seen in profile.

6 *Blanchard*, loc. cit. p. 1384.

7 *Van der Hoeven*, Tijdschrift. loc. cit. IX. 1842, p. 68, and X. 1843, p. 369.

8 The nervous system had already been partially described by *Treviranus* (Verm. Schrift. I. p. 38, Taf. IV. fig. 24) ; but especially, and with full details, by *Tulk*, loc. cit. p. 324, Pl. V. fig. 31.

9 For the nervous system of the Scorpionidae, see *Treviranus* (Bau. d. Arach. p. 14, Taf. I. fig. 13, and Zeitschrift f. Physiol. IV. p. 89, Taf. VI. fig. 1–3, and *Muller*, loc. cit. p. 60, Taf. I. fig. 5, 7) ; but especially *Newport's* excellent description (Philos. Trans. 1843, p. 260, Pl. XII.) ; he has traced, with *Androctonus*, the nerves of the extremities even into the tarsal articles and terminal hooks.*

* [§ 301, note 9.] See also *Dufour* (Ann. d. Sc. Nat. XV. 1851, p. 250). This anatomist has found a fourth abdominal ganglion, situated just behind the thoracic mass, from which passes off a pair of nerves to the pulmonary organs (*Scorpio occitanus*). — ED.

§ 302.

A Splanchnic nervous system has been observed with only the higher Arachnoidae ; but here it is highly developed. The odd stomachic nerve has been the part most difficult to discover ; it is observed, however, with some Araneae, — the posterior border of the brain sending off two small filaments which traverse the central opening of the stomach but unite on its dorsal surface.[1] The Scorpionidae have a similar stomachic nerve which also arises from the brain by two filaments which have a small ganglion at the point of their union.[2]

With the Phalangidae, Araneae, Galeodea, and Phrynidae, the splanchnic nerves are very distinct. They arise from the posterior border of the ventral nervous mass situated in the cephalothorax, and are distributed to the digestive, respiratory, circulatory, and genital organs, and have, sometimes, ganglia on their course. With *Phalangium*, there are three of these nerves arising from the posterior border of the transverse portion of the ventral mass. The middle one of these nerves divides into two branches, which dilate into two ganglia connected together by a transverse anastomosis. From these two ganglia arises a nervous plexus, which is distributed to the internal genital organs, and to the corium. The lateral nerves, directly after their origin, likewise divide into two branches, each of which forms a ganglion ; the external nerves after a shorter, and the internal after a longer course. The two external ganglia thus formed send filaments to the terminal portion of the genital organs, while those of the two internal ganglia are distributed to the digestive tube and neighboring organs.[3] With the Araneae, the Galeodea, and Phrynidae, the posterior extremity of the principal ventral ganglion sends off two considerable nervous cords, contiguous, which pass into the abdominal cavity where they are distributed, radiatingly, to the digestive organs, to the pulmonary sacs, to the genital organs, and to other abdominal viscera. Sometimes, before dividing, they unite in a common ganglion.[4]

1 This *Nervus sympathicus recurrens* was discovered by *Brandt*, with *Epeira* ; see Mediz. Zool. II. p. 90, Taf. XV. fig. 4, d., and fig. 6, e., or in the Isis, 1831, p. 1105, Taf. VII. fig. 6, b., and Bemerk. üb. d. Mundmagennerven, loc. cit. p. 15, or Ann. d. Sc. Nat. V. p. 94, and XIII. p. 185, Pl. IV. fig. 2, c. This same nerve has been refound by *Grube* (loc. cit. p. 302), with other indigenous Araneae. With *Mygale*, according to *Dugès* (Ann. d. Sc. Nat. VI. p. 175), there are, instead of two simple filaments, two lateral ganglionic net-works, from the brain to the stomach.

2 See *Newport*, loc. cit.*

3 See *Treviranus*, Verm. Schrift. I. p. 38, Taf.

IV. fig. 24, and *Tulk*, loc. cit. p. 325, Pl. V. fig. 31, 33.

4 This ganglion has been observed by *Treviranus* (Bau d. Arach. p. 45, Taf. V. fig. 45), with the indigenous Araneae, and by *Dugès* (Ann. d. Sc. Nat. VI. p. 175), with *Mygale*. According to *Brandt* (Mediz. Zool. II. Taf. XV. fig. 3, and Ann. d. Sc. Nat. XIII. p. 185, Pl. IV. fig. 4), this ganglion is wanting with *Epeira*, and *Treviranus* (Zeitsch. f. Physiol. IV. p. 95), has vainly sought for it in a Brazilian spider. *Blanchard* (loc. cit. p. 1384), has found it with *Galeodea*, and *Van der Hoeven* (Tijdsch. X. p. 370), with *Thelyphonus*.

* [§ 302, note 2.] See also *Dufour*, loc. cit. p. 251. — ED.

CHAPTER IV.

ORGANS OF SENSE.

§ 303.

The multi-articulated antennae with which the Crustacea and Insecta are endowed, are absent with the Arachnoidae, or, more properly speaking, they are changed into prehensile and masticatory organs.[1]

The palpi, which are absent with only a few Arachnoidae,[2] must be regarded as the principal seat of the sense of Touch. These tactile organs always receive two considerable nerves arising from the anterior extremity of the ventral ganglionic mass.[3] A very delicate sense of touch exists, also, in the extremity of the feet, which are well supplied with nerves; and, for this object, the feet of the Opilionina and Phrynidae have the form of multi-articulated antennae.

With the Araneae, this point admits of no doubt, for these organs (the feet) are especially used in the formation of the web.

§ 304.

Although we must grant to the Arachnoidae the sense of Taste, and that of Smell; and although many facts show that they have the sense of Hearing highly developed, yet, at present, nothing satisfactory has been discovered either as to the locality or the structure of the organs which are the seat of these senses.[1]

§ 305.

The organs of Vision of the Arachnoidae consist always of simple eyes (*Stemmata*); but among the lower Arachnoidae, there is a complete series, namely, the parasitic Mites, and allied groups, which are entirely deficient in these organs.[1]

The stemmata of the Arachnoidae have exactly the same organization as the simple eyes of the Crustacea. They are composed of a simple and convex cornea, of a spherical lens, and of a concavo-convex, vitreous body, which is surrounded by a Retina. Each of these eyes is enveloped, before and behind, by a pigment tunic corresponding to the Chorioïdea; its color

---

1 See § 306. *Latreille* (Règne anim. IV. 1829 p. 207), has regarded these mandibles as transformed antennae, but usually they have been considered as the first pair of maxillae. This view of *Latreille* is the correct one, since the nerves of these organs do not arise from the abdominal ganglia, but directly from the brain, as those of the antennae of Crustacea and Insecta.

2 These palpi are wanting with *Pycnogonum, Phoxichilus, Phoxichilidium* and *Pallene*; see *Savigny*, Mém. loc. cit. I. Pl. V. fig. 3; *Johnston*, Mag. of Zool. and Bot. I. Pl. XIII. fig. 1–8; *Milne Edwards*, Hist. Nat. d. Crust. Pl. XLI. fig. 6. With the Scorpionidae, as well as with

Obisium, Chelifer, Phrynus, and *Thelyphonus,* the palpi are forficulate, and are used as prehensile organs.

3 See *Treviranus*, Zeitsch. f. Phys. IV. p. 94, Taf. VI. fig. 4, No. 4 (a Brazilian spider), and *Doyère*, loc. cit. p. 349, Pl. XVII. fig. 1, n. a. (*Milnesium*).

1 According to analogy, the sense of taste, with the Arachnoidae, is seated probably at the entrance of the œsophagus.

1 The eyes are wanting with *Demodex, Sarcoptes, Pteroptus, Dermanyssus, Gamasus, Thyroglyphus, Glycyphagus, Acarus, Argas, Ixodes* &c.

is very variable, and, in front, it terminates between the lens and the vitreous body by a ring which resembles an Iris. When two of these stemmata are contiguous, the pigment tunic is common between them.[2]

The number, the situation, the disposition, and the direction of the eyes, present so many variations, that they have been used by zoologists to characterize the genera. *Chelifer, Erythraeus, Smaridia, Tetranychus, Arrenurus*, and the Tardigrada, have two of these organs on the anterior portion of the back, while with many Oribatea, they are lateral and anterior. With *Trombidium*, there are two eyes also, but they are situated directly above the first pair of legs, on clavate peduncles.[3] With the Pycnogonidae, and with *Obisium*, there are four eyes situated on the first segment of the body; there are the same number, also, with *Bdella, Rhyncholophus, Eylais, Atax, Diplodontus, Hydrachna*, and *Limnochares*, situated on the anterior part of the back.[4]

With the Opilionina, there are two median, larger, and two lateral, smaller eyes. The first of these are situated on a tubercle, and their corneae face right and left.[5] With *Galeodes*, there are six eyes on the anterior border of the first segment of the body; of these, the middle or largest pair is directed upwards; another, situated in front of these last, forwards; and the remaining pair, inserted above the anterior legs, laterally.[6] The Araneae have, usually, eight eyes; only a few have but six.[7] These eyes, always situated on the cephalothorax, are generally of different sizes with the same individual, and are either grouped symmetrically upon the anterior median line of the cephalothorax, or scattered on its lateral border.[8] The dorsal eyes are directed upwards, and the marginal ones, forwards or laterally.

The disposition and direction of these organs are conformable with the animal's mode of life; some species watch their prey in crevices, fissures, or tubes; while others remain motionless in the centre of their webs, or lurk from side to side, — a kind of life requiring them to look in all directions. The color of the pigment of the eyes is based also upon the same relations; for, with the diurnal species, it is green, reddish, or of a

2 For the structure of the eyes of Arachnoidae, see *Soemmering*, De ocul. hom. animal. sect. horizont. p. 74, Tab. III.; and *Garde*, Nov. Act. Nat. Cur. XI. p. 338 (*Mygale*); but especially *Muller*, Zur. vergleich. Physiol. d. Gesicht-sinn. p. 316, Taf. VII. fig. 8–11, or Ann. d. Sc. Nat. XVII. 1829, p. 234, Pl. XII. fig. 1–4 (*Andractonus* and *Galeodes*). *Brants* (Tijds. &c. V., or Ann. d. Sc. Nat. IX. 1838, p. 308) has confirmed *Muller's* observations for the eyes of *Buthus* and *Mygale*; but he observed, also, tubes situated behind the vitreous body, and analogous to those of the eyes of Crustacea and Insecta. *Muller*, however (Arch. 1838, p. 139), has been unable to find them, but he observed that the fibres of the optic nerve, after having entered the eye, are separated by the long filamentoid pigment bodies; and he adds, that these fibres should not be confounded with the vitreous cones of the faceted eyes, the first becoming opaque in alcohol, while the second preserve their transparency.

3 These pedunculated eyes, already figured by *Degeer* (loc. cit. p. 57, Taf. VIII. fig. 15, y. y.), have been described by *Hermann* (loc. cit. p. 19, Pl. III. fig. E. G.), as *Oculi inferi*; see, also, *Treviranus*, Verm. Schrift. I. p. 49, fig. 31, 33, 34, o. o.

4 With *Bdella*, the eyes are wholly lateral. With the Hydrachna, above named, they are united in pairs, so that each pair would easily be taken for a single eye. With *Atax, Diplodontus*, and *Hydrachna*, the two pairs of eyes are widely separated. But with *Eylais*, and *Limnochares*, they are closely approximated. With the young of these aquatic mites, their position is often different (*Dugès*, Ann. d. Sc. Nat. I. p. 144, Pl. IX. X.).

Wagner's attributing (Lehrb. d. vergleich. Anat. p. 431) compound eyes to certain Hydrachnea, is due, without doubt, to his regarding as such the approximated simple eyes. *Dujardin* (Ann. d. Sc. Nat. III. p. 19), however, affirms that *Penthaleus* has a single eye, composed of eight to ten facets, while some of the species of *Oribates* and *Molgus* have only a single stemma situated on the back.

5 See *Treviranus*, Verm. Schrift. I. p. 24, Taf. II. fig. 10. The two lateral eyes are wanting with many Opilionina. According to *Tulk* (loc. cit. p. 326, Pl. V. fig. 32), there is a pair of muscles inserted on the two middle eyes, by which their contents can be displaced.

6 See *Muller*, Zur vergleich. Physiol. &c. p. 332, Taf. VII. fig. 11.

7 There are six eyes with *Scytodes, Segestria, Dysdera*, and *Uptiotes*.

8 See *Savigny*, Descript. de l'Égypte, loc. cit. Pl. I.–VII. and *Walckenaer*, loc. cit. Pl. I.–IV., &c.

brownish black, as with the other Arachnoidae ; but with the nocturnal
spiders, it is replaced by a membrane which has a splendid lustre.[9] With
the Phrynidae, there are also eight stemmata, of which two are situated on
the middle of the cephalothorax, and the remaining six form a triangle
composed of three on each of its sides.

With the Scorpionidae, the eyes are the most numerous. There are two
large eyes on the middle of the cephalothorax, then a row of from two to
five smaller on each side of its anterior border.

The number of optic nerves depends, usually, upon that of the eyes.
But the Scorpionidae form an exception in this respect ; for their brain
sends off, at the side of the two median optic nerves, two other nerves,
common, and belonging to the two rows of marginal stemmata, but which
do not divide until they have reached these organs.[10] On account of the
usually deep position of the brain, the optic nerve is generally of considerable length ; but the Pycnogonidae alone differ in this respect from the
other Arachnoidae, for, with *Phoxichilus*, the four eyes are situated directly
on the brain, and, with *Ammothea*, this last sends off, as a common optic
nerve to the four eyes, a large, short prolongation.[11]

CHAPTER V.

DIGESTIVE APPARATUS.

§ 306.

The entrance of the digestive canal is surrounded by very variable
organs, but, with all, the Mandibles are always wanting. The organs usually
called such are only antennae metamorphosed into prehensile and masticatory parts. This is shown not only from the cerebral origin of their
nerves, but by the fact that they, or more properly the Cheliceres, never
act, like the mandibles of the other Arthropoda, in a horizontal direction.
Most of the Arachnoidae live on liquid food, and, therefore, the basilar
article of the maxillae is more or less abortive, and is rarely used in mastication, while the succeeding articles are changed into a usually very
large tactile or prehensile palpus.

In general, the organization of the parts of the mouth with the Arachnoidae may be divided into the following five types:

1. With the Tardigrada, there are real organs for suction. These consist of a kind of sucker, situated on the end of a fleshy proboscis which can
be retracted into the head. On each side of this proboscis there are two
stylets (teeth) which, by means of a special muscular apparatus, can be
protruded into the former.[1]

2. With most of the Acarina, the two cheliceres are sometimes forficulate or unciform, sometimes cultrate or styliform, and by their use, these

9 *Dugès*, Ann. d. Sc. Nat. VI. p. 175.
10 *Treviranus*, Zeitsch. f. Physiol. IV. p. 92,
Taf. VI. fig. 3 ; and *Muller*, Zur vergleich. Physiol. &c. p. 321, Taf. VII. fig. 10, or Ann. d. Sc.
Nat. XVII. p. 258, Pl. XVII. fig. 3.

11 *Quatrefages*, loc. cit. p. 77, Pl. I. fig. 1ᵃ,
2ᵃ.
1 See *Doyère*, loc. cit. p. 319, Pl. XIII.-XV.

small animals can pierce or cut as may be required. These cheliceres are free, or lodged in a sheath out of which they may be protruded ; sometimes they are covered, above or below, by a frontal or chin-like process. In a few instances, these processes are united, forming a proboscis out of which the cheliceres may be protruded.[2] The first pair of maxillae, which are inserted on the sides of the cheliceres, are wholly unfit for masticatory organs, and, being destined for tactile parts, they have the form of palpi. These palpi are sometimes multi-articulated, sometimes uni-articulated, and, from their various modifications, have received the names of *Palpi rapaces, anchorarii, fusiformes, filiformes, antenniformes, valvaeformes,* and *adnati.*[3]

3. The Oribatea, which, from their herbivorous nature, hold a distinct place, not only among the Acarina, but also among the Arachnoidae in general, are distinguished also for the organization of their buccal organs. Their cheliceres are protractile, and the first pair of maxillae, situated under them, forms a complete masticatory apparatus, their basilar article being developed at the expense of the rest into a large denticulated piece. The other articles form only a very short palpus.[4]

4. The Pycnogonidae, Opilionina, Pseudoscorpii, Galeodea, and Scorpionidae, all, have tri-articulated cheliceres. Under these last are situated the first pair of maxillae which have no masticatory character.[5] With the Scorpionidae, and Pseudoscorpii, they are long-forficulate, while, with the Galeodea, the Pycnogonidae and Opilionina, they are antenniform. With the Phalangidae, only, there is observed on their basilar article, a hairy, obtuse appendage, comparable to a rudimentary maxilla.[6] With the Scorpionidae, the two basilar articles of the pincers are so approximated by their flattened internal surfaces, that they may well be used for the bruising of soft animal substances.[7]

5. With the Phrynidae, and Araneae, the cheliceres have the form of bi-articulated, unciform antennae. The basilar article of these so-called mandibles is always very thick, and the terminal article consists of a small, very sharp hook.[8] When at rest, this last lies folded on the inter-

---

2 For the cheliceres of the Acarina, see the descriptions and figures of *Hermann, Dugès,* and *Dujardin,* (loc. cit.). These organs are forficulate with the Acarcea, Gamasea and Bdellea ; see *Dujardin,* Observ. au Microsc. Pl. XVII. fig. 10, 11 (*Acarus*). They are unguiculate with *Trombidium, Erythraeus, Smaridia, Atax,* and *Eylais* ; see *Treviranus,* Verm. Schrift. 1. Taf. V. fig. 29 (*Trombidium*). They are styliform with the *Ixodes, Tetranychus, Rhyncholophus, Rhaphignathus,* and *Hydrachna.* The frontal prolongation is regarded by some authors as an under lip. It is found with *Dermanyssus* and *Rhaphignathus* ; while, with *Ixodes,* it belongs to the chin, and thus forms an under lip. With *Smaridia,* and *Sarcoptes,* the cheliceres are encompassed by a kind of tube ; see *Dujardin* (Observat. &c. Pl. XVII. fig. 1-4 (*Sarcoptes*). With *Ixodes,* the cheliceres are cultrate and denticulate on their external borders ; see *Savigny,* Descript. de l'Égypte, Pl. IX. and *Audouin,* Ann. d. Sc. Nat. XXV. Pl. XIV. The brevity and inequality of these organs, as noticed by *Audouin* with *Ixodes erinacei,* were due to the circumstance that they were imperfectly and unequally protruded from their sheath.

3 This classification of the palpi belongs to *Dugès* ; see Ann. d. Sc. Nat. I. p. 11.

4 I have satisfied myself of the presence of horny denticulated maxillae, fitted for mastication, with *Hoplophora, Pelops, Zetes, Oribates, Damaeus,* and with other Oribatea.

5 Some Pycnogonidae form the only exception in this respect. With *Pariboea,* the cheliceres are simple, bi-articulate and clavate ; but with *Endeis, Pycnogonum,* and *Phoxichilus,* they are wholly wanting ; see *Philippi,* in *Wiegmann's* Arch. 1843, I. Taf. IX. fig. 1-3 ; also *Savigny, Johnston,* and *Milne Edwards,* loc. cit.

6 *Savigny,* Mem. &c. I. Pl. VI. fig. 2, d.

7 It is well known that the Scorpionidae and the other rapacious Arachnoidae, merely suck their prey ; but it is said that *Galeodes* devours completely the insects which it has caught, seizing them with their cheliceres, and eating them piece by piece. During these processes each chelicere acts separately (*Hutton,* Ann. of Nat. Hist. XII. 1843, p. 81, or *Froriep's* neue Not. XXVIII. p. 49). The Phalangidae have probably the same habits, for fragments of insects which they have eaten are found in their digestive canal (*Tulk,* loc. cit. p. 248).

8 See *Roesel,* loc. cit. Taf. XXXVII., and *Savigny,* Descript. de l'Égypte, Pl. I.-VIII., also *Lyonet,* loc. cit. Pl. XIX. XXI.

nal side of, or underneath the basilar article. It is erected when the animal, for defence, or for the seizure of its prey, inflicts a poisonous wound; and, for this purpose, the excretory duct of a poison-gland opens at the apex of each of these hooks.[9] The first pair of maxillae is changed, with the Araneae, into very long tactile, and with the Phrynidae, into prehensile organs. Their basilar articles form two upwardly directed prominences, which are contiguous at their bristly, internal borders, and thereby cover the entrance of the oral cavity.[10] As the Araneae bruise, by means of these prominences, their prey which they have seized and taken into their mouth, these parts may be regarded as rudimentary maxillae.

The entrance of the Oral cavity is surrounded, with most Arachnoidae, by a soft, unequal border. This may be regarded, in part, as an upper and under lip, and partly as a tongue.[11] The orifice and cavity of the mouth are often provided with small hairs pointing inwards, among which are sometimes observed horny ridges, which serve, probably, as teeth. The Araneae have this peculiarity, that their large oral cavity has a groove on the median line of the palate, which is continuous into the œsophagus.[12] Its lateral borders may be so approximated that it is changed into a canal. This apparatus is certainly very serviceable to these animals in sucking their prey, after it has been punctured repeatedly, and taken into the mouth.

With very many Arachnoidae, the food, before reaching the proper digestive tube, traverses a very short œsophagus.[13]

With the Araneae, this canal is geniculate, of a horny consistence, and, at the point where it enters the stomach, it presents a prismatic muscular enlargement on which is inserted a large muscle arising from the centre of the dorsal shield and passing through the central opening of the stomach.[14] This serves probably as a sucking apparatus during the prehension and deglutition of food.[15] With the Tardigrada, the œsophagus terminates also by a muscular apparatus of this kind, which, with *Macrobiotus*, and *Emydium*, is spheroidal, and with *Milnesium*, cylindrical.[16]

§ 307.

The Intestinal canal of the Arachnoidae is formed after two different types.

1. With the Tardigrada, Acarina, Pycnogonidae, Opilionina, Solpugidae, and Araneae, the stomach has a greater or less number of caecal

---

9 See § 315.

10 See *Treviranus*, Bau d. Arach. Taf. II. fig. 14–16, r., and *Brandt*, Mediz. Zool. 11. Taf. XV. fig. 9, 18, b.

11 With the Araneae, and Scorpionidae, the entrance of the mouth has a tumid, pilose upper lip. With the Opilionina, there are several such tumefactions, but with the Pycnogonidae, the oral orifice is prolonged, snout-like, between the maxillae.

12 See *Lyonet*, loc. cit. p. 401, Pl. XXI. fig. 4, 5, and *Dugès*, Ann. d. Sc. Nat. VI. p. 178.

13 With the Acarina, Pycnogonidae, and Araneae. *Quatrefages* (Compt. rend. XIX. 1844, p. 1152) thinks he has observed a ciliated epithelium in the

œsophagus of the Pycnogonidae; but, subsequently, he found that he was deceived, and that vibratile organs were wanting here as with all the Arthropoda.

14 *Brandt*, Mediz. Zool. I. p. 89, Taf. XV. fig. 6, b., or Ann. d. Sc. Nat. XIII. p. 183, Pl. IV. fig. 2. b.

15 This suctorial apparatus appears to have been well described and understood by *Wasmann* (loc. cit. p. 10, fig. 13, i. m.); but, already before this, *Lyonet* (loc. cit. p. 402, Pl. XXI. fig. 4, C D E.) had rightly perceived it; while *Brandt* (Med. Zool. 11. p. 87) had taken it for an *os hyoïdes*.

16 *Doyère*, loc. cit. p. 322, Pl XIII.–XV.

diverticuli, of the most varied form and size. It is continuous into a short, small intestine, which passes, in a straight line, to the anus situated usually at the posterior extremity of the body. Before reaching this point, the intestine has, usually, a dilatation bounded by a constriction, which may be regarded as a rectum, or better, perhaps, as a cloaca. With the Tardigrada, the stomach is oblong and occupies a large portion of the body. It is divided throughout by numerous constrictions into many irregularly disposed caeca.[1]

With the Acarina, whose anus is placed nearer the middle of the belly, there are, nearly always, three short caeca at the anterior part of the stomach, and two, longer and more or less constricted, in the lateral regions of the abdomen. With some species of parasitic Mites, these appendages of the stomach are bifurcated.[2] With the Pycnogonidae, the stomach is short, but has five pairs of very long caeca, some of which penetrate into the two cheliceres, and others into the eight long legs, even to the extremity of the tibiae.[3] With Galeodes, also, these appendages penetrate the legs, and the base of the cheliceres and palpi.[4] With the Phalangidae, the stomach is spacious and has thirty appendages of varied size. Thus, at its upper part, there are four rows of short caeca, and, upon the sides, three pairs, very long and extending over nearly the whole length of the visceral cavity; the middle pair of these last has, moreover, short sacculi.[5] With the Araneae, the stomach is situated in the cephalothorax, and presents a very remarkable disposition. At the posterior extremity of the thoracic cavity, and directly behind the sucking apparatus, it is divided into lateral halves which extend arcuately in front, and, uniting, form a ring from which are given off laterally five pairs of caeca extending towards the points of insertion of the legs and palpi.

The intestine arises from this annular stomach, opposite the sucking apparatus. It traverses the abdomen on the median line, and terminates, before reaching the anus, in a cloacal dilatation.[6]

2. With the Phrynidae,[7] and Scorpionidae,[8] the intestinal canal is very simple compared with that just described. It consists of a straight

1 Doyère, Ibid. p. 324, Pl. XV.

2 See Lyonet, loc. cit. Pl. XIII. fig. 11, 12; Dugès, loc. cit. I. Pl. I. fig. 27. II. Pl. VII. (Erythraeus, Dermanyssus and Ixodes); also, Treviranus, Zeitsch. f. Physiol. IV. p. 189, Taf. XVI. Ixodes has dichotomous stomachic appendages, of which the posterior, at the extremity of the body, curve first downwards, then forwards with a long course. These various caeca of the Acarina often appear, especially when filled with food, clearly defined, through the skin. But when empty, they are frequently overlooked in the small species, from the tenuity of their walls. However, I have always succeeded, even with the smallest Oribata, in distinguishing the walls of the intestine, especially when it contained food. I must, therefore, consider as wholly erroneous, the opinion recently advanced by Dujardin (Ann. d. Sc. Nat. III. p. 14, or Compt. rend. loc. cit. p. 1159), that the food eaten by the Acarina does not pass through a distinct digestive tube, but is freely effused in the interstices of the viscera.

3 Milne Edwards, Hist. Nat. d. Crust. III. p. 531, and Quatrefages, loc. cit. p. 72, Pl. I. II.

4 Blanchard, loc. cit. p. 1384.

5 Ramdohr, Abhandl. üb. d. Verdauungswerk. p. 205, Taf. XXIX.; Treviranus, Verm. Schrift. I. p. 29, Taf. III., and Tulk, loc. cit. p. 246, Pl. IV.

6 For the annular stomach of the Araneae, and on which, with Tegenaria, Treviranus (Bau d. Arach. p. 30, Taf. II. fig. 24, v. b.) has found only four caeca, see Brandt, Mediz. Zool. II. p. 89, Taf. XV. fig. 6, or Ann. d. Sc. Nat. XIII. p. 182, Pl. IV. fig. 2, or Isis, 1831, p. 1105, Taf. VII. fig. 6; also Owen, Lectures, &c., p. 257, fig. 110; and Wasmann, loc. cit. p. 11, fig. 17, 18. According to this last observer, the four pairs of stomachic caeca, with Mygale, bend downwards to the base of the eight legs, in order to pass into the thorax where they ramify and interanastomose.

With Argyroneta, and some species of Epeira, according to Grube (Müller's Arch. 1842, p. 208), the lateral halves of the stomach are not united in a ring at their anterior extremity, but are only contiguous.

With the Araneae, the walls of the stomach contain finely-granular cells which, by reflected light, have a milky aspect, and secrete perhaps a kind of gastric juice.

7 Van der Hoeven, Tijdschr. &c. IX. p. 68 (Phrynus).

8 Meckel, Beiträge, loc. cit. p. 107, Taf. VII. fig. 13; Treviranus, Bau d. Arach. p. 6, Taf. I. fig. 6, and Muller, loc. cit. p. 45, Taf. II. fig. 22.

tube, of nearly equal size throughout, without a stomachic dilatation and
without caeca, which opens by an anus at the posterior extremity of the
body.[9]

§ 308.

The Salivary glands exist with, perhaps, all the Arachnoidae; for, they
are found even in many of the lower forms, where their presence would be
least expected. With the Tardigrada, there are on each side of the sucking apparatus, large, lobulated glandular tubes, which appear to be organs
of this nature, although their outlets have not yet been distinctly traced.[1]
With the Oribatea, there is at the anterior extremity of the body, a pair
of similar tubes, but simple and colorless, which extend to the mouth,
and have undoubtedly a salivary function.[2]
With *Ixodes*, these organs are extraordinarily developed, consisting of
two large masses of vesicles situated on the sides of the anterior part of the
body, and opening by short ducts into two multiramose excretory canals.
These last, whose walls are traversed by a solid spiral filament, open into
the buccal cavity at the base of the lip-like process.[3]
With the Araneae, a slit in the upper lip leads into a cavity situated
above the palate, and at the base of this cavity is a transparent, glandular
mass, which, very probably, secretes the saliva; this flows up through the
slit in question, and moistens the substances from which the animal extracts its food.[4] As salivary organs should also be considered the two
pairs of glandular tubes, which, with the Scorpionidae, are situated on the
sides of the anterior part of the body, and extend forwards to open into
the œsophagus.[5]
With the Araneae, and Scorpionidae, there is a Liver distinct from the
digestive tube, which, for a long time was regarded as an adipose mass.
With the Tardigrada, Acarina, Pycnogonidae, and Opilionina, the walls of
the stomachic appendages are of this nature, for they are glandular and composed of granular and usually yellowish-brown cells.[6] With the Araneae,
the brown or dirty-yellow liver is very voluminous, filling a large portion
of the abdominal cavity, and enveloping most of the other viscera.
At first sight, it appears to be a compact mass, but, further examined,
it is found composed of numerous multiramose, closely-aggregated caeca.
The walls of these are thick, and crowded with hepatic cells, and they open
into the digestive canal near its middle by four short hepatic ducts.[7]

---

9 With the Scorpionidae, the anus is situated on
the penultimate caudal segment.

1 See *Doyère*, loc. cit. p. 321, Pl. XIII.–XV.

2 I have seen these glandular tubes with *Hoplophora*, *Zetes*, and *Oribates*.

3 The salivary glands of *Ixodes ricinus* resemble exactly the botryoidal ones of many of the Insecta.

The secretory vesicles of the saliva are filled with
transparent nucleated cells and surrounded by
numerous ramified tracheae with which it is impossible to confound the excretory ducts of these
glands; for with these last the spiral turns of thin
filament are very wide apart, while, in the tracheae,
the spiral windings are very close together.

4 This glandular apparatus has been seen by
Wasmann (loc. cit. p. 8, fig. 10) with *Mygale*;
I have found it also with other Araneae.

5 See *Müller*, loc. cit. p. 52, and *Newport*,
Philosoph. Trans. 1843, Pl. XV. fig. 39.

6 With the Tardigrada, Acarina, and Opilionina,
at least, I have seen, distinctly, hepatic cells in
the walls of the stomachic appendages. See also
Doyère, loc. cit. p. 327, Pl. XV.

7 *Treviranus* (Bau d. Arachnid. p. 30, 47, Taf.
II. fig. 24, dd., and Taf. V. fig. 47) had already
observed the communication between the liver and
the digestive organs. The remaining points in the
structure of this organ have been rightly estimated
by *Dugès* (Ann. d. Sc. Nat. VI. p. 179), *Grube*
(loc. cit. p. 239), and *Wasmann* (loc. cit. p. 13, fig.
17, m. n., 20–22). See also *Owen*, Lectures, &c.,
p. 258, fig. 110, l. i.

With the Scorpionidae, the liver is also very large, and composed of many lobes. It occupies the two sides of the abdominal cavity even to the base of the tail, and closely encompasses the intestine, the heart, and the genital organs. The ramifications of the biliary canals traverse, in groups, the parenchyma of this liver, and the bile is poured into the intestine by five pairs of short, excretory ducts, equally, but very widely separated from each other.[8] *

CHAPTER VI.

CIRCULATORY SYSTEM.

§ 309.

With many Arachnoidae, the circulatory system consists only of a Heart or an articulated dorsal vessel. With the higher forms, there is, in addition, a system of more or less developed blood-vessels; while with the lower species, such as the Tardigrada, the Acarina and the Pycnogonidae, not only all these vessels, but the heart, also, is absent. There is, therefore, in these last, no regular circulation, but the nutritive fluid fills all the interstices of the body, and, by the aid of the muscular movements and the contractions of the intestinal canal, is transferred in an irregular manner hither and thither in the visceral cavity and in the extremities.[1]

The Blood of the Arachnoidae is entirely colorless, and has a slightly milky aspect only when in considerable quantities. It contains a few granular blood-cells of a pretty regular, spheroidal form, and some very small, isolated granules, derived perhaps from broken blood-cells.[2]

8 See *Meckel*, Beitr. &c. p. 107, Taf. VII. fig. 13, 15; this author has seen four pairs of hepatic ducts. See, also, *Treviranus*, Bau d. Arachn. p. 8, Taf. I. fig. 6, A. v., and *Muller*, loc. cit. p. 35, 46, Taf. II. fig. 22, D. D.; finally *Newport*, Philosoph. Trans. 1843, Pl. XIV. fig. 32.

1 C. A. S. *Schultze* (in his memoir " Macrobiotus Hufelandii ") thinks he has observed blood-vessels in the Tardigrada; but neither *Doyère* (loc. cit. p. 310) nor I have been able to find them. For the interstitial circulation of the Pycnogonidae, see *Quatrefages*, loc. cit. p. 76. *Van*

Beneden has observed, in the extremities of these animals, regular blood-currents produced apparently by contractile membranes at the base of the legs; see Institut. No. 627, or *Froriep's* neue Notiz. XXXVII. p. 72.

2 For the blood of the Arachnoidae, see *Wagner*, Zur vergleich. Physiol. d. Blutes, Heft. I. p. 27, fig. 11 (*Scorpio europaeus*); *Horn*, Das Leben des Blutes, p. 10, Taf. I. fig. 12 (*Tegenaria domestica*), and *Doyère*, loc. cit. p. 309, Pl. XV. fig. 5. (*Tardigrada*.)

* [§ 308, end.] See, for some researches upon the hepatic organs of the Arachnoidae by means of chemical agents, and the positive determination thereby of the nature of the alleged hepatic appendages of the alimentary canal of these animals, *Will*, *Muller's* Arch. 1848, p. 50". — ED.

§ 310.

With the Arachnoidae, the circulatory organs, when present, are disposed in the following manner :

With the Phalangidae, they consist only of a Dorsal Vessel, which is three-chambered, and attenuated at both extremities.[1]

With the Araneae, the dorsal vessel is fusiform, and has many constrictions. It is situated principally in the abdomen, being attached to its dorsal wall by triangular transverse muscles. This heart, which extends also into the cephalothorax, sends off from each extremity and from its sides, many ramified, vascular canals, which are certainly Arteries.

The two of these last arising directly behind the peduncle of the abdomen, are distributed to the pulmonary sacs, while those following penetrate chiefly the liver. All these vessels gradually disappear in the parenchyma of the body, and the blood, after its effusion, continues to circulate in the lacunae, and, without the intervention of veins, is returned to the heart, or more properly into the blood-reservoir which corresponds to the dorsal sinus of the Crustacea. Thence it enters the heart through its lateral, valvular openings.[2]

The vascular system is most highly developed with the Scorpionidae. For, here, not only is there an articulated Heart and Arteries, but also a Venous system.[3] The cylindrical heart whose walls contain transverse and longitudinal muscular fibres, is retained in place between the diaphragm of the cephalothorax and the last abdominal segment, by several transverse triangular muscles. It has eight chambers whose size diminishes from before backwards. At each extremity it is prolonged into an

---

1 See *Tulk*, loc. cit. p. 249, Pl. IV. fig. 17, II., and *Treviranus*, Verm. Schrift. I. p. 31, Taf. III. fig. 16, k., and fig. 18.*

2 For the vascular system of the Araneae, see *Meckel*, in his translation of *Cuvier's* Leçons d'Anat. comp. Th. IV. p. 261; *Treviranus*, Bau d. Arach. p. 28, Taf. III. fig. 28–31, also his Verm. Schrift. I. p. 4, Taf. I. fig. 1; *Gaede*, Nov. Act. Nat. Cur. XI. p. 335, Tab. XLIV. fig. 3 (*Mygale*), and *Brandt*, Mediz. Zool. II. p. 89, Taf. XV. fig. 16, 17. See also *Dugès* (loc. cit. p. 181), who has been unable to find the venous system with the Araneae, but, at the same time, traced the heart even into the cephalothorax. *Wasmann* (loc. cit. p. 16, fig. 24), on the other hand, affirms that he has observed, with *Mygale*, venous trunks which entered the

heart above the points of origin of the arteries. The analogy between the heart of Crustacea and of Araneae has been especially pointed out by *Straus* (Considérat. &c. p. 345, and Traité d'Anat. comp. II. p. 251), and since confirmed by *Grant* (Outlines, &c., p. 452) and *Grube* (loc. cit. p. 300).†

3 *Treviranus* (Bau d. Arachn. p. 9, Taf. I. fig. 7), and *Muller* (loc. cit. p. 38, Taf. II. fig. 22), were acquainted with only the heart and larger vascular trunks of the Scorpionidae ; but *Newport* has given of the blood system of these Arachnoidae a complete and masterly description accompanied with very beautiful figures ; see Philos. Trans. 1843, p. 286, Pl. XIV. XV., or *Froriep's* neue Notiz. XXXIX. p. 51, fig. 38–40.

* [§ 310, note 1.] *Blanchard* (loc. cit. Ann. d. Sc. Nat. XII. 1849, p. 333) has extended our knowledge of the circulatory system of this family. The dorsal vessel terminates behind in a small vessel which runs to the extremity of the body. In front it passes into an artery of considerable size, which passes under the brain and sends off small branches to the œsophagus. At the base of this aorta the ophthalmic artery is given off, which bifurcates behind the eyes. From this portion of the heart also pass off branches to the stomach. This naturalist declares the existence here of his peritrachean system, which, together with the heart, he says he has injected through the lacunae. — ED.

† [§ 310, note 2.] According to *Blanchard* (loc. cit.), the blood, in the Araneae, passes to the respiratory organs, which it penetrates by a kind of infiltration ; from the lacunae of the walls of the lungs it is taken to the heart by means of the pulmono-cardiac vessels which have hitherto been taken for arteries. There are six pairs with *Epeira diadema*. But with those Araneae which have both lungs and tracheae, such as *Segestria, Dysdera*, &c., there is some modification, although the arterial system resembles that of the Araneae essentially pulmonary ; the heart is smaller and has fewer chambers, and the true arteries seem to lose their importance and give place to the peritrachean system of circulation. — ED.

arterial trunk. The anterior of these arteries very soon ramifies, and distributes blood to the feet, the pincers, the cheliceres, and to all the organs in the cephalic extremity. Two of its branches, bending downwards, embrace the œsophagus, and then join in a large common vessel called the Supra-spinal artery, which lies upon the ventral cord and accompanies it to the caudal extremity, giving off, in its course, numerous lateral branches.[4] The posterior arterial trunk is distributed in like manner to the posterior extremity, and gives off, right and left, numerous branches. The middle chambers of the heart send off, each, laterally, shorter arteries, which are distributed to the neighboring organs. Beside these arteries of the muscles and viscera, these animals have, also, a special Visceral artery, arising from the anterior arterial trunk before it divides into the two branches which form the supra-spinal artery. The visceral artery runs backwards towards the digestive tube, and sends branches to the liver.[5] The terminal ramifications of these various arteries are directly continuous, it is said, with a venous system.[6] In this last may be noticed, especially, a Sub-spinal vein, by which the blood is carried to the pulmonary sacs; thence to be borne to the heart by special vessels. These last open, probably, into a sinus, from which the blood passes into the heart through lateral openings, two of which exist in each of its chambers.†

4 This supra-spinal artery had been seen, it would appear, by *Müller* (loc. cit. p. 62, Taf. I. fig. 5, r. r.), but he took it for a ligament.

5 According to *Newport*, this visceral artery, which is simple with *Androctonus*, is divided into two trunks with *Buthus*.

6 *Newport* speaks in his memoir of various anastomoses occurring between the arteries and veins with *Scorpio*. But, as he nowhere describes precisely this point, and has not distinctly indicated it in his plates otherwise so beautiful, I demur admitting that, with the Scorpionidae, the arteries pass directly into the veins, and therefore, that these animals have a system of capillary vessels. This direct communication between these two systems does not exist with the other Arachnoidae, neither with all the other Arthropoda in general.*

* [§ 310, note 6.] In regard to the question of capillaries with the Scorpionidae, a remark of *Blanchard* (loc. cit.) may be given. He says, " I have proved with an entire certainty that the blood is distributed in all the cavities of the body, as with all the Articulata, and that it is conveyed to the lungs simply by means of the lacunae. Most of the vessels which arise from the sides of several of the chambers of the heart have appeared to me to be pulmono-cardiac vessels, wholly analogous to those we have described with the Araneae." — ED.

† [§ 310, end.] For further details on the circulatory system of the Arachnoidae, see the memoir quoted above of *Blanchard*. This naturalist has sought to extend his doctrine of the peritracheal circulation, to the different sections of the Arachnoidae. — ED.

CHAPTER VII.

RESPIRATORY SYSTEM.

§ 311.

The higher Arachnoidae respire by tracheae, or by lungs ; but in the lower, namely, the Tardigrada,[1] the Pycnogonidae,[2] and some parasitic Acarina,[3] no traces of respiratory organs have yet been found. With these animals therefore the respiration must be cutaneous.

Many Acarina, the Opilionina, the Pseudoscorpii and the Solpugidae, breathe by tracheae, while the Araneae, the Phrynidae and the Scorpionidae breathe by lungs. On this account, these animals have been divided, in zoological systems, into the *Arachnidae tracheariae* and *pulmonariae.* But this classification is valueless, since it has been shown that the Araneae possess both lungs and tracheae.

§ 312.

With the Acarina, the Tracheae are exceedingly tenuous, and it is only in the larger species that the spiral filament of these organs can be observed. They arise usually by a simple tuft from two stigmata which are sometimes concealed between the anterior feet, as with the Hydrachnea, the Oribatea, and the Trombidina, sometimes very apparent above the third pair of legs, as with the Gamasea, and sometimes behind the last pair of legs, as with the Ixodea.[1]

With the Hydrachnea, which live in the water and never come to the surface to take in air, the tracheae possess, probably, the power to extract from the water the air necessary for respiration.[2]

With the Pseudoscorpii, there is, on the ventral surface of the two first abdominal segments, a pair of lateral stigmata, with four short but large tracheal trunks from which arise numerous unbranched tracheae spreading through the entire body.[3] With the Solpugidae, whose tracheae

1 See *Doyère*, loc. cit. p. 316.
2 See *Quatrefages*, loc. cit. p. 76.
3 *Demodex, Sarcoptes, Acarus,* &c.
1 With *Trombidium*, there arise two simple and very distinct tracheal tufts from the two stigmata situated behind the second pair of legs (*Treviranus*, Verm. Schrift. I. p. 47, Taf. VI. fig. 32, t. t.). These tracheae do not proceed directly from the stigmata, but from two large, short trunks unobserved by *Treviranus*.
With *Gamasus*, and *Uropoda*, there are given off, from the two ramified tracheal tufts, two unbranched tracheae which, remaining of the same size, describe a slightly arcuate course along the lateral borders of the cephalothorax and terminate in caeca at the base of the parts of the mouth. The two lateral stigmata of *Ixodes* have been described by *Lyonet* (loc. cit. p. 288, Pl. XIV. fig. 3, 5), *Treviranus* (Zeitsch. f. Physiol. IV. p. 187, Taf. XV. fig. 2, f. f.), and *Audouin* (Ann. d. Sc. Nat. XXV. p. 419, Pl. XIV. fig. 2, q. r. s.). For the

tracheae of the Acarina, see, moreover, *Dujardin* (Ann. d. Sc. Nat. III. p. 16, or Compt. rend. loc. cit. p. 1160). It will be difficult, I think, to prove the assertion of *Dujardin*, that, with these animals, the tracheal system serves exclusively for the act of expiration, inspiration being performed wholly by the skin.
2 *Dugès* (Traité d. Physiol. II. p. 549) is certainly right in placing the tracheae of the Hydrachnea in the category of *Branchiae trachrales,* which are so widely spread with the aquatic larvae of Insecta (see below).
3 According to *Audouin* (Ann. d. Sc. Nat. XXVII. 1832, p. 62), the tracheae of *Obisium* are ramified, a statement which I have been unable to verify. It has already been stated that the scar-like fossae on the abdomen of *Chelifer* have been erroneously taken for stigmata (§ 298, note 4). The tracheae of the Pseudoscorpii are so easily seen by the microscope that it is incomprehensible how anatomists should have remained so long

ramify through the whole body like those of insects, there are three pairs of stigmata.[4] With the Phalangidae, the trachean system is highly developed, arising from two stigmata concealed under the coxae of the posterior legs, each of which has a horny valve. The two large trunks given off from these stigmata, run obliquely to the cephalic extremity; they intercommunicate by a transverse anastomosis, and give off, in all directions, numerous branches which are spread over the abdominal viscera, and penetrate even the palpi and legs.[5]

With many of the Araneae, there are, on the under surface of the abdomen, two orifices which lead into two pulmonary sacs, beside two other openings belonging to the trachean system. With *Segestria, Dysdera*,[6] and *Argyroneta*,[7] there arise from these two stigmata two large trunks surrounded by a kind of horny trellis-work. From the extremity of these trunks are given off innumerable, very small tracheae, which are unbranched and without the spiral filament. They are disposed in tufts, and are distributed, some in the abdomen, and others in the cephalothorax, penetrating even 'to the extremity of its members. With *Salticus*, and *Micryphantes*,[8] the two stigmata are situated at the posterior extremity of the body, far removed from the pulmonary sacs, and send off, directly, two tufts of unbranched tracheae, which are distributed exclusively to the abdominal viscera.[9] There is, with the other Araneae, a trachean system, very imperfect it is true, which has hitherto been overlooked by anatomists. Directly in front of the spinnerets, there is, with most species, a transverse fissure difficult to be seen, which leads into a very short trachean trunk. From this trunk are given off four simple tracheae which, singularly, are not cylindrical, but are flattened, riband-like, and without a trace of a spiral filament; these extend, with a gradual attenuation, to the base of the abdomen. These riband-like, silvery tracheae are composed of a thin, but solid, homogeneous membrane, which is enveloped by a soft, transparent pellicle corresponding to a peritoneum. The air received into these organs is separated into as fine portions as that of the lungs. These tracheae differ therefore, prominently, from those of the other Arachnoidae.[10]

ignorant of their existence, and even lately, that *Tulk* (Annals of Nat. Hist. XV. p. 57) should have failed to see them with *Obisium*.

4 See *Muller*, Isis, 1828, p. 711, and *Milne Edwards*, Règne anim. Illustr. Arachnides, Pl. II.

5 *Treviranus*, Verm. Schrift. I. p. 32, Taf. IV. fig. 19, and *Tulk*, loc. cit. p. 327, Pl. V. fig. 33.

6 *Dugès*, in "Le Temps," 1835, No. 1942, Feuilleton, Acad. d. Sc. Séance du 9 Février, or *Froriep's* neue Notiz. XLIII. p. 231, or Ann. d. Sc. Nat. VI. p. 183, and Règne animal, Arachnides, Pl. III. fig. 4, V. fig. 4. See also *Owen*, Lectures, &c., p. 259, fig. 112.

7 *Grube*, loc. cit. p. 300, and *Menge*, loc. cit. p. 22, Taf. I. fig. 6-14.

8 *Menge*, loc. cit. p. 23, Taf. I. fig. 15.

9 I have had an opportunity to satisfy myself of the existence of this interesting trachean system with *Segestria, Argyroneta, Salticus*, and *Micryphantes*. I should also add that the principal trunks are flattened, and that the contained air is finely divided, while that in the cylindrical tracheae given off from these trunks, forms a continuous column.

10 I have found this trachean system with *Epeira, Tetragnathus, Drassus, Clubiona, Theridion, Lycosa, Diomedes* and several others. I have been unable to perceive it in individuals escaping from the egg. *Thomisus viaticus* is the only species in which the four flattened trachean trunks are ramified, and thus serves as the passage to the most highly developed trachean system of *Salticus*. By direct light, they appear black, and thus it is possible that they may have sometimes been taken for urinary canals. But this error is unnecessary, for these last vessels burst from the slightest pressure and effuse granular contents, while the tracheae under such treatment become transparent, their contained air making its escape, and when the pressure is withdrawn they resume their black color."

* [§ 312, note 10.] See also for these anomalous tracheae, *Blanchard* (loc. cit. Ann. d. Sc. Nat. XII. 1849, p. 345), who regards them as only elongated pulmonary sacs; but especially *Leuckart* (Ueber den Bau und die Bendeutung der sog. Lungen bei den Arachniden, in *Siebold* and *Kölli-*

§ 313.

The Lungs of the Arachnoidae consist of round sacs situated near the lower surface of the abdomen and communicating, externally, by transverse fissures. Their internal surface has numerous thin solid lamellae, triangular or rhomboidal, and connected together like the leaves of a book. By reflected light these lamellae have the same silvery lustre as the tracheae, although, seen by direct light, they appear of a deep-violet, nearly black color. Each of these is formed by a membranous fold, between the two leaves of which the air enters from the general cavity of the lung and is divided into very minute portions. No traces of blood-vessels have been found in these Pulmonary lamellae. It is therefore very probable that the blood of the pulmonary arteries is effused into the parts surrounding the lungs, and in this way bathes the lamellae.[1]

With the Scorpionidae, the four anterior segments have, each, on their under surface, a pair of stigmata. These animals have eight pulmonary sacs, in each of which there are twenty fan-shaped lamellae.[2] The genus *Phrynus* has only two pairs of pulmonary sacs, the stigmata of which are placed between the first and second, and the second and third abdominal segments. But each sac has eighty lamellae.[3] With the Araneae, there are only two lungs occupying the base of the abdomen. The number of their lamellae is considerably less than in the preceding groups. But with the Mygalidae only, there is a second pair of lungs directly behind the first. The place occupied by these organs, is indicated, with the Araneae, by a triangular horny plate, at the posterior border of which is a stigma.[4]

1 These organs, with which no motions have been discovered, have been called Branchiae by many Zootomists. But the name of Lungs is very appropriate since the respiration is aërial and not aquatic.
2 For the lungs of the Scorpionidae, see *Meckel*, Translat. of Leçons d. Anat. comp. of *Cuvier*, Th. IV. p. 291; *Treviranus*, Bau d. Arach. p. 7, Taf. I., and Beobacht. aus d. Physiol. p. 25, fig. 40–42; *Müller*, Isis, 1829, p. 708, Taf. X. fig. 1–3, and in *Meckel's* Arch. loc. cit. p. 39, Taf. II. fig. 11–13. *Müller* has very correctly, and in the above-mentioned manner, understood this respiratory apparatus, while, on the other hand, *Treviranus* and other anatomists, think that the air, instead of entering between the leaves of the pulmonary lamellae, passes over their external surface, and that the blood penetrates between the two plates composing the leaves. *Newport* (Philos. Transact. 1843, p. 295, Pl. XIV.) is probably mistaken in saying that unnucleated cells and a very fine capillary net-work exist between these plates, and that the net-work arises from a branch of the pulmonary artery situated on the free border of each lamella.
3 See *Van der Hoeven*, Tijdsch. loc. cit.
4 The lungs of the Araneae have been studied by *Meckel* (Translat. Leçons d'Anat. comp. of *Cuvier*, loc. cit. p. 290), *Treviranus* (Bau d. Arachn. p. 24, Taf. II. and Beobacht. &c. p. 29, fig. 43–47), *Gaede*, Nov. Act. Nat. Cur. XI. p. 335, (*Mygale*); but especially by *Müller* (Isis, 1828, p. 700, Taf. X. fig. 4–6). See also *Menge*, loc. cit. p. 21, Taf. I. fig. 6–9. I am unable to say by what means the blood returns to the heart, whether by a direct course, or, more or less circuitously, through the interstices of the parenchyma, for there are no veins.

ker's Zeitsch. 1849, I. p. 246) who, contrary to *Blanchard*, advances the view that these organs are only a form of tracheae, infra-formed, and which are without the spiral filament, because their simple, unbranched condition does not require, like the ramose tracheae, a spring-like structure, to prevent them from collapsing. This view put forth together with the general doctrine that the pulmonary sacs of the Arachnoidae are, likewise, but modifications of the tracheal type, has many facts deserving the attention of anatomists, and especially the developmental relations of the spiral thread as observed in the embryos of these animals. — ED.

CHAPTER VIII.

ORGANS OF SECRETION.

I. *Urinary Organs.*

§ 314.

With most Arachnoidae, there are small, usually multiramose, glandular tubes, which open into the cloaca. By their structure and the nature of the fluid they secrete, they exactly resemble the Malpighian vessels of the Insecta, and like them, also, they have, for a long time, been regarded as hepatic organs; but now, they are known to be positively those of an urinary nature. The urine is usually accumulated in the cloaca, and consists of a troubled, dirty-white liquid, rarely reddish; and, by direct light, is found to hold in suspension innumerable dark molecules.

These organs appear to be absent with the Tardigrada, and Pycnogonidae. But, on the other hand, they are easily observed with many Acarina, where they consist of simple or ramose white tubes, situated between the appendages of the stomach.[1] With the Phalangidae, there are two pairs of urinary canals which wind between the stomachic caeca.[2] With the Araneae, these organs are numerous, multiramose, and of a white or reddish color. Their very small branches penetrate between the different portions of the liver, and end in two principal trunks or ureters, which open into a cloaca provided with a kind of diverticulum.[3] With the Scorpionidae, the organization in this respect is quite similar, and the canals, ramified in various ways, enter, some the interstices of the hepatic lobes, while others surround the digestive canal. They pour their product into the cloaca by two ureters which are situated back of the biliary canals.[4]

---

[1] I have discovered without trouble, these canals with the Hydrachnea, Gamasea, Trombidina, and Ixodea. *Treviranus* (Zeitsch. f. Physiol. IV. p. 189, Taf. XVI. fig. 8, n. n.) had already observed their insertion into the cloaca with *Ixodes*. With *Ixodes ricinus*, where they are simple and flexuous, I have seen them ascend even to the anterior extremity of the cephalothorax; this is entirely so with *Ixodes americanus*. The canals, which with *Nigua, Treviranus* (loc. cit. fig. 7, g. g.) has regarded as salivary organs, are certainly only the anterior extremities of the urinary vessels. The two species of *Ixodes* just mentioned have their cloaca filled with a white urine.

[2] See *Treviranus*, Verm. Schrift. I. p. 31, Taf. III. fig. 16, 17. *Tulk* (loc. cit. p. 249, Pl. IV. fig. 17) who has been unable to trace these canals to their points of insertion on the intestine, has taken a portion of them for salivary organs.

[3] *Ramdohr* (loc. cit. p. 208, Taf. XXX. fig. 2), and *Treviranus* (Bau d. Arach. p. 30, Taf. II. fig. 24) were only imperfectly acquainted with the urinary canals of the Araneae. They have been more exactly described by *Brandt* (Mediz. Zool. II. p. 89, Taf. XV. fig. 6, 17, or Ann. d. Sc. Nat. XIII. p. 183, Pl. IV. fig. 2, 3); but see, especially, *Wasmann*, loc. cit. p. 17, fig. 17, 21–23 (*Mygale*). In most species, the urine is of a dirty-white color; but with *Mygale*, it is reddish. In several individuals of a large species of *Mygale* preserved in alcohol, I have found, in the ureters, hard, reddish concretions which *Dugès* (Ann. d. Sc. Nat. VI. p. 180) had already observed. Treated with nitric acid and ammonia, I obtained purpuric acid.

[4] See *Treviranus*, Bau d. Arach. p. 6, Taf. I. fig. 6, and *Muller*, loc. cit. p. 47, Taf. II. fig. 22. This last anatomist says that these glandular canals communicate with the heart, but he has probably confounded them with the blood-vessels.

II. *Organs of Special Secretions.*

§ 315.

Very many Arachnoidae have Poison-glands, the product of which is excreted through the extremity of a hollow claw. With the Phrynidae, the Araneae, and some Acarina, there are two such glands in communication with the terminal hooks of the chelicères. They have been often taken for salivary organs. With the Trombidina, there are, on each side of the cephalothorax, two small, flexuous, colorless, glandular tubes, which, at their anterior extremity, are dilated, each, into a cylindrical, thin-walled poison-reservoir. From this reservoir arises a long, narrow canal, which runs to the chelicères.[1] With the Araneae, the poison-apparatus consists of two tubes, often a little curved, and surrounded by a layer of flattened, spiral, muscular fasciculi.[2] These two glands are situated at the base of the chelicères, extend more or less into the cephalothorax,[3] and, in front, become suddenly attenuated, forming a narrow excretory duct which terminates at the apex of the hollow claw of the chelicères.[4] With the Scorpionidae, this apparatus is situated in the last caudal segment; it consists of two oval vesicles, whose excretory ducts open at the apex of the sting situated on the end of the tail. These two glands are surrounded by a layer of flat, circular, smooth, muscular fasciculi.[5]

With the Araneae, there is another and very remarkable secretory apparatus, — the Silk organs. Its product is a viscous, transparent liquid which hardens quickly on exposure to the air, forming threads. It escapes by three, rarely by two pairs of spinnerets, situated behind the anus.[6] The glands which secrete it are composed of transparent nucleolated cells, and are of very variable form and disposition, but always situated in the midst of the abdominal viscera. About five kinds of these glands may be distinguished, although not always simultaneously in the same individual. The threads have probably different qualities, according to the glands from which they are secreted.

The genus *Epeira*, containing all these five kinds of glands, will serve as the type for their description. There are observed :[7] 1. Small pyriform follicles, aggregated in groups of hundreds, and having short excretory

---

1 The two poison-glands of the *Trombidium holosericeum*, and *Rhyncholophus phalangioides*, have the form of a ring with a small opening. *Treviranus* (Verm. Schrift. I. p. 48, Taf. VI. fig. 34) has described only very imperfectly these glands with the first mentioned of these animals ; and not having seen their excretory ducts, he took them for salivary glands. *Dugès* (Ann. d. Sc. Nat. III. p. 10), on the contrary, perceived their true relation to the chelicères.

2 It is remarkable that these muscular fasciculi present such different histological characters. I have seen them distinctly striated with *Lycosa*, *Drassus*, *Tegenaria* and *Micryphantes*. They are smooth with *Epeira*, *Thomisus*, *Clubiona* and *Mygale* ; with *Salticus*, they present obscure transverse lines, so that I am undecided whether they belong to the first or to the second of these categories.

3 With *Mygale*, these glands are entirely concealed in the basilar article of the chelicères.

4 *Treviranus*, Bau d. Arachn. p. 31, Taf. II. fig. 21, 22 ; *Lyonet*, loc. cit. p. 397, Pl. XX. fig.

16, 17 ; *Brandt*, Mediz. Zool. II. Taf. XV. fig. 6, or Ann. d. Sc. Nat. XIII. Pl. IV. fig. 2 ; and *Wasmann*, loc. cit. p. 19, fig. 25, 26. For the intimate structure of these glands, see *Meckel*, in *Muller's* Arch. 1846, p. 35.

5 *Muller*, in *Meckel's* Arch. loc. cit. p. 52, Taf. I. fig. 7, 8. *Serres* (loc. cit. p. 90) regards the portion of these glands which is surrounded by muscular fibres, as a reservoir of poison, and that this last is secreted by innumerable glandular follicles enveloping the muscular layer. In fact, with *Scorpio europaeus*, I have seen this layer covered, externally, with a stratum of cylindrical cells.

6 The Mygalidae have two pairs of these papillae, or spinnerets, instead of six, the usual number.

7 I speak here upon the careful investigations of *H. Meckel* (*Muller's* Arch. 1845, p. 50, Taf. III. fig. 40-49). For the older descriptions, see *Treviranus*, Bau d. Arach. p. 41, Taf. IV. V., and Verm. Schrift. I. p. 11, Taf. I. fig. 4 ; and *Brandt*, Mediz. Zool. II. p. 83, Taf. XV. fig. 5, or Ann d. Sc. Nat. XIII. p. 184, Pl. IV. fig. 5.

canals, which are interlaced in a screw-like manner, and open at the six spinnerets;[8] 2. Six long, flexuous tubes, which gradually enlarge into as many pouches, and are then continuous, each, into an equally long excretory duct which forms a double loop; 3. Three pairs of glandular tubes similar to the preceding, but which open externally through short excretory ducts; 4. Two groups of multiramose follicles, whose pretty long excretory ducts run to the two upper spinnerets; 5. Two slightly ramified cacca, varicose at intervals, and which terminate, by two short excretory ducts, in the middle spinnerets.

Most Araneae have three pairs of spinnerets, that is, papillae in the form of an obtuse cone; the middle pair of these is composed of two, and the anterior and posterior pairs, of three articles. The apex of these papillae defines the passage of the thread, and is surrounded by stiff bristles and hairs, and dotted with numerous small, horny tubes, which are only prolongations of the excretory ducts. Each of these tubes is composed of two pieces; one, basilar and thick, the other, terminal and very small, and through the orifice of which the web-liquid escapes in the form of a very delicate thread.[9] The number of these tubes varies according to the species, the age, and the sex.[10] Those belonging to the unbranched glands are distinguished from the others by their size. With some species of *Clubiona* and *Drassus*, there are, beside the usual six spinnerets, two others, composed of a single article and joined together. This fourth pair is situated on the belly, forward of the others, and is connected with a kind of comb (*Calamistrum*) attached to the metatarsus of the two posterior legs.[11]

With *Phalangium*, there is an S-shaped glandular tube situated on the digestive canal, and ending at both extremities by a narrow duct. Its nature is yet unknown; and although the outlet of these excretory ducts has not been discovered, yet as this apparatus is found only with males, it may well be supposed to have some connection with the genital functions.[12]

With some Acarina, there are certain phenomena indicating that these animals have special secretory organs, whose product, like the web-liquid, is hardened on its evacuation. Thus, with some species of the genus *Uropoda*, there is formed, by a substance of this kind, a peduncle situated at the posterior part of the abdomen, and by which these animals fix themselves to insects. This stalk, dilated disc-like, was taken formerly for an organ of suction.[13] Many species of *Hydrachna* fix, by a kind of glue,

---

8 *Mygale* has only this one kind of glands; they form four groups, situated immediately at the base of the spinnerets.

9 The spinnerets and terminal tubes have already been very exactly figured by *Leeuwenhoek* (loc. cit. p. 326, fig. 5, 6), and by *Roesel* (loc. cit. Taf. XXXVIII. fig. 4). See also *Lyonet*, loc. cit. p. 387, Pl. XIX. fig. 6–12; *Wasmann*, loc. cit. p. 20, fig. 31–34, and *H. Meckel*, loc. cit. p. 54, Taf. III. fig. 43–45.

10 There are more than a thousand of these tubes on the spinnerets of *Epeira*; with *Tegenaria*, there are about four hundred; with *Clubiona* and *Lycosa*, three hundred; with *Segestria*, one hundred, and their number is even less with the small spiders; see *Blackwell*, Transact. of the Linn. Soc. XVIII. 1841, p. 219, and Ann. of Nat. Hist. XV. p. 221, and *Menge*, loc. cit. p. 24.

11 This pair of accessory spinnerets is found, ac-

cording to *Blackwell* (loc. cit.), with *Clubiona atrox*, *Drassus viridis, imus, parculus* and *triguus*.

12 See *Treviranus*, Verm. Schrift. I. p. 37, Taf. III. fig. 17, h., and *Tulk*, loc. cit. p. 252, Pl. IV. fig. 21.

13 See *Degeer*, loc. cit. p. 52, Taf. VII. fig. 16, and *Dugès*, Ann. d. Sc. Nat. II. p. 30 (*Uropoda vegetans*). The peduncle is more or less long and often attached to the hardest parts of the Coleoptera. Its formation is connected with some metamorphosis of these animals, and is without doubt due to a secretion produced by some glandular apparatus opening near the anus. This view appears, at least, more natural than that advanced by *Dugès* (loc. cit. p. 30), and adopted by *Dujardin* (Compt. rend. loc. cit. p. 1160), that this peduncle is formed by the feces hardening after their escape from the anus.

the anterior portion of their body on aquatic plants, and in this position, wait the completion of their moulting.[14] The secreting organs of this substance have not yet been discovered.

CHAPTER IX.

ORGANS OF GENERATION.

§ 316.

All the Arachnoidae reproduce by a sexual generation, and their male and female genital organs are situated upon different individuals. The eggs are fecundated in the genital organs of the females, and the males have often copulatory organs of a very singular character. The Tardigrada form an exception in this respect, being hermaphrodites, and wanting the copulatory organs.

In general, the genital organs of the Arachnoidae are composed of the following parts. The ovaries or testicles are always double, but sometimes blended together on the median line. They are situated in the abdomen, and have two excretory ducts, which usually open at a common genital orifice at the base of the abdomen, or under the thorax. The ovaries, when filled with eggs, have always a botryoidal aspect. Only a few species have an ovipositor or a penis. The excretory ducts of both the ovaries and the testicles sometimes have appendages which, with the females, serve to receive the sperm, or to secrete a viscous substance for enveloping the eggs ; and which, with the males, represent an epididymis or the seminal vesicles. Quite often, the males differ from the females in a special modification of their cheliceres, their palpi, or some of their legs. When this is the case, these organs serve, during copulation, to hold the females, or play the part of a penis.

The eggs of the Arachnoidae are spheroidal, rarely oval,[1] and composed of a smooth chorion enclosing a vitellus consisting of vesicles filled with a colorless and, also, often highly-colored fat, in the midst of which is concealed the germinative vesicle. The germinative dot is sometimes simple, sometimes composed of a group of small granules.[2] The eggs of *Lycosa*, *Thomisus*, *Diomedes*, *Salticus* and *Tegenaria*, are remarkable ; for, beside the germinative vesicle, they contain, before being filled with the vitellus, a peculiar, round, finely-granular, solid nucleus.[3]

---

14 According to *Dugès* (Ann. d. Sc. Nat. I. p. 170), *Hydrachna cruenta*, adult, before moulting, bores into aquatic plants by means of its oral organs. But I have seen it fixed, also, upon smooth glass walls, with the parts of its mouth enveloped in a kind of cement.

1 The eggs are oval with the Oribatea and Scorpionidae.

2 The germinative dot is simple and flattened with *Scorpio*, *Thomisus*, *Theridion*, *Micryphantes*, *Lycosa*, *Phalangium*, *Obisium*, *Trombidium*, *Hydrachna*, *Ixodes*, *Oribates*, *Bdella*, &c. It is composed of a group of granules with *Epeira*,

Clubiona and *Salticus* ; see *Wagner*, Prodom. &c. loc. cit. p. 8, Tab. I. fig. 11 (*Epeira*).

3 This nucleus which appears to contain a central nucleolus, is distinguished, with direct light, by its dirty-yellow color, and it has always appeared to me that there were detached successively with mixed with the albumen, without the nucleus diminishing in size. At all events, this nucleus plays an important part in the development of the eggs, for it appears very early, and does not disappear until quite late. It has also been observed

The lower Arachnoidae produce only a small number of eggs at a time, but these are often of a size disproportionately large to that of the animal.[1]

As yet, only very incomplete researches have been made on the elements of the Sperm. It appears, however, that the spermatic particles differ considerably in the various groups. Those of the Tardigrada have the cercarian form; those of the Scorpionidae, on the contrary, are simply filamentoid. But both kinds have very active movements which are suspended by the contact of water.[5] The Sperm of the Araneae always contains spherical or reniform motionless corpuscles.[6] With the Acarina, the spermatic particles are motionless and of most varied forms.[7]

---

by *Wittich* (Observ. quaed. de Aranearum ex ovo evolut. Dissert. Halis, 1845, fig. 1, A.).*

4 With the Tardigrada, the eggs are very large, as are also those of *Oribates, Sarcoptes* and *Demodex*.

5 See *Doyère*, loc. cit. p. 354, Pl. XVI. fig. 5 (*Macrobiotus*), and *Külliker*, Schwelz. Denkschr. VIII. loc. cit. p. 25, Taf. II. fig. 16 (*Scorpio europaeus*). I have observed that the characteristic movements of the spermatic particles ceased instantly from contact with water, and that the particles themselves became twisted and doubled.

6 With *Tegenaria, Salticus, Lycosa* and *Theridion*, the spermatic particles have the form of round cells, while those of *Micryphantes* and *Clubiona* are reniform or semilunar. They are formed in groups in the mother-cells. With *Tegenaria*, a round nucleus is easily distinguished in the spermatic particles. With *Lycosa*, this nucleus is oblong, curved and attached to the wall of the cell; and this led me at first to think these cells were the spermatic particles in their first stages of de-

velopment, and that their definite form would be cercarian. But I quickly abandoned this idea when I found the same form in the seminal receptacle of the females, where, evidently, the spermatic particles cannot be present except in their perfect state.

7 With *Trombidium, Zetes, Oribates,* and *Hoplophora*, the spermatic particles are developed, as I have satisfied myself, under the from of very small, rigid corpuscles, in very large cells. With *Bdella*, they are produced in a similar manner, but are fusiform. With other Acarina they are found of remarkable forms. Thus, in the testicles of the *Hydrachnen* and *Gamasea*, I have observed round masses of cuneiform bodies, at the larger extremity of which there was an oblong granular spot. I have also satisfied myself that these motionless spermatic particles of such large size are preceded in their development by round nucleated cells. In the testicles of *Ixodes ricinus*, I have seen countless transparent staff-like bodies, pretty long and large, motionless, but swollen at one of their extremities when placed in water.†

---

* [§ 316, note 3.] The development and structure of the eggs of Araneae have recently been carefully studied by *Wittich* (Die Enstehung des Arachnideneles im Eierstocke, die ersten Vorgänge in demselben nach seinem Verlassen des Mutterkörpers; in *Müller's* Arch. 1849, p. 113), and by *J. V. Carus* (Ueber die Entwickelung des Spinneseies, in *Siebold* and *Külliker's* Zeitsch. II. 1850, p. 97). The structure of the ovary of these animals is no less beautiful than singular; it resembles a bunch of grapes enclosed in a common capsule. The eggs are developed, each, on the extremity of a pedicle which is attached to the main stem or rachis. The details of the development of the ova are briefly as follows: On the extremity of the pedicle appears a delicate vesicle, or cell, which contains a nucleated cell. This nucleated cell is the germinative vesicle, with its dot, and does not increase so rapidly in size as the vesicle in which it is contained; but this last dilates and expands, and minute cells appear in the liquid, lying between its membrane and the germinative vesicle. These newly-formed cells constitute the vitellus; and when the ovum is completely formed, it consists of vitellus in which is concealed the germinative vesicle with its dot. In a word, the ovum is here formed as elsewhere, except that it is developed on the extremity of a pedicle. In regard to the peculiar bodies mentioned above by *Siebold*, as found in the vitellus, their presence and struc-

ture have been observed by both *Wittich* and *Carus;* they are composed of concentric layers around a nucleus. Of their nature and function nothing is known. — ED.

† [§ 316, note 7.] I have studied the development and nature of the spermatic particles of the Araneae and Acarina, but with results different from those above mentioned. With the first of these, they are developed, as usual, in special daughter-cells, and invariably consist of an arcuate staff, to which is attached a short but very, very delicate tail; indeed, this tail is so tenuous that only the best and highest microscopic powers can bring it out. It escaped the watchful eyes of *Wagner* and *Leuckart*, and led them to adopt erroneous views of the formation of these bodies (see Art. *Semen*, Cyclop. Anat. & Phys. fig. 374). With the Acarina, the particles have the same form and character, but are much more minute and difficult of examination. It would appear from the description given above by *Siebold*, that he must have taken for spermatic particles the peculiar granule-like bodies found in the sperm of the Araneae. These bodies are very hydroscopic, but are homogeneous, and although I could make out nothing further as to their structure, yet it is evident that they are wholly different from the true spermatic particles, and cannot be considered as either undeveloped or modified forms of these last. — ED.

I. *Hermaphrodite Arachnoidae.*

§ 317.

The Tardigrada have only a single, but large, ovarian tube, applied on the posterior half of the digestive canal and opening into the cloaca. This last which is only a dilatation of the rectum, receives, also, two lateral, narrower, seminiferous tubes, together with the excretory orifices of a pyriform seminal vesicle. With *Milnesium, Emydium,* and *Macrobiotus ursellus,* the eggs are surrounded by a smooth chorion, and deposited in a solid epidermis which is detached during the moulting, — so that all the eggs are finally contained in this envelope. But the other species of *Macrobiotus* shield their eggs in another manner, by surrounding each with a very solid, granular capsule.[1]

II. *Female Arachnoidae.*

§ 318.

The female organs of the Acarina consist of two ovarian sacs, the oviducts of which open in a common vulva situated in the middle of the belly, or further forwards on the thorax, sometimes between, sometimes behind the last two pairs of legs.[1] With many of these animals, the oviduct opens into a protractile ovipositor by the use of which the eggs are lodged under the epidermis of plants or animals.[2] A great number of Mites

---

1 For the genital organs of the Tardigrada, see *Doyère,* loc. cit. p. 350, Pl. XIII. XIV. XVI. *Gorze (Bonnet,* Abhandl. aus d. Insekt. 1773, p. 374), and *O. F. Müller* (in *Fuessly,* Arch. d. Insektenkunde, Hft. VI. p. 27, Taf. XXXVI. fig. 4, 5) had already observed that the Tardigrada deposit their eggs in their cutaneous envelope.*

1 With the Gamasea, and Ixodea, the vulva is situated on the thorax ; while, with the Trombidina, Bdella, Hydrachnea, and Oribatea, it is upon the belly ; see *Treviranus,* Verm. Schrift. Taf. V. *(Hydrachna* and *Trombidium); Audouin,* Ann. d. Sc. Nat. XXV. Pl. XIV.; *J. Müller,* Nov. Act. Nat. Cur. XV. Tab. LXVII.; and *Treviranus,* Zeitsch. f. Physiol. IV. Taf. XVI. fig. 2 *(Irodes).* For the ovaries and oviducts of the Acarina, we have only the works of *Treviranus,*Verm. Schrift. I. p. 47 Taf. VI. fig. 32 E. q. *(Trombidium),* and Zeitsch. f. Physiol. IV. p. 100, Taf. XVI. fig. 7, 8, 10, λ. λ. *(Ixodes).* I have observed with *Ixodes ricinus* the following peculiarities, which were probably overlooked by *Treviranus* with *Ixodes americanus.* The two long ovaries anastomose arcuately at the posterior extremity of the abdomen. The two oviducts, here given off, open right and left into a pyriform uterus whose neck communicates laterally with a large caecum coming from the vulva. This caecum is divided by a septum into a posterior, or larger, and an anterior, or smaller, portion. The first receives the sperm which flows from the second during copulation, and thence passes into the uterus and even into the oviducts. The anterior portion represents the vagina properly speaking, and is in communi-

cation with two short cylindrical glands filled with transparent cells, and which secrete probably a substance for enveloping the eggs. I have, moreover, found with other Acarina (for example, with the Hydrachnea, Gamasea, and Oribatea) various organs belonging to the genital apparatus, but without perceiving their relations as clearly as with *Ixodes.* However this may be, I am convinced that *Dujardin* (Ann. d. Sc. Nat. III. p. 20) goes too far in saying that, with most Acarina, the eggs are developed loosely in the parenchyma of the body, without the necessity of an ovary with proper walls. According to this same naturalist (Ibid.), the Oribatea are viviparous and have a large vulva which can be closed by two lateral alae, and before which is an orifice closed also by a similar apparatus. This last orifice belongs to a tube which *Dujardin* regards as a penis ; so that the Oribatea would be hermaphrodites. As to the first point, — the viviparity of these animals, I have verified it for *Hoplophora, Zetes* and *Oribates ;* but I cannot say as much of the second point, for, as I have satisfied myself, the posterior orifice is an anus, and the anterior a vulva having an ovipositor.

2 For example, *Hydrachna ;* see *Dugès* Ann. d. Sc. Nat. I. p. 165. A parasitic mite long known under the name of *Hydrachna concharum* or *Limnochares anodontae,* and which lives in the cavity of the mantle of *Anodontae,* buries its eggs deeply in the skin of that organ ; see *Pfeiffer,* Naturg. deutsch. Land u. Süsswasser-Mollusk. Abth. II. p. 27, Taf. I.; and *Baer,* Nov. Act. Nat. Cur. XIII. p. 590, Tab. XXIX.

---

* [§ 317, note 1.] See also *Kaufmann* (loc. cit. in *Siebold* and *Kölliker's* Zeitsch. III. 1851, p. 230), who has studied the development of the

eggs from their first stages, in the ovary of *Macro bius Dujardin.* — ED.

surround their eggs, grouped together, with a tough coagulable substance, and glue them to various bodies. It is, therefore, very probable that there are special organs for the secretion of this substance.[3]

As yet, we have no observations on the internal genital organs of the Pycnogonidae, although, for a long time, the females have been recognized by their filiform oöphores, composed of nine to ten articles, and situated in front of the first pair of legs.[4]

With the Phalangidae, the two ovaries are blended together, and form a flexuous tube occupying a large portion of the abdomen and continuous anteriorly into two short oviducts. These last unite in a large oviduct situated, loop-like, in the posterior extremity of the abdomen, between the convolutions of the ovaries. Its anterior extremity receives a second oviduct, which, after describing numerous convolutions, opens in a horny, articulated oviposit̄or. This last can be protruded between the posterior legs by means of a special muscular apparatus, on the under side of which are two caeca opening into the oviduct at the base of the ovipositor. These organs are either seminal receptacles, or the secretory organs of a viscous substance.[5]

With most Araneae, the two oblong ovaries are concealed between the hepatic lobes, and open by the intervention of two short oviducts, into a vagina situated between the two pulmonary sacs. This vagina is supported by a horny plate, and opens externally through a transverse fissure, after having previously received the excretory ducts of the two contiguous *Receptacula seminis*. These last are pyriform and nearly always composed of a deep-brown, horny substance; they are attached to the cutaneous envelope, and have, each, an equally horny excretory duct which is more or less long and interlaced with the corresponding one on the other side.[6] The females surround their eggs in groups, with a web, so that they have no organs for secreting a viscid substance.

The Epeiridae offer a remarkable modification in their external genital organs. The entrance to their vagina is covered by a horny process, directed from before backwards, and at the base of which there are pyriform, pedunculated, seminal reservoirs.[7] It is yet undetermined whether this process is connected with the act of copulation, or with the deposition of the eggs.

The Scorpionidae have three ovaries consisting of as many longitudinal tubes united by four pairs of transverse ones. The two external of the former tubes are continuous anteriorly as oviducts, and unite in a short vagina which opens at the base of the abdomen. Before their junction, the oviducts dilate into a round pouch, which, as it sometimes contains the sperm, may be regarded as a *Receptaculum seminis*.[8]

---

3 For example, *Eylais*, *Limnochares*, and *Diplodontus*.

4 See *Johnston*, Magaz. of Zool. I. p. 370 Pl. XIII.; *Milne Edwards*, Hist. Nat. d. Crust. Pl. XLI. fig. 7; and *Philippi*, in *Wiegmann's* Arch. 1843, I. p. 177, Taf. IX. With *Phoxichilidium*, the two oöphores are only five articled.

5 See *Treviranus*, Verm. Schrift. I. p. 34, Taf. IV. fig. 20, 23; also *Tulk*, loc. cit. p. 318, Pl. V. fig. 26-29.

6 *Treviranus* (Bau d. Arachn. p. 37, Taf. IV. fig. 32) has figured very correctly the ovaries and their oviducts. He has even seen the *Receptacula seminis*, but he mistook them for cartilaginous bodies (Ibid. p. 38, Taf. II. fig. 20, o. and Taf. IV. fig. 40, o. 41). The anatomists who succeeded him

pahl no attention to these organs. The seminal receptacles are short and pyriform with *Lycosa*, *Theridion*, and *Micryphantes*; but they have a long excretory duct entwined with its opposite, with *Drassus*, *Salticus*, and *Thomisus*.

7 This process is S-shaped with *Epeira diadema*. It has been described and figured by *Leeuwenhoek*, loc. cit. p. 356, fig. 8; *Roesel*, loc. cit. p. 253, Taf. XXXVII. fig. 1, b. and Taf. XXXVIII. fig. 1, 3; and by *Degeer*, loc. cit. p. 85, Taf. XII. fig. 10. See, also, *Treviranus*, Bau d. Arachn. p. 39, Taf. II. fig. 18, c.; and *Savigny*, Descript. de l'Egypte, loc. cit. Pl. II. fig. 8 m. With *Nephila fasciata*, this process is tongue-shaped.

8 The female organs of the Scorpionidae have been described by *Meckel* (Beitr. loc. cit. p. 113,

III. *Male Arachnoidae.*

§ 319.

From the few observations hitherto made upon the male organs of the Acarina, it appears that they are formed after very different types. With *Trombidium*, there are twenty red, testicular vesicles, attached by short peduncles to the annular *Vas deferens* which opens between the posterior legs. This last, before its termination, receives also two brown, long-pedunculated vesicles, whose nature is yet unknown.[1]

With *Ixodes*, the testicles consist of a group of four to five pairs of longer or shorter follicles, which unite in the middle of the abdomen, and send off two small *Vasa deferentia* to the base of the chin-like process. This last, together with the chelicercs, these animals introduce deep into the vagina during copulation, while their two palpi, separated at a right angle, are applied upon the thorax of the female.[2] With *Gamasus*, there appear to be only two simple, isolated, testicular follicles, each having a deferent canal. With many Acarina, there is a short penis situated at a point corresponding to that of the vulva of the females, and sometimes concealed within the body.[3] With other Acarina, the males are distinguishable from the females, by the larger size of their chelicercs, and some of the legs which serve to retain the females during copulation.[4]

With the Phalangidae, the testicles consist of numerous small caeca, all united at one point into a long, flexuous deferent canal. This last is continuous into a *Ductus ejaculatorius* which traverses a muscular penis; this terminates with a hook-like gland, and its body is horny and surrounded by a muscular sheath out of which it can be protruded under the thorax.[5] With many Opilionina, the posterior legs have remarkable spines and excrescences which, undoubtedly, are used during copulation.[6]

With the Araneae, the testicles consist of two long, simple, interlaced caeca, concealed between the hepatic lobes.[7] From them pass off two

Taf. VII. fig. 18–20); *Treviranus* (Bau. d. Arachn. p. 12, Taf. I. fig. 12), and *Muller* (loc. cit. p. 53 Taf. II. fig. 14–19). This last naturalist has found, with the large African scorpions, in the lateral long, varicose and caecal appendages of the ovaries. These appendages do not increase in size except in proportion as the eggs are developed; while, with the small European scorpion, the eggs produce only simple pyriform folds on the ovarian tubes. I have found sperm and very active spermatic particles in the seminal receptacles of living females of *Scorpio europaeus*.

1 I have proved this complicated disposition of the male organs with *Trombidium holosericeum*, where it had been wholly misapprehended by *Treviranus* (Verm. Schrift. I. p. 48, Taf. VI. fig. 35).

2 This singular mode of coition had already been observed by *Degeer* (loc. cit. p. 45, Taf. VI. fig. 6) with *Ixodes ricinus*, and subsequently by *Ph. W. J. Muller* (Germar's Magaz. d. Entomol. II. 1817, p. 281); but it remained wholly unobserved by the other entomologists. It appears that the male of *Ixodes ricinus*, which differs considerably from the female, has been mistaken for a different species and named *Ixodes reduvius* (Audouin, loc. cit. XXV. p. 422, Pl. XIV. fig. 4), or *Ixodes marginalis* (Hahn, Die Arachn. II. p. 63, fig. 153). The characters peculiar to this sex are, a

dorsal shield covering the whole body, palpi shorter, and teeth less numerous upon the also shorter chin like process.

3 The penis is sub-ventral with *Bdella*, sub-thoracic with *Gamasus*, and behind the genital orifice with *Oribates*. With *Arrenurus*, it is inserted on a tuberosity of the abdomen, giving the males a very singular appearance (Dugès, loc. cit. 1. p. 155. Pl. X. fig. 20).

4 With the males of certain species of *Gamasus*, the two chelicercs are perforated, and the second pair of legs is very stout and provided with spines and excrescences. With *Dermaleichus*, it is the third pair of these organs which is sometimes very large and armed with robust nails. With *Sarcoptes*, the posterior legs are long and armed with nails and discs, while with the females these same legs are abortive.

5 *Treviranus*, Verm. Schrift. I. p. 36, Taf. IV. fig. 21, 22; and *Tulk*, loc. cit. p. 250, Pl. IV. fig. 21–24.

6 With *Eusarcus, Gonyleptes*, &c. The very large chelicercs of the Phalangidae are not used in the act itself of copulation, but are employed to fight with on these jealous occasions; see *Latreille*, Hist. Nat. d. Fourmis, p. 380.

7 For the testicles of the Araneae, see *Treviranus*, Bau d. Arachn. p. 37, Taf IV. fig. 33, and *Brandt*, Mediz. Zool. II. p. 89, Taf XV. fig. 7.

deferent canals to the base of the abdomen where, between the two pulmonary sacs, there is a genital opening in a small horny plate. This opening is only a simple transverse fissure which, in copulation, does not come in contact with the vulva of the female. These animals always use their hollow, spoon-shaped palpi, which often have a very complicated structure. They are filled with sperm and applied to the entrance of the vulva. For this purpose, the last article of the palpi, which is always hollow and much enlarged, contains a soft spiral body terminated by a curved, gutter-like, horny process. Beside this, there is an arched, horny filament, and several hooks and other appendages of the most varied forms. These appendages are protractile, and serve, some to seize the female, and others as conductors of the sperm.[8]

With the Scorpionidae, each of the two testicular tubes forms a loop enveloped by the substance of the liver, and connected with its mate by two transverse canals. The anterior border of each of these loops sends off a short *Vas deferens* which opens at the base of the abdomen, receiving in its track two caeca of unequal length. Of these, the longer contains a granulo-vesicular substance, and is, perhaps, an accessory gland ; while the shorter, from the character of its contents, is evidently a *Vesicula seminalis*.[9] A deeply crenated, small papilla projects out of the genital orifice, and, as it is wanting with the females, may be regarded as a

---

8 For a long time, the excretory point of the *Vasa deferentia* was undetermined, because the two sexes of these animals had never been seen to place in contact these genital orifices during copulation. But when it was discovered that only the palpi of the males touch the vulva, the excretory ducts of the testicles were sought in these palpi. It is only recently, however, that it has been perceived, that with these animals, as with the Libellulidae (see my memoir in *Germar's* Zeitsch. f. d. Entomol. II. p. 423), the copulatory organ and the *Vesicula seminalis* are entirely removed from the male genital orifice. In order to be convinced that the application of the male palpi against the female vulva constitutes really the act of copulation, it is only necessary to examine the palpi under the microscope and compress them. From the last article a large quantity of sperm will be seen to escape. Then again, after copulation, the *Receptacula seminis* of the females will be found filled with the fluid. The form of the palpi with the males varies almost infinitely according to the genera and species. They are very simple and slightly swollen with *Clubiona* and *Lycosa*, while, with *Epeira, Tegenaria, Linyphia, Micryphantes, Salticus, Argyroneta*, &c., their last article is so complicated that the most minute description would be inadequate in giving an exact idea of it ; see the figures of *Lyonet*, loc. cit. p. 383, Pl. XIX. XX.; *Trevi-*

ranus, Bau d. Arachn. p. 37, Taf. IV. fig. 35–37 ; *Brandt*, Mediz. Zool. II. p. 87, Taf. XV. fig. 1 ; *Savigny*, Descript de l'Egypte, Pl. I.–VII.; *Menge*, loc. cit. p. 35, Taf. III. fig. 13–27.

Treviranus (Bau d. Arach. p. 33) has made an exposition of the older opinions of *Lister, Lyonet, Clerk*, and *Degeer*, on the copulation of these animals ; but he regarded the act as only a prelude for exciting the sexual desires, and which would be followed by a real copulation, consisting of the contact of the male and female genital orifices. Moreover, *Treviranus* had never observed this last act. The more recent observers, such as *Dugès* (Ann. d. Sc. Nat. VI. p. 187), *Menge* (loc. cit. p. 36), and *Blackwall* (Annals of Nat. Hist. XV.) have naturally only confirmed the views of the older naturalists. According to *Menge*, the males of *Linyphia* and *Agelena* evacuate a drop of sperm from their genital orifice, which is then received and absorbed by the last article of their palpi.*

9 For the male organs of the Scorpions, which have been very imperfectly described by *Treviranus* (Bau d. Arachn. p. 22, Taf. I. fig. 11), see *Meckel* (Beiträg. loc. cit. p. 114, Taf. VII. fig. 14), *Serres* (loc. cit. p. 89) and *Muller* (loc. cit. p. 59, Taf. I. fig 8). I have seen distinctly with living individuals of *Scorpio europaeus*, spermatic particles in motion in the small caeca which I have called the seminal vesicles.

---

* [§ 319, note 8.] I have made a microscopic examination of this curious palpus-structure in connection with the general structural relations of the internal genital organs, in some of the common Araneae (*Tegenaria, Agelena*, &c.) where this formation is most marked. In *Agelena*, the peculiar, corkscrew-like, horny process, situated in the last, spoon-shaped article of the palpus, contains a canal throughout, which commences in a kind of receptaculum at the base of the process. This receptacle is filled with the peculiar granule-like bodies mentioned above (Note to §

316, note 7). As the most repeated and careful examinations showed no spermatic particles in this palpus-capsule, I was led to advance the view that the palpi were only excitatory and not intromittent organs, in the copulatory act (see Proceed. Boston Soc. Nat. Hist. IV. 1851, p. 106). But the question is still open, and especially as some recent investigators of the economy of these animals have observed facts that would indicate the intromittent function of these organs ; see particularly, *Blackwall*, Ann. Nat. Hist. passim, for several years past. — Ed.

rudimentary penis. It is yet undetermined if the two external, lateral, pectiniform appendages situated near the genital orifice in both sexes, serve any purpose during copulation.[10]

§ 320.

We have, as yet, only insufficient observations on the Development of the Arachnoidae. However, those that we possess upon its first stages,[1] show that, with few exceptions,[2] here as with the Crustacea, the disappearance of the germinative vesicle is followed by a superficial and partial segmentation of the vitellus. There is thereby formed a thin embryonic layer, composed of molecular corpuscles retained in a transparent viscid liquid, and distinguished from the rest of the vitellus by its white color. While this oval blastoderma, whose longitudinal axis corresponds to the ventral or nervous side of the embryo, extends towards the sides and the back, it divides into an external or serous, and an internal or mucous lamella, the last of which gradually covers the remaining portion of the vitellus, and becomes changed into the digestive tube and its appendages. In the mean while, there appear, on the external surface of the serous lamella, various symmetrical prominences and projections, which in time become the segments of the body, the parts of the mouth, the tactile, and the locomotive organs.

With the exception of the Scorpionidae, and Oribatea, which are viviparous, the embryos of all the Arachnoidae are developed subsequent to the deposition of the eggs.[3]

With the majority of Arachnoidae, the embryos, at their escape from the egg, have the form of the adult.[4] The lower orders, only, form an exception in this respect, for they acquire their definite form after several moultings, and a true metamorphosis. With the Pycnogonidae, these

---

[10] *Tulk* (Ann. of Nat. Hist. XV. p. 56) has lately expressed the opinion that these combs serve to clean the palpi, the tarsi, and the extremity of the tail. He adduces, as proof, the presence of transparent combs of exactly identical form, with *Obisium*, between the pincers of the cheliceres, and which are used for this purpose.

[1] The first phases of development have not been observed as yet except with the Araneae and Scorpionidae; see *Herold*, De Generat. Aranear. in ovo, 1824; *Rathké*, Zur Morphol. Reisebemerk. aus Taurien, 1837, p. 17, and in *Burdach's* Physiol. fl. 1837, p. 242; the same in *Froriep's* neue Notiz. XXIV. 1842, p. 165 (*Lycosa saccata*); also *Kölliker*, in *Müller's* Arch. 1843, p. 139 (*Scorpio europaeus*); finally, *Wittich*, Observ. quaed. de Aranear. ex ovo evolut., Halis. 1845, fig. 1, A. As to the development of the Acarina, I have satisfied myself upon the eggs of the Oribatea,

that the same phenomena occur as with most of the other Arachnoidae.*

[2] I have distinctly seen, in the eggs of *Macrobiotus Hufelandii*, the segmentation involve the entire vitellus. *Kölliker* (*Müller's* Arch. 1843, p. 136) has made the same observation with *Pycnogonum*.†

[3] The relations of the Scorpionidae in this respect are very remarkable, for their embryos are developed in the ovaries at the spot even where the eggs are formed; see *Muller*, loc. cit. p. 55, and *Rathké*, Zur Morphol. loc. cit. It is evident that the sperm must ascend from the two seminal reservoirs into the ovaries to fecundate the eggs. With the Oribatea, the embryos appear to be developed in a kind of uterus situated immediately behind the ovipositor.

[4] With the Araneae, the sexual differences which are so striking do not appear until after the first moulting.

---

* [§ 320, note 1.] See, also, *Wilson*, Researches into the structure and develop. of a newly-discovered parasitic Animalcule, &c., in the Philos. Trans. 1844, p. 305 (*Entozoon (Demodex) folliculorum*), and *Van Beneden*, Recherches sur l'Atax ypsilophora, in the Mém. de l'Acad. Roy. de Bruxell. XXIV. *Wilson's* details are imperfect, and throw but little light on the real character of the development of the follicle-parasite. It would appear, however, to be truly one of the Arthropoda. — ED.

† [§ 320, note 2.] For the embryology of the

Tardigrada, see *Kaufmann*, loc. cit. in *Siebold* and *Kölliker's* Zeitsch. III. 1851, p. 220. The type of development is like that of the Articulata in general, and this would seem to clearly settle the position of these animals in this class. *Kaufmann* confirms the observation of *Siebold* as to the segmentation of the entire vitellus. After this process has occurred, the mulberry like mass is changed into the embryo, exactly as is observed with the eggs of the Arachnoidae in general. — ED.

changes should be most prominent, for their embryos have a short unarticulated body, and, beside the cheliceres, are provided with only four bi- or
tri-articulated feet.   There is, however, a very long lash, attached, sometimes to the two cheliceres, sometimes to each of the four legs.   It is not
until after successive moultings, that the other legs, the divisions of the body,
and the extremities, appear.[(5)]

With *Emydium*,[(6)] and most of the Acarina, the embryos have only six
legs, when hatched ; but as they otherwise resemble the adults, their
metamorphosis consists only in the appearance of another pair of legs.   A
true metamorphosis is observed only with *Hydrachna*.   Here the embryos
have a very long and large snout which might easily be taken for a head
distinct from the trunk.[(7)]   This disproportion between these two parts is
subsequently reversed, when the young pierce with their snout the bodies
of insects, while their own bodies, gorged with food, become of a monstrous
size.   These young have six legs, and, during their parasitic life, were for a
long time described as distinct species under the generic name *Achlysia*,
until it was discovered that they possessed eight legs after their first moulting.[(8)]   Similar metamorphoses occur with the *Trombidia* which, as red,
hexapod larvae, are attached to flies, grasshoppers, plant-lice and various
other terrestrial insects.   These, also, have been formed into proper genera
under the names of *Astoma, Leptus,* and *Ocypeta.*[(9)] *

5 See *Kröyer*, Naturhist. Tids. III. 1840, p.
299, or Isis, 1841, p. 713, Taf. III. Tab. III. or
Ann. d. Sc. Nat. XVII. p. 288, Pl. XIII. B.
6 See *Doyère*, loc. cit. p. 358. The embryos of
the other Tardigrada have four pairs of legs.
7 See *Dugès*, Ann. d. Sc. Nat. I. p. 166, Pl. XI.
fig. 47. It follows clearly from the position of the
eyes, which are situated not upon this snout, but
upon the cephalothorax, that this snout is only a
support of the parts of the mouth, and not a head.
8 See *Audouin*, Mém. sur l'Achlysie in the Mém de
la Soc. d'Hist. Nat. de Paris, I. p.98, Pl. V. No. 2). He

found these red *Achlysiae* with monstrous bodies on
the dorsal surface of the abdomen of *Dytiscus* and
Hydrophilus. Others, smaller, were observed even
more frequently upon the segments of the body and
the articles of the extremities, with *Nepa*, and *Ranatra*. The true nature of these epizoa has been
cleared up by *Burmeister* (Isis, 1834, p. 138, Taf.
I. fig. 1–6), and by *Dugès* (Ann. d. Sc. Nat. I.
1834, p. 166, Pl. XI. fig. 49–55).
9 See *Gervais*, in *Walckenaër*, Aptères, &c., III.
p. 178.

* [§ 320, end.] There is some ground for the
opinion that alternation of generation, so called,
occurs with some of the Arachnoidae. *Dujardin*
(Ann. d. Sc. Nat. 1849, XI. p. 243) has examined
the wall-mite found on the house and other flies —

the *Acarus muscarum* of *Degeer*, and the *Hypo-
pus* of *Dugès*. It has neither mouth nor digestive
apparatus, but simply adheres to the animal on
which it lives, by a sucker. It may be the nurse
of a *Gamasus ?* — ED.

BOOK FOURTEENTH.

INSECTA.

CLASSIFICATION.

§ 321.

As anatomists have been able to examine, with few exceptions, nearly all the orders and families of the Insecta, their anatomical researches have not been restricted, as in the preceding classes, to isolated genera or species, but have embraced entire families. With such abundant materials, we should go beyond the limits of our work in enumerating here all the genera, or even all the families, whose organization has been studied.

A. INSECTS WITHOUT METAMORPHOSIS.
(*Insecta ametabola.*)

ORDER I. APTERA.
FAMILIES: PEDICULIDAE, NIRMIDAE, PODURIDAE, LEPISMIDAE.

B. INSECTS WITH INCOMPLETE METAMORPHOSIS.
(*Insecta hemimetabola.*)

1. MOUTH SUCTORIAL.

ORDER II. HEMIPTERA.
FAMILIES: COCCIDAE, APHIDIDAE, PSYLLIDAE, CICADIDAE, CERCOPIDAE, NAUCORIDAE, NEPIDAE, COREIDAE, PENTATOMIDAE.

2. MOUTH MANDIBULATE.

ORDER III. ORTHOPTERA.
FAMILIES: PHYSOPODA, FORFICULIDAE, PSOCIDAE, PERLIDAE, EPHEMERIDAE, LIBELLULIDAE, TERMITIDAE, ACRIDIDAE, LOCUSTIDAE, ACHETIDAE, PHASMIDAE, MANTIDAE, BLATTIDAE.

C. Insects with complete Metamorphosis.

(*Insecta holometabola.*)

1. Mouth Suctorial.

a. Two wings. Under lip changed into a suctorial organ.

ORDER IV. DIPTERA.

Families: Pulicidae, Nycteribidae, Hippoboscidae, Muscidae, Oestridae, Syrphidae, Conopidae, Stomoxydae, Bombylidae, Anthracidae, Leptidae, Henopidae, Asilidae, Stratiomydae, Tabanidae, Tipulidae, Culicidae.

b. Four scaly wings. Maxillae changed into a suctorial organ.

ORDER V. LEPIDOPTERA.

Families: Tineidae, Pyralidae, Geometridae, Noctuidae, Bombycidae, Hepiolidae, Zygaenidae, Sphingidae, Papilionidae.

c. Four naked wings. Tongue changed into a suctorial organ.

ORDER VI. HYMENOPTERA.

Families: Apidae, Andrenidae, Vespidae, Formicidae, Scoliadae, Mutillidae, Pompilidae, Crabonidae, Bembecidae, Chrysididae, Cynipidae, Ichneumonidae, Siricidae, Tenthredinidae.

2. Mouth Mandibulate.

a. Two posterior wings only.

ORDER VII. STREPSIPTERA.

b. Four membranous wings.

ORDER VIII. NEUROPTERA.

Families: Phryganidae, Sialidae, Hemerobidae, Myrmeleonidae, Rhaphidiadae, Panorpidae, Mantispidae.

c. Two upper wings corneous, and two under membranous.

ORDER IX. COLEOPTERA.

Families: Pselaphidae, Coccinellidae, Chrysomelidae, Cerambycidae, Curculionidae, Cistelidae, Meloidae, Tenebrionidae, Pyrochroidae, Elateridae, Lamellicornes, Clavicornes, Hydrophilidae, Hydrocanthari, Staphylinidae, Carabidae.

BIBLIOGRAPHY.

Réaumur. Mémoires pour servir à l'histoire des Insectes. 1734.

Roesel. Insekten-Belustigungen. 1746.

Swammerdamm. Bibel der Natur. 1752.

Lyonet. Traité anatomique de la chenille, que ronge le bois de saule, 1762, and Anatomie de différentes espèces d'Insectes, in the Mém. du Muséum, &c., XVIII.-XX. 1829-32.

Degeer. Abhandlungen zur Geschichte der Insekten. 1776.

Meckel. Beiträge zur vergleichenden Anatomie. 1808.

Gaede. Beiträge zu der Anatomie der Insekten. 1815.

Suckow. Anatomisch-physiologische Untersuchungen der Insekten und Krustenthiere. 1818.

Straus-Dürckheim. Considérations générales sur l'Anatomie comparée des animaux articulés. 1828.

Kirby and *Spence.* Introduction to Entomology, 1816-28, or its translation into German by *Oken,* 1823-33.

Burmeister. Handbuch der Entomologie. 1832.

Brandt and *Ratzeburg.* Medizinische Zoologie. Bd. II. 1833.

Ratzeburg. Die Forstinsekten. 1837.

Lacordaire. Introduction à l'entomologie. 1834-38.

Westwood. An introduction to the modern classification of Insects. 1839.

Newport. Article *Insecta,* in the Cyclopaedia of Anatomy and Physiology, II. 1839.

Léon Dufour. Recherches anatomiques et physiologiques sur les Hémiptères, 1833; and, Recherches anat. et physiol. sur les Orthoptères, les Hyménoptères et les Neuroptères, 1841. Both of these are in the Mém. à l'Acad. royale d. Sc. de l'Institut. de France, IV. VII.

Nicolet. Recherches pour servir à l'histoire des Podurelles, 1841, in the Neue Denkschr. der allg. schweizer. Gesellschaft, &c. Bd. VI.

ADDITIONAL BIBLIOGRAPHY.

Dufour. Études anatomiques et physiologiques, sur une Mouche, in the Mém. de l'Instit. IX. 1846, p. 545.

——— Recherches anatomiques et physiologiques sur les Diptères, in Ibid. XI. 1851, p. 171. See also his various communications on the anatomy and metamorphosis of different Insecta, in the Ann. d. Sc. Nat. VII. 1847, p. 5, and p. 14; VIII. 1847, p. 341; IX. 1848, p. 91, 199, 205, 344; XIV. 1850, p. 179; XVII. 1852, p. 65.

Blanchard. De la Circulation dans les Insectes, in the Ann. d. Sc. Nat. IX. 1848; p. 359 — also in extract in the Comp. rend. XXIV. 1847, p. 870, or in the Ann. Nat. Hist. XX. 1847, p. 112, or in *Schleiden* and *Froriep's* Not. LXVI. 1847, p. 342.

——— Nouvelles observations sur la circulation du sang et la Nutrition chez les Insectes, in the Ann. d. Sc. Nat. 1851, XV. p. 371.

Stein. Vergleichende Anatomie und Physiologie der Insekten. In Monographieen, Erste Monographie: Die weiblichen Geschlechtsorgane der Käfer. Mib. 9 Kupertafeln. Berlin, 1847.

Siebold. Ueber die Fortpflanzung von Psyche : Ein Beitrag zur Natur-geschichte der Schmetterlinge — in *Siebold* and *Külliker's* Zeitsch. I. 1848, p. 93 ; see, also, for further researches on the subject, his Bericht üb. die entomol. Arbeiten d. schles. Gesellsch. in J. 1850, or its transl. in the Transact. of the Ent. Soc. I. 1851, p. 234.

Leydig. Die Dotterfurchung nach ihrem Vorkommen in der Thierwelt und nach ihrer Bedeutung, in the Isis, 1848, Hft. 3.

—————— Einige Bemerkungen über dei Entwickelung der Blattläuse, in *Siebold* and *K üliker's* Zeitsch. 1850, II. p. 62.

—————— Anatomisches und Histologisches über die Larve von Corethra plumicornis, in Ibid. III. 1852, p. 435.

—————— Zur Anatomie von Coccus hesperidum, in Ibid. V. 1853, p. 1.

Meyer. Ueber die Entwickelung des Fettkörpers, der Tracheen und der Keimbereitenden Geschlechtstheile bei den Lepidopteren, in *Siebold* and *K üliker's* Zeitsch. I. 1849, p. 175 ; see also the Mitth. d. naturf. Ges. in Zürich, Hft. 2, p. 206.

See, also, the various writings referred to in my notes. — ED.

CHAPTER I.

EXTERNAL ENVELOPE AND CUTANEOUS SKELETON.

§ 322.

The cutaneous envelope of the multi-articulate body of the Insecta con-sists, as with the other Arthropoda, of a kind of external skeleton, of a con-sistence sometimes leathery and soft, sometimes horny and solid. Its elasticity and flexibility is limited to the points of junction of the segments of the body, and of the articles of the extremities. Its characteristic chemical substance is likewise chitine, a peculiar azotic matter insoluble in caustic potass, and with which highly-colored pigments are often chemically combined.[1] Chitine enters also into the composition of the hairs and the scales of the skin, and the internal processes which may be regarded as an Internal Skeleton.

§ 323.

Histologically, the cutaneous envelope is so variously and often so ex-

---

1 See *Odier*, Mém. de la Soc. d'Hist. Nat. de Paris, 1. loc. cit. ; *Lassaigne*, Compt. rend. XVI. 1843, p. 1087, or *Froriep's* neue Notiz. XXVII. p. 7, and *Schmidt*, Zur vergleich. Physiol. &c. p. 32. *Lassaigne* has proposed for this substance the name *Entomoderm*. The coloration of the cutaneous skeleton is probably due to an oil with which the chitine is impregnated, especially with the Coleoptera.

See *Bernard-Deschamps*, Sur les Elytres des Coléoptères, in the Ann. d. Sc. Nat. III. 1845, p. 354.*

* [§ 322, note 1.] Recent researches have shown that the peculiar substance Chitine is not limited in its distribution to the Arthropoda, for it has been found in nearly every class of the Inver-tebrata. See *Grube*, *Müller's* Arch. 1848, p. 461, and *Wiegmann's* Arch. 1850, p. 253 ; *Schultze*, Beitr. zur Naturgesch. d. Turbellarien, p. 33 ; and *Leuckart*, Morphol. der wirbellosen Thiere, p. 49, in *Siebold* and *Külliker's* Zeitsch. 1851, p. 192, and in *Wiegmann's* Arch. 1852, p. 22. — ED.

traordinarily complicated, that it is very difficult to recognize its elements. When horny, there can always be distinguished an epidermis composed of unnucleated, lamellated cells intimately blended together. These cells, however, are often polyhedral, and so disposed as to form a simple layer; in other cases, they are more or less blended together, giving rise to undulating or imbricated lines in the epidermis. In order to study the subjacent layer, or dermis, the cutaneous envelope must be macerated and decolored in caustic potass. This layer will then be found to be composed usually of several lamellae superposed in various ways and thereby often producing very elegant markings. In many instances, these reticulated or radiated markings would indicate the presence here of intercellular passages, and porous canals.[1] In the thin, membranous portions of the skeleton, for instance, the wings, the structure usually appears wholly homogeneous.

On the external surface of this envelope there are o en numerous excrescences, such as tubercles, spines and hairs, which are usually hollow. The hairs are sometimes simple and smooth, sometimes set with small hairs or barbellate.[2] Many of these cutaneous formations are inserted by a small peduncle in small fossae, to which they loosely adhere, and from which they are very readily detached. Usually, they are flattened, scale-

---

1 Histological researches upon the cutaneous skeleton have, as yet, been extended over only a few species. I am able to cite only the works of *H. Meyer* (*Müller's* Arch. 1842, p. 12 (*Lucanus cervus*), and of *Platner* (Ibid. p. 38, Taf. III. (*Bombyx mori*).

2 These barbellate hairs are found with the larvae of all the Bombycidae (*Réaumur*, Mém. &c. Tom. I. Pl. VI., and *Degeer*, Abhandl. I. Taf. IX.-XIII.).

They are easily rubbed off, and when brought in contact with our skin, they insinuate themselves by the truncated extremity, and thereby often provoke an insupportable itching or even an inflammation. The processionary moths are so much feared in this respect as to pass for being poisonous ; see *Nicolai*, Die Wander-oder Prozessionsraupe, 1833, p. 21, and *Ratzeburg*, Du Forstinsekten, II. p. 127, Taf. I. fig. 11, 12, and Taf. VIII.* The pains which these hairs can produce with man, may be judged by the disease which *Ratzeburg* suffered, and of which he has given an account (Entom. Zeit. 1846, p. 35).

The symptoms spoken of by this excellent entomologist may be explained without attributing any specific poisonous property to these hairs, if it be considered that, like a fine powder, they rest on the skin and may enter the respiratory organs by inhalation, and penetrating the tissues encounter a multitude of nervous fibres. Their passage into the tissues is the more easy, since they are fusiform, very sharp at both extremities, the free

one of which is provided with denticulations pointing upwards, while the opposite one is loosely inserted in a small fossa, so that they are detached without breaking from their fastenings by the least contact. The deep-colored spots observed on the back of the processionary moths, and which are divided into four parts by crucial lines (*Ratzeburg*, Die Forstinsekt. loc. cit. Taf. VIII. . fig. 1.ᴸ and 1. ʲ), consist of callosities on which are situated thousands of these small fossae from which arise an infinite number of hairs. With many birds and insectivorous reptiles, the hairs of the moths which these animals have eaten, traverse the mucous membrane of the stomach and enter the tissues. I should not have thus mentioned this subject, since for a long time the true nature of the hairy stomachs of old cuckoos has been understood (see the discussion on this subject between *Brehm*, *Richter*, *Carus*, *Oken*, and *Bruch*, in the Isis, 1823, p. 222, and 666, Taf. VIII., also, 1825, p. 579, Taf. IV.), if, recently the passage of hairs from the digestive tube into the mesentery of frogs had not given rise to a similar error. The mesentery of these reptiles very often contains fragments of hairs and the spines of insects, surrounded by concentric layers of connected tissue and thus arrested in their course. These encysted hairs have been described by *Remak* (*Müller's* Arch. 1841, p. 451) under the name of parasitic enigmatical horny fibres, while *Mayer* at Bonn has gone so far as to take them for Pacinian corpuscles (Die Pacinischen Körperchen, 1844, p. 14, fig. 2).*

---

* [§ 323, note 2.] *Will* (*Schleiden* and *Froriep's* Not. 1848, Aug. p. 145) has made chemico-microscopical investigations upon the nature of this peculiar poisoning power manifested in the processionary moths ; his researches were upon *Bombyx processionea*. The poisonous material was found to be formic acid in a free and highly-concentrated state ; it was met with in all parts of the caterpillar, but especially in the faeces, in the greenish-yellow liquid emitted by these animals when di-

vided, and in the hairs. These hairs were mostly hollow, and their cavity was not closed at their base, but passed through the skin and appeared connected with glands below. These observations are the more interesting since this same observer has shown that the poisonous material of the poison-apparatus of the Hymenoptera, consists likewise of formic acid. See my note under § 347, note 11. — Ed.

like, and colored. Their forms vary infinitely not only according to the species, but also according to the regions of the body. They are often ribbed longitudinally, and denticulated or deeply serrated on their borders. These scales are often inserted on the skin perpendicularly, thus forming a pelt-like covering easily wiped off; sometimes they are imbricated and exactly fitted to each other.[3]

§ 324.

The various parts of the cutaneous skeleton of the Insecta have been so carefully studied in Zoology, that they may well be passed over here. As the internal surface of this skeleton furnishes points of insertion to the voluntary muscles, the segments of the body on which these last are attached, would naturally be developed in proportion to their volume. Thus with those Insecta which have powerful masticatory organs, the head is remarkably large on account of the prominent development of the masticatory muscles; in the same way likewise other fossorial, rapacious, or saltatory Insecta, indicate their habits of life by the size of their legs, which are endowed with great muscular power. It is for the same reason, also, that with the species which fly, the mesothorax and metathorax are so largely developed; for these contain not only the muscles of the last two pairs of legs, but also those of the wings. These relations are especially distinct in those families or genera which embrace both winged and unwinged species.

On the internal surface of the cutaneous skeleton, are found, in the head and thorax, processes which may be regarded as an internal skeleton. Some of these serve as septa, which separate certain organs, and others furnish points of insertion to muscles, and then are often bifurcated.[1]

CHAPTER II.

MUSCULAR SYSTEM AND LOCOMOTIVE AND SONIFEROUS ORGANS.

§ 325.

The muscular fibres of the Insecta are striated, not only in the voluntary muscles, but often also in those of organic life, as in the stomach and intestine.[1] All are colorless or of a dirty yellow color. This last is especially observed with the muscles of the thorax belonging to the wings, which differ,

---

3 This covering has a velvety aspect with the Lepidoptera, Anthracidae, and Bombylidae ; it is scaly on the bodies of many of the Curculionidae, Melolonthidae, Clavicornes, Lepismidae, Poduridae and on the wings of the Culicidae, and Lepidoptera. It has always excited the attention of naturalists, who have figured it in their works. See also Réaumur, and Degeer, loc. cit., also Lyonet, Mém. du Muséum, XX. p. 82, Pl. VI.-XI. ; Bernard-Deschamps, Ann. d. Sc. Nat. III. 1835, Pl. III. IV. ; Ratzeburg Die Forstinsekt. II. Taf. I.;

Dujardin, Observ. au microscop. p. 121, Pl. VII. IX. XI. XII. ; Nicolet, loc. cit. p. 22, Pl. II. (Poduridae), and H. Fischer, Isis, 1846, p. 401, Taf. IV. (Coleoptera).

1 For the internal skeleton of the Insecta, see Audouin, Ann. génér. d. Sc. physiq. VII. p. 182, or Meckel's deutsch. Archiv VII. p. 435 ; Eschscholtz, Isis, 1822, p. 52 ; Burmeister, Handb. &c. I. p. 251, and Newport, Cyclopaed. loc. cit. II. p. 909.

1 Necrophorus.

furthermore, from the others, in their transverse striae being less distinct, and their fibrillae being more easily observed ; indeed, with most species, these last show a tendency to separate even from the least pressure, while those of the other voluntary muscles are very compactly united together.[2]

The muscles are attached directly upon the cutaneous skeleton, as with the Crustacea. In the extremities, only, do there appear to be tendons ; but these are merely very long, flattened processes of the skeleton, situated in the axis of the articles of the extremities. They serve as points of attachment to short muscular fasciculi which are there obliquely inserted, after having arisen from the inner surface of the articles. With adult insects, the segments of the body are only slightly movable, and have but few muscles. But with the larvae, whose extremities are rudimentary, or even wholly wanting, there is a very prominent muscular system situated directly beneath the skin, and composed often of several layers of flattened fasciculi.[3]

§ 326.

The Locomotive organs, properly speaking, of the Insecta, are the legs and the wings. The true legs never exceed three pairs, and are inserted upon the first three segments back of the head — *Prothorax, Mesothorax, Metathorax.* Each leg is divided into a *Coxa*, a *Trochanter*, a *Femur*, a *Tibia*, and a *Tarsus.* The tarsus is divided into several articles, the number of which reaches even five. The form of these legs varies infinitely according to the mode of life of each species. The most common are those to which are usually given the names : *Pedes cursorii, ambulatorii, gressorii,*

2 This is so with the Diptera, Hemiptera, and Hymenoptera, with the muscles of whose wings the elements may be very easily separated.*
3 *Lyonet* (Traité, &c., p. 114, Pl. VI.–VIII.) has given a very detailed account of the muscles of the larva of *Cossus.* See also *Newport*, Philos. Transact. 1836, p. 537, Pl. XXVII. (the cutaneous muscles of the larva of *Sphinx ligustri*), and

Straus, Consid. &c. p. 140, Pl. III. IV. (*Melolon-tha vulgaris*).
Reference may be also made to the works of *Meckel* (System, &c., III. p. 22), *Cuvier* (Leçons, &c. II. p. 64), *Burmeister* (Handb. &c. I. p. 267), *Lacordaire* (Introduct. à l'Entomol. II. p. 249), and of *Newport* (Cyclopaed. &c. loc. cit. p. 954).

* [§ 325, note 2.] The delicate and beautiful structure of the thoracic muscles of some of the Insecta, has been carefully studied of late, and has aided not a little in the elucidation of the histology of the muscular tissue. See *Lebert*, Recherches sur la formation des muscles, &c., in the Ann. d Sc. Nat. XIII. 1850, p. 182–195, Pl. VII. fig. 18–27 ; but especially *Aubert*, Ueber die eigenthümliche structur der Thoraxmuskeln der Insekten, in *Siebold* and *Kölliker's* Zeitsch. IV. 1853, p. 388.
This last-mentioned naturalist states that he has observed a new form of muscle-element in the thorax of the Libellulidae, consisting of primitive muscular bands by means of which, with a beaker-shaped apparatus, the wings are moved. These fibres consist of flattened riband-like bands, striated on their borders as well as on their flat surface.
I have recently studied the elements of the thoracic muscles of many Diptera (*Culex*, espec-

ially) with the highest and best microscopic powers. The discs composing the primitive fibrillae, easily separate and may be studied by themselves. But, with the best powers, I have been unable to observe in them anything but that each disc is a solid, homogeneous light-refracting body. The fibrilla appears to be formed by the aggregation of these discs in a linear series and with regular inter-spaces ; no sheath, by which these discs would be retained in a row, was observed. Contraction of the fibrilla takes place by an approximation of the discs to each other, and the consequent greater or less disappearance of the interspaces. These phe-nomena can be easily observed, and, it may be added, there are often seen isolated fibrillae con-tracting and relaxing, thus showing that the real phenomenon of muscular contraction, does not de-pend upon the nervous system, however much the action of this last may serve as a stimulant.   Ed.

natatorii, saltatorii, raptorii, and *fossorii.*[1]   The tarsal articles are often enlarged, in which case they form, on their under surface, either a naked, fleshy sole, or a thickly-pilose ball of the foot, which is usually cordiform. With the Dytiscidae, several of these tarsal joints are changed into a disc provided with suckers.   The last article of these organs bears usually two movable hooks, which are sometimes deeply bifid or denticulate on their concave border.[2]   It is rare that there is a third hook between them.[3]   But with the Strepsiptera, and Physopoda, all the hooks are wanting.   The Diptera and many Hymenoptera, have, moreover, under these hooks, soft lobules (*Arolia*) provided with numerous small papillae, by means of which these insects can fix themselves to objects.[4]   With a great number of larvae, the six legs are very short or abortive, or even completely wanting.   In the first case, they consist, usually, of merely a hook, but these larvae have also many short obtuse processes on the other segments of the body, and by means of the hairy soles on the extremity of these, they can fix themselves on bodies and thereby move along.[5]   With many entirely apodal larvae, these processes are replaced by simple tubercles, or by belts of backwardly-pointing bristles or spines, which serve as points of support in their locomotion.[6]

The organs of flying are the anterior and the posterior Wings.   The first are inserted on the mesothorax, and the second on the metathorax; but in nearly all the orders, there are genera with which these organs are wholly wanting.[7]   In other genera, the females alone are wingless,[8] which is also true of the neuters of certain families.[9]   It is more common still, to see the posterior wings changed into balancers;[10] and this same transformation occurs also, but very rarely, with the elytra.[11]   The wings, properly speaking, are only prolongations of the cutaneous skeleton traversed by tracheae and blood-canals.   Their forms, their nervures, their folds, &c., as well as their sometimes complete abortive condition, may here be passed over, for all these points belong to the domain of Zoology.*

1 For the marching, leaping, and swimming of insects, see *Straus,* Consider. &c. p. 180.

2 Thus, the hooks are bifid with *Meloë,* and denticulate with the Pompilidae, Hippoboscidae, Cistelidae, and with *Taphria, Dolichus, Calathus,* and *Pristonychus,* of the Carabidae.

3 This third hook is found, for example, with *Lucanus cervus.* With larvae of the Meloidae, the legs are terminated by three very remarkable, straight, lanceolate hooks, known by the name of *Triungulinus.* Many of the Curculionidae can grapple objects by means of the immovable hooks on their tibiae.

4 The Tenthredinidae have a lobule of this kind on each of their legs, and the Diptera have even two to three. For these lobules, as well as for the tarsi of Insecta in general, see the beautiful figures given by *Everard Home* (Lectur. on Comp. Anat. IV. Pl. LXXXI.-LXXXIV.). According to *Blackwall* (Trans. of the Linn. Soc. XVI. p. 487, 767; and Ann. of Nat. Hist. XV. p. 115), the papillae of the *Arolia* secrete a viscid substance which enables the Insecta having these organs to walk on

steep and smooth surfaces.   But this assertion requires further proof, although it is admitted by *Spence* (Trans. of the Entomol. Soc. IV. p. 18).

5 With the Lepidoptera, and Tenthredinidae.

6 With many Diptera.

7 Among the Orthoptera, the wings are wanting with some Blattidae, Acrididae, Phasmidae, and Psocidae; among the Hemiptera, with *Acanthia,* and *Rhizobius;* and among the Diptera, with *Melophagus, Phthiridium,* and *Pulex flagellos.*

8 With *Lampyris,* some Blattidae, Coccidae, Bombycidae, Geometridae, also with the Mutillidae and the Strepsiptera.

9 With the Formicidae and the Termitidae.

10 With the Coccidae and the Diptera. The posterior wings are entirely wanting with some Ephemeridae.

11 The two singularly distorted balancers of the Strepsiptera, situated in front of the wings and in rapid and unceasing motion, are, from their insertion, only abortive elytra.

* [§ 326.] *Leidy* (Proceed. Acad. Sc. Philad. III. 1846, p. 104) has described a peculiar mechanism by which the membranous wings of *Locusta* are closed in a plicated manner like a fan.   This mechanism consists of spiral ligamentous bands, wound, like the thread of a screw, around the transverse or connecting veins, which latter are also flexible.   By this arrangement, upon the contraction of the alary extensors, the spring-like ligaments, or ligamenta spiralia, are stretched in the expansion of the wings, and upon the relaxation or cessation of the action of the muscles, the physical properties alone of the ligamenta spiralia, in resuming their unstretched state, close the wings.— ED.

The movements of the wings are produced by two extensor and several smaller flexor muscles, which arise from the middle and posterior thoracic segments, and are inserted on a tendinous process at the base of each wing. The size of these muscles is proportionate to the size of the wings and their mode of use in flight. They are, consequently, all equally developed when the four wings participate equally in the act of flying, as is the case with the Lepidoptera, Hymenoptera, the majority of the Neuroptera, the Libelludidae, Perlidae, and finally, the Cicadidae, and the Aphididae.

The muscles of the anterior wings are comparatively smaller than those of the posterior, when the first are not used, properly speaking, except to cover the latter, as is the case with the Coleoptera, the Bugs, and many of the Orthoptera.[12]

With most of the Poduridae, and with the young larvae of *Xenos sphecidarum*, there is a peculiar, fork-shaped, saltatory organ. It is inserted on the posterior extremity of the body or under the abdomen, and, when at rest, points horizontally backwards. By means of a special muscular apparatus, this fork is bent forwards and applied against the abdomen; it is then returned violently into its original position, thereby tossing the animal a considerable distance.[13]

§ 327.

Many of the Insecta produce sounds which we perceive partly as clear tones, and partly as confused noises. These are due, sometimes to particular soniferous organs, but more often to vibrations of the cutaneous skeleton produced by special muscular organs, or to the rubbing of certain parts against each other. But in every case, without exception, the sound is due to the action of voluntary muscles, and has no connection with the respiratory organs.[1]

The buzzing produced by many Diptera, and Hymenoptera, during flight, is due, without doubt, to vibrations of the thorax produced by the rapid and successive contractions of the muscles of the wings in this act.[2] Further researches are necessary to show if the sounds produced by certain butterflies are due to rubbings of some parts of the cutaneous skeleton, or to a special soniferous apparatus.[3] But the sharp sounds of many Coleoptera

12 For the flight of Insecta, see *Straus*, Considér. &c. p. 200; but especially the extensive work, illustrated with many figures, of *Chabrier*, Mém. du Muséum, VI.–VIII.

13 See *Nicolet*, loc. cit. p. 39, Pl. III.; and my Beiträg. zur Naturg. d. wirbell. Thiere. p. 84, Taf. III. fig. 70. This fork is wanting in the genera *Achorutes* and *Anurophorus*, as well as with the larvae of *Stylops* among the Strepsiptera.

1 Some Insecta produce sounds by striking or rubbing certain parts of their cutaneous skeleton against the body. The males of *Mycterus curculioides* knock with such violence the extremity of their body against the boughs on which they have alighted, that they produce a pretty loud sound, designed, probably, to call the females. The larvae of *Vespa crabro*, when hungry, scratch the walls of their cells with the point of their jaws, and thus call the attention of the parents to their condition.

2 The buzzing of these Insecta has been attempted to be explained in various ways. At all events, it cannot be due alone to the movements of the wings, for it persists sometimes after the removal of these last. Neither can it be attributed to the rapid passage of air through the stigmata of the thorax, causing vibrations in that part of the body. *Burmeister*, who has advocated this last opinion (Handb. &c. I. p. 508, and in *Poggendorf's* Ann. d. Physik. XXXVIII. 1836, p. 283, Taf. III. fig. 7–9), compares these sounds to those produced by a siren; but this theory has been fully refuted by *Goureau*, *Solier*, and *Erichson* (see *Silbermann's* Revue Entomol. III. p. 105, and Ann. de la Soc. Entom. de France, VI. 1837, p. 31, and *Wiegmann's* Arch. 1838, II. p. 193). The various sounds of flies and bees may be imitated, moreover, by placing a vibrating tuning-fork in contact with a band of stretched paper, — an experiment in which the vibrations of the air take no part.

3 It is said that, with *Euprepia pudica*, the peculiar sounds are produced by the rubbing of a callosity of the two posterior hips against the middle hips (*Solier*, Ann. de la Soc. entom. loc. cit.). The cry of *Acherontia atropos* has been attempted.

are caused by the rubbing of their prothorax against the peduncle of the mesothorax, or by grating the ridges of the abdomen against the internal surface of the elytra. *Reduvius stridulus, Mutilla europaea,* and *Mantis religiosa,* produce also certain sounds by the friction of particular portions of their skeleton.[1] The males of many Acrididae produce their creaking sounds by playing, as with the bow of a violin, upon the lateral borders of their Elytra, by their posterior thighs which have a longitudinal granular ridge on their internal surface.[5] The peculiar cry of the male Locustidae and Achetidae is produced by the base of their elytra. The very hard and sharp internal border of one of these elytra rubs against a horny ridge upon the under surface of the other, close to the tympanitic disc.[6]

With the males of the musical Cicadidae, there is a very remarkable soniferous apparatus, situated on the under surface of the first abdominal segment. It consists of two spacious drums at the base of which is a dry, plicated membrane, to which is attached a large muscle of conical form, arising from a median, bifurcated process of the second abdominal segment. The entrance of each of these drums is more or less covered by a rounding operculum which is free behind. The muscle draws the membrane inwards, then relaxing, this last returns by its own elasticity, producing, as from the bending up and down of a metallic plate, a loud, clanging sound. This sound is undoubtedly considerably increased by the resonance of the air in the drums and in the neighboring vesicular tracheae.[7]

to be explained in various ways, but none of the causes yet assigned are satisfactory ; see *Passerini,* Ann. d. Sc. Nat. XIII. 1828, p. 332 ; *R. Wagner,* in *Muller's* Arch. 1836, p. 60 ; *Goureau, Nordmann,* and *Duponchel,* Ann. de la Soc. entom. VI. IX., or *Wiegmann's* Archiv. 1839–41.*

4 See *Burmeister,* Handb. &c. I. p. 507, and *Goureau,* in *Silberman's* Revue Entom. III. p. 101.

5 See my observations in *Wiegmann's* Arch. 1844, I. p. 53. This fiddling movement may be easily observed with the males of *Gomphocerus* and *Oedipoda. Pneumora maculata* has, upon the sides of the second abdominal segment, a very strongly denticulate, oblique ridge, against which is rubbed, probably, a horny process situated on the internal surface of the posterior thighs. I am yet unable to account for the way in which the males of *Oedipoda stridula* produce their hoarse buzzing, during flight.

6 For this soniferous apparatus, see *Goureau* and *Solier,* Ann. d. la Soc. Entom. 1837, p. 31 ; *Newport,* Cyclopaed. loc. cit. II. p. 928, fig. 394–396 ; *Goldfuss,* Symb. ad. Orthopt. quorund. oeconomiam, Bonn. Diss. 1843, p. 5, fig. 1–10 ; and my observations in *Wiegmann's* Arch. loo. cit. p. 60. *Burmeister* (Handb. &c. I. p. 511) has sought to explain this sound by referring it to the powerful escape of the air from the stigmata of the Locustidae and Acrididae ; but this is unsatisfactory.

7 For the soniferous apparatus of the musical Cicadidae, see *Réaumur,* Mém. V. 4th mém. Pl. XVII. ; *Burmeister,* Handb. &c. I. p. 513 ; *Ratzeburg,* Mediz. Zool. II. p. 208, Taf. XXVII. and especially *Carus,* Analekt. zur Naturwiss. p. 142, fig. 1–18.†

* [§ 327, note 3.] *Haldeman* (*Silliman's* Jour. May, 1848) states that *Lithosia miniata,* Kirby, or an allied species, produces an audible stridulation by vibrating the pleura beneath the wings, this part being marked in recent specimens by parallel lines, apparently indicating the position of the muscles. According to him, it is possible that the European *Acherontia atropos* may produce its peculiar sound in a similar manner. — ED.

† [§ 327, note 7.] See also my investigations upon this apparatus of the *Cicada septendecim* in the Proceed. Boston Soc. Nat. Hist. 1851, p. 72 — ED.

CHAPTER III

NERVOUS SYSTEM.

§ 328.

The central parts of the nervous system consist, with the Insecta, as with the other Arthropoda, of a Brain and a Ventral Cord.[1]

The brain is situated in the cephalic segment, and is composed of a *Ganglion supraœsophageum*, connecting with a *Ganglion infraœsophageum* which is smaller, by two lateral commissures which embrace the œsophagus. The first of these ganglia corresponds to the cerebrum of the Vertebrata; and the second is comparable perhaps to the cerebellum or spinal cord.

The ventral cord succeeds upon the sub-œsophageal ganglion, and consists, sometimes of a single ganglionic mass, sometimes of a chain of ganglia more or less approximated and connected by double, longitudinal commissures.[2] The number of the ventral ganglia, which is never greater than that of the segments of the body, as well as the presence and length of the longitudinal commissures, depend often upon the number, the size, and the mobility of the segments of the body to which they belong. With those Insecta whose segments are very short and rigid, the ganglia are closely approximated or even entirely blended together; while, in most larvae, where the segments of the body are equally developed and flexible, the ganglia are separate, nearly equal in size, and connected by pretty long commissures. These last are rarely united into a single cord, although the ganglia, not only those of the ventral cord, but also those of the brain, appear, nearly always, to be composed each of two united ganglia.

Aside from the differences presented according as the insect may be a larvae, a pupa, or an imago, the nervous system varies so much even in the same group, that it may be quite dissimilar in species which, in other respects, are very closely allied. These modifications refer to the number of the ganglia, the length of their commissures, and the more or less complete fusion between certain ganglia.

The superior cerebral ganglion, which is often composed of two hemispheres more or less fused together, gives off the two antennal and the two optic nerves. The simple eyes or stemmata, either when alone, or when coëxistent with compound or faceted eyes, always receive their nerves from the same ganglion; although these last are sometimes given off from a trunk in common with the optic nerve. The sub-œsophageal ganglion furnishes nerves chiefly to the mandibles, to the maxillae and their palpi. With the perfect Insecta, the three thoracic are much more voluminous than the abdominal ganglia. They send nerves not only to the legs, but also to the muscles of the wings.

---

1 For the nervous system of the Insecta in general, see *Burmeister*, Handb. &c. I. p. 290; *Lacordaire*, Introd. &c. II. p. 183 , *Newport*, Cyclopaed. II. p. 942, and *Blanchard*, Ann. d. Sc. Nat. V. 1846, p. 273. *L. Dufour* (Mém. prés. à l'Acad. d. Sc. IX. 1846, p. 562, Pl. I. fig. 16) has given a very exact description of this system in *Sarcophaga haemorrhoidalis*.

2 Entomologists are not agreed as to the number of the abdominal ganglia, for the sub-œsophageal ganglion is sometimes regarded as the first of the ventral cord. For the nervous system of the larva and pupa of *Sarcophaga*, see *L. Dufour*, loc. cit. Pl. I. fig. 12-15.

The ventral ganglia are usually small, co-equal, and give off no nerves except to the segments of the abdomen. The last ventral ganglion, alone, is larger, for it furnishes, in addition, nervous filaments to the rectum, and to the excretory ducts of the genital organs.

Usually, the nerves arise in the ganglia by two or three principal roots. Some nerves, however, arise from the interganglionic cord. In those species where the ventral ganglia are entirely fused together, the nerves arise close together, but immediately diverge in different directions.

§ 329.

As to the Intimate Structure of the nervous system of the Insecta, both the nerves and the ganglia are always surrounded by a fibrous neurolemma, and, according to carefully-made researches, are never wanting in the two usual anatomical components. Between the extremely tenuous primitive fibres, are interposed, in the ganglia, very small globules. These last, nucleolated, usually contain also a finely-granular substance, colorless, though sometimes reddish or brownish.[1]

There are, with the Insecta, as with the Crustacea, two modes of the disposition of the fibres in the ganglia. The first, which form nervous, inferior cords, are disseminated in the ganglia; while the others, which belong to the superior nerves, simply pass through or over these ganglia. These two kinds of fibres give off, laterally, nervous filaments, which, uniting, form peripheric nerves of a mixed character. The superior cords correspond, probably, to the motor nerves, and the inferior to the sensitive nerves of the Vertebrata.[2]

§ 330.

The Disposition of the nervous system in the various orders of Insecta presents the following differences:[1]

Among the Aptera, the ventral cord of the Pediculidae is composed of three contiguous ganglia situated in the thorax. The prothoracic ganglion connects with the brain, and the metathoracic sends nerves to the abdomen.[2] The nervous system of the Poduridae differs from this, in their three thoracic ganglia being separated, and their interganglionic longitudinal commissures being wide apart.[3] More widely different still, is the ventral cord of the Lepismidae; it is composed of eleven ganglia connected by double longitudinal commissures.[4]

With the Hemiptera, this system is limited to two thoracic ganglia, of which the anterior is the smaller. With *Pentatoma*, and *Cicada*, these two ganglia are not separated except by a constriction; while, with *Nepa*,

1 It was undoubtedly from their extreme delicateness that these ganglionic globules were overlooked by *Treviranus* (Beitr. g. zur Anfklär. d. Erscheinung. u. Gesetze d. organisch. Lebens. l. Hft. 2, p. 62). They have been distinctly seen by *Ehrenberg* (Unerk. Struct. &c. p. 56, Taf. VI. fig. 6 (*Geotrupes*)), *Puppenheim* (Die specielle Gewebelehre d. Gehörorg. p. 51), *Helmholtz* (De fabric. Syst. nat. &c. p. 21), *Hannover* (Recherch. microscop. &c. p. 71, Pl. VI. fig. 81, 82 (*Aeschna*)) and *Will* (*Müller's* Arch. 1844, p. 81).

2 This difference of the nervous cords was first pointed out by *Newport* with the pupa and imago of *Sphinx ligustri* (Philos. Trans. 1834, part II.

p. 389, Pl. XIII.–XVII. and Cyclopaed. loc. cit. p. 946). *Hagen* (Entom. Zeit. 1844, p. 364) has since observed it with *Aeschna grandis* and *Gryllotalpa vulgaris*.

1 Various and special accounts of the general disposition of the nervous system of insects may be found in *Curier*, Lecons, &c., III. 1845, p. 334.
2 *Swammerdamm*, Bib. der Natur. p. 36, Taf. II. fig. 7.
3 *Nicolet*, loc. cit. p. 44, Pl. IV. fig. 1 (*Smynthurus*).
4 *Treviranus*, Verm. Schrift. II. Hft. 1, p. 17, Taf. IV. fig. 3.

they are connected by two long commissures. From the posterior of these ganglia pass off, in a ventral cord, two main trunks, approximated, which send off, in their course, lateral branches towards the periphery; with *Pentatoma*, these two main trunks are fused into one.[5]

With the Diptera, the ganglionic chain is always connected by simple commissures. The number of ganglia varies with the families, and, usually, is proportionate to the length of the segments of the body.[6] The ventral cord is most concentrated with the Hippoboscidae,[7] the Oestridae, and the Muscidae calypterae; it consists of only a single thoracic ganglion, from which pass off nerves in various directions. The Muscidae acalypterae, on the contrary, the Syrphidae[8] and the Conopidae, have, beside this thoracic ganglion, one or two ventral ganglia; while the Scenopinidae have five, and the Tabanidae, Stratiomydae,[9] Therevidae, Leptidae, Asilidae, and Bombylidae, have six. Their number is still larger with the Empidae, Culicidae, and Tipulidae;—there being, in the first, three thoracic and five abdominal ganglia; and in the last two, three thoracic and six abdominal ganglia. The larvae of the Diptera usually have one more pair of ganglia than the adults. It is only in those species whose ventral cord is fused into a single mass, that the same concentration is observed with the larvae.[10] The larvae of the Diptera have either a moniliform ventral cord, composed of ten approximated ganglia, or a chain of eleven of these masses, connected by long commissures, which are often double.[11]

With the Strepsiptera, alike in the three states of larva, pupa and imago, the ventral cord consists only of a large thoracic ganglion, from which pass off nerves in various directions.[12]

With the adult Lepidoptera, the ventral cord consists of seven ganglia, of which the first two are the largest and belong to the thorax. The connecting commissures are not double except between the thoracic ganglia; those of the others being more or less fused into a single cord. In the Caterpillars, the ventral cord consists of eleven nearly equal ganglia; the two commissures between the first three of these, are quite wide apart;

5 *Treviranus*, Beitr. zur Anat. u. Physiol. d. Sinneswerk. lift. 1, Taf. 11. fig. 24 (*Cicada*), and *L. Dufour*, Recherch. sur les Hémiptères, p. 259, Pl. XIX. fig. 801-803 (*Pentatoma*, *Nepa* and *Cicada*).*
6 For the nervous system of the Diptera, see *L. Dufour*, Ann. d. Sc. Nat. 1844, p. 245.
7 *L. Dufour*, Ibid. III. 1845, p. 64, Pl. II. fig. 12.
8 *Burmeister*, Handb. d. Entomol. I. 307, Taf. XVI. fig. 11 (*Eristalis tenax*).
9 *Swammerdamm*, Bib. d. Nat. p. 270, Taf. XLI. fig. 7 (*Stratiomys*).
10 I have found the ventral cord of the larva of *Oestrus bovis* concentrated into a single large ganglion, situated at the extremity of the thorax. I think, therefore, that the description of the nervous system of the larva of *Oestrus ovis* given by J. L. Fischer (Observ. de Oestro ovino atque bovino. Diss. Lips. 1787, p. 32, or in *Werneri*, Vermium

intestin. exposit. contin. tertia. p. 28, Taf. III. fig. 4), and according to which, two long ganglionic cords, connected by transversal anastomoses, extend the whole length of the body,—is based on inexact observations. In the larvae of *Piophila* and *Eristalis*, several ventral ganglia fused together form a single abdominal cord ; see *Swammerdamm*, Bib. der Nat. p. 279, Taf. XLIII. fig. 7, and *Burmeister*, Handb. loc. cit. Taf. XVI. fig. 10.
11 The larvae of *Stratiomys* have an abdominal cord composed of ten contiguous ganglia (*Swammerdamm*, Bib. der Nat. p. 264, Taf. XL. fig. 5). With those of *Culex*, *Chironomus*, *Simulia*, and other Tipulidae, the ten ganglia are wide apart, and connected by double longitudinal commissures.†
12 In the apodal larvae and the larvae-like females of *Xenos Rossii*, I have found this nervous mass in the first segment of the body which corresponds to the cephalothorax.

* [§ 330, note 5.] For the nervous system of *Belostoma*, with all its details, see *Leidy*, History and Anatomy of the hemipterous Genus Belostoma, In the Jour. Acad. Nat. Sc. Philad. I. 1847, p. 65, Pl. X. fig. 13. — ED.
† [§ 330, note 11.] In the larva of Corethra

plumicornis, *Leydig* found the ventral cord composed of eleven, instead of ten ganglia ; see Anatomisches und Histologisches üb. d. Larve von Corethra plumicornis in *Siebold* and *Kölliker's* Zeitsch. III. 1852, p. 438. — ED.

while those of the others are usually fused together.[13] During the pupa-state, a remarkable change takes place. The commissures between the first and second, and the third and fourth ganglia, are gradually shortened. The ganglia are thereby gradually approximated, and, in the end, are fused together, forming the two thoracic ganglia of the adult, which send off nerves to the legs, and to the muscles of the wings. At the same time, the fifth and sixth ganglia entirely disappear or are fused into one.[14]

With the Hymenoptera, the ventral cord is composed of seven to eight ganglia connected by double commissures. The first of these, smaller than the second, is, like it, produced by the fusion of several ganglia; and both are situated in the thorax. Of the remaining five or six abdominal ganglia, the last two are closely approximated, or fused into one.[15] Here, as with the Lepidoptera, the number of ganglia in the ventral cord of the larvae, is eleven, as has been specially shown in the false caterpillars of the Tenthredinidae.[16]

With the Orthoptera, and Neuroptera, the nervous system is nearly always composed, in their various states, of three thoracic and six to seven abdominal ganglia connected by double commissures and forming a chain as long as the body.[17]

With the Coleoptera, the number and disposition of the ventral ganglia present the widest variations of all. The longitudinal commissures, always double, are shortened or even wholly wanting at certain points. The ganglionic chain is, therefore, more or less abbreviated, and sometimes the ganglia are almost fused into a single mass. In this respect this system here presents two principal types, the limits between which, however, have

13 For the nervous system of the larvae of *Vanessa urticae* and *Bombyx mori*, see the figures of *Swammerdamm*, loc. cit. p. 387, 230, Taf. XXVIII. fig. 3, and Taf. XXXIV. fig. 7 ; also for that of the larva and imago of *Cossus ligniperda*, the works of *Lyonet*, Traité, &c., p. 190, Pl. IX., and in the Mém. du Mus. loc. cit. p. 191, Pl. LI. (17). For that of *Gastropacha pini*, pupa and imago, see *Suckow*, Anat. physiol. Untersuch. p. 40, Taf. VII. fig. 37, 38 ; but see especially the excellent description of that of the larva, pupa, and imago of *Sphinx ligustri*, for which we must thank *Newport*. Philos. Trans. 1832, p. 383, Pl. XII. XIII.; also, 1834, p. 389, Pl. XIII–XVIII., and Cyclop. &c. loc. cit. p. 943, fig. 406, 414, 415.

14 This metamorphosis of the nervous system was first observed by *Herold* (Entwickelungsgesch. d. Schmetterlinge, loc. cit. Taf. II.) with *Pontia brassicae*, and has since been confirmed by *Newport* with *Sphinx ligustri* and *Vanessa urticae* ; see Philos. Trans. 1834, Pl. XV–XVI. fig. 20–30, and Cyclop. loc. cit. p. 962, fig. 420–423.

15 See *Swammerdamm*, Bib. der Nat. p. 207, Taf. XXII. fig. 6 (*Apis mellifica*) ; *Treviranus*, Biologie, V. Taf. I. (*Bombus muscorum*), and *Brandt* and *Ratzeburg*, Medizin. Zool. II. p. 203, Taf. XXV. fig. 31 (*Apis mellifica*). For the disposition of the ventral chain of the Lepidoptera in general, see, moreover, *L. Dufour*, Recherch. sur les Orthopt., Hymenopt. &c. p. 381. According to this last-mentioned naturalist, the number of ventral ganglia is five with *Vespa, Scolia*, and with most of the Apidae and Andrenidae ; six with *Odynerus, Sphex, Pompilus, Chrysis*, the Ichneumonidae,

Bembecidae, with *Larra*, and *Tiphia* ; four with *Tripoxylon*, and three with *Eucera*.

16 The ventral chain of the Tenthredinidae, Apidae, Vespidae and other Hymenoptera, undergoes, undoubtedly, with the pupae, a metamorphosis similar to that occuring with the Lepidoptera.

17 *Swammerdamm*, loc. cit. p. 108, Taf. XIV. (a pupa of *Ephemera*) ; *Marcel de Serres*, Mém. du Mus. IV. 1818, Pl. VIII. (I.) fig. 1 (*Acridium*); *J. Muller*, Nov. Act. Nat. Cur. XIV. Tab. IX. fig. 4, and XII. p. 568, Tab. L. fig. 1 (*Acridium* and *Bacteria*); *Newport*, Cyclop. II. p. 950, fig. 409, 410 (*Forficula* and *Locusta*); finally, *L. Dufour*, Ann. d. Sc. Nat. XIII. 1828, p. 361, Pl. XXII. fig. 4 (*Forficula*), Recherch. sur les Orthopt. &c. p. 281, Pl. II. fig. 7 (*Oedipoda*), and p. 561, Pl. XI. fig. 160 (*Libellula*). According to *L. Dufour*, there are seven ventral ganglia with *Libellula* and *Ephemera*, while there are only six with *Perla* and *Phryganea*. But *Pictet* (Recherch. pour servir à l'hist. et à l'anat. des Phryganides. Pl. II. fig. 33–36) and *Burmeister* (Handb. &c. II. p. 895, 898) assign to these Insecta, in both their larva and their perfect state, eight ventral ganglia. There are even nine of these ganglia with the Ephemeridae, according to *Burmeister* (loc. cit. p. 763). In the very clubbed larvae of *Myrmeleon* there are eight contiguous ventral ganglia beside two thoracic ones (*Cuvier*, Leçons, &c. III. p. 341). *Loew* (Germar's Zeitsch. IV. p. 424) remarks that the proper Neuroptera are distinguished by the separation of their last two abdominal ganglia, while, with all the Orthoptera, they are fused together.*

* [§ 330, note 17.] See also for a description and figures of the nervous system of *Spectrum*

femoratum, Leidy, Proceed. Acad. Sc. Philad. 1846, III. p. 83. — ED.

not yet been definitely fixe l.[18]   The first type consists of an absence of
all the longitudinal commissures, as is the case with most of the Lamellicornes, the Curculionidae, and the Scolytidae.   Here, the ventral cord is
limited to three ganglia connected together ; of these, the first corresponds
to the prothoracic, and the second, the larger, to the second and third
thoracic ganglia.   This last is succeeded by an oblong, ganglionic mass,
representing the concentrated abdominal portion of the cord, and from
which arise the nerves of the muscles of the abdomen.[19]   In the second
type, the abdominal portion of the cord occupies the entire length of the
body.   This is the case with the Cistelidae, Oedemeridae and Cerambycidae, which have five ganglia in the abdomen.[20]   With the larvae of the
Coleoptera, these two types are more clearly defined, there being no intermediate forms.[21]

§ 331.

The Splanchnic nervous system consists, with the Insecta, in all their
states, of a single and a double nervous cord.   Sometimes the first, sometimes the second of these is the more developed.

The single Stomato-gastric nerve arises from the anterior border of the
cerebral hemispheres, by two short filaments, which, directly in front
of the brain, meet in a ganglion (*Ganglion frontale*) lying upon the œsophagus.   From this ganglion are given off several nervous filaments which
go to the upper lip ; while, from the opposite side, arises a simple nerve

18 There has recently appeared a very detailed
memoir, accompanied with beautiful figures, on the
nervous system of the Coleoptera, by *Blanchard* ;
see Ann. d. Sc. Nat. V. 1846, p. 273, Pl. VIII.–XV.,
and Règne animal, illustr. Insectes, Pl. III. III.
bis. and IV. (*Melolontha, Carabus, Otiorhyn-
chus, Cerambyx*).
19 See *Straus*, Considér. &c. p. 391, Pl. IX. fig.
1 (*Melolontha vulgaris*), and *Blanchard*, loc. cit.
An analogous concentration of the nervous system
occurs in the families of Histeridae, Gyrinidae,
Nitidulidae, and Scaphididae, where the ventral portion forms a single oblong ganglion, while the three
thoracic ganglia are connected by double longitudinal commissures.   In most of the other families, the
three thoracic ganglia are more or less separated,
and the abdominal portion is modified in various
ways.   With the Endomychidae, Meloïdae and
Chrysomelidae, there are only four abdominal
ganglia connected by very short double commissures ; see *Audouin*, Ann. d. Sc. Nat. IX. 1826,
p. 36, Pl. XLII. fig. 16 (*Lytta*) ; *Brandt*, Mediz.
Zool. II. p. 103, Taf. XVII. fig. 2, Taf. XIX. fig.
15 (*Meloe* and *Lytta*) ; *Newport*, Cyclopaed. loc.
cit II. p. 950, fig. 408 (*Timarcha*) and *Joly*,
Ann. d. Sc. Nat. II. 1844, p. 24, Pl. IV. fig. 16
(*Colaspis*).   With the Dytiscidae, and with *Hytu-
rus*, there are six abdominal ganglia, and the commissures are also very short ; see *Burmeister*,
Handb. loc. cit. Taf. XVI. fig. 9 (*Dytiscus*).   This
figure, however, is not fully exact, if compared
with that of *Blanchard* (loc. cit. p. 343, Pl. X. fig.
1).   With the Staphylinidae, Silphidae, and Hydrophilidae, the abdominal portion, although composed of eight ganglia, is not prolonged much
into the abdomen ; it is longer and composed of
six to seven ganglia with the Carabidae, Lucanidae, and Pyrochroïdae ; see *L. Dufour*,
Ann. d. Sc. Nat. VIII. 1826, p. 27, Pl. XXI. bis,
fig. 2 (*Carabus*), and Ibid. XIII. 1840, p. 332,
Pl. VI. fig. 9 (*Pyrochroa*).   It is even still longer

and composed of eight ganglia with the Elateridae,
Cleridae and Telephoridae.
20 See *Blanchard*, loc. cit.
21 With those species of the Lamellicornes, and
Curculionidae, whose ventral cord is very much
concentrated, the eleven large component ganglia
are, with the larvae, united into one knotty mass,
without any trace of commissures ; see *Swammerdamm*, loc. cit. p. 131, Taf. XXVIII. fig. 1 (*Oryctes*) ; *L. Dufour*, Ann. d. Sc. Nat. XVIII. 1842,
p. 170, Pl. IV. fig. 11 (*Cetonia*) ; *Burmeister*,
Zur Naturgesch. d. Calandra, p. 13, fig. 13, 14 ;
Blanchard, Ann. d. Sc. Nat. loc. cit. Pl. XIV. fig.
1 (*Calandra*).   With the larvae of the Meloïdae,
Pyrochroïdae, Lucanidae, Chrysomelidae, Tenebrionidae, as well as of most of the other families
of the Coleoptera, the ventral chain occupies nearly
the entire length of the body, and is composed of
eleven ganglia having double commissures ; the
thoracic ganglia exceed but little in size those of
the abdomen ; see *Brandt*, Mediz. Zool. II. p. 105,
Taf. XVII. fig. 20, Taf. XIX. fig. 31 (*Meloe* and
Lytta) ; *L. Dufour*, Ann. d. Sc. Nat. XIII. 1840,
p. 327, Pl. V. fig. 8 (*Pyrochroa*), and XVIII.
1842, p. 172, Pl. V. fig. 17 (*Dorcus*) ; *Newport*,
Cyclopaed. loc. cit. p. 943, fig. 404 (*Timarcha*) ;
Joly, Ann. d. Sc. Nat. II. 1844, p. 24, Pl. IV. fig.
14 (*Colaspis*), and *Blanchard*, Ibid. Pl. XV. fig.
7, Pl. X. fig. 5 (*Chrysomela* and *Tenebrio*).   It
is only with the larvae of the Carabidae, Silphidae,
Staphylinidae and Diaperidae, that the ventral
cord, although composed of eight ganglia, does not
extend into the last abdominal segments ; while
that of the larvae of the Dytiscidae, composed of
seven ganglia, does not reach beyond the middle
of the abdomen ; see *Burmeister*, Trans. of the
Entomol. Soc. Lond. I. p. 239, Pl. XXIV. fig. 9
(*Calosoma*) ; *Blanchard*, Ann. d. Sc. Nat. loc. cit.
Pl. IX. fig. 3, 5, Pl. XI. fig. 4, Pl. X. fig. 2 (*Silpha, Staphylinus, Diaperis*, and *Dytiscus*).

(*Nervus recurrens*) which passes over the œsophagus to the stomach, giving off branches right and left. Reaching the stomach, it divides, after having formed a ganglionic enlargement, into two principal branches.

The double Stomato-gastric nerve consists of one, two, or three pairs of small ganglia, situated behind the brain, on each side of the œsophagus, and communicating with each other, with the posterior extremity of the brain, and with the *Nervus recurrens*, by delicate filaments. These filaments send fine threads to the œsophagus, and, at certain points, anastomose with the single nerves.[1]

With the Hemiptera, a single Splanchnic nerve has been observed, and, for the double system, there has been seen, on each side of the œsophagus, two small ganglia, one behind the other.[2]

With the Diptera, the splanchnic system appears to be present; at least, there has been observed on the Chyliferous stomach of the Hippoboscidae, a pair of filaments belonging, probably, to the double system.[3]

The Lepidoptera have a highly-developed *Nervus recurrens*, which often forms, with the caterpillars, several small ganglia lying behind each other on each side of the brain, and connected together by a double nervous arch. The double system arises on each side of the œsophagus, from two ganglia, situated one behind the other, which, with the caterpillars and pupae, are often approximated to a blending together, and which send off, beside the filaments anastomosing with the recurrent nerve, threads to the dorsal vessel.[4] The Hymenoptera,[5] Neuroptera, and Orthoptera, also, have the two kinds of splanchnic systems. The double trunks are highly developed with the Acrididae, and the Gryllotalpida, and have two pairs of ganglia at their upper extremity, beside one or two on their course; while, with the Libellulidae, Blattidae, and especially the Phasmidae, the single nerve is the most developed.[6]

<hr/>

1 For the Splanchnic nervous system of the Insecta, of which *Swammerdamm* had already observed the recurrent nerve, see, beside the general works of *Burmeister* (Handb. &c. I. p. 308), and *Lacordaire* (Introduct. &c. II. p. 214), especially *J. Müller*, Nov. Act. Acad. Nat. Cur. XIV. 1828, p. 73; *Brandt* (Isis, 1831, p. 1103, also his Bemerk. über die Mundmagen-oder Eingeweidenerven d. Evertebr. 1835, p. 16, or Ann. d. Sc. Nat. V. 1836, p. 95), and *Newport*, Cyclop. &c. loc. cit. II. p. 957).

2 *Meckel* (Beitr. zur vergleich. Anat. I. p. 4) has observed the *Nervus recurrens* in the common *Cicada*, and *Brandt* (Bemerk. &c. p. 23, Taf. II. fig. 1, 2) has observed the same with *Lygaeus*, and at the same time the ganglia of the double system.

3 See *L. Dufour*, Ann. d. Sc. Nat. III. 1845, p. 67.

4 The recurrent nerve was first discovered in the silk-worm by *Swammerdamm* (Bib. der Nat. p. 132, Taf. XXVIII. fig. 3, g.). Subsequently, *Lyonet* (Traité, &c., p. 577, Pl. XII. fig. 1, Pl. XIII. fig. 1, Pl. XVI. fig. 14, Pl. XVIII. fig. 1) described with the larva of the Goat-moth, the double system and its relations with the dorsal vessel. Since then, the two systems have been observed in the larvae, pupae, and imagines of various Lepidoptera; see *Suckow* (Anatom. physiol. Untersuch. 40, Taf. VII. fig. 33–38, (pupa and imago of *Gastropacha pini*)), who has described the double system and the cardiac nerve. See, also, *J. Müller* (Nov. Act. Nat. Cur. loc. cit. p. 97 (the recurrent nerve of a larva of *Sphinx*)), and *Brandt* (Isis, loc. cit. p. 1104, Taf. VII. fig. 3, 4, and Bemerk. &c. p. 20), who has described the two systems with the imago and larva of *Bombyx mori*. The works of *Newport* (Philos. Trans. 1832, p. 383, Pl. XII. XIII., and 1834, p. 389, Pl. XIII. XIV.) on the larva and imago of *Sphinx ligustri*, are very distinguished.

5 See *Treviranus* (Verm. Schrift. III. p. 59), who thinks he has observed the *Nervus recurrens* with *Apis mellifica*; *Brandt*, also (Medizin. Zool. II. p. 203, Taf. XXV. fig. 32, and his Bemerk. &c. p. 22), has described the two systems in this species, and in the Bumble-bee (*Apis terrestris*).

6 According to *Burmeister* (Handb. &c. I. p. 310, Taf. XVI. fig. 6 (*Gryllus migratorius*)), the recurrent nerve leaving the frontal ganglion, runs backwards and ends, after a short course, in a ganglion which connects by two filaments with the internal ganglia of the double system. These last send off several branches to the œsophagus, and connect, through two filaments, with the external ganglia of the same system. From these external ganglia arise two lateral trunks which run along the œsophagus and are distributed to the gizzard, forming a nervous plexus having four ganglia. See, also, for the same species, *Brandt*, in the Isis, 1831, p. 1104, Taf. VII. fig. 5. According to this last author (Bemerk. &c. p. 23, Taf. II. fig. 7–9), the double system of *Gryllotalpa* is similarly disposed, only the nervous plexus of the gizzard arises from two posterior ganglia of the two trunks. See, also, for that of *Gryllotalpa*, *L. Dufour*, Recherch. sur les Orthopt. &c. p. 285, Pl. III. fig. 22. With *Phasma ferula*, the four anterior ganglia of the single system are small, but, for compensation, the double system is very com-

The Coleoptera have, in both their larval and their perfect states, a feebly-developed double nervous system arising from two pairs of ganglia, and a highly-developed *Nervus recurrens* which, with a few species, forms, directly behind the *Ganglion frontale*, a second ganglion. It runs along the œsophagus, and usually forms, posteriorly, still another ganglion, and then divides dichotomously.[7]

A great number of the Insecta have, in all their states, another system of nerves, called Respiratory nerves, which, in view of their functions, ought very properly to be classed among the mixed nerves, for they contain not only motor, but also vegetative fibres. This system arises by several single roots from the longitudinal commissures of the ventral cord. Each of these roots divides into two *Nervi transversii* which deviate from each other at right angles, and anastomose with the ganglia of the ventral chain and with its peripheric nerves, receiving at the same time organic fibres from the ganglia of the double splanchnic system. These respiratory nerves are distributed to the large tracheau trunks, and especially to the muscles of the stigmata. The respiratory movements of Insecta cannot, therefore, be regarded as properly of a voluntary nature.[8]

CHAPTER IV.

ORGANS OF SENSE.

§ 332.

The sense of Touch appears to be seated, with Insecta, in very different parts of the body.[1] It is chiefly located in the palpi of the mouth, which, for this purpose, are usually terminated by a soft surface.[2] The antennae, also, serve as tactile organs, but in a very variable manner, according to their forms, the degree of their development, and the habits of

---

plete ; see *Brandt*, Bemerk. &c. p. 27, Taf. III. fig. 1–5, and *J. Muller*, Nov. Act. Nat. Cur. loc. cit. p. 85, Tab. VIII. fig. 1, 3. These two anatomists have given, moreover, details with figures on the splanchnic nerves of *Libellula, Blatta, Mantis* and *Gryllus*.

7 *Swammerdamm* (Bib. der Nat. p. 132, Taf. XXVIII. fig. 2) has observed the *Nervus recurrens* in the larva of *Oryctes nasicornis*. *Muller* (Nov. Act. Nat. Cur. loc. cit. p. 94, Tab. VII. fig. 4, 5) has figured it with *Lucanus* and *Dytiscus*. *Straus* (Consid. &c. p. 406, 391, Pl. IX.) has observed with *Melolontha*, not only the single nerve, but also the double system which, however, he mistook for the accessory ganglia of the brain ; *Brandt* (Mediz. Zool. 11. p. 103, 118, Taf. XVII. fig. 3, 4, Taf. XIX. fig. 20) was the first to understand the true nature of this system with *Meloë* and *Lytta*. See, moreover, *Burmeister* (Handb. &c. Taf. XVI. fig. 8 (a larva of *Calosoma*)), *Newport* (Philos. Trans. 1834, Pl. XIII. fig. 4, 5 (imago and larva of *Timarcha tenebriosa*), and *Cyclopaed*. &c. fig. 405, 412, 416–418 (*Timarcha, Meloë* and *Lucanus*)) ; also *Schiödte*, in *Kroyer's* Naturh. Tidskrift. IV. p. 101, Pl. I. *Acilius*.

8 Although *Lyonet* (Traité, &c., p. 98, 201, Pl.

IV. fig. 5, Pl. IX. fig. 1) had already described this respiratory system with the larva of the goat-moth, under the name of *brides épinières*, it is *Newport* who has recently called the attention of anatomists to this subject, by furnishing, with admirable details, the disposition of this respiratory plexus, in the larva, pupa and imago of *Sphinx ligustri* (Philos. Trans. 1832, Pl. XII. fig. 4, 1834, Pl. XIII. &c., and 1836, Pl. XXXVI., also Cyclopaed. loc. cit. p. 947, fig. 400). See, also, *Muller's* ideas (Archiv, 1835, p. 82) on the nature of this nervous system. With various Coleoptera and Orthoptera, with *Locusta, Gryllotalpa*, and *Carabus*, the single roots arise, according to *Newport*, from small ganglia, at the points where are given off the *Nervi transversii*.

1 For the senses of the Insecta in general, beside the works of *Spence* and *Kirby, Burmeister* and *Lacordaire*, see *Scheleer's* Versuch einer Naturgesch. d. Sinneswerkz. bei d. Insekten u. Würmern, 1798, a work in which are related the opinions of the older naturalists on this subject.

2 The tactile sense of the palpi is of great service to Insecta when they eat ; for these organs are used not only to feel the food but also to retain, and convey it between the jaws.

the species. These organs receive, each, directly from the superior cerebral mass, a nerve; these nerves perceive the slightest disturbances occurring in the antennal teguments, which are solid and often provided with hairs and bristles. With those Insecta with which these organs are very long, filiform, and movable in various directions, they serve, like the vibrissae of many mammalia, to announce the presence of external bodies. With very many other Insecta, they are very movable, and are distinctly used as tactile organs, like the fingers of the human hand.[3]   It is also by means of these organs, that insects perceive the various conditions of the atmosphere, especially the temperature, and thereby regulate their movements and actions.

With those Insecta whereof the parts of the mouth are changed into organs of suction, it is quite evident that the extremity of the snout or proboscis is the seat of a very delicate sense of touch. Also with those female insects having an ovipositor, which is used to deposit their eggs in holes of various depth, the apex of this organ must be endowed with the same power. Finally, this sense must be ascribed to the extremities of the legs of many Insecta, which, in either their larval or in their perfect state, use these organs for the performance of labors of a special nature.[4]   With the Poduridae, there is, upon the ventral surface of the first abdominal segment, a singular organ which is soft, protractile, bifurcated or bi-lobed, and probably of a tactile nature.[5]

§ 333.

Undoubtedly the sense of taste, with Insecta, is seated in the tongue, when this organ is present. The tongue, of a soft consistence, is particularly developed with the Carabidae, Locustidae, Acrididae, Libellulidae, and Vespidae, which are all mandibulated; and with the Apidae, and Muscidae, which lick up their food. With the suctorial Insecta, the tongue is either wanting, or changed into a horny bristle; — a transformation met with, also, in certain species having masticatory organs.

§ 334.

The organs of Olfaction with Insecta, have not yet been satisfactorily determined, although most of these animals by their aid, can perceive in a most wonderful manner, the food proper either for themselves or their young. The various hypotheses upon this subject are unsatisfactory, and especially those by which this sense is located in the hard and dry parts of the body, which are quite unfit to recognize odoriferous substances.[1]

---

3 This may be especially observed with the Hymenoptera.

4 As such I recollect only the Ateuchidae and *Rhynchites* among the Coleoptera, the fossorial Hymenoptera, and the larvae of the Phryganidae among the Neuroptera.

5 With *Smynthurus*, these organs consist of two long contractile cylinders; see *Degeer*, Abhandl. &c. VII. p. 20, Taf. III. fig. 10, and *Nicolet*, Re-

cherch. &c. p. 42, Pl. III. fig. 5, 19–22. I am not determined whether or not should be placed in the same category the soft protractile organs, often of a beautiful red or orange color, possessed by *Malachius* on the lateral portions of the body, by *Stenus* at the extremity of the abdomen, and by the larvae of various Lepidoptera (*Papilio machaon* and *podalirius*, *Harpyia vinula*, &c.) on the neck or back.*

1 According to *Rosenthal* (Reil's Arch. X. p.

* [§ 332, note 5.] See upon the protractile, tentacular organs of the larvae of the Papiliones, *Karsten* (*Muller's* Arch. 1848, p. 375). I have carefully and microscopically examined these organs

with *Papilio asterias*; I regard them as odoriferous and defensive, rather than tactile organs. — Ed.

§ 335.

There is the same uncertainty concerning the organs of Audition. Experience having long shown that most Insecta perceive sounds, this sense has been located sometimes in this, and sometimes in that organ. But in these opinions, it often seems to have been forgotten or unthought of, that there can be no auditive organ, without a special auditive nerve which connects directly with an acoustic apparatus capable of receiving, conducting, and concentrating the sonorous undulations.[1]

Certain Orthoptera are the only Insecta with which there has been discovered, in these later times, a single organ having the conditions essential to an auditive apparatus. This organ consists, with the Acrididae, of two fossae or conchs, surrounded by a projecting horny ring, and at the base of which is stretched a membrane resembling a Tympanum. On the internal surface of this membrane, are two horny processes to which is attached an extremely delicate vesicle filled with a transparent fluid and representing a membranous labyrinth. This vesicle is in connection with an auditory nerve which arises from the third thoracic ganglion, forms a ganglion upon the tympanum, and terminates in the immediate neighborhood of the labyrinth by a collection of cuneiform, staff-like bodies with very finely-pointed extremities (primitive nerve-fibres?), which are surrounded by loosely-aggregated, ganglionic globules.[2]

The Locustidae and Achetidae have a similar organ, situated in the

---

136, Taf. VIII. fig. 5, 6), the olfactory organ of the Muscidae is a double, oblong fossa, situated under the antennae, and covered by a plicated membrane formed by the cutaneous envelope, which is otherwheres solid and dry. Until lately, from the time of *Reaumur*, the sense of smell has been located in the antennae, although they present no trace of a humid surface, and have none of the anatomical and physiological conditions requisite for being the seat of this function. See *Lefebvre*, Ann. d. la Soc. entom. d. France, VII. p. 395, or Ann. d. Sc. Nat. XI. 1839, p. 191 ; and *Kuster*, Isis, 1844, p. 647. The same objections might be raised against the opinion of *Marcel de Serres* (Ann. du Mus. XVII. p. 426), who locates this sense, with the Orthoptera, in the palpi. Equally groundless appears the view of *Baster* quoted by *Straus* (Consider. &c. p. 420), that this sense is seated in the stigmata of the tracheae. *Treviranus* seeks to avoid the difficulty in supposing that the entire buccal cavity, which is humid, can receive odorous impressions. *Erichson* (Diss. de fabr. et usu antenn. in Insect., Berlin, 1847) has recently appeared anew in favor of the antennae. According to him, the numerous small fossae of these organs are covered internally with a delicate membrane sensible to odors.*

1 The author who has erred most widely in this respect, is *L. W. Clarke* (Magaz. of Nat. Hist. September, 1838, or in *Froriep's* neue Notiz. IX. p. 4, fig. 12, a–n), who has described at the base of the antennae of *Carabus nemoralis*, Illig. an auditive apparatus, composed of au *Auricula*, a

Meatus auditorius externus and internus, a Tympanum, and a Labyrinthus, of all of which there is not the least trace. The two white convex spots at the base of the antennae of *Blatta orientalis*, and which *Treviranus* (Annal. d. Wettermuisch. Gesellsch. f. d. gesammte Naturkunde, I. Hft. 2, p. 169, Taf. V. fig. 1–3) has described as auditory organs, are, as *Burmeister* has correctly stated (Handb. II. p. 469), only rudimentary accessory eyes. *Newport* (Trans. of the Entom. Soc. II. p. 229) and *Goureau* (Ann. d. l. Soc. ent. X. p. 10) think that the antennae serve both as tactile and as auditory organs. But this view is inadmissible, as *Erichson* (*Wiegmann's* Arch. 1839, II. p. 285) has already stated, except in the sense that the antennae, like all solid bodies, may conduct sonorous vibrations of the air ; but, even admitting this view, where is the auditory nerve? for it is not at all supposable that the antennal nerve can serve at the same time the function of two distinct senses.

2 This organ has been taken for a soniferous apparatus by *Latreille* (Mém. du Mus. VIII. p. 123) and *Burmeister* (Handb. I. p. 512). *J. Muller* (Zur vergleich. Physiol. d. Gesichtssinn. p. 439, and Nov. Act. Nat. Cur. XIV. Tab. IX.) was the first who fortunately conceived that with (*Gryllus hieroglyphus*, this was an auditory organ. He gave, however, this interpretation only as hypothetical ; but I have placed it beyond doubt by careful researches made on (*Gomphoceros*, *Oedipoda*, *Podisma*, *Caloptenus* and *Truxalis* (*Wiegmann's* Arch. 1844, I. p. 56, Taf. I. fig. 1–7).

---

* [§ 334, note 1.] See also *Burmeister* (Zeit. für Zool., Zoot., und Paläontol. von *D'Alton* und *Burmeister*, No. 5, p. 49, Taf. I. fig. 25–29), who likewise advocates the auditory function of the antennae. But *Burmeister* and *Erichson* differ

somewhat in their statements upon the intimate auditory structure of these organs, and, therefore, as to the exact mode by which audition occurs. — ED.

fore-legs directly below the coxo-tibial articulation.[3] With a part of the Locustidae,[4] there is, on each side at this point, a fossa; while with another portion of this family,[5] there are, at this same place, two more or less spacious cavities (Auditive capsules) provided with orifices opening forwards.[6] These fossae and these cavities have each on their internal surface, a long-oval tympanum. The principal trachean trunk of the leg passes between the two tympanums, and dilates, at this point, into a vesicle whose upper extremity is in connection with a ganglion of the auditory nerve. This last arises from the first thoracic ganglion, and accompanies the principal nerve of the leg. From this ganglion in question passes off a band of nervous substance which stretches along the slightly excavated anterior side of the trachean vesicle. Upon this band is situated a row of transparent vesicles containing the same kind of cuneiform, staff-like bodies, mentioned as occurring with the Acrididae. The two large trachean trunks of the fore-legs open by two wide, infundibuliform orifices on the posterior border of the prothorax, so that here, as with the Acrididae, a part of this trachean apparatus may be compared to a *Tuba Eustachii*.[7] With the Achetidae, there is on the external side of the tibia of the fore-legs, an orifice closed by a white, silvery membrane (Tympanum), behind which is an auditory organ like that just described.[8]

§ 336.

The organs of Vision consist of simple, or of compound eyes.[1] The first occur chiefly with the larvae of holometabolic Insecta; and the second with Insecta in their perfect state. There are, however, many species which have both kinds of eyes in their imago state. These organs are wanting with only a few adult Insecta,[2] but are wholly absent with many larvae and pupae of the holometabolic species.[3]

1. The Simple eyes (*Ocelli, Stemmata*) are composed of a convex, spheroidal, or elliptical cornea, behind which is situated a spherical or cylindrical lens, lodged in a kind of calyx formed by an expansion of the optic nerve, and which is surrounded by a variously colored pigment-layer, as by a Chorioidea.[4] These stemmata are sometimes so closely situated

---

3 See my researches in *Wiegmann's* Arch. loc. cit. p. 72, Taf. I. fig. 8–17.

4 *Meconema, Barbitistes, Phaneroptera, Phylloptera.*

5 *Decticus, Locusta, Xiphidium, Ephippigera, Saga, Conocephalus, Callinemus, Acanthodis, Pseudophyllus,* &c.

6 In his classification of the Locustidae, *Burmeister* (Handb. &c. II. p. 675) has made use of the different forms of these orifices;—differences, however, which had before been pointed out by *Degeer* (Abhandl. Th. III. p. 285, Taf. XXXVII. fig. 5 and 6) *Lansdown Guilding* (Linn. Trans. XV. 1827. p. 153).

7 These two infundibuliform orifices of the tracheae, which *L. Dufour* (Recherch. sur les Orthopt. &c. p. 279, Pl. I. fig. 2) has called *vessies acrostatiques,* have generally been regarded as the stigmata of the prothorax, although the true stigmata, of the ordinary form and size, are situated in front of the orifices in question.

8 With *Acheta achatina* and *italica,* there is a tympanum of the same size, on the internal surface of the legs in question; but it is scarcely observable with *Acheta sylvestris, domestica* and *campestris*

1 For the eyes of the Insecta, see *Marcel de Serres,* Mém. sur les yeux comp. et les yeux lisses d. Ins.; *Treviranus,* Verm. Schrift. III. p. 147, and Beit. zur Anat. u. Physiol. d. Sinneswerkz. IIft. I. p. 84; finally, *J. Müller,* Zur vergleich. Physiol. des Gesichtssinn, p. 326, or in Ann. d. Sc. Nat. XVII. 1829, p. 242 (in extract), and his Memoir in *Meckel's* Arch. 1829, p. 38.

2 The eyes are wanting in many species of *Ptilium* which live under the bark of trees (*Erichson,* Naturgesch. d. Insekt. Deutschl. III. p 32); with *Anophthalmus,* which live in caverns (*Sturm,* Deutschl. Fauna Abth. V. Bd. XV.), and with *Claviger,* which live in ant-nests.

3 As such may be cited the larvae of Hymenoptera, excepting, however, those of the Tenthredinidae; those of the Diptera, which live in decomposing animal and vegetable substances; those of the Elateridae, Histeridae, Lamellicornia, Tenebrionidae, and in general the apodal larvae of Coleoptera; finally, the parasitic larvae of the Strepsiptera, whose females are also blind in the imago state.

4 For the simple eyes of *Dytiscus,* see *Müller,* in *Meckel's* Arch. loc. cit. p. 39, Taf. III. fig. 1, 2; for those of *Cicada, Vespa, Bombus,* and *Libel-*

on the brain that their optic nerves consist only of small papillae on this last;[5] but, when further removed from the brain and grouped together, the optic nerves arise by a common trunk which divides into as many branches as there are eyes.[6]

The number and disposition of the stemmata vary very much in the different orders. When they alone constitute the visual organs, they are always situated on the lateral parts of the head, — where they may be disposed either, as one on each side, or as several irregularly grouped together (*Ocelli gregati*), or regularly arranged in rows (*Ocelli seriati*). There is only one simple eye on each side with the Pediculidae, Nirmidae, Coccidae, the larvae of the Phryganidae and Tenthredinidae, and the aquatic ones of very many Diptera. These organs are in groups of four to eight with the Poduridae,[7] with the larvae of Lepidoptera, the hexopod larvae of the Strepsiptera, the larvae of the Hemerobidae, Mymeleonidae, Raphididae, and with the hexapod ones of the Coleoptera.[8] The winged males of the Strepsiptera have the largest number of stemmata aggregated in groups; they here form two lateral, globe-like projections, and constitute the transitionary form to the faceted eyes, for there are fifty to seventy on each side, separated from each other only by hairs.[9] Very many Insecta with two, faceted eyes, have, also, on their front, three stemmata disposed in a triangle.[10]

2. The Compound eyes, or those whereof the cornea is faceted, are composed of simple eyes so thickly set together that their more or less thick, slightly convex, quadrangular, or hexagonal corneae are contiguous.[11]

The size of these facets is not uniform even in the same eye, for sometimes those above, or those in the centre, are the larger.[12] Behind each cornea is situated, in place of a lens, a transparent pyramid the apex of which is directed inwards and received into a kind of transparent calyx corresponding to a *Corpus vitreum*. This last is surrounded by another calyx formed by the expansion of a nervous filament arising from the

lula, see *Treviranus*, Beitr. &c. p. 84, Taf. II. fig. 25–35.

5 *Bombus, Apis, Vespa*; see *Treviranus*, Biologie, V. Taf. II., and his Beitr. &c. Taf. II. fig. 29; and *Brandt* and *Ratzeburg* Medizin. Zool. II. Taf. XXV. fig. 31, 32.

6 With many of the larvae of the Lepidoptera and the Coleoptera, the optic nerves arise by two more or less long roots; see *Lyonet*, Traité. &c. p. 581, Pl. XVIII. fig. 1, No. 1, and fig. 6 (larva of the goat-moth); *Suckow*, Anat. physiol. Untersuch. p. 41, Taf. III. fig. 34 (pine caterpillar), and *Burmeister*, Trans. Entom. Soc. I. p. 239, Pl. XXIII. fig. 7 (larva of a *Calosoma*). The three stemmata of *Cicada* receive their nerves from a common trunk arising from the middle of the brain; see *Treviranus*, Beitr. Taf. II. fig. 24, and *L. Dufour*, Recherch. sur les Hémiptères, &c., Pl. XIX. fig. 203.

7 See *Nicolet*, Recherch. sur les Podurelles, loc. cit. p. 28, Pl. II. III.

8 Such are the carnivorous larvae of the Carabidae, Staphylinidae, Dytiscidae, Dermestidae, Silphidae, &c., and the herbiferous larvae of the Chrysomelidae. Those of *Cicindela* have only two large stemmata on each side of the head, and those of *Lycus, Meloë, Lampyris* and *Cantharis*, have only one.

9 See *Templeton*, Trans. Entom. Soc. III. p. 54, Pl. IV.

10 There are three frontal stemmata with many of the Orthoptera (Mantidae, Acrididae, Libelluli-

dae, Perlidae, Psocidae, Ephemeridae and some Phasmidae); with some Neuroptera (*Hemerobius, Panorpa, Phryganea*), and Hemiptera (*Pentatoma, Coreus, Berytus, Cicada*). This is the case also with many Diptera, such as the Muscidae, Syrphidae, Stomoxidae, Bombylidae, Anthracidae, Oestridae, Asilidae, Empidae, &c.; they are wanting with *Tabanus, Haematopota, Conops, Hippobosca, Metophagus*, and many of the Tipulidae. With the Hymenoptera, they are constantly present except with the neuter ants and with the females of *Mutilla* and *Myrmosa*; there are only two of these eyes with most of *Gryllus; Sciophila, Mycetobia* and *Loja*, of the Diptera; *Sesia, Euprepia, Pyralis* and a great number of the Noctuidae, of the Lepidoptera; *Gryllotalpa, Blatta* and *Termes*, of the Orthoptera; and *Omalium* and *Anthophagus*, of the Coleoptera.

11 For the intimate structure of the compound eyes, see *Straus* (Consid. &c. p. 411, Pl. IX.), *Doyès* (Ann. d. Sc. Nat. XX. 1830, p. 341, Pl. XII., or in *Froriep's* Not. XXIX. p. 257), *R. Wagner* (*Wiegmann's* Archiv, 1835, I. p. 372, Taf. V.), and especially *Will* (Beitrag. zur Anat. d. zusammengesezten Augen mit facettirt. Hornhaut. 1840).

12 These differences in the size of the facets had been observed by *Marcel de Serres* (loc. cit. p. 45) with *Libellula*. They exist also in the eyes of *Lagria flava, gibbosa, atra, Tabanus rusticus*, and some other Diptera; see *Ashton*, Trans. Entom. Soc. II. p. 253, Pl. XXI.

ganglion on the extremity of the optic nerve, a short distance from the brain.[13] Each lens-like pyramid with its vitreous body and nervous filament is enveloped by a Chorioidea usually of a brown color, which forms, behind the cornea, a kind of pupil,[14] but to which are due, by no means, the beautiful colors so often observed in the eyes of these animals.[15]

The size and form of the compound eyes, as also the number of their facets, are very varied.[16] The larvae and pupae of the hemimetabolic Insecta have, usually, a less number of facets and consequently smaller eyes, than the perfect forms. With the Libellulidae, and Diptera, the eyes are very large;[17] while with the Formicidae, they are perhaps the smallest of all. With many Diptera, and some Hymenoptera, those of the males are much larger than those of the females, and are often contiguous in front or above.[18] With some Hymenoptera, and Diptera, they are pilose, — the hairs being inserted in the angles of the facets.[19]

The compound eyes are usually spherical or oblong; and, with many Cerambycidae, and with the Vespidae, they are deeply emarginate in front, or on their internal border. With *Diopsis*,[20] they have a very singular appearance, being supported on two very long, rigid, frontal processes, and their direction cannot, as with other Insecta, be changed without a turning of the head.[21]

CHAPTER V.

DIGESTIVE APPARATUS.

§ 337.

The Insecta very often use their labial and maxillary palpi to seize and to convey food to the mouth, and even to introduce it wholly within this last. With many species, the fore-legs are used to seize and retain the food, and the first pair is sometimes changed for this purpose even into rapacious organs.[1] With the larvae and pupae of the Libellulidae, there

---

13 According to *Müller* (Arch. 1835, p. 613), these retinae are formed only by a prolongation of the neurilemma, while the proper nervous substance does not extend beyond the extremity of the vitreous body ; but *Will* denies this (*Müller's* Arch. 1843, p. 349).

14 Each of these pupils, according to *Will* (*Müller's* Arch. 1843, p. 350), is moved by thirty to thirty-five delicate fibres which arise on the four transparent cylinders surrounding the pyramidal lenses ; but *Brants* (Tijdsk. voor natuurhjke geschied. en physiologie. 1844, II.) regards them as tracheal branches and not contractile fibres.

15 The beautiful emerald color of the eyes of many Libellulidae, Tabanidae, Hemerobiidae, &c., is due to the cornea ; for the chorioideae are of the same dead color as those of other Insecta.

16 There are sometimes several thousands of these facets in the eyes of large size ; see *Müller*, Zur vergleich. Physiol. d. Gesichtssinn, &c., p. 340 ; and *Will*, Beitrag. &c. p. 10.

17 The largest eyes are observed with the Henopidae, where they cover nearly the whole head ; see *Erichson*, Entomographien. Hft. I. p. 132, Taf. 1.

18 Among the Hymenoptera, the genera *Astata*, *Larra*, *Tachytes*, *Apis* ; and among the Diptera, the Muscidae, Syrphidae, Leptidae, Tabanidae, Stratiomydae, and many other families.

19 With *Apis*, *Tabanus*, *Anthomyia*, *Eristalis*, *Volucella*, and other *Diptera*.

20 See *Linné*, Amoenitates academicae. VIII. Tab. VI. and *Dalman*, in *Fuessly's* Archiv d. Insekt. Hft. 1, Taf. VI. or Isis, 1820, p. 501, Taf. V.

21 The Insecta scarcely move their head when they look in different directions. This renders very singular the extended mobility of the head with *Mantis religiosa*, which, in watching for its prey, looks on all sides.

1 For example, with *Syrtis*, *Naucoris*, *Nepa*, *Ranatra*, *Hemerodromia*, *Mantis*, *Mantispa*, &c.

is, attached to the under lip, a peculiar prehensile organ which covers, like a mask, the masticatory organs, and, by means of a double articulation, can be let down and then returned with the utmost quickness. During this manœuvre, the prey is seized by two acute hooks inserted on the anterior border of this lip, and carried to the mouth.[2]

The parts of the mouth of the Insecta may be divided into Masticatory and Suctorial organs, between which, however, there are many intermediate forms. The second are, properly speaking, modifications of the first, and for this reason, the last should be described first; the special details of these organs, however, belong to the domain of Zoology.

These masticatory organs [3] consist of a pair of Mandibles and a pair of Maxillae, which move laterally and are more or less covered by an upper (*Labrum*), and an under (*Labium*) lip. The upper jaws (*Mandibulae*) exceed in hardness all the other parts of the masticatory apparatus, and consist of two simple, horny organs, often denticulated at their extremity. The under jaws (*Maxillae*) are, usually, softer, and composed of several pieces, — of which the most essential are : *Palpi maxillares*, composed of from one to six articles, and directed outwards ; and the stipule, usually denticulated or ciliated, and divided into a *Lobus externus* and *internus*. The under lip, which supports two *Palpi labiales* composed of from one to four articles, may thus be considered as another pair of maxillae the lateral halves of which are more or less fused together on the median line.[4] Such are the oral organs with the Coleoptera, the Neuroptera, and the Orthoptera. It is interesting to remark that the Orthoptera, in the widest acceptation of the term, have in common, this character, that their under lip is divided by a deep fissure into lateral halves, while that of the Neuroptera and Coleoptera consists of a single piece.[5]

At the base of the under lip is attached the tongue, which, either fleshy or horny, is single or cleft. Often it is completely abortive, but in other cases, on the contrary, it is very long and changed into a suctorial organ. This last form is most prominent with the Hymenoptera, where the two jaws have, at the same time, ceased to be masticatory organs, and form a sheath enveloping the tongue and labial palpi.[6]

The oral parts are changed into suctorial organs with the Diptera, Hemiptera, and Lepidoptera. The first have a *Proboscis*, formed by the under lip transformed into a suctorial tube (*Theca*) which is often geniculate. At its base are from four to six bristles which may be regarded, some as maxillae and mandibles, and others as representing the tongue.[7]　With

2 See *Roesel*, Insektenbelustigungen, II. Insectorum aquatil. Classis II. p. 12, Taf. III. IX., and *Suckow*, in *Heusinger's* Zeitsch. d. organ. Physik. II. Taf. I.

3 Beside the so often cited writings of *Straus*, *Kirby* and *Spence*, *Brandt* and *Ratzeburg*, *Burmeister*, *Lacordaire*, *Newport*, and *Westwood*, see *Savigny*, Mém. sur les anim. sans vertèbres, I. p. 1, Pl. I.–IV.; also, Isis, 1818, p. 1405, Taf. XVIII. *Nees von Esenbeck*, Isis, 1818, p. 1386, and *Suckow*, in Heusinger's Zeitsch. &c. III. Taf. I.–IX.

4 This opinion, before advanced by *Oken*, *Savigny*, and *Leach*, has been sustained with very many details by *Brullé* (Ann. d. Sc. Nat. II. 1844. p. 324).

5 On account of these modifications of the under lip, to which *Erichson* (Entomograph. Hft. 1, p. 6, and in *Germar's* Zeitsch. I. p. 150, Taf. II.) has especially called the attention, we can distinguish, in their perfect state, the hemimetabolic from the holometabolic Neuroptera. This justifies the separation we have made of the first whose pupae take food and are active, from the second whose pupae are inactive and do not eat. We have placed these last among the Orthoptera, because, like them, they have in all their states a bifid under lip. The differences between the under lip of the Orthoptera and that of the Neuroptera are well shown in *Savigny's* excellent figures of the buccal organs of these insects (Descript. d. l'Egypte, Orthoptères, Pl. I.–VII. and Neuroptères, Pl. I.–III.

6 See *Swammerdamm*, Bib. der Nat. Taf. XVII. fig. 5 ; *Treviranus*, Verm. Schrift. II. Hft. 2, p. 112, Taf. XII.–XIV.; *Brandt* and *Ratzeburg*, Mediz. Zool. II. Taf. XXV. fig. 8–16; *Newport*, Cyclop. loc. cit. p. 897, fig. 375, 376 ; but especially *Savigny*, Descript. de l'Egypte, Hymenoptères, Pl. I.–XX.

7 See *Savigny*, Mém. sur les anim. sans vertè-

the Hemiptera, the suctorial apparatus is lengthened into a *Rostrum*, by the under lip being changed into two quadri-articulate grooves united so as to form a tube, and enclosing the setiform mandibles and maxillae.[8] With the Lepidoptera, the changes are still greater, for the mandibles are only very small appendages, while the maxillae are transformed, each, into a semi-canal which can be rolled up spirally, and when united form an organ of suction (*Lingua spiralis*). At the base of this last are two very short maxillary palpi, bi- or tri-articulate, while the two tri-articulate and very hairy labial palpi consist of two pretty large appendages between which the suctorial tube retreats when rolled up.[9]

The buccal organs begin to atrophy with the Aptera. The four palpi present with the Lepismidae, are already wanting with the Poduridae;[10] and with the Nirmidae, they, as well as the maxillae, are very small, while the mandibles are quite large.

With the Pediculidae, there are still wider modifications; for here there is a protractile suctorial tube composed of four stiff bristles (rudimentary jaws) which are enclosed in a soft and equally protractile sheath (under lip.)[11]

With the Larvae of Insecta, the buccal organs are most usually masticatory; for, not only the larvae of the Coleoptera, the Orthoptera, and many of the Neuroptera and Hymenoptera, have the same organs of this kind (masticatory) as the perfect insects,[12] but also the larvae of the suctorial Lepidoptera,[13] and those with a distinct head of certain Diptera [14] with which, however, the maxillae and palpi are very frequently wanting.[15] But with the acephalous larvae of Diptera, those of the Strepsiptera, as also with the parasitic ones of some Hymenoptera, the mouth is formed rather for sucking than for masticating the food; for, on the inner side of the soft tumid lips, either the horny organs are wholly wanting,[16] or the mouth is armed with two parallel hooks, which are used partly to grapple and partly to puncture the bodies these animals attack.[17]

bres, I. Pl. IV. fig. 1, and *Newport*, Cyclopaed. loc. cit. fig. 379–381.*

8 *Savigny*, Mém. &c. I. Pl. IV. fig. 2, 3; *Ratzeburg*, Mediz. Zool. II. Taf. XXVII.; and *Burmeister*, Handb. &c. II. Taf. I.

9 This suctorial tube is pretty long with the Papilionidae and the Sphingidae; it is very short with many Bombycidae and Pyralidae; see *Savigny*, Mém. &c. I. p. 1, Pl. I.–III.; *Ratzeburg*, Die Forstinsekt. II. p. 2, Taf. I.; and *Newport*, Cyclopaed. loc. cit. p. 900, fig. 377, 378.

10 *Nicolet*, Recherch. p. 34, Pl. IV.

11 *Burmeister*, Linnaea entomologica. II. p. 569, Tab. I.

12 See *Ratzeburg*, Die Forstinsekt. I. III.; *Hartig*, Die Aderflüger Deutschlands, Taf. I.–VIII.; *Burmeister*, Trans. Entom. Soc. I. Pl. XXIII. XXIV. (*Calosoma*), and Naturgesch. d. Calandra, fig. 10–12; *Waterhouse*, Trans. entom. Soc. I. Pl. III.–V. (*Rhaphidia*, and various Coleoptera).

13 *Lyonet*, Traité, &c., Pl. II.; and *Ratzeburg*, Die Forstinsekt. II. Taf. I.

14 Such are the larvae of *Culex*, *Chironomus*, *Corethra* and *Simulia*, and many other of the aquatic Tipulariae.

15 In the larvae of *Sciara*, *Mycetophila*, *Sciophila*, *Ceroplatus*, &c., which live in rotten wood or in fungi; see *L. Dufour*, Ann. d. Sc. Nat. XI. 1839, p. 204, Pl. V. fig. 23, XII. p. 10.

16 The mouth of the apodal larvae of the Strepsiptera (see my researches in *Wiegmann's* Arch. 1843, I. p. 159, Taf. VII. fig. 14), and of the young larvae of *Microgaster* (*Ratzeburg*, Die Ichneumon. d. Forstinsekt. p. 13, Taf. IX.) has, in place of jaws, soft papillae which, as these larvae approach the end of their development, are changed into horny jaws by means of which these Insecta make a passage into the skin of the animals in which they live.

17 With the Muscidae, Oestridae, Syrphidae, and other Diptera; see *Swammerdamm*, Bib. der Nat. Taf. XLIII. fig. 5, and *L. Dufour*, Ann. d. Sc. Nat. 1. 1844, p. 372, Pl. XVI. fig. 8, 10, XII. 1839, p. 4, Pl. II. III.

* [§ 337, note 7.] See in this connection the memoir of *Blanchard* (De la Composition de la bouche dans les Insectes de l'ordre des Diptères, in the Compt. rend. 1850, XXXI. p. 424), who shows that the mouth of the Diptera presents appendages wholly comparable to those of the other Insecta, except that these appendages are modified in a special manner. — Ed.

The mouth of the larvae of the Myrmeleonidae, Hemerobidae, and Dytiscidae, is of a very peculiar construction. There is no oral orifice, properly speaking, and the maxillae and mandibles are wholly unfit for mastication, the latter being changed into two curved hooks, hollow and with a narrow fissure at their extremity. These larvae bury these hooks in the insects they have seized, and through the cavity of these organs, which communicates at its base with the œsophagus, suck the blood.[18]

A considerable number of the Insecta take no food during their perfect state, the object of their existence being only to accomplish the act of reproduction. Their jaws are often very rudimentary and are fit neither for sucking nor for masticating.[19] In some cases, indeed, not only are these organs wanting, but the oral orifice is closed as with all inactive pupae.[20]

§ 338.

The Digestive Canal of Insecta and their larvae, is more or less long, sometimes extending from the mouth directly to the anus upon the median line; sometimes forming in the abdomen loops and convolutions. It is retained in place not by a mesentery, but by numerous fine tracheae, which envelop its entire extent. It is always wholly invested by a homogeneous peritoneal envelope under which lies a muscular tunic, composed of longitudinal and circular fibres, which are especially developed about the mouth and anus. Internally, it is lined throughout by an epithelium which is extremely thin at the middle portion of this canal, but very solid and composed of chitine at its two extremities. In the middle portion just mentioned, there is a layer of aggregated cells, evidently of a glandular nature, between the epithelium and the muscular tunic.

The different parts of this canal in the Insecta may be properly distinguished in the following manner. The first portion is the *Oesophagus*, muscular, occupying the three thoracic segments and often dilated at its posterior part into a crop (*Ingluvies*) and muscular gizzard (*Proventriculus*). Sometimes there is appended to the œsophagus a sucking stomach consisting of a more or less pedunculated, thin-walled vesicle, which is multiplicated on itself when empty.

The second portion consists of a *stomach* (*Ventriculus*), in which the chyle is formed, and which is continuous at the point of insertion of the Malpighian vessels, with the third portion of the digestive canal. This third portion commences by a small and usually short *Ileum*, which is followed by a *Colon*, larger and of variable length. This last often has a *Caecum* at its anterior extremity and terminates posteriorly in a short muscular *Rectum*.[1]

---

[18] See *Roesel*, Insektenbelust. III. Taf. XVII. XVIII. (*Myrmeleon*), 11. Insect. aquit. classis I. Taf. I.–III. (*Dytiscus*); *Ratzeburg*, Forstinsekt. III. Taf. XVI. (*Hemerobius*). With the larva of *Dytiscus*, the body of the maxillae is wholly abortive, but always provided with palpi. With those of *Hemerobius*, the maxillae are small, deficient in palpi, and play in a groove on the concave side of the mandibles; finally, with those of *Myrmeleon*, these organs are wholly enclosed in the cavity of the mandibles.

[19] The maxillae are rudimentary and very soft with the Ephemeridae, and Phryganidae, in the last stages of their development. The very short proboscis of many Bombycidae and Hepiolidae, appears equally unfit to receive food. Finally, the two small, intercrucial maxillae of the males of the Strepsiptera, are wholly inadequate for the functions of masticatory organs.

[20] Movable oral organs and an oral orifice are wanting with many Oestridae, and Henopidae, as well as with the male Coccidae.

[1] The functions of these different portions of the digestive canal do not always correspond to those of those parts having the same names with Vertebrata. *Burmeister* (Zur Naturgesch. d. Calandra p. 9) is certainly correct in saying that the stomach is the chylopoietic part, thus combining the func-

With nearly all Insecta in their perfect state, this colon or large intestine contains from four to six organs of a peculiar structure and doubtful function. These consist of transparent protuberances, disposed in successive pairs, or forming a transverse series. They are round, ovoid, or oblong, their base being sometimes surrounded by a horny ring, and they are traversed by numerous tuft-like tracheae.[2] The Lepidoptera, especially, are remarkable for their numerous organs of this kind.[3] It is singular that they are wanting in all insects during their larval and pupa states.

The Anus of Insecta, in all their states, is invariably situated on the last segment of the body. With the quiescent and non-feeding pupae, both the anus and the mouth are wanting, but with the larvae of only the Strepsiptera, the Apidae, and the Vespidae, are both ileum and colon wanting at the same time.[4]

The form and disposition of the different parts of the digestive canal vary infinitely, according to the habits of life and the states of development of the Insecta in which they are observed. On this account it is very difficult to make any general statement of the various structural relations.[5] But that condition may be taken as the fundamental type which belongs to those perfect insects whose life is pretty long and which have masticatory organs. Such, therefore, will receive our first consideration.

With the Coleoptera,[6] the oesophagus is nearly always terminated by a

---

tions of the stomach and small intestines of the Mammalia. The crop and gizzard correspond to parts of the same names with birds. The ileum, which is usually regarded as analogous to the small intestine of the Vertebrata, probably plays a very subordinate part in the act of digestion. *Burmeister* thinks that it serves only to conduct the chyme or chyle, but with certain species where it is very long, it is probably the seat of a second digestion. The caecum often serves to receive the secretory product of the Malpighian vessels, and therefore belongs rather to the urinary than to the Chylopoietic apparatus (see § 346).

2 It is hardly comprehensible how organs so common with the Insecta, should, as yet, be so little known.

Swammerdamm, however, observed them with *Apis mellifica* (Bib. der Nat. Taf. XVIII. fig. 1), and *Suckow* (*Heusinger's* Zeitsch. III. p. 21, Taf. VI. fig. 121, 128) has mentioned them with *Vespa crabro*, and *Apis mellifica*, under the name of callous swellings. *Brandt* and *Ratzeburg*, Mediz. Zool. II. Taf. XXV. fig. 29 (*Apis mellifica*), as well as *Burmeister* (Handb. &c. I. p. 149) speak of them very slightly. *L. Dufour* (Recherch. sur les Orthopt. &c. p. 396, 427), has figured them with various Orthoptera, Neuroptera and Hymenoptera under the name of *Boutons charnus*; finally, *Newport* (Cyclopaed. &c. II. p. 970, fig. 424, (*Carabus monilis*)) has designated them as Glandular protuberances. All the figures above cited give the external form of these organs but not

their internal structure. They are especially apparent and four in number with the Muscidae ; see *Ramdohr*, Abhandl. üb. d. Verdauungswerkz. &c. Taf. XIX. fig. 2, M. M. ; and *Suckow*, loc. cit. Taf. IX. fig. 153. The four with *Melophagus* are very singular and different from those of the other pupiparous Diptera, in that their external surface is covered with small solid scales ; see *L. Dufour*, Ann. d. Sc. Nat. III. 1845, p. 71, Pl. II. fig. 13–15.

3 I have counted, with the Zygaenidae, thirty of these swellings, and nearly a hundred with the Papilionidae, Noctuidae and Geometridae. *Hepialus*, *Tinea*, and *Adela*, have, by exception, only six. *Treviranus* (Verm. Schrift. II. p. 106, Taf. XII. fig. 4), and *Lyonet* (Mém. du Mus. &c. XX. p. 184, Pl. XVIII. fig. 6) have taken these organs for glands with *Papilio*.

4 The digestive canal is probably organized in a similar manner with the larvae of the Hymenoptera and the Diptera, which are parasitic in the bodies of other Insecta.[*]

5 For the digestive tube of the Insecta, beside the works already cited of *Swammerdamm*, *Gaede*, *Burmeister*, *Lacordaire*, and *Newport*, see, especially, *Ramdohr*, Abhandl. üb. d. Verdauungswerkz. &c. ; *Marcel de Serres*, Ann. du Mus. XX. p. 48 ; and *Suckow*, in *Heusinger's* Zeitsch. III. p. 1.

6 The digestive organs of the Coleoptera have been especially studied by *L. Dufour* (Ann. d. Sc. Nat. II. III. 1824, and I. 1834). See, moreover,

---

* [§ 384, note 4.] See, for the intestinal canal of the larvae of Hymenoptera, *Ed.* Grube (*Müller's* Arch. 1849, p. 50), who, from examinations of the larvae of wasps and hornets, concludes that a straight alimentary canal opening at the posterior extremity is always present, but that only the muscular tunic forms the continuous tube, — the lining membrane of the stomach ending caecally, and the

same membrane of the intestine commencing caecally, and, finally, that the intestine serves, during the larval state, only to receive the secretion of the Malpighian vessels which are urinary organs. But it is doubtful if the contents of the stomach are expelled by mouth during the larval state. This closed pyloric end of the stomach is opened during the transition to the pupa state. — ED.

crop-like dilatation,[7] which, with the Cicindelidae, Carabidae, Dytiscidae, and Gyrinidae, is followed by an ovoid gizzard. This last is longitudinally plicated internally, and these folds are usually armed on their borders with cilia or horny hooks. The intestinoid stomach is of median length with the carnivorous Coleoptera, but very long and more or less flexuous with those which are herbivorous.[8] Nearly always, its whole external surface is numerously constricted, and covered with small caeca.[9] The ileum and colon are, usually, rather short.[10]

Among the Orthoptera, the families Forficulidae, Termitidae, Blattidae, Achetidae, Locustidae, Acrididae, and Mantidae, are distinguished for their large crop,[11] which, with *Gryllotalpa* is completely constricted from the œsophagus. The gizzard is of variable length, and covered, internally, with rows of horny denticulated plates.[12] The stomach is tubular, of equal calibre, median length, and rarely makes a half or an entire turn.[13] In most of the families just mentioned, its upper extremity has two, six, or eight caeca,[14] and its posterior part is continuous into an often somewhat flexuous ileum upon which succeeds a short colon. With the Perlidae, the gizzard is wanting, but the upper extremity of the stomach has from four to eight caeca, pointing forwards.[15] With the Phasmidae, and the Libellulidae, the oesophagus is long and large, and protrudes somewhat into the straight, oblong, constricted stomach, which is without caeca and is succeeded by a very short ileum and colon.[16] The digestive tube of the Ephemeridae, which, in their perfect state, take no food, is feebly developed. Its walls are very thin throughout, and the oesophagus is directly continuous with the stomach which is a bladder-like dilation and succeeded by a short, straight intestine.[17]

Ramdohr, Magaz. d. naturf. Freunde zu Berlin, 1807, p. 207, Taf. IV. (*Carabus*); *Brandt*, Mediz. Zool. II. Taf. XVII. XIX. (*Meloë* and *Lytta*); *Straus*, Considér. &c. Pl. V. (*Melolontha*).*

7 With *Oedemera*, this crop is constricted from the stomach; see *L. Dufour*, loc. cit. III. Pl. XXX. fig. 7, 8.

8 The stomach is of the greatest length with the Melolonthidae and Hydrophilidae; see *Straus*, loc. cit. Pl. V., and *Suckow*, loc. cit. II. Taf. III. IV.

9 This constricted stomach is especially observed with the herbivorous Coleoptera, as, with *Meloë*, *Lytta*, and *Cantharis*; but is wholly wanting with *Lycus*, *Telephorus*, *Malachius*, and *Cistela*. With the Elateridae, the stomach is smooth, but, at its upper extremity there are two caecal folds, which, with the Buprestidae, are very long; see *L. Dufour*, loc. cit. III. Pl. XI. fig. 1, 3, 4; *Meckel*, Beitr. &c. I. lfft. 2, p. 129, Taf. VIII. fig. 5; and *Gaede*, Nov. Act. Nat. Cur. XI. p. 330, Tab. XLIV. fig. 1.

10 With the Dytiscidae, a pretty long and small caecum extends forwards from the rectum; see *Ramdohr*, Abhandl. &c. Taf. II.; *L. Dufour*, loc. cit. III. Pl. X. fig. 3, and *Burmeister*, Handbuch, &c., I. Taf. X. fig. 4.

11 See *Ramdohr*, Abhandl. &c. Taf. I.; *Marcel de Serres*, loc. cit. Pl. I.–III.; *Gaede*, Beitr. &c. Taf. I. II. (*Blatta* and *Acheta*); *Suckow*, loc. cit. III. Taf. VII. fig. 134–136 (*Gryllotalpa*); *Burmeister*, Handb. &c. I. Taf. XI. fig. 1–6; and *L. Dufour*, loc. cit. XIII. 1828, p. 350, Pl. XX.

(*Forficula*), and his Recherch. sur les Orthopt. &c. loc. cit. Pl. I.–V. XIII.

12 See the figures cited in the preceding note. According to *L. Dufour* (Recherch. &c. p. 608, Pl. XIII. fig. 196), this gizzard with its dental apparatus is wanting with *Termes*; but, according to *Burmeister* (Handb. I. p. 137, Taf. XI. fig. 8–10), it is present being concealed at the base of the oesophagus.

13 *Gryllotalpa* and *Ephippigera*.

14 These caeca are wanting with *Forficula* and *Termes*. There are only two with *Acheta*, *Gryllotalpa*, *Locusta*, and *Ephippigera*; six to eight with the Mantidae, Blattidae, and Acrididae. In these last, each of these caeca sends off two deverticula, one forwards, and the other backwards.

15 See *Suckow*, in *Heusinger's* Zeitsch. II. p. 267, Taf. XVI. fig. 7; *L. Dufour*, loc. cit. Pl. XIII. fig. 198; and *Pictet*, Hist. Nat. des Névroptères. Famille des Perlides. These caeca are wanting with *Nemura*.

16 See *Ramdohr*, Abhandl. &c. Taf. XV. (*Libellula* and *Agrion*); *Suckow*, loc. cit. II. Taf. II. fig. 14 (*Aeschna*); *L. Dufour*, Recherch. &c. p. 568, Pl. XI. (*Aeschna* and *Libellula*); and *Muller*, Nov. Act. Nat. Cur. XII. p. 571, Tab. L. (*Bacteria*). These stomachic appendages are wanting with *Psocus* also; see *Nitzsch*, in *Germar's* Mag. IV. p. 277, Taf. II. fig. 1.

17 See *L. Dufour*, Recherch. &c. Pl. XI. fig. 167, and *Pictet*, Hist. Nat. des Insect. Névropt. Famille des Éphémérines.

---

* [§ 338, note 6.] See, also, *Leidy*, loc. cit., Flora and Fauna within Animals, &c., for full details of the intimate anatomy of the alimentary canal of *Passalus cornutus*. — Ed.

The predatory Panorpidae, which are rapacious, differ notably from the other Neuroptera, and resemble rather the preceding order. Their œsophagus is short and straight, and, in the thorax, is succeeded by a spherical muscular gizzard which is lined internally with a brown chitinous membrane covered with stiff hairs. The stomach is tubular and straight; the ileum makes two convolutions before passing into the long colon.[18] With the other Neuroptera, namely, the Myrmeleonidae, Hemerobidae, Sialidae, and Phryganidae, the œsophagus is long, and dilated, posteriorly, into a kind of pouch; and often there is a long, thin-walled, sucking stomach inserted on one of its sides. The proper stomach is of a median length, and is more or less transversely constricted.[19] The two other portions of the digestive canal are very small and straight.

The Hymenoptera, which often sip up their fluid flood, have a long œsophagus which dilates into a thin-walled, sucking stomach.[20] With the Vespidae, Apidae, and Andrenidae, this stomach is often only a lateral fold of the œsophagus, and with many Crabonidae, it is attached solely by a short and narrow peduncle.[21] Many species of this order have a rudimentary, callous gizzard, enveloped by the base of the stomach. In the genera *Formica*, *Cynips*, *Leucospis* and *Xyphidria*, it is very apparent, and consists of a globular, uncurved organ. Those Hymenoptera which are engaged during a long and active life[22] in labors for the raising and support of their young, have a pretty long and flexuous stomach and intestine, and the first has, usually, many constrictions. The Cynipidae, Ichneumonidae, and Tenthredinidae, which, after copulation and the deposition of their eggs, take no further care in the act of reproduction, have only a very short small stomach and intestine.

But the modifications of each of the various portions of the digestive tubes are most prominent with the sucking Insecta, especially with the Hemiptera.[23] The œsophagus of these last is usually short and small, while the stomach is generally very long, and describes many convolutions in the abdominal cavity. This stomach, as to form and structure, may be divided into two or three distinctly-defined portions. The first consists of a glandular ante-stomach which is straight, large, and divided by several constrictions. The second has the form of a long, flexuous canal, whose walls are glandular, and which dilates, at its posterior extremity, into an oval pouch.[24] With the Cicadidae, it forms a kind of loop, its posterior extremity being attached to the ante-stomach with *Tettigonia*, *Cercopis*, and

18 *Ramdohr*, Abhandl. &c. p. 150, Taf. XXVI. fig. I, and *L. Dufour*, Recherch. &c. p. 582, Pl. XI. fig. 169.

19 *Ramdohr*, Abhandl. &c. Taf. XVI. fig. 2, Taf. XVII. fig. 2, 6; *L. Dufour*, Recherch. &c. Pl. XII. XIII.; and *Pictet*, Recherch. pour servir à l'hist. et à l'anat. des Phryganides.
The Myrmeleonidae and Hemerobidae, alone, have a spherical callous gizzard situated between the stomach and œsophagus.

20 See *Swammerdamm*, Bib. der Nat. Taf. XVIII. fig. 1; *Treviranus*, Verm. Schrift. II. Taf. XIV. XVI.; *Brandt* and *Ratzeburg*, Mediz. Zool. II. Taf. XXV. fig. 29; *Ramdohr*, loc. cit. Taf. XII.–XIV.; *Suckow*, loc. cit. III. Taf. VI. VII. VIII.; finally, *L. Dufour*, Recherch. &c. p. 380, Pl. V.–X.

21 With *Chrysis*, and *Hedychrum*, this sucking stomach consists of two lateral cæca situated at the lower end of the œsophagus; see *Suckow*, loc. cit. III. Taf. IX. fig. 155, and *L. Dufour*, loc. cit. Pl. IX. fig. 113, 116.

22 The Apidae, Andrenidae, Vespidae, and Larridae.

23 For the digestive apparatus of the Hemiptera, see *Ramdohr*, Abhandl. &c. Taf. XXII. XXIII.; *Suckow*, loc. cit. III. Taf. VII. VIII.; *L. Dufour*, Recherch. sur les Hémiptères. p. 20, Pl. I.–IX.*

24 *Notonecta*, *Naucoris*, *Velia*, *Lygaeus*, *Coreus*, *Pyrrhocoris*, *Pentatoma*, *Tetyra*, *Syromastes*, &c.

* [§ 338, note 23.] For the digestive apparatus with all its details of *Belostoma*, see *Leidy*,

Proc. Acad. Nat. Sc. Philad. I. 1847, p. 62. — Ed.

Ledra ; and with *Cicada,* it penetrates even under the muscular tunic of this ante-stomach.[25] With the Pentatomidae, and some Coreidae, there is even a third stomach, quite remarkable, consisting of a very narrow, slightly-flexuous canal, on which are inserted two or four rows of closely-aggregated glandular tubes.[26] The ileum and colon are nearly always fused into a pyriform pouch, upon which is sometimes inserted a kind of lateral caecum.[27] With the Cicadidae, however, the ileum is distinct, narrow, and nearly always very long and flexuous.

The Diptera have a sucking stomach with a more or less long peduncle, inserted upon one of the sides of the short, small œsophagus. This peduncle accompanies the stomach even into the abdominal cavity, where it terminates in a pouch whose thin walls are composed solely of delicate muscular fibres. This pouch is oblong or round, and often divided, heart-shaped, by a deep fissure.[28] The proper stomach is always long and intestinoid, except at its anterior extremity, where it is often dilated. It is situated in the abdominal cavity and makes many convolutions. In some families, only, there are two lateral caeca inserted near its cardiac extremity.[29] The ileum is small, of median length, and is succeeded by a pyriform colon.

The Lepidoptera, which, in their perfect state, live only upon the juices of flowers, suck up this kind of food by means of a thin-walled, sucking stomach, situated at the anterior extremity of the abdominal cavity, and opening by a short peduncle into the posterior extremity of the small, long oesophagus.[30] The stomach is pretty long and large, often varicose, and always straight. The ileum is long, small, and nearly always forms several

25 See *Ramdohr,* loc. cit. Taf. XXIII. fig. 3 ; *Suckow,* loc. cit. Taf. VII. fig. 138 ; and *L. Dufour,* loc. cit. Pl. VIII. or Ann. d. Sc. Nat. V. 1825, p. 157, Pl. IV. It was formerly thought that the second stomach of *Cicada* opened into the ante-stomach, but the true relation of these organs has been pointed out by *Doyère* (Ann. d. Sc. Nat. XI. 1839, p. 81, Pl. I.) and confirmed by *L. Dufour* (Ibid. XII. p. 287). The annular stomach of *Dorthesia* and *Psylla* is probably arranged in the same manner ; see *L. Dufour,* Recherch. &c. loc. cit. Pl. IX. fig. 108, 110.

26 There are four rows of these glands with *Pentatoma,* and *Tetyra,* and two, only, with *Syromastes,* and *Corcus* ; see *Ramdohr,* loc. cit. p. 189, Taf. XXII. fig. 3, 4 ; *L. Dufour,* Recherch. &c. p. 21, Pl. I. II. These two authors have taken these rows of glands for transverse-plicated semi-canals. *Treviranus* (Annal. d. Wetterauisch. Gesellch. &c. I. Hft. 2, p. 175, Taf. V. fig. 4) is still more mistaken in taking the four rows in *Pentatoma rufipes,* for as many adjacent, but distinct intestinal tubes.

27 *Corcus, Pelozunus, Ranatra,* and *Nepa.*

28 The sucking stomach is simple with the Tipulidae, and Leptidae ; it is cordate with the Tabanidae, Syrphidae, and Muscidae ; see *Ramdohr,* and *Suckow,* loc. cit. ; *Treviranus,* Verm. Schrift. II. p. 142, Taf. XVII. ; and *L. Dufour,* Ann. d. Sc. Nat. I. 1844, p. 376, Pl. XVI. fig. 12.

29 With the Tabanidae, these two caeca point forwards ; but with the Leptidae and Bombylidae, backwards ; and with the Syrphidae, there are four of them, varicose, two pointing forwards

and two backwards. The Diptera fill this sucking stomach with liquid (honey, blood, &c.), or solid (pollen-grains) substances, but which, certainly, are only there deposited without being changed, for the walls of this organ do not present the least traces of a glandular structure. It is, moreover, remarkable that the Pulicidae and the Hippoboscidae, which feed exclusively on animal juices, have a kind of crop at the posterior extremity of the oesophagus, but no trace of a sucking stomach ; see *Ramdohr,* loc. cit. Taf. XXI. XXIII. (*Melaphagus* and *Pulex*), also *L. Dufour,* Ann. d. Sc. Nat. VI. 1825, p. 303, Pl. XIII. fig. 1, and III. 1845, p. 69, Pl. II. fig. 13 (*Hippobosca* and *Melophagus*). With *Pulex,* the crop is provided with large cilia on its internal surface, and thus resembles a gizzard.

30 See *Swammerdamm* Bib. der Nat. Taf. XXXVI. fig. 1 (*Vanessa urticae*) ; *Treviranus,* Verm. Schrift. II. p. 103, Taf. XI., and Annal. d. Wetterauisch. Gesellsch. III. Hft. 1, p. 147, Taf. XVI. (*Vanessa, Sphinx,* and *Deilephila*) ; *Suckow,* loc. cit. Taf. IX. fig. 161 (*Yponomeuta*) ; and *Newport,* Cyclop. loc. cit. fig. 430, 431 (*Sphinx* and *Pontia*). This sucking stomach is double with the Zygaenidae (*Ramdohr,* loc. cit. Taf. XVIII. fig. I) ; it is wholly wanting with the Hepialidae, Bombycidae, and in general all the imago Lepidoptera which do not eat. See *Treviranus,* Verm. Schrift. loc. cit. p. 107, and Annal. d. Wetterauisch. Gesellsch. loc. cit. p. 158, Taf. XVII.; and *Lyonet,* Mém. du Mus. XX. p. 208, Pl. XIX. fig. 10.*

* [§ 338, note 30.] See, also, for the intimate structure of the intestinal canal and its appendages of *Bombyx mori* (both larva and imago), *Filippi,* Annali della R. Accad. d'agricoltura di Torino, V., or *Wiegmann's* Arch. 1851, Th. II. p. 217. — ED.

convolutions. The colon is constantly of a large size, and is often dilated into a caecum at its anterior portion.[31]

Among the Aptera, the Nirmidae, Poduridae, and Lepismidae, have, at the posterior extremity of the œsophagus, a kind of crop, which, with *Lepisma*, is succeeded by a globular gizzard provided with six teeth. The proper stomach has the form of a long tube, and is not flexuous as with the Pediculidae. With these last, and with the Nirmidae, which are parasites, it has, at its anterior extremity, two caeca directed forwards. But the intestine which succeeds it, is very short with all the Aptera.[32]

With all the hemimetabolic Insecta, or the Orthoptera and Hemiptera, the digestive canal of the larvae and pupae differs but little from that of the perfect insects.[33] With the Coleoptera, the larvae likewise resemble the perfect insects in this respect, — their mode of life being generally the same, as has already been evinced by the structure of their oral organs. The stomach is usually shorter and larger, and the number of its appendages less, than with the perfect forms.[34]

The larvae of the remaining holometabolic Insecta, which differ essentially from the imagines as to their oral organs, beside living upon different food, have also a digestive canal so different, that it must undergo a constant and gradual change during the quiescent pupa state.[35] Most of these larvae have powerful masticatory organs, — such are those of the Lepidoptera, the Tenthredinidae, the Siricidae, Phryganidae, Sialidae, and the cephalous ones of the Culicidae and Tipulidae. The digestive canal here is straight and rarely longer than the body; its greater portion consists of a large and usually varicose stomach, while the ileum and colon are pretty short. With the larvae of the Lepidoptera, the cylindrical ileum is large and divided into six lateral pouches, by as many longitudinal septa.[36] But with the cephalous larvae of the Mycetophilidae and Sciaridae, and the acephalous ones of the Diptera, the digestive canal is formed upon a wholly different

31 This caecum is found with *Hipparchia, Pontia, Sphinx, Gastropacha, Euprepia, Acidalia, Cabera, Adela, Chilo*, and *Tinea*. It is wanting with *Vanessa, Zygaena, Hepiolus, Cossus, Yponomeuta*, and *Pterophorus*.

32 See *Nitzsch*, in *Germar's* Magaz. d. Entom. III. p. 280 (Nirmidae); *Nicolet*, loc. cit. p. 46, Pl. IV. fig. 2 (Poduridae); *Swammerdamm*, Bib. der Nat. p. 33, Taf. II. fig. 3; *Ramdohr*, loc. cit. p. 185, Taf. XVI. fig. 3, and Taf. XXV. fig. 4, and *Treviranus*, Verm. Schrift. II. p. 13, Taf. III. fig. 1–6 (*Pediculus* and *Lepisma*).

33 See *Suckow*, in *Heusinger's* Zeitsch. II. Taf. I. fig. 8 (*Aeschna*), und *Rathké*, in *Müller's* Arch. 1844, p. 35, Taf. II. fig. 4 (*Gryllotalpa*).

34 With the larvae of *Calosoma*, the stomach is straight and without caeca (*Burmeister*, Trans. of the Entom. Soc. I. p. 236, Pl. XXIV. fig. 10, 11). With *Hydrophilus piceus*, and *Dytiscus marginalis*, it is varicose, slightly tortuous, and without caeca (*Suckow*, in *Heusinger's* Zeitsch. II. Taf. IV. fig. 26, and *Burmeister*, Handb. I. Taf. X. fig. 3). The larvae of the Lampyridae, Pyrochroidae, Mordellidae, and Curculionidae, differ but little from the imagines as to their digestive canal (*L. Dufour*, Ann. d. Sc. Nat. III. 1824, Pl. XI. fig. 7 (*Lampyris*); Ibid. XIII. 1840, Pl. V. fig. 5 (*Pyrochroa*); XIV. 1840, Pl. XI. fig. 9 (*Mordella*); and *Burmeister*, Zur Naturg. d. Calandra, p. 8, fig. 3.) The most marked difference between the larvae and the imagines, is observed with the Lamellicornes. The first have a very spacious,

straight stomach, which, at both extremities and sometimes also in the middle, has a circle of simple or varicose, thickly-set caeca; the ileum is very short, and the large intestine extremely large and always bent forwards; see *Roesel*, Insektenbelust. II. Taf. VIII. IX.; *Suckow*, loc. cit. III. Taf. III. fig. 87 (*Melolontha*), *L. Dufour*, Ann. d. Sc. Nat. XVIII. 1842, Pl. IV. fig. 8, Pl. V. fig. 18 (*Cetonia* and *Dorcus*); finally, the excellent work of *De Haan*, Sur les métamorphoses des Coléoptères, Mém. I. les Lamellicornes, in the Nouv. Ann. du Mus. IV. 1835, p. 153, Pl. XVI.–XIX.

35 For this metamorphosis of the intestinal canal, see *Dutrochet*, Jour. de Physique, &c., LXXXVI. 1818, p. 130, or *Meckel's* deutsch. Archiv IV. p. 285, Taf. III. (*Bombyx, Myrmeleon, Apis, Polistes, Tenthredo* and *Eristalis*). This metamorphosis with *Sarcophaga haemorrhoidalis* has been described and figured by *L. Dufour*, Mém. présentés, &c., IX. p. 580, Pl. III.

36 See *Swammerdamm*, Bib. der Nat. Taf. XXXIV. fig. 4; *Lyonet*, Traité, &c., Pl. XIII.; *Ramdohr*, loc. cit. Taf. XVIII. fig. 5. Many naturalists have carefully observed the metamorphoses of the digestive canal with the Lepidoptera; see *Herold*, Entwickelungsgeschichte d. Schmetterl. Taf. III. fig. 1–12 (*Pontia brassicae*); *Suckow*, Anat. physiol. Untersuch. p. 24, Taf. II. fig. 1–10 (*Gastropacha pini*); and *Newport*, Philos. Trans. 1834, Pl. XIV. fig. 11–13 (*Sphinx ligustri*). This last author has figured the digestive canal *in situ* in all the three states.

plan. It exceeds more or less the length of the body, and there is a crop
at the posterior extremity of the œsophagus upon which succeeds a long
and tortuous stomach. Upon the cardiac portion of this last there are
inserted two to four cæca directed either forwards or backwards, and with
some larvæ of the Muscidæ, there is also a long, sucking stomach upon
one of the sides of the œsophagus.[37]

With the larvæ of the Neuroptera, which suck up their liquid food through
tubular mandibles, the posterior extremity of the œsophagus is dilated into
a pyriform sucking stomach, which is followed by the proper stomach,
large, of median length, and slightly flexuous. The extremely small ileum
is long and makes several· convolutions, while the colon is large, vesiculiform, and continuous into a horny tubular rectum.[38]

§ 339.

As to the glandular appendages of the digestive canal of the Insecta,
the Salivary Organs are quite widely distributed, as well with the Imagines
as with the Larvæ and feeding Pupæ. These organs consist of one, or
two, rarely three pairs of colorless tubes of unequal length. These are
sometimes prolonged into the thorax, while in other cases they accompany
the digestive canal into the abdominal cavity where it makes many convolutions. Their excretory ducts are composed of a solid membrane, and are
distinctly separated from the glandular portion.[1] This last is composed
of three layers, namely: an external, homogeneous envelope, — an intimate
tunic accompanying the excretory duct, — and a middle layer composed of
colorless, glandular, nucleated cells, which often form very fine excretory
tubes opening into the common duct. Frequently, also, these ducts contain
a spiral filament like the tracheae; they open, each, at the base of the oral
cavity by a distinct orifice, and it is rare [2] that they unite, forming a
common duct; sometimes they have, near their excretory openings, special
salivary reservoirs.[3] With very many Aptera,[4] Diptera, Lepidoptera,
and Coleoptera,[5] the salivary organs consist of two simple tubes, which,
with the larvæ of the second and third of these orders, often extend a considerable way into the abdominal cavity.[6] With the Cerambycidæ, Te-

---

37 See *Swammerdamm*, Bib. der Nat. Taf. XLI.
fig. 6, Tab. XLIII. fig. 5 (*Stratiomys* and *Piophila*); *Ramdohr*, loc. cit. Taf. XIX. fig. 1
(*Musca*); *L. Dufour*, Ann. d. Sc. Nat. Xl. 1839,
p. 212, Pl. V. fig. 23, XII. p. 13, 18, Pl. I. fig. 1,
4, and I. 1844, p. 372, Pl. XVI. fig. 8 (*Ceroplatus*,
Sapromyza, *Piophila*).

The metamorphosis of this digestive canal, in
the pupa of *Sarcophaga carnaria*, is represented
in a suite of figures, published by *Suckow*, in
Heusinger's Zeitsch. III. Taf. IX. fig. 147-153.
But *Suckow* has fallen into the same error as
Ramdohr (loc. cit. p. 171) in regarding the cæcal
appendages of the stomach of the larvæ as four
tubes connecting the stomach with the salivary
canals.

38 See *Ramdohr*, loc. cit. p. 154, Taf. XVII. fig.
1; and *L. Dufour*, Recherch. &c. p. 589, Pl. XII.
fig. 175 (*Myrmeleon*). The large intestine together
with the rectum, does not serve, with this larva, as
a defecating organ, but, as is very extraordinary,
has the function of a Spinneret (see § 347).

1 For the intimate structure of these organs, see
H. Meckel, in *Müller's* Arch. 1846, p. 25, Taf. I.
II.

2 *Piophila*, *Musca*, *Sarcophaga*, *Tabanus*,
Hippobosca, *Oestrus*, *Mordella*, *Mantis*, and
Forficula.

3 With *Forficula*, *Musca*, *Sarcophaga*, and
Hippobosca, each of these excretory ducts is dilated
into a roundish reservoir; but with the Termitidæ,
Acrididæ, Achetidæ, and Mantidæ, there is an
oblong, pedunculated reservoir common to both
ducts. See, for the figures, the various memoirs of
L. Dufour.

4 With the Nirmidæ.

5 *Pyrochroa*, *Lixus*, *Phyllobius*, *Diaperis*,
Lema, *Oedemera*, *Chrysomela*, *Coccinella*. In
this last genus, the two salivary vessels are torose.

6 See the figures in the works of *Swammerdamm*,
Lyonet, *Ramdohr*, *Suckow*, *Herold*, and *L. Dufour*.

nebrionidae, Mordellidae,[7] and most of the Hymenoptera,[8] they consist of two rather short, ramified tufts, often contained entirely in the head.[9] Among the Neuroptera, the Myrmeleonidae and Sialidae have two simple short salivary tubes, while, with the Phryganidae and Hemerobidae, they are ramified and highly developed.[10] It is quite remarkable that there is, in this respect, a sexual difference with the Panorpidae; the males have three pairs of very long, tortuous tubes, while, with the females, the only vestiges of this apparatus are two indistinct vesicles.[11] Among the Orthoptera, the salivary organs are entirely absent with the Libellulidae, and Ephemeridae. On the other hand, they are highly developed with the Achetidae, Acrididae, Locustidae, Mantidae, Blattidae, Termitidae, and Perlidae, where they consist of two, four, or six botryoidal masses of vesicles, situated in the thorax, and having long, excretory ducts, beside, also, often long-pedunculated pyriform reservoirs.[12] Among the Hemiptera,[13] these organs are absent with the Aphididae and the Psyllidae; but, on the other hand, they are very large and of a remarkable structure with the Bugs and Cicadidae. Here they are nearly always lobulated, and are divided by a constriction into two portions, of which the upper is much smaller than the lower, and often both have long digitiform processes. The excretory duct divides, immediately after its origin, into two special canals of equal or very unequal length, which extend, serpentinely, first, into the abdominal cavity, and then ascend to the mouth.[14] Beside these two constricted glands, many Bugs have, also, one, rarely two pairs of simple salivary tubes,[15] which are often dilated, vesiculiform, at their extremity.[16] The salivary organs of the musical Cicadidae differ in many respects from those of the others of this family; for, beside the two simple tortuous tubes, there is, in the head, another pair of glands, composed, each, of two tufts of short, cylindrical caeca, situated one behind the other.[17] It is yet undetermined

7 See L. Dufour, Ann. d. Sc. Nat. IV. 1824, Pl. XXIX. fig. 4, 5, XIV. 1840, Pl. XI. fig. 16.
8 See L. Dufour, Recherch. &c. p. 390, fig. 48, 72, 109, 148 (Apis, Andrena, Philanthus, and Xyphidria.
9 With the Coleoptera, the ramified glands end in long, tortuous caeca; while with the Hymenoptera, their extremities are vesiculiform, thereby giving the whole gland a botryoidal aspect.
10 See L. Dufour, Recherch. &c. p. 563, fig. 179, 184, 191, 192, 208, 209 (Myrmeleon, Sialis, Hemerobius, and Phryganea).
11 See Brants, Tijdschr. voor naturl. Geschid. en Pfysiologie, 1839, p. 173; and L. Dufour, Recherch. &c. p. 582, fig. 169 (Panorpa).
12 See L. Dufour, Recherch. &c. p. 296, Pl. I.-V. XIII. (Tridactylus, Oedipoda, Gryllotalpa, Ephippigera, Mantis, Blatta, Termes, and Perla).*
13 For the salivary organs of the Hemiptera, see, beside Ramdohr, loc. cit. Taf. XXII. XXIII. especially L. Dufour, Recherch. sur les Hémiptères, p. 118, Pl. I. IX.
14 The two excretory ducts are of the same length

with Ranatra, Nepa, Naucoris, Corixa, Reduvius, and Syrtis. One is very long and the other very short with Tetyra, Pentatoma, Syromastes, Coreus, Lygaeus, Aphrophora and Cercopis With the Hydrocorisae, above cited, the two salivary glands are, moreover, composed of numerous round secretory vesicles. In general, these glands have been regarded as composed of two vesicles each of which has a proper excretory duct; but this view is incorrect. The two excretory ducts are always the result of the division of a common trunk which arises at the constricted point of the gland. With Ranatra, alone, the anterior is entirely separated from the posterior portion of the gland.†
15 There is only one pair of simple salivary glands with Tetyra, Pentatoma, Pyrrhocoris, Lygaeus, Naucoris, Nepa and Ranatra; two pairs with Coreus and Alydus. With Nepa and Ranatra, they dilate into an oval reservoir.
16 Syrtis, Reduvius, Pelrogonus, and Corixa.
17 See L. Dufour, Ann. d. Sc. Nat. V. 1825, p. 158, Pl. IV. and Recherch. &c. Pl. VIII.

* [§ 339, note 12.] See also Leidy, loc. cit. p. 82 (Spectrum femoratum). — ED.
† [§ 339, note 14.] With Belostoma, the salivary glands are four in number, are of conglomerate structure and situated on each side of the oesophagus into the commencement of which they empty. Two of them are long and extend backwards as far as the commencement of the ab-

domen; while the other two are about one-fourth as long. Beside these, on each side of the oesophagus, there is situated a sigmoid caecal pouch which opens by a narrow duct into the commencement of the oesophagus in the vicinity of the termination of the salivary ducts; these are perhaps reservoirs of the saliva; see Leidy, loc. cit. p. 63. — ED.

whether one of the pairs of these glands with these Hemiptera, may not be
a poison apparatus.

The Insecta have no distinct Hepatic Organs, but the function of a Liver
is performed by the walls of the stomach, the internal tunic of which is
composed of closely-aggregated hepatic cells. With many species whose
stomach has caecal appendages, the walls of these last have a similar hepatic
structure, and must secrete, therefore, a bile-like fluid.[18]

With some Insecta, the ileum has glandular appendages, whose product
is perhaps analogous to a pancreatic fluid. The two or four rows of follicles which, as before mentioned, are situated on the ileum of the Pentatomidae and some Coreidae, would, in the same manner, be regarded as a
Pancreas. The same remark applies to the ramified appendages, which,
with *Gryllotalpa,* open into the stomach below the two caeca, as well, also, as
to the two or three follicles which, with *Pyrrhocoris,* are inserted, laterally,
on the posterior part of the ileum.[19]

There is found, with all Insecta, a *Corpus adiposum,* — a tissue, composed
of adipose cells, which is intimately connected with the functions of digestion
and assimilation. This body is especially developed towards the end of
the larval state, and it disappears, for the most part, during the pupa
period, so that only a few traces of it are found with Insecta in their perfect state. It is usually of a white, or a dirty-yellow color, but is also
observed of a green, red, or orange hue. In the larvae, the fat cells generally form pretty large, lamelliform lobes, sometimes ramified or reticulated, sometimes plicated, spread through the abdominal cavity in all the
intervals of the viscera. These lobes are always traversed and retained
in place by numerous tracheau branches. With the perfect Insecta, the
remains of this body are not usually found except in the posterior portion
of the abdominal cavity, where they consist of fat-cells loosely scattered,
and not retained by the tracheae.[20]

---

18 For these biliary organs, see *J. Müller,* De
Gland. struct. p. 67. The Malpighian vessels
which were formerly regarded as biliary tubes,
will be treated of in future (§ 346).*

19 See *L. Dufour,* Recherch. sur les Orthopt. p.
332, Pl. II. fig. 19 (*Gryllotalpa*), and Recherch.

sur les Hémiptères, p. 44, Pl. II. fig. 19, 21 (*Pyrrhocoris*).†

20 See *L. Dufour,* Recherch. sur les Carabiq., in
the Ann. d. Sc. Nat. VIII. 1826, p. 29 ; Recherch.
sur les Hémipt. p. 141, and Recherch. sur les
Orthopt. p. 291, 385, 562.‡

* [§ 339, note 18.] The liver of the Insecta, as
well as that of the Invertebrata generally, has been
investigated by *Will* (*Müller's* Arch. 1848, p. 502)
who has applied the same chemical mode of inquiry, as that of *Brugnatelli* and *Wurzer* upon
the Malpighian vessels showing their urinary character (see infra § 345, note 2). With the Insecta,
he regards as hepatic the caecal and other glandular appendages which, when present, lie upon the
so-called *Ventriculus,* thus confirming the supposition expressed in the foregoing note. For the intimate microscopic structure of the liver of the Insecta, see *Leidy,* Amer. Jour. Med. Sc. XV. 1848,
p. 1. — ED.

† [§ 339, note 19.] For the hepatic organs of
Belostoma, see *Leidy,* loc. cit. p. 63, Pl. X. fig.
4, i. They consist of four long very tortuous tubes

closely applied to the intestinum tenue ; they join
the intestine at the junction of the duodenum and
ileum. — ED.

‡ [§ 339, note 20.] See, upon the *Corpus adiposum, Mayer* (Ueber die Entwickelung. des Fettkörpers, &c. bei den Lepidopteren, in *Siebold*
and *Kölliker's* Zeitsch. I. p. 175) who has traced
its development.

These adipose bodies are formed from a great
number of separate, flattened, usually many-
pointed lobes. These lobes consist of pouches with
structureless walls, and filled with fat-globules.
Each pouch is originally a simple cell with a large
nucleus attached to its wall. In this cell are
formed daughter-cells, which, when filled with fat,
burst, and thereby the mother-cell becomes the fat-
containing sac. — ED.

CHAPTER VI.

CIRCULATORY ORGANS.

§ 340.

The Circulatory System is feebly developed with Insecta, consisting of a contractile, articulated *Vas dorsale*, and a cephalic *Aorta*. The first serves as a heart, and the second is a simple conductor of the blood from the heart into the body. In both of these vessels, the blood moves from behind forwards, and, at its escape from the aorta, traverses the body in all directions, forming regular currents which have, however, no vascular walls. In this way, it penetrates the antennae, the extremities, the wings, and the other appendages of the body, by arterial currents, and is returned by those of a venous nature. All the venous currents empty into two lateral ones running towards the posterior extremity of the body, and which enter, through lateral orifices, the dorsal vessel.[1]

---

1 *Swammerdamm, Malpighi,* and others of the older anatomists, had already formed a pretty exact idea of the circulation of the Insecta. But, subsequently, it was entirely abandoned when it was observed that the dorsal vessel was a closed tube, and served only as a simple reservoir of the nutritive juices. *Carus* was the first to demonstrate anew the existence of a circulation which has since been confirmed with all the three stages of insects. See *Carus,* Entdeck. eines einfachen, vom Herzen aus beschleunigt. Blutkreisl. in den Larven netzflüglich. Insekt. 1827 ; Nov. Act. Acad. Nat. Cur. XV. part II. p. 8, Taf. LI. ; and Lehrb. d. vergleich. Zo t. 1834, p. 687 ; *R. Wagner,* Isis, 1832, p. 320, 778 ; *Burmeister,* Handb. &c. I. p. 164, 436 ; *Bowerbank,* Entom. Mag. I. 1833, p. 239, IV. 1835, p. 179 (also in *Froriep's* neue Notiz. XXXIX. p. 349) ; *Tyrrell,* Philosoph. Trans. 1835, p. 317 ; *Newport,* Cyclop. &c. II. p. 980 ; *Milne Edwards,* Ann. d. Sc. Nat. III. 1845, p. 278 ; and *Quatrefages,* Instit. 1845, p. 305. This circulation carried on by the dorsal vessel, having been observed by so many distinguished naturalists, it is truly incomprehensible that *L. Dufour* (Recherch. sur les Hémipt. p. 272 ; Recherch. sur les Orthopt. p. 287 ; Ann. d. Sc. Nat. XVI. 1841, p. 10 ; Mém. presentes à l'Inst. IX. p. 595, 601) can persist in denying that the dorsal vessel is anything but a secretory organ which, according to him, has no opening and therefore nothing in common with a heart. He cites the authority of *Cuvier* who was unwilling to accord to the *Vas dorsale* either the name or the functions of a heart (Cuvier, Mém. sur la manière dont se fait la nutrition dans les Insectes, in the Mém. d. l. S sc. d'Hist. Nat. de Paris, VII. 1798, p. 34, or *Reil's* Arch. V. p. 97). *L. Dufour* adduces, moreover, in support of his erroneous view, the following remark of Carus (Erläu-

terungst. IIft. VI. p. 8), "In the perfect Insecta, whose respiration is performed by a system of tracheae traversing the entire body, the circulation of blood would be useless." But to this it may be replied, that *Carus,* by these words, has contradicted his proper observations ; for he has shown that there is a circulation in many perfect insects, as is stated not only in the Nov. Act. Nat. Cur., loc. cit., but also in the Erläuterungstafeln from which the above citation was taken. At all events, the proposition of *Carus* is correct, " that in insects, the blood must come in contact with the atmospheric air, which is accomplished by means of the tracheam system." But this applies only to the small portion of the circulation connected with the respiratory process ; whereas, the larger portion, destined for the general nutrition of the tissues, does not evidently require the presence of tracheae. The presence of a real blood-circulation by means of the *Vas dorsale,* is so easily observed, that the injections of *Blanchard* are scarcely necessary (Compt. rend. XXIV. 1847, p. 870).

If, in certain species, although transparent, these phenomena are not observable, we must not be too hasty in denying its real existence, for the blood, which is not visible except through its globules, is often so poor in these last, as to elude our observation. *Verloren* has recently given a very complete résumé of what has been done on this subject, and has added new and confirmatory observations ; see Holländische Beitr. zu den anat. und. physiol. Wissenschaft. 1. Hft. 2, p. 220 ; and Mémoire en réponse à la question suivante : éclaircir par des observations nouvelles le phénomène de la circulation dans les Insectes, en recherchant si peut la reconnaître dans les larves des différents ordres de ces animaux, in the Mém. couronn. par l'Acad de Belgique, XIX. 1847.*

---

* [§ 340, note 1.] The results obtained by *Blanchard* have been very satisfactorily confirmed by *Agassiz* (Proceed. Amer. Assoc. Advancem. Sc. 1849, p. 140, also in its translation into French in the Ann. d. Sc. Nat. 1851, XV. p. 358), who has succeeded in distinctly injecting the tracheae by the dorsal vessel. These experiments I have had the

good fortune to witness, and their character was such as to leave with me no doubt as to the peritrachean circulation. See, also, the additional evidence which *Blanchard* (Compt. rend. Oct. 6, 1851) has recently furnished of a peritracheam circulation, which is very important and weighty. He took advantage of the well-known fact that silk-

The Blood of the Insecta is usually a colorless liquid, though sometimes yellowish, but rarely red.[2]   In this liquid are suspended a few very small, oval, or spheroidal corpuscles, which are always colorless, have a granular aspect, and are sometimes nucleated.[3]

The Dorsal Vessel, which is constricted at regular intervals, is always situated on the median line of the abdomen, being attached to the dorsal wall of its segments by several triangular muscles whose apices point outwards.   Its walls contain both longitudinal and transverse fibres, and, externally, are covered by a thin peritoneal tunic.   Internally, it is lined by another very fine membrane, which, at the points of these constrictions, forms valvular folds, so that the organ is divided into as many chambers as there are constrictions.   Each of these chambers has, at the anterior extremity on each side, a valvular orifice which can be inwardly closed.[4] The returning blood is accumulated about the heart and enters into it during the diastole of each of its chambers, through the lateral orifices.[5] It then passes, by the regularly successive contractions of the heart, from behind forwards into the aorta which is only a prolongation of the anterior chamber.   This aorta consists of a simple, small vessel, situated on the dorsal surface of the thorax, and extending even to the cephalic ganglion, where it either ends in an open extremity, or divides into several short branches which terminate in a like manner.[6]   The length of the dorsal vessel depends, in all the three states of insects, upon that of the abdomen. The number of its chambers is very variable, but is, most usually, eight.[7]

The blood, after leaving the aorta, traverses the body in currents which

2 The blood is red in many larvae of *Chironomus.*

3 For the blood of Insecta, see *Wagner*, Zur vergleich. Physiol. d. Blutes, Hft. I, p. 26, Hft. 2, p. 59, and Isis, 1832, p. 323; *Horn*, Das Leben d. Blutes, p. 9, Taf. I. and *Newport*, Institut. 1845, p. 241, or Ann. d. Sc. Nat. III. 1845, p. 364, or *Froriep's* neue Notiz. XXXIV. p. 9.

4 For the structure of the dorsal vessel, see *Straus*, Consid. &c. p. 356, Pl. VIII. (*Melolontha vulgaris*); *Wagner*, Isis, 1832, loc. cit. Taf. II. (larvae of Diptera and Ephemeridae), and in *Müller's* Arch. 1835, p. 311, Taf. V. (larva of *Corethra plumicornis*); *Newport*, Philos. Trans. 1843, p. 272, and Cyclop. loc. cit. p. 976, fig. 433, A. and 434 (*Lucanus cervus* and *Asilus crabriformis*); finally *Verloren*, Mém. loc. cit. p. 31, Pl. III.–VII. (*Chironomus, Sphinx, Rhynchophorus, Pompilus, Syrphus,* and *Vespa*). The

constrictions of the dorsal vessel are feebly marked with the larvae of the Diptera and Hymenoptera.*

5 According to *Newport* (Cyclop. loc. cit. p. 977), the space in which the blood accumulates about the heart is surrounded by a very thin membrane, and may therefore be regarded as a true auricle.

6 The Aorta is divided at its extremity with *Meloë, Blaps, Timarcha, Vanessa,* and *Sphinx ;* see *Newport*, Cyclop. loc. cit. p. 978.

7 With the Orthoptera, Lepidoptera, and their larvae, as also with various larvae of Diptera. It is rare that the number of chambers exceeds eight, as, for example, with the Poduridae (*Nicolet*, loc. cit. p. 50, Pl. IV. fig. 3).   More commonly there are seven, as with *Lucanus* and *Dytiscus* (*Newport*, Cyclop. loc. cit. fig. 433, A., and *Wagner*, Icon. Zoot. Taf. XXIII. fig. 2).   *Burmeister* (Handb. I. p. 165) has observed only four with the larva of a *Calosoma.*

worms fed on different artificially-colored leaves produced correspondingly colored cocoons.   He therefore fed, in the same manner, various larvae, and, upon dissection, found not only their blood but also their tracheae colored like the color used. With the tracheae, this color was deepest at the base, but gradually paled away towards their extremity.   What adds a corroborating value to these experiments is the fact that the muscles here remained uncolored, thus showing that this special tracheal coloration was not due to a bathing of the general fluids of the body.   Compare also the recent various notes and papers of *Blanchard*, in Ann. d. Sc. Nat. — Ed.

* [§ 340, note 4.]   See also, for histological de-

tails upon the heart, *Leydig, Siebold* and *Kölliker's* Zeitsch. 1852, III. p. 446 (larva of *Corethra plumicornis*).   This naturalist has here described a new and peculiar kind of valves, which deserve particular notice.   In the last chamber of the heart, there are six or eight pairs of roundish, clear bodies, attached to the inner surface of the heart by a peduncle.   They alternate in their position, one beyond the other, so that, during the systole, two of them are so opposed that the calibre of the chamber is completely closed at that point. Each of these curious valves is only a pedunculated nucleated cell · see loc. cit. Taf. XVI. fig. 2, c. — Ed.

are always extravascular, and in this way bathes all the organs." The
newly-prepared nutritive fluid passes through the walls of the digestive
canal in which it is found, into the visceral cavity, and thence directly into
the blood. Latterly, this extravascular circulation has been called in question, but its presence may be easily and directly observed with very many
perfect Insecta and their larvae. The vascular walls supposed to have
been seen at certain points, are, undoubtedly, the result of some error of
observation or interpretation.[9] This is also true of the pulsatile organs
supposed to have been observed in the legs of many water-bugs, and which
were thought to affect the circulation.[10]

CHAPTER VII.

RESPIRATORY SYSTEM.

§ 341.

The Insecta respire, in all their conditions of life, by means of a system
of Tracheae which are spread through the entire body and penetrate all the
organs. This system of air-vessels either opens externally by stigmata
through which the atmospheric air is introduced directly, or they have no
external communication, but derive the air from the water by means of
lamelliform or tubular prolongations with which the tracheae terminate,
and which have often been compared to branchiae.[1] In the first case, they
are called Pulmonary tracheae, and in the second, Branchial tracheae.

---

8 In the antennae, the legs, the filaments of the tail, and other appendages, the arterial and venous currents are contiguous. But in the wings they are are isolated; and although they may be observed in the nervures of the wings, yet these last should not therefore be regarded as true blood-vessels, for their cavities are only prolongations of the visceral cavity, as is shown by the fact that they are sometimes traversed at the same time by branches of tracheae. In the memoir of *Verloren* (loc. cit. p. 76) will be found a very complete account of all the reasons opposing the presence of vascular walls in Insecta.

9 The same should probably be said about the thin walls which *Bowerbank*, and *Newport* (loc. cit.) think they have observed with *Ephemera* concerning the two lateral currents which run towards the posterior extremity of the abdomen. Another vessel which, according to *Treviranus* (Zeitsch. f. Physiol. IV. p. 182, Taf. XIV. fig. 13) and *Newport* (Philos. Trans. 1834, p. 395, Pl. XIV. fig. 9, and Cyclop. loc. cit. p. 980), is found in the larvae and imagines of Lepidoptera above the ganglionic chain, and is the analogue of the supra-spiral artery of the Myriapoda (§ 284), requires further research, for it may be questioned if such an organ, found only in certain groups of Insecta, is really a vessel.

10 Very dissimilar and contradictory opinions have been published on these pulsatory organs. *Behn* (*Müller's* Arch. 1835, p. 554, Taf. XIII. fig. 13, 14, or Ann. d. Sc. Nat. IV. 1835, p. 5) has described them with *Corixa, Ploa, Naucoris, Nepa,* and *Ranatra*, as thin, movable lamellae attached to the inner wall of the tibiae. *Verloren* (Mém. loc. cit. p. 82, Pl. VI. fig. 24, 25) has confirmed these observations with the Cicadidae, although neither *L. Dufour* (Ann. d. Sc. Nat. IV. 1835. p. 313) nor *Wesmael* (Bullet. de l'Acad. de Bruxell. III. p. 158) has been able to discover them in the water-bugs above cited. It is possible that these apparent pulsations are produced simply by the contractions of neighboring muscular fibres.

1 See *Burmeister* (Handb. &c. I. p. 179; *Lacordaire*, Introduct. &c. II. p. 89; and *Newport*, Cyclop. loc. cit. p. 985). These organs have not the structure of true branchiae, and the blood is not subjected in their interior to the respiratory act, as is shown by the small quantity of this fluid which traverses them. These false branchiae are evidently designed to receive air, or, to speak more properly, to act, through endosmosis and exosmosis, in the transference of air from the water into the tracheal system. *Dugès* (Traité de Physiol. II. p. 519) is therefore correct in terming them *Branchies tracheales*.

The tracheae are cylindrical tubes of variable size, which often form, in their course, vesicular dilatations and numerous anastomoses. They divide, like blood-vessels, into many branches which gradually decrease in size, ending, at last, caecally, so that the expired air passes out by the same way that it entered.

The intimate structure of these organs is remarkable, and has always attracted the attention of anatomists.[2] When filled with air they present a beautiful, silver appearance. Externally, they are invested with a thin transparent, colorless, or very rarely brownish membrane, corresponding to a peritoneal envelope.[3] Internally, they are lined with another membrane still finer, which presents a lamellated epithelial structure.[4] Between these two membranes is situated a solid spiral filament whose turns are usually near together. This filament is sometimes cylindrical, sometimes flattened, usually transparent and colorless, and in a few instances only, of a dark color.[5] Often, its course is unbroken for a long distance, and rarely is its extremity forked. The new threads always begin between the turns of the preceding one, as may be easily observed at the commencement of each trachean ramification. In the ultimate trachean branches, these threads gradually decrease in size, and at last become indistinct. In the vesicular dilatations of the tracheae, with many Insecta, the spiral thread is often wholly wanting.[6] *

§ 342.

The Branchial tracheae are found only in certain aquatic larvae and pupae, and never in the perfect Insecta. The absence of stigmata here is compensated by the existence of false branchiae (*Branchiae spuriae seu tracheales*), which are cylindrical, or riband-like organs covered by a very

---

2 For the internal structure of the tracheae, see, beside the works of *Burmeister*, *Lacordaire*, and *Newport*, that of *C. Sprengel*, Comment. de partib. quibus Insect. spiritus ducunt, 1815 ; *Suckow*, in *Heusinger's* Zeitsch. II. p. 24, Taf. 1. fig. 10 ; *Straus*, Consid. &c. p. 315, Pl. VI. fig. 5 ; *Newport*, Philos. Trans. 1836, p. 529 ; and *Platner*, in *Müller's* Arch. 1844, p. 38, Taf. III.

3 This membrane is brown in the Libellulidae and Locustidae ; this coloration is due to a finely-granular substance contained in the membrane.

4 See *Platner*, loc. cit. Most anatomists regard this internal membrane as mucous. This being admitted, it was very natural to suppose that it, like that of the lungs of the Vertebrata, is covered with cilia. But here, as well as in other regions of the

body of insects, there is no trace of ciliated epithelium, which, indeed, would be incompatible with the presence of chitine. *Peters* (*Müller's* Arch. 1841, p. 233) was certainly deceived when he thought he observed ciliary movements in the tracheae of *Lampyris*, *Coccinella*, *Musca*, and other Insecta. He has himself admitted that he was not able to distinguish the cilia. For my part, I have sought in vain for this movement in the tracheae, and *Stein* (Vergleich. Anat. u. Physiol. d. Insekt. 1847, p. 105) has been equally unsuccessful.

5 The tracheae of the larvae of the Dytiscidae owe their black color to the spiral filaments.

6 With the Muscidae, Syrphidae, Vespidae, Apidae, and Melolonthidae.

---

* [§ 341, end.] See, also, for investigations upon the intimate structure of the tracheae, *Dujardin* (Comp. rend. 1849, p. 674), and *Mayer* (Ueber die Entwickelung, des Fettkörpers, der Tracheen, &c. &c., bei den Lepidopteren, in *Siebold* and *Kölliker's* Zeitsch. I. p. 175). The views of *Dujardin* are different from those usually received, for he regards the spiral thread not as a special formation, but only a fold like thickening of the internal membrane, — which membrane is not composed of cells but is a structure analogous to the wing-membrane, and is covered with hairs and points. On

the other hand, *Mayer*, who has studied the embryonic development of these organs, states that the spiral thread is originally a homogeneous membrane, which ultimately splits up into the threads.

This subject of the structure of trachea has now an additional point of interest, from its relations to *Blanchard's* views of a peritrachean circulation in the Insecta. In this connection see especially *Filippi* (Annali della R. Acad. d'agricoltxro di Torino, V., also *Wiegmann's* Arch. 1851, Th. II. p. 145). — ED.

thin cutaneous membrane, and containing one or several finely-divided trachean trunks. These trachean branchiae are either isolated, or fasciculated; in this last case, they are often digitiform, or penniform, and their ultimate ramifications are usually deficient in the spiral filament. All the air-vessels which these branchiae contain, arise from the larger trachean trunks. These branchiae occur with various Tipulidae, with a *Nymphula*, with the Phryganidae, Sialidae, Ephemeridae, Perlidae, Libellulidae, and with the Gyrinidae.

This trachean system is most simple with the larvae of Tipulidae of the genera *Chironomus, Tanypus, Corethra* and *Simulia,* as also with some larvae of the Phryganidae, of the genera *Rhyacophila* and *Hydropsyche,* where the tracheae, instead of forming cutaneous appendages, are subcutaneous and can therefore extract air from the water. The larvae of *Corethra* are distinguished for having in the thorax and abdomen, directly beneath the skin, two adjacent trachean vesicles, by means of which, very probably, the necessary renewal of air takes place.[1] With the pupae of *Simulia,* there are two branchial tufts on the sides of the prothorax, composed, each, of six to eight long caecal tubes, which contain each a single simple trachea deficient in the spiral thread.[2] Of the various larvae of the Lepidoptera, living under the water, that of *Nymphula stratiotalis,* alone, has trachean branchiae. These consist of fasciculate filaments situated on the sides of the abdominal segments.[3] With the larvae of *Sialis,* each of the six, seven or eight abdominal segments has upon its sides an articulated, filiform thread, containing a trachean vessel, and which may, therefore, be regarded as a trachean branchia.[4] Most of the larvae and pupae of the Phryganidae, have, at the same points, one or two filiform, trachean branchiae, rarely ramified, and united in groups of from two to five, which stand out towards the back.[5] With those of the Ephemeridae, each of the anterior abdominal segments has a pair of these branchiae which are sometimes ramified in the most varied manner, and sometimes consist of two kinds, some being lamelliform and alternating with the others which are fasciculate.[6] With all the Ephemeridae, these organs have movements which are sometimes slow and rhythmical, and sometimes rapid and oscillatory.

With the Perlidae, the branchiae are filiform, ramified, and situated on the three thoracic segments of the larvae and pupae, or bound together in several short fasciculi which cover the base of the legs.[7]

Among the Libellulidae, the larvae and pupae of *Agrion* and *Calo-*

1 See *Reaumur,* Mém. loc. cit. V. Pl. VI. fig. 7, or *Lyonet,* Mém. du Mus. XIX. Pl. IX. fig. 14, 15.

2 See *Verdat* and *Fries,* in *Thon's* Entom. Arch. II. p. 66, 61, Taf. III. One must be careful, and not confound, as has sometimes occurred, the hairy tufts of these larvae for the tufts of tracheae.

3 See *Degeer,* Abhandl. I. Abth. III. p. 85, Taf. XXXVII. fig. 5, 6.

4 See *Roesel,* Insektenbelust. II. Insecta aquat. Class. II. Taf. XXIII.; *Degeer,* Abhandl. II. Taf. XXIII.; *Suckow,* in *Heusinger's* Zeitsch. II. Taf. III. fig. 23, 24; and *Pictet,* Ann. d. Sc. Nat. V. 1836, Pl. III. During the passage into the pupa state, which occurs with *Sialis* out of the water, these tracheae are cast off.

5 See the figures of *Pictet,* Recherch. pour servir à l'hist. et à l'anat. d. Phryganides, Pl. II. &c., and *Degeer,* Abhandl. II. Taf. XII. The bran-

chiae are ramose with *Hydropsyche,* and *Rhyacophila.*

6 See *Swammerdam,* Bib. der Nat. Taf. XIII.-XV.; *Reaumur,* Mém. VI. Pl. XLII.-XLVI.; *Degeer,* Abhandl. II. Taf. XVI.-XVIII.; *Suckow,* in *Heusinger's* Zeitsch. II. Taf. III. fig. 21, 22; *Carus,* Entdeck. eines Blutkreisl. loc. cit. Taf. III.; and the figures of *Pictet,* Hist. d. Insect. Névropt. Ephémérines.

7 See the figures of *Pictet,* Hist. d. Névropt., Perlides. According to *Newport* (Ann. of Nat. Hist. XIII. p. 21, or *Froriep's* neue Notiz. XXX. p. 179, or Ann. d. Sc. Nat. I. 1844, p. 183), these branchial tufts persist, with *Pteronarcys regalis,* to the imago state. This would be a very extraordinary anomaly, and should be confirmed, for, from the observations of *Newport,* it does not appear that the tufts of hair situated on the thorax of this Perlide really preserve the structure of branchial tufts.

pteryx are distinguished for having three long, lamelliform branchiae, with a rounded extremity, and situated vertically upon the posterior part of the abdomen.[8]

The trachean branchiae of *Aeschna*, *Libellula*, and the other Libellulidae, are formed upon a wholly different plan. They are situated in the very large rectum, and consist of numerous epithelial folds which are traversed by a great number of very fine branches of many large trachean trunks. The rectum is, moreover, invested by a very highly-developed muscular tunic, and its orifice has three pyramidal valves which regulate the entrance and the escape of the water required for respiration.[9] Finally, the larvae of *Gyrinus* have a pair of long branchiae upon the sides of each of the first seven abdominal segments, and two pairs on those of the eight.[10]

§ 343.

The tracheae most universal with Insecta are those termed Pulmonary, which are characterized by the presence of stigmata (*Spiracula*). These last are round orifices or narrow two-lipped openings, situated at various points on the external surface of the body, and which, with many soft-skinned Insecta, are surrounded by a horny ring. Usually, their borders are fringed with small, short, simple or pinnate hairs,[1] and can be opened and shut by means of an internal muscular apparatus; this last is sometimes attached to two inwardly-projecting horny plates. By these means, many Insecta have well-marked respiratory motions, especially of the abdomen.[2]

With the larvae of the Lamellicornes, the stigmata have a peculiar organization. They are closed by a horny membrane whose semilunar borders are cribriform for the free passage of air.[3]

The larvae of the Oestridae have two large stigmata, covered each by a similar plate or membrane, at the extremity of the abdomen; and with some larvae of the Muscidae, the posterior stigmata are closed in the same manner, excepting that the membrane is perforated by three very distinct openings.

Each stigma is usually the entrance of only a single trachean trunk

8 *Roesel*, Insectenbelust. II. Insecta aquatica, Class. II. Taf. IX. XI.; and *Carus*, Entdeck. &c. Taf. I.

9. *Roesel*, loc. cit. Taf. III.–VIII. and *Suckow*, in *Heusinger's* Zeitsch. II. p. 35, Taf. I. II.

10 *Roesel*, loc. cit. III. Taf. XXXI. and *Degeer*, Abhandl. IV. Taf. XIII. Further researches are required to decide if the penniform appendages, situated on the sides of the abdominal segments of certain larvae of the Hydrophilidae, are really trachean branchiae. But it appears to me that, with these larvae, the pulmonary and trachean branchiae are confounded; see *Roesel*, Insectenbelust. II. Insect. aquat. Class. I. Taf. IV. and *Lyonet*, Mém. du Mus. XVIII. Pl. XXIII. (12), fig. 47 (*Hydrophilus caraboides*.)

1 *L. Dufour*, Ann. d. Sc. Nat. VIII. 1826, p 20, Pl. XXI.

2 The Locustidae, Libellulidae, and other Orthoptera, make true movements of inspiration and expiration, by alternately dilating and contracting the abdominal segments. With the Apidae, Vespidae, and other Hymenoptera, the alternate contractions and dilatations of the abdominal cavity are due to the protractile and retractile movements of the abdominal segments. Many Lamellicornes make these respiratory movements before flying, probably that they may fill their trachean system with air.

3 *Sprengel* (loc. cit. p. 9, Tab. I.) has described very correctly the stigmata of the Lamellicornes. *Treviranus* (Die Erschein. und Gesetze d. Organ. Lebens, I. p. 258) thinks that these lamellae are not perforated and that the air enters these tracheae by endosmose, although *Burmeister* (Handb. &c. I. p. 172) says he has observed a single central opening. I have been unable to confirm the statement of *Sprengel*, and think that these perforations might easily elude the observation, from their being concealed beneath a kind of net-work on the external surface of these lamellae. *L. Dufour* (Ann. d. Sc. Nat. XVIII. 1842, p. 173, Pl. IV. fig. 7) has also misapprehended the stigmata of the larvae of *Cetonia*; for that which he has described as a transverse fissure is only a fold, due to a pressure exercised during the manipulation, on the horny lamella which more lly is convex and imperforate in its centre.

which ramifies more or less directly; sometimes, however, several trunks arise from the same stigma.[1]

With perfect Insecta, the stigmata are nearly always situated on the sides of the body in the membrane connecting the two segments, being always wanting, however, in the membrane which unites the head and prothorax, and that between the last two abdominal segments. In many cases, they are covered by the borders of the segments. With the Coleoptera, the stigmata are often situated so high upon the back as to be concealed by the elytra.[2] The number and position of the stigmata vary infinitely, and are not invariable in the different conditions (larva, pupa and imago) of even the same species. These variations are the least with the hemimetabolic Insecta. But among the Hemiptera, the Naucoridae and Nepidae form a remarkable exception in this respect. They have, excepting those of the thorax, only two stigmata at the posterior extremity of the abdomen, and which alone serve, probably, for respiration when these insects are in the water; with Nepa, and Ranatra, these anal stigmata are situated at the base of a long tube formed by the union of two semicanals.[6] The smallest number of stigmata, consisting of two situated adjacently at the posterior extremity of the abdomen, occurs with the larvae of the Dytiscidae, Stratiomydae, Conopidae, and some Tipulidae and Tachinariae. Sometimes these two stigmata are situated at the extremity of a longer or shorter Respiratory tube (Sipho), surrounded by a circle of stiff or penniform bristles. In some cases this siphon is very long and articulated, and can be intussuscepted like the tubes of a telescope.[7] When these Insecta become pupae, these stigmata are sometimes remarkably modified. The pupae of Culex lose their anal siphon, and acquire, instead, two others which are infundibuliform and situated laterally between the prothorax and mesothorax.[8] The pupae of Ptychoptera respire by means of a flexible siphon situated in the neck.[9] With the Strepsiptera, the male, as well as the apodal female

4 In the larvae of the Lamellicornes; see *Sprengel*, loc. cit. Tab. I. fig. 1 (larvae of *Geotrupes*). Some Capricornes present, in their perfect state, a very singular organization in this respect. Their thoracic stigmata send off not only several large tracheal trunks, but also an infinite number of small branches; see *Pictet*, Mém. d. l. Soc. d. phys. &c. de Genéve, VII., 1836, p. 393, fig. 5, 6 (*Hammaticherus heros*), or Ann. d. Sc. Nat. VII. 1837, p. 63.

5 From this arrangement, the Dytiscidae and Gyrinidae, which live in the water, must, in order to breathe, emerge the posterior part of their body to draw fresh air under their elytra, whence it is taken into the tracheae. The Notonectidae, Hydrophilidae, Parnidae, and other aquatic Coleoptera, respire under the water by means of a provision of air which, after their immersion, adheres to the hairs of the legs. With *Hydrophilus*, the renewing of this air occurs in a very remarkable manner. They protrude only their antennae out of the water, and, bending them backwards, thus establish a communication between the external air and that adhering to the under surface of the body; see *Nitzsch*, in *Reil's* Arch. II. p. 440, Taf. IX.

6 See *Roesel*, Insectenbelust. III. Taf. XXII. XXIII.; and *L. Dufour*, Recherch. sur les Hémipt. p. 244, Pl. XVII. fig. 195, Pl. XVIII. With *Nepa*, it is true there are stigma-like rings on the other abdominal segments, but they are closed, and *L. Dufour* has properly called them false stig-

mata. In the young age of these insects; these false stigmata are open and situated in two pilose grooves located under the belly at some distance from the lateral borders, and which are prolonged even to the end of the siphon, where they blend into one. The air is conducted by these grooves into the stigmata.

7 By this disposition of the stigmata, the larvae of the Dytiscidae, Culicidae and Stratiomydae, are obliged, in order to breathe, to rise to the surface of the water, where they emerge only the stigmatic orifices, and the air then adheres to the coronets of hairs on the stigmata. Many Tipulidae, such as *Ptychoptera*, communicate even more easily with the air by means of their long, articulated, siphontube; see *Swammerdamm*, Bib. der Nat. Taf. XXXI. fig. 5, Taf. XXXIX. (*Culex* and *Stratiomys*); *Lyonet*, Mém. du Mus. XIX. Pl. XVIII. (10) fig. 1–3 (*Ptychoptera*). The parasitic larvae of the Conopidae, and of *Ocyptera* of the Tachinariae, which live in the cavity of the body of *Cassida*, *Pentatoma*, *Bombus* and *Andrena*, obtain the necessary air for their respiration by placing the posterior extremity of their body, which has two stigmata, in contact with a stigma or tracheal trunk of the insect in which they live; see *L. Dufour*, Ann. d. Sc. Nat. X. 1827, p. 255, VII. 1837, p. 16, Pl. I. fig. 13.

8 See *Swammerdamm*, loc. cit.

9 See *Lyonet*, loc. cit. p. 4, 5.

37*

pupae, respire by two stigmata situated on the sides of the cephalo-thorax.[10]

Most of the acephalous larvae of Diptera have only four stigmata, of which two are situated on the truncated extremity of the abdomen, and the two others, smaller, upon the sides of the second segment of the body. These last have sometimes a tubular form, and with some species, are even divided digitiformly at their extremity.[11] With the larvae of many Syr-phidae and Tachinariae, the two posterior stigmata consist of two siphons, which are often fused into one.[12] The larvae of the Coccidae have only four stigmata situated on the under side of the middle portion of their body.[13] Most of the larvae of the Coleoptera, Hymenoptera, Lepidoptera, as well as the cephalous ones of the Diptera, have numerous stigmata situ-ated on each side in the middle of the segments of the body, and which are never wanting, constantly, except with the second and third thoracic, and the last abdominal segments.

§ 344.

The numerous differences of the trachean system [1] in the various families of the Insecta may be classed under two principal forms.

1. With the first and most common, there are two large lateral trunks upon the sides of which open trunks which arise from the stigmata. From these lateral trunks branch off tracheae to the various parts of the body.

2. With the second form, the trunks which arise from the stigmata or trachean branchiae, directly ramify over the organs, but give off, both forwards and backwards, branches of communication to the neighboring trunks. The branches of one and the same segment frequently inter-anastomose by transverse trunks.

Often these two forms of tracheae coëxist in the same individual. In many cases, the secondary tracheae, in opening into the main trunks, are dilated into a large vesicle, or have upon their course numerous similar vesicles which give the whole system a varicose aspect.

Among the Aptera, the trachean system is of the first form with the Pediculidae, Nirmidae, and Poduridae.[2] But the Lepismidae form an exception in this respect, each of their stigmata opening into a trunk, which, without anastomosing with the neighboring trunks, is isolatedly ramified.[3]

With the Hemiptera, the trachean system presents many modifications. The trunks arising from the stigmata, sometimes ramify without anastomos-ing, and sometimes open into two lateral trunks. The musical Cicadidae

---

10 See my Memoir in *Wiegmann's* Arch. 1843, I. Taf. VII.

11 See *Bouché*, Naturgesch. d. Insekt. Taf. V. VI.; *L. Dufour*, Ann. d. Sc. Nat. XII. 1839, Pl. II. III., and XIII. 1840, Pl. III. and I. 1844, Pl.XVI.(*Tachina, Anthomyia, Helomyza, Sapro-myza, Piophila,* &c.).

12 The siphon is very long, articulated, and situ-ated at the extremity of the body with the larvae of *Eristalis* ; see *Reaumur*, Mém. loc. cit. IV. Pl. XXX. XXXII.

13 See *Burmeister*, Handb. &c. II. Taf. I. fig. 10-12.

1 See, beside the works of *Burmeister* and *La-cordaire*, the work of *Marcel de Serres*, in the Mém. du Mus. IV. p. 313.

2 With the Poduridae, the six trachean branches given off from the two main trunks, have each an oval dilatation ; see *Nicolet*, loc. cit. p. 47, Pl. IV. fig. 3.

3 *Guérin* (Ann. d. Sc. Nat. V. 1836, p. 374) thinks that the trachean system is wanting with *Machilis* ; but this must be incorrect, for *Bur-meister* (Isis, 1834, p. 137) has observed this sys-tem with *Lepisma*, with which it had for a long time before been sought in vain. I have very dis-tinctly observed it in *Machilis*, as well as in *Le-pisma*, and its organization is the same in both. The vesicles which, with *Machilis*, are situated on the sides of the abdominal appendages, and which *Guérin* thinks are respiratory organs, must have another function.

and the Pentatomidae, have varicose tracheae. With *Cicada*, there are two of these vesicles situated at the base of the abdomen distinguished for their very large size.[1] With *Nepa*, the primary trunks pass into the two lateral trunks, and form transverse anastomoses which extend from one side of the body to the other. In the thorax, the two lateral trunks form several large vesicles, between which arise, upon the sides, two other trunks which send an infinite number of very fine branches to the thoracic muscles.[5]

With the Diptera, this system is of the first form. It often presents, especially with those having a large and short abdomen, vesicular dilatations of which there are two, situated at the base of the abdomen, very large and distinct, sometimes filling nearly the whole abdominal cavity.[6] It is, moreover, with the larvae of this order, that this form of tracheae is most completely represented. The two lateral trunks are connected by the same number of transverse anastomoses as there are segments of the body.[7]

With the Lepidoptera in all their states, this system is also of the first form.[8] With the imagines of some Sphingidae, Bombycidae and Noctuidae, whose flight is continual, there are numerous vesicular dilatations and appendages of the tracheae.[9]

The tracheae of the Hymenoptera, which, throughout, are of the first form, send off from their two principal trunks numerous transverse anastomoses, and usually present vesicular dilatations at many points.[10] Of these last, those situated on the abdominal portion of the two trunks are very large, and often contiguous, so that the trunk to which they belong appears like a large sac constricted from point to point.[11] Sometimes there are only two of these vesicles, which are distinguished from the rest by their enormous volume, situated at the base of the abdomen.[12] With the larvae, there are found, pretty commonly, two main trunks connected by transverse communicating tubes.[15]

---

4 See *Burmeister*, Handb. &c. II. Taf. 1. fig. 10-12 (Coccidae) ; *L. Dufour*, Recherch. loc. cit. Pl. XVII. fig. 194 (*Tetyra*), and *Carus*, Analekt. &c. p. 156 (*Cicada*).

5 See *L. Dufour*, Recherch. &c. p. 244, Pl. XVIII.

6 With the Muscidae, Syrphidae, Tabanidae, Asilidae, Leptidae, &c. For the trachean system of the larva and pupa of *Sarcophaga haemorrhoidalis*, see *L. Dufour*, Mém. présentés, &c., IX. p. 572, Pl. II.

7 *Swammerdamm*, Bib. der Nat. Taf. XL. fig. 1 (larva of a *Stratiomys*) ; *Bouché*, Naturgesch. d. Insekt. Taf. VI. fig. 1 (larva of an *Anthomyia*), and *L. Dufour*, Ann. d. Sc. Nat. XII. 1839, Pl. I.-III.

8 *Lyonet*, Traité, Pl. X. XI. (larva of *Cossus ligniperda*). While the Syrphidae and Muscidae are passing into their pupa-state, the posterior stigmata disappear, the two anterior ones alone remaining active. With the Syrphidae these last often appear as two short tubes inserted on the cervical region.

9 See *Sprengel*, loc. cit. Tab. III. fig. 24 (*Sphinx ligustri*). Sometimes the number of these append-

ages is reduced to two large aëriferous reservoirs situated in the thorax ; see *Suckow*, Anat. physiol. Untersuch. p. 36, Taf. VII. fig. 30 (*Gastropacha pini*).

10 These dilatations are wanting with the Cynipidae, Chalcididae, and some Ichneumonidae. For the trachean system of the Hymenoptera in general, see *L. Dufour*, Recherch. sur les Orthopt. p. 374.[*]

11 With the Apidae, Andrenidae, Vespidae and Bembecidae ; see *Brandt* and *Ratzeburg*, Mediz. Zool. II. Taf. XXV. fig. 30 (*Apis mellifica*), and *Newport*, Philos. Trans. 1836, Pl. XXXVI. or *Cyclop*. &c. II. fig. 436 (*Bombus terrestris*).

12 With many of the Tenthredinidae, with *Myrmosa, Scolia, Crabro, Pompilus, Sphex,* &c.

13 See *Swammerdamm*, Bib. der Nat. Taf. XXIV. fig. 1 (larva of a bee). According to the observations of *Ratzeburg* (Die Ichneumon. d. Forstinsekt. p. 63, 81, Taf. IX.), the parasitic larvae of *Microgaster* and *Anomalon* are very singular. When young, they have no traces of tracheae, and respire, perhaps, by means of a caudal appendage enveloped by a thin membrane.

---

* [§ 344, note 10.] See *Newport* (On the formation and use of the air-sacs and dilated tracheae in Insects, Trans. Linn. Soc. June, 1847) ; these sacs are formed during the metamorphoses of the insect, and he adopts the view of Hunter, that the vesicles serve chiefly to enable the insect to alter its specific gravity at pleasure during flight, and thus diminish the muscular exertion required during these movements. — ED.

The true Neuroptera, in all their states, have a pretty simple trachean system provided with two lateral trunks. But with the Orthoptera, on the contrary, this system is usually very complicated. It is, indeel, less so with the Blattidae, Forficulidae, Ephemeridae, and Perlidae ;[14] but with the Libellulidae, the two lateral trunks are very large and arise from the trachean branchiae together with two other trunks.[15] With the other Orthoptera, the tracheae are very numerous and disposed according to the second type or form, their trunks being connected by a multitude of voluminous, longitudinal, and transverse anastomoses, giving the whole a reticulated aspect.[16] With the Acrididae, most of the transverse anastomoses have large air-reservoirs on their course.[17]

With the Coleoptera, the tracheae are always highly developed, and disposed, with the larvae, after the first type, but with the imagines, after the second.[18] With these last, the anastomosing canals, which connect the primary trunks, are often double.[19] With the Palpicornes, and Lamellicornes, this system is most highly developed, — the fine as well as the larger tracheae having a multitude of terminal vesicles.[20]

CHAPTER VIII.

ORGANS OF SECRETION.

I. Urinary Organs.

§ 345.

The Malpighian vessels, which are widely spread among the Insecta in all their conditions,[1] must now, since uric acid has been detected in their secretion, be regarded as Kidneys.[2]

---

14 See Swammerdamm, Bib. der Nat. Taf. XIV. and Carus, Entdeck. &c. Taf. III. (larva and pupa of an Ephemera).

15 Suckow, in Heusinger's Zeitsch. II. Taf. I. II. (larva and imago of an Aeschna).

16 With the Locustidae, Achetidae and Mantidae ; see L. Dufour, Recherch. sur les Orthopt. &c. p. 269, Pl. 1. fig. 1 (Ordipoda), and Marcel de Serres, Mém. du Mus. IV. p. 331, Pl. IV. (16) (Mantis), also in Isis, 1819, p. 627, Taf. IX.

17 Marcel de Serres, loc. cit. Pl. III. (15) (Truxalis), and L. Dufour, loc. cit. Pl. 1. (Oedipoda).

18 See Burmeister, Trans. Entom. Soc. I. Pl. XXIV. fig. 9 (larva of Calosoma sycophanta), and Audouin, Ann. d. Sc. Nat. IX. 1826, Pl. XLIII. fig. 3 (Lytta vesicatoria).

19 See L. Dufour, Ann. d. Sc. Nat. VIII. 1826, p. 23, Pl. XXI. bis. fig. 1, and Pictet, Mém. de Genève, VII. p. 397, fig. 6 (Hammaticherus heros).

20 Swammerdamm, Bib. der Nat. Taf. XXIX. fig. 9 (Geotrupes nasicornis), and Straus, Consid. &c. Pl. VII. (Melolontha vulgaris). See, also, for the Coleoptera in general, L. Dufour, Ann. d. Sc. Nat. VIII. 1826, p. 22.

1 As yet only Coccus, Chermes, and the Aphididae, have been found wanting the Malpighian ves-

sels ; see Ramdohr, Verdauungswerk. d. Insekt. p. 198, Taf. XXVI. and L. Dufour, Recherch. sur les Hemipt. p. 116, fig. 114. I have been unable to find them with the Strepsiptera in their various stages of development. The male imagines of Xenos Rossii, alone, have presented to me, at the extremity of the digestive canal, a singular glandular appendage resembling a cribriform lobe, and which serves, perhaps, as a urinary organ.

2 For a long time the Malpighian vessels were regarded as biliary organs, when Rengger expressed the opinion that they were urinary organs, without, however, having demonstrated the presence of uric acid in their secreted product (Physiol. Untersuch. uber die Haushalt. der Insekt. 1817, p. 27). This chemical proof was furnished by Brugnatelli and Wurzer (Meckel's Deutsch. Arch. II. 1816, p. 629, and IV. 1818, p. 213), with Bombyx mori. Subsequently, the existence of this acid has been confirmed by Chevreul with Melolontha vulgaris (Straus, Consid. &c. p. 251), and by Audouin with Lucanus cervus and Polites gallica (Ann. d. Sc. Nat. V. 1836, p. 129). See, also, Meckel, Ueb. die Gallen — und Harnorgane der Insekten, in his Arch. 1826, p. 21, and Groshans, De System. uropoët., quod est Radiat. Articulat. et Mollusc. Acephalorum. 1837, p. 39.

These always consist of several very long small tubes which, either separately, or by means of one or two common excretory ducts, are inserted upon the posterior or pyloric extremity of the stomach. These ducts are sometimes dilated, bladder-like, at their point of insertion. The opposite extremity of these uriniferous canals either terminates caecally, or passes arcuately into that of another. When, as is usual, they are very long, they embrace the digestive canal with numerous irregular convolutions. With certain species, they creep, by their anterior extremity, between the tunics of the stomach, or by their posterior between those of the colon ; this remarkable relation has often led to the opinion that these organs have two outlets into the digestive canal.[3]

These vessels are yellowish or brownish in color, and often slightly varicose.[4] They are composed of an external homogeneous tunic filled internally with cells. These last are very large, and are disposed rather in rows, than adjacently ; and nowhere can there be perceived in the interior of the vessels a glandular canal defined by a special epithelium. Each cell contains a clear, colorless nucleus, and a multitude of very fine granules which appear black by direct light, but by reflected light present a dirty-yellow or brown, rarely a green or red, aspect.[5] The granular contents of the cells, which give to these vessels their peculiar color, are scattered, when the cells are ruptured, through the intercellular spaces, and flow gradually into the digestive canal, Thus excreted, they accumulate in the colon or in its caecal appendage, and are evacuated with the faeces, or separately, as a troubled liquid of a color varying according to the species.[6]

§ 346.

The Malpighian vessels present numerous modifications as to their number, their length, their points of insertion, and their modes of grouping, in the different orders of the Insecta.[1]

With the Aptera, they are of median length ; with the parasitic species, and with the Lepismidae, they are four in number ; and six with the Poduridae.[2]

The Hemiptera have never more than four of these vessels, which are pretty long, whose extremities are looped with the Hydrocorisae and many

3 L. Dufour has clearly demonstrated the usual caecal terminations of these vessels ; see Ann. d. Sc. Nat. XIV. 1840, p. 231, Pl. XI. fig. 11 (larva of a Mordella), and XIX. 1843, p. 155, Pl. VI. fig. 9 (Hammaticherus heros).

4 The uriniferous canals of Melolontha vulgaris and Sphinx ligustri form, in this respect, a remarkable exception. In a great part of their course, they have on each side short caeca, pectinately disposed ; see Ramdohr, Abhandl. &c. Taf. VIII. fig. 1, 2 ; L. Dufour, Ann. d. Sc. Nat. III. 1823, Pl. XIV. fig. 4, 5 ; Straus, Consid. &c. Pl. V. fig. 6, 10 (Melolontha) ; and Newport, Cyclop. loc. cit. p. 974, fig. 432 (Sphinx).

5 For the intimate structure of these vessels, see H. Meckel, in Muller's Arch. 1846, p. 41, Taf. II.

6 With the holometabolic Insecta, the urine is evacuated isolately, especially when they approach the completion of their pupa-state. It is well known that the Lepidoptera, when bursting from their pupae, emit a considerable quantity of urine, of a variable color. In the larva and pupa of

Myrmeleon, it is gradually accumulated to a large quantity of a rose-color, in the digestive tube, and which the perfect insect immediately discharges on leaving the pupa-envelope, as a solid or elongate ovoid body. Réaumur (Mém. VI. 10 mém. Pl. XXXIV. fig. 12, 13) and Roesel (Insektenbelust. III. p. 123, Taf. XX. fig. 28, 29) have taken this urinary concretion for the egg of this insect. Sometimes there is precipitated in the urine, red crystals of a quadra-pyramidal form ; for example, with the larvae of Sphinx and Ephemera.

1 For these modifications in the different orders of Insecta, see the figures belonging to Ramdohr's work (Verdauungswerkz. &c.) ; those of Suckow, in Heusinger's Zeitsch. III. and L. Dufour, Sur les vaisseux biliares ou le foie des Insectes, in the Ann. d. Sc. Nat. XIX. 1843, p. 145, Pl. VII.-IX.

2 See Treviranus, Verm. Schrift. II. Taf. III. fig. 1 (Lepisma), Swammerdamm, Bib. der Nat Taf. II. fig. 2 (Pediculus), and Nicolet, loc. cit. Pl. IV. fig. 2 (Podura).

of the Geocorisae.[3]   With some species, their excretory ducts form one or two vesicular dilatations situated above the colon.[4]   It is only with a few Geocorisae, and with the Cicadidae, that the extremities of these canals are free.[5]   With this last group, and with the Cercopidae, they creep with a portion of the intestine, between the tunics of the ante-stomach, before opening into the lower extremity of the true stomach.[6]

With the Diptera, there are four long uriniferous vessels.   The Culici-dae and Psychodae, alone, by exception, have five.[7]   With very many species, these canals are united in twos, and open, by a common excretory duct, into the lower extremity of the stomach.[8]   Loop-like anastomoses occur only with the Tipulidae, Leptidae, and Bombylidae.[9]

With the Lepidoptera, there are nearly always six long, free, uriniferous tubes, which open into the stomach by two excretory ducts.[10]

The Hymenoptera are distinguished for their considerable number of these vessels, which are usually short and surround the pylorus in numbers of twenty to one hundred and fifty.[11]   With the Orthoptera, these vessels are inserted in a similar manner,[12] but are often much more numerous.[13]   The Termitidae, alone, form an exception,—having only six.[14]

The true Neuroptera are distinguished from the Orthoptera in that their vessels of this nature are long, flexuous, and only six to eight in number.[15]

With the Coleoptera, they are usually long, make numerous convolutions, and never exceed four or six in number.[16]   When four, they are nearly always joined by twos at their extremity; and when six, they are often attached by their extremities to the colon.[17]

The urinary vessels of the larvae and pupae resemble somewhat those of the perfect Insecta.[18]   With the larvae of certain Hymenoptera, and Orthoptera,

3 With the Naucoridae, Nepidae, with *Salda, Capsus* and *Reduvius*.   With *Dorthesia*, the four canals form also two short loops; see *L. Dufour,* Recherch. &c. p. 19, Pl. I.–IX.

4 *Pentatoma, Tetyra, Pyrrhocoris, Lygaeus, Gerris, Stenocephalus.*

5 *Cimex, Ploiaria, Miris, Alydus* and *Coreus.* In the last two genera the uriniferous canals, free, terminate at the pylorus in a common reservoir. With *Alydus, Aradus, Aneurus; Cixius, Issus,* and *Asiraca,* they unite in twos in a common excretory duct.   With *Psylla,* they consist only of four rudimentary caeca; see *L. Dufour,* Recherch. loc. cit.

6 It was a long time before there was an exact idea of the canals with *Cicada. Doyère* (Ann. d. Sc. Nat. XI. 1839, p. 81, Pl. I.) was the first who perceived that they penetrated between the tunics of the stomach; but he supposed they reappeared on its surface after a short course. He did not, therefore, attribute to these insects, only two uriniferous vessels. This last error has been rectified by *L. Dufour* (Ibid. XII. p. 287).

7 See *L. Dufour,* Ann. d. Sc. Nat. XIX. loc. cit. Pl. VIII. fig. 26 (*Anopheles*).

8 With the Muscidae, Oestridae, Conopidae, Syrphidae, and Hippoboscidae.   With the Stratiomydae, the four canals unite into one excretory duct; see *Swammerdamm,* Bib. der Nat. Taf. XLI. fig. 6 (*Stratiomys*); *L. Dufour,* loc. cit. Pl. VIII. fig. 28 (*Sargus*).

9 *Ramdohr,* loc. cit. Taf. XX.

10 With *Pterophorus* and *Yponomeuta, Suckow* has found only four uriniferous vessels (loc. cit. Taf. IX. fig. 159, 161).

11 See *L. Dufour,* Recherch. sur les Orthopt. Pl. III.–X.   The smallest number of these canals is found with the Formicidae, Cynipidae, and Ichneumonidae.

12 With the Ephemeridae, alone, the form of these canals is somewhat different, in that their free extremities are nearly always thickened, and that the excretory ducts take one or two spiral turns.

13 See *L. Dufour,* Recherch. sur les Orthopt. &c. Pl. I.–IV. XI. XIII.   *Gryllotalpa* is distinguished from the other Orthoptera in that the urinary canals are disposed fasciculate and terminate in a single excretory duct.

14 *L. Dufour,* Recherch. loc. cit. Pl. XIII. fig. 196.

15 *L. Dufour,* Ibid. Pl. XI.–XIII.   There are six of these vessels with the Phryganidae, Sialidae, Panorpidae, Rhaphidiidae: and eight with the Myrmeleonidae and Hemerobidae.

16 There are four urinary vessels with the Carabidae, Staphylinidae, Gyrinidae, Palpicornes, Lamellicornes, Cantharidae, and Buprestidae; six with the Byrrhidae, Nitidulidae, Dermestidae, Cleridae, Meloidae, Pyrochroidae, Bruchidae, Bostricidae, Capricornes, Chrysomelidae, and Coccinellidae.

17 For the uriniferous canals of the Coleoptera, see, beside *Ramdohr,* and *Suckow,* loc. cit., *L. Dufour,* Ann. d. Sc. Nat. 1824, II.–IV.; 1834, I. Pl. II. III.; 1840, XIII. Pl. V. VI.; XIV. Pl. XI.; XIX. Pl. VI.   With *Donacia,* the six vessels have a very peculiar aspect.   Two pairs unite loop-like at their posterior extremities, and their anterior ends unite in a common reservoir; while the third pair are free and open isolately at the pylorus; see *L. Dufour,* Ann. d. Sc. Nat. 1824, IV. Pl. VII. fig. 7, 8, and 1844, XIX. Pl. VII. fig. 10.

18 Beside *Ramdohr,* and *Suckow,* loc cit., see *L. Dufour,* Ann. d. Sc. Nat. XII. 1839, Pl. I.; XIII. Pl. V.; and XVIII. Pl. IV. (larva of a Tipulide, a *Sapromyza,* a *Pyrochroa,* and of a *Cetonia,* &c.); *De Haan,* Nouv. Ann. du Mus. IV. Pl. XVI.–XIX.

alone, their number is smaller,[19] and with those of the Lepidoptera, the extremities of the six tubes of this kind are insinuated between the tunics of the colon; while, with the imagines they are free.[20] With the Buprestidae, the larvae have six, but the imagines only four, of these vessels.[21]

II. Organs of Peculiar Secretions.

§ 347.

A great number of the Insecta, in both their larval and their perfect state, have glandular organs which secrete very varied products remarkable for their specific properties.

Many species have a secretory apparatus analogous to the cutaneous glands of the Vertebrata, which have received the name of *Glandulae odoriferae*. These consist of round follicles situated under the skin, whose very short excretory ducts open between the segments of the body, or between the articulations of its extremities. Their product emits a powerful odor, and, with some species, is evacuated in the form of droplets,[1] or, with others, covers the whole surface of the body, being perceived only by its odor.[2] The disagreeable odor emitted by the Bugs is due to a fluid secreted by a single, yellow, or red pyriform gland, situated in the centre of the metathorax, and opening between the posterior legs.[3] With other Insecta, there are analogous secretory organs, concealed in the posterior extremity of the abdomen, which copiously emit a fetid, troubled liquid, through an orifice situated by the side of the anus. These Anal Glands are usually double, and consist of simple follicles whose secretory product accumulates in round, or oblong contractile reservoirs.[4] With many

---

(larva of various Lamellicornes); and *Burmeister*, Trans. of the Entom. Soc 1. Pl. XXIV. fig. 10 (larva of a *Calosoma*), and his Abhandl. z. Naturgeschichte d. Calandra, loc. cit. fig. 3.

19 The larvae of the Aphidae and Vespidae have only four uriniferous vessels ; see *Swammerdamm*, Bib. der Nat. Taf. XXIV. fig. 6 (larva of a bee) ; *Suckow*, in *Heusinger's* Zeitsch. III. Taf. VI. fig. 180, and *Ramdohr*, loc. cit. Taf. XII. (larva of a *Vespa*); finally, *Rathké*, in *Müller's* Arch. 1844, p. 36, Taf. II. (larva of a *Gryllotalpa*).

20 See *Lyonet*, Traité, &c., Pl. XIII., and *Suckow*, Anat. u. physiol. Untersuch. Taf II.

21 See *L. Dufour*, Ann. d. Sc. Nat. XIV. 1840, p. 114. *Loew* (Entom. Zeit. 1841, p. 37, fig. 3) did not, probably, observe these canals in the larva of *Buprestis mariana* ; for, otherwise, he would not have regarded as such the two caecal appendages at the upper extremity of the stomach, and which the Buprestidae have also in their imago-state (see § 338).

1 With *Euprepia*, and *Zygaena*, a fluid of this kind, yellowishly transparent, exudes under the collar ; and with many Meloidae, Chrysomelidae, and Coccinellidae, it escapes from the knee-joints. The larvae of these last Coleoptera, as well as these

of many Tenthredinidae, emit droplets of fluid from the surface of their skin from the least touch. Very often the odor of this fluid reminds one of fresh poppy-juice. The fluid emitted from the cephaloprothoracic articulation, with *Colymbetes* and *Dytiscus*, has a very nauseating order. I am unable to decide whether or not the transparent liquid which escapes with various Aphididae through two tubes on their abdomen, belongs to this same category of secretions.

2 Certain Phryganidae, Hemerobidae, Crabronidae, Scoliadae, Ichneumonidae, &c., emit specific odors without the secretion of their *Glandulae odoriferae* being visible.

3 See *L. Dufour*, Recherch. loc. cit. p. 266, Pl. XVII. fig. 194. Moreover, the opinion that all the Bugs emit a bad odor is incorrect ; for with many, as for example *Syromastes*, the *Glandulae odoriferae* exhales a very agreeable odor resembling that of a fine bergamot pear.*

4 These anal glands, which *Burmeister* (Handb. I. p. 157), *Grant* (Outlines, &c., loc. cit. p. 554) and other anatomists have mistaken for urinary organs, consist, with the Dytiscidae and Gyrinidae, of two simple, long and flexuous caeca, whose reservoirs, having two short excretory ducts situated

---

* [§ 347, note 3.] With *Belostoma*, the odoriferous glands consist of two pretty long caecal tubes situated in the metathorax, beneath the other viscera, and extending into the anterior part of the

abdomen. They open externally between the coxae of the posterior legs. See *Leidy*, loc. cit. p. 64. — ED.

Coleoptera, these anal glands secrete a caustic fluid which has a penetrating and more or less aromatic odor. They are somewhat ramified, or composed of vesicles disposed botryoidally, and open into one or several long, excretory ducts.[5] These last open into two pyriform, muscular reservoirs, whose powerful contractions expel, as a means of defence, the secreted fluid.[6] The Formicidae, also, have, in the anal region, a glandular apparatus from which they eject a caustic, acid fluid. This apparatus is single and composed of one reservoir whose neck opens into a simple tube.[7] The larvae of *Harpyia*, also, defend themselves by ejecting an irritating liquid secreted by a glandular sac, which opens directly back of the head on the under surface of the first segment of the body.

Among the Hymenoptera, the females of the Vespidae, Fossores, Andrenidae, and Apidae, have, in the anal region, a glandular apparatus which secretes a poisonous fluid introduced by means of a hollow sting into the tissues of their prey or enemies.[8] This Poison-apparatus is composed of two long tubes which are sometimes very ramose.[9] The intimate structure of these tubes resembles that of the salivary glands.[10] The two poison-glands are sometimes isolated, sometimes united into a common canal, and their product is poured into a pyriform reservoir, which has thin but contractile walls, whose longer or shorter excretory duct opens into the sting.[11] This sting is formed by the intimate union of two lateral pieces, and plays in a cleft horny sheath. Often, its extremity is covered with backwardly-pointing denticles.[12] Both the sheath and the sting have, at their base, a peculiar muscular apparatus by which they are protruded and withdrawn.

near the arms, ejaculate a highly stinking liquid. With the Silphidae, where this apparatus is single, the reservoir opens laterally into the rectum ; see *H. Meckel,* in *Muller's* Arch. 1846, p. 47, and *L. Dufour,* Ann. d. Sc. Nat. VIII. 1826, p. 15, III. Pl. X. fig. 3, 4, 5, Pl. XIII. fig. 5, 7 (*Dytiscus, Gyrinus* and *Silpha*). With *Gyrinotalpa,* the anal glands consist of small lobular bodies inserted on the reservoir which receives their product ; see *L. Dufour,* Recherch. sur les Orthopt, &c., p. 346, Pl. II. fig. 19.
5 With the Carabidae, and Staphylinidae ; see *L. Dufour,* Ann. d. Sc. Nat. VIII. 1826, p. 6 ; II. Pl. XX. XXI.; III. Pl. X.; and VII. Pl. XIX. XX.; *J. Muller,* De Glandul. Struct. &c. Tab. I. fig. 13–18 ; and *Stein,* Vergl. Anat. u. Physiol. d. Insekt. 1847, Taf. I. fig. 4, g. g. (*Dianous*) and Taf. III. fig. 3, l. n. (*Oxytelus*).
6 With *Brachinus,* as is well known, this product is so volatile as to immediately become gaseous on its ejection.*
7 See *L. Dufour,* Recherch. sur les Orthopt. &c. p. 415, Pl. VII. fig. 86.
8 The Bees, which have a poison-apparatus of this kind, ought to be regarded as females whose genital organs are undeveloped. Many fossorial Hymenoptera, which feed their young with insects,

wound these last with their sting, that they may be mastered, and conveyed the more easily to the nest. Indeed, some carry their prey into their nests transfixed with the sting. (See my Observ. quaed. de Oxybelo atque Miltogramma, 1841, p. 11.) The wound does not always kill the insect, but simply disables it, so that they remain fresh for several days by the side of the larvae for whose food they are to serve.
9 There are two simple tubes with *Vespa, Scolia, Crabro, Halictus, Apis,* &c. ; but they are ramified with *Pompilus, Philanthus, Larra, Bombus,* &c.
10 For the intimate structure of these poison-glands, see *H. Meckel,* in *Muller's* Arch. 1846, p. 45, Taf. III.
11 This poison-apparatus is described more in detail in the works of *Swammerdamm,* Bib. der Nat. p. 183, Taf. VIII. (*Apis*); *Brandt* and *Ratzeburg,* Mediz. Zool. II. p. 203, Taf. XXV. fig. 39–42 ; *Ramdohr,* Abhandl. üb. d. Verdauungswerkz. &c. Taf. XIV. fig. 5 (*Pompilus*), and *Suckow,* in *Heusinger's* Zeitsch. II. Taf. XIV. fig. 38, 46 (*Apis* and *Crabro*).†
12 See *Swammerdamm,* loc. cit. Taf. XVIII. fig. 3.

* [§ 347, note 6.] For the peculiar glandular apparatus for this purpose, with *Brachinus,* see *Karsten,* in *Muller's* Arch. 1848, p. 367. Contrary to other Zootomists, this observer regards this apparatus as of a urinary nature, for he states that an analysis of its secretion furnishes a product analogous to urea. — ED.

† [§ 347, note 11.] The poison of the poison apparatus in the Hymenoptera has been investigated by *Will* (*Schleiden* and *Froriep's* Not. 1848, Sept. p. 17) who found, with Ants, Bees, and Wasps, that this product consisted of formic acid and a whitish, fatty, sharp residuum, the former being the poisonous substance. — ED.

There is another category of secretory organs which, with many females, open at the base of the ovipositor, but as they are intimately connected with the act of oviposition, they will be most properly described with the genital organs.[13]

A very large majority of the holometabolic Insecta have, in their larvae-state, silk-organs, the secretion of which they use, some, to weave a cocoon when about to pass into the pupa-state, or to close a hollow refuge they have sought; others to fasten together foreign bodies for the fabrication of their retreat. These organs are, therefore, most developed at the period when these insects approach their pupa-state; but with the larvae of the Psychidae, Tortricidae, and Lasiocampadae, they are already active during the first epochs of life. The silk-secreting portion of this glandular apparatus consists of two long, somewhat flexuous, thick-walled caeca, situated on the sides of the body, and continuous, in front, into two small excretory ducts, whose common orifice is on the under lip, and usually at the extremity of a short tubular protuberance.[14] With the larvae of *Myrmeleon*, the silk-apparatus is very remarkable, for the rectum itself is changed into a large sac and secretes this substance, which escapes through an articulated spinneret projecting from the opening of the anus.[15]

With the Apidae, there is a very remarkable Wax-secreting apparatus. This wax is elaborated by the Workers under the form of thin discs, which are formed between the imbricated posterior legs, without there having been discovered, as yet, in this region, the orifices of any special glands. It must therefore be supposed that it is produced by an exudation from the thin membranes which connect the different parts of the legs.[16] Moreover, many other Insecta have secretory products which transude through the skin without the existence of any special glandular apparatus, and which are hardened by the air like wax. These products are usually whitish, pulverulent, filamentous, or flocculent substances, which catch upon the surfaces of bodies.[17]

---

13 See § 350.

14 See *Roesel*, Insektenbelust. III. Class. I. Papilionum nocturnorum. Tat. IX. (*Bombyx*); *Lyonet*, Traité, &c., p. 498, Pl. XIV. XV. (*Cossus*); *Suckow*, Anat. u. physiol. Untersuch. p. 29, Taf. VII. fig. 31 (*Gastropacha*); *Pictet*, Recherch. pour servir à l'hist. d. Phryganides, Pl. III. fig. 1 (*Phryganea*). The decrease of these organs during the pupa-state has been very carefully detailed by *Herold*, Entwickelungsgesch. d. Schmetterl., Taf. III. and by *Suckow*, loc. cit. Taf. II. (*Pontia, Gastropacha*).

15 See *Réaumur*, Mém. &c. VI. Pl. XXXII. fig. 7, 8; *Ramdohr*, Abhandl. &c. Taf. XVII. fig. 1.

16 For the intimate structure of the wax-secreting portions of the skin with the workers of bees, see *Treviranus*, Zeitsch. f. Physiol. III. p. 62, 225; and *Brandt* and *Ratzeburg*, II. p. 179, Taf. XXV. fig. 18. The production of wax with bees has lately been the subject of much research among French naturalists. *Milne Edwards* has advocated the opinion before rejected by him, that this substance is secreted by special glands. But *L. Dufour*, after carefully-made researches, failed to discover them. See the various memoirs on this question in the Compt. Rend. XVII. and in the Institut. 1843, also in *Froriep's* neue Not. XXVIII. XXIX.

It is, moreover, easy to be convinced of the absence of these glands with the bee-workers; but if certain Andrenidae are examined, there will be found, on each side of their posterior tibiae, a small pyriform follicle with an excretory duct, and which secretes an oily substance.

17 These cutaneous secretions are observed with various Coccidae and Aphididae, whose entire bodies they cover with a powdery or woolly substance. With the females of *Dorthesia*, not only the entire body is covered with a substance which forms a solid white crust, but also the eggs after their deposition are invested with a similar envelope and thereby glued to the abdomen of the mother. With many male Coccidae, this secretion forms, at the posterior extremity of the abdomen, a bundle of very diverging, long, white and perishable hairs. With some Cicadidae (*Lustra* and *Flata*), the thorax and abdomen are covered, in places, by a kind of mould of a similar origin. The larvae of many Tenthredinidae (for example, *Tenthredo ovata*), as well as those of certain Coccinellidae (*Scymnus*), exude a liquid which, upon drying, forms white flocci."

---

* [§ 347, note 17.] See upon the subject of these secretions *Dujardin* (Mém. sur l'étude microscopique de la cire, in the Ann. d. Sc. Nat. XII. 1849, p. 250); his observations were made upon

The Phosphorescent Organs of the Lampyridae and certain Elateridae,[18] consist of a mass of spherical cells, filled with a finely-granular substance, and surrounded by many numerous trachean branches.[19] This substance which, by day-light, appears of a yellow, sulphur-like aspect, fills, with the Lampyridae, a portion of the abdominal cavity, and shines on the ventral surface through the last abdominal segments, which are covered with a very thin skin ; while, with the Elateridae, the illumination occurs through two transparent spots situated on the dorsal surface of the prothorax. The light produced by these organs so remarkably rich in tracheae, is undoubtedly the result of a combustion kept up by the oxygen of the air of these vessels. This combustion explains the remission of this phosphorescence observed with the brilliant fireflies, and which coincides, not with the movements of the heart, but with those of inspiration and expiration.[20]

CHAPTER IX.

ORGANS OF GENERATION.

§ 348.

The Insecta always multiply by means of genital organs situated in different individuals,[1] and, invariably, are provided with copulatory organs.‡ With certain species, namely, with the Apidae, and Termitidae, the females

18 For the phosphorescene of these Coleoptera, see *Carus*, Analekt. &c. p. 168 ; *Burmeister*, Handb. I. p. 534, and *Lacordaire*, Introduct. &c. II. p. 140.
19 The intimate structure of these organs has been studied with *Lampyris italica* by *Peters* (*Müller's* Arch. 1841, p. 229), and by *Morren* (*Isis*, 1843, p. 412). This last author says that this insect contains phosphorus, but adduces no fact in support of this assertion.
20 *Matteuci* has made numerous experiments on the phosphorescence of *Lampyris italica* ; from which it appears that the phosphorescent substance burns by means of the oxygen contained in the tracheae, without any increase of the temperature, and without any indication of the presence of phosphorus ; see *Matteuci*, Leçons sur les phénom. phys. d. corps vivants, Paris, 1847, p. 151, and Compt. Rend. XVII. 1843, p. 309, also in *Fro-*

various Insecta, among which were *Dorthesia*, *Alaerodes*, &c. The wax consists of fibres which are perpendicular to the secreting surface, and is a true product of the integument independent of any special glandular apparatus. — ED.

* [§ 347, note 20.] See, also, a note by me upon the intimate structure of the phosphorescent organs in *Pyrophorus phosphorus*, Proceed. Boston Soc. Nat. Hist. 1850, p. 290. — ED.

† [§ 348, note 1.] See, also, for cases of true hermaphroditism in the Insecta, *Wing* (Trans. of the Ent. Soc. London, V. p. 119) and *Westmael* (Bull. de l'Acad. d. Brux. 1849, II. p. 378). — ED.

riep's neue Not. No. 583, p. 168, and in *Schleiden* and *Frorirp's* Not. No. 9, p. 135.*
‡ *Hartig* has declared that certain species of *Cynips* are hermaphrodites ; but *Ratzeburg* and I have shown that this assertion is based on an erroneous interpretation of the organization of the females of *Cynips* ; see *Germar's* Zeitsch. f. Ent. III. p. 322, Taf. I.; and IV. p. 380, 396.
The true hermaphrodites which have as yet been found in the other orders of Insects, notably among the Lepidoptera, ought to be regarded as monsters. *Klug* (Verhandl. d. Gessellsch. naturf. Freunde in Berlin, I. p. 363, and Jahrb. d. Inseckt. I. p. 254), *Ochsenheimer* (Die Schmetterl. von Europa IV. p. 185) and *Lefebvre* (Ann. d. l. Soc. Entom. IV. 1835, p. 145) have given a list of the cases of hermaphroditism with insects. See also *Burmeister*, Handb. I. p. 338.†

‡ [§ 348.] The copulatory organs of the Insecta present wide and manifold variations, as has been shown especially by the recent researches of *Lacaze Duthiers*, Recherches sur l'Armure génitale des Insectes, in the Ann. d. Sc. Nat. 1849, XII. p. 353, 1850, XIV. p. 17 ; also his Recherches sur l'Armure génitale femelle des Insectes Orthoptères, id Ibid. XVII. 1852, p. 207, and Recherches sur l'Armure génitale femelle des Insectes Hémiptères, Ibid. XVIII. 1852, p. 337, finally the same of the Insectes Névroptères, Coléoptères, Diptères, in Ibid. XIX. 1853, p. 25, et seq. — ED

are much less numerous than the males. In the colonies of Bees, Termites, and Ants, there are, beside the males and females, a multitude of neuter individuals known as the Workers or Soldiers.

The sexual parts of insects are developed chiefly during the pupa-state; but their rudiments exist already in the youngest larvae, with which the sexes may then be distinguished.[2]

The female genital organs persist in a rudimentary germ-like condition with many larvae of Bees, probably owing to the influence of nourishment, for by increasing that of the workers these last may be raised to the rank of females or Queens.[3]

The Aphididae are very remarkable in that they produce, for several successive generations, only females which, in their turn reproduce, but viviparously and without the direct influence of the males.[4]

The genital organs of the Insecta are composed in general, of two symmetrical Ovaries, or Testicles, situated in the abdominal cavity, and of two oviducts, or Deferent canals (*Tubae*, or *Vasa deferentia*) which unite in a common excretory duct (*Vagina*, or *Ductus ejaculatorius*) opening back of

2 *Herold* (Entwickelungsgesch. d. Schmetterl.) has made very interesting researches on this premature development of the genital organs with *Pontia brassicae*, and which accord with the observations of *Suckow* (Anat. u. physiol. Untersuch. p. 31, Taf. III. V.) on those of *Gastropacha pini*. See, also, *Herold*, Disquisit. de Animal.Vertebr. carent. in ovo format. Tab. I. fig. 9, or Ann. d. Sc. Nat. XII. 1839, p. 186, Pl. VII. fig. 8. To be convinced that in the other orders of Insecta the genital organs are also developed at a very early period, it is only necessary to cast a glance over the figure which *Suckow* (*Heusinger's* Zeitsch. II. Taf. X. fig. 9) has given of *Aphrophora spumaria*, and *L. Dufour* (Ann. d. Sc. Nat. XIII. 1840, Pl. III. fig. 5) of *Pyrochroa coccinea*.

3 For the origin of the neuters with the Hymenoptera, see *Treviranus*, Zeitsch. f. Physiol. III. p. 220. In all the bee-workers there are found vestiges of the ovaries and of the seminal receptacle. See *Ratzeburg*, Nov. Act. Nat. Cur. XV. part II. p. 613, Tab. XLVII. and my observations in *Germar's* Zeitsch. IV. p. 375.

4 This mode of generation of the Aphididae (see § 350) quickly reminds one of that which *Steenstrup* has called Alternate Generation. Certain species of *Cynips* belong probably to the same category, for their males have yet been undiscovered. *Hartig* (*Germar's* Zeitsch. IV. p. 398) has been unable to find any individuals of this sex among thousands of *Cynips folii* and *divisa*. Similar observations have been made by *L. Dufour* (Recherch. sur les Orthopt. &c. p. 527). It is to me probable, also, that the capacity which many entomologists attribute to *Psyche* of laying eggs without a previous copulation is an example of alternate generation.*

* [§ 348, note 4.] The peculiar economical relations of certain Hymenoptera (*Cynips*) referred to above have received some explanation by the researches of *Frauendorf* (Hardinger Berichte üb. d. Mittheil. v. Freunden d. Naturwiss. in Wien. IV. p. 247, or *Wiegmann's* Arch. 1840, Th. II. p. 118), upon *Gastropacha lanestris*. He gathered two nests of the larvae at the end of June, 1836; by the middle of August the caterpillars had spun up, and on Sept. 18, the first imago appeared, and the second on Dec. 14; both of these were males; in the spring of 1837, some twenty individuals of both sexes appeared; others, likewise, in the autumn of 1837; others still in the following year, and the last of them on the 4th of March, 1842. The pupa-state of the last of the brood was therefore five and a half years, while that of the first was only as many weeks.

In regard to the alleged anomalous reproductive relations of *Psyche*, they have received the special attention of *Siebold*, who has quite cleared up the subject (Ueber die Fortpflanzung von Psyche : Ein Beitrag z. Naturgeschichte der Schmetterlinge, in *Siebold* and *Kölliker's* Zeitsch. I. 1848, p. 93; also in his Bericht üb. die entomol. Arbeiten d. schles. Gesellsch. im J. 1850, or its transl. in the Transact. of the Entom. Soc. London, I. 1851, p. 234. In the first of these researches made upon the genera *Psyche* and *Fumea*, there was no evidence that, with the individuals of these genera, reproduction occurs in an anomalous manner, that is, without the aid of the male; on the other hand, the facts of the well-developed character of the internal genital organs of the females, and of the capacity of the male to impregnate the female while she is concealed deeply in her case — these precluded the hypothesis of *Lucina sine concubitu*. But subsequent researches made upon *Talaeporia* have shown him that, with the individuals of this genus non-sexual reproduction does occur, presenting similar phenomena and conditions as the generation of the viviparous Aphides. It is proper to remark, however, that the carefully-made researches and experiments of *Speyer* upon the genital organs and mode of reproduction of *Talaeporia lichenella*, several years before, had shown that two successive generations here occur without the presence of males; see his paper in the Entom. Zeit. 1847, p. 18. For the phenomena and their interpretation of the development of the viviparous Aphididae, see my note at § 355, end. — ED.

the anus. This duct has several double or single appendages, of which one with the females serves as a seminal receptacle (*Receptaculum seminis*), or as a copulatory organ (*Bursa copulatrix*), while the others, in both sexes, are true secretory organs. The vagina is often prolonged into a horny ovipositor, and this same organ modified, with the males, is the *Penis*.

The Eggs of Insecta are very varied in their forms and colors. Externally, they are frequently marked by prominences and raised lines, forming a very varied, and often a very elegant design.[5] Those of some Cynipidae, Ichneumonidae, and Siricidae, have one of their ends prolonged into a long, straight or curved thread.[6] With some Hydrocorisae they are oblong and their posterior extremity is covered with long, stiff bristles.[7] They have, usually, a very solid chorion, and a thin vitelline membrane. The vitellus is composed of fat-vesicles more or less colored, which communicate their color to the entire egg. The germinative vesicle contains a germinative dot which is often composed of several parts.[8]

These eggs are formed after two different types.

1. With the Orthoptera, and various Coleoptera, the germinative vesicle is formed in the posterior extremity of the tubular ovaries, and is gradually surrounded by a mass of granular vitelline substance. This vitelline mass continues to increase until, at last, there is formed on its surface a chorion, at first soft, but which finally becomes solid. During the course of this development, the eggs succeed each other in a row, and in this way advance towards the opening of the ovarian tube.[9]

2. With the Lepidoptera, Diptera, Hymenoptera, Neuroptera, Cicindelidae, Carabidae, and Hydrocanthari, the mode of formation is wholly different. The vitelline mass which is disposed around the germinative vesicle, increases in the following manner: Between each two vitelline masses, there appear a group of large vitelline cells whose contents are blended with the subjacent vitelline mass; while, the chorion is developed from a layer of vitelline cells, commencing by its inner portion. It gradually extends over the vitelline mass and cells, and finally, when the vitellus has reached a certain volume, closes at the upper portion of this last. The epoch at which the eggs reach their maturity coincides, with the Lepidoptera, Tipulidae, and Ephemeridae, with the end of their pupa state, so that these insects are able to deposit their eggs as soon as they have cast off their pupa envelope; while, with the Libellulidae, the Locustidae, and especially the Apidae, the eggs are not matured in the ovaries until a long time after.[10]

With all Insecta, the sperm contains very active filiform spermatic particles which become immediately stiff and looped when put in water. These particles are developed in large cells whose involucrum finally dis-

5 *Kirby* and *Spence* (Einleitung, &c., p. 100, Taf. XV.) have figured a great number of eggs of insects of various forms. See, also, *Burmeister*, Handb. &c. Taf. 1. and *Lacordaire*, Introduction, &c., Pl. 1.

6 See *L. Dufour*, Recherch. sur les Orthopt. fig. 128, 149 (*Cynips* and *Xiphydria*); *Hartig*, in *Wiegmann's* Arch. 1837, I. p. 151, Taf. IV. (*Tryphon, Paniscus*, and other Ichneumonidae), and in *Germar's* Zeitsch. f. Entom. p. 327, Taf. I. fig. 5, 6 (*Cynips*).

7 The eggs of *Ranatra* have two long bristles; while, with those of *Nepa*, these last form a cornet; see *Rorsel*, Insektenbelust. III. Taf. XXII. XXIII., and *L. Dufour*, Recherch. sur les Hémipt. Pl. XVI.

8 See *Wagner*, Prodromus, &c., p. 9, Tab. II. fig. 18-22.

9 See *Wagner*, Abhandl. d. physical. mathemat. Klasse. der Akad. zu München. II. 1837, p. 554, Taf. II. fig. 1 (*Agrion*), and *Stein*, Vergl. Anat. u. Physiol. d. Insekt. I. p. 47, Taf. IX. fig. 4, 8 (*Telephorus* and *Acheta*).

10 *Herold* was the first who observed this remarkable mode of the formation of the eggs with the Lepidoptera; see his Disquisit. &c. Tab. I. fig. 11-18, or Ann. d. Sc. Nat. XII. 1839, p. 195, Pl. VII. fig. 13-18. Researches even still more detailed have been made by *Stein*, Vergl. Anat. &c. p. 52, Taf IX fig. 2, 9, 13 (*Pontia* and *Pterostichus*)

appears, while the spermatic particles thus formed remain together for some time and finally are united in fasciculi of variable forms.[11]

With many species, these bundles are disposed one after another, and then united forming long, vermicular bodies.[12] Only gradually, as the sperm mass passes along the deferent canals, are the spermatic particles separated to unite again under new and remarkable forms. These last consist of long, penniform bodies, having very singular movements, for their free extremities oscillate to and fro without cessation.[13] A kind of spermatophore is also observed in the female organs of many species belonging to the Lepidoptera, Orthoptera and Coleoptera. It consists of a peculiar hollow body, usually somewhat pedunculate, with pretty solid albumen-like walls, and filled with spermatic particles.[14]

I. Female Genital Organs.

§ 349.

The two Ovaries are always composed of a larger or smaller number of tubes, whose free extremities are extremely small, but which gradually increase in size to their point of insertion on the oviducts. From their caecal terminations is prolonged a delicate thread, which, bound together with the others, serves to attach the two ovaries to the thorax.[1] The

---

11 For the spermatic particles of the Insecta and their development, see my memoir in *Müller's* Arch. 1836, p. 30 ; and *Kölliker*, in the Neue schweiz. Denkschrift. VIII. p. 24.*

12 See my memoir in *Müller's* Arch. loc. cit. p. 38, Taf. III., fig. 16–18 (*Pontia*). These vermiform bundles are observed not only with all the Lepidoptera, but also with certain Diptera and Coleoptera ; see *Loew*, Horae anatom. Hft. 1, 1841, p. 26, Taf. II. (*Scatopse*), and *Hammerschmidt*, Isis, 1838, p. 358, Taf. IV. (*Cleonus* and various Lepidoptera). This last mentioned naturalist has, however, taken these cords for gigantic spermatic particles, to which he has given the name of *Pagiura, Spirilura* and *Cincinnura*.

13 I have discovered these penniform bodies composed of spermatic particles in the *Receptaculum seminis* of *Locusta* and *Decticus* ; see Nov. Act. Nat. Cur. XXI. 1845, p. 251, Tab. XIV. XV. *Dujardin* (Observ. au Microscop. 1842, Pl. XI. fig. 18, 19) had already perceived similar bodies in the male organs of *Tettigonia plebeja* and *Sphodrus terricola*. *Stein* (Vergl. Anat. &c. p. 106, Taf. I. fig. 19 (*Loricera*) has also found them in the seminal receptacles of the females of various Carabidae.

14 Pyriform, short-pedunculated spermatophores are found in the *Receptaculum seminis* of the

fecundated females of *Locusta* and *Decticus* (see my memoir in the Nov. Act. Nat. Cur. loc. cit. p. 262, Tab. XVI. fig. 14, 15), while those observed in the *Bursa copulatrix* of many Lepidoptera are round and long-pedunculated. With the Coleoptera, spermatophores are often found, also, in the copulatory pouch of the fecundated females. Their forms vary considerably, and I may mention specially those of *Chrina fossor*, which are elongate and remarkable for their very long and twisted peduncle ; see *Stein*, loc. cit. p. 91, Taf. I. VII. VIII. The older entomologists took these spermatophores for the penis which was detached in the copulatory act — an opinion which I myself formerly entertained (*Müller's* Arch. 1837, p. 399, 410) ; but, since, I have learned the true nature of these bodies, and the rectification of this error, made by *Stein* (loc. cit. p. 86), is perfectly correct.

1 For the different dispositions of these tubes, see *J. Müller*, Nov. Act. Nat. Cur. XII. p. 585 ; *Burmeister*, Handb. I. p. 199 ; and *Lacordaire*, Introduct. &c. II. p. 329. The ovaries of the Strepsiptera are organized after a wholly different type. The simplicity of the female organs here is very remarkable, and in this respect they hold an exceptional position. The two ovaries are, at first, two long bodies, composed of innumerable germs. When these last have matured, they are disengaged

---

*[§ 348, note 11.] The spermatic particles of the Insecta are described above, as well also by *Wagner* and *Leuckart* (loc. cit. Cyclop. Anat. and Phys.), as being invariably filiform. This is incorrect : it is true they are generally so ; as, for instance, with all the Coleoptera, Lepidoptera, Diptera, Aptera, Hemiptera ; but with some families of the other orders (the Hymenoptera, Neuroptera and Orthoptera) their form is quite different, and I am only surprised that it has not before been noticed. Thus, with the Libellulidae, Ephemeri-

dae, Andrenidae, Vespidae, &c., these particles have the form of those of the Araneae — an arcuate staff, to which is attached a delicate tail ; while, with the Phasmidae, they consist of a spoon-shaped head with a very conspicuous tail — indeed, quite resembling those of many of the Rodentia. It is scarcely necessary to add that in this class these particles are, as is the case with all the other classes of animals, developed in special cells. The whole subject is deeply interesting, in both a histological as well as a zoological point of view. — ED.

38*

ovarian tubes are, moreover, always enveloped by numerous trachean networks.[2]   Upon their length, which is very variable, depends the number of the eggs or germs which are disposed in a single file; and in this way, they may be distinguished as uni-, bi- and multi-locular. The two Oviducts are usually short and often dilated into a kind of calyx at their upper extremity, if there are numerous ovarian tubes meeting at this point.

The Seminal receptacle (*Receptaculum seminis*) is a double or single, solid capsule (*Capsula seminalis*) of variable form and surrounded by a muscular layer. It opens into the vagina below the point of junction of the two oviducts, by means of a canal of variable length (*Ductus seminalis*). This duct has sometimes a simple, or a bifurcated appendage (*Glandula appendicularis*).[3]   The seminal receptacle never contains spermatic particles with those females which have not rejected their pupa covering, or especially with those still in a virgin state; but after copulation it always contains a multitude of these particles moving very actively, and these movements are kept up for a long period, as may be observed with those females which live over the winter.[4]

The Copulatory pouch (*Bursa copulatrix*) consists nearly always, of a spacious, pyriform reservoir, which, with only a few exceptions, opens into the vagina below the seminal receptacle. During copulation, it receives the penis, and often, also, the sperm which enters either by portions contained in the spermatophores, or enveloped by a shapeless gelatinous substance.[5]

The secretory organs situated at the lower end of the vagina, consist, usually, of two rather long, glandular tubes on each side of the vagina, into which they open, either directly, or through two small special excretory ducts. They often have, on their course, two vesiculiform reservoirs. In most cases, these glandular organs appear to form a Sebaceous or

and scattered through the cavity of the body between the fat-cells. The females are apodal, and the ventral surface of their body, which resembles that of the larvae, is occupied by a shallow canal (Incubatory canal) which terminates caecally in the penultimate segment of the body, and opens upon the cephalothorax by a semilunar orifice (Genital opening). From this canal pass off into the visceral cavity three to five forward-bent tubes. The eggs are developed in the visceral cavity, and by these tubes the young larvae make their exit therefrom; see my Beitr. zur Naturgesch. d. wirbell. Thiere. p. 75, Taf. III. fig. 62, 67; and *Wiegmann's* Arch. 1848, I. p. 147. Formerly, I erred in taking the ventral for the dorsal surface with these insects.

[2] *J. Müller* has taken these filaments for vessels communicating between the ovaries and the dorsal vessel; see Nov. Act. Cur. XII. p. 580.

[3] For a long time this *Receptaculum seminis* remained wholly unobserved, or was taken for a *Bursa copulatrix*, or an organ secreting a viscous substance for gluing the eggs together and to foreign objects. The older descriptions and figures give, therefore, only an imperfect idea. It is only lately that the constant presence and true nature of this organ have been recognized (see my memoir in *Müller's* Arch. 1837, p. 592, and *Stein*, Vergl. Anat. &c. 1847, p. 96). Yet, at this day, the copulatory pouch and seminal receptacle are frequently confounded together; and *L. Dufour*, in particular, persists in his old error in designating this seminal receptacle as a *Glande sébifique*.

[4] See my observations made upon *Vespa* (*Wiegmann's* Arch. 1839, I. p. 107) and *Culex* (*Ger-*

mar's Zeitsch. II. 1840, p. 442). *Stein*, also (loc. cit. p. 112), has shown that the spermatic particles remain alive a long time in the seminal receptacles of the Coleoptera. The liquid secreted by the accessory gland serves, probably, to keep the spermatic particles fresh, and to prevent them from desiccation. The fecundation takes place undoubtedly when the eggs pass in front of the orifice of the seminal receptacle, which is then probably compressed by an investing muscular apparatus. This long preservation of sperm in the seminal receptacle explains how the females of certain species can lay eggs so long after copulation, and at a time when the males have all disappeared. The time of the full maturity of the eggs in the ovary, moreover, does not always coincide with that of the heat and copulation with the male. The observations which have been made on this last point have been collected by *Müller* (Nov. Act. Nat. Cur. XII. p. 624).

[5] This copulatory pouch, which, from its large size, was first perceived by entomologists, is even now often taken for a fecundating sac, or a seminal reservoir (*Spermatheca*). The spermatic particles are carried, undoubtedly by their own movements, from this copulatory pouch into the *Receptaculum seminis*; and very probably they begin to travel shortly after copulation, for, a long sojourn in the *Bursa copulatrix* does not appear advantageous, since those that remain over become stiff and dead-like in the midst of the seminal fluid, which is granulous and viscid. *J. Hunter* (Philos. Trans. 1774), in his experiments on artificial fecundation, was successful only when he took the sperm from the copulatory pouch of the females which had just come from copulation.

Mucous apparatus (*Glandulae sebaceae* or *colleteriae*), for they secrete a viscous, coagulable substance, which serves to envelop and glue the eggs together, and to fix them to foreign bodies. With the females of the Ichneumonidae, this apparatus secretes a kind of cement with which these insects close the wounds they have made in the bodies of the Insecta in which they have deposited their eggs. It is probable, also, that, with those Insecta which deposit their eggs by means of an ovipositor in the tissues of plants, thereby producing galls, these same organs serve as a kind of Poison-apparatus causing this diseased formation of the vegetable parenchyma.

§ 350.

The different parts of the female genital apparatus present, in the various orders and families, countless modifications as to number, form and disposition. The most important of these are the following:

With the Aptera, the two ovaries consist each of only four to five tubes, which, with the Pediculidae, open, all, at the top of the corresponding oviduct; while with the Lepismidae, they are separately inserted on the side of the moderately long oviduct. In both of these families, there are two short varicose caeca, which enter laterally the lower end of the vagina, and are probably sebaceous or viscous organs.[1] There appears to be here no seminal receptacle or copulatory pouch.

With the Hemiptera, the ovaries consist of four to eight tubes of variable length, disposed verticillate at the extremity of the short oviducts. The Psyllidae and Cicadidae, alone, form an exception in this respect. With the first, the ovaries are composed of ten to thirty unilocular tubes, and with the second, twenty to seventy bilocular ones. These last, moreover, are distinguished by their oviducts being divided into several branches, on the extremity of each of which is a tuft of ovarian tubes.[2] Their *Receptaculum seminis* consists of two small caeca.[3] The other Hemiptera have only a single seminal receptacle, which is pyriform with the Psyllidae and oviparous Aphididae;[4] is a long, slightly flexuous caecum with the Naucoridae, and Nepidae; and a very long, somewhat flexuous caecum with the Hydrometridae. With many Capsidae, and other Geocorisae, also, it is a pretty long and flexuous caecum, while, with the Pentatomidae, the rather short *Ductus seminalis* terminates in a brownish, horny, pyriform *Capsula seminalis*, the constrictions and protuberances of which often present a peculiar appearance. Sometimes this tube is dilated into a second vesicle, at whose base is a horny tube containing a second tube which is a direct prolongation of the *Capsula seminis*.[5] Most Hemiptera have no copulatory pouch, — the Cicadidae, alone, having one which consists of a narrow-necked, pyriform vesicle.[6] With the oviparous Aphididae,

1 See *Swammerdamm*, Bib. der Nat. p. 37, Taf. II. fig. 8 (*Pediculus*), and *Treviranus*, Verm. Schrift. II. p. 15, Taf. III. fig. 8, 9 (*Lepisma*).

2 See *L. Dufour*, Recherch. sur les Hémipt. Pl. XIV.–XVII., and Ann. d. Sc. Nat. V. 1825, p. 168, Pl. IV. (*Cicada*); and *Suckow*, in *Heusinger's* Zeitsch. II. Taf. XV. fig. 55, 57 (*Nepa* and *Cercopis*).

3 See *Meckel*, Beitr. &c. I. Ht. I. T d. I. fig. 6, i. 1.; *L. Dufour*, Ann. d. Sc. Nat. V. 1825, Pl. IV. fig. 5, 1. i., and fig. 8, d. d.; and *Doyère*, Ibi. VII. 1837, Pl. VIII. fig. 3–7, s. s (*Ledra* and *Cicada*).

4 See my memoir on the internal genital organs

of the oviparous and viviparous Aphididae, in *Froriep's* neue N tiz. XII. p. 308.

5 For the seminal receptacle of the Pentatomidae, see *L. Dufour*, Recherch. &c. loc. cit. Pl. XIV.–XVI., and *Siebold*, in *Müller's* Arch. 1837, p. 410, Taf. XX. fig. 4–6.*

6 See *Meckel* and *L. Dufour*, loc. cit. According to *Doyère* (loc. cit. p. 203, Pl. VIII. fig. 3), there is, with the female Cicadidae, a special orifice by the side of the oviduct, which is continuous with the ovipositor, and through which the penis protrudes into the copulatory pouch.

* [§ 350, note 5.] For the female organs of *Belostoma*, see *Leidy*, loc. cit. p. 64. — ED.

and many Geocorisae, the secretory apparatus consists of two round glandular sacs,[7] while, with the Cicadidae, it is a single, long flexuous tube.[8] The viviparous Aphididae differ from those which are oviparous, in that their eight ovarian tubes are multilocular, and their oviducts entirely without appendages; while with the second or oviparous, these eight tubes are unilocular, and there is a seminal receptacle and two sebaceous glands.[9]

With the Diptera,[10] the ovaries consist, usually, of numerous short, three or four chambered tubes. With only a few species, these tubes are long and have eighteen to twenty chambers.[11] The disposition of these tubes varies considerably. With some, they are simply terminal to the short oviduct; while with others they form one or more series on the sides of these organs, which, then, are longer.

The *Receptaculum seminis* presents the most varied forms,[12] it is usually, triple, rarely simple or double,[13] and is lined with a horny, brown substance. It has a round, pyriform, or oblong shape, and, in this last case, is often flexuous or spiral. The seminal ducts, which lead from the receptacles to the vagina, are sometimes isolated, and sometimes united into one or two common ducts before entering the vagina. Directly below them on each side, are the points of junction of the two secretory organs, which, always present with the Diptera, consist of two simple, rarely ramose tubes, whose very small excretory ducts have, exceptionally only, a vesiculiform dilatation.[14] The *Bursa copulatrix* appears to be wanting with all the Diptera. But, with many Muscidae, the vagina has, as a seminal receptacle or uterus, a spacious, and sometimes two-lobed reservoir in which the fecundated eggs are accumulated in great numbers, and remain until the larvae are sufficiently developed to be hatched, making these animals viviparous.[15] With certain species of *Tachina*, this uterus presents a remarkable form; the vagina is very long, spiral, and of equal size throughout; and, at certain periods, is crowded with larvae or small eggs.[16] With the pupiparous Hippoboscidae, the female organs are formed on an entirely special type, corresponding with the remarkable mode of the reproduction

7 See *L. Dufour*, Recherch. loc. cit. Pl. XIV. XV.

8 See *Meckel, Suckow, L. Dufour, Doyère*, loc. cit.

9 See my researches in *Froriep's* neue Notiz. XII. p. 307. *Dutrochet* (Ann. d. Sc. Nat. XXX. 1833, p. 204, Pl. XVII, C. fig. 1), it would appear, has unwittingly figured the genital organs of an oviparous *Aphis*, by taking the seminal receptacle for a sperm-secreting organ. In this way he was led to regard the viviparous Aphididae as hermaphrodites.

10 For the internal female organs of the Diptera, see *L. Dufour*, Ann. d. Sc. Nat. I. 1844, p. 253, and especially *Loew*, Horae anatom. p. 61.

11 *Ephydra* and *Tachina*; see *Loew*, loc. cit. Taf. IV. fig. 3, 10.

12 For the *Receptaculum seminis* of the Diptera, see *Siebold*, in *Muller's* Arch. 1837, p. 414, Taf. XX. fig. 7–10; and especially *Loew*, loc. cit. p. 89, Taf. IV.–VI., and in *Germar's* Zeitsch. III. p. 386, Taf. III.; the numerous figures of this author will give some idea of the inexhaustible variety of forms of these organs. When *L. Dufour* (Ann. d. Sc. Nat. I. 1844, p. 262) would regard the seminal receptacle as a reservoir of the neighboring secretory organs, it is evident that this distinguished entomologist must have entirely omitted a microscopical analysis of the substances found in the various glands and other organs of Insecta.

13 The *Receptaculum seminis* is simple with *Pulex, Empis, Dolichopus*, and *Bibara*; and double with *Piophila, Stomoxys* and *Borborus*.

14 See *Siebold*, and *Loew*, loc. cit. These glandular appendages secrete with certain Tipulidae a considerable quantity of gelatinous substance which envelops the eggs and binds them in a kind of collar. These collars, which are deposited in the water, have for a long time been figured by botanists among the algae under the name of *Glotonema*.

15 There are viviparous species in the genera *Musca, Anthomyia, Sarcophaga, Tachina, Dexia, Miltogramma*, &c.; see my memoir in *Froriep's* neue Notiz. III. p. 337, and in *Wiegmann's* Arch. 1838, I. p. 197; also my Observat. quaed. Entom. &c. p. 18. *L. Dufour* (Ann. d. Sc. Nat. I. 1844, p. 261) has designated this reservoir as *Reservoir ovolarrigère*; see also his Hist. d. Métamorph. et d. l'Anat. d. la Piophila petasionis, Ibid. p. 382, Pl. XVI. fig. 16, g. *Loew* (Horae anatom. Tab. IV. fig. 9, 11, 14, Tab. V. fig. 13) has figured analogous uteriform reservoirs with *Musca, Dexia, Piophila* and *Psila*.

16 This long spiral-form vagina, which was formerly described as an *ovarium spirale*, is found in *Tachina fera, tessellata, grossa, vulpina, haemorrhoidalis*, &c.; see my memoir in *Wiegmann's* Arch. loc. cit. p. 194, and *Réaumur* Mém. IV. 10 mém. p. 412, Pl. XXIX. fig. 7, 8

of these animals. The two ovaries are unilocular pouches of unequal size, inserted laterally, by means of a short oviduct, upon the vagina. The upper extremity of this vagina contains sperm, after copulation, and may, therefore, be regarded as a *Receptaculum seminis;* while the lower portion is widely dilated, and may, therefore, be considered as an uterus. The upper or narrower portion of the vagina receives two small, simple, or somewhat ramose glandular tubes (*Glandulae sebaceae*).[17] Below these glands are situated the two excretory ducts of a double glandular apparatus, very voluminous and multiramose, whose product serves, without doubt, to nourish the larvae which are provisionally developed in the uterus.[18]

With the Lepidoptera, each ovary is composed of four very long, spiral, multilocular tubes. The *Receptaculum seminis*[19] is pyriform, and often has a long, spiral *Ductus seminalis.*[20] At its base opens a simple or bifurcated accessory gland, and underneath it there is always a large, double, sebaceous gland, consisting of two rather long, flexuous, simple caeca. These last open into the vagina, by means of a short common excretory duct, and each, at their point of union, is usually dilated into a vesiculiform reservoir.[21] Some Lepidoptera have, moreover, two smaller ramose glands, situated near the orifice of the vagina, which secrete, perhaps, an odorous substance that excites the copulatory act.[22] The copulatory pouch, finally, is very remarkable in all the species of this order. It consists of a .arge, pyriform reservoir, sometimes constricted in its middle, and having for the reception of the penis, a canal which opens externally by a special orifice situated below the vulva. In its course this canal sends off a small, flexuous, lateral duct, which passes into the vagina opposite the mouth of the *Receptaculum seminis,* and thus forms a communication between this last and the copulatory pouch.[23]

With the Hymenoptera, the ovaries[24] vary very much as to the number of their component tubes, of which there are sometimes four to six, sometimes eight to ten, and with some species they range from twenty to a hundred.[25] These tubes are always multilocular, and never very long. The

---

17 These tubes are simple with *Melophagus*, and ramose with *Hippobosca.*

18 See my researches in *Müller's* Arch. 1837, p. 435, and those of *L. Dufour*, Ann. d. Sc. Nat. VI. 1825, p. 308, Pl. XIII., and III. 1845, p. 76, Pl. III. This last-mentioned naturalist has very well figured the female organs of *Hippobosca* and *Melophagus* ; only he is deceived relative to the glandular appendages of the vagina, in regarding the upper pair as a *Receptaculum seminis,* but which never contain spermatic particles.

19 For the appendages of the female organs of the Lepidoptera, see *Siebold*, in *Müller's* Arch. 1837, p. 417.

20 The seminal receptacle has been figured in its various stages of development by *Herold* (Entwickelungsgesch. d. Schmetterl. Taf. IV. fig. 1, u. y. p. and Taf. XXV.) as a unicornous secreting organ. See also *Suckow*, Anat. u. physiol. Untersuch. Taf. VI. g. g.

21 See *Herold*, loc. cit. Taf. III. fig. 1, t. z. and the following plates ; also *Suckow*, loc. cit. Taf. VI. l. l.

22 *Melitaea, Argynnis, Zygaena,* &c.

23 See *Herold*, loc. cit. Taf. III. fig. 1, x. f. g. and the plates following ; also *Suckow*, loc. cit.

Taf. VI. K. (indistinct). Moreover, *Malpighi* (De Bombyce, 1669, p. 81, Tab. XII. fig. 1, J. K. M.) had already perceived, with the silk-worm, all the appendages of the vagina, and specially the copulatory pouch with its canal of lateral communication. With *Euprepia Hebe* this canal has a pyriform deverticulum.*

24 For the female genital, organs of the Hymenoptera, see *L. Dufour*, Recherch. sur les Orthopt. &c. p. 406.

25 Each ovary is composed of three or four ovigerous tubes with *Xylocopa, Bombus, Anthophora, Chrysis ;* of five to six with *Nomada, Sapyga, Chalcis, Vespa ;* of eight to ten with *Pimpla, Paniscus ;* of ten to twelve with the Tenthredinidae ; of twenty to twenty-five with *Myrmica, Xiphydria* and *Banchus ;* and of more than one hundred with *Apis.* With *Chelonus* the ovaries present a remarkable exception ; they consist each of two long flexuous tubes, which are very widely dilated at their lower extremity. *L. Dufour* (loc. cit. p. 541, Pl. X. fig. 143) regards these swellings as a kind of uterus in which are developed the larvae of these Ichneumonidae ; but this assertion cannot be admitted without further research.

* [§ 350, note 23.] See also for the internal female genital organs, and especially their development, of the Lepidoptera, *Meyer*, loc. cit. *Siebold*

and *Kölliker's* Zeitsch. I. 1849, p. 192. This memoir contains many new details. -- ED.

Receptacula seminis is nearly always simple, round, or ovoid, and necked, and is continuous into a usually short, seminal duct.[26] A *Glandula appendicularis* is never absent, and consists, usually, of a bifurcate tube, which opens into the *Ductus seminalis*, and only rarely into the *Capsula seminalis* itself.[27]

With the Tenthredinidae this apparatus is, moreover, formed after a different type; the seminal vesicle is a simple deverticulum of the vagina, and more or less distinct from it, beside, it is deficient in the accessory gland.[28] The copulatory pouch is absent with all the Hymenoptera as are also the *Glandulae sebaceae* with those females which have a sting and a poison-gland; but these sebaceous glands are highly developed with those species having an ovipositor, into which last they open, and probably serve some purpose connected with the oviposition, partly as sebaceous, and partly as excitatory organs. This secretory apparatus consists of a simple or a double ramose gland, whose excretory duct receives the neck of a pyriform receptacle, or, sometimes, is itself dilated into a vesicular reservoir.[29]

With the Orthoptera, the two ovaries are nearly always composed of numerous, multilocular tubes, which usually open in a single row upon the internal or external side of two large and sometimes very long ovaries.[30] The seminal receptacle often consists of a simple longer or shorter pedunculated vesicle, whose closed extremity is dilated into a pyriform vesicle with the Psocidae, Forficulidae, Locustidae, Phasmidae and Mantidae.[31] A similar *Capsula seminis* is often found with the Acrididae on one of the sides of the *Ductus seminalis* and removed from its extremity.[32] Most of the Blat-

---

26 For the *Receptaculum seminis* see Siebold, Observ. quaed. Entom. loc. cit. p. 6, and in *Germar's* Zeitsch. IV. p. 362, Taf. II. With those females which, at short intervals, lay very many eggs, the seminal receptacle is very large; see *Swammerdamm*, Bib. der Nat. Taf. XIX. fig. 3, t. n. u., where the *Receptaculum seminis* of a honey-bee is very well represented.*

27 The *Glandula appendicularis* is simple and inserted on the *Ductus seminalis* with the Pteromalini and Cynipidae; it is double, and opens directly into the *Capsula seminis*, with *Vespa crabo* and *Tiphia femorata*.

28 The seminal receptacle is double, exceptionally, with *Lyda*.

29 This glandular apparatus is simple and has a lateral pyriform reservoir with various Ichneumonidae; see *L. Dufour*, Recherch. Pl. X. fig. 137–142 (*Pimpla* and *Bracon*). This naturalist calls this apparatus *Glande sérifique*, as distinguishing it from the *Glande sébifique*. With *Sirex*, I have observed the excretory duct of this single and multiramose gland dilated into a large reservoir. With the Tenthredinidae, it is also ramose, but double as well as its vesicular reservoir; see *L. Dufour*, loc. cit. Pl. X. fig. 155–157 (*Tenthredo* and *Cimber*).

30 With the Locustidae, Acrididae, Mantidae and Libellulidae, the ovarian tubes are inserted upon the internal side, and with the Phasmidae and Ephemeridae, on the outer side of the two oviducts. *Forficula gigantea* has, moreover, only five internal multilocular tubes, while with *Forficula auricularis*, the very long oviducts have on all sides a multitude of unilocular tubes. With *Mantis*, the ovarian tubes are unilateral, but united together in several bundles. With *Oedipoda cerulescens* and *Truxalis nasuta*, the two

long, flexuous, caecal oviducts, have tubes only at their lower extremity. The oviducts of *Perla bicaudata* are still more remarkable; they are very long, flexuous, and have ovarian tubes only on one side of their upper extremity, and anastomose in a loop-like manner. For all these differences, see *L. Dufour*, Recherch. sur les Orthopt. &c. Pl. II.–V. and Pl. XI. fig. 165, Pl. XIII. fig. 206, and in the Ann. d. Sc. Nat. XIII. 1828, Pl. XXI. XXII. (*Forficula*).

31 With *Forficula*, and *Acheta*, the seminal receptacle has a long and flexuous peduncle, which, with the Psocidae, and Locustidae, is shorter. That of *Psocus pulsatorius* contains several long-pedunculated, glandular bodies (*Nitzsch* in *Germar's* Magaz. IV. p. 282, Taf. II. fig. 3–5), which I formerly regarded as *Capsulae seminales* (*Müller's* Arch. 1837, p. 410,) but which are probably spermatophores. With *Perla*, the seminal receptacle is a simple caecum, twisted like a ram's horn, and the base of which supports several short glandular follicles (*Glandulae appendiculares ?*). For the seminal receptacle of the Orthoptera cited in the text, see especially *Roesel*, Insektenb. Th. II. Heuschrecken-und Grillen Sammlung. Taf. IX. fig. 3, k. (*Dicticus*); *L. Dufour*, Recherch. sur les Orthopt. Pl. III. fig. 31, Pl. IV. fig. 43 (*Acheta* and *Mantis*); &c. § etc. d. Fev. A.t Fat. Cur. XXI. part 2; tt. fig. 24 XIV fig 1 c. (*Locusta*).

32 See *Hegetschweiler*, De insect. genitalibus dissert., fig. VII. f. c.; and *Siebold*, in *Müller's* Arch. 1837, p. 409, Taf. XX. fig. 3 (*Gryllus*). The *Ductus seminalis* is usually very long and intertwisted, as, for example, with *Gryllus*, *Truxalis*, &c.

* [§ 350, note 26.] See also *Longstreth* (Proc. Acad. Sc. Philad. 1852, VI. p. 49) for some observations on the impregnation of the common honey-bee, as due to a *Receptaculum seminis*. — ED.

tidae[33] and Libellulidae[34] have a short, double, seminal receptacle, which, however, appears to be wholly wanting with the Ephemeridae. There is a round *Bursa copulatrix* only with the Libellulidae.[35] The glandular appendages of the vagina are also not found with all the Orthoptera. They are wanting with the Forficulidae, Phasmidae, Perlidae, Ephemeridae, Libellulidae and Acrididae, but with *Decticus* and *Locusta*, there is a sebaceous organ consisting of a simple, pretty long tube,[36] which, with the Achetidae, is more or less ramose, and with the Blattidae and Mantidae is composed of a considerable number of partly simple, partly ramose follicles.[37]

With the Neuroptera, the ovaries consist always of multilocular tubes. With the Hemerobidae, and Myrmeleonidae, there are ten inserted on the external side of the two large oviducts, and with the Phryganidae, their number is quite large, but their insertion on the oviducts is the same.[38] The ten with *Panorpa*, and the much larger number with *Sialis*, are disposed verticillate at the extremity of the oviducts. With *Myrmeleon* and *Panorpa*, the seminal receptacle is a long, pedunculated sac; and has, with *Hemerobius*, a single, and with *Raphidia*, a double *Glandula appendicularis*.[39] With the Phryganidae, this receptacle is still more complicated, for, beside a long, tortuous accessory gland, which is inserted on the neck, or at the base of the *Capsula seminis*, there is, at the lower extremity of the *Ductus seminalis*, another and flexuous glandular tube, and a short-pedunculated reservoir which corresponds perhaps to a copulatory pouch.[40] With *Sialis*, beside two lateral deverticula serving, probably, as copulatory pouches, the vagina has numerous vesicular appendages filled with a dark liquid, but the nature of these is still not understood.[41] With *Myrmeleon*, *Hemerobius*, and *Panorpa*, the vagina receives two simple, more or less flexuous, glandular tubes,[42] which are probably sebaceous organs, and with the Phryganidae, consist of six digitiform follicles.[43]

With the Coleoptera, the ovaries consist of trilocular, rarely multilocular tubes,[44] which are inserted on the calyciform upper extremity of the oviducts, in groups of five to ten or even of fifteen to thirty and forty.[45] Beside

33 *Blatta orientalis* has two short and flexuous seminal receptacles; but *Blatta germanica* has two large and two small ones; see *Siebold*, in *Muller's* Arch. 1837, p. 408.

34 The seminal receptacles of *Libellula*, *Aeschna* and *Diastatomma* consist of two small caeca, which, with *Calopteryx*, open into the vagina through a common duct; while, with *Agrion*, there is only a single long receptacle; see *Rathké*, De Libellular. partibus genital. Tab. I. fig. 11–13, Tab. II. fig. 12–14, and Tab. III. fig. 9–11, c., and *L. Dufour*, loc. cit. Pl. XI. fig. 165, d. d. (*Libellula*, *Aeschna* and *Agrion*). See also my memoir on the generation of the Libellulidae, in *Germar's* Zeitsch. I. p. 433.

35 See *Rathké*, loc. cit. Tab. I. fig. 11–13, Tab. II. fig. 12, 13, and Tab. III. fig. 9–11, b.

36 See *Roesel*, loc. cit. Taf. IX. fig. 3, i., and *Siebold*, Nov. Act. Nat. Cur. loc. cit. p. 255, Tab. XIV. fig. 1, e.

37 See *L. Dufour*, Recherch. &c. Pl. III. fig. 31, d. (*Oecanthus*), Pl. IV. fig. 43 (*Mantis*). It is not surprising that this wax-apparatus is so highly developed with the Blattidae and Mantidae, for, as is known, the females of these insects surround their eggs with very spacious, multilocular capsules, which they carry about with them, or fasten to foreign bodies; see *Garde*, Beitr. &c. Taf. I. fig. 13, 14 (*Blatta orientalis*), and Roesel, loc. cit. Th. IV. Taf. XII. (*Mantis*).

38 For the female organs of the Neuroptera, see *L. Dufour*, Recherch. sur les Orthopt. &c. Pl. XII. XIII.

39 *L. Dufour*, loc. cit. Pl. XII. fig. 174, d. (*Panorpa*).

40 *L. Dufour*, Ibid. Pl. XIII. fig. 211, 212.

41 *L. Dufour*, Ibid. Pl. XII. fig. 188, b.; and *Suckow*, in *Heusinger's* Zeitsch. II. Taf. XVI. fig. 16, d.

42 *L. Dufour*, Ibid. Pl. XII. fig. 174, 194, c. c.

43 *L. Dufour*, Ibid. Pl. XIII. fig. 211. By means of these glands the females of *Phryganea* envelop their eggs with a gelatinous substance which swells in water and often sticks to stones or aquatic plants, presenting the appearance of an annular spawn.

44 The ovarian tubes are multilocular with the Carabidae, Hydrocanthari, Cyphonidae, Telephoridae, and Curculionidae; in general they are bilocular with the Staphylinidae; see *Stein*, Vergl. Anat. &c. p. 29.

45 The ovaries are multitubular with the Carabidae, Hydrocanthari, Hydrophilidae, Elateridae, Chrysomelidae, and Coccinellidae; while with *Apion*, *Lixus*, and *Hylesinus*, there are only two on each side; see *L. Dufour*, Ann. d. Sc. Nat. VI. 1825, Pl. XVII.–XX.; *Suckow*, in *Heusinger's* Zeitsch. II. Taf. XIII., and *Stein*, loc. cit. Taf. III.–VIII.

these fasciculate, there are, also, here and there, botryoidal ovaries, in which there are numerous imbricated tubes inserted on a large calyx of each of the oviducts.[45] When these tubes are few in number, they are but rarely disposed in simple or double regular series.[47] With most species, the *Receptaculum seminis* is cuneiform and often arcuate; its internal walls are brown, solid and horny, and it communicates with the vagina or copulatory pouch by means of a long, flexuous, spiral *Ductus seminalis*. With many species, this receptacle is invested with a muscular apparatus, composed of striated fibres, and which undoubtedly is a compressor. Usually, there is, attached to the base of the receptacle, a simple, rarely bifurcate or multiramose, *Glandula appendicularis*, which is sometimes provided with a long, flexuous excretory duct.[48] Sometimes the entire *Receptaculum seminis* is composed of only a simple, rarely bifurcate, somewhat long caecum.[49] Most of the Hydrocanthari, and some Carabidae, with which the *Ductus seminalis* is inserted on the copulatory pouch, have the peculiarity that there arises from the *Receptaculum seminis* a special Fecundatory canal which opens into the upper portion of the vagina.[50] A *Bursa copulatrix* exists, generally, in this order. With only a few species, it consists of a simple dilatation of the vagina,[51] but, usually, it is a rather long, muscular caecum, separated from the upper wall of the vagina, and sometimes even flexuous when its length is considerable.[52] Very often, the vagina is quite long, curved S-shaped, and passes with the rectum into a cloaca-like canal. It has a complicated special muscular apparatus.[53] The glandular appendages of the vagina are wanting with the Coleoptera, but, with the Hydrophilidae, there are two multiramose appendages on the oviducts, which are probably sebaceous organs.[54] The same function may, perhaps, be attributed to the glandular walls of the upper extremity of the oviducts of the Staphylinidae and Histeridae.[55]

46 With the Meloidae; see *Brandt* and *Ratzeburg*, Mediz. Zool. II. Taf. XVII. fig. 2.

47 The ovaries are in single rows with *Macronychus, Oxytelus, Silpha*, and *Byrrhus*; but they are in two rows with *Stenelmis, Lycus, Oedemera*, and *Hydrobius*; see L. Dufour, Ann. d. Sc. Nat. III. 1835, Pl. VII. fig. 25, 27; and *Stein*, loc. cit. Taf. III. fig. 3, 16, Taf. IV. fig. 3, 4, and Taf. VI. fig. 8. There is a very remarkable disposition, according to Stein (loc. cit. p. 30, Taf. I. fig. 4), with *Dianous caerulescens, Myrmedonia canaliculata, Homalota canaliculata*, and a species of *Trichopteryx*, which, alone among all known Insecta, have only a single ovary and a single oviduct, the first being composed of ten to twelve tubes disposed in two rows.

48 For the different forms of the *Receptaculum seminis* of the Coleoptera, see L. Dufour, Ann. d. Sc. Nat. VI. 1825, and III. 1835, &c.; *Siebold*, in *Müller's* Arch. 1837, p. 404, Taf. XX. fig. I, and especially Stein, loc. cit. p. 96, with the corresponding figures. With the Elateridae, the accessory gland is distinguished by a very complicated structure and numerous ramifications; see L. Dufour, Ann. d. Sc. Nat. VI. 1825, Pl. XVII. fig. 8-10, and Stein, loc. cit. p. 129, Taf. V. The seminal receptacle is wholly wanting with *Xantholinus punctatus, Lathridius porcatus, Notoxus monoceros*, and *Lagria hirta*; see Stein, loc. cit. p. 93.

49 With the Carabidae, and some Staphylinidae, the seminal receptacle is double with *Stenus* and *Parderus*; see Stein, loc. cit. p. 97, Taf. I. III. fig. 6.

50 With the Hydrocanthari and some Cara-

bidae; see Stein, loc. cit. p. 99, Taf. I. fig. 12, Taf. II.

51 *Silpha, Dromius, Calosoma*, and other Carabidae.

52 See Straus, Consid. &c. Pl. VI. fig. 2, o. n. (*Melolontha*); *Brandt* and *Ratzeburg*, Mediz. Zool. II. Taf. XVII. fig. 2, n. m. (*Meloe*); *Suckow*, in *Heusinger's* Zeitsch. II. Taf. XIII.; *Siebold*, in *Müller's* Arch. 1837, p. 405, but especially Stein, loc. cit. p. 69, and the corresponding figures.

53 There is a long, flexuous, muscular vagina with the Cerambycidae, Curculionidae, Elateridae, Buprestidae, and most of the Heteromera; also, with the Histeridae, Dermestidae, Parnidae, &c.; see Stein's exact descriptions, loc. cit. p. 71, Taf. VI.-VIII.

54 See Stein, loc. cit. p. 33, Taf. IV. fig. 3 (*Hydrobius fuscipes*). With *Hydrobius piceus*, and *caraboides*, there are even two kinds of analogous appendages. One consists of eight bifurcate follicles, the other of simple tubes inserted on the calyx of the oviducts; see L. Dufour, Ann. d. Sc. Nat. VI. 1825, p. 445, Pl. XVIII. fig. 5, and *Suckow*, in *Heusinger's* Zeitsch. II. Taf. XIII. fig. 34. The bifurcated appendages were overlooked by this last naturalist. It is well known that the females of the Hydrophilidae enclose their eggs by groups in a cocoon (*Lyonet*, Mém. du Mus. &c. XVIII. p. 454, Pl. XXIV.) which those of *Spercheus* carry about attached to their posterior legs.

55 Stein, loc. cit. p 35.

§ 351.

The External Genital Organs of the females are pretty simple with the Aptera, Hemiptera, Lepidoptera, Coleoptera, with many of the Diptera, Orthoptera, and Neuroptera, and with some Hymenoptera. The orifice of the vagina is supported by an upper, and two lateral horny plates, whose size and form vary according to the species. With only some Coleoptera, Diptera, and Hymenoptera, the end of the vagina is protractile, appearing as a more or less articulated *Vagina tubiformis*.[1] These horny plates about the vaginal orifice serve to support the penis during copulation, and to facilitate the escape of the eggs during oviposition.[2] With the Acrididae, these plates are conical, and in two pairs, one upper, and one under, which may be opened and shut in a pincer-like manner. With several genera of the Tipulidae, and Asilidae, the two lateral plates are very long, and form a simple ovipositor (*Vagina bivalvis*).[3] With *Boreus*, and *Acheta*, this ovipositor is long, and with *Raphidia*, it is long and acinaciform. The Locustidae have also a similar and very prominent ovipositous sabre, but more complicated in that each of its plates is divided into three pieces, which are so disposed that the two internal, soft, are surrounded in a sheath-like manner by the four others, which are horny. With the Tenthredinidae, and with *Aeschna*, *Agrion*, and *Calopteryx*, there is an analogous apparatus situated at the posterior extremity of the abdomen, and covered by two valves, only that its pieces are denticulated in a saw-like manner, and therefore is called saw-ovipositor.[4] With the Siricidae, the ovipositing apparatus is likewise composed of two horny, denticulate plates; but is more auger-like in its form, and, with some species, projects far beyond the short lateral valves.[5]

The Ichneumonidae, Cynipidae, and Cicadidae have a more or less long ovipositor (*Terebra*), composed of two lateral groove-like sheaths, between which plays a kind of sting composed of two intimately-united horny shafts. This sting serves, partly to pierce the substance in which the eggs are to be deposited, and partly to push the eggs along the sheath formed by the groove-like valves.[6] All these different ovipositors have a muscular apparatus at their base, by which their component pieces are moved.

With some Libellulidae, there is a peculiar groove-like appendage on the penultimate abdominal segment. It serves to receive the eggs at the

---

1 The ovipositor is unarticulated and protractile with the Cerambycidae, while it is articulated with the Chrysididae and many of the Muscidae. In this last case, its pieces are movable, like the tubes of a telescope. They are only the terminal abdominal segments modified; see *L. Dufour*, Ann. d. Sc. Nat. I. 1844, p. 383, Pl. XVI. fig. 16 (*Piophila*).

2 For the ovipositor of Insecta, see *Burmeister*, Handb. &c. I. p. 203, Taf. XII., and *Lacordaire*, Introduct. &c. II. p. 353.

3 *Limnobia, Ptychoptera, Tipula, Ctenophora, Asilus, Iphria*. Among these Diptera, *Ctenophora ruficornis* is particularly distinguished by the length of the horny plates composing the ovipositor.

4 For the structure of this saw-like ovipositor, see *Lyonet*, Mé. du Mus. XIX. p. 57, Pl. VI.-VIII. (14–16) (5 inches à scie); and *Hartig*, Die Adlerflügler Deutschl. p. 37, Taf. I. u. d. f.; also, *Réaumur*, Mém VI. 11 mémoire, Pl. XL. fig. 6–

9 (*Agrion*). It is well known that these Insecta use this ovipositor to pierce the epidermis of plants, and to introduce therein their eggs. The deposition of the eggs with the Tenthredinidae has been described with details by *Dahlbom* (Isis, 1837, p. 76) and by *Ratzeburg* (Forstinsekten, Th. III. p. 65). I have, also, observed this act with *Agrion forcipula* (*Wiegmann's* Arch. 1841, 1. p. 205).

5 *Hartig* and *Ratzeburg* have given a detailed description of the auger of the Siricidae; it is particularly long with *Xiphydria* and *Sirex*.

6 For the ovipositor of the Hymenoptera, see *Hartig*, Die Adlerflügler Deutschl. p. 16; in *Wiegmann's* Archiv, 1837, 1. p. 151, and in *Germar's* Zeitsch. III p. 326; *Ratzeburg*, Mediz. Zool. II. p. 145, Taf. XXIII. (*Cynips*). For that of the Cicadidae, see *Réaumur*, Mém. V. 4 mémoire, Pl. XVIII.; and *Doyère*, Ann. d. Sc. Nat. VII. 1837, p. 195.

moment of their escape from the vagina, and in this way the eggs are collected in masses to be deposited in places fit for their incubation.[7]

II. Male Genital Organs.

§ 352.

The Testicles, which are double like the ovaries, consist, sometimes of two simple caeca, which are more or less long and torose, and sometimes of many caeca, very variable as to their forms and disposition. Their mode of grouping resembles that of the ovaries; indeed, their whole appearance and contour, and the number and composition of their various parts resemble remarkably those of the female organs. With many species, these organs are covered by a lively-colored pigment layer, or enveloped by a special membrane (*Tunica raginalis*).

The two *Vasa deferentia* are of variable length, often exceeding that of the body, and therefore making several convolutions in the abdominal cavity. When the testicles are composed of many caeca, there are often the same number of these canals; but they often unite, on each side, into a common duct. Sometimes they have, each, at their lower extremity, a vesicular dilatation which may be regarded as a *Vesicula seminalis*. At their point of junction on the *Ductus ejaculatorius*, there are usually situated two, longer or shorter, simple *Glandulae mucosae*, which secrete a quickly coagulating, granular mucus, which serves, during the copulatory act, partly to fill and distend the *Bursa copulatrix* together with the penis, and partly to surround portions of the sperm, and thereby form spermatophores.[1]

§ 353.

The principal modifications observed with the internal male organs of the Insecta, are the following.

Among the Aptera, *Lepisma* is distinguished in having numerous oval, testicular follicles, whose *Vasa deferentia*, after forming irregular ramifications, unite in two common excretory ducts, which, gradually enlarging, terminate in a *Ductus ejaculatorius* at the point of insertion of two arcuate accessory glands.[1]

With the Hemiptera, the internal genital organs are of very variable form.[2] The Pentatomidae have only two simple, pyriform testicles, often of a beautiful red color; at their free extremity they sometimes have several constrictions, and thus form the passage to the form proper to many Geocorisae, which have seven long testicular tubes united in a fan-like

7 The ovigerous groove is short and triangular with *Libellula vulgata* and *cancellata* ; long, acuminate, and perpendicular with *Cordulia metallica* ; long and cordately emarginate and closely applied against the abdomen with *Epitheca bimaculata*. A remarkable appendage, deeply excavated, situated to the exterior of the female genital organs of *Doritis Apollo* and *mnemosyne*, and upon which, as yet, no lepidopterist has given any details, is probably an ovigerous sac.

1 For the various forms of the simple and compound testicles, as well as for the male organs of the Insecta in general, see *Burmeister*, Handb. &c. I. p. 217, and *Lacordaire*, Introduct. &c. II. p. 305.

1 See *Treviranus*, Verm. Schrift. II. p. 15, Taf. IV. fig. 2. The Pediculidae have only two pairs of testicles.

2 See *L. Dufour*, Recherch. sur les Hémipt. Pl. X.-XIII.

manner.[7] Sometimes these seven tubes are grouped into a bundle at the upper extremity of each of the two deferent canals.[4] With the Cicadidae, the testicular tubes are extremely numerous and fasciculate in the same manner;[5] while with *Psylla*, there are only four, and with *Aphis*, only three on each side.[6] The Hydrometridae have only two or four long testicular follicles, on the sides of which arise the deferent canals. With *Pelogonus*, and *Notonecta*, there are two pairs of long, spiral tubes, while with *Nepa*, and *Ranatra*, there are five on each side, long and flexuous. The *Vasa deferentia* are short with most of the Geocorisae, the Psyllidae and the Aphididae; but with the Hydrocorisae, and the Cicadidae, they are long and intertwisted. The glandular appendages are highly developed with most Hemiptera and often open into the two deferent canals above the *Ductus ejaculatorius*.[7] But when these glands appear to be wanting, the deferent canals have upon their course, or at their extremity, vesicular dilatations which, perhaps, take their place.[8] With the Pentatomidae, the glandular appendages consist of two to four multiramose fasciculate tubes. The *Ductus ejaculatorius* is then dilated at its base into a kind of vesicle divided into two or three lobes, which serve probably as mucous reservoirs.[9]

With the Diptera, the male organs are much more simple, there never being but two simple testicles,[10] whose external envelope is often brown or yellow. These organs are usually pyriform or oval, but sometimes long or hooked or twisted in various ways.[11] The *Vasa deferentia* are usually of considerable length,[12] and open in the upper end of the *Ductus ejaculatorius*,[13] always in common with two simple and pretty long accessory glands.

With the Lepidoptera, the testicles are always composed of two round or oval follicles, often surrounded by a beautifully colored pigment.[14] Very often, also, they are so approximated on the median line of the abdomen, as to appear fused into a single round body.[15] The two deferent canals, after a short course, unite with two simple, long and very flexuous accessory glands, and then form a very long and torose *Ductus ejaculatorius*.[16]

---

3 *Coreus, Alydus, Pyrrhocoris, Acanthia.*

4 *Capsus, Miris, Aradus.*

5 *L. Dufour*, Ann. d. Sc. Nat. V. 1825, Pl. VI. fig. 6, 7, and in his Recherch. sur les Hémipt. Pl. XIII. fig. 152–155 (*Cicada, Aphrophora,* and *Issus*).

6 With *Aphis lonicerae*, the six testicular tubes are concentrated on the median line of the abdominal cavity, so that they might easily be taken for a single body; see my observations in *Froriep's* neue Notiz. XII. p. 307. According to *Morren's* description (Ann. d. Sc. Nat. 1836, p. 87, Pl. VI.), there would appear to be a real fusion of the testicular tubes with *Aphis persicae*.

7 With *Aradrus, Nepa, Cicada, Aphrophora*, the two simple glandular appendages, which are extraordinarily long and flexuous with the Cicadidae, are inserted on the sides of the deferent canals; while with *Aphis*, which has two, and with *Notonecta, Miris* and *Capsus*, which have four, the glandular tubes open into the *Ductus ejaculatorius*, conjointly with the deferent canals.

8 *Psylla, Pyrrhocoris, Velia* and *Gerris*. *L. Dufour* (Recherch. &c.) unhesitatingly calls these dilatations of the deferent canals *Vesiculae seminales*.

9 *L. Dufour* (Recherch. &c. Pl. X.) also regards this reservoir as a *Vesicula seminalis*.

10 The male organs of the Diptera have been described by *L. Dufour* (Ann. d. Sc. Nat. I. 1844, p. 250), and by *Loew* (Horae Anat. p. 9, Tab. I.–III.), whose account is very detailed and exact.

11 The testicles are long and regularly flexuous with *Myopa*, spiral-form with *Asilus* and *Dasypogon*, while those of the Hippoboscidae are extremely long and very torose; see *L. Dufour*, Ann. d. Sc. Nat. loc. cit.

12 *Stratiomys*, alone, has very long and torose deferent canals.

13 These two glands are very long with *Hippobosca, Dolichopus, Asilus,* and *Stratiomys*; ramose with *Trypeta* and *Psila*; with *Leptis*, they are wanting, being replaced, probably, by two swellings situated at the lower extremity of the two deferent canals. *Empis* and *Scatopse* have two pairs of glands, one above, the other below.

14 The testicles are carmen-red with *Argynnis, Hipparchia, Pontia* and *Liparis*; green with *Lycaena* and *Sphinx*.

15 *Suckow (Heusinger's* Zeitsch. II. Taf. X. fig. 10) has found two separated testicles with *Yponomeuta*. The fusion of these organs is complete with the Papilionidae, Sphingidae, Bombycidae, &c.

16 See *Herold*, Entwickelungsgesch. d. Schmetterl. Taf. IV. XXXII. (*Pontia brassicae*), and

With the Hymenoptera,[17] the testicles present many different forms. Beside two simple ovoid testicular follicles,[18] there are, not unfrequently, also two testicles composed of several long follicles, fasciculate, and surrounded, together with a portion of the torose deferent canal, by a common envelope; but, more commonly, these two testicles are contained in a capsule situated on the median line of the body.[19]

With the Tenthredinidae and the Siricidae, the testicles are separate and distinct, without capsules, and composed of round follicles disposed botryoidally.[20] The two deferent canals are usually pretty long, and have, sometimes, at their lower extremity, two vesicular dilatations which, containing sperm, may be regarded as seminal vesicles.[21] The deferent canals with the Hymenoptera have, usually, two pyriform accessory glands, whose excretory ducts unite into a short *Ductus ejaculatorius*.[22]

With the winged Strepsiptera, there are two pyriform testicles provided with very short deferent canals, which dilate above the *Ductus ejaculatorius* into two seminal vesicles; but nowhere has an accessory gland been observed.

With the Orthoptera, the two testicles are nearly always composed of a greater or less number of follicles. With the Acrididae, Locustidae, Achetidae, Blattidae and Mantidae, they are composed of long fasciculated or imbricated caeca, which, as with the Hymenoptera, are very often surrounded by a common envelope. In some species the two groups of testicular follicles are united into a common mass on the median line of the abdomen, by this *Tunica vaginalis*.[23] On the other hand, the Phasmidae, Libellulidae, Perlidae and Ephemeridae, have a multitude of round follicles, disposed botryoidally around a long dilated portion of each of the deferent canals.[24] These last are usually very short, and with the Achetidae and Locustidae, only, they are quite long, and spiral from beginning to end.[25] Many Orthoptera have highly-developed accessory glands surrounding a short *Ductus ejaculatorius*, on which they are sometimes disposed in successive groups.[26] A part of this apparatus, in which are

Suckow, Anat. u. physiol. Untersuch. Taf. IV. (*Gastropacha pini*).*

17 *L. Dufour* (Recherch. sur les Orthopt. p. 394, Pl. V.–X.) has furnished observations accompanied with very many figures on the male organs of the Hymenoptera.

18 The testicles are simple with *Parnopes, Cynips, Diploliepis* and *Chelonus*.

19 There are two unicapsular testicular bundles with *Apis, Xylocopa* and *Bombus*; see *L. Dufour*, loc. cit. fig. 54–62. The two testicular fasciculi are enclosed in a common capsule with *Anthophora, Anthidium, Odynerus, Tiphia, Scolia, Pompilus* and *Crabro*; see *L. Dufour*, loc. cit. Pl. VI.–IX.

20 *L. Dufour*, loc. cit. fig. 150–154 (*Tenthredo, Hylotoma* and *Cephus*).

21 The deferent canals terminate each with a seminal vesicle with *Cynips, Chelonus, Apis* and *Xylocopa*.

22 See *Brandt* and *Ratzeburg*, Mediz. Zool. Taf. XXV. fig. 35 (*Apis*), and *L. Dufour*, loc. cit.

* [§ 353, note 16.] See, also, for histological details on the internal male organs and their development, of the Lepidoptera, *Meyer*, loc. cit. *Siebold* and *Kölliker's* Zeitsch. I. 1849, p. 182. The formula of the development of the testicles is, of course, the same as that of the development of the

23 See *L. Dufour*, Recherch. sur les Orthopt. Pl. I.–V. There are two distinct fasciculate testicles with *Gryllotalpa, Oecanthus, Ephippigera*, and two groups of long, imbricated follicles with *Tetrix, Locusta* and *Decticus*. The testicles are fused into one body with *Oedipoda* and *Blatta*.]

24 See *Suckow*, in *Heusinger's* Zeitsch. II. Taf. XII. fig. 25, Taf. X. fig. 8; *Rathké*, De Libellur. partibus genital. Tab. I. fig. 3, and *L. Dufour*, loc. cit. Pl. II. fig. 164, and Pl. XII. fig. 204 (*Perla* and *Libellula*).

25 See *L. Dufour*, loc. cit. fig. 25, 36 (*Gryllotalpa* and *Ephippigera*).

26 The Perlidae have only two testicular follicles inserted on the deferent canals. *Tetrix*, the Acrididae, Achetidae and Blattidae, have two long and large fasciculi; finally, with the Mantidae and Locustidae, there are, besides these fasciculi, one or two pairs of shorter bundles; see *L. Dufour*, loc. cit. Pl. III.–V.

ovaries; but this observer shows that the spermatic particles are formed, like the ova, while the insect is in the pupa-state. — ED.

† [§ 353, note 23.] See also *Leidy*, Proceed. Acad. Sc. Philad. 1846, III. p. 80 (*Spectrum femoratum*). — ED.

situated here and there vesicular reservoirs, secretes, undoubtedly, with the Locustidae, a substance used in the formation of the spermatophores. But with the Phasmidae, Libellulidae and Ephemeridae, the *Ductus ejaculatorius* is wholly deficient in all kinds of glandular appendages.

With the Neuroptera, the various genera present only few modifications in their male genital organs. With *Panorpa*, the two testicles are very simple and ovoid;[27] but with the other species they consist of two tufts of long or round follicles.[28] With *Myrmeleon*, and *Hemerobius*, they are oval and surrounded by a distinct envelope. The two deferent canals are short, and always have on their lower extremity two long or ovoid accessory follicles.[29]

With the Coleoptera, the male organs vary very much.[30] With the Carabidae, Hydrocanthari, and Lucanidae, the testicles consist of two extremely long, torose caeca,[31] of which each is sometimes enclosed in two special envelopes.[32] The Elateridae, Tillidae, Cantharidae, very many Heteromera and Coccinellidae, have, on the other hand, a multitude of round or oblong, short follicles, fasciculate, composing the two testicles, which,[33] in some genera, are here also invested by a capsule.[34] With the Hydrophilidae, and Pyrochroïdae, these organs are composed of numerous short, aggregated follicles, situated laterally over a wide extent of the posterior extremity of the deferent canals.[35] With the Staphylinidae, and Silphidae, the testicular follicles are pyriform and inserted botryoidally on the posterior extremity of the simple or multiramose *vasa deferentia*.[36] With the Lamellicornes, Cerambycidae, Curculionidae, and Crioceridae, these organs are formed after a wholly different type, their number being two, six, or even twelve on each side. They are usually round follicles, flattened disc-like, and from which pass off pretty short excretory ducts to the extremity of the two common deferent canals.[37]

The *Vasa deferentia*, with the Coleoptera, are usually pretty long ; but with the Carabidae, Hydrocanthari and Cerambycidae they are very long, spiral or torose.[38] With a few species, only, is each of them dilated in its course into a *Vesicula seminalis*.[39] The accessory glands are never wanting in this order, and they either open, together with the deferent canals, into the upper extremity of the *Ductus ejaculatorius*, or they pass into these canals before they reach this duct. In very many species this gland-

---

27 *L. Dufour*, loc. cit. fig. 172.

28 *Sialis* and *Phryganea*.

29 See *L. Dufour*, loc. cit. Pl. XII. fig. 172-210 (*Panorpa*, *Myrmeleon*, *Sialis*, *Phryganea*), and *Suckow*, in *Heusinger's* Zeitsch. II. Taf. XVI. fig. 15 (*Sialis*).

30 For the male organs of the Coleoptera in general, see especially *L. Dufour*, Ann. d. Sc. Nat. VI. 1825, p. 152, Pl. IV.-IX. and I. 1834, p. 76, Pl. III. IV.

31 With *Harpalus*, the two caeca are united into a single clew.

32 *Cybister*, *Scarites*, and *Clivina*, have two testicles invested by a capsule.

33 Each testicular fasciculus is composed of from three to seven follicles with *Dermestes*, *Heteroceras*, *Anthrenus*, *Oedemera*, *Helops*, *Diaperis*, *Tenebrio*; while with *Blaps*, *Pimelia*, *Mylabris*, *Telephorus*, *Bostrichus*, the Elateridae and Coccinellidae, their number is much larger.

34 There is a *Tunica vaginalis* with *Clerus*, *Trichodes*, *Mylabris*, and which, with *Galeruca*, is even common to both testicles.

35 See *Swammerdamm*, Bib. der Nat. Taf. XXII. fig. 5; *Suckow*, in *Heusinger's* Zeitsch. II. Taf. X. fig. 1, 2 (*Hydrophilus*); *L. Dufour*, Ann. d. Sc. Nat. XIII. 1840, Pl. VI. A. fig. 18 (*Pyrochroa*).

36 The two testicles are multiramose with *Silpha*; see *L. Dufour*, Ann. d. Sc. Nat. VI. 1825, Pl. VI fig. 6.

37 *Hammaticherus*, *Anthribus*, *Lixus* and *Donacia* have two pairs of testicles ; *Melolontha* and *Prionus* six, *Trichius* nine, and *Cetonia* twelve. Beside *L. Dufour* (loc. cit.), see *Suckow*, in *Heusinger's* Zeitsch. II. Taf. XI. and *Straus*, Considérat. &c. Pl. VI.

38 These torosities are even surrounded with a capsule with *Cybister* ; see *L. Dufour*, Ann. d. Sc. Nat. VI. 1825, Pl. V. fig. 1.

39 With the Hydrophilidae, there is a vesicular dilatation at the lower extremity of the deferent canals ; but with *Anthribus*, and *Lixus*, it is situated at its opposite extremity.

ular apparatus consists of only two simple, longer or shorter caeca,[40] which are sometimes quite long and torose.[41] Another series of Coleoptera have four to eight caecal appendages, disposed in pairs, and variable as to length and volume. One of these pairs is probably only a reservoir for the secreted product of the others.[42] The *Ductus ejaculatorius* is always very muscular, and with very many species, quite long and flexuous, and the penis therefore can be widely protruded during copulation.

§ 354.

The Copulatory organs of the male Insecta are valve-like or forficulate, horny appendages,[1] which are so variable in their form that the most allied species differ, in this respect, widely and constantly.[2]

Beside these proper copulatory organs, situated at the posterior extremity of the abdomen, there are often on the antennae, the parts of the mouth, the legs and other regions of the body, auxiliary organs used for seizing and retaining the female, and which have long been objects of careful description in zoology.

With most Hemiptera, the posterior extremity of the abdomen conceals a horny capsule which contains a protractile, tubular penis. With very many Diptera, the copulatory organs project prominently in the same region of the body, and consist often of two horny valves of different forms which envelop a rather long penis.[3] The Lepidoptera, Hymenoptera, Orthoptera, and Neuroptera, have two pairs of valves, one internal, the other external, which enclose a tubular or groove-like penis.[4]

The Ephemeridae and the Strepsiptera, only, are distinguished by their very simple copulatory organs; for with the first there is only a simple penis without a valvular apparatus. This last is replaced by two long, small, triarticulated stylets, situated on the penultimate abdominal segment and curved inwardly; while with the Strepsiptera, the penis, also naked and horny, is so articulated that it can be applied laterally against the abdomen, like the blade of a knife in its handle.

With the Libellulidae, however, the orifice of the *Ductus ejaculatorius* is most simple, being covered only by two very small oval valves. But the penis is not wanting with these Insecta; it is singularly concealed, together with a horny-walled seminal vesicle, in a fossa situated at the base of the

40 With the Carabidae, Hydrocanthari, and with *Mordella, Anthribus, Galeruca* and *Coccinella.*
41 *Melolontha, Cetonia* and *Lucanus ;* see *L. Dufour, Straus,* and *Suckow,* loc. cit.
42 With the Staphylinidae, Cantharidae, Byrrhidae, Elateridae, Tillidae, Melolidae, Tenebrionidae, Pyrochroidae, Dermestidae, Cerambycidae, with *Donacia, Heteroecrus,* &c. ; see *L. Dufour, Suckow,* loc. cit. and *Brandt,* Mediz. Zool. II. Taf. XVII. XIX. This glandular apparatus is specially developed with *Hydrophilus piceus,* where, of the four pairs, one is distinguished for its length and thickness, and is composed at its extremity of numerous small follicles ; see *Swammerdamm,* Bib. der Nat. Taf. XXII. fig. 4 ; *L. Dufour,* l c. cit. VI. Pl. VI. fig. 7, and *Suckow,* loc. cit. Taf. X. fig. 1, 3.
1 See *Burmeister,* Handb. &c. I. p. 227, Taf. XIII.
2 As yet, these differences in form of the external male organs have been of little service to entomologists in the distinction of species, although, had

they been well understood, the formation of many bad species might have been prevented. They prevent allied species from producing bastards by adulterous connections, for the hard parts of the male correspond so exactly with those of the female, that the organs of one species cannot fit those of another. *L. Dufour* has, therefore, properly termed these copulatory organs as "*la garantie de la conservation des types, et la sauvegarde de la légitimité de l'espèce.*"
This horny apparatus, from its large and often tumid lateral valves is quite prominent with the Dolichopidae, Empidae, with *Asilus, Laphria, Ctenophora, Nematocera,* and other Tipulidae. See *Schummel,* Beitr. zur Entomol. Taf. I.–III. (*Tipula*).
4 With the Panorpidae, these copulatory organs are changed into very large pincers ; while with *Psyche,* the very long penis is protractile like the tubes of a telescope, thus enabling these butterflies to copulate with their females which remain concealed in sacks.

abdomen.[5] This penis is composed of three articles with *Aeschna, Libellula,* and *Gomphus ;* but of one only with *Calopteryx,* and *Agrion,* with which it is not directly adherent to the seminal vesicle. The male Libellulidae are obliged, before copulation, to fill their *vesicula seminalis,* which is situated at the base of the abdomen. This they accomplish by bending the posterior extremity of the abdomen, so as to meet and empty the semen into this vesicle. They then seize the female by the neck, by means of their anal pincers, and she places her genital orifice in contact with the copulatory apparatus of the male.[6] These anal pincers of the males have very distinct specific characteristics, while the females, on their part, have, in the separate species, equally specific sculptured markings on the prothorax.[7]

With the Coleoptera, the copulatory organs consist of a more or less horny sheath enveloped by a membranous prepuce, and containing a broadly-flattened penis which consists of a canal supported by two lateral horny ridges. At rest, these organs are entirely withdrawn into the abdominal cavity, but can be widely protruded out of it by means of a very remarkable muscular apparatus.[8] With the male individuals of *Dermestes,* there is a median orifice on the third and fourth abdominal segments, from which projects a brush of stiff bristles connected with a round muscular body situated on the internal surface of each of these segments. This brush is undoubtedly some way connected with the act of copulation.[9]

§ 355.

The development of the larvae of Insecta in the egg, occurs in the same manner as with most of the other Arthropoda. After the unusually early disappearance of the germinative vesicle,[1] there is formed, from a superficial and partial segmentation, a round or oblong-oval blastoderma, whose hyaline aspect contrasts with that of the rest of the vitellus.[2] This blas-

5 For the copulatory organs of the Libellulidae, see *Rathke,* De Libellar. partibus genital., and my researches in *Germar's* Zeitsch. II. p. 421.

6 The act of copulation of the Libellulidae has been represented by *Swammerdamm,* Bib. der Nat. Taf. XII. fig. 3 ; *Reaumur,* Mem. &c. VI. Pl. XL. XLI. ; and *Roesel,* Insectenbelust. Th. II. Insect. aquat. Class. II. Tab. X.

7 The different forms of these pincers have been figured in *Charpentier,* Horae Entomol. Tab. I., and *Selys Longchamps,* Monogr. des Libellul. d'Europe, Pl. I.–IV.

8 See *Straus,* Considér. &c. Pl. III. V.

9 See my observations in the Entomol. Zeitung. 1840, p. 137, and *Brullé,* Ann. d. l. Soc. Entom. VII. 1838, p. LIII. The golden-colored tuft of hairs situated at the base of the abdomen with the males of *Blaps,* does not correspond to that of *Dermestes,* because it is only external and does not project into the interior of the body.

1 The germinative vesicle is never observed in eggs that have been layed ; it has disappeared even in those still in the oviduct ; this disappearance would not appear, therefore, to depend upon the act of fecundation.

2 The first phases of the development of Insecta have been studied by *Herold* (Disquisit. de Anim. vertebr. carent. in ovo format. 1835–38) with *Spinx ligustri* and *Musca vomitoria;* and by *Kölliker* (Observ. de prima Insect. genesi, 1842, or Ann. d. Sc. Nat. XX. 1843, Pl. X.–XII.) with *Chironomus, Simulia,* and *Donacia.*

The ulterior phases have been traced by *Rathke* (*Meckel's* Arch. 1832, p. 371, Taf. IV. and *Müller's* Arch. 1844, p. 27, Taf. II.) with *Blatta orientalis* and *Gryllotalpa vulgaris ;* and by *Nicolet* (Recherch. &c. p. 18, Pl. I.) with the Poduridae.*

* [§ 355, note 2.] I am not aware that the numerous researches upon the embryology of the Insecta made within a few years, have added any new phases to the general type of development of these animals as brought out by the earlier investigators. The type of development with the Arthropoda is essentially the same in all of the classes of this section. What late observers have done, therefore, is the tracing of some of the secondary conditions of formation belonging to the different groups, and the observation of the details of development of different internal and external organs. These anomalies of development and reproduction, which continued research shows to be far from uncommon with the Insecta, will ultimately be found, probably, referable all to the phenomena, we have discussed below, of the Aphididae. — ED.

toderma, which corresponds to the ventral side of the future embryo, extends gradually in all directions and at last encompasses the whole vitellus, — its borders meeting on the dorsal surface. It may be divided into an external or serous, and an internal or mucous layer. In the first of these is developed, on the median abdominal line, the ventral cord ; while the second forms a semi-canal which gradually surrounds the vitellus and at last completely enveloping it, is changed into the digestive canal. The various appendages of this canal are subsequently formed by simple constrictions or deverticula from its cavity; while the other abdominal viscera are directly developed from a special blastoderma.

Upon the external surface of the serous layer are formed the parts of the mouth, the tactile organs, the legs, and the other appendages of the body, whose articulations, like those of the body itself, are produced by constrictions.

The dorsal vessel is formed between the two blastodermic layers on the side opposite that of the ventral cord. This development of the embryo takes place at the expense of the vitellus, which, enclosed in the digestive canal, is gradually consumed.*

* [End of § 355.] The subject, which has been frequently alluded to in this book, — the singular mode of reproduction of the viviparous Aphididae, is one of so much interest and importance in physiology, that I propose to give it something more than a brief mention. Moreover, I have enjoyed excellent opportunities for the study of these phenomena in question, and have advanced an interpretation of them, and their like elsewhere, quite different from that usually received.

My observations were made upon *Aphis caryae* (probably *Lachnus* of Illiger, or *Cinara* of Curtis), one of the largest and most favorable species for these investigations. This was in the spring of 1853. The first colony, on their appearance from their winter quarters were of mature size, and contained, in their interior, the developing forms of the second colony quite far advanced in formation. On this account it was the embryology of the third series or colony, that I was able to first trace. A few days after the appearance of the first colony (A), the second colony (B), still within the former, had reached two-thirds of their full embryonic size ; the arches of the segments had begun to close on the dorsal surface, and the various appendages of the embryo were becoming prominent ; the alimentary canal was more or less completely formed, although distinct abdominal organs of any kind belonging to the digestive system were not apparent.

At this time, and while the individuals B. were not only in the abdomen of their parents A., but were also enclosed each in its primitive egg-like capsule ; at this time, I repeat, appear the first traces of the germs of the third colony, C. Their first traces consisted of small egg-like bodies, arranged two, three, or four in a row, and attached at the locality where are situated the ovaries in the oviparous forms of the Aphididae. These egg-like bodies were either single nucleated cells of one three-thousandth of an inch in diameter, or a small

number of such cells enclosed in a simple sac. These are the germs of the third generation or colony, and they increase *pari passu* with the development of the embryo in which they are formed, and this increase of size takes place not by the segmentation of the primitive cells, but by the endogenous formation of new cells within the sac. After this increase has continued for a certain time, these bodies appear like little oval bags of cells, — all the component cells being of the same size and shape, — there being no one particular cell which is larger and more prominent than the others, and which could be comparable to a germinative vesicle. While these germs are thus constituted the formation of new ones is continually taking place. This occurs by a kind of constriction-process of the first germs ; one of the ends of these last being pinched off, as it were, and so, what was before a single body or sac, becomes two which are attached in a moniliform manner. The new germs thus formed may consist each of a single cell only, as I have often seen ; but they soon attain a more uniform size by the endogenous formation of new cells within the sac in which it is enclosed. In this way the germs are multiplied to a considerable number, the nutritive material for their growth being, apparently, a fatty liquid in which they are bathed, contained in the abdomen, and which is thence derived from the abdomen of the first parent. When these germs have reached the size of about one three-hundredth of an inch in diameter, there appears on each, near the inner pole, a yellowish, vitellus-looking mass or spot, composed of yellowish cells, which, in size and general aspect, are different from those constituting the germ proper. This yellow mass increases after this period, *pari passu* with the germ, and at last lies like a cloud over and partially concealing one of its poles. I would, moreover, insist upon the point that it does not gradually extend itself over the whole germ mass, and is, therefore, quite unlike a proligerous disc.

When these egg-like germs have attained the size of one one-hundred-and-fiftieth of an inch in diameter, there begins to appear distinctly the sketching or marking out of the future embryo. This sketching consists at first of delicately-marked retreatings of the cells here and there ; but these last soon become more prominent from sulcations, and, at last, the form of an articulated embryo is quite prominent.

During this time, the yellowish, vitellus-looking mass has not changed its place, and although it is somewhat increased in size, yet it appears otherwise the same. When the development has proceeded a little further, and the embryo has assumed a pretty definite form, the arches of the segments, which have hitherto remained gapingly open, appear to close together on the dorsal surface, thereby enclosing the vitellus-looking mass within the abdominal cavity. It is this same vitelloid mass thus enclosed, which furnishes the development of the new germs (which in this case would be those of the fourth colony, or D), and this germ development here commences with the closing up of the abdominal cavity, and then the same processes we have just described are repeated.

The details of the development subsequent to this time, — the formation of the different systems of organs, &c., are precisely like those of the development of true oviparous Arthropoda in general ; and although the ovoid germ has, at no time, the structural peculiarities of a true ovum,— such as a real vitellus, germinative vesicle and dot, yet if we allow a little latitude in our comparison and regard the vitellus-looking mass as the *mucous*, and the germ-mass proper as the *serous* fold of the germinating tissue, as in true ova ; if this comparison of parts can be admitted, then the analogy of the secondary phases of development between these forms, and true ova of the Arthropoda, can be traced to a considerable extent.

These secondary phases of development need not here be detailed, for they correspond to those described by *Herold*, *Kölliker*, of the true ovum in other Insecta, and which, too, I have often traced in various species of the Arthropoda in general.

When the embryo is fully formed and ready to burst from its capsule in which it has been developed, it is about one-sixteenth of an inch in length, or more than eight times the size of the germ, when the first traces of development in it were seen. From this last-mentioned fact, it is evident that, even admitting that these germ-masses are true eggs, the conditions of development are quite different from those of the truly viviparous animals, for, in these last, the egg is merely hatched in the body instead of out of it, and, moreover, it is formed exactly as though it was to be deposited, and its vitellus contains all the nutritive material required for the development of the embryo until hatched. With the Aphididae, on the other hand, the developing germ derives its nutritive material from the fatty liquid in which it is bathed, and which fills the abdomen of the parent. The conditions of development in this respect, are here, therefore, more like those of the Mammalia and the whole parent animal may be regarded in one sense as an individualized uterus filled with germs, — for the digestive canal with its appendages seems to serve only as a kind of laboratory for the conversion of the succulent liquids this animal extracts from the tree on which it lives, into this fatty liquid which is the nutritive material of the germs.

Omitting the curious and interesting details of the further history of the economy of these Insecta, as irrelevant to the point in discussion, we will now turn to see what view we should take of these processes, and what is their physiological interpretation. In the first place it is evident that the germs which develop these viviparous Aphides are not true eggs ; they have none of the structural characteristics of these last, — such as a vitellus, a germinative vesicle and dot ; on the other hand they are at first simple collections, in oval masses, of nucleated cells. Then again, they receive no special fecundating power from the male, which is the necessary preliminary condition of all true eggs ; and furthermore the appearance of the new individual is not preceded by the phenomena of segmentation, as is also the case with all true eggs. Therefore, their primitive formation, their development and the preparatory changes they undergo for the evolution of the new individual, are all different from those of real ova.

Another point of equal importance is these viviparous individuals of the Aphides have no proper ovaries and oviducts. Distinct organs of this kind I have never been able to make out. The germs, as we have before seen, are situated in moniliform rows, like the successive joints of confervoid plants, and are not enclosed in a special tube. These rows of germs commence, each, from a single germ-mass which sprouts from the inner surface of the animal, and increases in length and the number of its component parts by the successive formation of new germs by the constriction process as already described. Moreover, these rows of germs which, at one period, closely resemble in general form, the ovaries of some true Insecta, are not continuous with any uterine or other female organ, and therefore do not at all communicate with the external world, on the other hand, they are simply attached to the inner surface of the animal, and their component germs are detached into the abdominal cavity as fast as they are developed, and thence escape outwards through a *Porus genitalis*.

With these data, the question arises, what is the proper interpretation to be put upon these reproductive phenomena we have just described ? My answer would be that the whole constitutes only a rather anomalous form of gemmiparity ; as already shown, the viviparous Aphididae are sexless ; they are not females, for they have no female organs, they are simply *gemmiparous*, and the budding is internal, instead of external as with the Polypi and Acalephae ; moreover this budding takes on some of the morphological peculiarities of oviparity but these peculiarities are economical and extrinsic, and do not touch the intrinsic nature of the processes therein concerned. Viewed in this way, the different broods or colonies of Aphididae cannot be said to constitute as many true generations, any more than the different branches of a tree can be said to constitute as many trees ; on the other hand the whole suite, from the first to the last, constitute but a single true generation. I would insist upon this point as illustrative of the distinction to be drawn between *sexual* and *gemmiparous* reproduction. Morphologically, these two forms of reproduction, have, it is true, many points of close resemblance, but there is a grand physiological difference, the perception of which is deeply connected with our highest appreciation of individual animal life.

A true generation must be regarded as resulting only from the conjugation of two opposite sexes, — from a sexual process in which the potential representatives (spermatic particle and ovum) of two opposite sexes are united for the elimination of one germ. The germ power thus formed may be extended by gemmation or fission, but it can be formed only by the act of generation, and its play of extension by budding or by division must always be within a certain cycle, which cycle is recommenced by the new act of the conjugation again of the two sexes. In this way the dignity of the ovum as the primordium of all true individuality, is maintained.

I have thus treated this subject in some detail, not only from its wide bearing in the physiology of reproduction, but also from its direct relation to many phenomena alluded to in the preceding pages. In the memoir from which I have made this extract (read before the Amer. Acad. Arts and Sc., Oct. 11, 1853) I have entered into a full

discussion of those many points suggested by these studies. One of these, is, the relation of this subject to some of the various doctrines of development, which have been advanced in late years, such as that of *Alternation of Generation*, by *Steenstrup*, and that of *Parthenogenesis* by *Owen*. I have there attempted to show that the phenomena of these doctrines, as advanced by their respective advocates, all belong to those of gemmiparity, and that therefore Alternation of Generation and Parthenogenesis in their implied sense, are misnomers in physiology. Another point there treated in extenso, is the identity of this mode of reproduction we have just described in the Aphididae, with that observed in the so-called hibernating eggs of the Entomostraca (see above, § 200) and the like phenomena observed in nearly every class of the Invertebrata. They are all referable, in my opinion, to the conditions of gemmation, modified in each particular case, perhaps, by the economical relations of the animal.

See for some recent writings on this peculiar form of reproduction with the Insecta, and which contain many interesting physiological remarks, *Leydig*, Die Dotterfurchung nach ihrem Vorkommen in der Thierwelt und nach ihrer Bedeutung, in the Isis, 1848, Hft. 3 ; also, Einige Bemerkungen über die Entwickelung der Blattläuse, in *Siebold* and *Kölliker's* Zeitsch. II. 1850, p. 62 ; also Zur Anatomie von Coccus hesperidum, in Ibid. 1853, V. p. 1 ; *Victor Carus*, Zur näheren Kenntniss des Generationswechsels, Leipsig, 1849 ; and *Siebold*, as referred to in my note under § 348 note 4. I cannot here discuss the often similar and dissimilar views to those of my own above detailed, expressed by these different investigators.— ED.

INDEX.

ABBREVIATIONS

Acal., Acalephae.
Aceph., Acephala.
Ann., Annelides.
Arach., Arachnoidae.

Ceph'd., Cephalopoda.
Ceph'r., Cephalophora.
Crus., Crustacea.
Ech., Echinodermata.

Hel., Helminthes.
Inf., Infusoria.
Ins., Insecta.
Pol., Polypi.

Rhiz., Rhizopoda.
Rot., Rotatoria.
Turb., Turbellaria.

N. B. The Numbers refer to the Paragraphs.

A.

ACALEPHAE, 53.
ACEPHALA, 170.
Aciculi, Ann. 145.
Air cavity or reservoir, Acal. 65.
Ambulacra, Ech. 77, 91.
Ampulla, of Poli, Ech. 92.
ANNELIDES, 142.
Antennae, Ann. 149, 152. Arach. 306. Ins. 332.
Anus, Inf. 15. Ech. 82. Hel. 107. Rot. 136. Aceph. 188, 189. Ceph'r. 214. Ceph'd. 249. Crus. 279. Arach. 307. Ins. 338.
Apparatus, ceraceous, Ins. 347.
—— digestive. Inf. and Rhiz. 11. Pol. 35. Acal. 61. Ech. 82. Hel. 106. Turb. 125. Rot. 136. Ann. 152. Aceph. 188. Ceph'r. 213. Ceph'd. 248. Crus. 278. Arach. 306. Ins. 337.
—— ejaculatory, of sperm. Ceph'd. 259.
—— masticatory. Ech. 84. Hel. 108. Rot. 136. Ann. 154. Ceph'r. 213. Ceph'd. 248. Crus. 279. Arach. 306. Ins. 337.
—— mucous. Ins. 349.
—— respiratory. Inf. and Rhiz. 18. Pol. 41. Acal. 63. Ech. 89. Hel. 112. Turb. 126. Rot. 138. Ann. 158. Aceph. 193. Ceph'r. 219. Ceph'd. 253. Crus. 285. Arach. 311. Ins. 341.
—— rotatory. Rot. 133.
—— sebaceous. Ins. 349.
—— sericeous. Arach. 315. Ins. 347.
—— suctorial. Hel. 103. Arach. 306. Ins. 337, 338.
—— tentacular. Ech. 83.
—— venomous. Arach. 315. Ins. 347, 349.
Appendices, caecal. See Caecum.
—— cutaneous. Ann. 143.
ARACHNOIDAE, 295.
Arms. Aceph. 185. Ceph'd. 237.
Arolium. Arach. 299. Ins. 326.
Arteries. See Circulatory System.
Audition. See Auditive Organs.

B.

Balancers. Ins. 326.
Bladders, natatory. Acal. 58. Ceph'd. 256.
Blood. Pol. 39, 41. Hel. 110. Ann. 156. Aceph 191. Ceph'r. 216. Ceph'd. 251. Crus. 282. Arach. 309. Ins. 340.
Brain. Ann. 146, 147. Ceph'r. 206. Ceph'd. 240. Crus. 270. Arach. 300. Ins. 328.
Branchiae. Ech. 89. Ann. 159, 160. Aceph. 193. Ceph'r. 229. Ceph'd. 253. Crus. 285. Arach. 312. Ins. 342.
Bristles. Ann. 145. Aceph. 185.
Bursa Needhami. Ceph'd. 259.
Byssus. Aceph. 179.

C.

Caecum. Ech. 85, 86. Hel. 108. Ann. 154. Aceph. 189. Ceph'd. 249. Crus. 279. Arach. 307. Ins. 338.
Caeca, hepatic. See Liver.
Calamistrum. Arach. 315.
Canal, intestinal. Ann. 154. Aceph. 189. Ceph'r. 214. Ceph'd. 249. Crus. 279. Arach 307. Ins. 338.
—— stony. Ech. 75.
Canals, aquiferous. See Aquiferous System.
—— lateral. Hel. 110.
—— longitudinal. Hel. 111.
Capsule, genital. Ceph'd. 260.
—— egg. Ceph'd. 258.
—— ovarian. Ceph'd. 258.
Cardo. See Hinge.
Cavity, branchial. Ceph'r. 220.
—— incubatory. Crus. 292.
—— pulsatile. Inf. 16.
—— respiratory. Aceph. 194. Crus. 285.
Cell, lateral. Hel. 110.
Cells, chromatophoric. Ceph'd. 233.
—— hepatic. See Liver.
—— vitelline. Hel. 115. Turb. 129.